AUCTORES BRITANNICI MEDII AEVI • 26

GEOFFREY OF ASPALL QUESTIONS ON ARISTOTLE'S *PHYSICS*

Part 1

Edited by

SILVIA DONATI &
CECILIA TRIFOGLI

English Translation by

E. JENNIFER ASHWORTH &
CECILIA TRIFOGLI

Published for THE BRITISH ACADEMY
by OXFORD UNIVERSITY PRESS

Oxford University Press, Great Clarendon Street, Oxford OX2 6DP

First edition published in 2017

British Library Cataloguing in Publication Data
Data available

Library of Congress Cataloging in Publication Data
Data available

Typeset by
New Leaf Design, Scarborough, North Yorkshire
Printed in Great Britain
on acid-free paper by
CPI Antony Rowe Ltd
Chippenham, Wiltshire

ISBN 978–0–19–726599–4

CONTENTS

PART 1

GEOFFREY OF ASPALL
QUESTIONS ON ARISTOTLE'S *PHYSICS*
Part 1

PART 2

ACKNOWLEDGEMENTS

We acknowledge with gratitude the support that we have received from a number of institutions. Silvia Donati started her investigation of Geoffrey of Aspall's Questions on the *Physics* as part of a research project on the medieval reception of Aristotle's *Physics* (1989–1993), financed by the Deutsche Forschungsgemeinschaft and hosted by the Thomas Institut of the University of Cologne. She then continued her research under the aegis of the Thomas Institut, which she wishes to thank for its friendly hospitality and support. Constant support was also offered by her home institution, the Albertus Magnus Institut in Bonn. Cecilia Trifogli greatly benefitted from the financial support of the Alexander von Humboldt Stiftung during the initial stages of her research (1995–1996). In more recent years, four terms of sabbatical leave (Michaelmas terms 2009, 2011, 2013, 2016) granted by the University of Oxford, her home institution, have allowed her to focus on this project. It was during those leaves that she carried out a substantial part of the work.

A number of people have helped us with specific issues relating to translation, sources, and textual matters: Charles Burnett, Giovanni Catapano, Iacopo Costa, Matteo Di Giovanni, Sten Ebbesen, Gabriele Galluzzo, Lauge Nielsen, Giorgio Pini, Justin Stover, Martin Tracey. We warmly thank them all. Concetta Luna merits special mention. It is on her philological expertise that we have relied on a great number of occasions.

Finally, we owe a deep debt of gratitude to the late Francesco Del Punta, who taught us the skills of editing medieval philosophical texts and constantly encouraged us to pursue this kind of work. It is to his memory that this book is dedicated.

INTRODUCTION

1. Life and works of Geoffrey of Aspall

Geoffrey of Aspall was an eminent Aristotelian commentator, active at the Faculty of Arts of Oxford around the middle of the thirteenth century. The exact dates of his academic life are not known with certainty. In all probability his scholarly activity and production ended in 1263, when Aspall started a very busy career at the court of Edward I as financial clerk, which lasted until his death on 11 June 1287. The beginning of his academic life is more difficult to establish. In her pioneering study of Aspall's commentaries on Aristotle, Macrae suggested a date as early as 1243, and thus fixed the limits of Aspall's scholastic activity within the twenty years between 1243 and 1263.[1] On the basis of the evidence provided by Aspall's relationships with some of his contemporaries (John Peckham, Albert the Great, Robert Kilwardby, Adam of Buckfield) Plevano has revised Macrae's suggestion and pushed forward the beginning of Aspall's teaching to around 1250.[2] There is no direct evidence that Aspall's academic career took place in Oxford rather than in Paris, but a geographical example used by Aspall in one of his works saying that "many roads lead to London" allows one to infer that in all probability he taught at Oxford.[3]

Aspall lectured extensively on the Aristotelian *corpus* and in particular on the so-called *libri naturales* and on the *Metaphysics*. The list of works that can be ascribed either with reasonable certainty or with probability to Aspall includes commentaries on the following works of Aristotle: *Physics*, *De caelo et mundo*, *De generatione et corruptione*, *De anima*, *De sensu et sensato*, *De memoria et reminiscentia*, *De somno et vigilia*, *De longitudine et brevitate vitae* and *Metaphysics*.[4] There is evidence that Aspall also commented on the *Meteors* and on *De vegetabilibus*, but his commentaries on these two works have not yet been traced.[5]

[1] Macrae, 1968, 94–97. For further biographical details, see Emden, 1957, 1, 60–61.

[2] Plevano, 1993, 193–198.

[3] Macrae, 1968, 97. See also Appendix, Book VI, q. 28, n. 14: 'ad oxoniam contingit pluribus viis accedere'.

[4] More detailed lists of Aspall's commentaries, including the manuscripts in which they are preserved, are found in Macrae, 1968, 112–113, and Plevano, 1993, 186–189.

[5] Macrae, 1968, 111–112. Donati has recently suggested that Aspall's questions on book I of the *Meteors* are those preserved anonymously in manuscripts Cambridge, Peterhouse, 157, fols. 23v–34v, and Siena, Biblioteca Comunale degli Intronati, L.III.21, fols. 196r–224v. See Donati, 2012a, 253.

The sheer number of Aspall's commentaries on Aristotle indicates that Aspall should be considered a major representative of Aristotelianism at Oxford in the thirteenth century. Most of these commentaries, however, are transmitted anonymously in the manuscript tradition. The attribution to Aspall has been the result of a long and patient work of comparison between the anonymous works and those explicitly ascribed to him in the manuscripts, a work mainly carried out by Macrae in the 1960s[6] and more recently by Donati.[7] When Callus published his seminal paper on the 'Introduction of Aristotelian learning to Oxford' in 1943, he could ascribe to Aspall only a much smaller number of commentaries and some of them only very tentatively.[8] As we shall see in the next section, the commentary on the *Physics* is one of those for which the question of authorship arises.

Notwithstanding the substantial progress on the issue of Aspall's authorship since the publication of Callus' paper, some problems still remain. In particular, our editorial work on Aspall's commentary on the *Physics* raises doubts about Aspall's authorship of the anonymous commentaries on the *Metaphysics* and *De anima* traditionally ascribed to him,[9] since in Aspall's *Physics* commentary we have found references to a commentary on the *Metaphysics* and a commentary on *De anima* that are not verified by the two anonymous commentaries.[10] Aspall's *Physics* commentary also contains several references to a commentary on the *De memoria et reminiscentia* that are not verified by the anonymous commentary on this treatise considered as a possible work by Aspall. In this case, however, the lack of correspondence between the two works provides no reason against Aspall's authorship, given that only a fragment of the anonymous commentary on the *De memoria et reminiscentia* is preserved.[11]

[6] Macrae, 1968, 97–116.

[7] Donati, 2012a, 2013a.

[8] Callus, 1943, 46–49.

[9] The *Metaphysics* commentary is preserved in MS Cambridge, Gonville and Caius College, 509, fols. 52ra–123rb; the *De anima* commentary is preserved in MS Oxford, Merton College, 272, fols. 191ra–236rb, 237va–241rb. For the critical edition of the *De anima* commentary see below, n. 16.

[10] For the references to a *Metaphysics* commentary in Aspall's *Physics* commentary , see below, Book I, q. 39.1, n.7; q. 39.2, n.3; q. 65, n. 6; Book II, q. 7, n. 13; q. 15, n. 6; q. 16, n. 12; Appendix, Book I, q. 15, n. 5; q. 22, n. 9; Book II, q. 33, n. 1; Book VI, q. 6, n. 26. For the references to a *De anima* commentary, see below, Book II, q. 27, n. 24; Book III, q. 7, n. 1*; Appendix, Book I, q. 31, n. 16; Book II, q. 27, n. 25; Book VI, q. 10, nn. 9, 13, 21; q. 17, n. 17; q. 28, n. 14; q. 40, n. 4; q. 43, n. 7. For a discussion of these references, see Donati, 2013a, 316–317, 357–367. Doubts about the grounds for the attribution of the two anonymous commentaries to Aspall have been raised in Donati, 2012a, 251–255, and Donati, 2013a, 294–335, 348–367. However, until a more conclusive assessment of Aspall's authorship of these two anonymous commentaries is achieved, we will follow the traditional attribution to Aspall in the present edition.

[11] The anonymous commentary on *De memoria et reminiscentia* is preserved in MS Oxford, New

Aspall's commentaries belong to what Callus has classified as the third and last phase of the reception of Aristotle's learning in Oxford. These phases reflect different degrees of assimilation of Aristotle's thought and correspondingly different literary forms of the works on Aristotle.[12] The first phase is that of the paraphrases, that is, treatises that expound the doctrine of Aristotle on a given subject by way of a paraphrase. The treatise of John Blund on the soul is a paradigmatic example of a work belonging to the first phase.[13] The second phase is that of the literal exegesis of Aristotle's texts, either in the form of marginal glosses or of commentaries *per modum commenti*. Adam of Buckfield's literal commentaries are paradigmatic cases of works belonging to this phase. The third phase is that of the questions on Aristotle's text, that is, commentaries *per modum quaestionis*, which do not contain a systematic exposition of Aristotle's text but consist of questions in which problems arising out of the text or connected with it are discussed. Almost all of Aspall's works on Aristotle, and in particular the *Physics* commentary edited here, are commentaries *per modum quaestionis* and thus belong to this third phase. In addition to the commentaries *per modum quaestionis* on the *Physics*, *De caelo et mundo*, and *De Generatione et corruptione*, the manuscript tradition also ascribes commentaries *per modum commenti* on these works to Aspall, but his authorship of the first and third of such commentaries is not certain.[14]

The structure of Aspall's questions on Aristotle's works has been described in detail by Macrae.[15] The general structure is the standard one: the problem under discussion is stated in the title of the question; its actual discussion in the body of the question begins with arguments in favour of the view that is ultimately to be rejected; these arguments are then followed by arguments in favour of the contrary view (*ad oppositum*). Then comes the *solutio*, which contains Aspall's own reply to the question, followed by the reply to the arguments in favour of the contrary view. Within this general

---

College, 285, fols. 189rb–193rb. For the references to a commentary on *De memoria et reminiscentia* in Aspall's *Physics* commentary, see below, Book I, q. 40.2, n. 11; Book III, q. 7, n. 18; Appendix, Book VI, q. 5, n. 6. For a discussion of these references, see Donati, 2013a, 266–269, 335–336, 367–368.

[12] Callus, 1943, 38–39. On the last two phases distinguished by Callus, see also the detailed presentation by Macrae, 1968, 116–126. For some remarks on the chronology of these phases, see Plevano, 1993, 181–183.

[13] Blund's treatise has been recently re-edited with an English translation in John Blund, *Treatise on the Soul*, (Auctores Britannici Medii Aevi II), ed. by D. A. Callus & R. W. Hunt, with a new Introduction and English translation by M. W. Dunne, British Academy (Oxford: 2013).

[14] Donati, 2012a, 252. Aspall's *Physics* commentary contains a possible reference to a literal commentary on the *Physics* by Aspall himself (see Book I, q. 14, n. 4) which is not verified by the one ascribed to him.

[15] Macrae, 1968, 126–127.

structure, however, there are many variations both in the specific form that an individual question may take as well as in its length. One important general characteristic of Aspall's questions concerns the *solutio*: while in later commentaries *per modum quaestionis* the *solutio* is the *locus classicus* where the commentator expands and argues for his own view, this is not so in Aspall's questions, where the *solutio* is almost never given at great length, but is either stated in one sentence or may simply be implied by the replies to the initial arguments. The crucial elements of Aspall's view on a given problem are then to be found in his replies to the initial arguments, in which he addresses the objections to his own view. As a result, Aspall's treatment of the problem raised in the question is fragmentary and the difficult task of reconstructing his view from the scattered evidence provided by his replies to the arguments is left to the reader. It is especially with this difficulty in mind that in the last section of this Introduction we give an extensive outline of Aspall's discussion of the major topics treated in his questions on the *Physics*.

Aspall's other commentaries on Aristotle are still unedited, with the exception of his questions on *De anima*, edited by Čizmić, and his questions on *De somno et vigilia*, edited by Ebbesen.[16]

2. The Questions on the Physics

The questions that in the current state of research can be ascribed with reasonable certainty to Aspall are not a complete set of questions on the eight books of Aristotle's *Physics*. They cover books I–IV, book VI, and book VIII, that is, all the books except books V and VII. Furthermore, the extant questions on book IV are just a fragment, containing questions on the first topic of book IV, that is, place, but no questions on the two topics that follow, that is, void and time. Here are the details:

(i) The questions on books I–IV are explicitly ascribed to Aspall in one of the manuscripts preserving these questions: Oxford, Merton College, 272, fols. 88r–112v, where in the top left-hand corner of fol. 88r an early hand writes "secundum magistrum Galfridum de Haspyl".[17]

[16] Galfridus de Aspall, *Questiones in De anima*, ed. V. Čizmić (Munich: 2010); S. Ebbesen, 'Geoffrey of Aspall, *Quaestiones super librum De somno et vigilia*. An Edition', *Cahiers de l'Institut du Moyen Âge Grecque et Latin*, 83 (2014), 257–341 (available on line). In addition to these two series of questions, a selection of 15 questions from Aspall's commentary on the *Metaphysics*, Book VII, have been edited in Galluzzo, 2013, 47–81; a question on the *Metaphysics* about the change between contradictory states has been edited in Plevano, 1993, 227–232; one question on *De Caelo* about the eternity of the world has been edited in Galle, 2012, 349–355.
[17] See below, 3.1.

(ii) The questions on books I–IV in the Oxford manuscript are followed by a set of questions on book VIII (fols. 113r–118v). The problem of Aspall's authorship of the latter set of questions had been left open by Callus.[18] Macrae has taken it up again and showed that they can be ascribed with reasonable certainty to Aspall.[19] Macrae's conclusion has been corroborated by Donati.[20]

(iii) As to the questions on book VI, it was already known to Macrae that Aspall's questions on the *Physics* should also have contained questions on this book,[21] but the actual discovery of these questions is much more recent and due to Donati, who has found them, preserved anonymously and in two distinct redactions, in manuscripts Todi, Biblioteca Comunale, 23, fols. 59v–74r, and London, Wellcome Historical Medical Library, 333, fols. 81r–92r. Donati has established Aspall's authorship on the basis of a very accurate evaluation of the references found in these two sets of questions to the questions on *Physics* I and of the references found in other commentaries that are certainly by Aspall (Questions on *De generatione*, *De caelo*, and *De longitudine et brevitate vitae*) to the questions on *Physics* VI, and also of textual parallelisms.[22]

(iv) Immediately before the questions on book VI, the Todi manuscript contains a set of anonymous questions on book V of the *Physics* at fols. 57r–59v. Donati has carefully investigated the hypothesis of Aspall's authorship for this set of questions but she has not found conclusive evidence for it.[23]

(v) The fact that a question in book VI on the continuity of time deals with topics that in the Aristotelian tradition are *ex professo* discussed in connection with Aristotle's treatment of time in *Physics* IV (such as the existence of time, the dependence of time on the soul, and the unicity of time)[24] can be taken as an indication that Aspall's questions on book IV were originally more complete, also containing some questions on time, and not just the fragment about place that is known at present. The comparison with the authentic fragment preserved by the Oxford manuscript, however, shows that the complete set of Aspall's questions on book IV should not be identified with those contained in the Todi manuscripts at fols. 39v–57r.[25]

[18] Callus, 1943, 47.
[19] Macrae, 1968, 103–104.
[20] Donati, 1991, 422–424.
[21] Macrae, 1968, 111.
[22] Donati, 2013a, 219–289
[23] Donati, 2013a, 289–293. These questions on *Physics* V have been edited in the Appendix, pp. 1243–1260.
[24] Book VI, q. 12.
[25] Trifogli, 1996, 67–72.

(vi) There are no indications that Aspall discussed questions on book VII.

The questions on the *Physics* have come down to us in different versions, that is, in texts transmitting an exegetical and doctrinal content that is essentially the same but in different linguistic formulations. As we shall see in the section on the manuscript tradition,[26] any two distinct manuscripts of Aspall's questions on the *Physics* to some extent represent distinct versions. The presence of different versions has also been found for other works by Aspall, such as his questions on *De generatione et corruptione*[27] and those on *De caelo*,[28] and it is likely that future research will discover many other cases. The problem of the nature and origin of the different versions has been investigated by Donati. In her view, one hypothesis that can safely be ruled out is that the different versions are different redactions of the same work by Aspall himself. Much more likely is the hypothesis that different versions of one and the same commentary are, at least in part, the result of different *reportationes*, so that the common origin of different versions is a series of lectures given by Aspall on the relevant Aristotelian treatise, but the original material presented by Aspall in his lectures has come down to us in different reports compiled by his students. In addition to the hypothesis of different *reportationes*, we must also assume that the material collected by Aspall's students in their reports has been subsequently subjected to a number of editorial interventions in the course of its transmission, without being finally revised by Aspall himself.[29]

This kind of transmission of Aspall's questions on the *Physics* makes it very hard to establish its relative chronology. The problem is that the internal references to other works by Aspall found in his questions on the *Physics* lose their evidential value since they cannot be ascribed with certainty to Aspall himself, but could well have been added by an editor eager to point out the connection between Aspall's treatments of a given topic in different works.[30] Furthermore, for the time being we have not found indications that can help to establish the absolute chronology of Aspall's questions on the *Physics*.

The main sources that Aspall explicitly acknowledges are Aristotle, Averroes, and Augustine, in this order. Aspall is familiar with a vast part of the Aristotelian *corpus*, which extends beyond the works on which he commented: in his questions there are references not only to the *Metaphysics*

[26] See below, 3.2.
[27] Donati, 2012a, 257–264.
[28] Galle, 2012a, 322–328.
[29] Donati, 2013a, 369–389.
[30] Donati, 2013a, 390–399.

and the *libri naturales* but also to the logical works (e.g., *Prior* and *Posterior Analytics*, *Categories*, *De interpretatione*), on which he did not comment.

The numerous references to Averroes' commentary on the *Physics* show that Aspall heavily relies on Averroes' detailed exegesis for his understanding of Aristotle's text. He reads Averroes carefully and clearly regards Averroes as the Aristotelian *auctoritas*, whose interpretations of Aristotle should be taken into account. This does not mean, however, that Aspall is willing to accept all Averroes' views. On the contrary, like many other English commentators of his time, Aspall's attitude towards Averroes is very critical.[31] He does not hesitate to use harsh expressions like "falsum dicit"[32] and "falsum intellexit"[33] with reference to Averroes. Aspall does not consider Averroes a good interpreter of Aristotle's thought. He is puzzled by Averroes' views on some basic issues regarding motion, time, and place, and rejects them.[34]

Aspall apparently regards Augustine as the 'theological' *auctoritas*, so to say, the author whose views in theological matters are authoritative and should be taken into account. Aspall's attitude towards Augustine too, however, tends to be critical. For example, he devotes a specific discussion to Augustine's view on fate and openly criticizes some crucial aspects of it.[35]

The identification of the anonymous sources of Aspall's questions on the *Physics* is a difficult task that has not been systematically undertaken for the present edition. Roger Bacon and the author of the *Physics* commentary ascribed to Richard Rufus[36] are likely to be included among them, since Aspall quotes their opinions about several issues,[37] but whether he has a direct acquaintance with their works on the *Physics* and to what extent he uses them is still to be investigated. A fine-grained comparison between Aspall's questions on book III and the questions on the same book in the other English commentaries *per modum quaestionis* of his time has shown that a good number of arguments and opinions quoted in Aspall's questions

[31] On the Anti-Averroism of the English commentators on the *Physics* at Aspall's time see Trifogli, 2000, 47–59, 167–171, 192–197, 219–230, 240–246, 271–273.

[32] Book VI, q. 12, n. 23.

[33] Book III, q. 5, n. 6.

[34] See below, 5.6.2, 5.9.5, 5.11.1.

[35] Book II, qq. 25–27. See also below, 5.5.2.2, 5.5.2.3.

[36] The authorship of this commentary, published in ABMA XVI (as well as that of the commentary on *De generatione et corruptione* published under Rufus' name in ABMA XXI) is a matter of controversy among scholars. For arguments in support of Rufus' authorship, see Richard Rufus of Cornwall, *In Phys.*, 41–74; Wood, 2009; Richard Rufus of Cornwall, *In De gen. et corr.*, 17–54. For a critical review of those arguments, see Donati, 2005.

[37] Parallel passages with those two commentaries have been found at Book I, q. 12, n. 4; q. 17, n. 6; q. 32, n. 5; q. 44, n. 6; q. 49, n. 22; q. 79, n. 8; Book II, q. 22, n. 4; Book VI, q. 5, nn. 1–3; q. 12, n. 21.

are also present in other commentaries of the group but has not revealed a direct dependence of Aspall's questions on any of them.[38]

An indirect but very pervasive presence in Aspall's questions is that of Robert Grosseteste and in particular his theory of light and of intentional species. Indeed, Aspall devotes many, sometimes very long, discussions to a variety of issues connected with light. The most likely remote sources of such discussions are Grosseteste's treatises *De Luce* and *De lineis, angulis, et figuris seu de fractionibus et reflexionibus radiorum*.[39] By Aspall's time, however, Grosseteste's ideas expressed in these treatises had become a fundamental ingredient of the exegesis of Aristotle's natural philosophy. In most of the opinions connected with light reported by Aspall Grosseteste's original ideas are applied to specific notions of Aristotle's *Physics* and are intended to provide an explanation of these notions. A very good example of this kind of reading of Aristotle's *Physics* is an opinion that identifies the nature that Aristotle defines in book II as an internal principle of the motion and rest of a natural thing with the light of the celestial bodies diffused throughout the sublunar world.[40] Another major example is the appeal to the theory of intentional species in the account of Aristotelian substantial changes.[41] Obviously, Grosseteste's metaphysics of light is extraneous to Aristotle's natural philosophy, and the mere fact that Aspall reports so many readings of Aristotle that were heavily influenced by Grosseteste's ideas clearly shows that his questions on the *Physics* represent an early phase of the reception of Aristotelian learning in which Aristotle's thought has not yet been fully assimilated.[42] Aspall himself, however, does not passively accept this extraneous material but critically assesses it and this leads him to distance himself from at least part of it. For example, while he is still fully immersed in the species-account of substantial changes, he rejects the opinion that identifies the Aristotelian nature with the light of the celestial bodies mentioned above.[43]

[38] Trifogli, 2004, 169–187. A fine-grained comparison between Aspall's questions on books I and II of the *Physics* and those of the other contemporary English commentaries will be carried out by Silvia Donati within her research project on the early English tradition of *Physics* I–II, which will lead to the edition of several commentaries.

[39] The two treatises are edited in *Die Philosophischen Werke des Robert Grosseteste, Bischofs von Lincoln*, ed. L. Baur, Aschendorff (Münster: 1912), 51–59, 59–65.

[40] See below, 5.4.1.

[41] See below, 5.4.1. For a comprehensive study of the application of intentional species to natural causality in the English commentaries on Aristotle's *libri naturales*, see Donati, 2015b.

[42] On the earliness of the exegetical tradition to which Aspall's questions on the *Physics* belong, see Trifogli, 2000, 263–268.

[43] See below, 5.4.1.

Aspall's approach to Aristotle's *Physics* is, so to say, 'interdisciplinary': in addition to dealing *ex professo* with topics specific to the *Physics*, he does not hesitate, when the occasion arises, to devote extensive discussions to topics that specifically belong to other parts of the Aristotelian *corpus*. Examples of topics not specific to the *Physics* to which Aspall pays great attention are the notions of part and whole,[44] separate substances,[45] the soul,[46] sensory and intellectual cognition,[47] animals,[48] and substantial change.[49] Furthermore, in spite of the fact that Aspall did not comment on Aristotle's logical works, significant treatments of logical issues are found in his questions.[50]

3. Manuscript tradition and editorial principles

3.1. Manuscripts

Aspall's questions on the *Physics* are preserved in four manuscripts: a Cambridge manuscript (C), a London manuscript (L), an Oxford manuscript (O), and a Todi manuscript (T), which are all rich collections of commentaries on Aristotle. None of these manuscripts is a complete copy, in the sense of preserving all the questions that have been ascribed to Aspall (books I–IV, VI, VIII). T is the most complete: it contains all the questions except the fragment of questions on book IV; next comes O, which contains the questions on books I–IV, VIII but not those on book VI; finally, C and L preserve complementary sets of questions: C only contains those on books I–IV and L only those on books VI and VIII. A small fragment of Aspall's questions on book I is preserved in another Oxford manuscript (N).[51]

[44] See Book I, qq. 36–37; Book VI, qq. 9–10.
[45] See Book VIII, qq. 39–41.
[46] See Book VIII, qq. 21–24.
[47] The prologue of the *Physics* (Book I, chapter 1) is the *locus classicus* for raising issues about cognition in *Physics* commentaries, and indeed Aspall's questions on the prologue (Book I, qq. 1–21) contain many such discussions. But there are also many other places in which such issues are raised. See, for example, Book II, qq. 28–30, on divine knowledge; Book III, q. 7, on the cognition of motion; Book VIII, qq. 18–19, on the formation of sentences, and q. 28, on science as a quality of the soul; Book VI, qq. 2–3, on the cognition of continua and of substances.
[48] See Book II, qq. 32–43.
[49] See Book VI, qq. 14–17.
[50] See Book I, q. 29, on the significate of concrete names, and q. 53, on the names for matter; Book II, q. 24, on the truth-conditions of propositions about future contingents; Book VI, q. 32, on the distinction between the antecedent and the consequent of an inference, qq. 45–46, on the verbs 'begins' and 'ceases', and q. 51, on the universal quantifier 'every'.
[51] Zimmermann (1971, 25–26) also lists manuscript London, British Library, Royal 12.E.25 (English origin, end of the thirteenth century) as a further witness of Aspall's commentary on the *Physics*, but this manuscript has turned out to be a witness of the treatise *De principiis naturae* by the English Dominican William Rothwell. Cf. Shooner, 1973, 244–246. Zimmermann,

C= Cambridge, Gonville and Caius College, MS 509/386[52]
English manuscript of the second half of the thirteenth century. On the front pastedown a table of contents from the fifteenth century. 302 leaves, 280 × 195 mm., 2 cols. of 55 lines. Written in small neat anglicana hands of scholastic type. Red or blue initials flourished in the other colour; red or blue paraphs. On the front pastedown: "Ex dono Pulham". On fol. 302v: "Magister Ricardus de Pulham dedit librum istum ad collegium".

Aspall's Questions on *Physics* I–IV are preserved anonymously at fols. 124ra–155rb. *Incipit*: "Primo quaeritur quid sit abstrahere". *Explicit*: "et hoc quia in caelo non est contraria indigentia loci".

This manuscript also preserves other works by Aspall: (1) Questions on *Metaphysics* I–X (fols. 52ra–123rb); (2) Questions on *De longitudine et brevitate vitae* (fols. 276ra–286rb); (3) Questions on *De sensu et sensato* (fols. 287ra–302rb).

L= London, Wellcome Historical Medical Library, MS 333[53]
English manuscript of the first half of the fourteenth century. Two booklets. 152 leaves. Fols. 1–47 and 57–152: 290 × 220 (205 × 157) mm., 2 cols., 50 lines. Fols. 48–56: c. 275 × 200 (205 × 145) mm., 2 cols., 52 lines. Small textura hands, with many abbreviations. Col. initials, red paraphs. Gold-tooled leather binding (sec. XIX). A later reader has numbered the *quaestiones* throughout and made marginal notes, as well as supplying the table of *quaestiones* at fols. 151v–152v. On fol. 150 (visible under UVL): <"Iste liber deputatus est ad usum fratris de [...] Iunioris">. Purchased 1910.

Aspall's questions on *Physics* VI are preserved anonymously at fols. 81ra–92ra. *Incipit*: "*Si autem est continuum* etc. Quaeritur incidenter an materia cum potentia activa sit tota substantia rei compositae". *Explicit*: "et ita non distribuit pro omnibus punctis. Praeterea cuius actus". Those on *Physics* VIII are preserved anonymously at fols. 98ra–105va. *Incipit*: "Quaeritur primo an sit motus primus". *Explicit*: "primum motum plus faciat ad generationem hic inferiorum quam orbes inferiores".

---

however, was right in pointing out the close relationship of the treatise in the London manuscript to Aspall's commentary, since most of this treatise consists of literal excerpts from the section of Aspall's commentary on *Physics* I devoted to the three principles of nature (matter, privation and form). Cf. Donati, 2011.

[52] We use the descriptions in Thomson, 2013, 85–86, and Zimmermann, 1971, 12–14, supplemented by Weijers, 1998, 31–35.

[53] We use the description in Robinson, forthcoming, supplemented by Donati, 1993, 54–75, and Weijers, 1998, 31–35.

N= Oxford, New College, MS 285[54]

English manuscript of the thirteenth century. Structurally some eight booklets, presumably together early. 255 leaves (ii+253). 270 × 185 mm., the edges savagely retrimmed and spattered with red. Written mainly in anglicana hands.

The fragment of Aspall's Questions on *Physics* I is contained anonymously at fols. 114ra–115vb. *Incipit*: "[Q]*uoniam quidem intelligere* etc. Quaeritur de unitate scientiae naturalis". *Explicit*: "nec etiam erit devenire ad materiam primam. Et dicunt".

This manuscript also contains other works by Aspall: (1) Questions on *De generatione et corruptione* (fols. 38ra–57vb); (2) Questions on *De sensu et sensato* (fols. 164ra–189rb). Possibly by Aspall is also the collection of Questions on *De memoria et reminiscentia* (fragm.) preserved anonymously at fols. 189rb–193rb.

O= Oxford, Merton College, MS 272/304[55]

English manuscript of the end of the thirteenth century. Constructed as booklets corresponding to single works or groups, nonetheless mostly if not entirely together from the first. 302 leaves (iv+298); i–ii, 299–301 paper binding leaves. On iii v is a table of contents (without authors' names) from around 1500, including three items now lacking from the end: Questions on *De caelo*, *De memoria*, and *Metaphysics*. 285 × 210 mm., the edges drastically retrimmed. 2 cols., 49–65 lines. Written mainly in small, neat early anglicana hands, lemmata in gothic rotunda, written larger. Red and blue flourished initials; blue initials flourished in red, red or blue paraphs.

Aspall's questions on *Physics* I–IV are contained at fols. 88ra–112vb. They are ascribed to Aspall: at the head of fol. 88r "secundum magistrum Galfridum de Haspyl" in an early hand. *Incipit*: "*Quoniam autem oportet intelligere*… Quaeritur de unitate scientiae naturalis". *Explicit*: "per comparationem ad locum erit unum praedicamentum et per comparationem ad locatum erit aliud". The questions on *Physics* VIII are preserved anonymously at fols. 113ra–118vb. *Incipit*: "*Utrum autem factus sit aliquando motus, cum non esset prius, et corrumpitur iterum sicque moveri non sit aut neque* etc. Quaeritur primo an sit motus primus". *Explicit*: "Et similiter patet quod ratio eorum fuit nulla".

[54] We use the descriptions in Thomson, 2011, 365–368, and Zimmermann, 1971, 33–34, supplemented by Weijers, 1998, 31–35.

[55] We use the descriptions in Thomson, 2011, 320–323 (but we follow the first folios numeration), and Zimmermann, 1971, 28–31, supplemented by Weijers, 1994, 24–30, and Weijers, 1998, 31–35.

This manuscript also preserves several other works by Aspall: (1) Questions on *De generatione et corruptione* (fols. 175ra–190Ara); (2) Questions on *De anima* I–III (fols.191ra–236rb, 237va–241rb); (3) Questions on *De sensu et sensato* (fols. 254ra–273vb) (4) Questions on *De somno et vigilia* (fols. 274va–281va); (5) Questions on *De longitudine et brevitate vitae* (fols. 282rb–294vb). Possibly by Aspall (or by Adam of Buckfield) also is the literal commentary on *De generatione et corruptione* preserved anonymously at fols. 25ra–36vb.

T= Todi, Biblioteca Comunale, MS 23 (163)[56]
Second half of the thirteenth century.[57] 274 leaves (I, 272, I'). 324 × 227 mm. Two hands: hand A at fols. 1r–167v; hand B at fols. 168r–272v.[58] In the lower margin of fol. 168r a contemporary hand writes: "Isti quaterni sunt fratris H. de B(er)h(a)m", and "Isti .ix. quaterni sunt in custodia fratris Henrici de B(er)h(a)m".[59]

Aspall's questions on *Physics* I–III are preserved anonymously at fols. 1ra–39vb. *Incipit*: "<Q>uaestio est quid sit abstrahere. Dicendum quod abstrahere non est absolvere aliquid". *Explicit*: "Item Aristoteles loquitur de abstractione et lux non abstrahitur sed multiplicatur". The questions on *Physics* VI are preserved anonymously at fols. 59va–74ra. *Incipit*: "*Si autem est continuum et quod tangitur*. Primo quaeritur de continuo". *Explicit*: "sed mobile motu recto ita se habet quod omnes partes eius aequaliter moventur et aeque velociter; ergo multo fortius sic erit in motis motu circulari". The questions on *Physics* VIII are preserved anonymously at fols. 74rb–82vb. *Incipit*: "<S>uper VIII Physicorum quaeritur an sit motus primus et videtur quod non". *Explicit*: "Item, planetae secundum quod ordinantur sub diversis signis per hoc attribuitur eis".

This manuscript also contains several other works by Aspall: (1) Questions on *De longitudine et brevitate vitae* (fols. 84ra–93va); (2) Questions on *De sensu et sensato* (fols. 99vb–123ra); (3) Questions on *De caelo et mundo* (fols. 168ra–240vb); (4) Questions on *De generatione et corruptione* (fols. 241ra–272va). Also possibly by Aspall is the fragment of a commentary on the *Meteora* (prologue) preserved anonymously at fols. 163ra–167va.

[56] We use the description in Menestò, 2008, 1516–1522, supplemented by Louis Jacques Bataillon in Macrae, 1968, 132–133, Weijers, 1998, 31–35, and Donati, 2012a, 246–257.

[57] Probably early fourteenth century, according to Bataillon in Macrae, 1968, 132.

[58] According to Menestò, 2008, 1517, both hands are probably French, whereas according to Bataillon in Macrae, 1968, 132, the manuscript was written in England.

[59] According to Bataillon in Macrae, 1968, 133, the reference is to the village of Barham, near Canterbury. Menestò, 2008, 1518, however, reads "B(er)h(ei)m".

3.2. Quality and relationships of the manuscripts

We discuss separately sets of questions preserved in distinct sets of manuscripts.

3.2.1. Books I–III (O, T, C)

Each of the three manuscripts has a very high number of isolated readings that are not errors but variants, that is, they all result in acceptable texts, which often transmit what is essentially the same conceptual content but formulated in slightly different ways. These variant readings affect significant portions of Aspall's questions on *Physics* I–III. We here offer two examples taken from the questions on *Physics* II.

Example 1:[60]

O: Consequenter determinat Aristoteles de causis. Et primo de causis per se, secundo de causis per accidens. De causis per se dicit primo quod quattuor sunt causae, et dat tres proprietates. Prima est quod multae possunt esse causae eiusdem causati. Secunda est quod idem potest esse causa contrariorum. Tertia est quod aliqua possunt esse causa sibi invicem.	T: Consequenter determinat Aristoteles de causis (et primo de causis) per se, secundo de causis per accidens. De causis per se dicit primo quod quattuor sunt causae et dat tres proprietates. Prima est quod multae possunt esse causae eiusdem causati. Secunda est quod idem potest esse causa contrariorum. Tertia quod aliqua possunt esse sibi invicem causae.	C: *Determinatis autem his* etc. Capitulum de causa in quo primo determinat quae et quot sunt causae dicens quod sunt quattuor.
Cum Aristoteles determinet de causis per se et de causis per accidens, primo quaeritur de causis per se, circa quas quaeruntur quattuor quaestiones. Prima quaestio est de numero earum, secunda autem quaestio et tertia et quarta sunt circa tres proprietates ipsarum quas assignat Aristoteles.		
Quaeritur ergo primo an quattuor sint causae sicut narrat Aristoteles. Et videtur quod non, quia principium et causa idem...	Circa primum sic: videtur quod tantum tres sunt causae, quia causa et principium idem...	Et quod sint tantum tres probatio: causa et principium idem...

This passage is the beginning of the series of questions about Aristotle's discussion of the causes. The initial section of O is present in essentially the same formulation in T (the part in brackets is a marginal addition of T, which gives the reading of O) but in a very different and abbreviated form in C,

[60] Book II, q. 17, nn. 1*–1. The text of T and C is printed in the Appendix.

where it is prefaced by the Aristotelian lemma (printed in italics), absent in O and T. The second section of O is omitted by both C and T. Finally, in the third section, the first question about the number of causes begins with different formulations in the three manuscripts.

Example 2:[61]

O: <1> Item, duplex est intellectus, scilicet agens et possibilis; aut ergo tota anima dicitur intellectus agens aut pars animae aut intelligentia.	T: <1> Item, intellectus est duplex in homine, scilicet agens et possibilis. Agens autem, secundum Aristotelem, est omnia intelligens in actu, possibilis autem nihil intelligit nisi prius illuminetur ab agente.	C: <1> Item, in homine est duplex intellectus: agens et possibilis. Agens autem, secundum Aristotelem, est omnia intelligens in actu, possibilis autem nihil intelligit nisi prius illuminetur ab agente.
<2> Sed non tota, <a> tum quia intellectus agens omnia intelligit, intellectus possibilis nihil -tunc, si tota anima esset etc., sequeretur quod anima simul omnia intelligeret et nihil, quod est inconveniens. <b> Tum quia tunc esset unum et idem in ratione dantis sive agentis et patientis simul et semel, quod est inconveniens. Consequentia patet, quia intellectus agens est in ratione dantis et agentis, quia illuminat species receptas in imaginativa, et intellectus possibilis est in ratione recipientis. Recipit enim species illuminatas ab agente. <c> Tum quia, cum semper praesens sit illud quod illuminat, si sit tota anima, tunc semper intelligeret, quod est inconveniens.	<2> Sed quod intellectus agens sit intelligentia, et ita habeo propositum, videtur, quia non est tota anima, <b> quia, si sic, tota anima est agens et patiens, dans et recipiens, et illuminans et iluminatum, quod est inconveniens. <d> Item, illuminans aliud est distans essentialiter ab eodem; sed tota non distat a tota; quare etc.	<2> Sed probatio quod intellectus agens sit intelligentia, et ita habeo propositum, quod intellectus agens non est tota anima. <a> Si enim tota esset agens et ita actu omnia intelligens, et tota possibilis et ita nihil intelligens, tunc idem esset omnia intelligens et nihil intelligens. <b> Et praeter hoc agens est in ratione dantis, agentis et illuminantis, possibilis autem est in ratione patientis et recipientis. Si ergo tota esset agens et tota possibilis, tunc idem esset agens, dans, recipiens, illuminans, illuminatum. <d> Et praeter hoc, illuminans aliud est distans essentialiter ab eodem; sed tota non distat a tota; quare etc.
<3> Et eadem ratione nec erit intellectus agens pars animae, quia, cum illa pars animae semper sit praesens, tunc semper intelligeret.	<3> Eadem ratione nec est intellectus agens pars animae, scilicet pars formalis, quia, si esset, cum agens illuminet possibilem, tunc, cum pars formalis semper sit praesens, semper illuminaret, et ita possibilis semper intelligeret, quod falsum est.	<3> Nec est dicere quod intellectus agens sit pars animae, quia, si sic, cum agens illuminet possibilem, cum pars formalis sit semper praesens, semper illuminaret, et ita possibilis semper intelligeret, quod falsum est.

This passage contains an argument that proves that the agent intellect is a thing separate from the human soul. The formulation of section <1> is essentially the same in T and C but different in O, which omits the explanation of the notions of agent and possible intellects. Unlike T and C, however,

[61] Book II, q. 27, n. 4. The text of T and C is printed in the Appendix.

O adds a claim about the three candidates for being the agent intellect. Section <2> has different formulations in each of the three manuscripts, but the formulation of C is much more similar than that of T to that of O. For section <3> the formulations of C and T are similar and more expanded than that of O, but there is a redactional difference between C and T at the opening of the section: T, like O, writes "(Et) eadem ratione nec" whereas C writes "Nec est dicere".

It is obvious that differences between the three manuscripts like those illustrated by these two examples cannot be explained by assuming that any two of them derive from a common model by a simple process of copying. We should rather think that they result from deliberate editorial interventions and that therefore each manuscript to some extent transmits a different 'version' of Aspall's questions. On the whole, however, C and T are closer to each other than to O. Moreover, while the degree of distinction between T and C remains essentially the same throughout books I–III, there are extensive portions of text in which both T and C differ to a much higher degree from O than in others. This happens in more than half of the questions on book I (qq. 1–38, 42–48) and around one third of the questions on book II (qq. 17–20, 22, 27–36). For these portions of text too, the three manuscripts present essentially the same conceptual content, articulated in the same way, but its presentation diverges more radically in the sense that the textual differences between O and T/C are present throughout a question. It should also be noted that the redactional differences between T and C are no less deep in these portions of text than in the others. More explicitly, it is not the case that T and C testify to precisely the same text against O; it is just the case that their divergence from O becomes wider in some portions of the text. Furthermore, if we take O as the version of reference and consider the whole extent of the questions on books I–III, we can note that the version transmitted by C differs from that transmitted by O more widely than that transmitted by T: redactional changes that result in a very different formulation of an argument in O are relatively more frequent in C than in T, as can be gathered from the *apparatus criticus* of our edition.

The problem of the origin of the three different versions of Aspall's questions is a complex one and difficult to answer in all its aspects. One safe assumption is that they ultimately originate from one or more *reportatio* of Aspall's lectures on the *Physics*, that is, written notes of Aspall's lectures taken by students rather than produced by Aspall himself. The first relevant question is whether there is only one or more than one series of lectures at the origin of the different versions. Although we cannot establish this point beyond any doubt, it seems very likely that only one series of lectures was involved. The reason for this is that the kinds of differences that would

justify the hypothesis of different series of lectures, such as those in the content, the order, and the internal articulations of the questions, are almost absent from the three versions. Granted this, the next question then is whether the three versions originate from only one or from more than one *reportatio* of the same series of lectures. To answer this question, we need a criterion to distinguish between linguistic differences that correspond to distinct *reportationes* and those that correspond to redactional interventions operated on one and the same *reportatio*. A reasonable criterion is based on the extent of such differences: (i) when these differences affect an extensive portion of text (one question or a series of questions) throughout so that they affect each single section of that text, then the differences correspond to distinct *reportationes*; on the contrary, (ii) when the differences affect only some isolated passages or sentences within a question, then the differences represent redactional interventions on one and the same *reportatio*. Now, our three versions are such that:

(1) In the whole of the extensive text that they have in common versions C and T only have isolated differences (case ii).

(2) In the extensive portion of text corresponding to Book I, qq. 1–38, 42–48 and Book II, qq. 17–20, 22, 27–36, version O differs from versions C and T throughout (case i).

(3) In the rest of the questions on Books I–III, version O has only isolated differences from versions C and T (case ii).

Given our criterion, from (1) we infer that there are no distinct *reportationes* corresponding to versions C and T respectively. From (2) we infer that there are two different *reportationes* involved in the productions of the three versions, call them *reportatio* A and *reportatio* B. From (3) we infer that it is not simply the case that one *reportatio*, say A, corresponds to version O and the other *reportatio*, B, correspond to versions C and T; for if these were the case, the differences between version O and versions CT should be present throughout, contrary to (3). When we consider the three versions in their entirety, therefore, what we also need to assume is that a conflation has occurred between *reportationes* A and B, so that either (a) versions C and T or (b) version O result from the conflation of materials originally contained in the two *reportationes*. Call this conflated text *y*. Moreover, given that throughout the entirety of their texts each of the three versions presents individual differences affecting isolated passages, we also need to assume that each of them comes from its remote source through a stratification of subsequent redactional interventions.

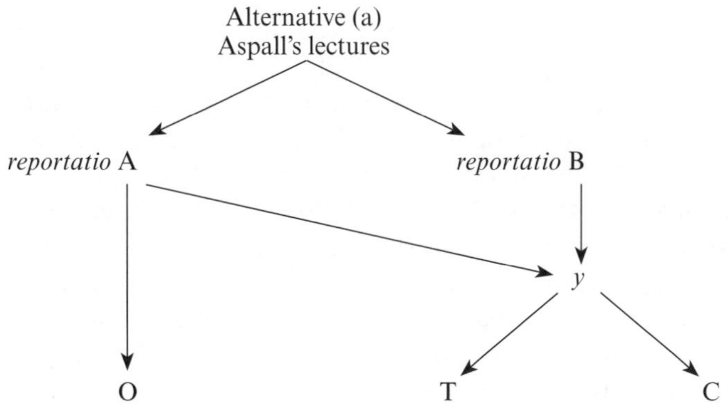

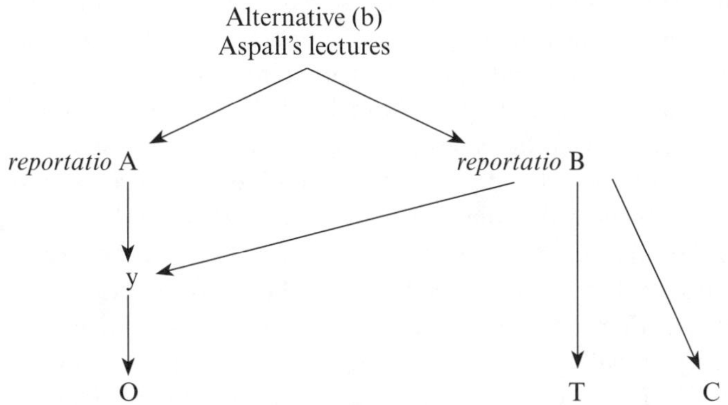

The further question of which of the two alternatives (a) and (b) holds is very difficult to answer and is here left open.

Both in alternative (a) and in alternative (b) there are extensive portions of text in which either the two version T and C or all the three versions derive from a remote common source. For these portions of texts, the further question arises of whether the versions deriving from a common source do so independently of one another or through the mediation of a common archetype, i.e., a copy serving as an intermediary between them and the original. This question too remains basically unanswered. On the one hand, the fact that the errors common to the relevant versions are few and scarcely significant suggests that there is no common archetype. On the other hand, it can well be the case that common mistakes have been corrected in the complex processes of transmission and redaction of our versions.

As to the quality of the three versions, the version transmitted by C is overall stylistically superior to that of both O and T: for example, some arguments are formulated in a more concise or clearer way, and explicit references to the relevant parts of Aristotle's discussion are more frequent and better articulated.[62] This stylistic superiority, however, seems largely due to a more pervasive editorial work so that a closer analysis reveals that version C is farther removed from the original than any of the other two versions, O and T. A clear indication of this is the fact that individual editorial interventions —typically, the reformulation of the original wording or the introduction of material from unknown sources—are much more frequent in version C than in versions O and T.

One atypical case of an individual intervention in version C is worth being singled out, because it seems to be a case of 'conflated' text, that is, a case in which the primary text of C originates from the combination of materials whose remote sources are two different *reportationes* of Aspall's lectures. This case occurs in question 44 of Book I, which belongs to the portion of text in which the differences between versions TC and version O are so extensive as to suggest the derivation from different *reportationes*. In q. 44 Aspall defends the view that matter began to exist. He considers a contra-argument based on the general principle that relatives are always mutual; this principle is then applied to the special case of the relatives 'beginning' and 'end' to conclude that, since matter does not have an end, it does not have a beginning either.[63] Aspall's reply to this argument is that not all kinds of relative are mutual; there are in fact some exceptions, and the relatives 'beginning' and 'end' are one such exception, so that something can have a beginning without having an end. Aspall's reply is reported in different formulations by versions TC and version O:[64]

TC: Ad secundum dicendum quod non sunt verissime relativa, sicut dicit Aristoteles in quinto *Metaphysicae* quod sensibile dicitur essentialiter ad sensum et non econverso, et sic non est ibi verissima relatio. Unde finis dicitur essentialiter ad principium, et ideo quod habet finem habet principium, et non econverso.	O: Secundae rationi dicendum quod quaedam sunt relativa quorum utrumque extremum relative dicitur ad aliud, ut pater et filius, quaedam quorum unum solum dicitur relative. Et sic est in proposito, quia finis dicitur in comparatione ad principium et non econverso, secundum Augustinum dicentem quod in quibusdam est accipere principium ubi non finem. Verbi gratia in numeris, quia, cum numerus crescat in infinitum per appositionem, non est sumere finem; est tamen ibi sumere primum aliquo modo, quia primum numerum.

[62] See, for example, the *apparatus criticus* at Book II, q. 9, nn. 2, 6, 7; q. 10, n. 8; q. 11, n. 2; q. 12, n. 4; q. 13, n. 9; q. 21, n. 1*; q. 24, n. 4; q. 25, n. 7; q. 37, n. 7; q. 39, n. 1; Book III, q. 1, nn. 2*, 1; q. 4, title; q. 8, n. 1*; q. 13, n. 1*; q. 13.2, nn. 2, 3, 7; q. 14.2, nn. 5–7.

[63] Book I, q. 44, n. 2, and Appendix, Book I, q. 44, n. 2.

[64] Book I, q. 44, n. 12, and Appendix, Book I, q. 44, n. 11.

The discussion of the argument about relatives then continues with a clarification about the different kinds of relatives invoked in the reply to that argument and other considerations. Furthermore, in this subsequent part of the discussion versions TC differ from version O. However, where the discussion ends in versions T and O, the conflation with version O occurs in version C; for C adds version O of the reply, presented as an alternative reply to that reported by TC in the text quoted above:[65]

> Aliter dicitur sic quod quaedam sunt relativa quorum utrumque extremum dicitur relative ad aliud, ut pater et filius, quaedam quorum unum solum dicitur relative. Et sic est in proposito, quia finis dicitur in comparatione ad principium et non econverso secundum Augustinum dicentem quod in quibusdam est accipere principium ubi non finem, verbi gratia in numeris, quia cum numerus crescat in infinitum per appositionem, non est sumere finem, est tamen sumere principium aliquo modo, quia ipsum (*pro* primum) numerum.

While the cases like that just presented in which the primary text of C is a conflation of version CT and version O are exceptional, there are a good number of cases in Book I in which passages corresponding to the version O are recorded in the margin of C.[66]

In addition to transmitting a version of Aspall's questions that is probably the result of extensive editorial work, C itself is a very bad copy of its model. It contains a high number of errors by omission and by substitution. On the contrary, O is largely correct. For example, in books I–III it contains only about twenty-one major errors by omission and about nine major errors by substitution. T is more correct than C but much more faulty than O. Furthermore, the original version of T contains many mistakes that have subsequently been corrected.

T also testifies to an editorial work in progress between the lines and in the margins: there is a significant number of passages in which stylistic changes occur, that is, where original readings of T that are correct are deleted and replaced with others or where the original text is made more complete by additions. The following passage contains examples of both kinds of editorial intervention:[67]

> Contra istud sic: Aristoteles dividit naturam in materiam et formam quae sunt partes compositi.

[65] Appendix, Book I, q. 44, n. 13 (*apparatus criticus*).

[66] See., e.g., C, fol. 124r, right mg. (cf. Book I, q. 1, n. 10); fol. 124v, left mg. (cf. q. 3.1, n. 2–3); fol. 125r, right mg. (cf. q. 6, n. 1); fol. 126v, lower mg. (cf. q. 15.2–16.2, n. 2–3); fol. 131r, right mg. (cf. q. 35, n. 29); fol. 134r, lower mg. (cf. q. 42.2, n. 6, 10); fol. 135v, lower mg. (cf. q. 46, n. 19); fol. 137v. left and lower mg. (cf. q. 46, n. 14, 15).

[67] Book II, q. 4, nn. 7–9.

> Item, dicit Aristoteles quod unumquodque magis est in sua natura cum est in actu quam cum est in potentia; intendit ergo quod illud sit natura quod prius est in potentia et postea in actu; *et huiusmodi sunt partes compositi.*
>
> Aliter dicitur sic quod *natura* nec est materia nec forma... *Unde debemus imaginari* quod...

All three manuscripts transmit this passage without relevant editorial variations. The parts printed in italics are where the subsequent editorial interventions of T occur: (i) the sentence "et huiusmodi sunt partes compositi" is deleted and replaced in the margin with "et tale est forma quae est compositi pars"; (ii) "de qua hic intendit" is added in the margin to "natura"; (iii) "Unde debemus imaginari" is deleted and replaced in the margin with "Si ergo imaginaremur". In this example, all three marginal interventions of T arguably improve the quality of the original text by making it more precise and clear. In many other cases, however, it is more difficult to find the rationale for the interlinear and marginal changes to the original text of T.

The fragment of Aspall's questions on book I preserved anonymously in N at fols. 114ra–115vb includes qq. 1–5, q. 42 and part of q. 43, ending *ex abrupto* at the beginning of the answers to the *quod non* arguments. Despite its fragmentary nature, N is important because the text it transmits is close to that of O and belongs to that portion of Book I in which the other two witnesses T and C are far removed from O. Thus, unlike T and C, N can be used to correct the mistakes of O (although the overall quality of O is better than that of N). Some mistakes common to O and N seem to indicate that they derive from a common model. However, not all the differences between O and N can be explained through the simple process of copying, but it is also necessary to assume that there were some redactional interventions in the transmission of the text. Most of these interventions do not affect the content of the text but only its style, in such a way that the style of O tends to be more elaborated than that of N.

3.2.2. Book IV (O, C)

The fragment of the questions on book IV transmitted by O is slightly more complete than that transmitted by C: C stops at the end of question 5, whereas O contains two other complete questions and a third incomplete one. There are editorial differences between the two versions but not as important and frequent as those in books I–III. Also, unlike in books I–III, in book IV O contains many interlinear and marginal corrections in which mistakes in the original text of O have been subsequently corrected.

3.2.3. Book VIII (O, L, T)

In all three manuscripts the questions cover topics up to the last chapter of book VIII but are left unfinished. Two main sets of questions about the heavens as self-movers are announced: one set about the intelligences moving the heavenly bodies and the other set about the heavenly bodies as moved by the intelligences.[68] The second set of questions is entirely absent from O, whereas L and T contain just one final question belonging to it.[69]

For book VIII too, each of the three manuscripts to some extent transmits a distinct redaction of Aspall's questions. Here is an example illustrating major editorial differences between any two of the three manuscripts:[70]

O: <1> Quaedam sunt animae virtutes quae uniuntur instrumento determinato, cuiusmodi sunt virtus visiva, auditiva et consimilia.	L: <1> Quaedam sunt animae virtutes quae insunt animae vel secundum partem essentiae eius vel secundum quod illae virtutes uniuntur instrumento determinato, cuiusmodi sunt virtus visiva, auditiva et consimilia.	T: <1> Quaedam autem insunt animae secundum partem essentiae eius et per determinatum respectum ad partes corporis, cuiusmodi et virtus visiva et auditiva etc.
<2> Et de istis non est verum secundum illos quod sunt in omni parte corporis...	<2> Et de istis non est verum secundum illos quod sunt in omni parte corporis...	<2> Et huiusmodi non sunt in omni parte corporis...

In L and T sentence <1> contains two claims about the powers of the soul under consideration, whereas in O it contains only one. The first claim in L and T—that these powers belong to a specific part of the essence of the soul (rather than to the whole essence of the soul)—is absent from O, but this omission results in an abbreviation rather than in an error. The second claim in L (i.e., these powers have a specific bodily organ) is the same as the single claim of O, but different from the second claim of T (that is, these powers have specific relations to parts of the body). Sentence <2> has the same formulation in O and L, but a more concise one in T.

In this example the distance between O and T is greater than that between O and L. This conclusion can be generalized to the whole extent of the questions on book VIII: the editorial differences between O and T are much more frequent and much more radical than those between O and L. Furthermore, while there is no dominant pattern for the differences between O and L, there is one such pattern for the differences between O and T: the text of T tends to be more abbreviated and concise than that of O, as the following example illustrates:[71]

[68] Book VIII, q. 34, n. 1*.

[69] This question is edited in the Appendix (Book VIII, q. 42).

[70] Book VIII, q. 31, n. 9.

[71] Book VIII, q. 20, title and n. 1.

O(L): <1> Quaeritur de hoc quod dicit consequenter quod gravia et levia non moventur ex se.	T: <1> Consequenter dicit quod grave et leve non moventur ex se.
<2> Contra: grave et leve moventur a natura; sed natura est principium etc.; ergo moventur a principio intra; sed quod movetur a principio intra movetur ex se; quare etc.	<2> Et videtur quod sic quia moventur a natura, et natura est principium motus etc.; ergo moventur a principio intra, et tale quid movetur ex se.

In sentence <1> containing the title of the question, T omits the initial part "Quaeritur de hoc quod". Similar abbreviations in the title of the questions in T occur in a relevant number of cases (qq. 6, 7, 8, 9, 12, 13, 14, 15, 16, 17, 18, 20). In sentence <2> T gives a more concise formulation of the minor premise of the second syllogism ("sed quod movetur... ex se") and entirely omits the reference to its conclusion ("quare etc."). Note also that the two manuscripts introduce the series of the contra-arguments with different formulas, but this happens rarely and should not be regarded as a distinctive feature of the two redactions.

Both L and T are considerably more faulty than O. For example, L has about twenty omissions by homoeoteleuton, T about twenty five whereas O has only two and both are corrected in the margins; both L and T have about ten errors of unconditioned long omissions (three or more words), whereas O does not have omissions of this kind. Compared to books I–III, however, the cases in which the initial text of O has been corrected are considerable more numerous. For example, the questions on book II and those on book VIII have approximately the same length in O, but the interlinear and marginal corrections of minor errors in the questions on book VIII are more than the double of those in the questions on book II.

3.2.4. Book VI (L, T)

There are some differences in the texts of L and T. The first two questions in L[72] are not present in T. L, however, covers topics only up to chapter 6 of book VI, whereas T covers up to the final chapter 10 (the last twelve questions found in T are absent in L).[73] Both L and T are incomplete (they stop *ex abrupto* in the course of their last question). Moreover, in the sections that are covered by both L and T, the two witnesses transmit two entirely distinct versions of the questions on book VI: although the same questions are raised and in the same order, linguistic differences are found almost in every single argument or even sentence of any question common to the two versions. A good illustration of this phenomenon is the following example of the different

[72] See Appendix, Book VI, qq. 1–2.
[73] Book VI, qq. 52–63.

formulations in which the two manuscripts present a conceptually identical argument against the existence of a continuum:[74]

T: Ab eo est continuitas a quo est rei unitas; sed a quo est rei entitas, ab eo est rei unitas; ergo a quo est rei entitas, ab eo est continuitas; sed entitas rei est a virtute caeli diffusa per universum; ergo si sit continuitas, hoc erit ab ista virtute, quod falsum est. Cum ista continuitatem non habeat, in aliis continuitatem causare non potest. Quod autem continuitatem non habeat probatio: quaecumque sunt continua per separationem possunt fieri contigua; sed impossibile est duo lumina secundum superficiem coniungi sive esse contigua; ergo etc. Minor patet per illud *De sensu et sensato*: "Duo lumina impossibile est secundum superficiem coniungi".	L: Ab eodem est unitas rei a quo est entitas, secundum Aristotelem in quinto *Metaphysicae*; sed entitas rei est a virtute caelesti diffusa a corporibus caelestibus; ergo et unitas; ergo et continuitas, cum sint ab eodem; sed continuitas non potest esse a virtute caelesti, quoniam ista virtus caelestis non est in se continua, quoniam, si sic, posset esse contigua; sed hoc est impossibile, quia dicit Aristoteles in *De sensu et sensato* quod lumen lumini non coniungitur secundum superficiem; ergo nec per separationem luminis a lumine fit lumen contiguum, si iterum uniatur.

The relationship linking L and T is compatible with the hypothesis that they are the result of two distinct *reportationes* of the same course of lectures. However, to explain the pecularities of the two texts we must also assume subsequent redactional interventions contributing to shape the original material of the two *reportationes* into the form in which they have come down to us.

A major editorial difference between the two versions is in their references to Aspall's questions on book I of *De generatione et corruptione*. In questions of book VI that have a parallel in questions on *De generatione et corruptione* T simply refers to the arguments in the parallel questions without formulating these arguments explicitly. For example, at the beginning of q. 14 in T we read: "Et patent argumenta in ista eadem quaestione super primum *De generatione*".[75] In the corresponding questions, however, L formulates these arguments explicitly and does not refer at all to the questions on *De generatione*.[76] This difference has been examined in great detail by Donati in her assessment of Aspall's authorship for the questions on *Physics* VI.[77]

Although we may safely assume that the remote sources of L and T are reports of Aspall's lectures on book VI, it is doubtful that all the material transmitted by the two versions originates from Aspall's lectures. For example, it is hard to believe that the first two questions in L (absent from T)—about the active potency of matter and the relationship between natural philosophy and metaphysics, respectively—could be the beginning of a course on Aristotle's treatise of the continuum. The inspection of the content of those two questions shows that they are compilations of material found elsewhere

[74] Book VI, q. 1, n. 1; Appendix, Book VI, q. 3, n. 1.
[75] Book VI, q. 14, n. 1.
[76] The questions in which this kind of difference appears are qq. 14–15 in T and qq. 18–20 in L.
[77] Donati, 2013a, 223–281.

in Aspall's questions.[78] Material that is conceptually extraneous to book VI but is actually almost a repetition of other sections of Aspall's questions on the *Physics* is also found in T.[79] Overall, it seems that the editorial process that, starting from Aspall's lectures on book VI, has resulted in the versions L and T has been much more radical and invasive than that relative to the other books.

The quality of each of the two manuscripts is rather poor. They both contain not only a very high number of minor errors of copying, which are often easy to correct, but also a good number of major omissions or corrupted passages, the sense of which has not been possible to restore. Other passages, although not evidently corrupt, are very obscure. These errors and textual problems affect L more seriously than T.

3.3. Criteria of edition

Given that the textual tradition of Aspall's questions on the *Physics* consists entirely of distinct versions, none of which completely coincides with any other, the only sensible criterion for their edition is to choose one of these versions and follow it consistently. We have chosen the version transmitted by O wherever it is available, that is, for books I–IV, and VIII.[80] For the remaining book VI we have chosen the version transmitted by T. The reasons for these choices are straightforward. The version transmitted by O (i) is more correct than those of the other manuscripts (that is, T and C for books I–III, C for book IV, and L and T for book VIII) and (ii) covers a larger portion of Aspall's questions. As to this second point, note that although manuscript T contains the questions on all books except the fragment of book IV, whereas O does not contain questions on book VI, there are strong doubts that the version of the questions on book VI and the version of the questions on books I–III, VIII transmitted by T belong to one and the same version of Aspall's questions. As we have just mentioned, Aspall's lectures on book VI seem to have been subject to a much more radical editorial work than those on the other books. On the contrary, the questions on books I–IV, VIII transmitted by O all seem to belong to one and the same version. Similar

[78] Q. 1 of Book VI in L is parallel to Book I, q. 35, nn. 28–29, and Appendix, Book I, q. 35, n. 29. Q. 2 of Book VI in L is a repetition of Book I, q. 2.

[79] For example, Book VI, q. 59 in T corresponds to Book III, qq. 4 and 12 (on the classification of motion in the categories and on action and passion as distinct categories, respectively). See also below, 5.6.1, 5.6.2, 5.6.3.3.

[80] Trifogli produced another edition of the questions on book III based exclusively on ms. O (without the help of the other witnesses to correct errors). It is available in a CD-Rom accompanying the volume Trifogli, 2007.

reasons motivate the choice of T instead of L for the questions on book VI: T is more correct than L and it covers a larger portion of book VI.

As to the edition of the questions on books I–IV, VIII, this is not a mere transcription of O. Despite the editorial differences, the other manuscripts containing these questions can be collated with O for large portions of their common text and have been effectively used to correct O. In those cases in which O testifies to a faulty text subsequently corrected between the lines or in the margins, we have compared the correction in O with the readings of the other manuscripts and when we judge that a reading of another manuscript is more reliable (for example, closer to the version O) than the interlinear and marginal correction in O, we have accepted it. On the other hand, coherently with the assumption of distinct versions, when we judge that the text of O is defensible, we do not accept the reading of another manuscript even when it is stylistically superior or gives a more complete text.

As to the edition of the questions on book VI, based on version T, the alternative version L is so different from T that it is not of immediate help in emending T. In most cases the text of L can simply give an idea of what the corresponding text of T can signify but it cannot also provide the particular readings required to emend it.

In addition to producing an edition of version O for books I–IV, VIII and of version T for book VI, we intend to make available the other versions too. For book VI the only way to achieve this is to present the whole version L separately from version T. We have thus printed version L in the Appendix. For books I–IV, VIII, the versions alternative to O can be collated with O for the whole of their common text with the exception of Book I, qq. 1–38, 42–48, and Book II, qq. 17–20, 22, 27–36 (the alternative versions for this set of questions are T and C). Accordingly, for the whole extent of books I–IV, VIII with the exception of the questions just mentioned, the alternative versions are documented in the *apparatus criticus* of version O, while for the remaining questions version T is printed in the Appendix and version C is documented in the *apparatus criticus* of version T. We have also printed in the Appendix some isolated questions (from Books I, III and VIII) contained in T (and/or C L) but not in O.

4. Presentation of the text

Latin text. With regard to orthography, we have chosen to normalize the spelling of this edition according to the conventions of classical Latin, because we think that it increases the clarity of the text. Similarly, we have not respected the paragraphing and punctuation of any of the manuscripts but employed our own with the aim of assisting the reader.

Italicization has been used for titles of works and for the Aristotelian *lemmata*. Citations of an author's words have been enclosed in double quotation marks, whereas single quotation marks have been used for terms or sentences in material supposition.

Angle brackets have been used for editorial additions and square brackets for editorial deletions. These editorial interventions have also been reported in the *apparatus criticus*.

To facilitate references, the questions belonging to each book have been numbered consecutively. Similarly, the paragraphs within each question have been numbered consecutively, starting from the one immediately following the title. All internal references are by book numbers, question numbers and paragraph numbers.

English translation. The aim of the translators was to produce a readable text that was neither overly free nor too grindingly literal. In some cases this has led to changes in the internal structure of the Latin sentences, and to the expansion of phrases that were too succinct. Where Aspall has introduced elaborate distinctions, we have numbered their different parts and used these numbers to refer to those parts in Aspall's subsequent references to them.

Apparatus criticus. For the questions on books I–IV, VIII (Version O) we give complete information about O except trivialities. In particular, we report on all the cases in which the printed text deviates from O; in such cases the readings of all the other available manuscripts are reported. We also report on the internal corrections of O. We have been very selective, however, about what to include in the *apparatus* from the other versions (when available). The general principle is that we report only on (i) major or otherwise significant redactional differences from O; (ii) major errors (and their corrections); (iii) variant readings common to two other manuscripts (except inversions that do not affect the sense and trivialities). In the case of T we have further restricted the *apparatus* (i) to the 'first hand' redactional differences of T from O, that is, we have not recorded the editorial work in progress that T is undergoing, typically in the margins.[81]

For the set of questions from books I and II in the version T (where also C is available) that we have printed in the Appendix, the *apparatus criticus* for T and C is not complete. It only reports (i) the cases when the printed text differs from that of T; in these cases the printed text is assumed to be that of C unless explicitly stated otherwise; (ii) marginal or interlinear corrections of major errors in T; (iii) major redactional differences of C; (iv) major errors of C (and their corrections).

[81] See the examples given above in 3.2.1.

The order of the sigla of the manuscripts in the *apparatus criticus* is not alphabetical but is that of the distance of the version transmitted by a manuscript from the version transmitted by O. Thus, for books I–III the order is OTC, for book IV OC, and for book VIII OLT.

For the questions on book VI transmitted by T we give complete information in the *apparatus criticus* about T (except trivialities).

The following abbreviations have been used:

add.	*addidit/addiderunt*
corr.	*correxit/correxerunt*
del.	*delevit/deleverunt*
fort.	*fortasse*
hom.	*per homoeoteleuton*
inv.	*invertit/inverterunt*
iter.	*iteravit/iteraverunt*
lac.	*lacuna*
mg.	*in margine*
om.	*omisit/omiserunt*
praem.	*praemisit/praemiserunt*
ras.	*rasura*
scrips.	*scripsit/scripserunt*
s.l.	*supra lineam*
suppl.	*supplevit/suppleverunt*
transp.	*transposuit/transposuerunt*

Apparatus fontium. We have identified the explicit sources and the internal references.

5. The main doctrinal points of Geoffrey of Aspall's Questions on the *Physics*

This section contains an outline of Aspall's discussion of the following major topics of Aristotle's *Physics*. From book I: the so-called principles of nature, that is, 5.1 Matter; 5.2 Privation; 5.3 Form. From book II: 5.4 Nature, and 5.5 Causes. From book III: 5.6 Motion, and 5.7 The infinite. From book IV: 5.8 Place. From book VI: 5.9 The continuum. From book VIII: 5.10 Eternity of the world, and 5.11 Self-motion.

The extensive investigation of the English commentaries on Aristotle's *Physics* of the thirteenth century carried out by Donati and Trifogli in the last twenty years, although not specifically focused on Aspall, has shed light on a good number of the doctrinal and exegetical aspects of the tradition to

which Aspall's Questions belong. We have added references to the studies relevant to a specific topic in a footnote at the beginning of the section devoted to that topic, but we have not explicitly used the results of those studies in our presentation of Aspall's discussion. We have tried instead to follow Aspall's text closely in order to reconstruct the salient points of his treatment of the topics listed above.[82]

5.1. Matter[83]

5.1.1. Ontological status of prime matter

Aspall's general characterization of prime matter is the standard one: prime matter is the component of a material substance that lacks any form. If we consider a material substance, for example, a man, and by means of a thought experiment remove all forms, such as humanity, corporeality, and substantiality, from him, what is left is prime matter (I, q. 43.2, n. 5; q. 48, n. 12). Aspall also considers an alternative hypothesis according to which prime matter should be identified with the ultimate genus of substance. Although he lists several arguments in favour of this hypothesis, he then rejects it (I, q. 48).

The most basic question about prime matter when defined as something absolutely formless is whether it exists (I, q. 42). An immediate objection to the existence of prime matter in this sense is based on the assumption that being presupposes a form so that only what has a form is a being. In Aspall's formulation of this objection, degrees of being correspond to degrees of form so that the more form something has the more being it has; therefore, what has no form at all has no being at all (I, q. 42.1, n. 1). In replying to this objection Aspall maintains that what requires a form is not a being as such but only a being of a certain kind, namely, a being in act. Accordingly, its lack of any form disqualifies prime matter, not from having being as such, but only from having actual being. Prime matter is a being but it is a being in potency (I, q. 42.2, nn. 1–2).

The description of prime matter as being absolutely formless and in potency is used by the vast majority of Aristotelian commentators but it is ambiguous and may conceal very different views about the ontological status of prime matter. One paradigmatic position in the medieval debate on matter

[82] We have not included two major topics from the first part of book I in our outline: (i) the methodological debate arising from the prologue of Aristotle's *Physics* and (ii) the metaphysical issues surrounding the analogy of being. Donati has devoted extended studies to the discussions of both topics in the English tradition (including Aspall). On the first topic, see Donati, 1995; on the second topic, see Donati, 2003, 2012b , 2014.

[83] See Donati, 1994, 1998b, 2002a, 2002b, 2007, 2012c, forthcoming.

is that of Aquinas, according to whom the being in potency of prime matter implies that prime matter possesses no actuality at all and is not a being in its own right independently of a form, so that prime matter exists only in so far as it is joined to a form. Another paradigmatic position is that of Scotus, who held on the contrary that although prime matter is in potency with respect to all forms, it does possess some actuality of its own and is a being in its own right, so that it can even exist without being joined to any form.[84]

Aspall does not devote a specific discussion to the ontological status of prime matter as potential being, but there are some indications that he ascribes to prime matter a certain degree of independent reality, so that anachronistically his position is more similar to that of Scotus than to that of Aquinas. Some clues to Aspall's view are found in the Question whether matter is a being in a qualified sense (*secundum quid*), that is, a diminished being (I, q. 57). Aspall raises this issue in connection with an opinion about the relationship between prime matter and form according to which prime matter absolutely speaking is neither the same as nor different from a form because the relations 'being the same' and 'being different' only hold between beings in an absolute sense. Matter, however, is not such a being, but a being in a qualified sense (I, q. 56, n. 14). An argument for the affirmative in this Question points out that the view that matter is a being in a qualified sense posits that matter as a being in potency is something intermediate between what is absolutely nothing and what is absolutely a being, that is, a being in act (I, q. 57, n. 1). Aspall rejects this view. He grants the claim that matter, being absolutely formless and potential, is not a being in an absolute sense if this is taken as equivalent to being a complete being. He maintains, however, that there is a more general notion of being in an absolute sense according to which both a being in act and a being in potency are beings in an absolute sense. As a result, matter as a being in potency is also a being in the absolute sense (I, q. 57, n. 7). To illustrate this, Aspall remarks that being in act and being in potency are not related to being in the way in which a man and a painting of a man are related to man, for 'man' is predicated equivocally of a man and of a painting of a man whereas being is predicated univocally of a being in act and a being in potency (I, q. 57, n. 4). A further argument against the view that matter is a being in a qualified sense makes the point that this is incompatible with the property universally ascribed to matter of being a component of a material substance. Since a material substance is a being in an absolute sense, it cannot have a being in a qualified sense as one of its components (I, q. 57, n. 6). The composition of matter and form in a material

[84] On Aquinas, see Wippel, 2000, 296–327. On Scotus, see Gilson, 1952, 432–444; Walter, 1965, 124–146; McCord Adams, 1987, II, 639–647, 671–695; Cross, 1998, 13–33.

substance is a composition of two distinct and irreducible things (*res*) (I, q. 56, n. 6).

Other indications of Aspall's view about the ontological status of prime matter are provided by his discussion of the coming into being of prime matter. Aspall discusses three Questions on this topic. The first two Questions deal with very general aspects of this problem, namely whether matter began to exist, and what efficient cause produced it. Aspall maintains that prime matter began to exist and was produced by the first cause, who created it from nothing. He also ascribes this view to Aristotle (I, qq. 44–45). The third Question deals with the more specific issue of whether prime matter was the immediate terminus of creation or whether it was created through a form, or in other words whether the creation of matter is an act distinct from and irreducible to the creation of a composite of matter and form (I, q. 46). This Question is the relevant one for our purposes, for the arguments in favour of the view that prime matter is created through a form are based on the assumption that prime matter is not a being in its own right but a being dependent on a form (I, q. 46, nn. 1–5), whereas the arguments in favour of the view that matter is the proper terminus of God's creative act are based on the assumption that matter is a being in its own right and is even prior by nature (although not temporally) to every form, being the subject of every form (I, q. 46, nn. 6–9). Aspall reports two opinions on the creation of matter, one of which posits that matter is the proper terminus of a creative act and the other of which posits that only the composite of matter and form is properly such a terminus (I, q. 46, nn. 10–11). Aspall sides with the first opinion and replies to the arguments showing that matter is not a proper terminus of a creative act (I, q. 46, nn. 12–19).

5.1.2. Essential properties of prime matter

Consistently with his view that prime matter is a being in an absolute sense, Aspall posits that prime matter has an essence. He says almost nothing, however, on this essence as such. His discussion is mainly about the essential properties of prime matter, namely those properties that belong to prime matter in virtue of its essence, and it shows that there were two major controversial issues in his time. One is whether prime matter in its essence is intelligible and knowable, and the other is whether it has parts. In what follows we shall concentrate on these two issues.

5.1.2.1. Intelligibility of prime matter Aspall maintains that since prime matter is a being in its own right independently of any form, it must also be intelligible and knowable in itself apart from any form (I, q. 46, nn. 8–9). He distinguishes, however, the case of God's intellect from that of the human intellect. While the divine intellect understands the nature of prime matter as

such, and not simply prime matter as related to a form, so that there is an idea of prime matter in God distinct from the idea of any form, the human intellect can understand prime matter only in relation to a form (I, q. 46, n. 12). This claim about the intelligibility of prime matter for the human intellect is certainly problematic. Things become intelligible for our intellect only in so far as they are abstracted from matter and only in so far as they have a form, but since prime matter cannot be abstracted from itself and does not have any form, it seems to follow that prime matter is not intelligible for us at all (I, q. 49, nn. 1–6). Aspall rejects this totally negative conclusion by appealing to the fact that there is a science dealing with prime matter, i.e., metaphysics, a fact that shows that we do have some understanding of prime matter (I, q. 49, n. 12). He concedes, however, that prime matter is only intelligible and knowable for us in a diminished way (I, q. 49, n. 16). There are no sense images (*phantasmata*) and intelligible species of prime matter as such but only of a composite of matter and form, so that prime matter is intelligible and knowable only in relation to a form (I, q. 49, nn. 22, 24–25). Aspall also lists some inferential ways (*per signa*) by which we come to grasp the nature of prime matter on the basis of its distinctive properties, such as being what persists in a substantial change, what carries a form, and what is left when all forms are removed from a composite (I, q. 49, n. 21).

Aspall devotes an important discussion to an alternative hypothesis according to which prime matter is absolutely intelligible, even for our intellect, because its ontological status is absolutely similar to that of our intellect. In particular, neither our intellect nor prime matter are beings in act before receiving a form, and both receive universal forms (I, q. 49, nn. 10–11). These similarities are strong enough to allow the conclusion that prime matter is not only absolutely intelligible but also that it itself understands, just as our intellect does (I, q. 51, nn. 1–4). Aspall rejects this hypothesis by showing that in fact the alleged strong similarities between prime matter and our intellect do not hold. While matter in itself is a being in pure potency, our intellect is something in act even before receiving forms from outside, that is, intelligible species, since even at the stage in which it lacks all intelligible species the intellect is the essential perfection of a human being which determines its specific being (I, q. 51, n. 11). Moreover, the intellect alone primarily and essentially receives universal forms, whereas any form received in prime matter is primarily received as singular (I, q. 52, nn. 12–15).

5.1.2.2. Parts of prime matter Granted that prime matter in its essence lacks any form, and in particular lacks any quantity, the controversial issue is whether it follows from this that prime matter in itself (i.e., apart from any form) is absolutely one and indivisible, without parts (I, qq. 54–55). The parts of prime matter under discussion here are not essential parts, that is, distinct

components of its essence in the way that matter and form are distinct essential components of a material substance. Since the essence of prime matter is absolutely undifferentiated and absolutely indivisible and one, prime matter does not have distinct essential parts. They are rather thought of as homogeneous integral parts, analogous to the parts of which a homogeneous continuum is composed and into which it is divisible (I, q. 43.2, n. 6).

The most compelling arguments in support of the indivisibility of prime matter rely on the general assumption that divisibility and parts can only derive from quantity so that only a quantum can be divisible and have parts (I, q. 54, nn. 1–3). The arguments for the opposite view are based on two main different kinds of consideration: metaphysical and physical.

(i) A general metaphysical principle in favour of the intrinsic divisibility of prime matter is that two distinct individuals have nothing that is numerically one in common. Since two individual material substances have prime matter in common, the prime matter existing in one of them cannot be numerically the same as the prime matter existing in the other, so that distinct individuals have distinct parts of prime matter. An example of the application of this principle is the following: consider a material continuum and divide it into two halves; then the matter in one of the two halves is numerically distinct from that of the other half, and to the two parts of the continuum there correspond two parts of prime matter. An objection to this is that although the parts of prime matter in a material substance are reducible to parts of quantity, so that in two parts of a material substance there are two distinct parts of quantity, and hence of a composite of prime matter and quantity, there are not also two distinct parts of prime matter as such. However, this objection is rejected by appealing to the principle that numerically distinct accidents inhere in numerically distinct subjects. Since the quantity of one of the two parts is numerically distinct from the quantity of the other part, these two quantities must have numerically distinct subjects; but their subject is prime matter (I, q. 54, nn. 18–20, 25).

(ii) Physical considerations in favour of the intrinsic divisibility of prime matter are based on Aristotle's claims about the reciprocal generations of elements, the subtle and the thick, and the rare and the dense. These claims show that the 'amount' of prime matter in distinct material substances does not always correspond to the amount of their quantity, so that the amount of prime matter must be distinct from that of quantity. For example, Aristotle maintains that ten handfuls of water are the minimum amount of water that can be transformed into one handful of earth. Given that matter is the persisting substrate of substantial change, this is taken to show that the amount of prime matter in ten handfuls of water is the same as the amount of prime matter in one handful of earth and that the amount of prime matter

in one handful of water is not the same as the amount of matter in one handful of earth (I, q. 54, nn. 14, 17, 21, 22).

Aspall reports an opinion which claims that prime matter is absolutely indivisible in its substance so that the whole of prime matter exists in each material substance. The effect of a form on prime matter is to multiply it as a whole, so that in two distinct material substances there are two distinct replicas of the whole of prime matter (I, q. 54, n. 24). Aspall does not follow this opinion. He thinks that a better opinion is the one positing that prime matter does have parts so that distinct individual substances and distinct parts of a continuum have distinct parts of prime matter as such and not only distinct parts of quantity or of a composite of prime matter and quantity. The parts of prime matter, however, exist only in potency (I, q. 54, n. 25). This qualification about the potential existence of parts in prime matter is not completely clarified. Aspall seems to understand it not in the strong sense that the parts of prime matter do not have any actual existence, but in the weaker sense that they do not have actual extension. Parts do exist in prime matter and their existence is naturally prior to quantity and independent from quantity, but they exist all together (*simul*) and not one outside the other. The intrinsic divisibility of prime matter is a divisibility into non-extended parts and not a divisibility into extended parts; it is only in virtue of quantity that the parts of matter become actually extended and spatially distinct. Distinct parts of prime matter are only numerically distinct but in no way spatially distinct (I, q. 54, nn. 26–30; q. 55, n. 9).

The nature of the intrinsic non-extended parts of prime matter is left without an adequate explanation. One issue that appears particularly problematic is how divisibility into non-extended parts can play the physical role ascribed to it by the arguments in support of the existence of parts of prime matter based on the mutual generation of the elements, the nature of the thick and of the subtle, and that of the rare and the dense. These arguments suggest that there must be something like a quantity or amount of prime matter, distinct and independent from that of a quantity in the proper sense, so that there can be more prime matter where there is less quantity and the other way around. It is not clear, however, how non-extended and spatially indistinct parts can play any role in defining a quantity of prime matter.

5.1.3. Matter as principle of nature

While Aspall ascribes a strong ontological status to prime matter as a thing in itself totally distinct from any form, he denies that prime matter as such can play the role of the material principle of nature discussed by Aristotle in *Physics* I. As he puts it, prime matter in its substance is a principle of a

composite substance *qua* substance but not also *qua* changeable substance (I, q. 58, n. 8). In another formulation, prime matter in its substance is the subject of the forms in a composite substance but it is not also the subject of generation, that is, of a substantial change of a composite substance (I, q. 60). More precisely, in Aspall's view, prime matter persists in a substantial change whereas no actual form persists; moreover, prime matter is the ultimate subject of any form. However, these two essential properties of prime matter are not sufficient to define the material principle of *Physics* I. Aspall maintains that the material principle must also be passive in a physical sense, that is, something on which the agent of a substantial change exerts its action by being in contact with it and which undergoes some changes in its substance as a result of this action. Furthermore, the material principle must not simply be the recipient of a form, but must be something that is transformed into that form through a substantial change and which has a physical inclination (*appetitus*) towards that form. Prime matter in its substance does not have these physical properties. It is indeed passive in the sense that it receives a form but it is not passive in a physical sense; and it does lack any form, but it does not have any physical inclination towards a form (I, q. 42.2, n. 9; q. 58, nn. 5–6; q. 60, nn. 8–9; q. 61, nn. 4–5). Accordingly, natural matter, that is, the material principle of *Physics* I, is not prime matter. What then is natural matter? In Aristotle's ontology it seems that the only plausible candidate for being natural matter with the physical properties ascribed to it by Aspall is some kind of composite of prime matter and a form. Aspall's ontology, however, includes non-Aristotelian entities, which he calls the diminished potencies of prime matter, and these entities define natural matter: natural matter is prime matter together with its diminished potencies (I, q. 58, n. 8; q. 60, nn. 12–13, 16).

5.1.4. Potencies of matter

Aspall does not devote a specific discussion to clarifying what kind of entities the diminished potencies of prime matter are exactly. What we learn from his Questions on the potencies of matter (I, qq. 62–65) is that the diminished potencies of prime matter are classified as active potencies and contrasted with passive potencies. The passive potency of prime matter is a receptive potency, i.e., a potential to receive a form. An active potency is that in virtue of which prime matter has a physical inclination towards a form and in virtue of which it actually receives that form through a physical change. While the receptive/passive potency is essential to prime matter, active potencies are not essential to it, although they can broadly speaking be said to be essential to prime matter in the sense that they are perfections that prime matter is apt by nature to have and towards which it has a natural inclination (I, q. 62, nn. 5–6; q. 64, n. 4).

A crucial question is whether an active potency of prime matter towards a form is in some way the same as that form, for example whether the active potency of matter towards the form of fire is in some way the same as the form of fire. Aspall's reply to this question is negative: the active potency of matter towards a form is something incomplete or diminished that cannot be called a form in actuality. Aspall, however, qualifies his negative reply by adding that an active diminished potency towards a form is the same in essence with that form. In our example, while the active potency towards the form of fire is in no way the same as the form of fire, it does belong to the essence of fire (I, q. 65, n. 3). While Aspall does not clarify his view about the essential link between an active potency and the corresponding form in this context, he is ready to point out some serious problems arising from it. One is that this view, when combined with his other view that the material principle is the composite of prime matter and of its diminished active potencies, undermines the essential distinction between the material principle and the formal principle of *Physics* I, a distinction which Aspall himself strongly supports (I, q. 65, n. 5). Another problem is that this view is difficult to reconcile with the assumption that something absolutely new comes to be as a result of a substantial change, given that the essence of the form induced at the end of a substantial change already existed in matter at the very beginning of the substantial change leading to that form (I, q. 65, n. 4).

Aspall leaves these problems open in the Question on this topic from book I, referring the reader to a Question on this topic in his *Metaphysics* commentary (I, q. 65, n. 6). In a Question on the same topic in book VI, however, Aspall clarifies some aspects of his position. He makes it clear that the diminished active potency to a form F is not form F in an incomplete or diminished degree, in the way in which darkness is a diminished incomplete whiteness. It can be said to be of the same essence as the corresponding form only in a broad sense, since it is 'towards' that form, that is, it is an ability or disposition towards that form but totally lacks that form. In particular, Aspall rejects an argument that proves that since the active potency is a diminished act, it is an incomplete act or form. He replies to the argument by distinguishing two senses of 'diminution': in one sense 'diminution' signifies a mere aptitude for or inclination towards an act together with the total privation of that act; in another sense it signifies an aptitude for or inclination towards an act together with only a partial privation of that act, that is, together with the privation of the completion of that act. The diminution involved in the active potency of matter is a diminution in the first sense, while that involved in an incomplete form like darkness is a diminution in the second sense (VI, q. 14, n. 3).

The interpretation of the relation between a diminished active potency and the corresponding form in terms of inclination and aptitude rather than in terms of essential identity provides an answer to the problems left open in the Question in book I.[85] It opens, however, another problem, namely, that of explaining in what sense an active diminished potency of matter is active, given that it is totally formless, as is prime matter itself. Aspall's solution to this problem is that a diminished active potency, unlike a form or something with a form, is not active by itself, but, unlike prime matter, it can be made active, that is, it can be activated so that it becomes an intrinsic active principle in the generation of a substance (I, q. 65, n. 7; VI, q. 14, n. 4). We shall deal with the question of the activation of the active diminished potencies of matter in the section devoted to nature.[86]

5.1.5. The matter of the celestial bodies

In his treatment of matter in *Physics* I Aristotle associates matter primarily with change and in particular with substantial change (generation and corruption) rather than with extension or locomotion. The celestial bodies are extended and subject to locomotion, but they are not subject to substantial change, since they are eternal. This gives rise to the standard questions in the Aristotelian tradition of whether the celestial bodies have matter and, if they do, whether their matter is of the same kind as the matter of the sublunar bodies. Aspall's discussion focuses on the latter issue: he asks whether the essence of prime matter is the same in the corruptible bodies here below and in the superior incorruptible bodies (I, q. 59). He reports an opinion, which he ascribes to master *William Lupus*, according to which the prime matter of celestial bodies is essentially different from that of the sublunar corruptible bodies. He sides, however, with an opposite opinion positing that prime matter in these two kinds of body has the same essence, so that there is only one essence of prime matter in all material things (I, q. 59, n. 22–24). The strongest arguments in support of the latter opinion appeal to the principle that if two material substances have univocal properties, then they also have univocal subjects of these properties. Since the ultimate subject of any property of a material substance is prime matter, if sublunar bodies and celestial bodies have some univocal properties, then on the basis of this principle one can conclude that they have the same kind of prime matter. It is then argued that locomotion, extension, and transparency are univocal properties of celestial and sublunar bodies (I, q. 59, nn. 10, 12–13).

Particularly relevant to the discussion about the nature of prime matter is Aspall's treatment of some arguments in favour of an essential difference

[85] See above, p. xlv.
[86] See below, pp. l–lii.

between the prime matter of the sublunar bodies and that of the celestial bodies. These arguments appeal to those properties of prime matter which fit it to be both a principle of contingency, namely that in virtue of which a thing can be and not be, and a passive principle, namely a principle in virtue of which a thing can be acted upon. Since, even granted that the celestial bodies have prime matter, their prime matter does not have these properties, it seems to follow that their prime matter is of a different kind from that of the sublunar bodies (I, q. 59, nn. 2–5). Aspall's reply appeals to the distinction between prime matter as such and natural matter. It is not prime matter in its essence which is a principle of contingency and passivity but natural matter, that is, prime matter together with some diminished potencies, which do not belong to its essence (I, q. 59, nn. 29–30). Accordingly, the prime matter in a celestial body and that in a corruptible body are essentially the same, but while a corruptible body also has natural matter, in a celestial body there is no natural matter.

5.2. Privation

Aspall's distinction between prime matter and natural matter is also related to his view about the ontological status of the privation of a form as a principle of nature. Aspall first argues that a privation is not the pure negation of a form, that is, the mere absence of a form, but the negation of a form together with an aptitude to that form. To illustrate this, Aspall remarks that the privation of a form is not analogous to non-seeing, which is the mere absence of sight, but to blindness, which is the absence of sight together with an aptitude to seeing. The point of this distinction is to establish that a privation is a thing. The mere negation of a form is not a thing, but the aptitude for a form is a thing, which Aspall also refers to as the thing of a privation (*res privationis*) or the thing corresponding to a privation (*res respondens privationi*). It is because privation is a thing that it is a principle of nature (I, q. 66, n. 8). The crucial question is now what this thing corresponding to the privation of a form is. The dominant view among later Aristotelian commentators is that this thing is simply prime matter, and that there is only a conceptual distinction between prime matter and the privation of a form. Thus prime matter is a thing that is apt by nature to receive a form and lacks that form. Aspall is not an exponent of this reductive view. He posits that the thing corresponding to the privation of a form is not prime matter as such but the diminished active potency towards that form (I, q. 67, n. 4); and, as we have seen above, a diminished active potency is a thing distinct from the essence of prime matter.[87] Thus, Aspall makes privation a thing in its own right, distinct from

[87] See above, p. xliv.

both prime matter and form. One motivation for this reification is to account for the standard claim that matter desires (*appetit*) a form. In Aspall's view, prime matter alone, being devoid of any form, is not adequate to account for this desire or inclination. Something else is needed, and privation is indeed such a thing, namely the thing existing in prime matter in virtue of which matter has a desire for or inclination towards a form (I, q. 72, n. 2).

5.3. Form

As is shown by Aspall's discussion, there was major uncertainty in his time about what kind of form it is that Aristotle posits as one of the three principles of nature, alongside with matter and privation. An opinion that seems to have enjoyed some popularity claims that such a form is the first cause, that is, translated into Aristotelian terms, the first mover of *Physics* VIII (I, q. 73, nn. 1–5, 10). Obviously, this opinion does not find any textual support in *Physics* I and it reflects basic misunderstandings about the overall aim of Aristotle's treatment of the principles of nature. Aristotle makes it sufficiently clear that the formal principle of *Physics* I is a component of a material substance and that it exists in a subject, that is, in matter, rather than being a thing existing over and above natural things and ultimately responsible for their existence and change (I, q. 73, nn. 6–8). Aspall rejects the opinion and points out an ambiguity in the notion of formal principle from which this opinion arises (I, q. 73, n. 12).

In Aspall's view, however, the requirement that the formal principle of *Physics* I be a component of a material substance is not enough to identify this principle. The reason for this is that, following the dominant view in his time, he posits a plurality of (substantial) forms in a composite substance. Any substance, be it of the most complex kind, like a human being, or of the most simple kind, like an element, contains many substantial forms, one for each subdivision of the Porphirian tree under which that substance falls. The question then arises of whether any and each of these substantial forms is the formal principle or rather only one or some of them. Aspall does not address this question in its generality. He explicitly considers only one candidate—the form of the most general genus in the category of substance, that is, the form of substantiality—and maintains that this form is the formal principle of nature. The rationale for this choice is that the form of the most general genus satisfies a priority-condition: it is the first form that comes to exist in the matter of a composite substance (I, q. 74, nn. 1–3). The idea, which is not spelled out in detail, seems to be that the multiplicity of substantial forms in a composite substance is ordered according to degrees of completeness in such a way that the first form is the least complete and the last form (i.e., that of the most special species) is the most complete. To this order of complete-

ness there corresponds an order of inherence in matter such that the most incomplete form is that which first comes to exist in matter and the most complete is the last one. The order of inherence in matter is not a temporal order, but an ontological one.

5.4. Nature[88]

5.4.1. The definition of nature

Aristotle defines nature as an internal principle of motion and rest in a natural thing. He arrives at this definition by taking it for granted that there is a distinction between natural things and non-natural things, and making it explicit what the ultimate difference between the two classes of things is. Thus, while a natural thing such as a plant has an intrinsic principle of change, a non-natural thing, for example an artefact like a bed, does not properly speaking have any such internal principle.[89] Aristotle then addresses the question of whether the nature of a thing is its matter or its form. His reply is that the nature of a thing is in a sense its matter and in another sense its form, but primarily and more appropriately its form.[90]

Aristotle's discussion makes it sufficiently clear that nature is not a single all-pervading principle existing over and above natural things and regulating their changes. This did not prevent 'cosmological' interpretations of Aristotle's notion of nature from arising. An opinion reported by Aspall posits that nature is a heavenly power, which is responsible for the total ordering of the universe and is identified with the heavenly light. According to this opinion, when Aristotle says that nature is (primarily) a form, he does not mean that it is the form that is an intrinsic component of a natural thing, but an external form (II, q. 4, n. 6).[91] Aspall quickly dismisses this opinion by pointing out that it is obvious that the matter and the form into which Aristotle 'divides' nature are the intrinsic components of a material substance (II, q. 4, nn. 7–8).

In Aspall's time, even the condition that the relevant form is an intrinsic component of a natural substance was not regarded as sufficient to settle the question of what nature is, since the common view, shared by Aspall, was that a material substance has many substantial forms, namely substantiality, corporeality, and so on, all the way down to the form of the most specific species. In particular, we have seen that when dealing with the question of

[88] See Donati, 2015a.

[89] Arist. *Phys.* II.1, 192b8–193a9.

[90] Arist. *Phys.* II.1, 193a9–194b15.

[91] For Aspall's discussion of some issues about the ontological status of light, see I, qq. 39–40. See also below, pp. lvi–lvii.

which form it is that Aristotle posits in *Physics* I as a principle of nature alongside matter and privation, Aspall himself identifies this 'first' form with the form of substantiality.[92] The connection between the form of substantiality as first form and nature as an internal principle of change is embodied in an opinion according to which the nature of a substance is the composite of matter and the first form of substantiality. What seems to have been regarded attractive in this opinion is that it allows one to give nature the status of a single all-pervading principle without violating the assumption that nature does not exist over and above natural things. The aggregate of matter and the form of substantiality exists in all material substances and is somehow something common to all of them, without intrinsic differentiation, given that only the more specific substantial forms are responsible for the differences between the various kinds of substances (II, q. 4, n. 9). Aspall rejects this opinion precisely because it posits nature as a single uniform principle. He rightly points out that specifically distinct natural substances have specifically distinct natural motions (e.g., fire goes up by nature, and earth goes down by nature); therefore, the nature responsible for these specifically distinct motions cannot be specifically the same (II, q. 4, n. 11).

While Aspall maintains that the nature responsible for the natural locomotion of an element is its most specific substantial form, he is reluctant to generalize this conclusion by positing that the nature responsible for any kind of change of a natural thing is such a form. One tacit assumption in his interpretation of Aristotle's definition of nature as an internal principle of change is that since Aristotle distinguishes different kinds of natural change, it may well be the case that the nature that is the internal principle of one kind of change is not the same as the nature that is the internal principle of another kind of change. In fact, Aspall maintains that the nature that is the principle of the substantial change that is the generation of a new substance is a different kind of entity from the nature that is the principle of locomotion. For example, while the nature responsible for the natural upward motion of fire is its specific substantial form, namely the form of fire, this form is not also responsible for the coming into being of fire. In Aspall's view, given that Aristotle identifies the nature of substance with a form rather than with matter, the nature of a substance is an active rather than a passive internal principle of any of its natural motions, that is, it is an internal mover. But how can the specific substantial form of a substance be an internal mover in its generation if it only exists at the end of its generation? (II, q. 3, nn. 1–6). Aspall's reply to this question is highly original as an interpretation of Aristotle's thought. As we have seen above in the section on matter,[93] Aspall

[92] See above, pp. xlviii–xlix.
[93] See above, pp. xliv–xlvi.

assumes that in addition to prime matter and a form, there is a third thing described as a diminished active potency towards a form. The role of this diminished potency is that of enabling matter to receive that form. Although the diminished potency towards a form is an active principle with respect to that form, it is not active by itself, so that taken by itself it does not qualify for being the internal mover in the generation of a substance. It becomes such a mover when it becomes active, and it does so when it is stimulated by the external agent of a generation. For example, suppose that some fire F1 is generated from air and that the external agent of this generation is some other fire F2. The diminished potency towards the form of fire existing in air needs to receive an external stimulus from fire F2 in order to activate itself and develop into the specific form of fire. This external stimulus is the intentional species of the form of fire. Accordingly, the nature that is an internal agent in the generation of fire is an aggregate of the diminished potency towards the form of fire together with the intentional species of the form of fire introduced by the fire that is the external agent in this generation. More generally, the nature involved in the generation of a substance is a diminished potency towards the substantial form of that substance together with the intentional species of the substantial form: it is this aggregate that develops by itself into the specific substantial form of the new substance (II, q. 4, n. 12; q. 5, nn. 12–13).

Many aspects of Aspall's account of nature as an internal principle of the generation of a substance are obscure and left without an adequate explanation. He seems to take this account for granted as reflecting an already established opinion (II, q. 4, n. 12). Indeed, his account is based on a view about natural agency that is very popular among English commentators of the mid-thirteenth century such as Roger Bacon.[94] The most striking aspect is the appeal to intentional species. In a Question specifically devoted to the activation of a diminished potency of matter in book I, Aspall takes it for granted that the external agent must introduce something into the patient in order to stimulate the diminished active potency towards a form and asks what this something is. The relevant candidates are (i) a real thing and (ii) the intentional species of a thing. His motivation for rejecting (i) is that a thing that is transferred from the agent to the patient could only be a composite of matter and form, that is, an integral part of the agent, with the unwanted consequence that the agent could eventually be extinguished as a result of its action. We are then left with (ii) but no positive arguments are given in favour of it. Aspall simply points out that an intentional species can produce specifically different effects depending on the nature of its recipients. So the intentional species of fire when it is received in the senses produces a sensation of

[94] See Donati, 2015b.

fire and when it is received in a physical subject in which a form contrary to that of fire exists, as happens in the case of the generation of fire from air, it stimulates the diminished potency towards the form of fire that exists in that subject (I, q. 77, nn. 1–2, 5).[95]

5.4.2. Some particular cases of natural things

Aspall singles out the following classes of natural things for a specific discussion: (1) the celestial bodies; (2) the earth; (3) animate bodies; (4) heavy and light bodies.

5.4.2.1. Celestial bodies The main issue about the nature of the celestial bodies is whether their eternal circular motion is natural to them, that is, in Aspall's understanding of Aristotle's definition of nature, whether the specific form of the heavens is what moves them with their eternal circular motion. Aspall reports the view of some people who think that the form of the heavens can certainly produce a single celestial revolution but not the endless cycle of celestial revolutions, and therefore posit that an intelligence or incorporeal mover is required to account for the eternity of celestial motion (II, q. 7, n. 7). Aspall rejects this view, pointing out that the form of the heavens remains absolutely unchanged throughout their eternal circular motion, so that if it can produce a single revolution, it can also produce the next one (II, q. 7, n. 8). He defends the view that the form of the heavens is their mover by appealing to the status of this form as the 'most noble' among the corporeal forms. According to Aspall, being the most noble implies that the form of the heavens produces a maximal degree of homogeneity in their parts; this in turn implies that each part of the heavens desires to be in the same place as any other part. Since the parts of the heavens are spatially extended and hence cannot be in the same place simultaneously, the only way one part can occupy the place of another part is through temporal succession, so that the place that is occupied by one part at a given time comes to be occupied by all the other parts at different times, and this is exactly what happens in a celestial revolution (II, q. 7, n. 9). Another argument for positing that the form of the heavens is their mover is based on the assumption that it belongs to the nature of the heavens that the celestial light emanated by some celestial bodies is responsible for the cycle of generation and corruption in the sublunar world, and this causal role requires the heavens to be eternally moved with a circular motion (II, q. 7, n. 10).

Aspall touches on the problem of whether we should posit an incorporeal mover of the heavens, that is, an intelligence, in addition to their corporeal form. He sides with a view according to which an intelligence is required to

[95] On Aspall's discussion of the activation of the diminished active potency, see also VI, q. 16.

account for the fact that the heavens in virtue of their circular motion are responsible for the generation of living things, like plants and animals, a causal role that the corporeal form of the heavens cannot have (II, q. 7, nn. 11–12). We will come back to this issue in the section on self-motion.[96]

Another issue concerning the nature of celestial bodies is how the condition expressed by Aristotle's definition that the nature of something is a principle not only of its motion but also of its rest is satisfied in the case of celestial bodies, given that they are never at rest. Aspall rejects an attempt to argue that in the celestial bodies there is a potentiality to be at rest, although this potentiality is never actualized (II, q. 7, nn. 15–17). He maintains instead that the heavens are indeed actually at rest in some sense: the rotation of the heavens is such that the heavens as a whole do not change place, though their parts do (II, q. 7, n. 18).

5.4.2.2. The earth Aspall supports Aristotle's view that the earth as a whole is naturally at rest in the centre of the universe by a cosmological argument that relates the being at rest of the earth to the motion of the heavens. He takes it as established by his discussion about the celestial bodies that the circular motion of the parts of the heavens is ordered by nature to the being at rest of the heavens as a whole. Since the earth is the centre around which the rotation of the heavens occurs, the being at rest of the heavens as a whole requires that the central earth is at rest, so that being directed by nature to the being at rest of the whole heavens, the circular motion of the parts of the heavens is also directed by nature to the being at rest of the earth (II, q. 9, n. 7). This argument, however, only shows that the being at rest of the earth is due to the nature of the heavens, and not also that it is due to the intrinsic nature of the earth itself. And the fact that the parts of the earth always do move downwards when they are not impeded, suggests that this motion is natural to them and therefore that their state of rest is violent, i.e., against their nature (II, q. 9, n. 1).

In reply to this objection to the naturality of the being at rest of the earth, Aspall makes the general point that a necessary condition for something's being at rest by violence is that it is at rest outside its natural place. This condition is not met by parts of earth which are not separated from the whole earth, for the natural place of a part of earth is the same as that of the whole earth, and this is the space around the centre of the universe actually occupied by the whole earth; therefore, even if one such part moves downwards when the parts below it are removed, it cannot be inferred from this that that part was at rest by violence, since it moves within its natural place (II, q. 9, n. 8). Some of Aspall's critical remarks seem to make the point that for this reply

[96] See below, pp. xcv–xcviii.

also to work in the case of parts of earth separated from the whole earth one needs to assume that the natural place of the earth is not a definite region of space (II, q. 9, nn. 9–11).

5.4.2.3. Animate bodies The controversial issue for this class of natural things is whether the nature of an animate body should be identified with its soul. The controversy concerns not only the human soul and its voluntary powers, but also the soul of plants and that of non-rational animals. Aspall reports an opinion that denies that the nature of an animate body is its soul because what is responsible for the distinctive natural motions of an animate body, such as nutrition, growth, and reproduction, is not its soul but a material principle, namely, heat of different kinds. The soul is only responsible for the boundaries of such natural motions, that is, for the fact that these motions do not occur indefinitely but have definite limits, so that, for instance, a plant does not grow beyond a definite size (II, q. 10, n. 6).

While Aspall does not explicitly reject this opinion, he seems to prefer a more nuanced one, which distinguishes a proper sense and a broad sense of nature, and which posits that the soul of an animate body is its nature taken in a broad sense and but not also its nature taken in a proper sense. Nature taken in a broad sense is any internal mover whereas nature in a proper sense is an internal mover with the additional property of not being essentially distinct from what is moved by it. For example, the form of a light body is its nature in a proper sense because it is a mover of that body but it is not essentially distinct from it. The soul of a living being, on the contrary, is its nature only in a broad sense because it is essentially distinct from the thing moved by it, that is, from the body. Properly speaking, then, the soul of an animate body is not its nature but simply its mover (II, q. 10, n. 7). This account assumes that there is a dualism of body and soul, which is not explained in this context. As we shall see in the section on self-motion, Aspall uses this dualism in support of Aristotle's view that animate bodies are self-movers.[97] Furthermore, according to this opinion, even in the natural non-voluntary motions of an animate body the soul is not properly speaking responsible for these motions as such but only for their limits (II, q. 10, n. 8).

5.4.2.4. Heavy and light bodies Although in the Aristotelian tradition the locomotion of a heavy body downwards and that of a light body upwards are regarded as the standard examples of natural processes, the 'naturality' of these motions is also open to a standard objection arising from Aristotle's discussion of self-motion in *Physics* VIII. Aristotle there denies that heavy and light bodies are self-movers, that is, bodies that are moved by themselves rather than by something else, and claims that self-motion is distinctive of

[97] See below, pp. xciii–xcv.

animate things.[98] But the claim that heavy and light bodies are not self-movers with respect to their characteristic locomotions seems to be incompatible with the naturality of these locomotions. If these locomotions were natural, then, according to the definition of a nature as an internal principle of motion and rest, they should derive from a source internal to these bodies, and so these bodies should be self-movers (II, q. 11, n. 1). The reason why heavy and light bodies are not self-movers is that for something to be a self-mover it is not sufficient that it contain within itself a mover and a moved thing, but it is also required that the mover and the moved thing be essentially distinct. This requirement is not satisfied by heavy and light bodies, for their only internal components are matter and their substantial form(s), and their substantial form is the mover; but matter cannot be the moved thing. Although matter is essentially distinct from form, it cannot have the status of the thing moved with a locomotion, precisely because it is formless and hence not an actual being. The thing moved with a locomotion is the whole heavy or light body, that is, the composite of matter and form, which is not essentially distinct from its form (II, q. 11, n. 2). Aspall accepts this explanation, which is very popular in the thirteenth century. He thinks, however, that this explanation also shows why the conclusion that the locomotions of heavy and lights bodies are not natural does not follow from the premise that they are not self-motions. This inference is based on the wrong assumption that the mover as a nature must not only be an internal source of motion but also essentially distinct from the thing moved by it. Consistently with his discussion of the case of animate bodies, Aspall points out that the essential distinction between the mover and the moved thing disqualifies the mover from being a nature properly speaking. The mover as a nature is not essentially distinct from the thing moved (II, q. 11, n. 8).

Other aspects of Aspall's discussion specific to self-motion will be examined in the section on that topic below.[99]

5.5. Causes

Aspall devotes two series of Questions to Aristotle's doctrine of the causes (*Physics* II.3–7). The first series is about the four per se causes (material, formal, efficient, and final causes), [100] and the second series is about the accidental causes (chance and fortune).[101] In addition, Aristotle's discussion of finality in nature (*Physics* II.8–9) gives rise to Questions about monsters.

[98] Arist. *Phys*. VIII.4, 255a5–7.
[99] See below, pp. xcii–xciii.
[100] Arist. *Phys*. II.3, 7
[101] Arist. *Phys.* II.4–6.

5.5.1. Per se causes

Aspall's discussion of Aristotle's doctrine of the four causes is short and mainly exegetical. His main contribution is that of showing why there are exactly four kinds of (per se) cause, a point that Aristotle leaves without proof. Aspall focuses on a material substance and assumes that there are two aspects of such a substance that require a causal explanation, namely its being and its becoming, that is, its changes. The being of a material substance is further divided into its potential and its actual being; its potential being is then ascribed to the material cause and its actual being to the formal cause, that is, to its matter and to its form respectively. The cause of the becoming or changes of a substance is the mover of this substance. Movers are of two kinds: a mover that moves only in so far as it is itself in motion, and a mover that moves without being itself in motion. In Aspall's view, a mover of the first kind is an efficient cause, while a mover of the second kind is a final cause (II, q. 17, n. 4). Aspall further clarifies the distinction between efficient and final causes in the case of a rational agent. An efficient cause is a mover that produces something, whereas a final cause is a mover that simply stimulates the efficient cause to produce something but does not produce something by itself. For example, in the case of the building of a house, the carpenter is the mover as efficient cause while the intentional species of the house to be built is the mover as final cause, since it stimulates the agent to produce a house. Thus, the carpenter moves as an efficient cause, that is, he produces something, only in so far as he is moved, that is, stimulated, by the intentional species of the thing to be produced (II, q. 20, n. 7).

Aristotle's claim that one and the same thing can be the cause of one effect by its presence and of the contrary effect by its absence, as when the pilot of a ship is the cause of its safety by his presence and of its wreck by his absence,[102] leads Aspall to discuss the counterexample of light. The absence of light does not cause any positive effect, for what results from the absence of light is darkness, which is not a thing but the mere absence of light. Aspall's explanation of this feature of the causality of light appeals to the ontological status of light as an external form, that is, a form that comes to exist in a material substance 'from outside'. The contrast here is between light and intrinsic forms, that is, forms that are educed from the potencies existing in the matter of a material substance, such as the forms of substantiality and that of corporeality. What Aspall means by a form's being educed from the potency of matter is that there is an active diminished potency in matter towards that form. Aspall further assumes that in the case of a form educed from a potency of matter, the matter has two kinds of potency, one towards

[102] Arist. *Phys.* II.3, 195a12–14.

that form and the other towards the contrary form. For example, since heat is a form educed from a potency of matter, there is in matter both a potency to heat and a potency to cold. Because of the presence in matter of both a potency towards form F and of a potency towards a form contrary to F, when form F no longer exists in matter, matter can acquire the form contrary to it. In the case of light, since light is not educed from a potency of matter and so there is no active diminished potency in matter towards light or its contrary, the absence of light does not result in any positive thing (II, q. 19, n. 5).

5.5.2. Accidental causes: chance and fortune. Fate

In the Questions on chance (*casus*) and fortune (*fortuna*), Aspall actually discusses two distinct views: that of Aristotle in *Physics* II.4–6 and that of Augustine in *De Civitate Dei* V. While Aristotle distinguishes between chance and fortune and posits that chance should be taken more widely than fortune,[103] Aspall almost exclusively uses the term 'chance' in connection with Aristotle's view. His Questions on chance are then about Aristotle's view, whereas those on fortune are about Augustine's view. Augustine's view also leads Aspall to discuss the notion of fate (*fatum*).

5.5.2.1. Chance Aspall presents the problem of what chance is as that of establishing whether chance is some kind of cause or rather some kind of effect. Authoritative advocates of these two views are Aristotle and Boethius respectively: Aristotle classifies chance as an accidental cause, whereas Boethius defines it as an unexpected event or effect (II, q. 21, nn. 1–8). Aspall's discussion shows that some commentators in his time thought that both Aristotle and Boethius were right and tried to reconcile their views in different ways. One way simply consists in postulating that the term 'chance' is equivocal and can signify both an accidental cause and an accidental effect equally well and primarily (II, q. 21, n. 9). Another more sophisticated way of reconciling the views consists in distinguishing a primary and a secondary significate of the term 'chance', so that its primary significate is an accidental event and its secondary significate is an accidental cause (II, q. 21, n. 12).

Aspall, however, is not satisfied with these attempts to identify chance either exclusively with a kind of cause or exclusively with a kind of effect or with both a cause and an effect. He maintains that chance should be conceived as a relation (*habitudo*) between a cause C and an effect E and that the relation should have the distinctive property that the cause C is not ordered to produce E. In other words, the accidentality distinctive of chance is neither the accidentality of a cause nor that of an effect but that of the causal relation between a cause and an effect. In Aristotle's example of the

[103] Arist. *Phys.* II.6, 197a36–b23.

man who goes to the market, meets his debtor there, and thus recovers his money, it is neither the man's going to the market (the cause) nor his meeting his debtor and thus recovering his money (the effect) that is by chance but the fact that his going to the market results in his meeting his debtor and thus recovering his money. Because chance is a relation holding between a cause and an effect the word 'chance' can denominate each of them: e.g., we can say that a man's going to the market is an accidental or chance cause of his meeting his debtor and that a man's meeting his debtor is an accidental or chance effect of his going to the market. In Aspall's interpretation, these two kinds of denomination are those used by Aristotle and Boethius respectively to describe chance (II, q. 21, n. 15).

Aspall offers the clarification that when Aristotle claims that a pale man is an accidental cause of a house[104] and that a man's going to the market is an accidental cause of his meeting his debtor[105] he is using two distinct notions of accidental cause. A pale man is an accidental cause of a house because being pale is an accident of the per se cause of a house, that is, of the builder, whereas a man's going to the market is an accidental cause of his meeting his debtor because it produces an effect that is not the one to which it is ordered (II, q. 21, nn. 5, 19). Aspall suggests that there is an important difference between the two kinds of accidental cause. Whereas an accidental cause of the first kind presupposes a per se cause because, for instance, reference to a pale man presupposes or is reducible to reference to a builder as the per se cause of a house, an accidental cause of the second kind does not presuppose a corresponding per se cause. In Aristotle's example, there is no genuine cause of meeting one's debtor other than going to the market, and the fact is that meeting one's debtor does not have a per se cause in the sense of a cause ordered to produce it (II, q. 22, n. 14).

Aspall accepts Aristotle's view that there are things that happen by chance. In reply to theological objections, he makes the point that the fact that each created thing or event is necessarily ordered to the first cause does not prevent it from happening by chance. If a thing is ordered to the first cause but is not intended by the first cause, then it does happen by chance, because in this case the first cause is not per se ordered to produce that thing. Similarly, the fact that something is produced by necessary causes does not prevent it from happening by chance. Thus, in an Aristotelian example, the falling of a stone on someone's head necessarily causes the fracture of his head, but this fracture comes about by chance because the falling of a stone is not intentionally ordered to this effect (II, q. 22, nn. 12–13, 16–18).

[104] Arist. *Phys.* II.5, 196b25–27.
[105] Arist. *Phys.* II.5, 197a14–17.

According to Aspall, the question whether there are future events that happen by chance merges with a question about the truth-value of propositions about the future. The latter is a major subject of controversy in the medieval debate about future contingents which arose from Aristotle's *De Interpretatione*.[106] Aspall reports two main opinions: (i) one says that every proposition about the future such as 'A will be' is false (II, q. 23, n. 1); (ii) the other says that every proposition about the future lacks a truth-value, being neither true nor false (II, q. 24, title).

Aspall rejects both (i) and (ii), and maintains that (iii) there are propositions about the future that are false and others that are true, and this depends on whether the event in question will actually happen or not: 'A will be' is true if A will be and false if A will not be (II, q. 24, n. 11).

The crucial problem that Aspall has to face in supporting (iii) is that of showing that (iii) is compatible with the distinction between contingent and necessary future events, that is, the problem of showing that to grant that a proposition about a future event is true does not imply that that event necessarily happens. This problem is spelled out in temporal terms as follows: suppose an event A happens at time t; then if the proposition 'A will be' is true, it is true at any time before t, so that it is eternally true (before t). From the hypothesis that 'A will be' is eternally true does it follow that 'A will be' is necessarily true? Aspall denies the validity of this inference by pointing out that the condition of being eternally true is not enough to make the proposition necessarily true; what is also required is that the proposition cannot be false for the whole time in which it is true. This additional condition is not satisfied in the case of contingent future events (II, q. 23, nn. 5–8).

Aspall also addresses the theological version of this problem, which arises from the assumption of God's foreknowledge of future events; and he reports a number of opinions about how God's foreknowledge can be reconciled with the contingency of future events (II, q. 23, nn. 9–11).

5.5.2.2. Fortune Aspall briefly reports Augustine's view in *De civitate Dei* that there are no fortuitous things and events properly speaking, in the sense of things that do not have genuine causes, since God and his providence are responsible for any thing or event. 'Fortune' is a term that people use when the causes of something are not known to them, but it is not part of a rigorous philosophical language, in which a sentence like 'Fortune wanted this' should be replaced with the sentence 'God wanted this' (II, q. 25, nn. 1–4).

According to Aspall's interpretation, Augustine (i) denies that there are things that happen by chance with respect to the first cause by positing that all things happen by divine necessity and (ii) maintains that God's foreknowledge

[106] Arist. *De Int.* 9.

of any future thing or event makes that thing or event necessary. Aspall is strongly against Augustine on this issue and defends the Aristotelian view that there are indeed things that happen by chance. He points out against (i) that the assumption that God is a cause of all things does not prevent some of them from happening by chance. God can cause something but yet he may not intend to cause it. Consistently with his definition of chance, Aspall maintains that a thing which is caused by God but not intended by him happens by chance, for the causal relation in this case has the accidental character distinctive of chance. Similarly, he argues against (ii) that God's foreknowledge of a future event as such is not enough to make it necessary; what is also required is that God has the intention to cause that event, that is, given Aspall's definitions, God's foreknowledge must be accompanied by his ordering and intending will. It is this ordering and intending will that is missing in God's foreknowledge of a chance event (II, q. 25, nn. 5–6).

5.5.2.3. Fate According to Aspall, Augustine associates the notion of fate with that of necessity, and maintains that things ordained by fate are those that have necessary causes of their coming into being, so that they are unavoidable. Augustine also distinguishes two notions of fate: (i) a theological one, according to which fate is the divine word or will, which orders all things, and (ii) a cosmological one, according to which fate is a causal order proceeding from the celestial bodies. Augustine then posits that all things are ordained by fate in the theological sense (i) but there are no things ordained by fate in the cosmological sense (ii) (II, q. 25, nn. 7–10; q. 27, n. 13).

Aspall thinks that Augustine's notion of fate is not an acceptable one because it does not take into account the fundamental distinction between things that come to be as intended by their causes (*cum intentione*) and things that come to be without being intended by their causes (*sine intentione*), that is, casual things. According to Aspall, the correct notion of fate only applies to things that are intended by their causes, so that things that happen by chance are not ordained by fate (II, q. 27, n. 14).

Aspall takes Augustine's view on fate as an occasion for a general discussion of necessity and contingency in natural and voluntary things or actions. Aspall examines (i) the case of voluntary things, understood as things that are produced by voluntary agents alone, and (ii) the case of natural things, those that have natural causes in addition to the divine will. (i) Among voluntary things, those caused by the divine will are all necessary. It is worthy of notice that Aspall includes in this class not only the celestial bodies, the intelligences and the rational souls, but also the elements. What he seems to have in mind is the first coming into being of an element, which is effected directly by God without any natural causes. Something caused by a created will can be either necessary or contingent. If the created will by which a thing is caused always

chooses what is good, then that thing is necessary, but if the created will chooses sometimes what is good and sometimes what is bad, then that thing is contingent. Accordingly, the voluntary actions of the separate intelligences are necessary because these intelligences always choose what is good, but the voluntary actions of men are contingent. (ii) For things produced by natural causes, Aspall distinguishes (a) the case in which a natural cause C is the immediate cause of a natural thing or event E from (b) the case in which C is a mediate cause of E. In case (a) the natural thing or event E necessarily happens, in case (b) it does not necessarily happen. This distinction is illustrated by some examples. (a) Rain necessarily happens whenever Saturn passes through the sun in summer, for although this celestial configuration occurs rarely, when it does occur, rain necessarily follows. (b) Grass, on the other hand, does not necessarily grow as a result of the relevant natural factors produced by the celestial bodies, such as heat and humidity, because these factors do not immediately produce the growth of grass but only do so through intermediate material factors, which sometimes offer resistance to the action of the celestial bodies or are not able to receive it (II, q. 27, nn. 15–18).

5.5.3. Monsters

In *Physics* II.8 Aristotle mentions monsters—living beings like man-headed oxen and olive-headed vines—in arguing for the existence of the final cause in nature.[107] Aspall points out that in his zoological treatises Aristotle also classifies a mule as a monster.[108] This leads Aspall to introduce a distinction between a broad sense and a proper sense of the term 'monster' so that a mule is a monster in a broad sense while living beings that belong in part to one species and in part to another, such as man-headed oxen and olive-headed vines, are monsters in the proper sense (II, q. 31, n. 1*). On Aspall's interpretation, a monster in the proper sense is a living being that is not intended by nature, whereas a monster in a broad sense is a living being that is intended by nature but the way in which it is generated deviates from the standard course of nature. Thus a mule is intended by nature but its generation violates the principle that like generates like, which regulates the natural productions of living beings (II, q. 31, nn. 1–2).

In the case of monsters in the proper sense, Aspall raises the question of whether such living beings exist by nature. The difficulty here seems to be that of reconciling the fact that these animals come to be through a natural process of generation with the fact that the outcome of this natural process is something that is classified as being against nature (II, q. 37, nn. 1–6). To deal with

[107] Arist. *Phys.* II.8, 199a32–b13.
[108] Arist. *De gen. an.* II.8.

this difficulty, Aspall introduces a distinction between a universal nature and a particular nature. This distinction is not found in Aristotle, who nowhere suggests that there is a universal nature in addition to the natures of the various kinds of natural substance. In Aspall's time, however, this distinction seems to be well-established and commonly accepted. Indeed, Aspall takes this distinction for granted, without feeling any pressure to justify or explain it. As we have seen, in his discussion of nature Aspall rejects the view that the nature defined by Aristotle in *Physics* II.1 is a universal cosmological principle;[109] it is likely, therefore, that in his understanding only a particular nature is a nature in Aristotle's sense. What is clear is that Aspall gives distinct roles to the two kinds of nature: a universal nature is a regulative or ordering principle (*ordinans*) whereas a particular nature is an operative principle (*operans*). This distinction allows Aspall to explain why it is not inconsistent to posit that monsters are both against nature and by nature: they are against nature if nature is taken as the universal ordering nature, since they are not intended by such a nature and are in fact contrary to its intention; they are by nature if nature is taken as the particular operating nature, since they are indeed produced through a natural process of generation (II, q. 37, n. 7).[110]

Aspall's analysis of the case of the monsters and his distinction between two kinds of nature seem to have major implications for natural teleology, but these are not developed by Aspall. Although his discussion about monsters is very extensive and covers many more topics than those presented here (II, qq. 38–43), they address very specific issues that more properly belong to the discussion of Aristotle's zoological treatises rather than to the general investigation of the *Physics*.

5.6. Motion[111]

5.6.1. Existence of motion

The main issue that Aspall discusses in the Question he devotes to the existence of motion is that of the different criteria for the existence of permanent things and successive things respectively. A permanent thing, for example, a material substance, is something whose parts can all exist at the same time and in particular in the present instant; a successive thing is something whose parts can only exist one after another in temporal succession and none of whose parts can exist in the present instant. Motion and time are commonly classified

[109] See above, p. xlix.

[110] For a discussion of closely related topics see VIII, qq. 6–7.

[111] A detailed paraphrase (in Italian) of the contents of each of Aspall's Questions on motion (*Physics* III.1–3) is given in Trifogli, 2004, 169–179. See also, Trifogli, 1991, 467–501; 2000, 37–86; 2004, 8–54.

as successive things. In his treatment of time in *Physics* IV.10, Aristotle presents some arguments against the existence of time that are based on its successive nature. The conclusion that time does not exist is inferred from the premise that no part of time exists: the only parts of time are past and future parts; but these do not exist because only what is present exists, and what is present in time is not a temporally extended part of time but an indivisible instant.[112] Like the vast majority of Medieval commentators, Aspall takes it that the arguments of *Physics* IV.10 also apply to motion, because the parts or phases of a motion like those of time are either past or future and no part of motion is present, since no part or phase of a motion can occur in the indivisible present instant. The main arguments presented by Aspall against the existence of motion are adaptations of the arguments found in *Physics* IV.10 (III, q. 1, nn. 1–3).

In his reply to these arguments, Aspall makes the preliminary point that they assume a criterion of existence that is valid for permanent things but does not apply to successive things. For a permanent thing to exist it is necessary and sufficient for all its parts to exist simultaneously in the present, but for the existence of a successive thing it is necessary and sufficient for its parts to exist one after another so that one is past and the other is future (III, q. 1, n. 5). Aspall thinks, however, that this preliminary point is not enough to settle the question of the existence of a successive thing. One may object that a motion (and likewise time) does not have past or future parts at all. This is because saying that a part of motion is past or future implies that that part is such that at some point it was present or will be present; but no part of a motion can ever be present (III, q. 1, n. 6). In reply to this objection, Aspall concedes that no part of a motion or of time can ever be present but he maintains that the argument that infers from this claim that motion and time do not exist should be rejected because it overlooks the important role of the present indivisible elements of motion and time, namely the present instantaneous change in motion and the present instant of time. These present indivisible elements are not simply boundaries between past and future parts of motion and of time but also have a causal role because they 'produce' motion and time respectively (III, q. 1, nn. 12–13). Like many thirteenth-century commentators, Aspall believes that Aristotle himself ascribes this causal role to the present instant of time in a difficult passage of *Physics* IV.11 where he suggests that the present instant is something that persists in time but is in some way diversified throughout time, in the same way in which the mobile body persists during its change but is in different states throughout its change.[113] According to Aspall's interpretation, Aristotle wants to say that the

[112] Arist. *Phys.* IV.10, 217b32–218a8.
[113] Arist. *Phys.* IV.11, 219b9–220a33.

present instant flows in time and produces time by its flow. He remarks that the present indivisible change has similar properties. He expands and tries to elucidate Aristotle's view by pointing out that both in the case of motion and of time, the uninterrupted succession of their parts is in some way present, that is, real, because 'it passes through' their indivisible present element, which is also described as what carries the whole of motion and time.

Aspall's attempt to explain the causal role of the present elements of motion and time is too short and obscure. It is clear that he thinks that the existence of a successive thing derives from the existence of its present indivisible element, but how exactly this derivation should be understood remains unexplained.[114]

5.6.2. Classification of motion in the categories

In *Physics* III Aristotle classifies four species of motion or change on the basis of the category to which the things with respect to which a change occurs belong. The four species are generation and corruption, which are changes in substance; alteration, which is a change in quality; increase and decrease, which are changes in quantity; and locomotion, which is a change in place or 'where'.[115]

Aspall assumes that the classification of the species of change given by Aristotle is based on considering motion as a way towards a thing belonging to a category (*via in rem praedicamenti*) and distinguishes it from a classification of motion itself considered as a species of a category (*species in genere*). The point of Aspall's distinction is the following: Aristotle's classification does not say what kind of thing a motion itself is but only what kinds of thing are those towards which a motion is directed. For example, saying that generation and corruption are changes in the category of substance does not mean that generation and corruption are themselves substances, but only that they are processes through which a substance comes to be or passes away. It is only the classification of a motion as a species of a category that says what kind of thing motion itself is. Aspall maintains that a motion is in different respects both a thing in the category of quality and a thing in the category of quantity. When a motion is considered as an accident of a mobile thing which denominates that thing, then it is a quality; when it is considered as involving a distance between its initial terminus and its final terminus, then it is a

[114] Aspall discusses the issue of the existence of motion again in Book VI, q. 58 in connection with Zeno's arguments. The two discussions have many elements in common but are not totally parallel.

[115] Arist. *Phys*. III.1, 200b32–34, 201a9–15.

quantity (III, q. 4, n. 10).[116] The question of how it is possible for a motion to be a species of two distinct categories is not elucidated by Aspall.

Aspall's main concern is to defend the assumption, which is implicit in his distinction between the two kinds of categorical classification of motion, that the relation of a motion to the final terminus towards which it is directed does not determine what kind of thing a motion is. Aspall is aware that this assumption is contrary to the authoritative opinion of Averroes, who maintains that a motion does not differ essentially but only in degrees of completeness from the final form towards which it is directed. For example, becoming white is not a thing distinct from whiteness, but it is whiteness in a diminished and incomplete state. According to Averroes then, the final terminus towards which a motion is directed determines the kind of thing that motion is because that motion is essentially the same as its final terminus.[117] It is clear that the explanation provided by Averroes of Aristotle's classification of the species of motion reflects a reductionist position about the ontological status of motion, according to which in order to account for the becoming white of a body, for example, it is not necessary to posit its becoming white as a thing by itself, conceived as an accident of the body and distinct from the qualities that that body successively acquires and loses when it becomes white, because its becoming white is essentially the same thing as one of these qualities, that is, the whiteness finally acquired by that body. Like many of his contemporaries, Aspall rejects this reductionist view. He argues against it by using assumptions that are typical of a realist view about motion. For example, one assumption is that motion is a successive thing inherent in the mobile body and irreducible to its permanent accidents, which are those with respect to which that body changes (III, q. 5, n. 1). Another assumption is that motion is the thing through which a permanent form comes to be and therefore it cannot be the same as that form (III, q. 5, n. 2). More generally, he rejects the view that the relation of a motion to its final terminus is so essential that it determines what kind of categorical thing motion is (III, q. 5, nn. 7–8). He argues that since motion is an accident of the mobile body, its classification in Aristotle's categories should reflect its status as an accident; but motion has the status of an accident when it is related to the mobile body and not when it is related to its final terminus. It is with respect to the mobile body that motion is an accident, whereas with respect to its final terminus it is a way towards it (III, q. 5, nn. 9–10).

Aspall often uses the expression 'way towards a form' to describe the kind of thing that motion is. This expression is borrowed from Averroes, who

[116] The same solution to the problem of the classification of motion in the categories is found in Book VI, q. 59, n. 5.

[117] Aver. *In Phys.* III, t. c. 4, fol. 87ra–b.

contrasts the view that motion is not essentially distinct from its final form, which he thinks is the true one, with the view that motion is a way towards the final form and really distinct from this form, which he regards as merely a common opinion.[118] Contrary to Averroes, Aspall maintains that it is Aristotle's genuine view that motion is a way towards a form. In his interpretation, when Aristotle defines motion as the act of what is potentially in so far as it is potentially,[119] the term 'act' does not stand for a form, not even for a form in an incomplete state, but for a way towards a form, whereas the potentiality in question is the potentiality to the form, so that Aristotle's definition should be read as saying that motion is the way through which a mobile thing acquires a form (Book III, q. 8, n. 5; q. 9, n. 4).

Aspall also raises the question of the sufficiency of Aristotle's classification of motion into four species, and lists three kinds of motion which apparently do not fall under any of these species: condensation and rarefaction; nutrition; and the process through which intentional species come to exist in the soul (III, q. 6). This last case is the most interesting. In Aspall's view this process is a motion because it occurs successively and not instantaneously; but it cannot be one of the four species of motion distinguished by Aristotle because it is not directed to a thing belonging to a category, but to an intention, which is not a thing belonging to a category (III, q. 6, nn. 7–8).

All three counterexamples are left without a solution. A reference to his commentary on the *De anima* may indicate that Aspall devotes a more extended discussion to the species of motion in that work (III, q. 7, n. 1*).

5.6.3. Action and passion

According to Aristotle, the existence of a motion depends not only on a thing that is moveable—the mobile thing or patient—but also on a thing that is capable of moving—the mover or agent. Furthermore, a motion involves not only an act of the patient, but also an act of the agent causing that motion. When a motion actually occurs, both what is capable of being in motion and what is capable of causing motion are actualized by becoming something actually in motion and something that actually causes a motion respectively. Aristotle defines passion as the act of the patient and action as the act of the agent. His main claim about action and passion so defined is that, like the way from Athens to Thebes and the way from Thebes to Athens, they are really the same thing and differ only in description. More explicitly, while the definition of motion makes it clear that motion is the act of the mobile thing or patient, Aristotle specifies that motion is also the act of the agent, so that action and

[118] Aver. *In Phys.* III, t. c. 4, fol. 87ra–b.
[119] Arist. *Phys.* III.1, 201a9–11.

passion are one and the same motion described in different ways. When a motion is described as related to the patient by being in the patient, then it is a passion, and when it is described as related to the agent by being produced by the agent, then it is an action. In the course of arguing for this main claim, Aristotle also establishes the point that the subject of both action and passion is the mobile thing or patient, for although action is the act of the agent, such an act does not exist in the agent itself but in the patient.[120]

Aspall focuses on the two main issues of Aristotle's treatment of action and passion in *Physics* III: the issue of the subject of motion, and that of the real identity of action and passion. He adds a third issue arising from Aristotle's discussion in the *Categories*, that of the categorical distinction between action and passion.

5.6.3.1. The subject of motion In support of Aristotle's position that the subject of motion is only the patient or mobile thing, Aspall points out that the subject of motion is the thing that primarily acquires something new through motion and this is the mobile thing rather than the mover. For example, it is the mobile thing that acquires whiteness through the motion of becoming white. The agent causing the becoming white does not acquire anything apart from a new relation to the mobile thing, that is, the relation of being the cause of its becoming white, but this acquisition is accidental, that is, dependent on the mobile thing's acquisition of whiteness (III, q. 10, nn. 9–10, 14–15).

Aspall also remarks that the abstract term 'motion' denominates both the mobile thing and the mover but in different ways: it denominates the mobile thing as the subject of motion and the mover as its efficient cause, just as 'health' denominates a body as its subject and a medicine as its efficient cause (III, q. 10, n. 12).

5.6.3.2. Real identity between action and passion In assessing Aristotle's view that action and passion are one and the same thing Aspall distinguishes two senses of 'action': the Aristotelian sense, according to which an action is a motion related to its agent, that is, a motion considered as being produced by an agent, and another sense according to which an action is the operation of the agent itself on the patient, that is, what the agent does in order to produce that motion or change in the patient. In the case of voluntary agents, such as the first cause, the intelligences and the human soul, the operation of the agent is specified as a voluntary conversion of the agent into a patient, whereas in the case of natural agents it is specified as the influence of the power of the agent on the patient. Given this distinction between two notions of action, Aspall maintains that Aristotle's view only holds for an action understood in

[120] Arist. *Phys*. III.3, 202a13–b22.

Aristotle's sense as a motion related to its agent. It is not valid, however, if action is taken as the operation itself of the agent. This operation is not the same as the motion that the agent causes in the patient but is rather the causal process by means of which the agent causes that motion: it is what the agent does in order to produce a motion in the patient (III, q. 11, nn. 8–9).

5.6.3.3. Action and passion as distinct categories Like many other Aristotelian commentators, Aspall finds the thesis of the real identity between action (in Aristotle's sense) and passion put forward in *Physics* III difficult to reconcile with Aristotle's position in the *Categories* that action and passion are distinct categories.[121] According to the authoritative interpretation of Boethius, the ten categories correspond to the first ten kinds of things, so that a distinction between things is required for a distinction between categories (III, q. 12, n. 1).

Aspall maintains that the opinion of Boethius is not as conclusive against the categorical distinction between action and passion as first appears. In Aspall's view, when Boethius claims that distinct categories signify distinct things, he uses the term 'things' in a broad sense to include both 'true' things and the 'circumstances' of things. For example, the same 'true' thing corresponds to time and to 'when', that is, a thing belonging to the category of quantity, but a 'when' adds to time the circumstance of being related to a temporal thing, and because of this additional circumstance it gives rise to a category distinct from quantity. The same remark applies to the case of place and 'where'. In the case of action and passion as well there are distinct circumstances added to the same motion: the circumstance of being from the agent, characteristic of an action, is distinct from that of being in the patient, characteristic of a passion (III, q. 12, n. 10).

Aspall, however, is not totally happy with this solution to the problem of the categorical distinction of action and passion. He thinks that it is open to the counterexample of accession and recession from a terminus. For instance, the motion from Athens to Thebes can be regarded both as a recession from Athens and an accession to Thebes. The different circumstances in this case seem to be completely analogous to those of being from the agent and of being in the patient, but accession and recession do not belong to distinct categories (III, q. 12, n. 11). This counterexample shows that not all distinct circumstances give rise to distinct categories and raises the problem of giving criteria for determining which distinct circumstances produce distinct categories and which do not. Aspall considers two distinct attempts to solve this problem, but in the end he leaves the question open (III, q. 12, nn. 12–15).[122]

[121] Arist. *Cat.* 4, 1b25–2a4.

[122] A different discussion of the categorical distinction between action and passion is found in Book VI, q. 59, where the problem is again left open.

5.7. The infinite[123]

5.7.1. The infinite in discrete quantities

In Aristotle's view there is no actual infinity in discrete quantities or numbers. He admits that a number can be increased to infinity but he denies that this process will eventually lead to a number that is infinite in act, that is, to a number greater than any finite number. Aristotle's main argument in support of this view is based on the assumption that the increase of a number depends on the division of a continuum, so that the infinity of a number is of the same kind as that associated with the number of the parts divided from a continuum. This is not an actual infinity, for the number of the parts so divided is always finite, although it can always be increased by dividing another part.[124]

Aspall reports and accepts Aristotle's view (III, q. 13.2, n. 1; q. 19.2, n. 2). However, he also presents a number of objections to it. Most of them appeal to the structure of a continuum and point out that Aristotle's argument against the actual infinity of number only considers the parts actually divided from a continuum and neglects the 'undivided' parts, that is, the parts of an undivided continuum. These parts are infinitely many, although according to Aristotle they exist only potentially as long as the continuum is undivided so that they do not form an actually infinite multitude. As an objection points out, however, the number corresponding to the undivided parts of a continuum exists in act and is infinite so that such a number is actually infinite (III, q. 19.1, n. 1). The relevant assumption here is that the actuality of a number does not depend on the actuality of the multitude to which it belongs, so that both the number of potentially existing things and that of actually existing things exist in act. Accordingly, although the undivided parts of a continuum only exist potentially, their number exists actually and it is infinite because the potential parts of a continuum are infinite, and therefore it is an infinite in act (III, q. 19.1, n. 6). Another version of this kind of argument considers true propositions associated to the potential parts or potential points of a continuum. For example, since there are infinitely many points existing potentially in a continuum, the proposition saying that a given point exists potentially in a continuum is actually true; but there are infinitely many true propositions of this kind, as many as there are potential points, so that there are infinitely many actual truths associated to a continuum (III, q. 13.1, n. 3). The same applies to propositions about the halves of an undivided

[123] A detailed paraphrase (in Italian) of the contents of each of Aspall's questions on the infinite (*Physics* III.4–8) is given in Trifogli, 2004, 179–187. See also Trifogli, 1993; 1994; 2000, 87–132; 2004, 54–86.

[124] Arist. *Phys*. III.7, 207b10–15.

continuum such as 'this half is greater than its half' (III, q. 19.1, n. 2). In this case too, the relevant assumption is that the actuality of a true proposition does not depend on the actual existence of the thing this proposition is about (III, q. 19.1, nn. 3–4).

Aspall thinks that these objections are not conclusive. In particular, he denies that the number of the potential parts of a continuum exists in act. In his view, there are only two proper ways in which a number exists in act: either because it is the number of a multitude of actually distinct things or because it consists in an actual mental process of counting a multitude of things. Since the potential parts of a continuum are not actually distinct one from the other and since they cannot actually be counted one by one, their number cannot be actual in either way (III, q. 19.2, n. 2). Similarly, Aspall maintains that the actuality of a true proposition presupposes that we actually conceive what the proposition is about, and since we cannot actually conceive infinitely many points or parts of a continuum, there are not infinitely many actual truths associated to them (III, q. 13.2, n. 8).

A different kind of counterexample to Aristotle's view that there is no actual infinity in numbers is that of the human soul. On Aspall's interpretation, Aristotle maintains not only that the world and the human species are eternal but also that the human soul is immortal so that it survives the death of the human being to which it belongs. It follows that there are now infinitely many human souls, so that the number of the human souls is actually infinite (III, q. 13.1, n. 1). Aspall presents more than one solution to this counterexample (III, q. 13.2, nn. 2–4) but does not agree with any of them. The most relevant one assumes that Aristotle's claim that no number is actually infinite should not be understood as applying to any kind of number but only to the number that results from the division of a continuum, that is, the number associated to the parts divided from a continuum. Accordingly, since the number of human souls is not that associated to a continuum, the actual infinity of this number is compatible with Aristotle's view (III, q. 13.2, n. 2).

5.7.2. The infinite in continuous quantities

Aristotle maintains that the only kind of infinity that exists in continuous quantities is the potential infinite by division, that is, the infinite associated with the infinite divisibility of a continuum into its parts. He argues that there are two other kinds of infinity that cannot exist in such quantities: an actual infinite, that is, a body of infinite extension, and a potential infinite by addition analogous to the one he allows in the case of numbers.[125] Aspall's most extensive discussion of the potential infinite by division is found in his

[125] Arist. *Phys.* III.6, 206b3–27.

Questions on book VI about the composition of a continuum,[126] whereas in his Questions on Aristotle's doctrine of the infinite in book III he concentrates on the two negative theses about the infinite by addition. While he accepts both of them, he also thinks that each of them involves its own problems, which he tries to solve.

(i) The main problem with the thesis that there is no actually infinite body or extension is that it seems to admit some counterexamples. The most important is that of light. Following an Aristotelian principle, Aspall maintains that an infinite power can only exist in an infinite body. According to Augustine, however, the power of sunlight to illuminate is infinite, that is, the sun can multiply its light indefinitely. It seems to follow, therefore, that the sun, having an infinite power, is an infinite body (III, q. 14.1, nn. 3–4).

Aspall considers two main solutions to this problem without explicitly adopting either of them. The first rejects Augustine's premise that sunlight has an infinite power to multiply itself (III, q. 14.2, n. 4). The second solution concedes that the sun can multiply its light indefinitely but denies that this infinite multiplication presupposes a single infinite power existing in the sun. Rather, there are infinitely many powers responsible for the multiplication of light in the medium: the light in one part of the medium has the power to generate the light in the adjacent part of the medium and this is how light can propagate indefinitely (III, q. 14.2, nn. 5–8).

Aspall specifies that the Aristotelian thesis that there is no infinite extension applies not only to physical extension or magnitude, like that in the counterexample of the sun, but to extension as such, so that it is the nature of extension itself that is incompatible with the infinite (III, q. 15, n. 8; also I, q. 33, n. 8). Hence, a geometrical line cannot be infinite. One of Aspall's arguments to show why this is so claims that the existence of an infinite line or more generally of an infinite extension violates the principle that a part is necessarily a part of a whole so that it is impossible to posit a part without positing a whole. Now, an infinite line cannot be a whole, according to Aspall, because it is not possible to encompass it as a whole; on the other hand, a line does contain parts, so it cannot be infinite (III, q. 15, n. 5).

(ii) The Aristotelian thesis that there is no potential infinite by addition in continuous quantities raises two main problems. One is that of explaining the asymmetry between the potential infinite by addition and the potential infinite by division in continuous quantities: why is it the case that a continuous quantity can be divided and so made smaller and smaller ad infinitum but it cannot be increased and so made bigger and bigger ad infinitum? (III, q. 15, n. 2). The second is that of explaining the asymmetry between the potential infinite by addition in continuous quantity and the potential infinite by

[126] See below, pp. lxxviii–lxxx.

addition in numbers: why is it the case that a number can be increased ad infinitum whereas such an increase is not possible for a continuous quantity? In reply to the first question, Aspall maintains that the infinite divisibility of a continuous quantity is accounted for by something intrinsic to its nature, that is, matter with its infinite parts.[127] But there is no such intrinsic cause for its infinite increase: the increase of a given magnitude presupposes that something extrinsic to it is added to it, so that it cannot be accounted from inside it, as it were (III, q. 15, n. 10). This explanation leads to the second question, that about numbers, because it seems that the process of increasing a number by addition also presupposes the addition of something external to that number. Aspall replies that the two cases are not similar. The difference is that there is indeed something intrinsic to the nature of a number that accounts for its increase by addition, while there is not such a cause in a continuous magnitude. The parts or units of a number are such that each of them or any aggregate of them do not form a whole—something self-contained to which nothing can be added—but are such that they are ordered or potential to another unit or number. Any continuous magnitude, on the contrary, is by its nature a whole and as such is not ordered or potential to another magnitude (III, q. 15, nn. 11–12).

Other differences between the structure of magnitudes and that of numbers relevant to the infinite are given by Aspall in his discussion of the 'comparability' of the infinites (III, q. 20).

5.8. Place[128]

Aspall's treatment of Aristotle's theory of place in *Physics* IV.1–5 is quite short. Although he touches on the main traditional topics, such as the immobility of place (IV, q. 4) and the place of the heavens (IV, q. 5), he devotes a significant discussion only to a few of them, namely those connected with the status of place as an accident of the containing body.

Aristotle defines the place of a body as the limit of the body containing it and in contact with it.[129] For example, the place of a liquid totally enclosed in a vessel is the limit of the vessel in contact with that liquid. This definition makes it clear that (i) place is an accident and not a substance, and that (ii) it is an accident of the containing or locating body. This is because the limit of a body is indeed its surface, which is one of its accidents. Moreover, a limit inheres in the thing of which it is the limit so that it is again one of its accidents.

[127] On the infinite parts of matter, see Book I, q. 34, nn. 5–6.
[128] See Trifogli, 1998, 179–208; 2000, 133–202; 2007, 8–55.
[129] Arist. *Phys.* IV.4, 212a2–6.

Both properties (i) and (ii) of Aristotelian place seem, however, to have been controversial at Aspall's time.

One major motivation for arguing against (i) comes from Aristotle's notion of natural place, according to which the natural place of a body has some active role with respect to the body located in it in that it has a power to preserve that body's nature.[130] It seems that this active role cannot be accounted for by place if place is simply an accident and in particular a quantitative accident, like a surface. The active role of place requires a homogeneity in nature between the located body and its place (IV, q. 1.1, nn. 6–7). In dealing with this kind of objection, Aspall concedes that the surface of the containing body is simply the mathematical place of the located body but not also its natural place. This surface as mathematical place is responsible only for the being spatially contained of the located body but not also for the preservation of its nature. Natural place is something more than mathematical place, for it is the combination of mathematical place and the preservative power exerted by the containing body on the located body. Aspall claims that while mathematical place is an accident, natural place is in some way a substance. What he seems to have in mind is not that natural place is properly speaking a natural substance, but that it requires a natural substance or substances from which its preservative power derives. Aspall thinks that these relevant substances are not only the natural bodies immediately containing the located bodies, but must also include the celestial bodies (IV, q. 1.2, nn. 2, 6; q. 7, nn. 6–7).

Another major source of objections against (i) the property of place of being an accident is Aristotle's requirement that the place of a body be immobile, that is, not subject to locomotion.[131] In the Aristotelian tradition, this insistence on the immobility of place is considered a serious problem, since it seems incompatible with the nature of place as the limit of the containing body. Since the containing body, being a natural body, is mobile, it must carry its limit along during its locomotion. Aspall's discussion shows that in his time some commentators entertained the idea that one way to save the immobility of place was to posit that place is a substance in its own right rather than an accident, so that it enjoys some kind of independent existence (IV, q. 1.1, nn. 4–5; q. 1.2, nn. 3–5). Aspall, however, is not attracted by this line of argument in support of the immobility of place (IV, q. 1.2, nn. 8–9). He also considers other attempts to save the immobility of place. The most successful, in his view, is one that tries to account for the immobility of place in terms of the immobility (that is, lack of translational motion) of the whole heavens (IV, q. 4, nn. 5–11).

[130] Arist. *Phys.* IV.1, 208b8–14; IV.5, 212b29–213a10.
[131] Arist. *Phys.* IV.4, 212a16–21.

(ii) The controversy about the subject of place arises from a popular view in Aspall's time that place is not only an external container of the located body but also spatially coexists with the located body. Place somehow has a third dimension—depth—in virtue of which it extends inside the located body. While the two dimensions that define place as an external container belong to the containing body, and thus place so considered exists in the containing body as its subject, the third dimension of place does not belong to the containing body, given that it is external to that body. But since Aspall accepts Aristotle's rejection of an incorporeal extension, the existence of the third dimension of place must depend on a body, and the only plausible candidate is the located body. Thus, the dependence of the depth of place on the located body leads Aspall to conclude that place has two subjects: both the containing body and the located body (IV, q. 3, n. 14). Aspall is aware that this is a difficult position, because it seems to violate the principle that one accident has only one subject, but he does not provide an adequate solution to this problem (IV, q. 3, nn. 2–6, 14).

Another problem that Aspall leaves without a clear solution is that of the relationship between the extension of place and that of the located body. He tries to reconcile the view that the extension of place is caused by the located body with the assumption that extension must belong to place prior to its receiving the located body, given that place can receive something with extension, such as the located body, only by being itself extended. Aspall's attempted reconciliation appeals to the distinction between the temporal order and the order of nature: in the temporal order extension comes to belong to place simultaneously with its receiving the located body, but in the order of nature extension belongs to place before it receives the located body (IV, q. 6, nn. 1–3, 5–7; VI, q. 40).

5.9. The continuum

5.9.1. Existence and coming into being of a continuum

While in his treatment of continuity in *Physics* VI Aristotle takes it for granted that a continuum exists and concentrates on the investigation of its nature, Aspall devotes a preliminary discussion to the existence and production of a continuum.

In the short Question on the existence of a continuum (VI, q. 1) Aspall focuses on two of its general properties: (i) being something divisible into parts and (ii) being something one. As to (i), Aspall reports the opinion of some eminent philosophers who maintain that matter has neither distinct parts nor any potency for extension. This opinion is used as a premise in an argument that concludes that no material substance is divisible, for if matter

is not in potency to parts and extension then no form that comes to exist in matter can give it parts and extension (VI, q. 1, n. 4). Consistently with his view about prime matter in the Questions on *Physics* I,[132] Aspall rejects this opinion and claims that matter has intrinsically distinct parts and a potentiality for extension. He also implicitly suggests that the divisibility into parts of matter is a necessary condition for the existence of a continuum (VI, q. 1, nn. 8, 5).

As to (ii), Aspall considers an argument against the existence of unity in things based on the non-Aristotelian assumption that the continuity and unity of a thing derives from a celestial power, identified with the celestial light; but since celestial light is not continuous, it cannot produce a continuum (VI, q. 1, n. 1). In another Question Aspall ascribes the non-Aristotelian assumption to Ovid, who invokes a celestial power diffused throughout the universe, the so-called 'golden chain', to account for the unity of the universe. Ovid's view is there openly contrasted with the view that the unity of the universe is a unity of order and that the relevant order is that of all created things to the first cause, a view ascribed by Aspall to Aristotle (VI, q. 5, nn. 6–8). In his discussion of the existence of a continuum, Aspall accepts the non-Aristotelian assumption in this argument only in the weak formulation that the celestial light is a cause that preserves the being of a thing and its unity, but not also a cause that produces the being of a thing and its unity. In addition, he draws a contrast between the celestial light as universal agent and a particular agent of the same nature as the thing produced by it, and he maintains that the principle that an agent must have properties of the same kind as the thing produced by it only holds for a particular agent (VI, q. 1, n. 6).

A more extensive discussion is devoted to the issue of the coming into being of a continuum, in particular to the question of whether a continuum comes to be as the result of a natural process of generation (VI, qq. 4–5). This discussion focuses on the paradigmatic case of a continuous magnitude. Arguments based on Aristotelian claims point out that quantity or magnitude remains numerically the same in a process of generation. For example, in the generation of fire from air, the quantity of the fire generated from air is numerically the same as the quantity of air from which fire is generated; hence, since quantity persists throughout a process of generation, it is not produced by generation (VI, q. 4, nn. 1, 3). Aspall rejects these arguments. He denies that different individual substances, like air and the fire generated from it, have 'common' accidents, that is, accidents that are numerically the same in both. He also denies that any substantial form remains numerically the same in a process of generation, and concludes from this that no accident remains numerically the same. Since an accident is caused by a substantial form in

[132] See above, pp. xli–xliii.

virtue of the being that a substantial form has in matter, the numerical identity of an accident also presupposes the numerical identity of the substantial form by which that accident is caused. In the example of the generation of fire from air, since the substantial form of fire and that of air are distinct substantial forms, the quantity caused by one of them cannot be numerically the same as that caused by the other. Aspall adds that specifically distinct substantial forms can cause accidents that are specifically the same (although numerically distinct). The reason for this is that specifically distinct substantial forms can have similar beings in matter, and it is the being in matter of a substantial form that is responsible for an accident (VI, q. 4, nn. 6–7, 9–10).

His positive view on this issue is that a magnitude considered in its essence in so far as it is a magnitude, that is, considered as a mathematical thing, is not subject to generation and corruption, whereas a magnitude considered in its actual existence in matter can be subject to generation and corruption. More precisely, while a magnitude existing in the unchangeable matter of a celestial body is not subject to generation and corruption, a magnitude existing in the changeable matter of substances here below is subject to these changes (VI, q. 4, n. 18). Aspall makes it clear that a magnitude considered in its essence as a mathematical thing, and thus not subject to change, is not a magnitude considered abstractly from matter as such, but only from the privation with which matter is joined. His idea is that a magnitude in its essence is caused by the union of matter and a substantial form. Since the coming to be of a substantial form in matter is prior by nature to the coming to be of a privation in matter, and since among the three principles of nature—matter, form, and privation—it is only privation that is a principle of change, a magnitude in its essence does depend on matter but at the same time it is not changeable (VI, q. 5, nn. 1–5).

Aspall specifies that the magnitude of a generable and corruptible substance here below is subject to generation only accidentally and not also per se. What is primarily generated is a substantial form along with the active and passive qualities accompanying a substantial form, whereas a magnitude is produced only secondarily, and in particular as a result of the generation of qualities (VI, q. 4, n. 11). He considers the objection that the dependence of the generation of quantities on that of qualities violates the commonly accepted assumption that quantities are ontologically prior to qualities. In reply to this objection Aspall makes it clear that a process of generation does not start from qualities, that is from the corruption of the qualities of the thing to be corrupted which are contrary to those of the thing to be generated, but from the stimulation of the active potency of matter caused by the introduction into matter of the intentional species of the substantial form of the

agent, and he tacitly assumes that the existence of quantities in matter results from such a stimulation (VI, q. 4, nn. 12–16).

5.9.2. Cognoscibility of a continuum

Another preliminary Question raised by Aspall but not addressed by Aristotle himself is that of whether there can be intellectual cognition of a continuum (VI, q. 2). The dominant issue discussed in this Question, however, is that of the sensory cognition of a continuum. Following Aristotle, Aspall maintains that our intellect can only know a thing via the sensory cognition of that thing, so that human intellectual cognition presupposes sensory cognition. Accordingly, if there is no sensory cognition of a continuum, there is no intellectual cognition of it either. But the assumption that there is a sensory cognition of a continuum is regarded as controversial. Aspall reports a number of arguments showing that a continuum cannot be a sensible thing properly speaking. Many of these arguments question the general assumption that the so-called common sensibles (e.g., a continuum or magnitude, motion, shape) are per se sensible. The account of sensory cognition at work in these arguments implies that, for a thing to be perceived by the senses, it is necessary for that thing to send its (sensible) species to the senses. The arguments then show that a continuum (and more generally a common sensible) cannot send its own species to the senses (VI, q. 2.1, nn. 1–5, 7).

Aspall rejects these arguments. He thinks that they are contrary to Aristotle's view, given that Aristotle maintains that not only the proper sensibles but also the common sensibles are per se sensibles.[133] He accepts the species-account of sensory cognition but he insists that a continuum does send its proper species to the relevant senses. This is shown, in his view, by the fact that the senses can discriminate between the colour of a thing and its magnitude or continuum; but in order to do so the senses need to have a sensible species of a continuum distinct from that of a colour (VI, q. 2.1, nn. 8–12). He admits, however, that the sensory cognition of a continuum is subordinate to that of a proper sensible: a proper sensible per se and primarily sends its species to the senses, whereas a continuum sends its species 'under' the species of a proper sensible (VI, q. 2.2, n. 2). The claim that the species of a continuum is sent to sight, for example, under the species of a colour, which is a proper object of sight, is left without an adequate explanation. In its support Aspall invokes a parallelism between the real properties and the intentional properties of sensible things: just as a colour and a continuum as real things are such that the continuum underlies the colour (being its proper subject), so a colour and a continuum as intentional species are such that the species of a continuum is latent under that of the colour (VI, q. 2.1, n. 11).

[133] Arist. *DA* II.6, 418a8–20.

Aspall also uses this parallelism to argue that a substance as distinct from its sensible accidents also sends its own species to the senses, albeit not primarily but under the species of some of its accidents, so that not only the colour of Socrates is properly perceived but also Socrates as a substance, albeit in a derivative way (VI, q. 3, nn. 3, 5–6). This account of the perception of a substance is discussed in more detail in a Question from book I.[134]

In a short Question on the intelligibility of motion in book III, Aspall considers the view that both motion and time as successive things are perceived by means of the species of the relevant permanent things, that is, of the mobile body and of the space traversed by it, and not by means of their own species as successive things. Another view considered there is that the perception of motion and of time depends on the perception of their indivisible elements, that is, the present instantaneous change and the present instant. Aspall does not express agreement with either of the two views (III, q. 7, nn. 15–16).

5.9.3. Composition of a continuum

Aristotle's major thesis about a continuum is that a continuum is divisible ad infinitum into divisible parts, so that a continuum is not composed of indivisibles. For example, a line, being a continuum, is composed of further lines and not of points.[135] Aspall accepts Aristotle's position and defends it against a variety of arguments intended to show that a continuum is composed of indivisibles (VI, q. 8):

(i) Some arguments point out that there are physical phenomena that seem to involve distinct bodies that occupy the same place. For example, a mixed body is such that it can be subject to both action and passion in any of its parts; but since action and passion are contrary, they cannot simultaneously occur in the same bodily part, and so it is necessary to posit that in each spatial portion of a mixed body there are two distinct bodily parts. Similarly, in a process of increase the body subject to increase and the body added to it come to occupy the same place. The principle is then invoked that two bodies can be in the same place only if one or both of them are indivisible, that is, altogether lacking extension, a principle which is closely related to that of the impenetrability of bodies (VI, q. 8, n. 2–4).

In replying to these arguments, Aspall distinguishes two ways in which the claim the two bodies are in the same place can be understood: either in the sense that (a) the total place occupied by the two bodies is greater than the place that one of them alone would occupy, or in the sense that (b) the total place occupied by the two bodies is equal to the place that one of them alone

[134] Book I, q. 40.
[135] Arist. *Phys.* VI.1, 231a21–26.

would occupy. Aspall accepts that if the claim is understood in sense (b), then it does follow from it that one or both of the two bodies is indivisible, but he denies that two bodies can be in the same place in sense (b). It is perfectly possible for two bodies to be in the same place in sense (a), since this does not imply that one or both of them are indivisible. Indeed, this is what happens in a mixed body or in a process of increase (VI, q. 8, n. 5).

(ii) Other arguments show that the assumption that a continuum is divisible to infinity leads to the admission of an actual infinity (VI, q. 8, nn. 9–10). For example, the infinite divisibility of a line implies that there are infinitely many points in a line. Aspall's treatment of these arguments (VI, q. 8, nn. 23–25) is parallel to that given in his discussion of the infinite in discrete quantities in book III.[136]

(iii) Another argument points out that if a continuum is divisible to infinity, any one of its parts is also divisible to infinity, and is therefore continuous. Accordingly, the form of continuity denominates both a whole and each of its parts. This gives rise to the problem of accounting for the difference between the form of continuity and other forms, such as the form of animal. The form of animal denominates the whole animal alone, and not any of its parts (VI, q. 8, n. 15). Aspall devotes a specific Question to the issue raised in this argument (VI, q. 9). In his view, the problem arises from failing to distinguish between the existence of a form in a part of a whole and the denomination associated with that form, and as a result assuming that the mere existence of a form in a part is sufficient for the part to be denominated by that form, so that for example, since the form of animal exists in a foot, the foot itself can be denominated 'animal'. Aspall insists that the existence of a form and the denomination associated with that form should be kept distinct; and it is this distinction that makes it possible to account for the difference between the form of animal and the form of continuity. Each of these two forms is a form of a whole composed of parts, and each exists in a whole and denominates that whole, but the major difference between the two forms is that while the form of animal denominates the whole animal alone and not its parts, even though it does exist in those parts, the form of continuity denominates both the whole and each of its parts. Aspall then specifies that while the form of continuity denominates each of its parts it does so only in so far as each part is considered as part of the whole continuum and not as a continuum in itself. He also maintains that while the whole form of animal exists in each part of the animal, the whole form of continuity does not exist as such in each part of a continuum. On the contrary, to the parts of the continuum there correspond parts of the form of continuity. However, even the partial existence of the form of continuity in a part of a continuum is sufficient for

[136] See above, pp. lxix–lxx.

denomination, whereas the total existence of the form of animal in a part of an animal is not sufficient to denominate that part, since the vital operations distinctive of the whole animal, such as breathing, cannot be ascribed to all its parts, given that these include such parts as hands and feet (VI, q. 9, nn. 7–11).[137]

5.9.4. Minima naturalia

It is a matter of debate in Aspall's time as well as later in the thirteenth century whether Aristotle's thesis in *Physics* VI that a continuum is infinitely divisible is valid only for a mathematical continuum, that is, a continuum considered simply as quantity, or is also valid for a natural continuum, that is, a natural body.[138] Aristotle himself provides a starting point for this debate, since in *Physics* I he claims that there is a minimum of flesh, thus implying that flesh is not infinitely divisible into parts that are flesh.[139] On other occasions Aristotle claims that there is a maximal size for the extension of natural things.[140] This claim can easily be used to support the existence of a minimum in natural things. For example, one of the arguments presented by Aspall against the infinite divisibility of a continuum claims that since there is a maximal magnitude, there must also be a minimal magnitude, given that being a maximum and being a minimum are contraries and so they are apt by nature to exist in the same thing (VI, q. 8, n. 1; also, I, q. 78, nn. 1–2). Aspall devotes specific discussions to the *minima naturalia* (I, qq. 78–79) and the *maxima naturalia* respectively (VI, q. 11). His discussion of the *maxima naturalia* is essentially parallel to that of the infinite in continuous quantities that we have presented above.[141] We concentrate here on the question of the *minima naturalia*.

From Aspall's report of contemporary opinions about the *minima naturalia*, it appears that the dominant distinction that framed the debate in his time is that between a virtual minimum (*minimum virtute*) and a minimum by action (*minimum actione*). The distinction between 'virtual' and 'by action' is ascribed to Aristotle in a passage of *On sense and sensibilia* where he says that something can be sensible, that is, able to produce a sensation, in two ways: either by action or virtually (I, q. 79, n. 9).[142] In Aspall's discussion this

[137] Aspall devotes specific discussions to other issues concerning parts and wholes in Book I, qq. 36–38, and Book VI, q. 10.

[138] The fact that there are a good number of Questions devoted to mathematical things in Book VI (qq. 18–26) suggests that Aristotle's treatment of the continuum is considered as the *locus classicus* for the discussion of mathematical things. These Questions, however, only deal very marginally with the issue of continuity. The main focus is on the general ontological properties of mathematical things.

[139] Arist. *Phys*. I.4, 187b18–21.

[140] Arist. *DA* II.4, 416a16–17.

[141] See above, pp. lxx–lxxii.

[142] Arist. *On sens*. 6, 445b29–446a15.

distinction is taken for granted and is left without explanation, but the idea behind it seems to be the following: a virtual minimum is a part of a natural body so small that if it were divided, the parts resulting from its division would not have the same nature as the natural body from which they are divided, whereas a minimum by action is a part so small that if it were divided, the parts resulting from its division would not have the kinds of action distinctive of the natural body from which they are divided. A clear assumption about this distinction is that the existence of a virtual minimum implies the existence of a minimum by action but not the other way around. Indeed, one important view in the debate reported by Aspall is that in a natural body there is no virtual minimum but there is a minimum by action (I, q. 79, n. 9).

Aspall too adopts a version of this view. He makes the preliminary point that a minimum does exist in those natural bodies such that any part of them, however big or small, when separated from its whole, fails to keep its nature, or, in Aspall's words, such a separated part can be said to be of the same nature as the whole only in an equivocal sense. This is the case of flesh: any part of the flesh of a living body, once separated from the whole flesh and hence from the living body to which it belongs, is no longer flesh, except equivocally, since it loses the property of being alive which is essential to flesh. Thus, according to Aspall, Aristotle is right when he claims that there is a minimum of flesh, but this should be understood in the sense that the minimum of flesh is actually the whole flesh of a living body and not any of its parts, however big or small (I, q. 79, n. 12). The really controversial question about the existence of a minimum then arises for those natural bodies such that they can be divided into parts having the same nature, as happens with the elements and mixed inanimate bodies. Aspall claims that there is no virtual minimum in a body of this kind. This means that, given any part of such a body, this part can be further divided into parts that have the nature of the whole from which they are divided (I, q. 79, n. 12).

As to the existence of a minimum by action, Aspall distinguishes between the internal action of a natural body (or of its form) of in-forming its own matter and external actions, like that of acting on another body and changing it into something of its own nature, as happens when fire acts on air and changes it into fire, and that of moving towards its natural place. He maintains that there is no minimum for the internal action, so that given any part of a natural body, however small, the form actually in-forms that part just as it in-forms the whole body. As to external actions, he maintains that there is a minimal part of a natural body that can actually produce the effect of that action, so that, for example, there is a minimal part of fire that is capable of actually transforming air into fire. However, there is no minimal part that has the inclination to perform the external actions proper to its species. For

example, any part of fire, however small, retains its inclination to move upwards and to convert air into its own nature, but because of its smallness it may not be able to actually achieve these effects (I, q. 79, n. 12).

5.9.5. Continuity of time[143]

Aspall's main concern in the Question devoted to this topic (VI, q. 12) is that of defending Aristotle's view in *Physics* VI that time is continuous[144] against some objections arising from Aristotle's discussion of time in *Physics* IV.10–14:

(i) The arguments against the existence of time in *Physics* IV.10 suggest the view that the only existing element of time is the indivisible present instant;[145] this view, in turn, seems to lead to the conclusion that time is composed of instants, given the plausible assumption that if time is a being, it must be composed of beings (VI, q. 12, n. 4). The idea that an instant can be part of time is also made plausible by the comparison between an instant and a unit: an instant is a more complex indivisible than a unit, given that, according to Avicenna, a unit only expresses a negation—the mere lack of division of a being—and so there is no nature underlying a unit, whereas there is a nature underlying an instant. But a unit, despite its absolute simplicity, can be part of something, that is, of a number; therefore, an instant too, being more complex than a unit, can be part of something, that is, of time (VI, q. 12, nn. 2–3).

Aspall concedes that an instant is more complex than a unit but he denies that this premise is enough to allow one to conclude that the instant can be a part of time, just as a unit is a part of a number. What is also required is the assumption that the divisibility of time is of the same kind as that of a number. This assumption, however, is clearly against Aristotle who argues that time, being a continuum, is divisible ad infinitum, which is not true of a number (VI, q. 12, n. 12).

While Aspall denies that time is composed of instants, he admits that the present instant has a crucial role in accounting for the existence of time as a successive thing (VI, q. 12, nn. 13–14). His remarks on the role of the present instant in accounting for the existence of time are parallel to those about the existence of motion as a successive thing, which we have presented above.[146]

(ii) In *Physics* IV.14 Aristotle claims that there is only one time for all simultaneous motions.[147] This claim gives rise to the traditional objection that

[143] See Trifogli, 1998, 221–256; 2000, 203–261; 2007, 74–120.
[144] Arist. *Phys.* VI.1, 231b18–232a22.
[145] Arist. *Phys.* IV.10, 218a3–8.
[146] See above, pp. lxii–lxiv.
[147] Arist. *Phys.* IV.14, 223a29–b12.

the unity of time is incompatible with the status of time as an accident of a motion. Since numerically distinct subjects have numerically distinct accidents, if time were an accident of motion, there would be as many numerically distinct times as there are simultaneous motions. But if motion cannot be the subject of time, it seems that the only alternative left is that the soul is its subject, a view that Aspall ascribes to Averroes. Since the soul is indivisible, time too, being a passion or accident of the soul, is indivisible (VI, q. 12, nn. 6–7).

Aspall rejects the view that the subject of time is the soul. He assumes that time is a real thing, whereas what exists in the soul is the intentional species of a thing (VI, q. 12, n. 20). He accepts Averroes' view that the perception of time depends on the soul, but he denies that its existence does too (VI, q. 12, n. 23). Aspall sides with the realist view that time is an extra-mental thing, something that exists in motion. On the problem of how to reconcile the existence of time in a motion with its unicity, he reports an opinion according to which all motions which occur simultaneously can be regarded as just one subject with respect to time, an opinion advocated by the author of the commentary ascribed to Rufus and by Bacon.[148] Aspall thinks that this solution is not a good one, and leaves the problem open (VI, q. 12, nn. 21–22).

Another traditional objection to the continuity of time derives from the assumption that there is a fastest motion, that is, the motion of the heavens, for since this is the fastest motion, it takes a minimal time (VI, q. 12, n. 1). Aspall reports the replies of Alexander and of Averroes. In particular, Averroes maintains that any motion in so far as it is a motion (in its nature as motion) can be made faster or slower, although in some cases it is the nature of the thing subject to that motion that prevents an increase or decrease in its speed (this is the case of the heavens) (VI, q. 12, nn. 9–11). In the Question on the continuity of time Aspall does not side with either of these two views, but when he addresses this issue in more general terms in a later Question from book VI that asks whether it is possible for any motion to be made faster or slower, he defends Averroes' view. In reply to some objections to this view, he suggests that a more accurate formulation of Averroes' distinction about motion is one between motion considered as a continuum and motion considered as something that arises from the principles of the mobile thing. A motion considered in the first way can be made faster or slower, but this is not always possible when a motion is considered in the second way (VI, q. 28, nn. 1–6).

[148] On this opinion see Trifogli, 2000, 246–256.

5.9.6. Divisibility of motion

In his discussion of the continuity of motion in *Physics* VI, Aristotle mainly has in mind the case of locomotion, which is described as a mobile body's traversal of a certain space in a certain time. This description suggests that the quantity and parts of motion can be defined in two ways, namely, with respect to space and with respect to time. That is, in Aristotelian terms, a motion can be divisible according to the division of space and according to the division of time.[149] Aspall tries to relate this abstract quantitative description of a motion to a more metaphysical analysis of motion. He maintains that a motion contains two main components: succession over a space, and its duration. He also refers to succession over a space as the substance itself of motion, thus implying that this is the more fundamental ingredient in the ontology of motion. These two metaphysical ingredients of a motion are those that provide the foundation of the divisibility of motion which is associated with space and the divisibility associated with time respectively. A motion's succession over a space is divided according to the division of that space, whereas a motion's duration is divided according to the division of time. Accordingly, two motions taking the same time are equal with respect to their duration, but they are not necessarily equal with respect to their succession over a space, since the spaces traversed in the two cases can be unequal. Similarly, two motions traversing the same space are equal with respect to succession over a space, and hence with respect to the substance of motion, but they are not necessarily equal with respect to the duration of motion, since one can be faster than the other (VI, q. 27, nn. 6–7).

Aspall considers an objection to his account of the divisibility of motion which arises from its ontological status as an accident of the mobile body. The objection appeals to the general principle that the only divisibility of an extended accident is that deriving from its subject. This principle is illustrated by the example of whiteness, for the parts and divisibility of whiteness exclusively derive from and are defined by those of the surface that is its subject (VI, q. 27, n. 3). Aspall's reply is that the principle invoked in this objection is not universally valid. More precisely, it is valid for absolute accidents, but not for relative accidents because the thing that constitutes the term of the relation defining a relative accident is also responsible for its divisibility. Motion is a relative accident and so it is an accident of a substance in so far as this substance is subject to a motion, and so in so far as it has a relation to a space traversed by it (VI, q. 27, n. 9).

The appeal to the status of motion as a relative accident of the mobile body is also used by Aspall to reply to an objection against the temporality of

[149] Arist. *Phys*. VI.1, 231b18–20.

motion. The objection is based on the assumption that the temporal conditions for the existence of an accident are defined by those of its subject. Accordingly, since the mobile body, as a permanent thing, can exist in an instant, it can also move in an instant (VI, q. 33, n. 4). Aspall maintains that the assumption made by this objection is not universally valid, since it is valid for absolute accidents but not necessarily for relative accidents. In particular, since motion is an accident of the mobile body in relation to a space to be traversed, and since a space can be traversed only in time and not in an instant, motion can only inhere in a mobile body in time (VI, q. 33, n. 13).

Furthermore, Aspall accepts Aristotle's view that something indivisible cannot per se be subject to motion,[150] so that a per se mobile thing is necessarily extended and divisible (VI, q. 37.2, n. 1). He denies, however, that the divisibility of the mobile body defines a quantitative kind of divisibility of its motion. While the divisibility of space and that of time define quantitative parts of a motion, the divisibility of the mobile body defines the divisibility of motion in its subjective parts, that is, it gives rise to numerically distinct motions. Thus, if a mobile body is divided into parts A and B, the motion of A and the motion of B are not two parts of the same motion, but two distinct motions. These two motions exist only potentially as long as the mobile body is simply divisible but not actually divided (VI, q. 39, n. 8).

5.9.7. Motion in an instant

In his discussion of Aristotle's thesis that every motion is continuous Aspall mainly focuses on the divisibility of a motion with respect to time, and in particular on the question of whether a motion can occur in an instant. Aristotle absolutely denies the possibility of an instantaneous motion and accordingly he denies the possibility that a motion is composed of temporally indivisible parts.[151] Following the standard medieval interpretation, Aspall assumes that Aristotle's denial only applies to motions properly speaking, that is, to qualitative motions (alterations), quantitative motions (increase and decrease), and locomotion, and does not also apply to substantial motions, that is, generation and corruption. Substantial forms are introduced and removed instantaneously, although the qualitative changes that lead to the introduction or removal of a substantial form do take time (VI, q. 52, n. 5).

Aristotelian commentators, however, also point to some exceptional cases of non-substantial motions that seem to occur instantaneously. Averroes, for example, admits that there are instantaneous motions, although he maintains that they do not exist in their own right but only as attached, so to say, to a locomotion, which is a temporal motion. Thus, an instantaneous motion can

[150] Arist. *Phys*. VI.10, 240b8–13.
[151] Arist. *Phys*. VI.3, 234a24–31.

immediately follow a locomotion, or in Averroes's own words, an instantaneous motion can be the end of a locomotion. This case is illustrated by the example of the illumination of a house by a candle which results from the locomotion through which the candle is carried into the house.[152]

Aspall reports Averroes' view and then rejects it. He admits that the illumination of an extended space, that is, the multiplication of light through a corporeal medium, occurs in an instant, but he denies that this is a motion properly speaking, and hence it is not one of those motions which Aristotle intends to debar from occurring instantaneously. He also remarks that Averroes' account is too restrictive, for illumination can occur not only at the end of a locomotion, but also at the end of a qualitative motion, as in the case of the illumination produced by a fire generated from burning coal (VI, q. 33, nn. 5–8).

Aspall devotes an extensive discussion to the multiplication of light through a medium in book I (I, q. 39.2, nn. 8–12; q. 40.2) but he does not deal explicitly with issues related to the temporality of a motion.

Another apparent counterexample to Aristotle's view that a motion cannot occur in an instant discussed by Aspall is that of the dissolution of a mixed body. When a mixed body is dissolved, it is dissolved into the elements that compose it in such a way that the elements leave the mixed body and move towards their natural places so that, for instance, fire moves up, and earth moves down. The separation of the elements from the mixed body and their locomotion occurs simultaneously with the last change through which the mixed body is corrupted, and so, since this last change, being a corruption, occurs in an instant, the locomotion of the composing elements also occurs in an instant (VI, q. 13, n. 1).

Aspall's reply to this argument is not completely clear. He denies the argument's assumption that a mixed body is dissolved into the four elements in their pure natures, so that the elements exist potentially in a mixed body and are actualized when the mixed body is dissolved. Empirical evidence shows that fire in its pure nature only results from the dissolution of a mixed body in the special case in which the mixed body is dissolved by being burnt. What happens more usually is that the fine-grained components of a mixed body are dissolved into something that has the nature of air, but not air understood as the pure element. Similarly, the gross components are dissolved into something that has the nature of earth, but not earth understood as pure element but rather a mixture of water and earth. As a result, the fine-grained components move towards the parts at the circumference of the mixed body, while the gross components tend towards its central parts (VI, q. 13, nn. 10–12). Aspall's account does not explicitly address the problem of the

[152] Aver. *In Phys.* VI, t. c. 32, fol. 266ra–b.

temporality of these kinds of locomotion as presented in the original objection. The implicit idea, however, seems to be that the locomotions of the elements that result from the dissolution of the mixed body do not occur simultaneously with the instantaneous dissolution of the mixed body, but after it. More precisely, since the immediate results of the dissolution of the mixed bodies are not pure elements but rather mixtures of elements, a process of purification of the elements is still necessary before they can move with their natural locomotions.

Aspall deals with two important general issues related to Aristotle's denial of instantaneous motion.

(i) One is that of establishing what is responsible for the temporal succession of a motion. Aspall discusses this issue in replying to an objection against the temporality of the motion of the heavens. The objection concludes that the celestial motion is instantaneous from the premise that the heavens do not offer resistance to their mover (VI, q. 33, n. 1; VIII, q. 2, n. 1). This objection has an Aristotelian origin, for in a famous argument against the possibility of motion in a void Aristotle assumes that the resistance offered by a corporeal medium to a motion through that medium is necessary for the temporality of a motion, so that any motion through a void would be instantaneous, given that the void does not offer any resistance.[153] Aspall rejects Aristotle's assumption. In his view, what accounts for the temporal succession of a motion is not the resistance offered by the mobile body to its mover or to the medium, but the existence of a multiplicity of intermediate states between the initial and the final terminus of a motion. More explicitly, in the case of a locomotion this multiplicity is that of the parts of the spatial extension traversed in this kind of motion. Thus, the motion of the heavens takes time and is not instantaneous, despite the fact that the heavens do not offer resistance to their mover, because the heavens (or their parts) have to traverse many spatially extended parts in order to complete their revolution (VI, q. 33, n. 9; VIII, q. 2, n. 4).

Aspall points out that this account of the temporality of a motion holds not only for locomotion but also for the other kinds of motion, that is, those involving a change with respect to forms rather than places (III, q. 3, n. 5). Consider the generation of fire starting from air. As we have seen,[154] Aspall maintains that even from the very beginning of this change, when matter is still under the form of air, there is an active diminished potency towards the form of fire existing in matter. In Aspall's view, it is this diminished potency that is responsible for the fact that the generation of fire takes time, and not the resistance offered by the form of air existing in matter. The generation of

[153] Arist. *Phys.* IV.8, 215a24–216a22.
[154] See above, pp. l–lii.

fire would take time even if the starting point of this process were simply matter without the form of air and with the diminished potency towards the form of fire, which does not offer any resistance to the agent producing fire. The reason for this is that there are degrees in the diminished potency towards the form of fire existing in matter, degrees corresponding to different states of actualization of the form of fire (VIII, q. 2, n. 4).

(ii) The other issue is a logical one about the truth-conditions of propositions about the existence of motion. The question here is whether the view that motion does not exist in an instant implies that the proposition 'something is in motion' is not true at an instant. Aspall reports an argument against the temporality of the celestial motion that assumes a positive reply to this question. The argument is built on the inference that if the proposition 'the heavens are in motion' is true, it is true at this instant; it then uses the further assumption that a proposition is true at an instant if and only if its predicate inheres in its subject in an instant, so that in our case, since 'the heavens are in motion' is true at the present instant, the heavens move in the present instant, and so their locomotion is instantaneous (I, q. 35, n. 23; VI, q. 13, n. 5)

Aspall reports a reply to this argument that denies that the proposition 'the heavens are in motion' is true at an instant, appealing to the general principle that a proposition in which the predicate stands for a successive thing can be true only in time and not at an instant. Aspall rejects this reply. He points out that the proposition 'the heavens are in motion' is equivalent to the proposition 'the heavens were in motion and the heavens will be in motion', and this latter proposition is indeed true at every instant, although in the case of a temporally finite motion the analogous proposition is true at all instants except the first and the last (VI, q. 13, n. 15).

Aspall's own reply to the argument consists in denying the assumption that in order for the proposition 'the heavens are in motion' to be true at an instant, it is necessary for the heavens to be in motion in an instant. Saying that the proposition 'the heavens are in motion' is true at an instant means that the true composition of the predicate with the subject of that proposition is performed by the human soul at that instant, but it does not follow from this that the predicate inheres in the subject in that instant. For example, both the proposition 'man is animal' and 'the heavens are in motion' are true at an instant in the sense that the mental composition of the predicate with the subject happens at an instant, but while the predicate 'animal' of the first proposition inheres in its subject 'man' in an instant, the predicate 'in motion' of the second proposition does not inhere in its subject 'the heavens' in an instant (I, q. 35, n. 24; VI, q. 13, n. 15).

5.10. Eternity of the world

Following Aristotle, Aspall maintains that the world as a whole is neither generable nor corruptible. He also claims, however, that it had a beginning and will have an end, thus implying that the physical world, understood as the world subject to change, started to exist as a result of a creative act of God rather than a process of generation, and it will cease to exist at some point, again as a result of a divine action rather than a process of corruption (VIII, q. 4, n. 11). Aspall devotes two distinct Questions to the issue of the ingenerability and incorruptibility of the world and to that of its temporal creation.

5.10.1. Ingenerability and incorruptibility of the world

Aspall's discussion focuses on the ingenerability and incorruptibility of the celestial bodies. In Aristotle's view, unlike the bodies here below, the heavens are not subject to generation and corruption, and because of their continuous motion they are responsible for the unending cycles of generation and corruption of the bodies here below.[155] Aspall reports a number of arguments that point out some relevant similarities between the structure of the heavens and of the bodies here below, in particular, the elements, and he concludes on the basis of these similarities that since the bodies here below are subject to generation and corruption, the heavens too must be subject to these substantial changes (VIII, q. 4, nn. 1–10).

The general strategy of Aspall in defending Aristotle's view against these arguments is to show that there are no such relevant similarities, for the heavens are of a totally different nature from the bodies here below. In particular, Aspall rejects the claims that in the heavens there are (i) the qualities of being hot and being cold, and (ii) those of being light and being heavy. As to (i), he specifies that the heavens produce heat and cold just as sublunary bodies do but they are not hot and cold as the sublunary bodies are. For example, the heat produced by the sun and that produced by fire are of the same kind but this does not imply that the qualities responsible for these univocal effects in the sun and in fire respectively are of the same kind. According to Aspall, the principle that if the effects are univocal, then their causes are univocal is not universally valid. Non-univocal efficient causes, such as the first cause and creatures, can indeed produce univocal effects (VIII, q. 4, n. 12). As to (ii), he concedes that a star would move upwards if it were placed here below but he argues that it cannot be inferred from this that the star is a light body. The star placed here below would move upwards simply in order to be joined once more with the whole heavens to which it

[155] Arist. *De gen. et corr.* II.10, 336a13–b24.

naturally belongs, and not because of its lightness, as happens in the case of fire (VIII, q. 4, n. 13).

A special problem arises in connection with the sensible properties of the heavens, such as being visible, since Aristotle makes an explicit claim that strongly suggests that a given sensible property is of only one kind in all the things having that property (VIII, q. 4, n. 5). Aspall wants to defend the view that the being visible of the heavens is of a different kind from the being visible of the bodies here below, and he reports different attempts to reconcile this view with that of Aristotle, without agreeing with any one of them (VIII, q. 4, nn. 15–17).

5.10.2. Temporal creation

Aspall claims that the Christian view that the world was created in time so that motion, time, and natural kinds all had a temporal beginning, is not contrary to Aristotle's view that the world is not subject to generation and corruption, but he does not devote a specific discussion to showing how the two views are reconcilable (VIII, q. 5, n. 18). He explicitly considers the alternative view whereby there is an eternal creation, so that the world is eternal, just as God himself is eternal, but the world is nevertheless created by God. He ascribes this view to the ancient philosophers and mentions the example they use to illustrate it: even if a footprint had existed in the dust from eternity, it would still have a foot as its cause (VIII, q. 5, n. 7). Aspall thinks, however, that the assumption that an eternal effect has an efficient cause is untenable. An eternal effect can have a cause of its conservation in being, but it cannot have an efficient cause because, in his view, it is an essential condition of something's being an efficient cause that it bring its effect into being. According to Aspall, therefore, the idea of eternal creation is incompatible with the assumption that the first cause is an efficient cause of the world (VIII, q. 5, n. 39).

Aspall devotes special attention to the 'logic' of temporal creation. Consider the case of time and suppose that it began to exist so that it passes from non-being to being. There is then a last instant of the non-being of time and a first instant of its being. Are these instants the same instant or two distinct instants? The difficulty here is that neither of these two alternatives seems tenable. (i) If the last instant of the non-being of time and the first instant of its being are the same instant, then contradictory states, like the non-being and the being of time, will exist in the same instant, that is, they will be simultaneous, which is against the principle of non-contradiction. (ii) If, on the other hand, the last instant of the non-being of time is a distinct instant from the first instant of its being, then according to Aristotle, there will be an intervening time between these two instants, which would imply that time exists before its beginning (VIII, q. 5, n. 1).

Aspall's solution consists in positing that the last instant of the non-being of time and the first instant of its being are distinct instants but in denying against (ii) that from this it follows that there is something intermediate between these two instants. He maintains that Aristotle's claim that between any two indivisibles there is something intermediate is only true for indivisibles belonging to the same continuum, such as two points belonging to the same line, but it is not also true for indivisibles belonging to distinct continua, For example, in two contiguous lines, the last point of the first line and the first point of the second line are distinct but nothing intermediate exists between them. Now, the last instant of the non-being of time and the first instant of its being belong to two distinct continua, that is, the last instant of the non-being of time belongs to eternity whereas the first instant of its being belongs to time. Thus the non-being of time is not simultaneous with its being but it is immediately followed by its being (VIII, q. 5, nn. 2, 19–23).

Aspall reports a view that tries to solve similar logical problems arising about the passing into non-being of an intelligence by positing that contradictories can be true in the same instant of time in virtue of the fact that there is a distinction of degrees within an instant, so that one degree corresponds to the being of that intelligence and another one to its non-being (VIII, q. 5, n. 25). Aspall rejects this view (VIII, q. 5, nn. 26–27).

Another question to which Aspall devotes some attention is whether temporal creation is compatible with the absolute immutability of God: does the passage from non-acting to acting correspond to some kind of change in God? Aspall's reply is negative. The passage from non-acting to acting involves a change in the agent only in the case of natural agents and does not apply to voluntary agents, whether created or eternal. A voluntary agent passes from non-willing to produce something to willing to produce it without a change in its will. For example, it is the same act of will by which I want to do something tomorrow and I do not want to do it today (VIII, q. 5, nn. 9–10, 31–35).[156]

Aspall also denies that the act of creation produces a relative change in God in the sense that God acquires a new relation to creatures. He holds that the relation between God and creatures is non-mutual: creatures come to be related to God as a result of God's act of creation but God does not come to be related to creatures (VIII, q. 5, nn. 36–37; also I, q. 22, n. 10).

[156] A different reply, which admits a distinction between divine will and human will, is found in Book I, q. 22, n. 11.

5.11. Self-motion

5.11.1. Heavy and light bodies

In *Physics* VIII.4 Aristotle maintains that heavy and light bodies are not self-movers with respect to their natural motions. For example, fire is not itself the cause of its natural upward motion and earth is not itself the cause of its natural downward motion. Thus, heavy and light bodies are subject to their natural motions but they are not also what causes these motions, so that they do not move themselves in accordance with their natural motions: they are not self-movers. Aristotle suggests that there are two kinds of external things that are in different ways responsible for the natural motion of heavy and light bodies: their generators, that is, what generates those bodies by introducing the relevant substantial form into their matter, and what removes the obstacle to their motion.[157]

An obvious problem with Aristotle's claim that heavy and lights bodies do not cause their own natural motions is that this claim seems incompatible with the notion of natural motion. According to the definition of *Physics* II.1,[158] nature is a principle of motion internal to the thing in motion. From this definition it follows that the natural motion of a thing is caused by something internal to that thing. But then it seems that a thing that is subject to a natural motion should also be the cause of that motion, so that that thing causes its own natural motion. This problem is clearly presented in Aspall's Question on the self-motion of heavy and light bodies. Some of the arguments in favour of Aristotle's view that heavy and light bodies are not self-movers derive this conclusion from the premise that the 'natural' motions of heavy and light bodies are in fact not natural (VIII, q. 20, nn. 6–8). Aspall rejects these arguments (VIII, q. 20, nn. 19–20). Like the majority of Aristotelian commentators, he accepts Aristotle's view that heavy and light bodies do not move themselves in accordance with their natural motions but he thinks that it is compatible with the view that these motions are natural.

To go into more detail, one can add that Aspall maintains that in a heavy or light body there is an intrinsic active principle of its natural motion, and that it is because of the existence of this internal active principle that the natural motion of a heavy or light body is said to be natural. This principle is the substantial form of the body together with an immaterial celestial power. Aspall specifies that the substantial form of a light or heavy body, or more generally, the substantial form of every material substance, is an active internal principle and that it has three distinctive actions: it gives perfection to matter, it acts on matter by removing a form contrary to itself and inducing a

[157] Arist. *Phys*. VIII.4, 255a1–256a3.
[158] Arist. *Phys*. II.1, 192b8–33.

form similar to it, and it moves the body towards its natural place. However, a material form cannot perform any of these three actions by itself but only by being itself moved by an immaterial celestial power (VIII, q. 20, n. 15). Aspall does not explain the role of this immaterial power in detail, but he rejects an opinion according to which this immaterial power is the only mover causing the natural motion of a heavy or light body, so that the substantial form of that body does not play any active role (VIII, q. 20, nn. 9–10). In his view, the substantial form of a heavy or light body is what causes its natural motion, although it also requires an immaterial power to bring this about.

On Aspall's interpretation, however, the existence of an internal active principle sufficient to move a heavy or light body in accordance with its natural motion is not enough for a heavy or light body to qualify as a self-mover in the proper sense. Following some suggestions by Aristotle himself,[159] Aspall posits that two other conditions are required for self-motion in the proper sense: (i) that the thing that moves itself contains two essentially distinct parts such that one is the mover and the other is the thing moved by the mover; (ii) that the thing that moves itself can offer resistance to its mover and so stop moving when it wants to do so. These two additional conditions are not met by a heavy or light body. In particular, a body of this kind fails to meet condition (i) because its only components are matter and substantial form, and while its substantial form is indeed the mover, matter as such cannot be the thing having locomotion (VIII, q. 20, n. 16). Aspall reports Averroes' view according to which the natural motion of a heavy and light body is some kind of self-motion, although not in the proper sense, and he rejects it (VIII, q. 20, nn. 11–14).

5.11.2. Animate bodies

Aristotle maintains that animate bodies are self-movers with respect to locomotion, so that they are the cause of their locomotion.[160] On Aspall's interpretation, Aristotle identifies the internal mover of an animate body with its soul and the internal moved thing with its body, so that it is the soul of an animate body that is responsible for its locomotion (VIII, q. 21.2, n. 1).

In connection with Aristotle's view so interpreted Aspall raises a question about the assumption that the soul moves the body. The main problem with this is that the soul is not a body, but an incorporeal substance, according to Aspall, and an incorporeal substance does not seem to be the kind of thing that can move a body, for it seems that only a body can produce a motion in a body. Aspall's discussion of this problem mainly focuses on the requirement that the mover and the thing moved by it should be in contact and, more

[159] Arist. *Phys.* VIII.4, 255a5–18.

[160] Arist. *Phys.* VIII.2, 253a11–15; VIII.4, 254b14–20, 27–33.

precisely, be in the same place.[161] From this requirement it follows that in order to move the whole body, the soul should be present in the whole body; but since the soul is indivisible (without extension) it cannot be present in the whole body, that is, in each of its extended parts. If the soul is present in one part of the body, then the whole of it is there, since the soul does not have parts, and hence it is not also present in another part of the body (VIII, q. 21.1, nn. 3–4). The question then arises of how it is possible for the soul to be everywhere in the body, given that the soul is indivisible, while the body is extended. Aspall reports an opinion that explains the ubiquitous presence of the soul in the body by appeal to the nature of light. This opinion is very obscure and Aspall quickly dismisses it (VIII, q. 21.2, nn. 4–5). In his view, the true explanation of the presence of the whole soul everywhere in the body should appeal to two properties of soul: (i) its lack of extension and (ii) its lack of a position in body and, more generally, its lack of quantitative properties. In particular, it is the second property of a soul that explains the different topological properties of a point and the soul. A point contained in a line is not everywhere in a line; but this is due to the fact that a point has a position in a line and is a quantity. If with a thought-experiment we remove position and quantity from a point, then that point will be everywhere in a line (VIII, q. 21.2, nn. 6–8).

Coherently with his view that the whole soul is in each part of the body, Aspall also maintains that each power of the soul is in each part of the body. For example, the visual power exists not only in the eyes but also in a hand, although the existence of such a power in a hand does not make the hand capable of seeing (VIII, q. 31, n. 11).

Another aspect of Aristotle's view that needs clarification is whether the soul and the body of an animate substance meet the condition of being essentially distinct, a condition that, according to Aspall, a self-mover in a proper sense must satisfy.[162] The main problem here is that soul and body are related as act and potency: the soul is the act of the body, and the body is in potency to the soul; but a potency and the corresponding act do not seem to be essentially distinct (VIII, q. 22.1, n. 1). In replying to this objection Aspall maintains that we should distinguish two kinds of act-potency relationship: (i) one in which the potency is simply something that receives the act without being transformed into the corresponding act, and (ii) one in which the potency is transformed into the corresponding act. When act and potency are related as in (i), act and potency can be essentially distinct. Of this kind is the relation between matter and form and that between body and soul: in being actualized by a form, matter does not become that form, but simply receives that form;

[161] Arist. *Phys.* VII.2, 243a32–35.
[162] See above, p. xciii.

similarly, in being actualized by a soul, the body does not become the soul. When act and potency are related as in (ii), however, they are not essentially distinct. This is the case of the potentiality of a genus with respect to each species: the essence of the genus becomes a species by the addition of a specific difference (VIII, q. 22.2, n. 2). Aspall specifies that the coming into being of a species from its genus is not a natural process of generation but a rational process, and he adds some clarifications about the similarities and differences between the two kinds of process (VIII, q. 22.2, nn. 4–5).

5.11.3. The heavens[163]

Aspall reports that the theologians and the philosophers have contrasting views on the self-motion of the heavens: according to the theologians, the heavens are not self-movers whereas according to the philosophers, they are self-movers. The contrast derives from a more fundamental disagreement about the movers of the heavens. The theologians posit that the only mover intrinsic to the heavens is their substantial form, whereas the philosophers posit that the heavens have two intrinsic movers: their substantial form, which is a corporeal form, and an intelligence, which is a spiritual form and in this respect similar to the soul of a living body. According to the theologians, then, the essential components of the heavens are exactly similar to those of an elementary body: matter and their corporeal substantial form; but although the substantial form is the mover, matter cannot be the thing moved by it, so that the heavens, just like elementary bodies, fail to qualify as self-movers in the proper sense. According to the philosophers, on the contrary, the heavens are self-movers in the proper sense. The condition that a per se self-mover must have two essentially distinct components such that one is the mover and the other the thing moved by it is met by the heavens: the composite of matter and the substantial corporeal form of the heavens (i.e., the body of the heavens) is perfectly suitable for being the thing moved by an intelligence (VIII, q. 34, nn. 3–5).

Aspall says that the opinion of the theologians is probably the true one (VIII, q. 34, n. 3), but he focuses on the opinion of the philosophers in his discussion and shows how it can be defended against some objections:

(i) One objection is that the motion produced by an intelligence would not be a natural motion for the heavens, since it does not derive from the 'natural' form of the heavens, which is their substantial corporeal form. Aspall replies that an intelligence is a nature in relation to the heavens, although not in the proper sense but only by analogy. The intelligence moves the heavens in its capacity as a voluntary agent which has an appetite for moving the heavens and this appetite is somehow natural (VIII, q. 34, n. 6).

[163] See Donati, 2015a.

(ii) Another objection points out that the fact that the motion of the heavens is eternal and so does not have a state of rest as its natural end shows that this motion cannot be produced by an intelligence. This objection assumes that producing a motion finalized to a state of rest is a distinctive feature of a perfect mover, which an intelligence clearly is. In particular, an intelligence is a much more perfect mover than the human soul; on the other hand, all the motions produced by the human soul are directed to a state of rest in a determinate place (VIII, q. 34, n. 8). Aspall replies that the motion of the heavens produced by an intelligence is directed to an end, but such an end is not a state of rest. Aspall reports two views about the end of the motion of the heavens. One is that of Augustine, who posits that the heavens move with their circular motion in order to become similar to the first cause; the other is that of Aristotle, who posits that the end of the celestial motions is the generation of things here below. He sides with Aristotle's view (VIII, q. 34, nn. 11–13).

(iii) Another objection to the view that an intelligence is a mover of the heavens is based on the medieval doctrine of the bodily spirits, which are thought of as very subtle material entities playing the role of instrumental principles in the operations of the soul upon the body.[164] In accordance with this theory, it is argued that the soul does not move the body immediately but by means of a spirit. The fact that a soul moves the body by means of a spirit is regarded as a mark of the perfection of the soul as a mover. Since an intelligence is a more perfect mover than the soul, if an intelligence were a mover of the heavens, it should move it by means of a spirit; but there are no such bodily spirits in the heavens (VIII, q. 34, n. 9). Aspall replies to the objection that it is not because of its perfection as a mover that a soul moves the body by means of the bodily spirits, but because of the physical constitution of the body moved by it. Since the body of a living substance here below is composed by parts of very different natures (its organic parts), a bodily spirit, which is a material entity in some sense similar to all of these parts, is then required for the soul to be able to move each part of the body. The heavens, however, are bodies of an absolutely uniform nature and so do not require a bodily spirit as an intermediate principle (VIII, q. 34, n. 15).

5.11.4. The intelligence as mover of the heavens

Aspall singles out for a specific discussion some issues arising in connection with the view of the philosophers that an intelligence is a mover of the heavens.

One general issue is whether the intelligence is simply a mover of the heavens or also its perfection, that is, a form inhering in the heavens and

[164] See Bono, 1984.

contributing to its nature. The relevant comparison here is that between the intelligence and the soul of a living substance here below: it is in virtue of the soul that a living substance is a self-mover and it is in virtue of an intelligence that the heavens are a self-mover; but since the soul is both a mover and a perfection of a living substance, this may suggest that for something to qualify as a self-mover it is necessary for its intrinsic mover also to be its perfection (VIII, q. 37, nn. 1–4). Aspall rejects this suggestion and maintains that the intelligence is not a perfection of the heavens (VIII, q. 37, n. 10). He specifies, however, that the relationship between the intelligence and the heavens is not merely that between the efficient cause of a locomotion and the thing moved by it with this motion, as is the relationship between a sailor and a ship. The difference is that in addition to moving the heavens the intelligence also sustains them in their existence, whereas a sailor does not have any causal role with respect to the existence of the ship moved by him. The aggregate of the sailor and the ship does not count as a self-mover because there is no relevant union between the two, but there is such a union between the intelligence and the heavens, since the intelligence is responsible for the existence of the heavens (VIII, q. 37, n. 13).

Another issue is whether the intelligence that is the mover of the first mobile body (the outermost celestial orb) is subject to a motion, either per se or accidentally, or is absolutely immobile. This question is connected to the exegetical issue of whether the immobile mover whose existence Aristotle proves in *Physics* VIII is the intelligence that moves the heavens or the first cause.[165] Aspall reports an opinion according to which we should distinguish between the conjoined mover of the heavens, that is, the intelligence, and its separate mover, that is, the first cause: the conjoined mover can be subject to a motion whereas the first cause is absolutely immobile. Therefore, the immobile mover of *Physics* VIII is the first cause (VIII, q. 36, n. 11). Aspall quickly dismisses this opinion because he thinks that it does not reflect Aristotle's genuine view and that it is also incompatible with the view of Aristotle's most authoritative commentator Averroes. According to Aspall, then, the immobile mover of *Physics* VIII is the intelligence that is the conjoined mover of the heavens (VIII, q. 36, nn. 12–14). Aspall's interpretation, however, is open to the major problem of accounting for the immobility of the intelligence. How can the intelligence be absolutely immobile if it is joined to the heavens, which are constantly subject to motion? It seems to be universally the case that if a thing A is contained in or joined to another thing B and B is subject to a motion, then also A moves together with B. This is illustrated by the case of animal-motion: since an animal is a self-mover, its

[165] Arist. *Phys*. VIII.5.

soul, which is the mover intrinsic to it, is accidentally subject to a motion, for the soul is carried along when the animal moves (VIII, q. 36, n. 1).

Aspall defends the view that the intelligence is absolutely immobile, despite being the conjoined mover of the heavens. He gives two explanations of the immobility of the intelligence. One is in terms of its location: the intelligence is located in the whole heavens but in such a way that it is not in anyone of their parts; the whole heavens as a whole, however, are not subject to a motion, given that in the rotation of the heavens only the parts and not the whole change place (VIII, q. 35, nn. 14, 20; q. 36, n. 15). The other explanation, which Aspall ascribes to Averroes, appeals to the special status of the intelligence that is the mover of the first heavens, according to which that intelligence is not supported as it were, or carried along, by the heavens. The relevant contrast here is with the case of the sailor moving a ship: although the sailor is only a mover of the ship and not also its perfection, the sailor is moved accidentally when the ship moves, because he is supported or carried by the ship. The intelligence, however, is not supported or carried by the heavens, rather the heavens are supported by the intelligence that moves them (VIII, q. 36, nn. 16–17).[166]

To his discussion of the physical role of the intelligences as movers of the heavens, Aspall appends a short discussion of some of their more metaphysical and general properties, such as their classification in a genus, and their distinction into species (VIII, qq. 39–41).

[166] On the special status of the intelligence as mover of the heavens, see also, Book VIII, q. 38, n. 10.

BIBLIOGRAPHY[167]

Bono, J. J., 1984, 'Medical Spirits and the Medieval Language of Life', *Traditio*, 40 (1984), 91–130.

Borgo, M., 2014, 'The Latin Medieval Translations of Aristotle's *Metaphysics*', in F. Amerini & G. Galluzzo (eds.), *A Companion to the Latin Medieval Commentaries on Aristotle's Metaphysics* (Leiden – London: 2014), 19–57.

Callus, D. A., 1943, 'Introduction of Aristotelian Learning to Oxford', *Proceedings of the British Academy*, 29 (1943), 229–281.

Cross, R., 1998, *The Physics of Duns Scotus. The Scientific Context of a Theological Vision* (Oxford: 1998).

Del Punta, F.-Donati, S.-Trifogli, C., 1996, 'Commentaries on Aristotle's *Physics* in Britain, ca. 1250–1270', in J. Marenbon (ed.), *Aristotle in Britain during the Middle Ages* (Turnhout: 1996), 265–283.

Donati, S., 1991, 'Per lo studio dei commenti alla *Fisica* del XIII secolo, I: Commenti di probabile origine inglese degli anni 1250–1270', Part 1, *Documenti e studi sulla tradizione filosofica medievale*, 2 (1991), 361–441.

Donati, S., 1993, 'Per lo studio dei commenti alla *Fisica* del XIII secolo, I: Commenti di probabile origine inglese degli anni 1250–1270', Part 2, *Documenti e studi sulla tradizione filosofica medievale*, 4 (1993), 25–133.

Donati, S., 1994, 'Wissenschaft und Glaube bei der Frage nach dem Ursprung der Materie in einigen ungedruckten Physikkommentaren aus dem 13. bis zum Anfang des 14. Jahrhunderts', in I. Craemer-Ruegenberg & A. Speer (eds.), *Scientia und ars im Hoch-und- Spätmittelalter*, Miscellanea Mediaevalia 22, 1 (Berlin: 1994), 399–420.

Donati, S., 1995, '*Physica* I, 1: l'interpretazione dei commentatori inglesi della *Translatio Vetus* e la loro recezione del commento di Averroè', *Medioevo*, 21 (1995), 75–255.

Donati, S., 1998a, 'Il commento alla *Fisica* di Adamo di Bocfeld e un commento anonimo della sua scuola', *Documenti e studi sulla tradizione filosofica medievale*, 9 (1998), 111–178.

Donati, S., 1998b, 'Materie und räumliche Ausdehnung in einigen ungedruckten englischen *Physikkommentaren* aus der Zeit von etwa 1250–1270', in J. Aertsen & A. Speer (eds.), *Raum und Raumvorstellungen im Mittelalter*, Miscellanea Mediaevalia 25, 1 (Berlin-New York: 1998), 17–51.

Donati, S., 2002a, 'The Doctrine of *dimensiones indeterminatae* in the Commentary Tradition of the *Physics* in the XIIIth and in the Early XIVth Century', in C. H. Leijnhorst, C. H. Lüthy & J. M. M. Thijssen (eds.), *The Dynamics of Aristotelian Natural Philosophy* (Leiden-Boston-Cologne: 2002), 188–223.

Donati, S., 2002b, 'La discussione sulla materia nella tradizione di commento della *Fisica*: commentatori inglesi degli anni 1240–1270 ca.', in G. Fioravanti,

[167] Here we give the complete references to the secondary literature mentioned in the Introduction and in the *Apparatus fontium* to Aspall's text.

C. Leonardi, & S. Perfetti (eds.), *Il commento filosofico nell'occidente latino (secc. XIII–XV)* (Turnhout: 2002), 185–232.

Donati, S., 2003, 'La discussione sull'unità del concetto di ente nella tradizione di commento della *Fisica*: commenti parigini degli anni 1270–1315 ca.', in M. Pickavé (ed.), *Die Logik des Transzendentalen* (Berlin-New York: 2003), 60–139.

Donati, S., 2005, 'The Anonymous Commentary on the *Physics* in Erfurt, Cod. Amplon. Q. 312 and Richard Rufus of Cornwall', *Recherches de Théologie et Philosophie médiévales*, 72 (2005), 232–362.

Donati, S., 2007, 'Materia e dimensioni tra XIII e XIV secolo: la dottrina delle *dimensiones indeterminatae*', *Quaestio*, 7 (2007), 361–393.

Donati, S., 2008, 'Guglielmo di Clifford (+ 1306) e alcuni commenti anonimi ai *libri naturales* del ms. Cambridge, Peterhouse, 157 (*De anima*, *De generatione et corruptione*, *Meteora*, *De somno et vigilia*, *De vegetabilibus*)', *Documenti e studi sulla tradizione filosofica medievale*, 19 (2008), 501–618.

Donati, S., 2011, 'Il *De principiis naturae* di Guglielmo di Rothwell O.P. e le questioni sulla *Fisica* di Goffredo di Aspall (+1287)', in G. Kapriev (ed.), *Nomina essentiant res* (Sofia: 2011), 283–305.

Donati, S., 2012a, 'Goffredo di Aspall (+ 1287) e alcuni commenti anonimi ai *Libri naturales* nei mss. London, Wellcome Hist. Med. Libr., 333 e Todi, BC, 23 (*Qq. Super I De gen. et corr.*, *Qq. super Phys.* V, VI)', Part 1, *Documenti e studi sulla tradizione filosofica medievale*, 23 (2012), 245–320.

Donati, S., 2012b, 'La doctrine de l'analogie de l'être dans la tradition des commentaires de la *Physique*. Quelques modèles interprétatifs (commentaires de la Faculté des arts, autour 1250–1300)', *Revue Thomiste,*112 (2012), 31–59.

Donati, S., 2012c, 'Robert Kilwardby on Matter', in H. Lagerlund & P. Thom (eds.), *A Companion to Robert Kilwardby (*Leiden: 2012), 239–273.

Donati, S., 2013a, 'Goffredo di Aspall (+ 1287) e alcuni commenti anonimi ai *Libri naturales* nei mss. London, Wellcome Hist. Med. Libr., 333 e Todi, BC, 23 (*Qq. Super I De gen. et corr.*, *Qq. super Phys.* V, VI)', Part 2, *Documenti e studi sulla tradizione filosofica medievale*, 24 (2013), 219–418.

Donati, S., 2013b, 'Pseudoepigrapha in the «Opera hactenus inedita Rogeri Baconi»? The Commentaries on the *Physics* and on the *Metaphysics*', in O. Weijers & J. Verger (eds.), *Les débuts de l'enseignement universitaire à Paris (1200–1245 environ)*, Studia Artistarum, 38 (Turnhout: 2013), 152–203.

Donati, S., 2014, 'English Commentaries before Scotus. A Case Study: The Discussion on the Unity of Being', in F. Amerini & G. Galluzzo (eds.), *A Companion to the Latin Medieval Commentaries on Aristotle's Metaphysics* (Leiden: 2014), 137–207.

Donati, S., 2015a, 'Is Celestial Motion a Natural Motion? Averroes' Position and Its Reception in the Thirteenth- and Early Fourteenth-Century Commentary Tradition of the *Physics*', in P. J. J. Bakker (ed.), *Averroes' Natural Philosophy and Its Reception in the Latin West* (Leuven: 2015), 89–126.

Donati, S., 2015b, 'Tra psicologia e filosofia della natura: la teoria delle *species* nella discussione sulla causalità naturale (Commenti inglesi ai *Libri naturales*, 1240–1270 ca.)', *Documenti e studi sulla tradizione filosofica medievale*, 26 (2015), 277–336.

Donati, S., forthcoming, 'Matter, Being and Intelligibility in the Commentary Tradition of the *Physics*: English Commentators, ca. 1240–1270'.

Ebbesen, S., 2014, 'Geoffrey of Aspall, *Quaestiones super librum De somno et vigilia*. An Edition', *Cahiers de l'Institut du Moyen Âge Grecque et Latin*, 83 (2014), 257–341.

Emden, A. B., 1957, *A Biographical Register of the University of Oxford to A. D. 1500,* 3 vols (Oxford: 1957–59).

Gál, G., 1953, 'Robert Kilwardby's Questions on the *Metaphysics* and *Physics* of Aristotle', *Franciscan Studies*, 13 (1953), 7–28.

Galle, G., 2012, 'Geoffrey of Aspall's Questions on Whether the World is Generable and Perishable. An Edition and Discussion', *Documenti e studi sulla tradizione filosofica medievale*, 23 (2012), 321–355.

Galluzzo, G., 2013, 'Geoffrey of Aspall's Commentary on Aristotle's *Metaphysics*, Book Zeta. Some Major Themes', *Documenti e studi sulla tradizione filosofica medievale*, 24 (2013), 1–81.

Gauthier, R. A., 1983, 'Notes sur Siger de Brabant. 1. Siger en 1265', *Revue des sciences philosophiques et théologiques*, 67 (1983), 201–232.

Gilson, E., 1952, *Jean Duns Scotus. Introduction a ses positions fondamentales* (Paris: 1952).

Glorieux, P., 1971, *La Faculté des Arts et ses Maitres au XIIIe siècle* (Paris: 1971).

Hamesse, J., 1974, *Les Auctoritates Aristotelis. Un florilège médiéval* (Louvain-Paris: 1974).

Ker, N. R., 1969, *Medieval Manuscripts in British Libraries* (Oxford: 1969).

Macrae, E., (with a note by Louis-Jacques Bataillon) 1968, 'Geoffrey of Aspall's Commentaries on Aristotle', *Mediaeval and Renaissance Studies*, 6 (1968), 94–134.

McCord Adams, M., 1987, *William Ockham* (Notre Dame, Ind.: 1987).

Menestò, E. & al., 2008, *I manoscritti medievali della biblioteca comunale "L. Leoni" di Todi* (Spoleto: 2008).

Moorat, S. A. J., 1962, *Catalogue of Western Manuscripts on Medicine and Science in the Wellcome Historical Medical Library* (London: 1962).

Plevano, R., 1993, 'Richard Rufus of Cornwall and Geoffrey of Aspall. Two Questions on the Instant of Change', *Medioevo*, 19 (1993), 167–232.

Powicke, F. M., (1931), *The Medieval Books of Merton College* (Oxford: 1931).

Robinson, P., forthcoming, *Catalogue of Medieval Manuscripts of Latin Commentaries on Aristotle in British Libraries*, Volume III.

Sharpe, R., 2001, *A Handlist of the Latin Writers of Great Britain and Ireland before 1540* (Turnhout: 2001).

Shooner, H.V., 1973, *Codices manuscripti Operum Thomae de Aquino* (Roma: 1973).

Thomson, R., 2009, *A Descriptive Catalogue of the Medieval Manuscripts of Merton College, Oxford, with a description of the Greek Manuscripts by N. G. Wilson* (Cambridge: 2009).

Thomson, R., 2011, *Catalogue of Medieval Manuscripts of Latin Commentaries on Aristotle in British Libraries*, Volume I: Oxford (Turnhout: 2011).

Thomson, R., 2013, *Catalogue of Medieval Manuscripts of Latin Commentaries on Aristotle in British Libraries*, Volume II: Cambridge (Turnhout: 2013).

Trifogli, C., 1991, 'Le questioni sul libro III della *Fisica* in alcuni commenti inglesi intorno alla metà del sec. XIII', Part 1, *Documenti e studi sulla tradizione filosofica medievale*, 2 (1991), 443–501.

Trifogli, C., 1993, 'Le questioni sul libro III della *Fisica* in alcuni commenti inglesi intorno alla metà del sec. XIII', Part 2, *Documenti e studi sulla tradizione filosofica medievale*, 4 (1993), 135–178.

Trifogli, C., 1994, 'An Aspect of Medieval Mathematics: Infinity in Number in Some English Commentaries of the XIIIth Century', in I. Craemer-Ruegenberg & A. Speer (eds.), *Scientia und ars im Hoch-und-Spätmittelalter*, Miscellanea Mediaevalia 22, 1 (Berlin: 1994), 343–353.

Trifogli, C., 1996, 'Le questioni sul lib. IV della *Fisica* in alcuni commenti inglesi intorno alla metà del sec. XIII', Part 1, *Documenti e studi sulla tradizione filosofica medievale*, 7 (1996), 39–114.

Trifogli, C., 1998, 'Le questioni sul lib. IV della *Fisica* in alcuni commenti inglesi intorno alla metà del sec. XIII', Part 2, *Documenti e studi sulla tradizione filosofica medievale*, 9 (1998), 179–260.

Trifogli, C., 2000, *Oxford Physics in the XIIIth Century (ca. 1250–1270): Motion, Infinity, Place and Time* (Leiden: 2000).

Trifogli, C., 2004, *Liber tertius Physicorum Aristotelis. Repertorio delle questioni. Commenti inglesi ca. 1250–1270* (Firenze: 2004).

Trifogli, C., 2007, *Liber quartus Physicorum Aristotelis. Repertorio delle questioni. Commenti inglesi ca. 1250–1270* (Firenze: 2007).

Walter, A. B., 1965, 'The Ockhamist Critique', in E. McMullin (ed.), *The Concept of Matter in Greek and Medieval Philosophy* (Notre Dame, Ind.: 1965), 124–146.

Weijers, O., 1994, *Le travail intellectuel à la Faculté des arts de Paris: textes et maîtres (ca. 1200–1500)*, fasc. 1 (Turnhout: 1994).

Weijers, O., 1998, *Le travail intellectuel à la Faculté des arts de Paris: textes et maîtres (ca. 1200–1500)*, fasc. 2 (Turnhout: 1998).

Wippel, J. F., 2000, *The Metaphysical Thought of Thomas Aquinas. From Finite Being to Uncreated Being* (Washington D.C.: 2000).

Wood, R., 2009, 'The Works of Richard Rufus of Cornwall: The State of the Question in 2008', *Recherches de Théologie et Philosophie médiévales*, 76 (2009), 1–73.

Zimmermann, A., 1971, *Verzeichnis ungedruckter Kommentare zur Metaphysik und Physik des Aristoteles aus der Zeit von etwa 1250–1350* (Leiden-Köln: 1971).

QUAESTIONES SUPER PHYSICAM

LIBRI I–IV, VIII

(RECENSIO O)

TABULA QUAESTIONUM

LIBER I

LIST OF QUESTIONS

BOOK I

Q. 26 Quare Aristoteles dixit 'quaedam'.
Q. 27 An haec divisio entis -ens aliud substantia, aliud accidens- sit divisio pure univoci vel aequivoci.
Q. 28 Utrum accidens sit unum genus ad novem praedicamenta.
Q. 29 Utrum nomen concretum det intelligere substantiam sive subiectum tamquam significatum suum.
Q. 30 Utrum accidens possit esse per se sine substantia.
Q. 31 Utrum accidens possit praedicari de substantia.
Q. 32 An illa consequentia teneat: 'substantia et accidens sunt; ergo multa sunt'.
Q. 33 Utrum infinitum quantitati congruat in se.
Q. 34 An infinitum reperiatur in alio a quanto, ut in alio praedicamento.
Q. 35 An sequatur: 'continuum est; ergo multa sunt'.
Q. 36 Utrum pars et totum sint idem.
Q. 37 An omnes partes eiusdem continui sint eaedem in substantia.
Q. 38 De partibus diversorum totorum, ut corruptibili et incorruptibili, utrum conveniant in aliqua natura.
Q. 39 Utrum unum et idem sit substantia et accidens.
Q. 40 An accidens, cum multiplicat speciem suam in medio vel ad sensum, simul cum specie accidentis commultiplicetur aliquid substantiae deferentis illud accidens.
Q. 41 De argumento Aristotelis: si omnia sunt unum, tunc idem est bonum et malum.
Q. 42 Utrum materia prima sit.
Q. 43 Supposito quod materia prima sit, utrum sit prima.
Q. 44 De materia prima an incepit esse.
Q. 45 Habito quod materia prima incepit esse, a quo incepit esse.
Q. 46 Habito quod materia prima incepit esse, et hoc a causa prima, qualiter incepit esse, an scilicet per formam vel per se ipsam.
Q. 47 Utrum materia prima sit substantia vel accidens.
Q. 48 Habito quod materia prima sit substantia, utrum sit substantia in actu vel non.
Q. 49 Utrum materia prima possit intelligi ab intellectu possibili.
Q. 50 An materia prima sit una numero.
Q. 51 Dato quod materia prima intelligitur, an materia prima intelligat.

LIBER II

BOOK II

Q. 29 Qualiter prima causa intelligat istas propositiones: 'homo est animal', 'homo est substantia'.
Q. 30 Qualiter Deus scit contingentia.
Q. 31 De monstris communiter dictis, quare sunt sive dicuntur animalia monstruosa.
Q. 32 Quare huiusmodi animal est sterile.
Q. 33 De convenientia asini cum equa.
Q. 34 De monstris proprie dictis, utrum possit esse aliquod tale animal.
Q. 35 De insertis, utrum ibi sint duae animae vegetativae vel una tantum.
Q. 36 Quid illorum convertit alterum in sui naturam.
Q. 37 Dato quod tale animal possit esse, utrum sit a natura.
Q. 38 An sit unum individuum vel duo.
Q. 39 De assimilatione prolis in lineationibus corporis et dispositione faciei et huiusmodi membrorum, utrum talis assimilatio habeat causam naturalem.
Q. 40 Quae est causa huius assimilationis.
Q. 41 Utrum aliqua pars quantitativa concepti sit a patre.
Q. 42 Utrum unus homo per plures accessus possit super unum generatum.
Q. 43 Utrum duo patres possunt super generationem unius filii.

LIBER III

Q. 1 Utrum motus sit.
Q. 2 De dicto Aristotelis quod est loqui de infinito propter continuitatem in motu.
Q. 3 Utrum omnis motus sit continuus.
Q. 4 In quo genere sit motus.
Q. 5 De opinione Commentatoris quod motus est ipsa forma inducenda non differens ab ipsa nisi secundum magis et minus.
Q. 6 Utrum sufficiant species enumeratae ab Aristotele.
Q. 7 Utrum motus possit intelligi.
Q. 8 Utrum motus sit actus.
Q. 9 Utrum motus sit actus entis in potentia secundum quod huiusmodi.
Q.10 In quo est motus ut in subiecto.
Q. 11 Utrum actio et passio sint motus unus.
Q. 12 Utrum actio et passio diversificent praedicamenta.
Q. 13 An sit infinitum in discretis.

BOOK III

LIBER IV

LIBER VIII

BOOK IV

BOOK VIII

Q. 6 About Aristotle's statement that nothing disordered comes from nature.

Q. 7 About Aristotle's statement that nature is the cause of every ordering.

Q. 8 About Anaxagoras' assumption that all the elements were at rest for an infinite time in some chaos and then intellect arrived and separated them.

Q. 9 About Aristotle's assumption that a man sometimes moves and sometimes does not.

Q. 10 Whether sound is absolutely one accident.

Q. 11 Whether a sound that is numerically one is carried from the struck thing to the hearing.

Q. 12 About Aristotle's statement that an animal is moved by itself only in locomotion.

Q. 13 About Aristotle's statement that between any two contrary motions an intermediate rest intervenes.

Q. 14 About Aristotle's assumption that contrary natural motions occur in equal times.

Q. 15 About Aristotle's statement that the parts of an erosion do not correspond to the parts of the falling of a drop <of water>, but the whole erosion occurs in the last instant of time.

Q. 16 About Aristotle's statement that the elements are naturally at rest in their places, and they move there only by violence.

Q. 17 About Aristotle's statement that in every motion there is a corruption of the terminus from which and a generation of the terminus towards which.

Q. 18 Whether judgment is a motion.

Q. 19 Whether the soul produces a composition in time.

Q. 20 About Aristotle's statement that heavy and light things are not moved by themselves.

Q. 21 Whether the soul moves the body.

Q. 22 Whether the soul moves the body by itself.

Q. 23 Whether the soul moves the body without being in-formed by a determinate species.

Q. 24 Whether the soul moves the body in accordance with every species of locomotion.

Q. 25 About Aristotle's statement that there is not a first among infinitely many things.

Q. 26 About Aristotle's way of arguing: one can find a mover and a moved thing together, and one can find a moved thing by itself; therefore, one can also find a mover by itself.

Q. 27 About Aristotle's assumption that it is impossible that a mover should be moved by the same species of motion with which it moves.

Q. 28 Whether the sciences acquired by the learner are one or more qualities of his soul.

Q. 29 About Aristotle's statement that the mover is in act and the mobile thing is in potency.

Q. 30 About Aristotle's statement that the generator has the same species as the thing generated by it.

Q. 31 Whether the soul moves the body more in one part than in another or whether more of the soul is in one part of the body than in another.

Q. 32 About Aristotle's statement that in locomotion each of the two <that is, the mover and the moved> is in contact with the other or alternatively that one of the two is in contact with the other.

Q. 33 Why Aristotle says: "Each of the two <is in contact> with the other or one of the two <is in contact> with the other".

Q. 34 Whether the heavens are moved by themselves.

Q. 35 Whether an intelligence is everywhere as a whole in the heavens.

Q. 36 Whether an intelligence is moved either per se or accidentally.

Q. 37 Whether the mover of the first orb is its perfection.

Q. 38 Whether the intelligence that moves the heavens moves them with the whole of its power.

Q. 39 Whether the movers of the heavens are in a genus.

Q. 40 Whether the intelligences differ in species.

Q. 41 Whether an intelligence and a rational soul are the same in species or not.

LIBER I

1*. *Quoniam autem oportet intelligere* [a]*et scire circa omnes scientias quarum sunt principia aut causae aut elementa* etc.

<Quaestio 1>
Quaeritur de unitate scientiae naturalis.

1. Et quod non sit una probatio: ab unitate subiecti dicitur scientia una; sed in naturali scientia non est aliquod subiectum unum [b]et univocum de quo traditur tota naturalis scientia; ergo etc. Minor patet per Commentatorem dicentem quod 'corpus' dicitur aequivoce de corpore hic inferius et de corpore supracaelesti; cum igitur in naturali scientia tractetur de utroque corpore, relinquitur quod subiectum non est univocum.

2. Item, scribitur in libro *De generatione* quod non habentia materiam et formam agunt impassibiliter entia. Et glossatur: non habentia materiam et formam, ut corpora supracaelestia. Sed illa propositio hoc modo falsa est, cum omne corpus habeat materiam et formam, nisi intelligatur quod non habeant materiam et formam univoce sicut et corpora hic inferiora. Ergo, ut videtur, corpora caelestia et hic inferiora aequivoce habent materiam et formam. Ergo subiectum naturalis philosophiae non est univocum.

3. Item, in libro de naturalibus tractatur de natura; sed 'natura', secundum Commentatorem, dicitur aequivoce de his quae semper transmutantur, ut corpora caelestia, et de his quae aliquando transmutantur et aliquando non[c], ut corpora inferiora; quare etc.

4. Item, de motu et de mutatione tractatur in naturali scientia; sed motus et mutatio non habent aliquod unum subiectum, ut patet, proprie sumendo subiectum; ergo de necessitate sequitur quod non est aliquod unum subiectum et univocum in naturali scientia.

[a] et ... elementa *om.* N
[b] et univocum *om.* N
[c] non *suppl. s.l.* O

BOOK I

1*. *When the objects of an inquiry, in any department, have principles, causes, or elements, <it is through acquaintance with these> that understanding and knowledge is attained.*[1]

<Question 1>
The question is about the unity of natural science.

1. Arguments that it is not one: a science is said to be one from the unity of its subject; but natural science does not have one univocal subject with which the whole natural science deals; therefore, <natural science is not one>. The minor premise is clear from the Commentator's statement that 'body' is said equivocally of a body here below and of a celestial body;[2] therefore, since natural science deals with both kinds of body, it follows that its subject is not univocal.

2. Furthermore, in *On the Generation* it is written that things that do not have matter and form act without being subject to passion.[3] And this claim is glossed as follows: things that do not have matter and form, namely the celestial bodies. But since every body has matter and form, the proposition is false when glossed in that way, unless it is understood in the sense that the celestial bodies do not have matter and form univocally with the bodies here below. Therefore, it seems that the celestial bodies and the things here below have matter and form equivocally. Therefore, the subject of natural philosophy is not univocal.

3. Furthermore, the book about natural things deals with nature; but according to the Commentator, 'nature' is said equivocally of the things that are always subject to change, namely the celestial bodies, and of those that are at some time subject to change and at another time are not, namely the bodies here below;[4] hence, <the subject of natural philosophy is not univocal>.

4. Furthermore, natural science deals with motion and change; but a motion and a change do not have one subject, if subject is taken in a proper sense, as is clear; therefore, it necessarily follows that natural science does not have one univocal subject.

[1] Arist. *Phys*. I.1, 184a10–12.
[2] Aver. *In Metaph.* X, t.c. 26, fol. 276va.
[3] Arist. *De gen. et corr*. I.7, 324b4–5.
[4] Aver. *In Phys*. II, t.c.1, fol. 48rb; *In Metaph.* II, t.c. 16, fol. 36ra.

5. Item, in libro de naturalibus tractatur de motu; 'motus' autem dicitur aequivoce de motu recto et circulari; ergo non est aliquod subiectum univocum.

6. Ad oppositum: in metaphysica tractatur de prima causa et substantiis creatis et de omni substantia in quantum substantia et de omni accidente in quantum accidens est et universaliter de omni ente in quantum ens est; et tamen metaphysica praeter[a] [b]totam illam communitatem est scientia una; ergo multo fortius naturalis scientia erit scientia una, cum de tanta communitate non tractetur in naturali scientia.

7. [c]Solutio problematis est quod naturalis scientia est scientia una.

8. Ad primum dicendum, secundum quosdam, et similiter ad problema quod duplex est univocatio, naturalis scilicet et logicalis. Loquendo de logicali univocatione, dicunt quod ista scientia est una, quia secundum logicos illa sunt univoca in linea praedicamentali quae in substantia[d] conveniunt. Sed loquendo de naturali univocatione, dicunt quod non, ut probant rationes.

9. [e]Sed contra sic: magnitudo et numerus conveniunt in hoc quod quantitates[f] sunt, et ita esset una scientia de numero et magnitudine, quod falsum est. Et hoc concedendum est.

10. Unde [g]solutio problematis talis est: subiectum naturalis scientiae dupliciter potest considerari. Vel in comparatione ad subsequentia ipsum subiectum, ut passiones et proprietates et alia; sic subiectum non est unum nec est necesse nec requiritur hoc ad unitatem scientiae. Vel in comparatione ad antecedentia ipsum; et sic subiectum est unum in proposito, quia illud quod antecedit subiectum naturalis scientiae et ad quod reducuntur subiectum et omnia pertinentia ad subiectum est unum et univocum, scilicet natura, sicut patet.

11. Sed contra: instans est minimum in tempore et est aliud a natura; ergo non ad naturam reducitur[h]. Et tamen in scientia naturali tractatur de tempore et instanti.

12. Sed dicendum quod, licet non immediate reducatur ad naturam, causaliter tamen ad ipsam reducitur.

[a] praeter *scripsimus*; propter ON
[b] totam illam] istam totam N
[c] Solutio problematis est] dico ad quaestionem N
[d] substantia] subiecto N
[e] Sed contra] contra hoc N
[f] quantitates *scripsimus*; ipsae res ON (quantitas praedicatur univoce de quantitate continua et discreta TC)
[g] solutio ... est] dicendum quod N
[h] reducitur] etc. *add. s.l.* O

5. Furthermore, the book on natural things deals with motion; but 'motion' is said equivocally of a rectilinear and a circular motion; therefore, there is no univocal subject.

6. To the contrary: metaphysics deals with the first cause, with created substances, with every substance in so far as it is a substance, with every accident in so far as it is an accident, and universally with every being in so far as it is a being; metaphysics, however, is one science despite such a great community <of subjects>; therefore, there is a much stronger reason to hold that natural science is one science, since natural science does not deal with such a great community <of subjects>.

7. The solution to this problem is that natural science is one science.

8. According to some people, one must reply to the first argument,[5] and also to the problem <as a whole>, that there are two kinds of univocity: natural and logical. They claim that when one speaks of logical univocity, natural science is one, because the logicians think that those things that agree in substance are univocal in the predicamental line. But they say that when one speaks of natural univocity, natural science is not one, as the arguments prove.

9. Against this: a magnitude and a number agree in being quantities, and thus there would be one science dealing with both numbers and magnitudes, which is false. And this must be conceded.

10. Hence the solution to the problem is as follows. The subject of natural science can be considered in two ways: either (i) in relation to the things posterior to the subject itself, such as its passions, its properties, and the like; in this way the subject is not one nor is this necessary or required for the unity of a science; or (ii) in relation to the things prior to it; and in this way the subject is one in the present case, because the thing prior to the subject of natural science and to which its subject and all the things relative to its subject are reduced is one and univocal, namely, nature, as is clear.

11. But against this: an instant is a minimum in time and it is something other than nature; therefore, an instant is not reduced to nature. But natural science deals with time and with instants.

12. It must be said, however, that although an instant is not reduced to nature immediately, it is reduced to nature causally.

[5] See above, n. 1.

13. Unde ad rationes Commentatoris dicendum quod Commentator intelligit in suis dictis [a]analogiam pro aequivocatione et quod utitur [b]aequivocatione pro analogia. Analogia autem non impedit univocationem quin etiam subiectum propter hoc possit esse unum.

<Quaestio 2>
Quaeritur de subalternatione scientiae naturalis.

1. Et quod subalternetur metaphysicae probatio, quia illa scientia subalternatur alii cuius subiectum est sub subiecto illius. Sed subiectum naturalis scientiae est sub subiecto metaphysicae; corpus enim mobile, quod est subiectum naturalis scientiae, est sub ente vel substantia, quod est subiectum metaphysicae; ergo etc.

2. Idem patet per expositionem nominis. Dicitur enim 'metaphysica' quasi transcendens naturam; sed excessum continetur sub excedente ipsum; ergo etc.

3. Ad hoc idem est ratio Alexandri talis: illa scientia quae non demonstrat sua principia, sed accipit [c]ea demostrata ab alio, est illi subalternata; sed naturalis philosophia non demonstrat sua principia, sed accipit ea[d] demonstrata ab alio, ut a metaphysica; ergo etc.

4. Ad oppositum: subalternans et subalternata non utuntur eodem medio nec eadem demonstratione; sed metaphysica in undecimo et naturalis in octavo *Physicorum* utuntur eadem demonstratione, ut demonstratione quia; quare etc. Per istam rationem respondet Commentator Alexandro.

5. [e]Unde ad problema dicendum quod non subalternatur [f]naturalis scientia metaphysicae. Cuius causa est quod[g] quattuor sunt condiciones requisitae ad hoc quod aliqua scientia alii subalternetur. Prima condicio est quod subalternata[h] habeat aliquid additum super subalternantem[i] quod sit diversum genere. Secunda condicio est quod subalternata[j] non egrediatur de principiis subalternantis[k]. Tertia condicio est quod ex illo[l] et addito possit

[a] analogiam pro aequivocatione] aequivocationem pro analogia N
[b] aequivocatione pro analogia *scripsimus cum mg.* C; analogia pro aequivocatione ON
[c] ea demonstrata] eadem O
[d] ea] eadem O
[e] Unde ... dicendum] dico ad quaestionem N
[f] naturalis scientia metaphysicae] neque etiam metaphysica subalternatur N
[g] quod] quia N
[h] subalternata *intellige* subiectum subalternatae
[i] subalternantem] subalternans O; *intellige* subiectum subalternantis (aliquid apponatur subiecto scientiae subalternantis TC)
[j] subalternata *intellige* subiectum subalternatae
[k] subalternantis *intellige* subiecti subalternantis
[l] illo *sc.* subiecto scientiae subalternantis (ex subiecto illo et ex addito fiat unum TC)

13. Thus, in reply to the arguments of the Commentator[6] it must be said that in his claims the Commentator understands equivocity to mean analogy and uses the term 'equivocity' in place of 'analogy'. But analogy does not rule out univocity or prevent the subject from being one.

<Question 2>
The question is about subalternation in natural science.

1. Arguments that natural science is subaltern to metaphysics: a science whose subject is included in the subject of another science is subaltern to it; but the subject of natural science is included in the subject of metaphysics, since mobile body, which is the subject of natural science, is included in being or substance, which is the subject of metaphysics; therefore, <natural science is subaltern to metaphysics>.

2. The same conclusion becomes clear through the exposition of the name; for the word 'metaphysics' suggests that metaphysics transcends nature; but what is exceeded is contained in what exceeds it; therefore, <natural science is contained in metaphysics>.

3. The following argument of Alexander proves the same conclusion:[7] a science that does not prove its principles, but takes them as proved by another science is subaltern to it; but natural philosophy does not prove its principles, but takes them as proved by another science, namely by metaphysics; therefore, <natural philosophy is subaltern to metaphysics>.

4. To the contrary: a subalternating science and a subaltern science do not use the same middle term or the same kind of demonstration; but metaphysics in book XI[8] and natural science in *Physics* VIII[9] use the same kind of demonstration, namely demonstration of the fact; therefore, <natural science is not subaltern to metaphysics>. The Commentator replies to Alexander with this argument.[10]

5. Thus, we should reply to this problem by saying that natural science is not subaltern to metaphysics. The reason for this is that four conditions are required for the subalternation of one science to another. The first condition is that the subaltern science should have something added to the subalternating science that is different in genus from <that subalternating science>. The second condition is that the subaltern science should not emerge from the principles of the subalternating science. The third condition is that from the subject of the subalternating science and what is added to it something

[6] See above, nn.1, 3.
[7] Cf. Aver. *In Metaph.* XII, t.c. 5, fol. 292vb.
[8] Arist. *Metaph.* XII.6, 1071b3–22.
[9] Arist. *Phys.* VIII.5–6, 10.
[10] Aver. *In Metaph.* XII, t.c. 5, fols. 292vb –293va.

fieri unum. Quarta condicio est ut non demonstrent[a] per idem medium[b]. Secunda condicio et quarta deficiunt in proposito, quia corpus mobile egreditur de principiis entis. Et similiter quarta, sicut patet.

6. Per hoc vero patet solutio ad rationes.

1*. Quaeritur de prima conclusione.

<Quaestio 3>

Et primo quaeritur de maiore, circa quam primo quaeritur [O 88rb] an contingat aliquid scire vel intelligere.

<Quaestio 3.1>

1. Et quod non videtur, quia, si contingeret aliquid scire, illud scitum moveret intellectum nostrum; sed nihil potest movere intellectum nostrum. Probatio, quia, si aliquid moveret intellectum nostrum, hoc maxime esset per doctrinam et disciplinam; sed doctrina et disciplina non possunt movere intellectum. Probatio: scribit Boethius istam propositionem "Non est universale intelligibile nisi prius fuerit singulare sensibile"; sed ista doctrina et ista disciplina sunt insensibilia per auctorem *Sex principiorum*; ergo etc.

2. Item, omne quod fit notum fit notum ex notioribus; ergo, si animae in sua prima creatione fieret[c] aliquid notum, hoc erit per aliquid notius. Sed non est ei aliquid notius, quia non est ei aliquid notum; ergo etc.

3. Item, scribitur[d] quod circa intellectivam partem animae non fit[e] alteratio; sed cum anima primo creatur, est non sciens, quia creatur sicut tabula nuda; ergo numquam fiet sciens, quia, si sic, esset alteratio [f]in ea, quod est inconveniens.

[a] demonstrent] demonstrant O
[b] medium] sed quod una demonstret per medium quia quod alia demonstrat per medium propter quid *add.* N
[c] fieret] fiat N
[d] scribitur] dicit Aristoteles in VII Physicorum N
[e] fit] sistit N
[f] in ea *om.* N

that is one can come to be. The fourth condition is that they should not use the same medium in demonstrations. The second and the fourth conditions are not met in the present case, because a mobile body emerges from the principles of being, and similarly the fourth condition <is not met>, as is clear.

6. With this the solution to the arguments[11] is perfectly clear.

1*. Questions about the first conclusion.[12]

<Question 3>
The first questions are about the major premise,[13] and the first of these is whether it is possible to know or understand something.

<Question 3.1>

1. Arguments for the negative: if it were possible to know something, the known thing would move our intellect; but nothing can move our intellect. Proof: if something moved our intellect, this would maximally happen in virtue of a doctrine and a discipline; but a doctrine and a discipline cannot move our intellect. Proof: Boethius wrote this proposition "Something is not universal and intelligible unless it was singular and sensible beforehand";[14] but this doctrine and this discipline are not sensible, according to the author of *Six principles*;[15] therefore, <a doctrine and a discipline are not universal and intelligible>.

2. Furthermore, everything that comes to be known comes to be known from things that are better known; therefore, if something came to be known by a soul in its first creation, this would be in virtue of something better known. But there is nothing better known by a soul <in its first creation> because there is nothing known by it; therefore, <nothing comes to be known by a soul>.

3. Furthermore, it is written that an alteration does not occur in the intellectual part of a soul;[16] but when a soul is first created, it has no knowledge, because it is created as a blank slate;[17] therefore, it will never come to have knowledge, because, if it did, there would be an alteration in it, which is absurd.

[11] See above, nn.1–3.

[12] That is, the science of nature must determine what relates to its principles. See Arist. *Phys.* I.1, 184a14–16.

[13] See above, q. 1, n. 1*.

[14] We could not find this claim in Boethius but a very similar one is in Arist. *De anima* III.7, 431a16–17.

[15] Anon. *Sex princ.* 12, AL I.6–7, 37.17–21 and critical apparatus.

[16] Arist. *Phys.* VII.3, 247b1–2.

[17] Cf. Arist. *De anima* III.4, 429b29–430a2.

4. Item, si intellectus intelligeret[a], intelligeret aut substantiam aut accidens; sed non substantiam[b], quia tunc intelligeret maxime substantiam, ut substantiam quae est generalissimum.

5. Sed contra: omne [c]recipiens denudatur a recepto; si ergo intellectus debeat intelligere et recipere substantiam quae est generalissimum[d], denudatur a forma generis generalissimi. Sed prout est in praedicamento substantiae, habet formam generalissimi; ergo non potest [e]eam[f] recipere nec intelligere.

6. Et quod accidens non intelligatur, probatio, quia, secundum Averroem[g], verius fit unum ex intellectu et intellecto quam ex materia et forma; sed ex substantia et accidente non potest vere fieri unum; ergo etc.

7. Item, si accidens intelligeretur, tunc moveret intellectum. Sed quod non potest movere intellectum nostrum, probatio: movens nobilius est moto; nullum accidens est nobilius substantia; ergo etc.

<Quaestio 4>
Quaeritur etiam si intellectus possit intelligere universale.

<Quaestio 4.1>

1. Et quod [h]non probatio, quia omne quod intelligitur intelligitur per speciem; sed universale non potest intelligi per speciem, quia sic sequeretur quod speciei esset species, cum universale sit species, quod est inconveniens et contra Aristotelem in quinto huius.

2. Item, si intelligeret universale, tunc intelligeret maxime universale quod est ens; sed ens non potest intelligi, quia sic intelligeretur per suam speciem; [i]sed non habet speciem; ergo etc. Probatio minoris, quia, si ens habeat speciem, aut est illa species ens aut[j] non ens; sed non est non ens; ergo erit ens. Cum ergo omnis species est per additionem supra suum genus, illud ens tunc erit per additionem supra ens simpliciter. Sed omnis species simplicior est eo cuius est species; sed illud ens, cum se habeat per additionem, erit compositius; ergo etc.

[a] intelligeret *suppl. mg.* O; aut *add. sed del.* N
[b] substantiam] substantia O
[c] recipiens ... recepto] receptum denudatur a recipiente N
[d] generalissimum] illa substantia *add.* ON *sed del.* O
[e] eam … intelligere] intelligi nunc recipi N
[f] eam *suppl. s.l.* O
[g] Averroem *scripsimus cum* TC; Aristotelem ON
[h] non probatio] videtur *scrips. sed suppl. s.l.* non N
[i] sed … speciem *om. hom.* N
[j] aut] vel O

4. Furthermore, if an intellect understood <something>, it would understand either a substance or an accident; but not a substance, because then it would understand what is maximally substance, namely the substance that is the most general genus.

5. But against this: every receiving thing is stripped of the received thing; therefore, if an intellect must understand and receive the substance that is the most general genus, that intellect is stripped of the form of the most general genus. But in so far as it is in the category of substance an intellect has the form of the most general genus; therefore, an intellect can neither receive nor understand that form.

6. And that an accident is not understood is proved because, according to Averroes, something that is more truly one comes to be from an intellect and what is understood than from matter and a form;[18] but something that is truly one cannot come to be from a substance and an accident; therefore, <an accident is not understood>.

7. Furthermore, if an accident were understood, then it would move an intellect. But that an accident cannot move our intellect is proved as follows: a mover is more noble than the thing moved; but no accident is more noble than a substance; therefore, <an accident cannot move our intellect>.

<Question 4>
Another question is whether an intellect can understand a universal.

<Question 4.1>

1. Arguments for the negative: everything that is understood is understood in virtue of its species; but a universal cannot be understood in virtue of its species, because if it were so understood, then, given that a universal is a species, it would follow that there would be a species of a species, which is absurd and contrary to Aristotle in *Physics* V.[19]

2. Furthermore, if an intellect understood a universal, then it would maximally understand the universal that is being; but being cannot be understood, because if it were, it would be understood in virtue of its species, but being does not have a species; therefore, <being is not understood>. Proof of the minor premise: if being has a species, that species is either a being or a non-being; but it is not a non-being; therefore, it will be a being. Since every species comes about through addition to its genus, that being will then come about through addition to being absolutely. But every species is more simple than the thing of which it is a species; but that being, since it comes about through addition, will be more composite; therefore, <being does not have a species>.

[18] Aver. *In De anima* III, t.c. 5, ed. Crawford, 404.503–508.
[19] Arist. *Phys.* V.2, 225b33–226a23.

<Quaestio 3.2>

1. Omnes istae rationes conclundunt falsum.

2. Ad [a]primam rationem dicendum quod auctor *Sex principiorum* intelligit quod haec doctrina et haec disciplina sunt insensibilia per se consideratae; si tamen considerentur sub suis signis, ut sub voce, sic sunt sensibilia. Et hoc modo doctrina et disciplina movent intellectum, scilicet prout sunt sub voce.

3. Ad secundam dicendum quod maior intelligenda est de cognitione conclusionum per praemissas, et hoc modo verum concludit.

4. Ad tertiam dicendum quod circa [b]animam intellectivam non cadit alteratio, quantum est a parte animae, nisi secundum quod anima unitur cum corpore.

5. Vel sic: duplex est alteratio, scilicet quaedam naturalis, [c]et talis alteratio[d] fit a contrario in contrarium, et talis alteratio non est circa animam intellectivam; et quaedam spiritualis, et talis [e]alteratio est[f] circa animam intellectivam, quoniam alteratio spiritualis est ad meliorationem[g].

6. Nota quod humanitas [h]per se non est sensibile nisi per sua accidentia.

7. Ad quartam sic, quod maior intelligenda est sic: [i]recipiens denudatur a recepto eo [j]modo quo est receptum; sed receptum est in praedicamento substantiae ut res est et recipitur ab intellectu ut species et[k] intentio.

8. Ad quintam dicendum quod Averroes[l] intelligit per 'verius', hoc est minus numerosum, quia materia et forma ponunt in numerum, et ideo possunt copulari per coniunctionem mediam; intellectus autem et intentio rei sive species non ponunt in numerum. Probatio: omne quod numeratur cum aliquo numeratur cum eo sub aliqua forma communi; sed intellectus et intentio rei in nullo communicant, quia non in communissimo, ut in ente, quia intellectus est ens, intentio autem est non ens; et ideo in nullo communicant. Unde ex hoc patet iam quod ista propositio falsa est 'Deus et angelus sunt duo' et omnes tales propositiones.

[a] primam rationem] primum N
[b] animam intellectivam] intellectivam partem N
[c] et ... contrarium *om. hom.* N
[d] alteratio *post* contrarium *transp. sed corr.* O
[e] alteratio est] est alteratio quae est N
[f] est *ex corr.* O
[g] meliorationem] melioritionem (*lectio incerta*) O; melius N
[h] per ... sensibile] non est per sensibilis N
[i] recipiens ... recepto] receptum denudatur a recipiente N
[j] modo... sed] quod N
[k] et] vel N
[l] Averroes *scripsimus*; Aristoteles ON

<Question 3.2>

1. All these arguments[20] reach false conclusions.

2. To the first argument[21] it must be said that the author of *Six principles* means that this doctrine and this discipline are not sensible considered by themselves, but if they are considered through their designations, namely as referred to by spoken words, in this way they are sensible. And a doctrine and a discipline move an intellect in this way, that is, in so far as they are referred to by spoken words.

3. To the second argument[22] it must be said that the major premise should be understood as being about the cognition of a conclusion through some premises, in which case the conclusion <of the argument> is true.

4. To the third argument[23] it must be said that an alteration does not occur in an intellectual soul in so far as it depends on the soul itself, although it does occur in so far as it depends on the soul's being joined to a body.

5. Another reply is that there are two kinds of alteration: (i) one is natural, and this kind of alteration is from a contrary to a contrary, and this kind of alteration does not occur in an intellectual soul; (ii) another is spiritual, and this kind of alteration occurs in an intellectual soul because a spiritual alteration is directed towards an improvement.

6. Note that humanity is not sensible per se but only through its accidents.

7. To the fourth argument[24] it must be said that the major premise should be understood in this way: a receiving thing is deprived of the nature of the received thing in the way in which it is received; but the received thing is in the category of substance in so far as it is a thing, and it is received by the intellect in so far as it is a species or intention.

8. To the fifth argument[25] it must be said that Averroes means by 'more truly' less numerous, because matter and a form produce a number, and therefore they can be joined by an intermediate conjunction; but an intellect and an intention or species of a thing do not produce a number. Proof: everything that is numbered with something is numbered with it under a common form; but an intellect and an intention of a thing have nothing in common, because they do not have the most common thing, namely being, in common, since an intellect is a being, but an intention is not a being, and therefore they have nothing in common. From this it is now clear that the proposition 'God and an angel are two' is false as are all propositions of this kind.

[20] See above, q. 3.1, nn. 1–7; q. 4.1, nn. 1–2.
[21] See above, q. 3.1, n. 1.
[22] See above, q. 3.1, n. 2.
[23] See above, q. 3.1, n. 3.
[24] See above, q. 3.1, nn. 4–5.
[25] See above, q. 3.1, n. 6.

<Quaestio 4.2>

1. Ad sextam dicendum quod universale dupliciter potest considerari: vel prout est quaedam species immediate recepta in anima intellectiva, et sic non intelligitur per suam speciem, quia tunc est ipsa species intellecta; vel prout est res quaedam extra animam, et tunc non est species, sed res, et tunc potest intelligi per speciem.

2. Ad septimam dicendum quod, sicut species coloris non est color, sic id[a] quod est species entis non est ens.

<Quaestio 5>

Quaeritur an omnia intelligamus et etiam quaeritur an contingat intelligere causam primam, quoniam, habito quod contingat intelligere causam primam, multo fortius sequitur quod omnia alia possumus intelligere et etiam quod intelligamus.

1. Quod autem intelligamus causam primam probatio: quandocumque obiectum sensus[b] est praesens sensui in debito situ, statim a sensu apprehenditur; ergo, quandocumque obiectum intellectus[c], ut intelligibile, est praesens intellectui, **[O 88va]** statim ab intellectu apprehendetur. Sed secundum Augustinum, Deus essentialiter, potentialiter et praesentialiter est ubique; ergo in intellectu humano; ergo etc.

2. Item, habetur in *Ethicis* quod omnis mens humana appetit summe[d] bonum cognoscere[e], cuiusmodi est prima causa; sed nihil appetimus nisi quod volumus et nihil volumus quod ignoramus; quod enim volo non ignoro; ergo etc.

3. Item, in coniunctione animae cum corpore corpus non amittit suas[f] proprietates, ut patet manifeste. Cum ergo multo potentior sit anima supra corpus quam corpus supra animam, sequitur quod multo fortius anima non amittet suas proprietates. Sed anima ante suam coniunctionem omnia intellexit, [g]et etiam primam causam; ergo et post unionem.

[a] id *scripsimus*; ens ON
[b] sensus *om.* ON *sed suppl. s.l.* O
[c] intellectus *om.* ON *sed suppl. s.l.* O
[d] summe] supremum N
[e] cognoscere] contingere N
[f] suas] partes *add. sed del.* O
[g] et ... causam *om.* N

<Question 4.2>

1. To the sixth argument[26] it must be said that a universal can be considered in two ways: either (i) in so far as it is a species immediately received in an intellectual soul, and in this way a universal is not understood in virtue of its species, because in this case it itself is an understood species; or (ii) in so far as it is a thing outside the soul, and then a universal is not a species, but a thing, and then it can be understood in virtue of its species.

2. To the seventh argument[27] it must be said that just as the species of a colour is not a colour so the thing that is a species of being is not a being.

<Question 5>
The question is whether we understand all things, and also whether it is possible to understand the first cause, because, once it has been established that it is possible to understand the first cause, there is a much stronger reason to conclude that we can understand all other things, and also that we do understand.

1. Arguments that we understand the first cause: whenever a sensible object is present to the senses in the appropriate location, it is immediately apprehended by the senses; therefore, whenever an intellectual object, that is, an intelligible thing, is present to an intellect, it is immediately apprehended by the intellect. But, according to Augustine, God is everywhere by his essence, his power, and his presence;[28] therefore, God is also present to the human intellect; therefore, <God can be apprehended by the intellect>.

2. Furthermore, it is stated in the *Ethics* that every human mind desires to know the supreme good,[29] which is the first cause; but we only desire what we will and we will nothing that we do not know, since I do not fail to know what I will; therefore, <a human mind knows the first cause>.

3. Furthermore, in the union of a soul with a body the body does not lose its properties, as is completely obvious. Therefore, since the soul has much more power over the body than the body has over the soul, it follows that there is a much stronger reason to hold that the soul will not lose its properties. But the soul understood all things including the first cause before its union; therefore, it also knows the first cause after its union.

[26] See above, q. 4.1, n. 1.
[27] See above, q. 4.1, n. 2.
[28] The correct reference is Peter Lombard, *Sententiae* I, dist. 37 c.1 § 5, ed. Coll. S. Bonav. IV, 265.2–4; based on *Glossa ord. in Cant*. 5.17, ibid., § 3, 264.6–8 and apparatus of the sources.
[29] Arist. *Eth. Nic*. X.7–9, 1177a12–1179b32.

4. Item, lapis, ubicumque ponatur, sive in intersticio ignis vel in aliquo alio loco, semper descendit ad locum proprium, ut deorsum; ergo, cum proprium animae sit omnia intelligere, et etiam primam causam, ubicumque ponatur et in quocumque loco, semper obtinebit suum proprium; ergo etc.

5. Item, anima cognoscit omnia principia, et qui cognoscit antecedens cognoscit et consequens; ergo anima cognoscit omnia principiata[a] et omnes conclusiones, et ita omnia intelligit.

6. Item, virtus visiva absque omni agente potest videre obiectum; ergo, cum intellectus sit potentior quam sensus, multo fortius percipiet intelligibile et apprehendet[b] sine aliquo alio agente, et ita non necesse est ponere intellectum agentem; ergo, ut videtur, intellectus erit in potentia accidentali ad intelligendum, et ita subito omnia intelliget.

7. Item, plus conveniunt anima sensitiva et intellectiva quam sensitiva et corpus; sed in tantum conveniunt anima sensitiva et corpus quod species corporis sine aliquo alio agente recipitur in anima sensitiva, ut patet; ergo multo fortius in tantum convenient anima sensitiva et intellectiva quod species recepta in virtute sensitiva sine aliquo intermedio agente vel coadiuvante[c] subito recipitur in intellectiva, et ita idem quod pridem.

8. [d]Solutio. Dicendum quod anima nostra intellectiva sive anima humana non omnia intelligit, nec etiam ipsum primum complete.

9. Unde ad secundam[e] rationem dicendum, secundum quosdam, quod omnis mens humana habet cognitionem de prima causa incompletam et eam cognoscere appetit, scilicet secundum cognitionem completam.

10. Sed contra: Deus, secundum Augustinum, est summe simplex, indivisibile, impartibile; ergo non habet partem et partem. Ergo qui cognoscit aliquid ipsius complete, cognoscit et totum complete, quod verum est.

[a] principiata *corr. ex* principia O
[b] apprehendet *lectio incerta* O
[c] coadiuvante] quoadiuvante O
[d] Solutio. Dicendum] dico N
[e] secundam *scripsimus*; propriam *(lectio incerta)* O; primam N

4. Furthermore, wherever a stone is located, either in a zone of the sphere of fire or in some other place, it is always moved downwards to its proper place, namely to the downward place; therefore, since it is proper to the soul to understand all things, including the first cause, wherever and in whatever place the soul is located, it will always achieve what is proper to it; therefore, <the soul knows the first cause>.

5. Furthermore, the soul knows all principles, and whoever knows an antecedent also knows the consequent; therefore, the soul knows all the things derived from the principles and all conclusions, and thus it understands all things.

6. Furthermore, the visual power can see its object without any other agent; therefore, since the intellect is more powerful than the senses, there is a much stronger reason to hold that the intellect perceives and apprehends an intelligible without another agent, and thus it is not necessary to assume an agent intellect; therefore, it seems, the intellect will be in accidental potency to understanding, and thus it will immediately understand all things.

7. Furthermore, a sensitive and an intellectual soul agree more than do a sensitive soul and a body; but a sensitive soul and a body agree to the extent that the species of a body is received in a sensitive soul without another agent, as is clear; therefore, there is a much stronger reason to hold that a sensitive and an intellectual soul agree to the extent that a species received in the sensitive power without an intermediate agent or co-agent is immediately received in the intellectual soul, and thus the same conclusion as before follows.

8. Solution. It must be said that our intellectual soul or the human soul does not understand all things, nor does it completely understand the first being itself.

9. Thus, according to some people, one should reply to the second argument[30] that every human mind has an incomplete cognition of the first cause and desires to know it with a complete cognition.

10. But against this: according to Augustine, God is absolutely simple, indivisible, inseparable into parts;[31] therefore, he does not have distinct parts. Therefore, anyone who knows something of God completely also knows the whole of him completely, and this is true.

[30] See above, n. 2.

[31] See, for example, Aug. *Conf.* XIII.3.4, CCSL 27, 244.10–12, and *De trin.* VI.6 (8), VII 1(2), CCSL 50, 236.1–3, 246.66–69. However, Aspall's formulation of Augustine's view is more similar to a claim about the immobile mover in Arist. *Phys.* VIII.10, 267b25–26 than to Augustine's own formulation. See *Physica* Transl. Vetus, AL VII 1.2, 340.19–21: "Manifestum itaque est quoniam indivisibile et inpartibile est nullam habens magnitudinem." Other contemporary commentators ascribe the view to Aristotle in the same context. See, for example, Anon., *Qq. in Phys.*, MS Cambridge, Gonville and Caius College, 367, fol. 120rb: "Aristoteles in fine octavi huius dicit quod est impartibile, indivisibile, nullam habens magnitudinem etc.".

11. Unde sic dicendum est quod[a] cognoscitur complete ab intellectu nostro, sed non cognoscitur adeo intensive[b] sicut cognoscitur post separationem animae a corpore. Et hoc patet sic: lux est indivisibilis; si vero appareat oculo meo, semper videtur; non tamen semper adeo intensive[c] videt oculus meus lucem, semper tamen videt adeo quantitative. Et sic patet responsio ad secundam[d] rationem.

12. Ad primam[e] dicendum quod causa prima dupliciter potest esse praesens apud intellectum nostrum: uno modo, secundum quod est praesens cuilibet rei, et sic prima causa non est materialis, sed est forma solum, quae est salvans omnem rem et omnem creaturam. Et si causa prima sic sit praesens apud intellectum possibilem, non est necesse ut ipsum apprehendat, sicut patet. Alio modo potest esse praesens secundum quod est apprehensibile et natum apprehendi. Et sic debet[f] intelligi ab intellectu possibili. Sed adhuc non semper intelligitur, sed solum quando intellectus noster possibilis non[g] convertit se ad naturam infimam nec mediam, sed solum ad naturam supremam, quae natura est solum ipsa prima causa. Unde homo, quando reliquerit omnia ista[h] inferiora et etiam omnia creata et etiam omnia media, ut caelum et angelos, et elevet suam cogitationem[i] usque ad ipsum primum summum bonum, tunc solum potest intellectus noster possibilis intelligere primam causam.

13. Ad tertiam dicendum quod, quando duo quorum unum est nobilius[j] et reliquum vilius adiunguntur ad invicem, nobilius amittit suas proprietates et debilitatur per vilius, vilius autem melioratur per nobilius, sicut patet in grosso exemplo: si argentum coniungatur cum stagno, stagnum nobilitatur et argentum vilioratur; argentum enim esset [k]magis nobile si per se poneretur, stagnum [l]autem non est ita nobile quando per se ponitur [m]sicut cum coniungitur[n] cum argento.

14. Et per hoc etiam patet responsio ad quartam.

[a] quod] non *add.* N
[b] intensive] insensitive O
[c] intensive] insensitive O
[d] secundam] primam N
[e] primam] secundam N
[f] debet] potest N
[g] non *suppl. s.l.* O
[h] ista] illa O
[i] cogitationem] cognitionem N
[j] nobilius] nobile N
[k] magis nobile] nobilius N
[l] autem ... nobile] tamen est ignobilius N
[m] sicut cum] quam quando N
[n] coniungitur *scripsimus cum* N; ordinatur O

11. Accordingly, it must be said that God is known by our intellect completely, but he is not known as intensively as he is known after the separation of the soul from the body. And this is made clear in this way: light is indivisible; and if light appears to my eye, it is always seen, but my eye does not always see light equally intensively, although it always sees the same quantity <of light>. And with this the reply to the second argument is clear.

12. To the first argument[32] it must be said that the first cause can be present to our intellect in two ways: (i) in one way as it is present to every other thing, and in this way the first cause is not material, but is only a form, which preserves every thing and every creature. And if the first cause is present in this way to the possible intellect, it is not necessary for the intellect to apprehend it, as is clear. (ii) In another way the first cause can be present in so far as it can be apprehended and is apt by nature to be apprehended. And it is in this way that it must be understood by the possible intellect. In this case too, however, the first cause is not always understood, but only when our possible intellect does not turn towards a low or an intermediate nature, but only towards the supreme nature, which nature alone is the first cause itself. Thus, when a man has abandoned all the things here below and all created things and all intermediate things, namely the heavens and the angels, and raises his thought up to the first supreme good itself, only then will our possible intellect be able to understand the first cause.

13. To the third argument[33] it must be said that, when two things one of which is nobler and the other less valuable are joined together, the nobler one loses its properties and is weakened by the less valuable one, whereas the less valuable one becomes better through the more noble, as is made clear by a rough example: if silver is joined with lead, lead becomes nobler and silver becomes less valuable; for silver would be nobler if it existed by itself, but lead is not so noble when it exists by itself as when it is joined with silver.

14. And from this the reply to the fourth argument[34] is also clear.

[32] See above, n. 1.
[33] See above, n. 3.
[34] See above, n. 4.

15. Ad quintam dicendum quod non sequitur, si cognoscat antecedens, quod cognoscat consequens, quia potest dubitare utrum sit consequens vel non; et etiam si sciat quod sit[a] consequens, potest dubitare quo modo et qua via sequatur. Unde non concludit.

16. Ad sextam dicendum et ad septimam quod species corporis quae est in corpore sub istis[b] condicionibus, 'hic' et 'nunc', est in virtute sensitiva sub eisdem condicionibus; sed species quae est in virtute sensitiva sub istis condicionibus, 'hic' et 'nunc', omnino denudatur et spoliatur ab istis condicionibus in intellectiva[c]. Et ita quoad hoc plus differunt intellectiva et sensitiva quam sensitiva et corpus. Et huiusmodi spoliationem[d] causat [e]intellectus agens ex parte animae intellectivae.

<Quaestio 6>
Consequenter quaeritur specialiter an intelligamus causam primam.

1. Et videtur quod sic per Aristotelem in libro *Posteriorum* sic: scire est causam rei cognoscere; ergo qui scit A scit ipsam per causas, ut per B; similiter, qui scit B scit per causam, ut per C. Et sic contingit procedere in infinitum usque ad causam primam; ergo, ut videtur, qui cognoscit A vel B cognoscit et primam causam.

2. Item, scribit Commentator quod, si aliquis[f] scit **[O 88vb]** aliquid, scit illud per causam primam; ergo, ut videtur, in minimum cognoscendo, cognoscitur et ipsa causa prima.

3. Item, Commentator: "Cognita quidditate substantiae sensibilis determinatae, cognoscitur et ipsa causa prima[g]".

4. Ad oppositum: intellectus ad intellectum est proportio; finiti ad infinitum non est proportio; sed intellectus noster est finitus, ipse Deus est infinitus; quare etc.

5. Item, infinitum non comprehenditur a finito; ergo etc.

6. Solutio problematis est quod contingit intelligere primam causam. Sed primae rationes concludunt falsum, scilicet quod in cognoscendo[h] istam rem, cognoscitur et ipsa causa prima.

[a] sit] scit O
[b] istis] his N
[c] intellectiva *lectio incerta* O
[d] spoliationem *scripsimus*; spoliatio ON
[e] intellectus agens] intellectum agentem N
[f] aliquis *scripsimus*; aliquid O
[g] prima] omnium *add. mg.* O
[h] cognoscendo] aliam *add. s.l.* O

15. To the fifth argument[35] it must be said that, if someone knows the antecedent, it does not follow that he knows the consequent, because he can doubt whether it is the consequent or not; and even if he knows that it is the consequent, he can doubt in what way and through which process it follows. Thus, the argument is not conclusive.

16. To the sixth and to the seventh arguments[36] it must be said that the species of a body that is in the body under the conditions 'here' and 'now' is in the sensitive power under the same conditions; but a species which is in the sensitive power under the conditions 'here' and 'now' is entirely stripped and deprived of these conditions in the intellectual soul. And thus in this respect an intellectual and a sensitive soul differ more than a sensitive soul and a body. And where the intellectual soul is concerned, the agent intellect is the cause of this deprivation.

<Question 6>
The next question specifically asks whether we understand the first cause.

1. Arguments for the affirmative: according to Aristotle in the *Posterior Analytics*, to know something is to know its cause;[37] therefore, whoever knows A knows it through its cause, for instance through B; similarly, whoever knows B knows it through its cause, for instance through C. And in this way one can proceed to infinity and reach the first cause; therefore, it seems that whoever knows A or B also knows the first cause.

2. Furthermore, the Commentator writes that, if someone knows something, he knows it through the first cause;[38] therefore, it seems that in knowing a minimal thing, one also knows the first cause itself.

3. Furthermore, the Commentator <says>: "Once the quiddity of a determinate sensible substance is known, the first cause itself is also known".[39]

4. To the contrary: there is a relation of an intellect to a thing understood; but there is no relation of a finite thing to an infinite thing; but our intellect is finite, and God is infinite; hence, <our intellect does not understand the first cause>.

5. Furthermore, an infinite thing is not comprehended by a finite thing; therefore, <the first cause is not comprehended by our intellect>.

6. The solution to this problem is that it is possible to understand the first cause. But the first series of arguments[40] reach a false conclusion, namely that in knowing this thing, the first cause itself is also known.

[35] See above, n. 5.
[36] See above, nn. 6–7.
[37] Arist. *An. post.* I.2, 71b9–12.
[38] Aver. *In Phys.* I, t.c. 1, fol. 6rb .
[39] Aver. *In Metaph.* VII, t.c. 5, fol. 156rb.
[40] See above, nn. 1–3.

7. Unde ad primam rationem dicendum quod cognoscere rem est dupliciter: aut per principia propria, et sic, cognita re aliqua, non necesse est ut cognoscatur prima causa, aut per totam suam causalitatem, et sic necesse est cognoscere causam primam.

8. Et similiter intelligendae sunt auctoritates Commentatoris. Qui enim scit aliquid per totam suam causalitatem cognoscit et primam causam, sicut patet.

9. Ad argumenta in oppositum dicendum quod infinitum potest esse duplex, scilicet infinitum in simplicitate, et talis infinitas non impedit infinitum comprehendi a finito, verbi gratia punctus, qui est in linea, est simplex et infinitus in simplicitate, et tamen recipitur in linea finita. Et sic est prima causa infinita, scilicet in simplicitate. Alio modo est infinitum, scilicet in quantitate, et hoc non potest comprehendi etc. nec habet aliquam proportionem ad finitum. Sed alia infinitas habet proportionem.

10. Vel sic: intellectus humanus duplicem habet potentiam, unam potentiam habet mediante qua operatur, et haec est finita, et aliam mediante qua recipit, et haec est infinita; unde intellectus noster uno modo est finitus, alio modo infinitus.

11. Vel sic: duplex est intellectus, scilicet practicus, qui intellectus est circa bonum, et talis intelligit primam causam, et talis intellectus est infinitus, et speculativus, qui est circa verum, et iste intellectus proprie non intelligit ipsum primum, sed potius ista inferiora.

12. Unde incidentaliter nota quod participare bonitate divina est dupliciter, vel formaliter et praedicative, et sic non participamus bonitate divina, vel exemplariter et imitative, et sic dicitur quod participamus. Unde haec propositio 'nos omnes participamus bonitate divina' est falsa nisi intelligatur.

1*. Ista quaesita quaeruntur a parte [a]de li 'intelligere'.

[a] de li *lectio incerta* O

7. Thus, to the first argument[41] it must be said that one can know a thing in two ways: either (i) through its proper principles, and in this way, when a thing is known, it is not necessary for the first cause to be known; or (ii) through all its causes, and in this way one necessarily knows the first cause.

8. And the authoritative texts of the Commentator[42] must be understood in the same way; for whoever knows something through all its causes also knows the first cause, as is clear.

9. To the arguments to the contrary[43] it must be said that there are two kinds of infinite: (i) the infinite in simplicity, and this kind of infinite does not prevent an infinite thing from being comprehended by a finite thing; for example, a point, which is in a line, is simple and infinite in simplicity, and yet it is received in a finite line. And it is in this way that the first cause is infinite, that is, in simplicity. (ii) The infinite in quantity, and this kind of infinite cannot be comprehended <by a finite thing> nor does it have a relation to a finite thing. But the other kind of infinity has a relation <to a finite thing>.

10. An alternative reply is that the human intellect has a twofold power: (i) one power by means of which it operates, and this power is finite, and (i) another by means of which it receives, and this power is infinite. Thus, our intellect is finite in one way and infinite in another way.

11. Another reply is that there are two kinds of intellect: (i) the practical intellect, which deals with the good, and this kind of intellect understands the first cause, and is infinite; (ii) the speculative intellect, which deals with the true, and this kind of intellect does not properly understand the first being itself, but rather the things here below.

12. Hence, we should note in passing that to participate in divine goodness can be understood in two ways: either (i) formally and by predication, and in this way we do not participate in divine goodness, or (ii) by example and imitation, and in this way it is said that we participate in divine goodness. Thus, the proposition 'we all participate in divine goodness' is false unless it is understood <as in (ii)>.

1*. All the questions raised so far are about the verb 'to understand' <as it appears in Aristotle's claim>.[44]

[41] See above, n. 1.
[42] See above, nn. 2–3.
[43] See above, n. 4.
[44] Arist. *Phys.* I.1, 184a10.

<Quaestio 7>

Gratia [a]de li 'scire' quaeratur an contingat scire per causam.

1. Et videtur quod non sic: dicit Aristoteles in septimo *Metaphysicae* quod scire cuiuslibet rei est scire suum 'quid est' per essentiam; sed quidditas sive essentia non est causa rei; ergo etc. Probatio minoris: dicit enim Aristoteles quod quidditas Socratis est idem cum Socrate; causa autem rei non est idem cum re, sed aliud.

2. Item, nos scimus res in intellectu possibili per delationem specierum ab obiecto usque ad intellectum possibilem; sed species rei non est eius causa, quia species sunt causatae a rebus, et causa praecedit ipsam rem; ergo etc.

3. Item, si cognoscimus aliquid per suam causam, cognoscimus illam causam in quantum causa est; sed omnes causae posteriores comparationem habent ad causam primam et in quantum huiusmodi sunt causata; ergo impossibile est cognoscere aliquam causam in quantum causa est, sed solum in quantum causata.

4. Quod autem illa comparatio sit substantialis et substantia rei causatae probatio: sit A prima causa, B autem causa posterior, C relatio inter ipsa. Quaero de C, per quod[b] importatur illa relatio: aut est idem cum B aut non. Si est idem, habetur propositum. Si non est idem, ergo diversum causatum; ergo habet relationem ad causam primam. Sit sua relatio D; quaero de D utrum sit idem cum C vel non. Si idem, habetur propositum, quia similiter tunc relatio alia erit idem cum re et substantia rei. Si non, contingit procedere sicut prius, et sic erit processus in infinitum, nisi concedatur quod illa relatio sit eadem cum illa re cuius est relatio, et hoc substantialiter.

5. Solutio. Ad primam rationem dicendum quod res et quidditas rei unum et[c] idem sunt, et ambo illa praecedit aliud scire quod[d] est scire per causam.

6. Vel sic: duplex est causa: completa[e], quae scilicet aggregat in se naturam omnium causarum, et haec causa est eadem cum quidditate rei, et causa incompleta[f], quae non aggregat in se nisi unam causam tantum sive aliquam[g] partem causarum, et haec non est eadem etc.

[a] de li *lectio incerta* O
[b] quod *scripsimus*; quam O
[c] et *suppl. s.l.* O
[d] quod *ex corr.* ? O
[e] completa *scripsimus*; completiva O
[f] incompleta *scripsimus*; incompletiva O
[g] aliquam *scripsimus*; alteram O

<Question 7>

Because of the verb 'to know',[45] the question arises whether it is possible to know through a cause.

1. Arguments for the negative: in *Metaphysics* VII Aristotle says that to know any thing is to know its 'what it is' in its essence;[46] but the quiddity or essence of a thing is not its cause; therefore, <to know something is not to know it through its cause>. Proof of the minor premise: Aristotle says that the quiddity of Socrates is the same as Socrates;[47] but the cause of a thing is not the same as that thing, but something else.

2. Furthermore, we know things in the possible intellect through the transfer of their species from the object to the possible intellect; but the species of a thing is not its cause, because species are caused by the things, whereas a cause precedes the thing itself; therefore, <to know something is not to know it through its cause>.

3. Furthermore, if we know something through its cause, we know its cause in so far as it is a cause; but all posterior causes have a comparative relation to the first cause and in so far as they are so related, they are caused; therefore, it is impossible to know a cause in so far as it is a cause but only in so far as it is caused.

4. Proof that that comparative relation is substantial and the substance of a caused thing: let A be the first cause, B a posterior cause, and C the relation between them. I ask about C, which stands for that relation, whether it is the same as B or not. If it is the same, we get the intended conclusion. If it is not the same, it is a different caused thing; therefore, it has a relation to the first cause. Let D be its relation <to the first cause>. I ask about D whether it is the same as C or not. If it is the same, we get the intended conclusion, because then the other relation will likewise be the same as that thing and the substance of that thing. If it is not the same, one can proceed as before, and so there will be an infinite regress, unless it is conceded that that relation is the same as the thing of which it is a relation, and is so substantially.

5. Solution. To the first argument[48] it must be said that a thing and the quiddity of that thing are one and the same, and that another kind of knowing, which is knowing through a cause, precedes both.

6. An alternative reply is the following. There are two kinds of cause: (i) a complete cause, that is, one which includes in itself the nature of all causes, and this cause is the same as the quiddity of a thing, and (ii) an incomplete cause, which includes in itself only one cause or one part of the causes, and this kind of cause is not the same <as the quiddity of a thing>.

[45] Arist. *Phys.* I.1, 184a10.
[46] Arist. *Metaph.* VII.6, 1031b20–21.
[47] Arist. *Metaph.* VII.6, 1032a6–8.
[48] See above, n. 1.

7. Ad secundam sic, quod duplex est cognitio intellectus: una quae est simplex notitia, et haec non est per causam, sed per speciem. Quaerit enim illa notitia in quo genere et in qua specie res est. Alia est cognitio quae est rei per suam causam. Et haec est duplex: quaedam enim fit per causas proprias et propinquas, et sic intelligitur illa littera: "Tunc arbitramur cognoscere etc."; aut per totalem suam causalitatem, et sic intelligendum est illud Commentatoris: "Qui cognoscit aliquid cognoscit omnes causas, tam propinquas quam remotas".

8. Ad tertiam dicendum quod est considerare rem sub duplici comparatione: aut sub illa comparatione quae[a] non addit aliquid supra suam essentiam, et ista comparatio est eadem cum sua essentia, et sic est causata; aut sub illa comparatione quae[b] aliquid addit supra essentiam, et haec non est eadem, et secundum hoc est causa.

9. Nota quod ista propositio est falsa 'verissima cognitio est per causam' —causa enim prima verissime cognoscitur, non tamen cognoscitur per suam causam, quia non habet causam—nisi intelligatur sic: in genere substantiarum causatarum, cuiusmodi non est causa prima.

10. Nota quod illa propositio est falsa 'verissima cognitio est per demonstrationem'—quia substantiae non possunt cognosci per demonstrationem nec etiam causa prima—nisi sic intelligatur: in genere accidentium. **[O 89ra]**

1*. *Prius temptandum est determinare ea quae sunt circa principia.*

<Quaestio 8>

1. Contra istam litteram arguitur sic: in naturali scientia incipiendum est ab eis quae sunt nobis magis nota; sed huiusmodi sunt composita ex principiis quam principia; ergo etc.

[a] quae *scripsimus cum* TC; qua O
[b] quae *scripsimus cum* TC; qua O

7. To the second argument[49] it must be said that there are two kinds of intellectual cognition: (i) one is a simple cognition, and this is not through a cause, but through a species; for that cognition asks in what genus and in what species a thing is. (ii) Another is a cognition of a thing through its cause. And this is of two kinds: (iia) one is through the proper and proximate causes of a thing, and the text "Then we judge that we know <a thing when we are acquainted with its primary causes or first principles>"[50] refers to this kind of cognition; or (iib) through all the causes of a thing, and the Commentator's claim, "Whoever knows something knows all its causes, both proximate and remote",[51] must be understood as being about this kind of cognition.

8. To the third argument[52] it must be said that it is possible to consider a thing under two kinds of comparative relation: either (i) under a comparative relation which does not add anything to its essence, and this comparative relation is the same as the essence of a thing, and in this way the thing is caused; or (ii) under a comparative relation which adds something to its essence, and this comparative relation is not the same as the essence of a thing, and in this way the thing is a cause.

9. Note that the proposition 'the truest cognition is through a cause' is false unless it is understood as referring to the genus of caused substances, to which the first cause does not belong; for the first cause is most truly known, but it is not known through its cause, since it does not have a cause.

10. Note also that the proposition 'the truest cognition is through a demonstration' is false unless it is understood as referring to the genus of accidents, because substances cannot be known through demonstration and neither can the first cause.

1*. *<Plainly, therefore, in the science of nature too> our first task will be to try to determine what relates to its principles.*[53]

<Question 8>

1. Arguments against this textual claim: in natural science we must begin with things better known to us; but such things are those composed of the principles rather than the principles themselves; therefore, <we must begin with the things composed of the principles>.

[49] See above, n. 2.
[50] Arist. *Phys.* I.1, 184a12–14.
[51] Aver. *In Phys.* I, t.c. 1, fol. 6rb.
[52] See above, nn. 3–4.
[53] Arist. *Phys.* I.1, 184a14–16.

2. Item, quae cadunt sub sensu nostro nobis sunt magis nota; sed principiata cadunt sub sensu nostro, principia vero non; quare ab eis est incipiendum.

3. Solutio. Duplex est intellectus: unus qui semper est in actu, et hic[a] semper cognoscit omnia, et principia et principiata, quia, si intelligeret alterum illorum tantum, tunc esset in potentia respectu cognitionis alterius. Et ideo, quando est in actu, intelligit tam principia quam principiata; sed per prius formas completas, ut principiata, cognoscit, quae in via naturae praecedunt principia. Alius est intellectus qui prius est in potentia et postea in actu, et hoc dupliciter: vel secundum eius simplicem notitiam, et haec[b] est nobis notior quoad principiata, vel secundum [[c]eius totalem] causalitatem. Et hoc dupliciter. Aut enim intelligit causas proprias, et sic principiata magis nota sunt, aut causas remotas, et sic principia sunt magis nota. Vel accipiendo cognitionem secundum totalem causalitatem, principia sunt magis nota. Et secundum hoc loquitur Aristoteles.

4. Vel dicendum quod, si consideretur[d] possibilitas nostra cognoscendi secundum quod exigit ordo naturae, sic principia sunt magis nota. Si autem dicatur solum secundum possibilitatem nostrae cognitionis et omittatur aliud membrum, tunc principiata sunt nobis magis nota.

<Quaestio 9>

Quaeritur de hoc quod dicit Commentator quod non omnis scientia considerat de omnibus[e], scilicet causis, sed etc.

1. Contra: quod mathematicus consideret de omnibus causis probatio: per Aristotelem omnis res habet suum 'quod[f] est' et suum 'quo est'; ergo in omni re est causa materialis et formalis. Item, omnis res causata est quoquo modo composita respectu causae primae et habet suam materiam cum aliquo quod

[a] hic *scripsimus*; haec O
[b] haec *scripsimus*; hic O
[c] eius totalem *delevimus*
[d] consideretur *ex corr.* O
[e] omnibus *suppl. s.l.* O
[f] quod *scripsimus*; quid O

2. Furthermore, the things that fall under our senses are better known to us; but the things derived from the principles fall under our senses, whereas the principles do not; hence, we must start with the things derived from the principles.

3. Solution. There are two kinds of intellect: (i) one which is always in act, and this kind of intellect always knows all things, both the principles and the things derived from the principles, because, if it understood only one of these two kinds of thing, then it would be in potency with respect to the cognition of the other. And therefore, when it is in act, it understands both the principles and the things derived from the principles; but it first knows complete forms, that is, the things derived from the principles, which precede the principles in the order of nature. (ii) Another kind of intellect is that which is first in potency and afterwards in act, and this comes about in two ways: either (iia) in accordance with the simple cognition of a thing, and with this kind of cognition the things derived from the principles are better known to us; or (iib) in accordance with the causes of a thing. And this comes about in two ways: (iib') <in the first way> either the intellect understands the proper causes of a thing, and in this way the things derived from the principles are better known, or <alternatively> it understands the remote causes, and in this way the principles are better known. (iib'') Or <in the second way> if we take the cognition <of a thing> in accordance with all its causes, the principles are better known. And this is what Aristotle's words mean.[54]

4. Alternatively, one might say that, if we consider the possibility that we know in the way that the order of nature requires, then in this way the principles are better known. If, however, only the possibility of our cognition is considered and the other condition is omitted, then the things derived from the principles are better known to us.

<Question 9>

The question is about the Commentator's claim that not every science considers all things, that is, all causes, but <some sciences, namely mathematical sciences, only consider the formal cause, and another science, namely divine science, considers three causes, that is, the efficient cause, the formal cause and the final cause, and another science, namely natural science, considers all four causes >.[55]

1. Against this: a mathematician considers all causes. Proof: according to Aristotle, each thing has its 'that which is' and its 'that by which it is';[56] therefore, in every thing there is a material cause and a formal cause. Furthermore, every thing that has been caused is in some way composite with

[54] See above, q. 8, n. 1*.

[55] Aver. *In Phys.* I, t.c. 1, fol. 6ra.

[56] This claim is not found in Aristotle or in Boethius, to whom it is more often attributed by medieval authors. On the origin of this medieval *auctoritas*, see Gauthier, 1983, 213–214.

est vere forma vel ei proportionatum[a]; ergo, ut videtur, omnis res habet in se causam materialem et formalem. Item, nihil se ipsum componit nec ad esse producit; ergo omnis res habet causam efficientem. Item, omne efficiens agit per aliquem finem; ergo omnis res causam finalem habet. Ergo, ut videtur, in omni re sunt illae quattuor causae; ergo, cum mathematicus considerat de rebus, necessario considerat de ipsis et eorum causis omnibus et non de una tantum.

2. Item, Aristoteles dicit in secundo *Posteriorum* quod ista conclusio, 'angulus in semicirculo est rectus' demonstratur per materiam; et ista conclusio est mathematica; quare etc.

3. Item, si illa materia quae magis grossa est potest ingredi demonstrationem, multo fortius ergo et illa materia quae magis spiritualis. Sed materia naturalis, quae materia est magis grossa, potest ingredi demonstrationem; ergo et materia mathematica, quae materia est magis spiritualis.

4. Quod etiam metaphysicus debet considerare de causa materiali, quod tamen negat Commentator, probatio, quia metaphysici est considerare omnem substantiam in quantum substantia est, et ita, cum materia sit substantia, debet considerare de materia.

5. Item, dicit Aristoteles in septimo quod quaedam est definitio quae fit per omnes causas, et talem potest metaphysicus facere; ergo etc.

6. Item, Aristoteles in primo libro docet sumere statum in causa materiali sicut et in aliis causis; ergo etc.

7. Ad rationes procedentes contra dictum Commentatoris de mathematica dicendum quod quaelibet res causata habet materiam et formam et omnes causas, vel vere vel proportionaliter. Unde, quando materia non est cum motu et transmutatione, tunc illa materia non est vera, sed tunc, quando sic denudatur, magis sapit nomen formae quam materiae. Et quia de tali materia

[a] proportionatum *scripsimus*; proportionato O

respect to the first cause and has its matter together with something that is truly a form or proportionate to it; therefore, it seems, every thing has in itself a material cause and a formal cause. Furthermore, "nothing is responsible for its own composition or brings itself into being;"[57] therefore, every thing has an efficient cause. Furthermore, every efficient cause acts for an end; therefore, every thing has a final cause. Therefore, it seems, the four causes are found in every thing; therefore, since a mathematician considers things, he necessarily considers them and all their causes and not just one of them.

2. Furthermore, in *Posterior Analytics* II Aristotle says that the conclusion 'an angle in a semicircle is a right angle' is proved through matter;[58] and this conclusion is mathematical; hence, <mathematical things have a material cause>.

3. Furthermore, if that matter which is thicker can enter a demonstration, there is a much stronger reason to hold that that matter which is more spiritual can also enter a demonstration. But natural matter, which is a thicker matter, can enter a demonstration; therefore, mathematical matter, which is a more spiritual matter, can also do this.

4. Proof that a metaphysician must also consider a material cause, a claim which the Commentator denies:[59] it pertains to a metaphysician to consider every substance in so far as it is a substance, and thus, since matter is a substance, he must consider matter.

5. Furthermore, in *Metaphysics* VII Aristotle says that there is a definition that involves all four causes,[60] and a metaphysician can produce this definition; therefore, <a metaphysician also considers matter>.

6. Furthermore, in *Metaphysics* I Aristotle teaches that it is possible to reach an end state in a material cause, just as <it is possible> in the other causes;[61] therefore, <a metaphysician considers matter>.

7. To the arguments against the Commentator's claim about mathematics,[62] it must be said that each caused thing has matter and a form and all <four> causes, either truly or by analogy. Thus, when matter is not found together with motion and change, that matter is not a true one, but when it is stripped of <motion and change> in this way, it is more suggestive of form than

[57] Cf. Nicolas of Amiens, *Ars catholicae fidei* I, prop. 3, ed. Dreyer, 80.11.

[58] Arist. *An. post.* II.11, 94a28–34.

[59] Aver. *In Phys.* I, t.c. 1, fol. 6ra.

[60] Arist. *Metaph.* VII.17, 1041a26–b9.

[61] The correct reference is Arist. *Metaph.* II.2, 994a1–11. Note that in the Arabic text of the *Metaphysics* and consequently in Averroes' commentary, Book I is incomplete (it begins at I.5, 987a9) and Book I and Book II are in the reverse order; this reversal is also found in some manuscripts of the Arabic-Latin translation of the *Metaphysics* and of the Latin translation of Averroes' commentary, in which therefore Book II may become Book I. Cf. Borgo, 2014, 50–51.

[62] See above, nn. 1–3.

considerat mathematicus, ideo dicit Commentator quod non habet causam materialem, et intelligit quod non veram. Similiter etiam ille efficiens verus est qui efficit rem per transmutationem et efficit motum; et talis efficiens non est in mathematica, et ideo Commentator etc. Similiter etiam verus finis quando aliquis movens sive efficiens producit aliquem finem per motum vel transmutationem; et quia talis finis non est in mathematica, ideo etc.

8. Ad rationes procedentes contra dictum Commentatoris de metaphysica dicendum quod, quia causa prima omnino est immaterialis, et causa prima principaliter est subiectum illius scientiae, ideo dicit Commentator quod caret causa materiali.

<Quaestio 10>
Quaeritur an naturalis possit probare etc.[a]

1. Et videtur quod non per Aristotelem in veteri *Metaphysica* dicentem quod male[b] est simul quaerere scientiam et modum scientiae; cum igitur Aristoteles quaerat hic scientiam naturalem et etiam eius modum per conclusiones, videtur quod vile facit probando conclusiones.

2. Solutio. Duplex est scientia. Est enim quaedam in qua demonstrantur speciales passiones de specialibus subiectis, et in talibus est illa propositio intelligenda. Unde Aristoteles in libro *Posteriorum* nullas facit demonstrationes nisi exempli gratia, et hoc quia dat ibi modum faciendi demonstrationes; sed postea in geometria et mathematica facit demonstrationes et non dat modum ibi faciendi demonstrationes. In scientiis tamen in quibus non demonstrantur speciales passiones de specialibus subiectis, cuiusmodi est naturalis scientia, ibi non est inconveniens.

3. Vel sic, quod naturalis in quantum naturalis non potest simul quaerere etc., naturalis tamen prout assumit formam metaphysici potest. Et hoc patet per ipsum Commentatorem in libro primo *Physicorum*.

[a] etc. *scripsimus*; t.$_{9}$. O
[b] male *scripsimus cum* TC; inutile O

matter. And since a mathematician considers this kind of matter, this is why the Commentator says that mathematical things do not have a material cause, and he means that they do not have a true material cause. Similarly, a true efficient cause is that which produces a thing through change and that produces motion; and this kind of efficient cause does not exist in mathematical things, and this is why the Commentator <says that mathematical things do not have an efficient cause>. Similarly, there is a true end when something that is a mover or an efficient cause produces an end through motion or change. And since such an end does not exist in mathematical things, therefore <the Commentator says that mathematical things do not have an end>.

8. To the arguments against the Commentator's statement about metaphysics[63] it must be replied that since the first cause is absolutely immaterial, and the first cause is the main subject of that science, this is why the Commentator says that metaphysics lacks a material cause.

<Question 10>
The question is whether a natural philosopher can prove <the method of science>.

1. Arguments for the negative: according to Aristotle in the old *Metaphysics*, it is wrong to inquire at one and the same time about science and the method of science;[64] therefore, since Aristotle inquires here[65] by way of conclusions both about natural science and about its method, it seems that he does something worthless in proving conclusions.

2. Solution. There are two kinds of science. (i) One kind of science is that in which special passions are proved of special subjects, and Aristotle's claim should be referred to these cases. Thus, in the *Posterior Analytics* Aristotle does not do any demonstrations except by way of example, and this is because he explains the way of doing demonstrations there; but afterwards in geometry and mathematics he does do demonstrations and does not explain the way of doing demonstrations there. But (ii) in the sciences in which special passions are not proved of special subjects, and natural science is of this kind, it is not absurd <to give both demonstrations and the way of doing demonstrations>.

3. Another solution is that a natural philosopher cannot at one and the same time investigate <both a science and the method of a science> in so far as he is a natural philosopher, but a natural philosopher can do that in so far as he takes on the role of a metaphysician. And this is clear from the Commentator himself in *Physics* I.[66]

[63] See above, nn. 4–6.
[64] Arist. *Metaph.* II.3 995a13–14 in the Arabic-Latin translation (see Aver. *In Metaph.* II, fol. 35ra).
[65] That is, in the *Physics.*
[66] Aver. *In Phys.* I, t.c. 12, fol. 11vb.

4. Item, omnis scientia et omnis philosophus supponit suum subiectum esse; sed naturalis suum subiectum esse non supponit, quia quaerit sua principia; ergo etc.

5. Solutio. Dicendum quod supponit principia sui subiecti, sed supponendo ea esse, quaerit quae et qualia sunt illa.

6. Item, nulla scientia facit considerationem ultra suum subiectum nec ad priora. Cum igitur principia sint priora subiecto naturalis philosophiae, non debet naturalis de illis principiis considerare nec inquirere.

7. Solutio. Dicendum quod naturalis non extendit suam considerationem ad principia ut reducat corpus mobile ad illa, et sic **[O 89rb]** esset inconveniens, sed ut reducat illa principia ad corpus mobile[a], et sic non est inconveniens.

8. Vel sic, quod naturalis in quantum naturalis non considerat de suis principiis nisi prout assumit formam metaphysici.

<Quaestio 11>

Quaeritur utrum Aristoteles loquatur de principiis scientiae vel rei, cum dicit quod "Incipiendum est a principiis etc.".

1. Et videtur quod de principiis scientiae, quia dicit in littera: "Scire et intelligere in omnibus scientiis quarum sunt principia". Et glossatur li 'quarum', scientiarum scilicet; ergo etc.

2. Item, dicit hic Aristoteles quod incipiendum est ab universalibus et in libro *Posteriorum*, quia universalia sunt principia artis et scientiae; ergo etc.

3. Ad oppositum: de eisdem principiis intendit in dando intentionem de quibus post pertractat in faciendo exsecutionem; sed exsequitur de principiis rei; ergo et de talibus intendit.

[a] mobile] esset primum et prius *add. sed del.* O

4. Furthermore, every science <presupposes> and every philosopher assumes that the subject of that science exists; but a natural philosopher does not assume that the subject of natural science exists, because he is investigating its principles; therefore, <it seems that he does something worthless in proving conclusions>.

5. Solution. It must be said that a natural philosopher assumes that the principles of his subject exist, and on the assumption that they exist he investigates what the principles are and of what kind they are.

6. Furthermore, no science considers something beyond its subject or prior to it. Therefore, since the principles are prior to the subject of natural philosophy, a natural philosopher must not consider or investigate those principles.

7. Solution. It must be said that a natural philosopher does not extend his consideration to the principles in order to reduce a mobile body to them, for it would be absurd <to consider the principles> in this way, but <he does so> in order to reduce those principles to a mobile body, and it is not absurd <to consider them> in this way.

8. Another solution is that a natural philosopher in so far as he is a natural philosopher does not inquire about the principles <of natural science> except in so far as he takes on the role of a metaphysician.

<Question 11>
The question is whether Aristotle speaks of the principles of a science or of things when he says that we must start from the principles.[67]

1. Arguments that he speaks of the principles of a science: in the text Aristotle says "to know and to understand in all sciences of which there are principles".[68] And the expression 'of which'[69] is glossed as referring to sciences; therefore, <Aristotle speaks of the principles of a science>.

2. Furthermore, here Aristotle says that we must start from universals,[70] and in the *Posterior Analytics*[71] <he says> that universals are the principles of art and of science; therefore, <Aristotle speaks of the principles of a science>.

3. To the contrary: the principles to which Aristotle refers when he expresses what he intends to do[72] are the same as those with which he later deals when he carries out his intention; but he deals with the principles of things; therefore, his intention is also to deal with such principles.

[67] Arist. *Phys*. I.1, 184a14–16.
[68] Arist. *Phys*. I.1, 184a10–11.
[69] Arist. *Phys*. I.1, 184a11.
[70] Arist. *Phys*. I.1, 184a23–24.
[71] Arist. *An. post*. II.19, 100a6–9.
[72] Arist. *Phys*. I.1, 184a14–16.

4. Solutio. Dicendum quod loquitur hic de utrisque, quia sunt eadem secundum rem, differentia solum secundum modum, secundum quod dicit Boethius quod omnia ea quae sensu percipimus, si ad rationem referamus sive ad intellectum, sunt universalia, si ad sensum, singularia. Ergo principia, secundum quod sunt signata sub istis condicionibus, 'hic' et 'nunc', et secundum quod sunt coniuncta cum transmutatione et motu, sunt principia rerum; secundum quod ab istis condicionibus denudantur eadem principia et similiter a motu et transmutatione, sic sunt principia scientiarum.

1*. *Innata est nobis via* etc.

<Quaestio 12>
Iuxta hoc quod dicit: "notioribus naturae" quaeritur utrum naturae sit aliquid notum.

1. Et videtur quod non, quia ubi est notitia, ibi est comprehensio, et ubi est comprehensio, ibi est tristitia vel delectatio; sed natura inferior neque tristatur neque gaudet; ergo etc.

2. Item, omnis notitia est alicuius vivi vel vitam habentis; sed natura inferior non est vivum nec alicuius vitam habentis; ergo etc.

3. Oppositum patet per Aristotelem dicentem quod naturae aliquid est notius; ergo et notum.

4. Solutio. Dicendum quod est considerare naturam dupliciter: vel in se, et sic non[a] habet aliquam notitiam, vel in comparatione ad causam primam sive ad suam causam moventem et efficientem. [b]Et sic habet notitiam, quia, si reducatur illa natura[c] ad suam causam moventem, ut ad intelligentiam[d], sic habet notitiam, videlicet ut in suo efficiente.

[a] non *suppl. s.l.* O
[b] Et sic *lectio incerta* O
[c] natura] movens *add. sed del.* O
[d] intelligentiam *scripsimus*; intellectivam O (mediantibus intelligentiis TC)

4. Solution. It must be said that Aristotle is speaking here of both kinds of principles, because they are the same in reality, and different only in mode, in accordance with Boethius's claim that all things that we perceive with the senses are universal in relation to reason or to the intellect, and are singular in relation to the senses.[73] Therefore, principles, in so far as they are designated under the conditions 'here' and 'now' and in so far as they are joined with change and motion, are the principles of things; <but> in so far as the same principles are stripped of these conditions and also of motion and change, they are the principles of sciences.

1*. *The natural way of doing this <is to start from the things which are more knowable and clear to us and proceed towards those which are clearer and more knowable by nature>*[74]

<Question 12>

In connection with Aristotle's expression "the things which are better known to nature"[75] the question arises whether something is known to nature.

1. Arguments for the negative: where there is cognition there is comprehension, and where there is comprehension there is unhappiness or happiness; but a nature here below is neither unhappy nor happy; therefore, <there is no cognition in a nature here below>.

2. Furthermore, every cognition belongs to something alive or having life; but a nature here below is not alive, and neither does it belong to something having life; therefore, <there is no cognition in a nature here below>.

3. The contrary is clear from Aristotle who says that something is better known to nature;[76] therefore, something is also known to nature.

4. Solution. It must be said that it is possible to consider a nature in two ways: either (i) in itself, and in this way it does not have cognition, or (ii) in relation to its first cause, that is to its moving and efficient cause; and in this way a nature has cognition, because, if a nature is reduced to its moving cause, namely to an intelligence, it has cognition in this way, that is, in its efficient cause.[77]

[73] Boethius, *De cons*. V.6.36, CCSL 94, 105.120–122.
[74] Arist. *Phys*. I.1, 184a16–18.
[75] Arist. *Phys*. I.1, 184a17–18.
[76] Arist. *Phys*. I.1, 184a17–21.
[77] The author of the commentary attributed to Richard Rufus holds a similar view; see Richard Rufus of Cornwall, *In Phys.*, 96.47–97.52.

5. Vel sic: duplex est notitia, una quae fit per principium apprehendens intrinsecum ipsius rei, et talis notitia non reperitur nisi in vivo etc. Alia est notitia naturalis secundum quod dicimus quod aliqua res magis scibilis est quam alia, et hoc quia habet aliquam proprietatem sive condicionem ei insitam a suo primo motore propter quam magis nata est movere intellectum simpliciter, et non intellectum obnubilatum per corpus, quam alia res. Et talem notitiam sive cognitionem potest natura inferior habere de rebus.

<Quaestio 13>

Quaeritur an eadem sint nota nobis et nota simpliciter.

1. Et videtur quod sic, quia omnis demonstratio procedit ex prioribus <[a]simpliciter; et procedit ex prioribus> quoad nos. Probatio: praemissae enim sunt nobis magis notae quam conclusio, et demonstratio procedit ex[b] praemissis ad conclusionem; ergo etc.

2. Item, homo est quodam modo finis omnium, quia scribitur sic a quodam philosopho: “Sumus enim quodam modo finis omnium”. Sed ordo rei maxime est in fine; ergo in homine; ergo maxime in perfecto hominis, ut in intellectiva virtute; ergo[c], [d]cum idem quod[e] est conservativum ordinis in uno erit et in alio, sequitur quod, si ordo rerum talis est quod sunt simpliciter nota, quod idem ordo servetur in fine; et sic, si res in se sint simpliciter notae, causant et similiter intelligibilia[f] in eodem ordine; ergo etc.

3. <Item[g]>, sicut est a parte sensibilium, similiter erit etiam a parte intelligibilium; sed virtus sensitiva citius apprehendit magis sensibile, ut patet comparando duo sensibilia eiusdem generis, licet non ita sit comparando diversa sensibilia diversorum sensuum[h]; ergo intellectus similiter apprehendet magis intelligibile, et [i]illud erit nobis magis notum; ergo etc.

4. Dicendum quod non sunt eadem etc.

[a] simpliciter … prioribus *hic aliqua deesse videntur, fort. per hom. omissa, quae ad sensum supplevimus*
[b] ex] prioribus *add. sed del* O
[c] ergo] etc. *add. mg.* O
[d] cum … etc. *scrips. sed del.* O
[e] quod *scripsimus*; idem O
[f] intelligibilia *scripsimus*; intellectiva O
[g] Item *supplevimus*
[h] sensuum *scripsimus*; sensibilium O
[i] illud erit nobis *suppl. s.l.* O

5. Another solution is that there are two kinds of cognition: (i) one that comes about through an intrinsic principle of apprehension in the thing itself, and this kind of cognition is only found in what is alive <and so not in a nature here below>. (ii) The other is natural cognition, as when we say that one thing is more knowable than another, and this because it has a property or condition conferred on it by its first mover in virtue of which it is more apt by nature to move an intellect taken absolutely—and not an intellect beclouded by a body—than another thing. And a nature here below can have this kind of cognition of things.

<Question 13>
The question is whether the same things are both known to us and known absolutely.

1. Arguments for the affirmative: every demonstration starts from what is prior <absolutely; but it also proceeds from what is prior> to us. Proof: the premises are better known to us than the conclusion, and a demonstration proceeds from the premises to the conclusion; therefore, <the same things are both known to us and known absolutely>.

2. Furthermore, man is in some way the end of all things, because a certain philosopher writes the following: "For we are in some way the end of all things".[78] But the order of a thing is maximally in its end; therefore, <it is> in man; therefore, the order of a thing is maximally in the perfect part of man, that is, in the intellectual power; therefore, since the same thing that preserves the order in one thing preserves it in another, it follows that, if the order of things is such that they are absolutely known, the same order is preserved in the end, and in this way if some things are in themselves known absolutely, they also cause notions that are intelligible in the same order; therefore, <the same things are both known to us and known absolutely>.

3. Furthermore, what holds for sensible things also holds for intelligible things; but the sensitive power more quickly apprehends what is more sensible, as is clear when one compares two sensible things of the same genus, although this is not true when one compares different sensible things relating to different senses; therefore, the intellect will apprehend what is more intelligible in a similar way, and such a thing will be better known to us; therefore, <the same things are both known to us and known absolutely>.

4. It must be said that <the things known to us and known absolutely> are not the same.

[78] Arist. *Phys*. II.2, 194a34–35.

5. Ad primam rationem in oppositum dicendum quod maior est falsa. Recipit enim instantiam in mathematicis, sicut patet.

6. Ad secundam dicendum quod ordo naturae talis est ut illud quod est magis notum naturae sit minus notum homini. Et hoc est quia anima ligata est corpori et recipit obnubilationes corporis et intelligit solum secundum possibilitatem corporis. Possibilitas vero cognitionis corporis est a posterioribus; unde homo habet maiorem et pleniorem sive potiorem cognitionem de posterioribus quam de prioribus, natura autem econverso; unde talis est ordo inter eos.

7. Ad tertiam rationem dicendum quod, quia sensitiva non potest agere nisi secundum instrumenta corporea nec recipere, ideo eius cognitio sive eius receptio non deterioratur per corpus; anima [a]vero <intellectiva> non intelligit per instrumenta corporea, sed per se ipsam; unde eius cognitio et receptio multum deterioratur per obnubilationem corporis. Unde non est simile.

<Quaestio 14>
Quaeritur utrum ea quae sunt magis[b] confusa sint nobis magis nota.

1. Et videtur quod non per Aristotelem dicentem quod definitum est magis notum definientibus, et tamen definientia sunt magis confusa definitis; animal enim, quod est definiens, est magis confusum quam homo, quod est definitum, et similiter rationale.

2. Item, universalia sunt magis confusa singularibus, secundum Aristotelem ibidem, et tamen non sunt nobis magis nota, immo <singularia[c]> sunt magis nota intellectui.

3. Solutio. Ad istam rationem dicendum quod tota definitio aggregata ex partibus est minus nota intellectui quam definitum; quaelibet tamen pars per se est notior, quia universalior.

4. Aliter dicitur ut patet in sententia[d].

[a] vero <intellectiva> non *scripsimus cum* TC; vero non *ex. corr.* O
[b] magis *suppl. s.l.* O
[c] singularia *supplevimus*
[d] sententia *scripsimus*; sensitiva O; *cf. app. font.*

5. To the first argument to the contrary[79] it must be said that the major premise is false; for it admits a counterexample in mathematical things, as is clear.

6. To the second argument[80] it must be said that the order of nature is such that what is better known to nature is less known to a man. And the reason for this is that the soul is bound up with the body and receives the darkness of the body and only understands in accordance with what is possible for the body; but the body's cognition can only start from what is posterior; thus, a man has a greater and fuller or more powerful cognition of what is posterior than of what is prior, but for nature it is the other way around; thus, this is the order between man and nature.

7. To the third argument[81] it must be said that since a sensitive soul can only act and receive in accordance with corporeal instruments, therefore its cognition or its reception is not made worse by the body; but a soul does not understand through corporeal instruments, but by itself; hence, its cognition and reception are made much worse by the darkness of the body. Therefore, the two cases are not similar.

<Question 14>
The question is whether the things that are more confused are better known to us.

1. Arguments for the negative: Aristotle says that the thing defined is better known than the things that define it;[82] but the things that define are more confused than the things defined by them, since animal, which is a thing that defines, is more confused than man, which is a thing defined by it, and similarly for rational.

2. Furthermore, universals are more confused than singulars, according to Aristotle in the same place,[83] but they are not better known to us, rather the singulars are better known to an intellect.

3. Solution. To this argument[84] it must be said that the whole definition taken as the aggregate of its parts is less known to the intellect than the thing defined; however, any of its parts taken by itself is better known, because it is more universal.

4. There is another reply, as is clear from the exposition.[85]

[79] See above, n. 1.
[80] See above, n. 2.
[81] See above, n. 3.
[82] Arist. *Phys.* I.1, 184a26–b12.
[83] Arist. *Phys.* I.1, 184a23–26.
[84] See above, n. 1.
[85] This is a reference to a literal commentary (*sententia*) on the *Physics* containing an alternative solution. This commentary, however, is not that ascribed to Aspall (cf. Donati, 2012, 252), but either that by Adam of Buckfield or one derived from it. The relevant passage from Buckfield's commentary is the following: "<1> Potest dici quod intelligit ipsum definitum esse notius

<Quaestio 15>

Iuxta istam quaestionem quaeritur utrum universale sit prius singulari vel non.

<Quaestio 15.1>

1. Et ista quaestio **[O 89va]** cedit in istam quaestionem. Habito enim quod universale sit prius, sequitur quod sit magis confusum.

2. Quod autem universale sit prius probatio: consequens per naturam prius est antecedente et notius; universale autem proprie est consequens. Quod enim sequitur ad multa proprie est consequens; sed universale semper sequitur ad multa.

3. Item[a], prius est illud a quo non convertitur etc.; sed ab universali[b] ad singulare non convertitur etc.[c]; ergo etc.

4. Item, secundum Aristotelem, prius est animal quam equus, et tamen animal est magis universale.

5. Item, definitio constat ex prioribus secundum naturam et ex illis quae sunt magis universalia et ex illis quae sunt notiora; ergo etc.

6. Ad oppositum: quod procedit ex alio, illud ex quo procedit est magis notum; sed universale procedit ex singularibus, ut patet per auctorem *Sex principiorum*; ergo etc.

7. Item, omne quod fit ab intellectu nostro, cum sit quoddam artificiatum, est posterius apud intellectum quam illud quod fit a natura et est res naturae. Sed universale est illud quod fit ab intellectu nostro quasi quoddam artificiatum, quia nos facimus universalia, sicut vult Aristoteles in libro *Posteriorum*; singulare autem est a natura et res naturae. Ergo universale erit posterior et ignotior, singulare vero prior et notior.

[a] Item] per *add.* O
[b] universali *corr. ex* universale O
[c] etc.] et per consequens *add. sed del.* O

<Question 15>
In connection with this question it is asked whether a universal is prior to a singular or not.

<Question 15.1>

1. And the former question[86] yields this question; for, once it has been established that a universal is prior, it follows that it is more confused.

2. Arguments that a universal is prior: a consequent is prior by nature and better known than an antecedent; but a universal is a consequent in the strict sense, because what follows from many things is a consequent in the strict sense; but a universal always follows from many things.

3. Furthermore, the prior is that from which there is no conversion <with respect to the implication of existence>;[87] but from a universal to a singular there is no conversion <with respect to the implication of existence>; therefore, <a universal is prior to a singular>.

4. Furthermore, according to Aristotle, animal is prior to horse,[88] and animal is more universal.

5. Furthermore, a definition consists of things that are prior in accordance with nature and of things that are more universal and of things that are better known; therefore, <a universal is prior to a singular>.

6. To the contrary: what proceeds from another thing is such that the thing from which it proceeds is better known; but a universal proceeds from singulars, as is made clear by the author of *Six principles*;[89] therefore, <a singular is better known than a universal>.

7. Furthermore, since anything that is produced by our intellect is some kind of artefact, in relation to the intellect it is posterior to what is produced by nature and is a thing of nature. But a universal is something produced by our intellect as if it were some kind of artefact, because we produce universals, as Aristotle claims in the *Posterior Analytics*;[90] but a singular is produced by nature and is a thing of nature. Therefore, a universal will be posterior and less known, whereas a singular will be prior and better known.

---

intellectui ipsa definitione aggregata ex genere et differentia. Et sic non est contrarietas; quamvis enim partes definientes sint universaliores ipso definito, ipsa tamen definitio aggregata ex partibus non est universalior ipso. <2> Vel aliter quod ipsum definitum uno modo notius est ipsis partibus definientibus, prout scilicet est totum confusum ex suis partibus, et hoc est in definitione. Alio autem modo est definitum minus notum, prout scilicet genus, quod est pars definitionis, continet plures species. Et hoc est in divisione. Unde definitum in definitione notius est intellectui quam genus, in divisione autem genus est notius quam species definita." (Donati, 1995, 222.150–161).

[86] See above, q. 14.

[87] Arist. *Cat.* 12, 14a29–30.

[88] Arist. *De gen. anim.* II.3, 736b2–3.

[89] Anon. *Sex princ.* 9, AL I.6–7, 37. 5–7.

[90] Arist. *An. post.* II.19, 100a3–b5.

8. Item, scribit Commentator quod intellectus possibilis est locus universalium.

<Quaestio 16>
Consequenter quaeritur utrum sit[a] notius etc.

<Quaestio 16.1>

1. Et quod singulare sit notius probatio: Aristoteles in libro *Posteriorum*: deficiente sensu deficit illa cognitio quae est a sensu propter hoc quod omnis cognitio nostra est a sensu; ergo etc.

2. Item, hoc patet per Aristotelem in libro *Posteriorum* dicentem quod ex[b] multis sensibus fit unum experimentum et ex multis experimentis fit una memoria et ex multis memoriis fit unum universale, quod est principium artis et scientiae. Et ita videtur quod cognitio universalis procedat ex cognitione singularium sive ex cognitione sensitiva.

3. Ad idem[c]: ad antecedens esse verum sequitur consequens esse verum; ergo similiter ad antecedens esse cognitum a nobis sequitur consequens esse cognitum a nobis; cum ergo ex cognitione universalis sive consequentis non sequitur[d] cognitio antecedentis[e], tunc illa cognitio[f] est prior et notior.

4. Contra[g]: scribitur a Boethio quod species est totum esse individui et non econverso; ergo, cum universale sit species, et tota cognitio singularis dependebit a cognitione universalis.

<Quaestio 15.2–16.2>

1. Solutio ad ista problemata talis est: duplex est comparatio universalis: aut ad suum singulare proprium aut ad aliud universale quod sit ipso magis commune. Verbi gratia primae comparationis est comparando hominem ad hunc hominem, comparationis secundae verbi gratia comparando hominem ad animal. Utraque autem comparatio potest fieri dupliciter: aut quoad esse naturale aut quoad esse intelligibile. Si primo modo, hoc potest esse dupliciter, quia aut quoad esse naturale naturae intendentis aut quoad esse naturae operantis. Si vero comparemus magis[h] commune ad commune minus vel universale ad proprium singulare quoad esse naturae intendentis, cum natura

[a] sit] intellectus *(lectio incerta) add. sed del.* O
[b] ex *suppl. s.l.* O
[c] idem *ex corr.* O
[d] sequitur *suppl. s.l.* O
[e] antecedentis] sed econverso *add. s.l.* O
[f] cognitio] antecedentis scilicet *add. s.l.* O
[g] Contra *ex corr.* O
[h] magis *scripsimus*; maius O

8. Furthermore, the Commentator writes that the possible intellect is the place of universals.[91]

<Question 16>
The next question is whether a universal is better known.

<Question 16.1>

1. Arguments that a singular is better known: in the *Posterior Analytics* Aristotle says that, when the senses are absent, the cognition that derives from the senses is also absent, because every cognition of ours derives from the senses;[92] therefore, <a singular is better known>.

2. Furthermore, this is clear from Aristotle's statement in the *Posterior Analytics* that from many sensory acts one experience comes about, and from many experiences one memory comes about, and from many memories one universal, which is a principle of art and of science, comes about.[93] And thus it seems that the cognition of a universal proceeds from the cognition of singulars, that is, from sensitive cognition.

3. To the same conclusion: the truth of a consequent follows from the truth of an antecedent; therefore, in the same way, a consequent's being known to us follows from an antecedent's being known to us; therefore, since the cognition of an antecedent does not follow from the cognition of a universal or of a consequent, the cognition of an antecedent is prior and better known.

4. Against this: Boethius writes that its species is the whole being of an individual and not the other way around;[94] therefore, since a universal is a species, the whole cognition of a singular will also depend on the cognition of a universal.

<Question 15.2–16.2>

1. The solution to these problems is the following. A universal admits two kinds of comparative relations: either (i) to one of its proper singulars or (ii) to another universal more common than it. An example of a comparative relation of kind (i) is when man is compared to this man; an example of a comparative relation of kind (ii) is when man is compared to animal. Both comparative relations can occur in two ways: either (a) with respect to natural being or (b) with respect to intelligible being. Case (a) is twofold: either (a') with respect to the natural being of the nature that intends <to produce> that

[91] Aver. *In De anima* III, t.c. 6, ed. Crawford, 416.91–417.108.
[92] Arist. *An. post.* I.18, 81a38-b9.
[93] Arist. *An. post.* II.19, 100a3–9.
[94] Boethius, *In Isag. Porph.* III.12, CSEL 48, 237.1–7. See also ibid., III.9, 228.21–22, 229.8–9.

intendat producere formam completam ex forma incompleta, sic minus commune et singulare sunt priora et magis[a] universale et species posteriora. Si vero comparentur ad invicem secundum esse naturae operantis, cum natura operetur propter formas completas, sic adhuc prius est iste homo quam homo et homo prius quam animal, quia natura non facit universalia, sed solum singularia. Et hoc quia nobilius quod potest natura facere est producere animam sensitivam; anima enim intellectiva non est a natura, sed a prima causa. Universale autem in sensitiva esse non habet, sed solum singulare; ideo etc.

2. Si secundo modo, hoc potest esse dupliciter: aut loquendo proprie de intelligere sive sumendo apprehensionem proprie, et sic animal est prius homine et homo prius isto homine. Et hoc patet per experimentum de homine stante a remotis; prius enim cognoscitur quod est corpus, deinde, cum appropinquat, cognoscitur quod sit animal, et deinde, cum magis appropinquat, cognoscitur quod est homo, et ultimo cognoscitur quod est iste homo, sive Socrates. Aut loquendo communiter, sumendo intelligere pro quacumque apprehensione, et sic dicitur quod singulare et minus commune sunt priora.

3. Vel aliter sic: duplex est scientia sive cognitio, scilicet una quae est experimentalis, et hac scientia sive ista cognitione est singulare notius et prius, et alia quae est causalis, quae scientia dicitur vere scientia; et hac scientia sive cognitione est universale prius et notius.

4. Per istas distinctiones ad omnes rationes patet solutio, quia ad rationes ad utramque partem.

5. Nota quod huiusmodi propositiones sunt falsae si per se intelligantur, 'universale facit notum' et 'per universalia cognoscimus substantiam' et quod cognitio nostra incipit ab universalibus, quia non[b] cognoscimus <per[c]> universale unde universale est, sed per suam quidditatem et essentiam; unde prima cognitio non est per universalia, sed per quidditatem rei, secundum quod dicit Aristoteles quod quidditas rei est eadem cum re. Per accidens autem cognoscimus per universalia; unde illae propositiones quas scribit Aristoteles in littera intelligendae sunt et habent veritatem per accidens et non per se.

[a] magis *scripsimus*; maius O
[b] non *suppl. s.l.* O
[c] per *supplevimus*

being or (a'') with respect to the natural being of the nature that produces that being. If we compare the more common to the less common or a universal to one of its proper singulars in way (a'), the less common and a singular are prior, and the more universal and a species are posterior, since nature intends to produce a complete form starting from an incomplete form. If they are compared one to the other in way (a''), since nature acts in virtue of complete forms, it is still the case that this man is prior to man and man is prior to animal, because nature does not produce universals, but only singulars. And the reason for this is that the noblest thing that nature can produce is a sensitive soul; for an intellectual soul is not produced by nature but by the first cause. However, a universal does not have being in a sensitive soul, but only a singular has being there; therefore, <nature does not produce universals>.

2. Case (b) is twofold. In one way (b') we either speak of understanding in the proper sense or we take apprehension in the proper sense, and in this way animal is prior to man and man is prior to this man. And this becomes clear through our experience of a man standing at a distance:[95] we first know that he is a body, then, when he comes closer, we know that he is an animal, and then, when he comes still closer, we know that he is a man, and finally we know that he is this man, for example, Socrates. Alternatively, (b'') we speak in the broad sense, taking understanding for any <kind of> apprehension, and in this way it is said that a singular and something less common are prior.[96]

3. Another reply is the following. There are two kinds of knowledge or cognition: (i) one is experimental, and through this kind of knowledge or cognition a singular is better known and prior; and (ii) another is causal. This knowledge is truly said to be knowledge; and through this kind of knowledge or cognition a universal is prior and better known.

4. With these distinctions the solution to all the arguments, both for and against,[97] is clear.

5. Note that such propositions as 'a universal produces what is known', 'we know a substance through universals' are false if they are understood per se, and the claim that our cognition starts from universals <is false> too, because we do not know through a universal in so far as it is universal, but through its quiddity and essence; thus, the first cognition is not through universals, but through the quiddity of a thing, in accordance with Aristotle's claim that the quiddity of a thing is the same as that thing. But we know through universals accidentally; thus, those propositions that Aristotle writes in the text[98] must be interpreted and they are true accidentally and not per se.

[95] Cf. Avic. *Liber primus naturalium*, tr. 1, c. 1, ed. Van Riet, 11.10–17.

[96] The solution in nn. 1–2 is very close to that contained in Adam of Buckfield's commentary on the *Physics*; the relevant text is published in Donati, 1995, 220.103–222.138.

[97] See above, q. 15.1, nn. 1–8; q. 16.1, nn. 1–4.

[98] Arist. *Phys.* I.1, 184a23–b14.

<Quaestio 17>

Quaeritur de dicto Commentatoris, qui distinguit triplicem compositionem, quarum unam dicit esse generis ex suis speciebus. Iuxta hoc quaeritur an genus componatur ex suis speciebus.

1. Et videtur quod non, quod est contra Commentatorem, quia, si sic, tunc genus nihil esset nisi aggregatio, et tunc unitas generis erit unitas aggregationis, et non convenientiae nec essentiae, quod est inconveniens.

2. Item, compositum **[O 89vb]** posterius est [a]componentibus; ergo, secundum Commentatorem, genus erit posterius speciebus; sed a posteriori ad prius tenet consequentia; sequetur ergo: 'animal; ergo homo', quod est inconveniens.

3. Item, quaecumque simul sunt per naturam et nec prius nec posterius, unum non est de[b] compositione alterius. Ista maior plana est. Sed genus et species simul sunt per naturam et non unum prius altero. Probatio: si genus esset prius[c], tunc in illo esse priori aut est genus aut non. Per te est genus; sed non est genus nisi respectu specierum; ergo sequetur quod, si genus sit prius speciebus, quod species erit prius speciebus, quod est inconveniens.

4. Ad problema dicendum quod unitas generis est unitas convenientiae et essentiae incompletae; et haec essentia completur in suis speciebus et reperitur secundum se totam et totaliter in qualibet sua specie.

5. Sed contra: quaecumque uni et eidem etc. Cum ergo duae species conveniant in tota essentia generis totaliter et ita sunt eadem respectu essentiae generis, sequetur quod duae species erunt [d]in genere essentialiter idem. Istam conclusionem concesserunt quidam propter hanc rationem. Et dicunt quod species non differunt essentialiter nisi [e]penes esse.

6. Sed contra eorum positionem sic: definitio dicit essentiam; ergo quaecumque differunt penes definitiones differunt penes essentiam. Sed species differunt penes definitiones; ergo etc. Quod est concedendum per Aristotelem dicentem quod, quando aliqua sunt in potentia in aliquo, maior et amplior est eorum differentia cum exierint de potentia; ergo, cum species et differentiae sint potestate in genere, maiorem et ampliorem differentiam habebunt cum exierint de potentia generis. Sed quando sunt potestate in genere, differunt

[a] componentibus ... Commentatorem *corr. in* secundum Commentatorem componentibus ergo O
[b] de *suppl. s.l.* O
[c] prius *corr. ex* prior O
[d] in genere *scrips. sed del.* O
[e] penes esse *inv. sed corr.* O

<Question 17>

About the Commentator's statement that there are three compositions, one of which he says is that of a genus from its species, the question arises whether a genus is composed from its species.[99]

1. Against the Commentator there are the following arguments: if this were the case, then a genus would be nothing other than an aggregate, and the unity of a genus would be the unity of an aggregate, and not the unity of agreement or of an essence, which is absurd.

2. Furthermore, a composite is posterior to its components; therefore, according to the Commentator, a genus will be posterior to its species; but a consequence from the posterior to the prior is valid; therefore 'animal; therefore, man' will follow, which is absurd.

3. Furthermore, all the things that are simultaneous by nature and neither prior nor posterior are such that one is not a component of another. This major premise is clear. But a genus and its species are simultaneous by nature and one is not prior to the other. Proof: if a genus were prior, then in that prior being there is either a genus or there is not. According to you there is a genus; but something is not a genus except in relation to its species; therefore, it will follow that, if a genus is prior to its species, a species will be prior to species, which is absurd.

4. To this problem it must be said that the unity of a genus is the unity of agreement and of an incomplete essence; this essence becomes complete in its species and is found as a whole and totally in any of its species.

5. Against this: all the things that <are the same> as one and the same thing <are the same as one another>. Therefore, since two species totally agree in the whole essence of a genus, and hence are the same with respect to the essence of a genus, it will follow that two species in a genus will be essentially the same. Some people conceded this conclusion because of this argument. And they say that the species do not differ essentially except in being.

6. One can argue against their position as follows: a definition expresses an essence; therefore, all the things that differ in definition differ in essence. But species differ in definition; therefore, <they differ in essence>. This must be granted, given what Aristotle says, namely that when some things are in potency in something, their difference is greater and wider when they exit from that potency;[100] therefore, since species and differences are in potency in a genus, they will have a greater and wider difference when they exit from the potency of a genus. But when they are in potency in a genus they differ in

[99] Aver. *In Phys.* I, t.c. 5, fol. 8rb.
[100] Not found.

penes esse; ergo[a], quando exierint de potentia ad actum, non solum differrent penes esse, sed etiam penes essentiam.

7. Sic respondendum est ad primam rationem.

8. Solutio vero problematis et secundae rationis[b] talis. Est duplex compositio: una quae est secundum excessum in praedicando et secundum continentiam, et alia quae est secundum excessum in existendo. Prima compositio est generis ex suis speciebus, quia genus excedit species in praedicando et etiam in continendo. Plura enim continet et de pluribus praedicatur genus quam species. Et in ista compositione compositum est prius componentibus, quia continet sua componentia in potentia et ipsum est in[c] actu. Secunda compositio est speciei ex genere et differentia, et tale compositum continet sua componentia in actu et est posterius suis componentibus. Et de tali composito intellexit Commentator et secundum hoc procedit secunda ratio.

9. Ad tertiam rationem dicendum quod genus dupliciter potest considerari: aut genus in ratione generis, et secundum hoc concludit ratio, aut id quod est genus sive secundum suam essentiam. Sed hoc est dupliciter: aut secundum viam intelligendi et considerandi aut secundum viam existendi. Si primo modo, hoc est dupliciter: aut loquendo de intellectu qui semper est in actu, et sic prius est species quam genus, aut de illo qui aliquando est actu, aliquando potentia, et sic prius est id quod est genus quam id quod est species. Si[d] secundo modo, hoc est dupliciter: aut complete, et sic simul sunt genus et species, aut incomplete, et sic id quod est genus est prius, sicut patet in quadam alia distinctione praeposita.

[a] ergo *suppl. mg.* O
[b] rationis] est *add. mg.* O
[c] in] suis componentibus *add. s.l.* O
[d] Si *lectio incerta* O

being; therefore, when they exit from potency to act, they will differ not only in being, but also in essence.[101]

7. One must reply to the first argument[102] in this way.

8. However, the solution to the problem and to the second argument[103] is the following. There are two kinds of composition: (i) one in accordance with an excess in predication and in containment, and (ii) the other in accordance with an excess in existence. Composition of kind (i) is that of a genus from its species, because a genus exceeds its species in predication and also in containment; for a genus contains more things and is predicated of more things than its species. And in this composition a composite is prior to its components, because it contains its components in potency and is itself in act. Composition of kind (ii) is that of a species from a genus and a difference, and this kind of composite contains its components in act and is posterior to its components. And the Commentator referred to this kind of composite[104] and the second argument applies to this kind of composite.

9. To the third argument[105] it must be said that a genus can be considered in two ways: either (i) as a genus in accordance with the *ratio* of genus, and in this way the argument is conclusive, or (ii) as the thing that is a genus, that is, in accordance with its essence. Case (ii) is twofold: <the genus is considered> either (iia) in accordance with the way of understanding and considering or (iib) in accordance with the way of existing. Case (iia) is twofold: either (iia') we speak of an intellect that is always in act, and in this way a species is prior to its genus, or (iia'') <we speak> of an intellect that is at some time in act and at another time in potency, and in this way the thing that is a genus is prior to that which is its species. Case (iia'') is twofold: either (iia''1) <the genus is considered> as complete, and in this way a genus and its species are simultaneous, or (iia''2) <it is considered> as incomplete, and in this way the thing that is a genus is prior, as is clear in another distinction made above.[106]

[101] The author of the commentary on the *Physics* attributed to Richard Rufus raises an objection similar to that in n. 6; see Richard Rufus of Cornwall, *In Phys.*, 213.306–215.355.
[102] See above, n. 1.
[103] See above, n. 2.
[104] Aver. *In Phys.* I, t.c. 5, fol. 8rb.
[105] See above, n. 3.
[106] See above, qq. 15.2–16.2, n. 2.

<Quaestio 18>
Quaeritur de hoc quod dicit Aristoteles quod totum est notius secundum sensum quam partes eius.

1. Quod autem hoc sit falsum probatio, quia nullum sensibile secundum se totum videtur unico aspectu; ergo pars sensibilis est notius apud sensum quam totum sensibile.

2. Item, nullum sensibile unico aspectu cognoscitur. Neque enim[a] videtur nisi exterior superficies; sed tota superficies exterior videri non potest, sed solum aliqua pars superficiei, sicut patet de corpore rotundo et similiter de aliis corporibus.

3. Item, virtus sensitiva, cum sit virtus materialis, operatur per instrumenta corporea, sed intellectiva non sic; ergo secundum possibilitatem sui instrumenti operatur. Sed unumquodque instrumentum corporeum facilius apprehendit minus quam maius; ergo etc.

4. Ad oppositum patet per Aristotelem in littera.

5. Solutio. Sentire totum sensibile, hoc est dupliciter: aut quamlibet partem in se aut quamlibet partem in toto. Primum membrum est impossibile, scilicet videre totum ita quod quamlibet partem [b]in se, nec sic intellexit Aristoteles. Si secundo modo, hoc est dupliciter: aut videre per totam superficiem exteriorem, et hoc adhuc est impossibile nec etiam sic intellexit Aristoteles, aut videre per totam superficiem unico aspectu determinato[c], et hoc est possibile et sic intelligit Aristoteles.

1*. De tertia ratione quae sumitur per naturam puerorum duo sunt quaerenda.

<Quaestio 19>
Primum est utrum sensitiva virtus potest super comprehensionem universalis.

1. Et videtur quod sic, quia virtus sensitiva potest super comprehensionem alicuius ex multis; ergo potest super universale. Probatio antecedentis: lupus, quam cito videtur ab ove, statim fugit ovis; et sic oves accipiunt illud universale de lupo quae est inimicitia; ergo etc.

[a] enim *suppl. s.l.* O
[b] in se *suppl. s.l.* O
[c] determinato *corr. s.l. in* indeterminato O (secundum partem determinatam obiectam visui T *mg.* C)

<Question 18>
The question is about Aristotle's claim that a whole is better known to the senses than are its parts.[107]

1. Arguments that this claim is false: no sensible thing is seen in its totality by a single visual act; therefore, a part of a sensible thing is better known to the senses than is the whole sensible thing.

2. Furthermore, no sensible thing is known by a single visual act, because only the exterior surface is seen, but the whole exterior surface cannot be seen <by a single visual act>, but only a part of that surface, as is clear in the cases of a round body and of other bodies.

3. Furthermore, since the sensitive power is a material power, it acts through corporeal instruments, but the intellectual power does not act in this way; therefore, the sensitive power acts in accordance with the possibility of its instrument. But any corporeal instrument more easily apprehends what is smaller than what is bigger; therefore, <a part is better known to the sensitive power than a whole>.

4. The contrary is clear from Aristotle in the text.[108]

5. Solution. Grasping a whole sensible thing by the senses can be understood in two ways: either (i) <grasping> every part in itself or (ii) <grasping> every part in the whole. Case (i) is impossible, that is, seeing a whole in such a way as to see every one of its parts in itself, and Aristotle did not mean this. Case (ii) is twofold: either (iia) seeing the whole exterior surface, and this is still impossible, nor did Aristotle mean this, or (iib) seeing the whole surface by just one determinate visual act, and this is possible, and Aristotle meant this.

1*. There are two questions about the third argument, which is taken from the behaviour of children.[109]

<Question 19>
The first question is whether the sensitive power can comprehend a universal.

1. Arguments for the affirmative: the sensitive power can comprehend something starting from many things; therefore, it can comprehend a universal. Proof of the antecedent: a sheep flees a wolf as soon it sees a wolf, and thus sheep grasp that universal feature of a wolf which is hostility; therefore, <the sensitive power can comprehend something starting from many things>.

[107] Arist. *Phys.* I.1, 184a24–25.
[108] Arist. *Phys.* I.1, 184a24–25.
[109] Arist. *Phys.* I.1, 184b12–14.

2. Item, sensus terminatur ad sensibile sicut intellectus ad intelligibile; ergo, sicut se habet intellectus ad sensibile in quantum sensibile, similiter se habebit sensus ad intelligibile in quantum intelligibile; sed intellectus quidam, ut intellectus angeli, potest super comprehensionem sensibilis in quantum sensibile ex quo potest super comprehensionem omnium; ergo et sensus potest super comprehensionem intelligibilis in quantum intelligibile. Universale est huiusmodi; ergo etc.

3. Item, quod potest in plus potest in minus; sed virtus sensitiva in plus potest quam unire sibi aliquod corporeum ut intelligibile, quia potest sibi unire aliquod incorporeum incorruptibile multo nobilius aliquo corpore, ut animam intellectivam, et multo plus est unire sibi aliquod incorruptibile secundum essentiam quam aliquod corporeum **[O 90ra]** secundum speciem et similitudinem solum; ergo etc.

4. Item, anima sensitiva et universale proportionantur quoad condiciones in existendo, quia, sicut universale est ubique et in omnibus suis partibus, similiter anima sensitiva ubique est in corpore; ergo, cum, sicut est in existendo, similiter sit in apprehendendo, tunc proportionantur ad invicem secundum condiciones apprehensionis; ergo etc.

5. Item, sunt quaedam res corruptibiles et quaedam res incorruptibiles; ergo, sicut se habent istae res, similiter se habebunt eorum apprehensiones; ergo erit quaedam anima incorruptibilis quae debet apprehendere res incorruptibiles, ut anima[a] intellectiva, et alia quae debet comprehendere res corruptibiles, quae debet esse corruptibilis; talem autem non est ponere nisi anima sensitiva; ergo etc.

6. Item, est maior mundus et minor mundus, qui dicitur homo; ergo, sicut est in maiori mundo, sic erit in minori. Sed ita est in maiori mundo quod corpora caelestia immittunt suas virtutes et sua lumina in corpora hic inferiora; ergo similiter erit in minori mundo quod anima intellectiva hominis species receptas in ipsa immittet et infundet ad inferiora hominis, ut ad virtutes sensitivas. Sed ea quae recipit anima sensitiva cognoscit; ergo etc.

7. Item[b], prius per naturam est quod intellectus agens abstrahat quam quod intellectus possibilis intelligat; in illo priori esse illa species non est in intellectu possibili, quia prius est per naturam etc., nec in intellectu agente,

[a] anima] quare non est ponere nisi anima sensitiva ergo etc. Item est maior mundus et minor mundus *add. sed del.* O

[b] Item] per *add.* O

2. Furthermore, the senses are terminated at a sensible thing just as an intellect is terminated at an intelligible thing; therefore, just as an intellect is related to a sensible thing in so far as it is sensible, so the senses will be related to an intelligible thing in so far as it is intelligible; but some intellects, for example angelic intellects, can comprehend a sensible thing in so far as it is sensible in virtue of the fact that they can comprehend all things; therefore, the senses can also comprehend an intelligible thing in so far as it is intelligible. A universal is that kind of thing. Therefore, <the senses can comprehend a universal>.

3. Furthermore, what has the power to do more has the power to do less; but the sensitive power can do more than join something corporeal, such as an intelligible thing, to itself, because it can <also> join something incorporeal and incorruptible and much nobler than a body, such as an intellectual soul, to itself, and to join something incorruptible in essence to itself is to do much more than to join something corporeal in species and likeness alone to itself; therefore, <the sensitive power can grasp an intelligible>.

4. Furthermore, a sensitive soul and a universal are proportional with regard to the conditions of existence, because just as a universal is everywhere and in all its parts, so a sensitive soul is similarly everywhere in a body; therefore, since just as things are related in existence, so they are similarly related in apprehension, it follows that a sensitive soul and a universal are proportional to one another with regard to the conditions of apprehension; therefore, <a sensitive soul can apprehend a universal>.

5. Furthermore, some things are corruptible and others are incorruptible; therefore, just as these two kinds of thing are related, so the apprehension of them will similarly be related; therefore, there will be an incorruptible soul that can apprehend incorruptible things, and this is the intellectual soul, and another soul that can comprehend corruptible things and should be corruptible; but no such soul can be posited other than the sensitive soul; therefore, <a sensitive soul can apprehend a universal>.

6. Furthermore, there is a major world and <there is> a minor world, which is said to be man; therefore, as it happens in the major world, so it will happen in the minor world. But in the major world it happens that the celestial bodies introduce their powers and their light into the bodies here below; therefore, it will similarly happen in the minor world that the intellectual soul of a man introduces and infuses the species received in it into the inferior parts of man, that is, into the sensitive powers. But the sensitive soul knows the things that it receives; therefore, <a sensitive soul knows a universal>.

7. Furthermore, abstraction by the agent intellect is prior by nature to the understanding of the possible intellect. In that prior state of being, the species in question is not in the possible intellect, because < abstraction by the agent intellect> is prior by nature <to the understanding of the possible intellect>,

quia iste intellectus illuminat solum et non recipit; ergo necesse est ponere istam speciem recipi et esse in illo priori esse in aliqua virtute sensitiva; ergo etc.

8. Ad oppositum: nullum universale manet in virtutibus affixis organis corporeis; sed virtus sensitiva operatur per instrumenta corporea; ergo nullum universale potest recipi in virtute sensitiva.

9. [a]Quod verum est et concedendum est, quod nullum universale comprehenditur[b] per virtutem sensitivam.

10. Ad primam rationem contra hoc dicendum quod, secundum Avicennam[c], amicitia et inimicitia sunt sensibilia, quia in sensibilibus posita, quia per colores tales vel tales et sic ordinatos vel sic cognoscitur inimicitia et amicitia. Unde dicendum quod ovis non comprehendit[d] aliquod universale, quando multotiens viderit lupos nec sic accidit istud universale, inimicitia, sed ex naturali instinctu fugit ovis lupum, non accipiendo sic aliquod commune de multis, quia, quando aliquis comprehendit[e] aliquod universale, accipit illud cum deliberatione, pertractando illud cum deliberatione ad multa. Unde, quia ovis non habet deliberationem talem, nec concipit inimicitiam cum deliberatione, sed in[f] instanti, viso lupo, concipit inimicitiam; ideo etc.

11. Ad secundam rationem dicendum quod angelum cognoscere sensibile in quantum sensibile est dupliciter: uno modo quod is sit sensus, quod intellectus angeli immutetur a sensibili, et iste sensus est impossibilis; alio modo, ut is sit sensus, quod percipiat[g] totam naturam sensibilis, et iste sensus verus est.

12. Ad tertiam rationem dicendum quod duae sunt condiciones ad hoc requisitae, ut aliquid possit sibi recipere aliquam speciem universalem: prima est quod non affigatur virtutibus affixis instrumentis corporeis; secunda condicio est quod post receptionem habeat potentiam convertendi se ad multa. Ambae istae condiciones deficiunt a parte virtutis sensitivae; ideo etc.

13. Ad quartam rationem dicendum quod non proportionantur[h] omnino quoad condiciones existendi, quia[i] universale, licet non secundum essentiam in suis suppositis numeretur, numeratur tamen secundum entitatem et esse; sensitiva autem non numeratur nec secundum essentiam nec secundum entitatem, sed solum secundum esse.

[a] Quod … nullum *iter. sed corr.* O
[b] comprehenditur *scripsimus*; componitur O
[c] Avicennam *scripsimus cum* TC; *lectio incerta* O
[d] comprehendit *scripsimus*; componit O
[e] comprehendit *scripsimus*; componit O
[f] in *suppl. s.l.* O
[g] percipiat *corr. s.l. ex* participet O
[h] proportionantur *scripsimus*; *lectio incerta* O
[i] quia] licet non secundum *add. sed del.* O

nor is it in the agent intellect, because this intellect only illuminates and does not receive; therefore, it is necessary to assume that in that prior state of being the species in question is received and exists in some sensitive power; therefore, <the sensitive power knows a universal>.

8. To the contrary: no universal remains in the powers joined to corporeal organs; but the sensitive power acts through corporeal instruments; therefore, no universal can be received in the sensitive power.

9. It is true and it must be conceded that no universal is apprehended through the sensitive power.

10. To the first argument[110] against this it must be said that, according to Avicenna, friendship and hostility are sensibles, because they are placed in sensible things, given that hostility and friendship are known through such and such colours ordered in this or that way.[111] Thus, it must be said that a sheep does not apprehend a universal when it sees wolves a large number of times, nor does the universal 'hostility' come about in this way, but a sheep flees a wolf by natural instinct, without extracting something common from many things. This is because when someone apprehends a universal, he deliberately considers it and deliberately relates it to many things. Thus, since a sheep does not have this kind of deliberation, it does not conceive hostility through deliberation, but once it sees a wolf, it immediately conceives hostility; therefore, <a sheep does not apprehend a universal>.

11. To the second argument[112] it must be said that the claim that an angel knows a sensible thing in so far as it is sensible can be taken in two ways: (i) in one way the sense is that an angelic intellect is modified by a sensible thing, and this sense is impossible; (ii) in another way the sense is that an angel perceives the whole nature of a sensible thing , and this sense is true.

12. To the third argument[113] it must be said that two conditions are required in order for something to be able to receive a universal species in itself: the first is that it is not joined to powers that are joined to corporeal instruments; the second condition is that after the reception it has the power of turning towards many things. Both these conditions are missing in the sensitive power; therefore, <the sensitive power cannot receive a universal species>.

13. To the fourth argument[114] it must be said that they are not altogether proportional with regard to the conditions of existence, because although a universal is not numbered in essence in its supposits, it is numbered, however, in its entity and being; but a sensitive soul is not numbered in its essence or in its entity, but only in its being.

[110] See above, n. 1.
[111] Avic. *De anima* I.5, ed. Van Riet, 85.88–86.6; IV.3, 38.33–39.38.
[112] See above, n. 2.
[113] See above, n. 3.
[114] See above, n. 4.

14. Ad quintam dicendum quod una virtus apprehensiva potest super utrumque, quia species utrobique habet easdem condiciones quoad receptum. Utraque enim denudatur ab istis condicionibus, esse 'hic' et 'nunc', et propterea una virtus potest apprehendere utramque.

15. Ad sextam dicendum quod aequivocatur 'reddere' sive 'infundere' etc. in praemissis, quia in maiori mundo sumitur 'reddere' pro redditione specierum in existendo, in minori mundo pro redditione specierum in cognoscendo; et ita aequivocatur 'reddere' etc.

16. Ad septimam[a] rationem dicendum quod simul secundum existentiam intellectus agens abstrahit et possibilis[b] recipit, prius tamen per naturam abstrahit quam recipiat. Verbi gratia, arguitur probando quod forma prima est eadem cum causa prima sic: prius per naturam est forma antequam materiae adveniat; in illo priori esse non est in materia; ergo, cum fuerit alicubi, necesse est ponere ipsam esse in causa prima. Et dicitur ad hoc sic, quia simul secundum existentiam sunt forma et materia, prius tamen per naturam est forma.

<Quaestio 20>

Secundum est utrum virtus sensitiva possit componere et utrum faciat aliquis compositionem per virtutem sensitivam.

1. Et videtur quod sic[c]: dicit Aristoteles, primo *De anima*, quod eiusdem virtutis est simplices voces formare et eas ad invicem componere; sed virtutis sensitivae est voces simplices formare, ut patet, quia avis, quae[d] utitur virtute sensitiva solum, informat voces significativas; ergo etc.

2. Item, canis, videndo dominum suum, applaudit et protendit[e] ei vultum dilectionis.

3. Item, si canis appelletur proprio nomine, adest; sed hoc non posset facere nisi faceret compositiones in sua virtute aestimativa; ergo etc.

[a] septimam] dicendum *add. sed del.* O
[b] possibilis *corr. ex* possibiles O
[c] sic *suppl. s.l.* O
[d] quae *scripsimus*; qui O
[e] protendit *scripsimus*; praetendit O

14. To the fifth argument[115] it must be said that one power of apprehension has power over both <corruptible and incorruptible things> because in both cases a species exhibits the same conditions with regard to what is received; for in both cases a species is stripped of the conditions 'being here and now', and because of this one power can apprehend both.

15. To the sixth argument[116] it must be said that 'to give' or 'to infuse' are used equivocally in the premises, because in the major world 'to give' is taken for the giving of a species in existence, but in the minor world <it is taken> for the giving of a species in knowing; and thus 'to give' is used equivocally <in the premises>.

16. To the seventh argument[117] it must be said that the agent intellect abstracts and the possible intellect receives simultaneously with regard to existence, but the abstraction is prior by nature to the reception. For example, the following argument can be presented to prove that the first form is the same as the first cause: the existence of a form is prior by nature to its coming to be in matter; in that prior state of being the form is not in matter; therefore, since it is somewhere, it is necessary to assume that it is in the first cause. And in reply to this argument it is said that a form and matter are simultaneous in existence, but a form is prior by nature.

<Question 20>
The second question is whether the sensitive power can compose and whether someone can make a composition through the sensitive power.

1. Arguments for the affirmative: in *On the Soul* I Aristotle says that it belongs to the same power to form simple vocal sounds and to compose them with one another;[118] but it belongs to the sensitive power to form simple vocal sounds, as is clear, because a bird, which only uses the sensitive power, forms significant vocal sounds; therefore, <it belongs to the sensitive power to compose vocal sounds>.

2. Furthermore, when a dog sees its owner, it shows its approval and affection by its expression.

3. Furthermore, when a dog is called by its proper name, it comes; but it could not do this unless it made compositions through its estimative power; therefore, <the sensitive power can make compositions>.

[115] See above, n. 5.

[116] See above, n. 6.

[117] See above, n. 7.

[118] This claim is not present in *De anima* I or in the passages of *De anima* II dealing specifically with voice (II.8, 420b5–421a6).

4. Item, si homo componat per sensitivam, multo fortius **[O 90rb]** et alia animalia, in quibus verius est anima sensitiva. Sed homo componit per sensitivam, faciendo compositionem talium propositionum 'Socrates est albus', 'lignum est album'; ergo etc.

5. Item, cum sit duplex compositio, scilicet realis et intentionalis, si aliquid a natura possit super compositionem realem, multo fortius et super compositionem intentionalem; difficilius enim est componere et producere[a] effectum quam effectum cognoscere, sicut patet de domo. Difficilius enim est componere domum quam domum factam cognoscere. Sed intellectiva non est a natura, sed a causa prima; ergo oportet quod illud[b] sit sensitiva, quia nobilius quod potest natura facere, hoc est producere sensitivam; ergo etc.

6. Item, in omni operatione vegetativae semper ipsa vegetativa unit obiectum[c] cum aliquo, quia, quando ipsa augmentat, unit obiectum[d] cum re augmentata et, quando nutrit, unit obiectum[e] cum re nutrita et, quando generat, unit superfluum cum superfluo. Ergo eadem ratione similiter erit a parte virtutis sensitivae; ergo etc.

7. Item, omne iudicans facit compositionem. Aliter enim non posset iudicare[f] nisi componeret. Ex hoc enim provenit iudicium. Sed sensus communis, per Aristotelem in libro *De anima,* iudicat de duobus sensibilibus etc.; ergo componit.

8. Item, eadem est proportio virtutis sensitivae ad species receptas in ipsa quae est proportio virtutis intellectivae ad species receptas in ipsa; sed intellectiva non solum[g] recipit species, sed etiam eas componit; ergo similiter sensitiva non solum recipiet species, sed etiam eas componit[h].

9. Item, aliqua virtus facit compositionem aliquorum ad invicem sine istis condicionibus, 'hic' et 'nunc', sicut virtus intellectiva; ergo erit aliqua virtus faciens compositionem aliquorum ad invicem sub istis condicionibus, 'hic' et 'nunc'. Et hoc non potest esse nisi sensitiva; relinquitur ergo quod sensitiva componat.

[a] producere *corr. s.l. ex* procedere O
[b] illud *sc.* quod potest super compositionem intentionalem
[c] obiectum *scripsimus cum* TC; ipsam O
[d] obiectum *scripsimus cum* TC; ipsam O
[e] obiectum *scripsimus cum* TC; ipsam O
[f] iudicare *scripsimus*; videre O
[g] solum *suppl. s.l.* O
[h] componit *scripsimus cum* TC; intelliget *scrips. sed del.* O

4. Furthermore, if a man composes through his sensitive power, there is a much stronger reason to hold that this also happens in other animals, in which a sensitive soul exists more truly. But a man makes compositions through his sensitive soul, as when he composes such propositions as 'Socrates is white', 'a piece of wood is white'; therefore, <the sensitive power can make compositions>.

5. Furthermore, since there are two kinds of composition, namely real and intentional, if something produced by nature can make a real composition, there is a much stronger reason to say that it can make an intentional composition; for it is more difficult to compose and produce an effect than to know an effect, as is clear in the case of a house. It is indeed more difficult to build a house than to know a house already built. But an intellectual soul does not derive from nature but from the first cause; therefore, it is necessary for the thing <that can make an intentional composition> to be a sensitive soul, because the noblest thing that nature can produce is a sensitive soul; therefore, <the sensitive power can make compositions>.

6. Furthermore, a vegetative soul always joins its object with something in all its operations, because when it produces an increase, it joins its object with the increased thing, and when it nourishes, it joins its object with the nourished thing, and when it generates, it joins something superfluous with something superfluous. Therefore, by the same argument the same will apply to the sensitive power; therefore, <the sensitive power can make compositions>.

7. Furthermore, everything that judges makes a composition; for it could not judge without composing, since judgment derives from composing. But according to Aristotle in *On the Soul*, the common sense judges about two sensible things;[119] therefore, the common sense makes a composition.

8. Furthermore, the relation of the sensitive power to a species received in it is the same as the relation of the intellectual power to a species received in it; but the intellectual power not only receives species but also composes them; therefore, the sensitive power will similarly not only receive species, but will also compose them.

9. Furthermore, there is a power that makes a composition of some things with one another without the conditions 'here' and 'now', namely the intellectual power; therefore, there will be a power that makes a composition of some things with one another under the conditions 'here' and 'now'. And this can only be the sensitive power; therefore, we are left with the conclusion that the sensitive power makes a composition.

[119] Arist. *De anima* III.2, 426b12–427a14.

10. Ad oppositum: dicit Aristoteles in libro *De anima* quod componere est propria operatio intellectivae, quia est quidam intellectus simplex qui accipit[a] et alius intellectus compositivus qui componit; ergo non erit actio sensitivae. Quod verum est, quia compositio intentionum quae [b]facit propositionem in mente vel voce sive pronuntiatione fit cum deliberatione in videndo utrum res unius intentionis conveniat rei alterius intentionis, et postea[c] coniungit ad invicem secundum convenientiam, ut in propositionibus affirmativis, secundum disconvenientiam, ut in propositionibus negativis. Unde, quia ista deliberatio[d] non potest reperiri in utentibus tantum sensitiva, ut in brutis, ideo etc.

11. Ad primam rationem dicendum quod vox avis non est significativa, quia nulla vox est significativa nisi sit ibi deliberatio et etiam quod proferens concipiat[e] vocem esse significativam, et etiam quod[f] intendat ipsam esse significativam, quia voces sunt notae earum passionum quae sunt in anima; quod ergo non concipit passionem, eius vox non est significativa. Sed avis non concipit passionem; ergo eius vox non est significativa.

12. Sed contra: vox laici in dicendo orationem dominicam est vox significativa et tamen non concipit significationem vocabulorum.

13. Sed dicendum quod, licet nesciat quid significet sua vox, intendit tamen suam vocem esse significativam; sed avis neque deliberat nec concipit neque intendit.

14. Ad secundam dicendum quod aut hoc faciunt ex naturali instinctu, sicut agnus in fugiendo lupum, aut ex assuetudine, et sic est de cane. Ex consuetudine enim applaudit domino suo et huiusmodi similia facit.

15. Ad tertiam rationem dicendum quod natura universalior est in operando quam in apprehendendo; et huius causa est quia limitatur virtus sensitiva, quae tantum est apprehensiva, secundum naturam.

16. Ad quartam rationem dicendum quod aequivoce se habet obiectum vegetativae et sensitivae, quia obiectum vegetativae est in operando, obiectum sensitivae autem in cognoscendo. Et similiter est de compositione, quia

[a] accipit *corr. s.l. ex* componit O (recipit TC)
[b] facit propositionem *scripsimus cum* TC; fit propositionum O
[c] postea] dicere ista *add. sed del.* O
[d] deliberatio] videtur sic dicere *add.* O
[e] concipiat *ex corr.* O
[f] quod] proferens *add. sed del.* O

10. To the contrary: in *On the Soul* Aristotle says that to make compositions is an operation proper to an intellectual soul,[120] because there is one simple intellect that receives and another composing intellect that makes compositions; therefore, making compositions will not be the action of a sensitive soul. And this is true, because the composition of intentions which produces a proposition in the mind or in utterance or expression, comes about through the deliberation involved in checking whether the thing of one intention agrees with the thing of another intention, and next by the joining of one intention with another in accordance with their agreement, namely in affirmative propositions, or in accordance with their disagreement, namely in negative propositions. Thus, since this deliberation cannot be found in things that only use a sensitive soul, namely brute animals, it follows that <a sensitive soul does not make compositions>.

11. To the first argument[121] it must be said that the vocal sound of a bird is not significant, because no vocal sound is significant unless it involves deliberation and unless what utters a vocal sound conceives and intends it as significant, because vocal sounds are signs of passions existing in the soul; therefore, if something does not conceive a passion, its vocal sound is not significant. But a bird does not conceive a passion; therefore, its vocal sound is not significant.

12. Against this: the vocal sound of a layman when he recites the Sunday prayer is a significant vocal sound, but he does not conceive the signification of the words.

13. It must be said that, although he does not know what his vocal sound signifies, he nonetheless intends it to be significant; but a bird neither deliberates nor conceives nor intends < its vocal sound to be significant>.

14. To the second argument[122] it must be said that they do this either by natural instinct, for example in the case of a lamb that flees a wolf, or out of habit, for example in the case of a dog, because a dog shows its approval to its owner and does similar things out of habit.

15. To the third argument[123] it must be said that nature is more universal in operating than in apprehending; and the reason for this is that the sensitive power, which is only apprehensive, is limited by nature.

16. To the fourth argument[124] it must be said that the objects of a vegetative soul and of a sensitive soul are to be taken equivocally, because the object of a vegetative soul consists in operating, whereas the object of a sensitive soul

[120] Arist. *De anima* III.6, 430a26–b6.
[121] See above, n. 1.
[122] See above, nn. 2–3.
[123] See above, nn. 4–5.
[124] See above, n. 6.

compositio vegetativae est compositio realis, compositio autem[a] sensibilis est compositio intentionalis.

17. Ad quintam rationem dicendum quod iudicare de aliqua re est dupliciter: uno modo apprehendere ipsam per suam speciem sine deliberatione, alio modo cum deliberatione; primo modo iudicat sensus communis, secundo vero modo non.

18. Ad sextam rationem dicendum quod non est eadem proportio, sicut patet alibi. Sensus enim non est ita potens supra sensibile sicut intellectus supra intelligibile; unde intellectus solum facit compositionem, sed imaginativa sub imperio rationis sive prout terminatur ad intellectivam potest facere compositiones.

<Quaestio 21>

Quaeritur utrum geometer debeat disputare contra sua principia destruentem.

1. Et videtur quod sic. Cuius est per intrumentum operari, eius est instrumentum observare et tueri; ergo cum geometer operetur per sua principia, debet ea observare et tueri. Sed hoc non potest facere nisi disputet contra destruentem sua principia; ergo etc.

2. Item, secundum Aristotelem in libro *Posteriorum,* nihil debet esse notius oppositis principiorum geometriae in geometria; ergo, sicut eius est disputare cum sustinente principia, similiter et erit eius disputare contra destruentem sua principia sive contra dicentem opposita suorum principiorum.

3. Oppositum[b] vult Aristoteles plane in littera.

1*. Quaeritur de ista consequentia: 'Si principium est, principiatum est'. Et primo quaeritur de ipsa in comparatione, secundo de ipsa consequentia in se sive absolute.

[a] autem] si sit *add. s.l.* O

[b] Oppositum] ad *praem. sed del.* O

consists in cognizing. And the same applies to composition, because the composition of a vegetative soul is a real composition, whereas the composition of a sensitive soul is an intentional composition.

17. To the fifth argument[125] it must be said that judging a thing can happen in two ways: (i) in one way by apprehending a thing through its species without deliberation, (ii) in another way <by doing that> with deliberation. The common sense judges in way (i) but not in way (ii).

18. To the sixth argument[126] it must be said that the relation is not the same, as is made clear elsewhere; for the senses do not have as much power over a sensible thing as the intellect does over an intelligible thing; thus, only the intellect makes a composition, but the imaginative power under the command of reason or in so far as it terminates at the intellectual soul can make compositions.

<Question 21>
The question is whether a geometer should dispute against someone who destroys his principles.

1. Arguments for the affirmative: those who act through an instrument should also preserve and protect their instrument; therefore, since a geometer acts through his principles, he must preserve and protect them. But he cannot do this unless he disputes against someone who destroys his principles; therefore, <a geometer should dispute against someone who destroys his principles>.

2. Furthermore, according to Aristotle in the *Posterior Analytics*, nothing should be better known in geometry than the opposites of the principles of geometry;[127] therefore, just as it belongs to a geometer to dispute with someone who supports his principles, so it will similarly belong to him to dispute against someone who destroys his principles or says the opposite of his principles.

3. Aristotle clearly sides with the contrary view in the text.[128]

1*. Questions about the consequence: 'if a principle exists, something derived from the principle exists'.[129] The first question is about the consequence considered in relation <to other things>, the second is about the consequence in itself or absolutely.

[125] See above, n. 7.
[126] See above, n. 8.
[127] Arist. *An. post.* I.2, 72a37–b3.
[128] Arist. *Phys.* I.2, 185a1–4.
[129] See Arist. *Phys.* I.2, 185a3–5.

<Quaestio 22>

1. Quod autem non sequatur probatio, quia tunc sequeretur quod, si principium est ab aeterno, ut causa prima, quod similiter principiatum, quod est falsum.

2. Ad[a] hoc dicunt quidam quod principium est ab aeterno, non tamen principium in ratione principii, et ideo non sequitur.

3. Sed contra hoc sic: si Deus nunc esset principium in ratione principii et prius non fuit sic, tunc in ipso facta est aliqua immutatio, quod est inconveniens.

4. Item, si non sit principium in ratione principii, sed debeat esse, tunc **[O 90va]** in ipso est potentia, quod est inconveniens.

5. Item, si in eo est potentia ad esse principium in ratione principii, ergo aliquid movet ipsum educi de potentia ad actum; sed hoc non est nisi sua voluntas, quia scribitur quod Deus nulla necessitatis condicione, sed voluntatis electione etc.[b]; ergo, cum sua voluntas sit uniformis, tunc numquam potest educi illa de potentia ad actum; et ita erit semper principium uniforme.

6. Item, scribit Augustinus: "Quia Deus bonus, ideo sumus"; ergo, cum Deus semper fuit bonus ab aeterno, et semper fuimus nos, et ita semper fuit principium in ratione principii.

7. Item, Aristoteles arguit in undecimo *Metaphysicae*, probando quod Deus omnia intellexit simul, sic: si non omnia simul intellexit, sed aliquid intelligit de novo, tunc facta est in eo aliqua nova dispositio; ergo, cum habitus praecessit actum, tunc factus est similis dormienti, quod est falsum. Cum eodem modo contingit arguere de creare; ergo, sicut ratio tenet ex una parte, similiter et ex alia.

8. Item, Aristoteles de intelligere arguit sic: si Deus non omnia simul intelligat, ergo successive intellexit; sed hoc non potest facere, quia sic esset ibi transmutatio et potentia, quod est inconveniens. Eodem modo contingit arguere de creare.

9. Item, ibi arguit Aristoteles de intelligere sic: si Deus aliquid novum intelligeret, tunc in ipso esset lassitudo, quod est inconveniens. Multo fortius

[a] Ad *iter. sed corr.* O
[b] etc. *suppl. s.l.* O

<Question 22>

1. Arguments that the consequence is not valid: if it were valid, it would follow that, if a principle exists from eternity, for example the first cause, something derived from it <would> similarly <exist from eternity>, which is false.

2. Some people reply to this that a principle exists from eternity, but not a principle under the *ratio* of principle, and therefore the conclusion does not follow.

3. Against this: if God were now a principle under the *ratio* of principle and he was not like this before, then some change would have occurred in him, which is absurd.

4. Furthermore, if God is not a principle under the *ratio* of principle, but he ought to be like that, then there is a potency in him, which is absurd.

5. Furthermore, if there is a potency in God for being a principle under the *ratio* of principle, then something moves him to be educed from that potency to act; but this can only be his will, because it is written that God <is moved> not by any condition of necessity but by the election of the will; therefore, since his will is uniform, it follows that his will can never be educed from potency to act; and thus God will always be a principle in a uniform way.

6. Furthermore, Augustine writes "Because God is good, hence we exist";[130] therefore, since God was always good from eternity, we always existed, and thus God was always a principle under the *ratio* of principle.

7. Furthermore, in *Metaphysics* XI Aristotle presents the following argument to prove that God understands all things simultaneously: if God did not understand all things simultaneously, but he <now> understands something for the first time, then a new disposition has come about in him; therefore, since a habit precedes an act, then God is similar to someone asleep, which is false.[131] Since it is possible to argue in the same way about creation, it follows that just as the argument is valid in one case, so it is <valid> in the other case.

8. Furthermore, Aristotle argues about understanding as follows: if God does not understand all things simultaneously, it follows that he understands them successively; but he cannot do this, because in that case there would be a change and a potency in God, which is absurd.[132] One can argue about creation in the same way.

9. Furthermore, in that text Aristotle argues about understanding as follows: if God were to understand something new, then God would experience weariness, which is absurd.[133] But if God did not create all souls

[130] Aug. *De doctr. christ.* I.32, CCSL 32, 26.3–4.
[131] Arist. *Metaph.* XII.9, 1074b17–18; cf. Aver. *In Metaph.* XII, t.c. 51, fol. 335rb.
[132] Arist. *Metaph.* XII.9, 1074b26–27.
[133] Arist. *Metaph.* XII.9, 1074b28–29.

sequitur tunc quod, si non creat omnes animas, quod in eo fuit lassitudo, quia difficilius est creare et facere aliquid quam intelligere.

10. Sustinendo istam responsionem, ad primam rationem dicendum quod non fit in eo aliqua transmutatio, quia, quando relatio fundatur super duo extrema, alteratio et transmutatio potest fieri in uno nulla transmutatione facta in altero. Verbi gratia: si sit Socrates albus et omnes alii sint nigri et nascatur Plato iam albus, fit transmutatio in Platone, nulla tamen facta transmutatione in Socrate. Similiter, si sol staret in meridie et non esset terra et ponamus quod non possit multiplicari nec lucere nisi habeat obstaculum; adveniat terra et fiat terra; statim lucet sol et tamen transmutatio facta est circa terram, nulla transmutatione facta circa solem. Ergo multo fortius ubi relatio fundatur tantum supra alterum, sicut est in proposito. Creator enim non refertur ad creaturam nisi quia creatura refertur ad creatorem.

11. Ad secundam rationem dicendum quod dissimiliter est de voluntate Dei et de voluntate nostra, quia scribitur quod Deus, aeterna voluntate non inchoata nec transmutata, creat; unde a parte voluntatis Dei nulla est transmutatio, quia non potest fieri de non volente volens, nos vero fimus[a] de non volentibus volentes. Similiter, sicut omnia scivit, praeterita, praesentia et futura, sic semper omnia ista voluit; unde non potest fieri de non volente volens.

12. Ad alias rationes quae sumuntur per auctoritatem Aristotelis in undecimo dicendum quod dissimiliter est de ista actione quae est intelligere et de ista actione quae est creare, quia non est cognitio in intellectu nisi quia fiat aliqua apprehensio ab extra, secundum Commentatorem super tertium *De anima* dicentem quod intellectui aestimatur accidere fatigatio per accidens, quia virtutibus materialibus, in quibus accidit imaginationem fieri, per se accidit fatigatio. Praeterea, ista operatio quae est intelligere est eadem cum causa prima; et ideo, quando non est aliud quod potest intelligere, intelligit se ipsum. Sic autem non est de creare, quia creare semper dicitur in comparatione ad aliud[b] a causa prima. Et ideo, in intelligendo se, cum omnia in se reluceant, intelligit omnia. Unde suus intellectus immediate terminatur[c] in se ipso et est idem cum sua[d] substantia.

[a] fimus *scripsimus*; fuimus *scrips. et add. s.l.* et fimus *(lectio incerta)* O
[b] aliud *scripsimus*; alium O
[c] terminatur *scripsimus cum* T; creatur O (causatur C)
[d] sua] forma *add. sed del.* O

<simultaneously>, there would be a much stronger reason to hold that God experiences weariness, because it is more difficult to create and produce something than to understand something.

10. If we accept this reply, to the first argument[134] it must be said that no change occurs in God because when a relation is founded on two extremes, an alteration and a change can occur in one of the two extremes when no change occurs in the other extreme. For example, if Socrates is white and all the other men are black and then Plato is born already white, a change occurs in Plato, but no change occurs in Socrates. Similarly, if the sun occupied its position at noon and the earth did not exist, and we assume that <the light of> the sun cannot be multiplied and it cannot illuminate unless there is an obstacle, then if the earth comes into being, the sun would immediately illuminate <it>, but a change has occurred in the earth while no change has occurred in the sun. Therefore, there is a much stronger reason to say that this is true when a relation is founded on only one of the two extremes, as happens in the present case; for the creator is related to a creature only because a creature is related to the creator.

11. To the second argument[135] it must be said that the cases of the will of God and of our wills are not similar because it is written that God creates by an eternal will which neither has a beginning nor is subject to change; thus, where the will of God is concerned there is no change, because he cannot start to will after not willing, but we start to will after not willing. Similarly, just as God knew all things—past, present, and future—so he always willed all these things; thus, he cannot start to will after not willing.

12. To the other arguments[136] that are based on the authoritative texts of Aristotle in *Metaphysics* XI it must be said that the cases of an action of understanding and that of an action of creating are not similar, because an act of cognition cannot occur in the intellect unless an apprehension from outside occurs. This is in accordance with the Commentator's claim in his commentary on *On the Soul* III that it is believed that weariness affects the intellect accidentally, because weariness per se affects the material powers in which imagination occurs.[137] Furthermore, the operation of understanding is the same as the first cause; therefore, when there is no other thing for the first cause to understand, the first cause understands itself. But this is not the case with the action of creating, because creating is always said in relation to a thing different from the first cause. Therefore, the first cause in understanding itself understands all things because all things shine in it. Thus, the intellect of God is immediately terminated at God himself and is the same as his substance.

[134] See above, nn. 3–4.
[135] See above, n. 5.
[136] See above, nn. 7–9.
[137] Aver. *In De anima* III, t.c. 7, ed. Crawford, 419.51–55.

13. Ad auctoritatem Augustini dicendum quod sic glossatur: quia sic vult sua voluntas, cui voluntati unitur bonitas. Vel sicut prius.

14. Sed contra: quod intelligere et creare sint idem probatio: causam primam intelligere nunc rem aliquam esse extra ipsam et creare aliquam creaturam extra ipsam sunt convertibilia. Sequuntur enim convertibiliter, sicut patet. Si enim intelligat nunc primo aliquam rem esse extra ipsum, nunc primo creat et econverso. Ergo, demptis eisdem utrobique, quae relinquuntur erunt eadem. Sed illa quae sunt eadem sunt illa: nunc primo rem aliquam extra ipsum esse et nunc primo creare aliquam rem. Ergo illud quod residuum est erit aequale sive ea quae residua sunt erunt aequalia, et hoc est: causam primam intelligere et causam primam creare sunt convertibilia.

15. Item, sequitur: si Deus incipit creare, ergo creatio incipit. Ista creatio aut est creata aut non. Si est creata, ergo creationis erit creatio, quod est inconveniens. Si non creatur, et omne quod non creatur est idem cum causa prima, quia solum increatum est causa prima, ergo creare erit idem cum causa prima sicut et intelligere; ergo etc.

16. Item, in omni praedicatione ubi praedicatum non est accidentale[a], praedicatum est idem cum subiecto; sed ista praedicatio vera est 'Deus creat', et praedicatum non est accidentale, quia nullum accidens praedicatur de Deo; ergo illud praedicatum erit substantiale; ergo etc.

17. Sustinendo istam responsionem[b], dicendum quod aliquid potest praedicari de Deo et per se dicere substantiam Dei, ut 'Deus est bonus'. Est aliquod praedicatum quod[c] cum alio dicit substantiam Dei, ut 'Deus vult facere'. Tertio modo est praedicatum quod neque per se neque cum **[O 90vb]** alio dicit substantiam primae cause, quia denominat effectum causatum in creaturis et prout ad creaturas terminatur, ut cum dicitur 'Deus creat'.

18. Ad primam rationem dicendum quod non demus idem, quia a parte una cadit esse, a parte alia non, quia hic est incongruitas 'Deus incipit creare esse animam extra se'.

[a] accidentale *scripsimus cum* TC; actuale O
[b] responsionem] rationem (*lectio incerta*) O
[c] quod *suppl. s.l.* O

13. To the authoritative text of Augustine[138] it must be said that it should be glossed as follows: because the will of God wills this, and goodness is joined with his will <hence we exist>. Or <one should reply> as before.[139]

14. But against this: a proof that to understand and to create are the same: the first cause's understanding that a thing now exists outside itself and the first cause's creating a creature outside itself are convertible, since one follows from the other reciprocally. This is clear; for if the first cause now first understands that a thing exists outside itself, it now first creates that thing and the other way around. Therefore, when the same things are subtracted from each case, the things left behind are the same. But the things that are the same are that a thing now first exists outside the first cause and that the first cause now first creates that thing. Therefore, the thing or the things left behind will be equivalent, that is, the first cause's understanding and the first cause's creating are convertible.

15. Furthermore, if God begins to create, it follows that creation begins. This creation is either (i) created or (ii) not created. If (i), it follows that there will be a creation of a creation, which is absurd. If (ii), and everything which is not created is the same as the first cause, because the only non-created thing is the first cause, it follows that to create will be the same as the first cause just as its understanding is as well; therefore, <the first cause's creating and its understanding are the same>.

16. Furthermore, in every predication where the predicate is not accidental, the predicate is the same as the subject; but the predication 'God creates' is true, and the predicate is not accidental, because no accident is predicated of God; therefore, that predicate will be substantial; therefore, <God's creating is the same as God>.

17. If we accept this reply, it must be said that (i) one kind of predicate of God expresses the substance of God per se, for example 'God is good'. (ii) Another kind of predicate expresses the substance of God together with something else, for example 'God wills to make something'. (iii) A third kind of predicate expresses the substance of the first cause neither per se nor together with something else, because it denominates an effect caused in creatures and in so far as the substance of the first cause is terminated at creatures, for example when we say 'God creates'.

18. To the first argument[140] it must be said that we do not subtract the same things, because in one case being is included, in the other case it is not, given that the following claim is inconsistent: 'God begins to create a soul that exists outside himself'.

[138] See above, n. 6.
[139] See above, n. 11.
[140] See above, n. 14.

19. Ad secundam rationem dicendum quod sicut est in motu, si quaeratur utrum motus incipiat esse vel utrum mutatio incipiat; aut ergo per mutationem aut per motum, dicendum quod per mutationem, et hoc non alia mutatione quae sit via in ipsam, sed per aliud efficiens. Sed si tunc obiciatur sic: ergo idem est causa sui ipsius, dicendum quod mutatio exit in esse per aliud quod sit causa sive agens, non tamen per aliud quod sit via in ipsam. Unde, cum sic dicitur: omne quod exit in esse exit per motum vel per mutationem, hoc intelligendum est de his quae sunt res actu existentes et non de via in rem. Isto eodem modo dicendum quod creatio exit in esse per creationem, non aliam a se, quae sit via in se ipsam, sed se ipsa; exigit tamen agens quod sit aliud a se.

20. Ad tertiam dicendum quod huiusmodi praedicata, 'creare', 'dominari' et huiusmodi, non praedicant divinam essentiam, sed solum praedicant effectum connotatum in creaturis; et huiusmodi non sunt[a] omnino essentialia, sicut intelligere et huiusmodi.

21. Aliter dicitur ad principale quod, sicut est in artificialiter agentibus quod, cum artifex debeat domum componere, prius habet similitudinem illius domus in mente sua quam actu componat, et dicitur quod verius dicitur principium respectu similitudinis quae est in anima quam respectu rei extra, unde sic est ex alia parte quod, sicut in primo ab aeterno fuerunt similitudines omnium rerum postea creandarum, sic ab aeterno respectu illarum idearum [[b]ab aeterno] fuit principium. Et sic esse principium est verius esse principium quam respectu rei[c] completae in esse specifico.

<Quaestio 23>
Quaeritur secundo de illa consequentia in se.

1. Et quod non sequatur probatio sic: principium est primum; sed, per Aristotelem, primum non est coaequaevum cum eo respectu cuius dicitur primum; cum ergo principium dicatur primum respectu principiati, principium non erit coaequaevum cum principiato.

[a] sunt *corr. s.l. ex* item O
[b] ab aeterno *delevimus*
[c] rei *scripsimus cum* TC; *om. et* esse *suppl. s.l.* O

19. To the second argument[141] it must be replied just as it is replied in the case of motion, if one asks whether a motion begins to exist or whether a change begins to exist; and hence whether this is through a change or through a motion. It must be replied that <a motion or a change begins to exist> through a change, but not through a change distinct <from that which begins to exist> and that is a way towards it, but through something else as its efficient cause. If someone then objects 'therefore, the same thing is the cause of itself', it must be said that a change comes to be through another thing which is its cause or agent, but not through another thing which is a way towards that change. Thus, the claim 'everything that comes to be comes <to be> either through a motion or through a change' must be referred to the case of things that exist in act and not to the case of the way towards a thing. In the same way it must be said that a creation comes to be through a creation, but not through a creation distinct from itself, which is a way towards itself, but by itself. However, it requires an agent distinct from itself.

20. To the third argument[142] it must be said that such predicates as 'to create', 'to dominate', and the like do not predicate the divine essence, but only the effect connoted in creatures; and these predicates are not absolutely essential, in the way that understanding and the like are.

21. Another reply is that, just as it happens in the case of artificial agents, that when an artisan must build a house, he has a likeness of that house in his mind before he actually builds it, and that artisan is more truly said to be the principle of the likeness in his soul than of the thing outside, so it happens in the other case that, just as the likenesses of all things to be created afterwards existed from eternity in the first being, so the first being was the principle with respect to those ideas that existed there from eternity. And to be a principle in this way is to be a principle more truly than <to be a principle> with respect to a thing complete in its specific being.

<Question 23>
The second question is about that consequence in itself.

1. Arguments that it does not follow: a principle is something that is first; but according to Aristotle, what is first is not coeval with the thing with respect to which it is said to be first; therefore, since a principle is said to be first with respect to the thing derived from that principle, a principle will not be coeval with the thing derived from that principle.

[141] See above, n. 15.
[142] See above, n. 16.

2. Item, licet simpliciter loquendo completum dicitur prius, loquendo tamen de eadem re, prius est incompletum; similiter prius est aliquis puer quam vir, <vir[a]> [b]tamen est prius si puer debeat fieri et generari. Cum principiatum debeat fieri ex principio, necessario principium erit prius; ergo etc.

3. Item, principium in quantum principium est pars essentiae principiati, et omnis talis pars prius est toto; ergo etc.

4. Ad oppositum: dicit Aristoteles in libro *De causis* quod materiam sine forma vel formam sine materia esse non est possibile. Ergo sequitur, si materia est, quod forma est[c]; et similiter sequitur: si forma est, materia est; ergo a primo, si materia est[d], compositum est. Et ita, si principium est, de necessitate principiatum erit.

5. Item, causa est ad cuius esse sequitur aliud esse; et causa et principium idem; ergo, si principium est[e], de necessitate oportet[f] ut principiatum sit.

6. Solutio. Ad illud problema respondendum per distinctionem talem: principium et principiatum possunt dupliciter considerari, vel quoad intentiones relativas vel quoad res quae subsunt intentionibus. Si primo modo sint, simul sunt sicut relativa. Si secundo modo, hoc est dupliciter: aut in via cognoscendi aut in via existendi. Si primo modo, hoc est dupliciter, quia aut in via experimentali sive quae est per simplicem notitiam, et sic causatum sive principiatum prius est[g], aut in via causali[h], et sic principium est prius quam principiatum. Si secundo modo, hoc est dupliciter: aut secundum actualem existentiam, et sic simul sunt principium et principiatum, aut secundum habitualem existentiam, et sic principium prius principiato.

[a] vir *supplevimus*
[b] tamen … generari *scrips. sed del. et* solutio *add. sed del.* O
[c] est *suppl. s.l.* O
[d] est *suppl. s.l.* O
[e] est *suppl. mg.* O
[f] oportet *ex corr. (lectio incerta)* O
[g] est *suppl. s.l.* O
[h] causali *corr. s.l. ex* casali O

2. Furthermore, although what is complete is said to be prior absolutely speaking, in the case of one and the same thing, however, what is incomplete is prior; in the same way, someone is a child before being a man, but a man is prior if a child must come to be and be generated. Since a thing derived from a principle must come to be from a principle, a principle will necessarily be prior to it; therefore, <the consequence does not follow>.

3. Furthermore, a principle in so far as it is a principle is part of the essence of a thing derived from it, and every such part is prior to the whole; therefore, <a principle is prior to the thing derived from a that principle>.

4. To the contrary: in the book *On Causes* Aristotle says that it is not possible for matter to be without a form or for a form to be without matter.[143] Therefore, if matter exists, it follows that a form exists; and similarly, if a form exists, it follows that matter exists; therefore, by the argument from first to last, if matter exists, the composite exists. And so, if a principle exists, a thing derived from a principle will necessarily exist.

5. Furthermore, a cause is that thing from whose being the existence of something else follows; and a cause and a principle are the same; therefore, if a principle exists, it is necessary that a thing derived from that principle exists.

6. Solution. To that problem it must be replied by distinguishing two ways of considering a principle and a thing derived from that principle: either (i) as to their relative intentions or (ii) as to the things that underlie these intentions. In way (i), a principle and a thing derived from that principle are simultaneous just as the relative things are. Case (ii) is twofold: either (iia) in the way of knowing or (iib) in the way of existing. Case (iia) is twofold: either (iia') in an experimental way, that is through a simple cognition, and in this way a thing caused or a thing derived from a principle is prior to that principle, or (iia'') in a causal way, and in this way a principle is prior to a thing derived from that principle. Case (iib) is twofold: either (iib') in accordance with actual existence, and in this way a principle and a thing derived from that principle are simultaneous, or (iib'') in accordance with dispositional existence, and in this way a principle is prior to a thing derived from that principle.

[143] The correct reference is Nicolas of Amiens, *Ars fidei catholicae* I, prop. 4, ed. Dreyer, 80.16–17. See also Appendix, Book I, q. 23, n. 5.

<Quaestio 24>
Quaeritur de dicto Commentatoris, qui dicit quod nihil differt dicere omnia esse unum quam esse unum hominem vel unum equum[a].

1. Quod falsum est[b], quia omnia reducuntur ad causam primam sicut ad unum, et tamen non omnia reducuntur ad unum hominem vel ad unum equum.

2. Item, tunc teneret fallacia consequentis, quia tunc sequeretur: 'unum; ergo homo vel equus', quod est inconveniens.

3. Item, contingit dicere quod duo denarii sunt unus numerus, non tamen contingit dicere, secundum Aristotelem, quod duo denarii sunt idem[c] denarius. Ecce per hoc vult significare quod aliqua possunt esse unum in[d] genere et duo in specie; ergo, ut videtur, potius diceret Commentator omnia esse unum animal vel substantiam unam[e] quam unum hominem vel equum. [f]Si differt dicere omnia esse unum et omnia esse unum animal, ergo et similiter differt dicere alia duo.

4. Solutio. Commentator intendit quod nihil differt sic dicere quoad inconveniens et aliter non. Et tunc patet solutio ad rationes, quia non dicit hoc simpliciter.

<Quaestio 25>
Quaeritur quare dicit omnia esse mobilia et non omnia quiescibilia.

1. Et dicunt quidam quod hoc est quia sequitur: 'quiescibile; ergo mobile', et non econverso, quia corpora caelestia moventur et numquam quiescunt. Et ideo, cum motus sit magis commune quam quies et etiam superius ad ipsam[g], sufficienter dedit ipsum intelligi nec etiam necesse fuit exprimere quietem, cum intelligeretur per suum commune.

2. Sed contra hoc sic: quod corpora caelestia quiescant probatio per Aristotelem dicentem quod aqua potest **[O 91ra]** moveri secundum suas partes, licet secundum se totam non moveatur; similiter corpora caelestia possunt moveri secundum partes suas, licet secundum se tota non moveantur. Et ita aliquo modo quiescunt.

[a] equum *suppl. s.l.* O
[b] est] ut videtur *add. s.l. (lectio incerta)* O
[c] idem *scripsimus cum* TC; *om. et illegibile suppl. s.l.* O
[d] in] specie *add. sed del.* O
[e] unam] quam unus *add. sed del.* O
[f] Si … duo *scrips. sed del.* O
[g] ipsam *scripsimus*; ipsum O

<Question 24>

The question is about the Commentator's statement that there is no difference between saying that all things are one and saying that there is one man or one horse.[144]

1. That claim is false, because all things are reduced to the first cause as to one thing, but it is not the case that all things are reduced to one man or to one horse.

2. Furthermore, then the fallacy of the consequent would be valid, because then 'one; therefore, a man or a horse' would follow, which is absurd.

3. Furthermore, it is true to say that two tens are one number, but, according to Aristotle, it is not true to say that two tens are the same ten.[145] With this Aristotle wants to signify that some things can be one in genus and two in species; therefore, it seems that the Commentator should say that all things are one animal or one substance rather than one man or one horse. If there is a difference between saying that all things are one and saying that all things are one animal, it follows that there is a similar difference in the other two cases <that is, between saying that all things are one and saying that all things are one man or one horse>.

4. Solution. The Commentator means that there is no difference in saying this where absurdity is concerned, but not in another respect. The solution to the arguments[146] is then clear, because the Commentator does not say this absolutely speaking.

<Question 25>

The question is why Aristotle says that all things are mobile and not that all things are capable of rest.[147]

1. Some people reply that this is because 'it is capable of rest; therefore, it is mobile' follows, but the converse does not, because the celestial bodies are subject to motion but are never at rest. Thus, since motion is more common than rest and also superior to it, it was sufficient for Aristotle's words to be understood of motion, and it was not necessary to mention rest explicitly, since rest would be understood through what is more common than it.

2. Against this: that the celestial bodies are at rest is proved from Aristotle's statement that water can be moved in its parts, even though it is not moved as a whole;[148] similarly, the celestial bodies can be moved in their parts, even though they are not moved as a whole. Thus the celestial bodies are in some way at rest.

[144] Aver. *In Phys.* I, t.c. 9, fol. 10ra.
[145] Arist. *Phys.* IV.14, 224a3–4.
[146] See above, nn. 1–3.
[147] Arist. *Phys.* I.2, 185a12–13.
[148] Arist. *Phys.* I.3, 186a16–18.

3. Item, corpora caelestia non quiescere est contra theologos, qui dicunt corpora caelestia quiescere quando mundus peribit, quod verum est forte.

4. Ideo dicendum est aliter et sic, quod, quia, per Aristotelem, quies est privatio motus, ideo posuit illud quod dicit positionem naturae potius quam illud quod dicit privationem.

<Quaestio 26>
Quaeritur quare dixit 'quaedam[a]'.

1. Cui respondet Commentator dicens quod hoc dixit propter animam intellectivam, quae non movetur per se, et propter animam sensitivam, quae est in loco per accidens.

1*. Quaeritur de conclusione illa 'tantum unum est'. Sed primo quaeratur de suis praemissis quas facit Aristoteles. Et primo de ente, ubi tria quaeruntur.

<Quaestio 27>
Primum est an haec divisio entis—ens aliud substantia, aliud accidens—sit divisio pure univoci vel aequivoci, quorum utrumque est falsum.

1. Et quod sit divisio pure aequivoci probatio: propter differentiam aliquorum sub eodem nomine dicimus ipsa esse aequivoce contenta sub illo nomine; sed inter omnia quaecumque continentur sub ente maxime differunt substantia et accidens; ergo maxime erunt aequivoca.

2. Item, plus illa differunt quae continentur sub eodem nomine quando ipsum non potest dici de utroque in recto, sed de uno in recto et de alio in obliquo, quam quando nomen ipsum potest dici de utroque in recto. Sed testante Aristotele in septimo *Metaphysicae*, accidentia non sunt entia nisi quia entis, per hoc[b] volens dicere quod ens non praedicatur de accidente in recto, sed[c] in obliquo.

[a] quaedam *post* Commentator (n. 1) *transp. sed corr.* O
[b] hoc] est Aristoteles *add. s.l.* O
[c] sed *scripsimus*; nisi O

3. Furthermore, <to hold> that the celestial bodies are not at rest is contrary to the theologians, who say that the celestial bodies will be at rest when the world perishes, which is perhaps true.

4. Therefore, another reply must be given, namely that since, according to Aristotle, rest is a privation of motion,[149] he therefore assumed what expresses the positing of a nature rather than what expresses a privation.

<Question 26>
The question is why Aristotle said 'some'.[150]

1. The Commentator's reply is that Aristotle said this because of the intellectual soul, which is not moved per se, and because of the sensitive soul, which is in a place accidentally.[151]

1*. Questions about the conclusion 'only one thing exists'.[152] The questions about the premises which Aristotle assumes,[153] however, come first. And first about being, about which there are three questions.

<Question 27>
The first question is whether the division of being into substance and accident is the division of something purely univocal or of something equivocal, each of which is false.

1. Arguments that it is the division of something purely equivocal: because of the difference of some things under the same name we say that those things are equivocally contained under that name; but of all the things contained under being, substance and accident are maximally different ; therefore, they will be maximally equivocal.

2. Furthermore, when some things are contained under the same name and are such that that name cannot be predicated of both of them in the nominative case but of one of them in the nominative case and of the other in an oblique case, those things are more different than the things such that that name can be predicated of both of them in the nominative case. But according to Aristotle in *Metaphysics* VII, accidents are beings only because they belong to a being,[154] and by this he means that being is not predicated of an accident in the nominative case, but in an oblique case.

[149] Arist. *Phys*. V.6, 229b25.
[150] Arist. *Phys*. I.2, 185a13.
[151] Aver. *In Phys.* I, t.c. 11, fol. 11ra–b.
[152] Arist. *Phys*. I.2, 185a22.
[153] Arist. *Phys*. I.2, 185a20–27.
[154] Arist. *Metaph*. VII.1, 1028a18–20.

3. Item, aequivoca, licet non conveniant in definitione aequivoci ad illa sive aequivocitatis, conveniunt tamen in eadem definitione et essentia alicuius alterius; verbi gratia: licet marina belua et latrabile animal non conveniant in definitione et essentia canis, conveniunt tamen in essentia superiori, ut in corpore, et habent eandem definitionem et essentiam corporis. Sed substantia et accidens in nulla definitione conveniunt; ergo etc.

4. Item, hoc patet per Porphyrium.

5. Quod autem sit[a] divisio pure[b] univoci probatio, quia praedicatur de pluribus differentibus specie in eo quod quid est; ergo etc. Prima pars patet. Secunda autem patet per hoc quod differentia praedicatur in quid et similiter genus et tamen non dicunt totam naturam rei; sed ens amplectitur totam naturam rei; ergo etc.

6. Item, species dicit totam naturam rei et similiter ens; ergo ens est species. Et species est maxime univocum; ergo etc.

7. Item, dicit Aristoteles in tertio *Metaphysicae* quod genus dicitur per prius et posterius de suis speciebus, et ita analogice; et ita non obstante analogia[c] est[d] genus.

8. Item, in septimo: "In genere multae latent aequivocationes", hoc est analogiae; ergo etc.

9. Item, dicit Commentator quod quaestio 'an sit' uno modo quaerit de accidente, alio modo de genere[e], quia, si ens sumatur communiter ad decem praedicamenta, sic sumitur pro genere[f], si pro vero, sic sumitur pro accidente. Per hoc vult Commentator quod ens sit genus ad decem praedicamenta.

10. Solutio. Ad primam rationem contra primam partem[g] dicendum quod tres sunt condiciones aequivocationis. Prima est ut attributum attribuatur per diversas rationes eis quibus attribuitur. Secunda condicio est ut definitio unius

[a] sit] in definitione *add. sed del.* O
[b] pure *scripsimus*; pur O
[c] analogia] ens *add. s.l.* O
[d] est *scripsimus*; etiam O
[e] genere *scripsimus cum* TC; substantia O; *cf. app. font.*
[f] genere *scripsimus cum* TC; substantia O; *cf. app. font.*
[g] partem] affirmativam rationem *add. sed del.* O

3. Furthermore, although equivocal things do not agree in the definition of what is equivocal with respect to them or in their equivocity, they do however agree in one and the same definition and essence of something else; for example, although a marine animal and a barking animal do not agree in the definition and in the essence of a dog, they do however agree in a superior essence, for example in that of a body, and they have the same definition and essence of a body. But substance and accident agree in no definition; therefore, <they are purely equivocal>.

4. Furthermore, the same conclusion is clear from Porphyry.[155]

5. Arguments that the division <of being into substance and accident> is the division of something purely univocal: being is predicated essentially of many things different in species; therefore, <being is predicated univocally>. The first part of this claim <that is, that being is predicated of many things different in species> is clear. The second part <that is, that being is predicated essentially of a thing> is clear because a difference is essentially predicated and so is a genus, but they do not express the whole nature of a thing; but being encompasses the whole nature of a thing; therefore, <being is essentially predicated of a thing>.

6. Furthermore, a species expresses the whole nature of a thing and so does being; therefore, being is a species. And a species is maximally univocal; therefore, <being is maximally univocal>.

7. Furthermore, in *Metaphysics* III Aristotle says that a genus is predicated of its species in accordance with the prior and the posterior,[156] and thus by analogy; and so it is a genus despite the analogy.

8. Furthermore, in *Physics* VII <Aristotle says that> many equivocities, i.e., analogies, are latent in a genus;[157] therefore, <a genus is predicated analogically>.

9. Furthermore, the Commentator says that the question 'whether a thing exists' is in one way about an accident and in another way about a genus,[158] because, if being is taken as something common to the ten categories, being is taken as a genus in this way, but if being is taken as the true, then it is taken as an accident. With this the Commentator wants to say that being is a genus with respect to the ten categories.

10. Solution. To the first argument against the first part[159] it must be said that there are three conditions for equivocity. The first is that an attribute is attributed to the things to which it is attributed in virtue of different *rationes*.

[155] Porph. *Isag.*, AL I.6–7, 12.2–6.
[156] Arist. *Metaph.* III.3, 999a6–12.
[157] Arist. *Phys.* VII.4, 249a22–23.
[158] Aver. *In Metaph.* V, t.c. 14, fol. 117rb–va.
[159] See above, n. 1.

non dependeat a definitione alterius. Tertia condicio est quod non attribuatur uni propter aliud. Prima condicio est in proposito, quia <[a]ens attribuitur> per diversas rationes[b], sed duae ultimae condiciones deficiunt. Unde minor primae rationis falsa est.

11. Ad secundam rationem dicendum quod ens sic dici de uno in recto et de alio [c]in obliquo facit ipsum non esse aequivocum. Dicitur enim de accidente mediante substantia, et ideo non potest praedicari de ipso in recto, sed in obliquo. Et ideo non est[d] aequivocum, quia deficit ultima condicio.

12. [e]Eodem modo dicendum ad tertiam rationem.

13. Ad quartam sic, quod logicus non requirit ad suam aequivocationem nisi quod nomen[f] dicatur de suis contentis secundum diversas rationes, et hoc quia solum respicit praedicationem[g] sive modum praedicandi. Et ad univocationem non requirit nisi quod dicatur de eis[h] secundum eandem definitionem. Et ideo, quia ens habet[i] diversas definitiones secundum quod dicitur de contentis[j], ideo logicus dicit ens dici[k] aequivoce. Naturalis autem et metaphysicus considerant de rebus non solum secundum praedicationem[l], sed secundum rei veritatem; et ideo superaddunt alias duas condiciones, quae deficiunt a parte entis. Et ideo non dicunt ens dici[m] aequivoce, sed analogice.

14. Praeterea, ad univocationem naturalem et metaphysicam exigitur eadem definitio simul cum hac condicione: aeque primo. Et ista ultima condicio deficit a parte entis.

15. Nota etiam quod duplex est analogia: una quae impedit naturam generis, et haec est illa quae non dicitur secundum eandem rationem de suis contentis; alia, quae non impedit, et haec est illa quae dicitur secundum eandem rationem.

16. Ad primam rationem in oppositum dicendum quod dicunt quidam quod ens dicit quid, non tamen praedicatur in eo quod quid. Et hoc quia li 'quid'[n] dicit quasi speciem et ens non est species.

[a] ens attribuitur *supplevimus*
[b] rationes] fit *add. s.l.* O
[c] in obliquo *suppl. s.l.* O
[d] est *suppl. s.l.* O
[e] Eodem … rationem *scrips. sed del.* O
[f] nomen *suppl. s.l.* O
[g] praedicationem *scripsimus*; praedicatum O
[h] eis *suppl. s.l.* O
[i] habet] istas *add. sed del.* O
[j] contentis] his scilicet substantia et accidente *add. s.l.* O
[k] dici] de eis *add. s.l.* O
[l] praedicationem *scripsimus*; praedicatum O
[m] dici *suppl. s.l.* O
[n] quid *scripsimus*; quod O

The second condition is that the definition of one thing does not depend on that of the other. The third condition is that an attribute is not attributed to one thing in virtue of another. The first condition holds in the present case, because being is attributed in virtue of different *rationes*, but the last two conditions are not met. Thus, the minor premise of the first argument is false.

11. To the second argument[160] it must be said that the fact that being is said of one thing in the nominative case and of another in an oblique case prevents it from being equivocal; for being is said of an accident by means of a substance, and this is why it cannot be predicated of an accident in the nominative case, but in an oblique case. Therefore, being is not equivocal, given that the last condition of equivocity is not met.

12. It must be replied to the third argument[161] in the same way.

13. To the fourth argument[162] it must be said that a logician only requires for equivocity that a name should be said of the things contained under it in accordance with different *rationes*—the reason for this is that he only considers predication or modes of predication—and he only requires for univocity that a name should be said of them in accordance with the same definition. Since being has different definitions when it is said of things contained under it, therefore a logician says that being is said equivocally. But a natural philosopher and a metaphysician consider things not only in accordance with predication but also in accordance with the truth of the matter, and therefore they add the other two conditions, which are not met by being. Thus, a natural philosopher and a metaphysician do not say that being is said equivocally, but analogically.

14. Moreover, the same definition along with the condition 'equally primarily' is required for natural and metaphysical univocity. And this last condition is not met by being.

15. Note also that there are two kinds of analogy: (i) one is that which rules out the nature of a genus, and this is that in which a name is not said of the things contained under it in accordance with the same *ratio*; (ii) the other is that which does not rule out <the nature of a genus>, and this is that in which a name is said in accordance with the same *ratio*.

16. To the first argument to the contrary[163] it must be said that some people say that being expresses the 'what it is', but is not predicated essentially. And the reason for this is that the 'what it is' expresses a species, as it were, but being is not a species.

[160] See above, n. 2.
[161] See above, n. 3.
[162] See above, n. 4.
[163] See above, nn. 5–6.

17. Vel sic: dicendum quod praedicatur in quid, sed non in idem quid; immo, praedicatur secundum definitionem diversam.

18. Sed contra: sicut homo de Socrate et Platone dicitur et est species eorum, similiter dicitur ens de omnibus hominibus; ergo, ut videtur, ens erit species homini.

19. Et dicendum quod conclusio forte vera est. Dicendum tamen quod, quia habet potentiam ut dicatur de aliis, ideo non erit species, quia hoc repugnat speciei specialissimae.

20. Ad alia quae sumuntur ab Aristotele per distinctionem patet analogiae[a].

21. Ad aliud quod sumitur a Commentatore dicendum quod non intelligit quod sit verum genus, sed quod se aliquo modo habet per **[O 91rb]** modum generis, quia per aliquam condicionem.

<Quaestio 28>

Quaeritur utrum accidens sit unum genus ad novem praedicamenta.

1. Et videtur quod sic: sicut 'substantia' dicitur a substando, similiter 'accidens' ab inhaerendo; ergo, sicut substantia dicitur genus propter hoc quod substat, similiter et accidens dicetur genus propter hoc quod inhaeret.

2. Item, quando duo dividunt aliquid[b], si unum est genus, et reliquum, sicut patet in divisione substantiae; ergo, si substantia est genus, similiter et accidens.

3. Item, praedicatur in quid de omnibus aliis accidentibus et ipsa participant eandem definitionem.

4. Item, cum sic dicitur, 'quantitas est accidens', istud praedicatum non inest ei[c] tamquam accidens nec tamquam proprium nec tamquam definitio, ergo tamquam genus.

5. Item, sicut est una potentia in materia prima respectu omnium formarum substantialium, similiter est ponere[d] respectu omnium formarum accidentalium; ergo, sicut potentia una transmutatur in genus, sic et alia potentia.

6. Ad oppositum: si sic esset, tunc esset ponere duo genera prima, et hoc est inconveniens.

7. Solutio. Dicendum quod accidens non est genus, et hoc quia omnia accidentia inhaerent, sed non omnia inhaerent eodem modo, quia quaedam

[a] analogiae *scripsimus*; analogice O
[b] aliquid] accidens. Item praedicatur in quid *add. sed del.* O
[c] ei] subiecto *add. s.l.* O
[d] ponere] unam *add. s.l.* O

17. Another reply is that being is predicated essentially, but not in accordance with the same essence; on the contrary, it is predicated in accordance with different definitions.

18. But against this: just as man is predicated of Socrates and of Plato and is their species, being is similarly predicated of all men; therefore, it seems that being will be the species of a man.

19. It must be said that the conclusion is perhaps true. On the other hand, since being has the potency of being predicated of other things, it will not be a species, because this is inconsistent with a most special species.

20. The distinction about analogy makes it clear how one should reply to the other arguments taken from Aristotle.[164]

21. To the other argument from the Commentator[165] it must be said that he does not mean that being is a true genus, but that it has the mode of a genus in some way, because it meets a condition for being a genus.

<Question 28>

The question is whether accident is one genus for the nine categories.

1. Arguments for the affirmative: just as the word 'substance' comes from subsisting so the word 'accident' comes from inhering; therefore, just as substance is said to be a genus because it subsists, so accident will also be said to be a genus because it inheres.

2. Furthermore, when two things divide something, if one of them is a genus, the other is as well, as is clear from the division of substance; therefore, if substance is a genus, accident is a genus too.

3. Furthermore, accident is predicated essentially of all the other accidents and all accidents participate in the same definition.

4. Furthermore, when it is said 'quantity is an accident', the predicate inheres in quantity neither as an accident nor as a *proprium* nor as a definition; therefore, <it inheres> as a genus.

5. Furthermore, just as in prime matter there is one potency with respect to all substantial forms, so we should assume that there is one potency with respect to all accidental forms; therefore, just as one of these two potencies is transformed into a genus, so the other potency is as well.

6. To the contrary: if this were the case, then we should assume two first genera, and this is absurd.

7. Solution. It must be said that accident is not a genus, and this is because all accidents inhere, but not all accidents inhere in the same way, because some accidents inhere as things and others as modes, for example, position,

[164] See above, nn. 7–8.
[165] See above, n. 9.

ut res et quaedam ut modus, [a]sicut positio[b], quae inhaeret ut circumstantia[c] et non ut res. Substantia autem uno modo et eodem substat; et ideo est unum genus.

<Quaestio 29>

Quaeritur utrum nomen concretum det intelligere substantiam sive subiectum tamquam significatum suum.

1. Et videtur quod sic, quia dat intelligere substantiam et accidens; sed subiectum est principalius et a principali[d] fit denominatio; ergo etc.

2. Item, secundum Aristotelem, concretum posterius est abstracto, et ratione accidentis non est posterius, quia accidens est commune utrobique; ergo oportet quod hoc sit ratione subiecti; ergo significat substantiam et subiectum. [e]Et quod posterius est formalius[f]; quare etc.

3. Item, hoc patet per utrumque commentatorem, quia unus commentator dicit quod per prius significat subiectum et per posterius accidens, alius commentator, ut videtur, dicit quod prius significat accidens et posterius subiectum.

4. Ad oppositum: dicit Commentator super quintum *Metaphysicae* quod accidens significat prius.

5. Item, dicit Aristoteles: "Album solam qualitatem significat"; ergo non subiectum significat.

6. Solutio. Dicendum quod non significat substantiam. Sed hoc potest esse dupliciter, aut quia nullo modo significat subiectum. Et huic consonat Aristoteles, scilicet quod non significat subiectum, sed solum dat ipsum intelligere tamquam terminans suum significatum, quia dicit Aristoteles in quarto *Metaphysicae* quod nihil differt dicere 'Socrates est sanus' et 'Socrates

[a] sicut … res *scrips. sed del.* O
[b] positio *scripsimus*; potentia O
[c] circumstantia *scripsimus*; instantia O (circumstantiae TC)
[d] principali] principari O
[e] Et … etc. *scrips. sed del.* O; *fort. ante* ergo significat *legendum*
[f] formalius *sc.* quoad significationem

which inheres as a circumstance and not as a thing. But substance subsists in one and the same way, and therefore it is one genus.

<Question 29>
The question is whether a concrete name brings to mind a substance or subject as its significate.

1. Arguments for the affirmative: a concrete name brings to mind a substance and an accident; but a subject is more principal and a denomination derives from what is principal; therefore, <a concrete name brings to mind a substance as its significate>.

2. Furthermore, according to Aristotle,[166] a concrete name is posterior to an abstract one, and it is not posterior because of an accident, because an accident is common to both; therefore, it is necessary that this is so because of a subject; therefore, a concrete name signifies a substance and a subject. And what is posterior is more formal <with regard to signification>; hence, <a concrete name brings to mind a substance as its significate>.

3. Furthermore, this conclusion is made clear by both commentators, because one commentator[167] says that a concrete name signifies a subject primarily and an accident secondarily, and the other commentator seems to say that a concrete name signifies an accident primarily and a subject secondarily.[168]

4. To the contrary: in his comment on *Metaphysics* V the Commentator says that a concrete name primarily signifies an accident.[169]

5. Furthermore, Aristotle says that 'white' only signifies a quality;[170] therefore, it does not signify a subject.

6. Solution. It must be said that a concrete name does not signify a substance. But this can come about in two ways: either (i) because a concrete name in no way signifies a subject. And <the view> that a concrete name does not signify a subject, but only brings that to mind as what terminates its significate is in accordance with Aristotle, because in *Metaphysics* IV he says that there is no difference between saying 'Socrates is healthy' and 'Socrates is in health'.[171] This makes it clear that a concrete name only signifies an accident. Because of the inclination of an accident to a subject, however, a concrete

[166] This is probably a reference to Arist. *Cat.* 1, 1a12–15.
[167] That is, Avicenna, according to Aver. *In Metaph.* V, t.c. 14, fol. 117ra. Cf. Avic. *Logica,* part 1, ed. Venice 1508, fol. 9va.
[168] Aver. *In Metaph.* V, t.c. 14, fol.117ra–b.
[169] Aver. *In Metaph.* V, t.c. 14, fol.117ra–b.
[170] Arist. *Cat.* 5, 3b19.
[171] The correct reference is Arist. *Metaph.* V.7, 1017a27–29, according to the Arabic-Latin translation (see Aver. *In Metaph.* V, fol. 116va–b).

est in sanitate'. Per hoc patet quod solum significat accidens. Tamen, per inclinationem accidentis ad subiectum dat intelligere subiectum, solum tamquam terminans inclinationem suam. Aut quia non significat subiectum primo, sed secundario, et huic consonat Commentator. Melius tamen est dicere cum Aristotele.

7. Unde ad primam rationem dicendum quod verum est quod dat intelligere ipsa, sed unum tamquam significatum et alterum tamquam terminans et non principale.

8. Ad secundam rationem dicendum quod diversus est modus significandi in albo quam in albedine, quia album significat accidens cum dependentia et per modum dependentis significat, albedo vero significat accidens cum dependentia, non tamen per modum dependentis; unde album ratione modi significandi est posterius albedine et non ratione subiecti.

9. Sed contra: quod non det intelligere subiectum tamquam terminans[a] probatio, quia sic sequetur quod idem erit abstractum et concretum, quia albedo significat accidens et per modum dependentis ad subiectum tamquam terminans, ita quod subiectum non sit de significato suo, sicut facit album; ergo etc.

10. Sed dicendum quod albedo significat sub dependentia, et non per modum dependentis, album vero significat et sic et sic, et ideo non concludit.

<Quaestio 30>
Quaeritur utrum accidens possit esse per se sine substantia.

1. Et videtur quod sic, quia dicit Aristoteles in quinto *Metaphysicae* quod ens dicitur esse per se secundum quod est commune ad omnia decem praedicamenta, quia omnia praedicamenta sunt per se; et ita videtur quod accidens sit per se sine substantia.

2. Item, species accidentium deferuntur ad organum visus, et illae species sunt ibi sine subiecto proprio et sunt illae species accidentia; ergo etc.

3. Sed dicitur quod non sunt illae species accidentia, quia species coloris non est color.

[a] terminans *corr. ex* determinans O

name also brings a subject to mind, but only as what terminates <the accident's> inclination. Or (ii) because a concrete name does not signify a subject primarily, although it signifies it secondarily, and this is in accordance with the Commentator. But it is better to follow Aristotle.

7. Thus, to the first argument[172] it must be said that it is true that a concrete name brings to mind a subject and an accident, but one <is understood> as the name's significate and the other <is understood> as what terminates the name's significate and not as what is principal.

8. To the second argument[173] it must be said that 'white' and 'whiteness' have different modes of signifying, because 'white' signifies an accident together with a dependence and does so in accordance with the mode of something dependent, whereas 'whiteness' signifies an accident together with a dependence, but does not do so in accordance with the mode of something dependent; thus, 'white' is posterior to 'whiteness' because of its mode of signifying and not because of its subject.

9. Against this, one can prove that a concrete name does not bring to mind a subject as what terminates <its significate>, because in that case it would follow that the same name would be abstract and concrete, because 'whiteness' signifies an accident and does so in accordance with the mode of something dependent on a subject as on its terminus, in such a way that the subject does not belong to its significate, and this is the same way in which 'white' would signify; therefore, <'whiteness' would be both abstract and concrete>.

10. But it must be said that 'whiteness' signifies under a dependence but not in accordance with the mode of something dependent, whereas 'white' signifies in both ways, and therefore the argument is not conclusive.

<Question 30>

The question is whether an accident can exist per se without a substance.

1. Arguments for the affirmative: in *Metaphysics* V Aristotle says that being is said per se in so far as it is common to all ten categories,[174] because all categories are per se; and thus it seems that an accident exists per se without a substance.

2. Furthermore, the species of accidents are carried to the organ of sight and exist there without their proper subject, and those species are accidents; therefore, <an accident can exist per se without a subject>.

3. It is replied that those species are not accidents, because the species of a colour is not a colour.

[172] See above, n. 1.
[173] See above, n. 2.
[174] Arist. *Metaph.* V.7, 1017a22–23.

4. Sed contra: Aristoteles dicit in secundo[a] *De anima* quod organum visus est non[b] color et non est coloratum, quia est receptivum cuiuslibet coloris; ergo, sicut negat ipsum esse coloratum, sic affirmat ipsum esse receptivum coloris; et sic, cum recipit speciem[c] coloris, illa species erit color.

5. Item, lux est accidens et tamen est sine subiecto in medio, quia, si A communicet B cum C nullam substantiam communicando, tunc B est accidens. Sit A corpus caeleste, B lucere[d], C autem illud quod recipit lucem[e]; sed corpus caeleste communicat lucere cum ista re, nullam substantiam suam communicando; ergo lux est accidens. Quod autem sit sine subiecto patet, quia nullum accidens transmutatur; ergo lux non transmutabitur; ergo, recedente una parte aeris, remanebit adhuc lux, et ita erit sine subiecto.

6. Item, virtus et scientia [f]non sunt in anima tamquam in subiecto, quia sic essent ibi ut res et non essent ibi tamquam in suo cognoscente, **[O 91va]** quod est inconveniens; ergo, cum non contingat ponere eas[g] esse in alio ut in subiecto, ut videtur, in nullo subiecto erunt.

7. Solutio. Ad primam rationem dicendum[h] quod esse per se est dupliciter: uno modo, quod possit per se existere, et sic sola substantia est ens per se, et sic non sumitur 'per se' in proposito in textu Aristotelis. Alio modo, quod[i] habet existentiam in alio vel cum alio, quod quidem habet esse per se, ut accidens habet[j] esse in substantia, <quae[k]> habet esse per se. Et hoc modo possumus dicere quod accidentia sunt per se, alio modo non.

8. Ad secundam[l] rationem dicendum quod virtus et vitium non compatiuntur[m] se in anima nec scientia et ignorantia. Et, si essent intentiones, compatiuntur se, quia intentiones oppositae compatiuntur se in anima. Et ideo virtus et scientia sunt in anima tamquam res et non tamquam intentiones. Cognoscuntur autem hoc modo, quia scientia non est per se obiectum virtutis

[a] secundo *scripsimus cum* (*post corr.* T) TC; tertio O (T *ante corr.*)
[b] non *scripsimus*; ut O
[c] speciem *scripsimus*; species O
[d] lucere] vel lux *add. mg.* O
[e] lucem *ex corr.* O
[f] non sunt *inv. sed corr.* O
[g] eas *scripsimus*; eos O
[h] dicendum *suppl. s.l.* O
[i] quod] non *add. sed del.* O
[j] habet *corr. in* habens O
[k] quae *supplevimus*
[l] secundam: *recte* quartam (*sed haec ratio secunda est in* TC); et ultimam *add. s.l.* O
[m] compatiuntur] in *add. sed del.* O

4. But against this: in *On the Soul* II Aristotle says that the organ of sight is not a colour and is not coloured, because it is receptive of any colour;[175] therefore, just as Aristotle denies that the organ of sight is coloured, so he affirms that it is receptive of a colour; thus, since the organ of sight receives the species of a colour, that species will be a colour.

5. Furthermore, light is an accident and yet it exists without a subject in the medium, because if A has B in common with C without having a substance in common, then B is an accident. Let A be a celestial body, B the action of illuminating, and C a thing that receives light; a celestial body has the action of illuminating in common with this thing C, without having its substance in common with it; therefore, light is an accident. That light exists without a subject is clear because no accident is changed; light then will not be changed; therefore, when one portion of air recedes, light will still remain, and so it will exist without a subject.

6. Furthermore, virtue and knowledge are not in the soul as their subject, because if they were, they would be there as things and they would not be there as in their knower, which is absurd; therefore, since it seems that it is not possible to assume that virtue and knowledge are in something other than <the soul> as subject, they will not be in a subject.

7. Solution. To the first argument[176] it must be said that 'being per se' can be taken in two ways: (i) in one way as something that can exist per se, and in this way only a substance is a being per se, and 'per se' is not taken in this way in the present text of Aristotle.[177] (ii) In another way as something that has existence in another thing or together with another thing, which itself has being per se, in the way that an accident has being in a substance, which has being per se. And in this way we can say that accidents exist per se, but not in the other way.

8. To the second argument[178] it must be said that virtue and vice are not compatible in the soul nor are knowledge and ignorance; but if they were intentions, they would be compatible, because opposite intentions are compatible in the soul; therefore, virtue and knowledge are in the soul as things and not as intentions. But they are known in this way because knowledge is not a per se object of the speculative power, but is its moving *ratio*, because a per se object of this power is something knowable. For example, something visible rather than vision is a per se object of sight, and vision is known in knowing something visible. Similarly in the present case, we know knowledge through knowing something knowable. Thus, knowledge is known neither per

[175] Arist. *De anima* II.7, 418b26–27.
[176] See above, n. 1.
[177] Arist. *Metaph.* V.7, 1017a22–23.
[178] See above, n. 6.

speculativae, sed est[a] ratio movens, quia per se obiectum[b] est cognoscibile. Verbi gratia: visio non est per se obiectum visus, sed visibile et, in cognoscendo visibile, cognoscitur visio. Similiter a parte illa, in cognoscendo scibile, cognoscimus nos scientiam. Unde scientia <nec[c]> per se nec per suam speciem cognoscitur, sed per scibile, quia numquam habuimus scientiam antequam scivimus scibile; unde scientiae erit scientia non per se, sed ex consequenti. Non sic autem est a parte virtutis, quia virtus non cognoscitur per se, quia non sequitur, si plus habeat de virtute, quod plus cognoscat virtutem, sed sequeretur, si esset per se scibile et cognoscibile[d] secundum quod res est. Item, non cognoscitur per suam speciem, quia nihil facit speciem quod non est corpus vel forma existens in corpore; sed virtus non est corpus nec forma existens in corpore, immo existit in anima, quod verum est. Immo cognoscitur per suum effectum et per signa, sicut cognoscimus iustitiam per suum effectum, ut, si aliquis reddat unicuique quod suum est et exerceat[e] opera iustitiae, cognoscimus quod talis est iustus. Et similiter[f] cognoscimus quod sit vitiosus.

9. Ad aliam[g] rationem dicendum quod illa species nec est substantia nec est accidens nec color. Et dicendum quod Aristoteles intendit sic quod caret quolibet colore <[h]non> ut sit receptivum cuiuslibet speciei coloris[i], <sed> quia <color est extremitas perspicui in corpore terminato. Sed> duplex est terminatio —una quae est terminatio corporeitatis, et sic terminatur aer et ignis quantum ad dimensiones corporales, alia quae est terminatio perspicuitatis, cuiusmodi est coloris[j] ad parietem. Et cum dicit Aristoteles: "Color est extremitas etc.", intelligit quod secundo[k] modo terminetur[l]. Visus autem non potest esse terminativum perspicuitatis, quia tunc staret illa species ibi et non deferretur[m] usque ad intellectum; et ideo dixit Aristoteles quod non est coloratum.

10. Ad tertiam rationem dicendum quod lux non est sola species, sicut dicunt quidam, quia species non facit denominationem, sed habitum solum, et lux facit denominationem; unde non est sola species, quod verum est, immo est substantia et similiter accidens, quia splendor lucis est accidens, res

[a] est] ibi *add. s.l.* O
[b] obiectum] eius *add. s.l.* O
[c] nec *supplevimus*
[d] cognoscibile] item non cognoscitur per suam speciem, quia nihil facit *add. sed del.* O
[e] exerceat *ex corr.* O
[f] similiter] per oppositum similiter *add. s.l. (lectio incerta)* O
[g] aliam *lectio incerta* O
[h] non ... Sed *hic aliqua omissa esse videntur quae ad sensum supplevimus; cf.* TC
[i] coloris] et non eius terminativum *add. s.l.* O
[j] coloris *suppl. s.l.* O
[k] secundo *scripsimus* (hic non sumitur terminatio pro terminatione corporeitatis, sed pro terminatione diaphaneitatis TC); utroque O
[l] terminetur *sc.* perspicuum
[m] deferretur *scripsimus*; deferetur O

se nor through its species, but through something knowable, because we never had knowledge before we knew something knowable; thus, there will be knowledge of knowledge not per se but derivatively. The case of virtue is different, because virtue is not known per se, because if someone has more virtue, it does not follow that he knows virtue better, but this would follow if virtue were knowable per se and in so far as it is a thing. Furthermore, virtue is not known through its species, because what is neither a body nor a form existing in a body does not produce a species; but virtue is neither a body nor a form existing in a body, on the contrary, it exists in the soul, which is true. Rather virtue is known through its effect and through its signs: for example we know justice through its effect, so that if a man gives to each what belongs to him and he performs acts of justice, we know that this man is just. In a similar way we know that a man is vicious.

9. To the other argument[179] it must be replied that a species is neither a substance nor an accident nor a colour. And it must be said that Aristotle means that the organ of sight lacks every colour, not in order to be receptive of every species of colour, but because colour is the limit of the transparent in determinately bounded body. But there are two kinds of termination—(i) one is the termination of corporeity, and air and fire are terminated in this way with regard to corporeal dimensions; (ii) another is the termination of transparency, and the termination of a colour at a wall is of this kind. When Aristotle says "colour is the limit <of the transparent in determinately bounded body>"[180] he means that colour is terminated in the second way. But sight cannot be a terminus of transparency, because then that species would remain in sight and would not be carried up to the intellect; and this is why Aristotle said that sight is not coloured.

10. To the third argument[181] it must be said that light is not only a species, as some people say, on the grounds that a species does not give rise to a denomination, but only to a habit, since light does give rise to a denomination. Thus, it is not only a species, which is true; rather, light is both a substance

[179] See above, nn. 2–4.
[180] Arist. *De sensu* 3, 439b11–12.
[181] See above, n. 5.

deferens splendorem est substantia. Et ita duo sunt in luce; unde lux est substantia cum accidente.

1*. *Si quidem enim erunt* etc.

<Quaestio 31>
Iuxta hoc quaeritur utrum accidens possit praedicari de substantia.

1. Et videtur quod non, quia in propositione affirmativa ubi praedicatum vere affirmatur de subiecto, hoc est propter identitatem et convenientiam praedicati cum subiecto; sed magis conveniunt Socrates et Plato quam Socrates et albus; ergo verior est praedicatio Platonis de Socrate quam albi de Socrate; sed primum est impossibile; ergo secundum. Minor patet, quia Socrates et Plato conveniunt in genere et specie, album et Socrates non conveniunt in genere nec in specie.

2. Item, album non dicitur de substantia nisi quia dat intelligere substantiam tamquam terminans[a], cum sit dependens. Sed illa causa nulla est, quia deficit alicubi, ut in relativis. Pater enim dat determinate intelligere filium terminantem suum significatum et in actu, et tamen praedicatio unius de altero est impossibilis.

3. Item, non contingit dicere quod unum potest praedicari de alio, quia sunt unum numero. Probatio enim quod non sunt unum numero, quia non sunt eadem specie; ergo multo fortius nec numero, quia quae differunt maiori diversitate differunt et minori diversitate.

4. Item, plus conveniunt qualitas et quale quam Socrates et album, quia sunt idem secundum subiectum et secundum formam et significatum, licet modo alio, sed Socrates et album non conveniunt nisi in subiecto et non in forma vel specie. Sed praedicare quale de qualitate est impossibile; ergo etc.

5. Item, intellectus possibilis sive informativus recipit intentiones sub abstractione, quia accipit eas per abstractionem intellectus agentis; ergo in compositione solum componet eas in abstractione et non in concretione. Ergo talis propositio falsa.

[a] terminans] sed illa causa nulla est *add. sed del.* O

and an accident, because the brightness of light is an accident, whereas the thing carrying its brightness is a substance. And thus two things are found in light with the result that light is a substance together with an accident.

1*. *For if <both substance and quantity and quality> are, then, <whether these exist independently of each other or not, what exists will be many>*[182]

<Question 31>
In connection with this the question arises whether an accident can be predicated of a substance.

1. Arguments for the negative: in an affirmative proposition the predicate is truly affirmed of the subject because of the identity and agreement of the predicate with the subject; but Socrates and Plato agree more than Socrates and white; therefore, the predication in which Plato is predicated of Socrates is truer than the predication in which white is predicated of Socrates; but the first predication is impossible; therefore, the second is too. The minor premise is clear because Socrates and Plato agree in genus and in species, whereas white and Socrates agree neither in genus nor in species.

2. Furthermore, white is only said of a substance because it brings a substance to mind as what terminates it, since it is something dependent. But this explanation is worth nothing, because it does not hold elsewhere, for example in the case of relatives; for father brings to mind the determinate understanding of son as what terminates its significate in act, but a predication in which one is predicated of the other is impossible.

3. Furthermore, it is not possible to say that one can be predicated of the other because they are one in number, since it is proved that they are not one in number: they are not the same in species; therefore, there is a much stronger reason to hold that they are not the same in number, because the things that differ by a greater diversity also differ by a minor diversity.

4. Furthermore, a quality and what has that quality agree more than Socrates and something white, because they are the same in subject and in form and in significate, although in different ways, whereas Socrates and something white agree only in subject and not in form or species; but it is impossible to predicate what has a quality of that quality; therefore, <it is also impossible to predicate white of Socrates>.

5. Furthermore, the possible or in-formable intellect receives intentions as abstract, because it receives them through the abstraction of the agent intellect; therefore, in producing a composition it composes intentions only as abstract and not also as concrete. Therefore, that proposition is false.

[182] Arist. *Phys.* I.2, 185a27–29.

6. Item, simplices intentiones rerum significant passiones et non ipsas res; ergo, sicut passiones sunt in anima, sic habebunt voces eis correspondentes. Sed passiones sunt in anima simplices et consistentes in abstractione; ergo etc.

7. Ad oppositum: accidentia 'sunt in' subiectis et 'dicuntur de', per Aristotelem **[O 91vb]**, libro *Praedicamentorum*.

8. Et similiter hoc patet in proposito, quod verum est.

9. Ad primam rationem in oppositum dicendum quod duplex est convenientia: quaedam in supposito, quaedam in forma et non in subiecto. Ad hoc quod dicatur unum vere de[a] alio non requiritur nisi convenientia in uno supposito et non in forma, et hoc patet per Aristotelem, tertio *De anima*. Unde licet plus conveniant in forma Socrates et Plato, minus tamen conveniunt in supposito, quae convenientia requiritur ad praedicationem[b].

10. Ad secundam dicendum quod unum dare intelligere aliud est dupliciter: aut actu aut potentia. Si primo modo, hoc est dupliciter: aut enim sunt idem secundum subiectum et suppositum aut non. Secundo modo, sic dant relativa intelligere se actu. Si primo modo, hoc est dupliciter. Aut enim dat intelligere illud gratia significati et modi significandi aut solum gratia significati. Primo modo dat concretum intelligere subiectum, secundo modo abstractum.

11. Ad quartam rationem dicendum quod idem est dupliciter: aut idem secundum suppositum et subiectum et secundum coexistentiam aut idem ita quod sint supposita eiusdem; primo modo Socrates et album sunt idem, secundo modo vero qualitas et quale. Et sic procedit ratio, primo modo non procedit.

12. Ad tertiam eodem modo[c] quo prius.

13. Ad alias duas rationes dicendum quod duplex est abstractio: una quae est a supposito signato, et ista fit per intellectum agentem, alia quae est abstractio in modo significandi. [d][e]Et talis abstractio est in proposito. Et cum tali abstractione simul <stat[f]> quod sit ibi aliquod suppositum signatum;

[a] de *suppl. s.l.* O

[b] praedicationem *scripsimus*; praedicatum O

[c] modo] in *add. sed del.* O

[d] Et … Socrate *immo, abstractio quae compatitur cum compositione cum supposito signato est abstractio a supposito signato (sc. universalis a particulari); cf.* TC

[e] Et … compositi *scrips. sed del.* O

[f] stat *supplevimus*

6. Furthermore, the simple intentions of things signify the passions of things and not the things themselves; therefore, just as these passions are <found> in the soul, so they will have spoken words corresponding to them. But the passions in the soul are simple and abstract; therefore, <they are not predicated of substances>.

7. To the contrary: accidents are in subjects and are said of subjects, according to Aristotle in the *Categories*.[183]

8. This is also clear in the present case, which is true.

9. To the first argument to the contrary[184] it must be said that there are two kinds of agreement: (i) one in supposit, and (ii) another in form and not in subject. For a true predication of one thing of another, nothing other than agreement in one supposit is required, and not that of a form, as is clear from Aristotle, *On the Soul* III.[185] Thus, although Socrates and Plato agree more in form, they agree less in supposit, and that agreement is required for predication.

10. To the second argument[186] it must be said that one thing can bring about the understanding of another in two ways: either (i) in act or (ii) in potency. Case (i) is twofold: either (ia) the two things are the same in subject and supposit or (ib) they are not. Relatives bring about the understanding of one another in act in way (ib). Case (ia) is twofold: either (ia') one thing brings about the understanding of the other in virtue of its significate and of its mode of signifying or (ia'') only in virtue of its significate. A concrete name brings about the understanding of its subject in way (ia'), and an abstract name in way (ia'').

11. To the fourth argument[187] it must be said that being the same is of two kinds: either (i) the same in supposit and in subject and in coexistence, or (ii) the same in such a way as to be supposits of the same nature. Socrates and something white are the same in way (i), whereas a quality and what has that quality are the same in way (ii). And the argument is conclusive if being the same is taken in way (ii) but it is not conclusive if it is taken as in way (i).

12. To the third argument[188] it must be said as has been said before.[189]

13. To the other two arguments[190] it must be said that there are two kinds of abstraction: (i) one is abstraction from a determinate supposit, and this is produced by the agent intellect; (ii) the other is abstraction in the mode of signifying, which is the kind of abstraction considered here. This kind of

[183] Arist. *Cat.* 1, 1a29–b3.
[184] See above, n. 1.
[185] This is probably a reference to Arist. *De anima* III.6, 430b26–27.
[186] See above, n. 2.
[187] See above, n. 4.
[188] See above, n. 3.
[189] See above, n. 11.
[190] See above, nn. 5–6.

unde simul stant, quod sit ibi abstractio in modo significandi et tamen compositio cum supposito signato, ut cum Socrate. Unde abstrahit non a materia quacumque, sed a materia subiecta universali et non a materia quae est altera pars compositi. Unde album abstrahitur ab hoc albo et ab illo, non tamen abstrahitur a subiecto.

<Quaestio 32>
Quaeritur an illa consequentia teneat: 'substantia et accidens sunt; ergo multa sunt'.

1. Et videtur quod non, quia subiectum et accidens non ponunt in numerum; ergo non sunt multa.

2. Item, nos non dicimus Socratem et suam albedinem esse duo nec Platonem cum sua nigredine quod sunt duo.

3. Item, per id capit res suam numerationem per quod capit suam individuationem; sed accidens capit suam individuationem a subiecto; ergo suam numerationem; ergo etc.

4. Ad oppositum: quaecumque differunt maiori diversitate differunt et minori; subiectum et suum accidens differunt genere; ergo et numero.

5. Solutio. Ponere in numerum est dupliciter: aut in numerum entium aut in numerum essentiarum, [a]et isto ultimo modo intelligit Aristoteles, quia ipsi dixerunt omnia esse unum secundum esse et secundum essentiam; [b]unde illo ultimo modo ponunt in numerum, primo modo non.

6. Sed contra: haec est prima divisio entis: aliud substantia, aliud accidens; ergo, cum dividentia debent ponere in numerum respectu divisi, tunc substantia et accidens ponunt in numerum respectu entium et sunt entia diversa.

7. Item, accidens non dicit essentiam, sed solum esse, quia accidens dicit esse substantiae; ergo substantia et accidens ponunt in numerum entium et non essentiarum.

8. Solutio. Duplex est essentia: una quae derivatur ab esse, alia est essentia a qua procedit ipsum esse. Primo modo habet accidens[c] essentiam, secundo modo [d]sive secundum quod derivatur ab ente, sic solae substantiae habent essentiam.

[a] et … essentiam *scrips. sed del.* O
[b] unde illo *ex corr.* O
[c] accidens] esse sive *add. sed del.* O
[d] sive ... sic *scrips. sed del.* O

abstraction is compatible with the existence of a determinate supposit, so that abstraction in the mode of signifying is compatible with a composition with a determinate supposit, for example with Socrates. Thus, this kind of abstraction does not abstract from every matter but from the matter that is the subject for a universal and not from the matter that is the other part of the composite. Thus, white is abstracted from this white thing and that white thing, but it is not abstracted from its subject.

<Question 32>
The question is whether the consequence 'substance and accident exist; therefore, many things exist'[191] is valid.

1. Arguments for the negative: a subject and an accident do not produce a number; therefore, they are not many.

2. Furthermore, we do not say that Socrates and his whiteness are two or that Plato and his blackness are two.

3. Furthermore, a thing receives its numeration from that from which it receives its individuation; but an accident receives its individuation from its subject; therefore, it also receives its numeration from its subject; therefore, <a subject and an accident are not many>.

4. To the contrary: all the things that differ by a greater diversity also differ by a minor one; but a subject and its accident differ in genus; therefore, they also differ in number.

5. Solution. Producing a number comes about in two ways: either (i) by numbering beings or (ii) by numbering essences. Aristotle[192] refers to way (ii), because those people said that all things are one in being and essence; thus, substance and accident produce a number in way (ii) but not in way (i).[193]

6. Against this: the first division of being is that into substance and accident; therefore, since the things that divide must produce a number with regard to the thing that is divided, it follows that substance and accident produce a number of beings and are different beings.

7. Furthermore, an accident does not express essence but only being, because an accident expresses the being of a substance; therefore, substance and accident produce a number of beings and not of essences.

8. Solution. It must be said that there are two kinds of essence: (i) one that derives from being, and (ii) another from which being itself proceeds. An accident has an essence of kind (i), whereas only substances have an essence of kind (ii), that is an essence that derives from a being.

[191] Arist. *Phys.* I.2, 185a27–29.
[192] Arist. *Phys.* I.2, 185a27–29.
[193] The author of the commentary on the *Physics* attributed to Richard Rufus and Adam of Buckfield hold a similar solution. See Richard Rufus of Cornwall, *In Phys.*, 99.98–100, and Donati, 1998, 138.

9. Contra: sicut sua entitas non ponit [a]in numerum cum entitate substantiae, [b]cum sit dependens [c]ab ea, sic nec sua essentia ponet in numerum cum essentia substantiae; ergo etc.

10. Item, 'quid est' per essentiam non reperitur in aliquo accidente, quia, secundum Aristotelem, non reperitur in carentibus genere, et accidentia carent genere, secundum Commentatorem.

11. Item, quae differunt secundum essentias differunt secundum naturam; si[d] accidens et substantia differunt secundum essentias, ergo secundum naturam. Sed hoc est falsum, quia accidens non habet aliquam naturam aliam a subiecto, cum natura dividitur in materiam et formam[e].

12. Solutio. Ipsi bene dixerunt. Sed ens[f] potest dupliciter considerari[g]: vel secundum quod dividitur in substantiam et accidens, et sic debemus dicere quod ponunt in numerum entium, vel ens secundum quod est subiectum existens, et sic non ponunt in numerum, quia omnia sunt in illo subiecto. Non enim est aliud subiectum in quo est humanitas et albedo, quia utrumque est in isto homine.

13. Ad rationes in oppositum dicendum quod duplex est essentia: una incompleta, et haec est essentia accidentis, quia causatur et dependens est, alia[h] est essentia completa, et haec est essentia substantiae. Quia[i] ex unione formae substantialis cum[j] materia causantur accidentia, inde[k] essentia accidentis causatur ex esse[l] [[m]quia ex unione formae substantialis in materia]. Unde ad primam rationem in oppositum dicendum quod, quia essentia unius est essentia incompleta et essentia alterius essentia completa[n], ideo non ponunt in numerum in aliqua essentia completa nec in aliqua essentia incompleta[o], et tamen numerantur.

14. Ad secundam rationem dicendum quod Aristoteles intelligit quod accidentia non sunt in genere, quia accidentia continent in se duas naturas vel quia non sunt aliquid completum nec essentiam completam habent.

15. Ad tertiam rationem dicendum quod maior est falsa.

[a] in numerum *ex corr.* O
[b] cum sit *suppl. s.l.* O
[c] ab ea *suppl. mg.* O
[d] si] ergo *add. s.l.* O
[e] formam] substantialem *add. s.l.* O
[f] ens] ponunt in numerum entium *add. sed del.* O
[g] considerari] vel sumi *add. sed del.* O
[h] alia *lectio incerta* O
[i] Quia] ergo *add. s.l.* O
[j] cum *scripsimus*; in O
[k] inde *scripsimus*; unde O
[l] esse] substantiae *add. s.l.* O
[m] quia ... materia *delevimus*
[n] completa] et *add. sed del.* O
[o] incompleta *corr. s.l. ex* completa O

9. Against this: just as the entity of an accident does not produce a number along with the entity of a substance, since it is dependent on <the entity of a substance>, so its essence will not produce a number along with the essence of a substance; therefore, <a substance and an accident do not produce a number of essences>.

10. Furthermore, the 'what it is' through essence is not found in an accident, because according to Aristotle it is not found in things that do not have a genus,[194] and according to the Commentator, accidents do not have a genus.[195]

11. Furthermore, things that differ in essence differ in nature; therefore, if accident and substance differ in essence, they also differ in nature. But this is false, because an accident does not have a nature distinct from its subject, since nature is divided into matter and form.

12. Solution. What these people said is correct.[196] But being can be considered in two ways: either (i) in so far as it is divided into substance and accident, and in this way we must say that substance and accident produce a number of beings, or (ii) in so far as it is an existing subject, and in this way they do not produce a number, because all the accidents are in that one subject; for the subject of humanity and that of whiteness are the same, since they both exist in this man.

13. To the arguments to the contrary[197] it must be said that there are two kinds of essence: (i) one is incomplete, and this is the essence of an accident, because it is caused and dependent, (ii) the other is complete, and this is the essence of a substance. Since accidents are caused by the union of a substantial form with matter, it follows that the essence of an accident is caused by being. Accordingly, to the first argument to the contrary[198] it must be replied that, since the essence of one of the two is incomplete and the essence of the other is complete, it follows that they do not produce a number in a complete essence or in an incomplete essence, and yet they are numbered.

14. To the second argument[199] it must be said that Aristotle means that accidents are not in a genus, either because accidents contain two natures in themselves or because they are not something complete and they do not have a complete essence.

15. To the third argument[200] it must be said that the major premise is false.

[194] Arist. *Metaph.* VII.4, 1030a11–12.
[195] Aver. *In Metaph.* VII, t.c. 13, fol. 163va–b.
[196] See above, n. 5.
[197] See above, nn. 9–11.
[198] See above, n. 9.
[199] See above, n. 10.
[200] See above, n. 11.

16. Quarta ratio contra responsionem adhuc: forma, quando consideratur sub illo priori esse per naturam antequam adveniat materiae, est essentia; ergo, cum accidens non potest sic considerari[a], non erit essentia.

17. Et dicendum quod conclusio vera est loquendo de essentia completa **[O 92ra]**, et sic procedit ratio, de essentia incompleta falsa est, et sic non procedit, quia forma, antequam adveniat materiae, habet essentiam completam, et debet concludere quod accidens non habet talem essentiam[b].

<Quaestio 33>
Quaeritur utrum infinitum quantitati congruat in se.

1. Et videtur quod non, quia, si sic, ita quod esset eius differentia, tunc conveniret ei aut per se aut per accidens. Non per se, quia omni quanto inesset tunc, et tunc esset reperire[c] in binario. Si per accidens, tunc esset reperire aliquid quod esset infinitum per se et mediante illo[d] infinito[e] conveniret [f]et quantitati, quod est inconveniens.

2. Item, non competit ei actu, sicut probatur inferius, nec potentia, quia frustra est potentia, cum non potest reduci ad actum, et potentia infiniti numquam potest reduci ad actum; ergo etc.

3. Ad oppositum: infinitum est cuius quantitatem etc. nihil est sumere extra quod sit aliquid sui, ut videtur accipiendo [g]spatium pedale: cum nihil est sumere extra, erit infinitum.

4. Illa ratio procedit accipiendo descriptionem per partem negativam. Accipiendo per partem affirmativam, arguitur sic: extra [h]spatium pedale semper est sumere aliquid; extra[i] ergo est infinitum; ergo infinitum est.

5. Item[j], 'a puncto extra signato super lineam infinitae quantitatis contingit perpendicularem ducere'; 'linea indefinitae quantitatis' est subiectum; ergo de eo oportet supponere quia est; et est infinitum; ergo etc.

6. Item, numerus est duplatus, triplatus, et sic in[k] infinitum; ergo etc.

[a] considerari] non habet essentiam et *add. mg.* O
[b] essentiam *suppl. mg.* O
[c] reperire] ipsum *add. s.l.* O
[d] illo *suppl. s.l.* O
[e] infinito] per se *add. s.l.* O
[f] et quantitati *scripsimus*; ei quantitas O
[g] spatium pedale *scripsimus*; sphaeram pedalem O
[h] spatium pedale *scripsimus*; sphaeram pedalem O
[i] extra *post* ergo est infinitum *transp. sed corr.* O
[j] Item] maxima est *add. s.l.* O
[k] in *suppl. s.l.* O

16. There is still a fourth argument against the reply: a form is an essence when it is considered under that being which is prior by nature to its coming to be in matter; therefore, since an accident cannot be considered in this way, it will not be an essence.

17. It must be said that this conclusion is true if we speak of a complete essence, and the argument succeeds in this way, but the conclusion is false if we speak of an incomplete essence, and in this way the argument does not succeed. This is because a form has a complete essence before it comes to be in matter, and the argument should conclude that an accident does not have this kind of essence.

<Question 33>
The question is whether the infinite is consistent with quantity in itself.

1. Arguments for the negative: if this were the case, the infinite would be a difference of quantity, and so the infinite would belong to it either per se or accidentally. But it does not belong to it per se, because <if it did> then an infinite would exist in every quantity and it would be possible to find an infinite in number two. If accidentally, then it would be possible to find something that is infinite per se and by means of that infinite the infinite would belong to quantity, which is absurd.

2. Furthermore, the infinite does not belong in act to quantity, as is proved below,[201] nor in potency, because when a potency cannot be actualised, that potency is in vain, and the potency of the infinite can never be actualized; therefore, <the potency of the infinite would be in vain>.

3. To the contrary: the infinite is such that <if one considers> its quantity, <one sees that> nothing belonging to it can be found outside it,[202] as is clear from the space of one foot: since nothing <belonging to that space> can be found outside it, it will be infinite.

4. That argument proceeds by taking the negative part of the description of the infinite. But by taking the affirmative part, one can argue as follows: it is always possible to find something outside the space of one foot; therefore, the infinite is outside; therefore, the infinite exists.[203]

5. Furthermore, in the claim 'it is possible to draw a line perpendicular to a line of an infinite quantity from a designated point external <to that line>' the subject is 'a line of indefinite quantity'; therefore, we must suppose that that line exists; and this line is infinite; therefore, <there exists an infinite line>.

6. Furthermore, a number can be doubled, tripled, and so on to infinity; therefore, <a number can be infinite>.

[201] See below, Book III, qq. 13–14, 19.
[202] Cf. Arist. *Phys.* III.6, 207a1.
[203] Cf. Arist. *Phys.* III.6, 207a1–2.

7. Ad quaestionem respondent quidam quod quantitas dupliciter potest sumi: uno modo, secundum quod quantitas est abstracta, et sic infinitas est differentia quantitatis, et sic contingit procedere in infinitum, et sic est quantitas in mathematicis; aut secundum quod concernit materiam naturalem et in ipsa radicatur, et sic infinitum non est differentia quantitatis, et sic procedunt rationes.

8. Sed contra arguit Aristoteles in quarto huius probando quod magnitudo infinitati non [a]solum secundum quid repugnat, sed etiam simpliciter repugnat.

9. Ideo aliter dicendum quod infinitum in potentia est differentia quantitatis et infinitas non est in rebus, sed in ratione proprie.

10. Unde ad secundam rationem dicendum quod duplex est potentia: una permanentis et alia successivi; potentia permanentis vana est nisi reducatur ad actum, potentia vero successivi non. Et talis est potentia infiniti, [b]quia successive una pars est in potentia et successive alia. Et est potentia permanentis, quae simul exit in esse quando tota primo excitatur[c] et tota recipit totam completionem, ita quod non una pars recipit completionem et non tota, immo tota recipit primo minorem completionem, secundo maiorem, tertio adhuc maiorem.

11. Aliter dicendum sic, quod duplex est actus[d]: unus completus, et talis potentia[e] debet reduci ad actum; alius actus est incompletus, sicut actus infiniti, et [f]talis potentia[g] non debet procedere in actum.

12. Ad primam rationem dicendum quod infinitas est quantitatis differentia per se, secundum quod est in potentia et non secundum quod est in actu. Et secundum hoc procedit ratio. Unde, sicut infinitum est, sic congruit quantitati, et hoc est tantum[h] potentia.

[a] solum secundum quid *suppl. s.l.* O
[b] quia … alia *scrips. sed del.* O
[c] excitatur *corr. ex* exitatur O
[d] actus] et similiter potentia *add. s.l.* O
[e] potentia] completa correspondens illi actui *add. s.l.* O
[f] talis potentia *sc.* actus incompleti
[g] potentia] sua *add. s.l.* O
[h] tantum] in *add. s.l.* O

7. Some people reply to this question that quantity can be taken in two ways: (i) in one way in so far as it is something abstract, and in this way infinity is a difference of quantity and it is possible to proceed to infinity, and it is in this way that quantity exists in mathematical things; or (ii) in so far as it concerns natural matter and is rooted in it, and in this way the infinite is not a difference of quantity, and on this interpretation the arguments succeed.

8. But Aristotle argues against this in *Physics* IV, proving that a magnitude is incompatible with the infinite not only in a qualified way but also absolutely.[204]

9. Therefore, another reply should be given, namely that the infinite in potency is a difference of quantity, and infinity does not properly exist in things, but in the mind.

10. Thus, to the second argument[205] it must be said that there are two kinds of potency: that of a permanent thing and that of a successive thing. The potency of a permanent thing is in vain unless it is actualized, but this is not true of the potency of a successive thing. The potency of the infinite is that of a successive thing, because there is a succession of potential parts <in the infinite>. The potency of a permanent thing is that which comes to be all at once when the whole of it is first stimulated and the whole of it becomes complete, so that it does not happen that just one part instead of the whole potency becomes complete. Rather, the whole potency first receives a minor degree of completion, secondly a greater one, <and> thirdly a still greater one.

11. Another reply is that there are two kinds of act: (i) one complete, and the potency <relative to this kind of act> must be reduced to act; (ii) the other incomplete, such as the act of the infinite, and the potency <relative to this kind of act> need not proceed to act.

12. To the first argument[206] it must be said that infinity is a per se difference of quantity, in so far as infinity is in potency and not in so far as it is in act. And the argument succeeds on this interpretation. Thus, given the way that the infinite exists, it is consistent with quantity, and this is only in potency.

[204] The correct reference is Arist. *Phys*. III.5, 204b1–7.

[205] See above, n. 2.

[206] See above, n. 1.

<Quaestio 34>

Quaeritur an infinitum reperiatur in alio a quanto, ut in alio praedicamento.

1. Et videtur quod sic: in materia prima est infinitas et tamen est substantia; et ita est in alio quam in quantitate[a].

2. Item, finitum reperitur in quolibet genere, sicut species sub genere, quae per differentiam terminatur et finitur; et ubi reperitur unum contrariorum et reliquum; ergo etc., cum finitum et infinitum sunt contraria.

3. Solutio. Est una infinitas quae potest tolli per aliquid finitum ei additum, et talis infinitas dicitur incompletio et indeterminatio; et sic est genus infinitum, quia indeterminate se habet ut in qualibet specie reperiatur et determinate se habet[b] per adiunctionem differentiarum. Et ista infinitas reperitur in quolibet genere. Alia, quae nullo modo potest tolli, et talis est differentia quantitatis.

4. Et sic patet quod aequivocatur 'infinitas' in secunda ratione.

5. Similiter[c] etiam in prima, quia materia prima est infinita infinitate quae est indeterminatio et incompletio formae, quia materia prima est in potentia ad omnes formas; et sic fit ibi aequivocatio.

6. Aliter dicendum quod duplex est infinitas: una quae habet partes infinitas in sua potentia secundum substantiam, ubi non est pars extra partem, sed pars sub parte, ita quod pro 'simul' sumatur li 'sub'; hoc est confundere partes sub indistinctione, et ita non est eadem pars in manu et capite. Et talis infinitas est in potentia materiae primae, quia partes sunt infinitae in materia quae sunt in potentia ad extensionem. Alia est infinitas quae est cum extensione, et sic non est materia prima infinita, quia habet in se partes infinitas quae sunt in potentia ad extensionem.

[a] quantitate] etc. *add. mg.* O
[b] habet] vel determinatur *add. s.l.* O
[c] Similiter] autem *add. sed del.* O

<Question 34>
The question is whether the infinite is found in something other than quantity, that is, in another category.

1. Arguments for the affirmative: infinity exists in prime matter, but prime matter is a substance; and thus infinity exists in something other than quantity.

2. Furthermore, the finite is found in every genus as a species under a genus, which is terminated and made finite through a difference; but where one of two contraries is found the other is also found; therefore, since finite and infinite are contraries, <the infinite is also found in every genus>.

3. Solution. (i) One kind of infinity is such that it can be removed by the addition of something finite to it, and this kind of infinity is said to be an incompleteness and an indetermination; a genus is infinite in this way because it is indeterminate in such a way that it is found in every species and is made determinate by the addition of differences. This kind of infinity is found in every genus. (ii) The other kind of infinity is such that it cannot be removed in any way, and this kind of infinity is a difference of quantity.

4. And so it is clear that in the second argument[207] 'infinity' is taken equivocally.

5. The same applies to the first argument,[208] because, since prime matter is in potency to all forms, it is infinite by the kind of infinity that is an indetermination and an incompleteness with respect to a form; therefore, there is an equivocity in this argument.

6. Another reply is that there are two kinds of infinity: (i) one that has infinitely many parts in accordance with the potency belonging to its substance, in which the parts are such that they are not one outside the other but one under the other, and here 'under' is taken for 'in the same place', that is, the parts are found mixed together with a lack of <spatial> distinction; and in this sense the same part is not found in a hand and in the head. And this kind of infinity exists in the potency of prime matter, because there are infinitely many parts in potency to extension in matter. (ii) Another kind of infinity is that which exists with extension, and in this way prime matter is not infinite, because it has in itself infinitely many parts <which are only> in potency to extension.

[207] See above, n. 2.
[208] See above, n. 1.

<Quaestio 35>
Quaeritur an sequatur: ‘continuum est; ergo multa sunt’.

1. Et videtur quod non, quia unum membrum[a] non infert[b] oppositum, et unum et multa sunt opposita, et continuum est unum [c]membrum unius; ergo non sequitur ex membro unius suum oppositum quod multa.

2. Solutio. Dicitur quod non intelligit quod multa sint actu, sed potentia tantum, quia continuum est divisibile in infinitum.

3. Sed contra: dicit Aristoteles quod intellectus intelligit magnitudinem in tempore indivisibili; tunc, cum tempus sit continuum, erit tunc quoddam continuum indivisibile.

4. Item, proportionanda sunt in creaturis potentia activa et passiva; ergo erit aliqua potentia activa in creaturis infinita, si[d] est aliqua passiva in illis infinita, et hoc est inconveniens[e].

5. Item, si continuum dividitur in infinitum[f], quod omne [g]duplum sit dimidium[h] sequitur[i] sic[j]: sic se habet duplum ad dimidium sicut dimidium ad duplum; cum ergo omne dimidium est duplum, si debeat fieri talis divisio in infinitum, sequitur ergo similiter quod omne duplum est dimidium. Ergo, cum diameter mundi sit duplum, tunc erit dimidium. Et hoc est falsum, quia tunc aliqua linea **[O 92rb]** esset extra ipsum cuius esset dimidia.

6. Item, diminutio et subtractio procedunt ad non esse, sicut augmentum et appositio ad esse; sicut ergo in augmento, quod[k] procedit ad esse, est ponere statum, multo fortius ergo in[l] diminutione, quae procedit ad non esse sive in subtractione, vel aequaliter.

7. Item, dicit Aristoteles quod, si sphaera aenea tangeret superficiem planam, tangeret in puncto, quia, si tangeret in duobus punctis, cum puncta non continuantur neque contiguantur, inter illa duo puncta cadit linea media; si ergo sphaera tangeret in illis duobus punctis, linea inter puncta applicaret se secundum se totam superficiei[m]. Sed quaelibet pars superficiei[n] recta est;

[a] membrum *corr. s.l. in* oppositum O
[b] infert] suum *add. s.l.* O
[c] membrum … multa *scrips. sed del.* O
[d] si *scripsimus*; sicut O
[e] inconveniens] ergo, cum in indivisibili esset potentia passiva, aliter posset dividi, sequitur quod nullo modo potest dividi *add. sed del.* O
[f] infinitum] sequitur *add. s.l.* O
[g] duplum sit dimidium *scripsimus*; dimidium sit duplum O
[h] dimidium] et econverso *add. s.l.* O
[i] sequitur *scrips. sed del.* O
[j] sic *ex corr.* O
[k] quod *scripsimus*; quae O
[l] in *suppl. s.l.* O
[m] superficiei] corporis circularis *add. s.l.* O
[n] superficiei] rectae *add. s.l.* O

<Question 35>

The question is whether the inference 'a continuum exists; therefore, many things exist'[209] is valid.

1. Arguments for the negative: a member of a thing does not imply the opposite of that thing; but the one and the many are opposite, and the continuum is a member of the one; therefore, from a member of the one its opposite, namely that many things <exist>, does not follow.

2. Solution. It is said that Aristotle[210] does not mean that there are many things in act, but only in potency, because a continuum is divisible to infinity.

3. Against this, Aristotle says that the intellect understands a magnitude in an indivisible time;[211] then, since time is continuous, a continuum will be indivisible.

4. Furthermore, active and passive potency must be proportional in creatures; therefore, if there were an infinite passive potency in creatures, there would be an infinite active potency in them, and this is absurd.

5. Furthermore, if a continuum is divided to infinity, that every double is a half follows like this: as a double is to its half so a half is to its double. Therefore, since if this kind of division must proceed to infinity, every half is a double, it also follows that every double is a half. Therefore, since the diameter of the world is a double, then it will be a half. But this is false, because then there would be a line outside the diameter of which the diameter would be a half.

6. Furthermore, decrease and subtraction are directed towards non-being, just as increase and addition <are directed> towards being; therefore, just as it is necessary to come to a halt in an increase, which is directed towards being, there is a much stronger or an equally strong reason to hold that this happens in a decrease or a subtraction, which are directed towards non-being.

7. Furthermore, Aristotle says that, if a brazen sphere touched a plane surface, it would touch it at one point,[212] because, if the sphere touched it at two points, then since points are neither continuous nor contiguous, there would be an intermediate line between those two points; therefore, if a sphere

[209] Arist. *Phys.* I.2, 185b9–10.
[210] Arist. *Phys.* I.2, 185b9–10.
[211] Arist. *De anima* III.6, 430b7–9.
[212] Arist. *De anima* I.1, 403a13–14.

ergo linea[a] se applicaret[b] et secundum rectum. Si ergo volvatur sphaera et tangat in duobus punctis, sequetur necessario quod sphaera componatur ex lineis rectis; sed hoc falsum est; ergo primum. Tangit ergo in puncto. Si ergo sphaera moveatur super planum, constat quod lineam describet. Arguo tunc sic: omne tactum per contactum aliquorum, illa contacta sunt partes eius; sed tota linea tangitur per contactum punctorum quae sunt in ea; ergo puncta sunt partes lineae. Sed dicit Aristoteles, libro *De generatione*, quod, si linea componatur ex punctis sive ex indivisibilibus, quod tunc stabit divisio lineae; ergo etc.

8. Item, plus excedit substantia incorporea substantiam corpoream quam tempus instans, quia instans est pars temporis, et sic non est ex alia parte. Ergo, cum substantia corporea agit in tempore, tunc substantia incorporea[c] multo fortius aget in instanti[d]. Sed incorporeum, ut intelligentia, movet caelum; ergo motus caeli est[e] in instanti. Et omnis motus est continuus; ergo etc.

9. Item, motus est continuus et motus est in instanti. Probatio: si ex aere debeat fieri ignis, in tota transmutatione est subiectum sub forma aeris quousque subito inducatur forma ignis. Sed, per Aristotelem, ex uno pugillo aeris fiunt decem ignis; tunc, cum subito inducatur forma ignis, tunc[f] subito rarefit aer in decuplum. Sed rarefactio est motus; subito[g] ergo sive in instanti fit motus. Et quod est in instanti est indivisibile; ergo etc.[h] <...>

10. Item, vera quantitas est in anima, quia eadem est forma in obiecto et medio et in anima, quia quod solum diversificatur per diversitatem recipientium et numeratur non[i] dividitur per essentiam; sed forma non[j] diversificatur in illis nisi per diversitatem recipientium; sed vera quantitas est in obiecto; ergo vera quantitas erit in medio et in anima. Et quod est[k] in anima est indivisibile; ergo etc.

[a] linea *scripsimus cum* TC; recta *add.* O
[b] applicaret] corpori circulari *add. s.l.* O
[c] incorporea *corr. s.l. ex* corporea O
[d] instanti] vel in minori *add.* O
[e] est *suppl. s.l.* O
[f] tunc *scrips. sed del.* O
[g] subito] fit *add. sed del.* O
[h] etc. *hic aliqua deesse videntur, fort. per hom. omissa; cf. n. 26 et TC n. 11*
[i] non] solum *add. sed del.* O
[j] non] sic *add. sed del.* O
[k] est *suppl. s.l.* O

touched <a plane surface> at two points, the line between those points would apply itself as a whole to that surface. But any part of that surface is straight; therefore, the line between those points would apply itself to what is straight. Therefore, if a sphere revolves and touches <a plane surface> at two points, it necessarily follows that a sphere is composed of straight lines; but this is false; therefore, the first premise is false. Therefore, a sphere touches <a plane surface> at one point. Therefore, if a sphere moves over a plane, it is apparent that it describes a line. Then I argue as follows: when something is in contact through contact with some things, the things that are in contact are parts of it; but a whole line is in contact through contact with the points contained in it; therefore, points are parts of a line. But in *On Generation* Aristotle says that, if a line is composed of points or of indivisibles, then the division of a line will come to a halt;[213] therefore, <the division of a continuum comes to a halt>.

8. Furthermore, an incorporeal substance exceeds a corporeal substance more than time exceeds an instant, because an instant is a part of time, and this is not so in the other case. Therefore, since a corporeal substance acts in time, then there is a much stronger reason to say that an incorporeal substance will act in an instant. But an incorporeal substance, for example an intelligence, moves the heavens; therefore, the motion of the heavens is in an instant. And every motion is continuous; therefore, <a continuum is indivisible>.

9. Furthermore, a motion is continuous, and a motion is in an instant. Proof: if fire must come to be from air, during the entire change the subject is under the form of air until the form of fire is induced instantaneously. But according to Aristotle, ten handfuls of fire come to be from one handful of air;[214] therefore, since the form of fire is induced instantaneously, air becomes ten times more rare instantaneously. But rarefaction is a motion; therefore, a motion occurs instantaneously or in an instant. And what exists in an instant is indivisible; therefore, <a motion is indivisible>. <...>

10. Furthermore, a true quantity exists in the soul, because the same form is in the object, in the medium, and in the soul, since what is diversified and numbered only in virtue of the diversity of the recipients is not divided in essence; but a form is diversified in those things only in virtue of the diversity of the recipients; but a true quantity is in the object; therefore, a true quantity will be in the medium and in the soul. And what exists in the soul is indivisible; therefore, <a true quantity is indivisible>.

[213] Although this claim about a line is not found in *De gen. et corr.*, it is likely that it originates from Aristotle's discussion there of the 'division through and through'. Cf. Arist. *De gen. et corr.* I.2, 316a15–317a17.

[214] Arist. *De gen. et corr.* II.6, 333a20–22.

11. Item, sicut substantia vere praedicatur de substantia universali et de substantia particulari, similiter et quantitas vere praedicabitur de quantitate universali et de quantitate particulari. Sed quantitas universalis non est nisi in anima, quia intellectus agens facit universalitatem in rebus, ut dicit Commentator; cum ergo faciat universalitatem in rebus, multo fortius faciet universalitatem in se; ergo etc. Et, ut prius, quod est in intellectu est indivisibile.

12. Item, forma multipliciter consideratur. Vel prout essentia vel prout forma vel prout quidditas vel prout universale. Sed solum est universale prout intellectus comparat formam communem ad plura et prout dicit potentiam multiplicandi, non tamen actu multiplicatam. Ergo[a] non habet esse actuale in rebus extra, quia in illis[b] actu multiplicatur; ergo oportet quod habeat esse actuale in anima. Sed quod est in indivisibili est indivisibile; ergo.

13. Item, scientia est in anima; ergo illud de quo est scientia sive demonstratio est in anima. Sed demonstratio non fit de hac linea, sed de illa quae per hanc intelligitur; ergo linea de qua fit demonstratio est in intellectu. Et illud de quo est scientia est universale; ergo etc. Et quod est in indivisibili est indivisibile; quare illud universale sive illa forma universalis quantitatis, ut linea, non erit divisibile, sed indivisibile.

14. Item, vere individuum et naturale est illud quod recipitur in sensu, quia illud per quod agens naturale corrumpit patiens naturale est[c] vera res naturalis. Sed visus corrumpitur aliquando propter excellentiam luminis; ergo etc. Ergo, cum eadem sit proportio sensibilis ad sensum et intelligibilis ad intellectum, sequetur tunc quod vere commune et vere universale est in intellectu, et[d] ita ut prius.

15. Item, quod plus quam similitudo sola recipiatur in sensu, et ita quod res habeat ibi suum verum esse probatio: sicut est in speculo materiali, sic est in speculo spirituali; sed ita est in speculo materiali quod, si res cuius similitudo recipitur in speculo recedat, non amplius manet similitudo; ergo similiter erit

[a] Ergo *scrips. sed del.* O
[b] illis] non *add.* O
[c] est *corr. ex* ergo (*lectio incerta*) O
[d] et] universaliter *add. sed del.* O

11. Furthermore, just as substance is truly predicated of a universal substance and of a particular substance, similarly quantity too is truly predicated of a universal quantity and of a particular quantity. But a universal quantity only exists in the soul, because the agent intellect produces universality in things, as the Commentator says;[215] therefore, since the agent intellect produces universality in things, there is a much stronger reason to hold that it will produce universality in itself; therefore, <a universal quantity exists in the intellect>, and we conclude as before that what exists in the intellect is indivisible.

12. Furthermore, form is considered in many ways: either as an essence or as a form or as a quiddity or as a universal. But it is considered as a universal only in so far as the intellect takes a common form in relation to many things and in so far as a form expresses a potency to multiplication but without being multiplied in act. Therefore, form considered as a universal does not have actual being in the things outside the soul, because it is multiplied in act in those things; therefore, it is necessary for it to have actual being in the soul. But what exists in an indivisible is indivisible; therefore <form considered as a universal is indivisible>.

13. Furthermore, knowledge exists in the soul; therefore, that about which there is knowledge or demonstration exists in the soul. But there is no demonstration about this line, but <only> about the line understood through this line; therefore, the line about which there is a demonstration exists in the intellect. And that of which there is knowledge is universal; therefore, <the line that exists in the intellect is universal>. And what exists in an indivisible is indivisible; hence, that universal or that universal form of a quantity, such as a line, will not be divisible, but indivisible.

14. Furthermore, a true and natural individual is that which is received in the senses, because a true and natural thing is that through which a natural agent produces the corruption of a natural patient. But in some cases sight is corrupted by an excess of light; therefore,<light is a true and natural individual received in the senses>. Thus, since the relations of something sensible to the senses and that of something intelligible to the intellect are the same, it follows that what is truly common and truly universal exists in the intellect, and so we get the same conclusion as before.[216]

15. Furthermore, here is a proof that what is received in the senses is something more than a mere likeness of a thing so that a thing itself has its true being in the senses: what happens in the case of a material mirror also happens in the case of a spiritual mirror; but if a thing the likeness of which is received in a material mirror recedes, its likeness no longer remains;

[215] Aver. *In De anima* I, t.c. 8, ed. Crawford, 12.25–26.
[216] See above, n. 13.

in speculo spirituali quod, si sola similitudo recipiatur, abeunte re non amplius manebit illa intentio et illa similitudo. Sed hoc est falsum. Dicit enim Aristoteles quod simulacra manent in organo post[a] absentiam rerum; ergo sic erit in intellectu, et[b] ita ut prius.

16. Solutio. Dicendum quod Aristoteles sufficienter arguit contra eos.

17. Ad primam rationem in oppositum dicendum quod tempus esse indivisibile est dupliciter, sicut possumus dicere quod[c] locus est indivisibilis dupliciter: uno modo[d] quia locus non habet partem et partem, et sic est oratio[e] impossibilis, quia, sicut quodlibet corpus locatum habet partes, similiter et locus; alio modo quia non habet partes distinctas, et sic possumus dicere quod aer et lux sunt in loco indivisibili. Eodem modo est[f] ex parte temporis, quia contingit intelligere 'in tempore indivisibili'[g], ita quod non sint partes distinctae respectu temporis, **[O 92va]** quia simul recipitur tota species in sensitiva et tota in omnibus virtutibus tam materialibus quam spiritualibus. Et ita partes non sunt distinctae respectu temporis.

18. Ad secundam rationem dicendum quod[h] in proficiendo est status[i], [j]in deficiendo vero non est status; et ideo est aliquod bonum summe et non est[k] summe malum. Et ideo, cum potentia passiva sit in deficiendo, potest esse infinita; potentia activa est in proficiendo; et ideo potentia activa non est infinita.

19. Ad tertiam rationem dicendum quod ratio[l] uno modo procedit, alio modo non, quia, si sumatur sic, scilicet quod reperiuntur semper in eodem continuo[m], si sic sumatur quod, sicut omne dimidium dicitur duplum, quod similiter omne duplum est dimidium, sic[n] tenet proportio. Sed hoc modo non[o] sumitur in arguendo. Et sic procedit ratio[p].

20. Ad quartam rationem dicendum sicut ad secundam.

[a] post *suppl. s.l.* O
[b] et] universaliter *add. sed del.* (*lectio incerta*) O
[c] quod *suppl. s.l.* O
[d] modo] aut *add. sed del.* O
[e] oratio *suppl. s.l.* O
[f] est *suppl. s.l.* O
[g] indivisibili *scripsimus*; insensibili O
[h] quod] non *add. sed del.* O
[i] status] non autem in deficiendo *add. s.l.* O
[j] in … status *scripsimus; post* malum *transp. sed del.* O
[k] est *suppl. s.l.* O
[l] ratio *scripsimus*; li sic O
[m] continuo] tenet similitudo *add. mg.* O
[n] sic] non *add. s.l.* O
[o] non *scrips. sed del.* O
[p] ratio] ex falso assumpto *add. s.l.* O

therefore, the same will happen in a spiritual mirror, so that, if only a likeness of a thing is received <in it>, when that thing is absent, its intention and its likeness will no longer remain. But this is false, because Aristotle says that sense images remain in the organ after the absence of the things <sensed>;[217] therefore, this will also happen in the intellect, and then we get the same conclusion as before.[218]

16. Solution. It must be said that Aristotle argues successfully against these people.[219]

17. To the first argument to the contrary[220] it must be said that the claim that time is indivisible can be understood in two ways, just as the claim that a place is indivisible can be understood in two ways: (i) in one way in the sense that a place does not have <one> part and <another distinct> part, and in this way the claim is impossible, because in the same way that every located body has parts, so does a place; (ii) in another way in the sense that a place does not have distinct parts, and in this way we can say that air and light are in an indivisible place. The same holds in the case of time, because it is possible to understand the locution 'in an indivisible time' in the sense that there are no distinct parts with respect to time, because the whole species is received simultaneously in the sensitive soul and in all the soul's powers, both material and spiritual. And thus the parts of a species are not distinct with respect to time.

18. To the second argument[221] it must be said that a process of improvement comes to a halt, but a process of deterioration does not come to a halt; and this is why there is something maximally good but there is nothing maximally bad. Therefore, since a passive power involves a deterioration, it can be infinite; but an active power involves an improvement, and this is why an active power is not infinite.

19. To the third argument[222] it must be said that the argument succeeds in one way but not in another, because the relation holds if it is assumed that a half and a double are always found in the same continuum, and if it is assumed that just as every half is said to be a double, so every double is similarly <said to be> a half. But this is not what the argument assumes. And it is in this <second> way that the argument succeeds.

20. To the fourth argument[223] it must be replied as it was replied to the second argument.[224]

[217] Arist. *De insomniis* 2, 460b2–3.
[218] See above, n. 13.
[219] Arist. *Phys.* I.2, 185b9–11.
[220] See above, n. 3.
[221] See above, n. 4.
[222] See above, n. 5.
[223] See above, n. 6.
[224] See above, n. 18.

21. Ad quintam sic, quod, etsi[a] forte punctus[b] per signationem non facit lineam, tamen punctus fluens potest facere lineam, et indivisibile sic se habens facit divisibile. Sic instans fluens facit tempus[c]. Et sic non procedit ratio.

22. Ad sextam sic, quod excedere est dupliciter, aut secundum naturam in essendo, et sic verum est quod plus excedit[d], aut secundum proportionem potentiae in operando, et sic falsum est.

23. [e]Ad septimam talem[f]:'caelum movetur', hoc praedicatum inest subiecto in instanti, quia omne quod est verum est verum in instanti. Et, si hoc, motus est in instanti. Quod vero praedicatum non insit subiecto in tempore probatio, quia, si sit vera in tempore ita quod non in hoc instanti, ergo in hoc instanti non est vera, sed statim post hoc erit vera et, si hoc, incipit esse vera. Sed hoc falsum est, cum prius fuit vera et nunc est vera et, si hoc, est vera in instanti. Et si hoc, res praedicati in instanti inest rei subiectae; aliter enim enuntiaret rem esse[g] aliter quam est. Et ita praedicatum[h] est indivisibile[i]. Ergo motus est indivisibilis. Et motus est continuus; ergo etc.

24. Sed dicendum quod haec propositio 'caelum movetur' est vera in instanti et non est necesse quod res praedicati insit rei subiectae in instanti, si sit praedicatum successivum, sed oportet quod anima, percipiens motum inesse caelo in tempore, in quolibet instanti motus sui componat, ita quod instans respondeat compositioni et non rei. Unde ad hoc quod propositio sit vera in instanti aut necesse est quod res praedicati in instanti insit rei subiectae, sicut haec[j]: 'homo est animal', vel quod res praedicati insit ante compositionem et post, et hoc si praedicatum sit successivum. Unde in principio motus, ut in ipsa mutatione a qua incipit motus, non potest dici quoniam movetur, nec similiter in ultimo; sed in omni parte sui motus, ita quod motus praecedat et motus subsequatur, est verum dicere quoniam movetur.

[a] etsi *corr. s.l. ex* si O
[b] punctus *suppl. s.l.* O
[c] tempus *corr. s.l. ex* lineam O
[d] excedit] etc. *add. s.l.* O
[e] Ad ... etc. *septima ratio ad oppositum hic introducitur*
[f] talem *corr. s.l. in* sic O
[g] esse *scripsimus*; in se O
[h] praedicatum *lectio incerta* O
[i] indivisibile] similiter et compositio *add. s.l.* O
[j] haec *scripsimus*; hic O

21. To the fifth argument[225] it must be replied that, although it is perhaps the case that a point does not produce a line by a process of designation, a flowing point, however, can produce a line, and an indivisible of this kind produces something divisible. It is in this way that a flowing instant produces time, and in this way the argument does not succeed.

22. To the sixth argument[226] it must be said that to exceed can be taken in two ways: either (i) in accordance with a nature in its existence, and in this way it is true that <an incorporeal substance> exceeds <a corporeal substance> more <than time exceeds an instant>, or (ii) in accordance with a comparison of powers in their operation, and in this way the claim is false.

23. To the seventh argument, which is the following: in the proposition 'the heavens are moved' the predicate inheres in the subject in an instant, because whatever is true is true at an instant. And if so, a motion exists in an instant. Proof that the predicate does not inhere in the subject in time: if the proposition is true in time in such a way that it is not true at this instant, therefore, it is not true at this instant, but immediately after this instant it will be true, and if so, it begins to be true. But this is false, since it was true before and it is true now, and if so, it is true at an instant. If so, the predicate-thing inheres in the subject-thing in an instant; for otherwise the proposition would express the claim that a thing is other than it is. Thus the predicate is indivisible. Therefore, a motion is indivisible. And a motion is continuous; therefore, <a continuum is indivisible>.

24. But it must be said that the proposition 'the heavens are moved' is true at an instant, but it is not necessary for the predicate-thing to inhere in the subject-thing in an instant, if the predicate is successive. However, it is necessary that while the soul is perceiving that a motion inheres in the heavens in time, it should produce a composition at every instant of their motion, in such a way that the instant corresponds to the composition and not to the thing. Thus, in order for a proposition to be true at an instant, either it is necessary for the predicate-thing to inhere in the subject-thing in an instant, for example in the proposition 'a man is an animal', or it is necessary for the predicate-thing to inhere <in the subject-thing> before and after the composition, and this is when the predicate is successive. Thus, at the beginning of a motion, that is, when that sudden change from which a motion begins occurs, it is not true to say that something is moved and similarly at the end of a motion; but it is true to say that something is moved in every part of its motion such that there is a motion before and after that part.

[225] See above, n. 7.

[226] See above, n. 8.

25. Ad aliud dicendum quod ex uno pugillo aeris fiunt decem ignis; sed hoc intelligendum est: [a]ex unico et[b] naturali sua[c] densitate [d]fiunt decem ignis. Unde, si ex aere fiat ignis, semper, dum est in rarefactione, est sub forma aeris. Et tunc demum, quando materia fit necessitas, totus aer est rarefactus, et tunc subito fit formae ignis immissio[e].

26. Ad aliud dicendum quod natura non permitteret superficies per aequidistantiam disiungi vel quod subito possit fieri multiplicatio.

27. Ad aliud dicendum quod non solum diversificatur per naturam recipientium, sed per lucem[f] agentem fit de re non res sive intentio; et ita in medio sive in anima non meretur nomen essentiae.

28. Ad aliud dicendum quod quadruplex est esse formae, quia prius per naturam est forma aliquid in se quam uniatur materiae, et tunc meretur nomen essentiae. Et posterius per naturam informat materiam, et tunc dicitur forma ab informando et est altera pars compositi. Et tertio per naturam perficit individuum, et tunc meretur nomen quidditatis. Et quarto per naturam perficit unum individuum, non ita tamen quin sit in potentia ad aliud perficiendum, et sic meretur nomen universalis quod est res.

29. Sed universale triplex habet esse: unum, quod est actuale. Et hoc est duplex: uno modo, si habeat actualem existentiam, et sic consideratur universale secundum quod est in individuis actualiter; et sic est unum de multis et in multis. Alio modo, si habeat esse actuale ita quod actu immutat[g] intellectum, et haec est eius ultima operatio, et tale esse habet in intellectu, quia universale est proprium obiectum intellectus. Unde simul stant quod universale habeat actuale esse in rebus extra et etiam quod habeat actuale esse in intellectu. Secundum[h] est esse habituale, et istud esse est illud esse quod debetur rei[i] in sua natura, cuiusmodi sumitur cum dicitur quaestio 'an sit', quia illud esse est

[a] ex unico et *scrips. sed del.* O
[b] et] in *add. s.l.* O
[c] sua] aeris *add. s.l.* O
[d] fiunt decem ignis *scrips. sed del.* O
[e] immissio] mutatio O (mutatio TC) *sed corr. s.l.* O
[f] lucem *corr. s.l. in* intellectum O
[g] immutat *corr. s.l. ex* mutat O
[h] Secundum] esse *add. s.l.* O
[i] rei *lectio incerta sed corr. s.l. in* ei O

25. To the other argument[227] it must be said that the claim that ten handfuls of fire come to be from one handful of air must be understood in the sense that ten handfuls of fire come to be from one handful <of air> in its natural density. Thus, if fire comes to be from air, as long as air is being rarefied, it is always under the form of air. And then finally, when matter becomes a necessity, the whole air is rarefied and then the introduction of the form of fire occurs instantaneously.

26. To the other argument[228] it must be said that nature would not allow a surface to be separated at the same distance or a multiplication to be able to occur instantaneously.

27. To the other argument[229] it must be said that a form is not only diversified through the nature of its recipients, but it also changes from being a thing to being a non-thing or intention through the agent light; thus, <when found> in the medium or in the soul a form does not deserve the name of essence.

28. To the other argument[230] it must be said that a form has four kinds of being, because (i) a form is something in itself prior by nature to being joined to matter, and at that stage it deserves the name of essence, and (ii) at a stage posterior by nature it in-forms matter, and at that stage a form is so-called from <its action of> in-forming and it is the other part of a composite. (iii) At a third stage by nature a form gives perfection to an individual, and at that stage it deserves the name of quiddity. (iv) At a fourth stage by nature a form gives perfection to one individual, but in such a way that it is in potency to give perfection to another individual, and in this way it deserves the name of the universal that is a thing.

29. But a universal has three kinds of being: (i) one is actual being, which is twofold: (ia) in one way, when a universal has actual existence, and in this way a universal is considered in so far as it actually exists in individuals, and it is one related to many and in many. (ib) In another way, when a universal has actual being in such a way that it actually modifies the intellect, and this is the last operation of a universal, and a universal has this kind of being in the intellect, because a universal is the proper object of the intellect. Thus, <the claims> that a universal has actual being in the things outside the soul and that it has actual being in the intellect are compatible. (ii) The second kind of being that a universal has is habitual being, and this kind of being is that owing to a thing in its nature, the being to which the question 'whether it exists' applies, because this is the being which a thing has in its nature. (iii) The third kind of being of a universal is its being in relation to a species, that is,

[227] See above, n. 9.
[228] This argument is missing in O; see Appendix, Book I, q. 35, n. 11.
[229] See above, n. 10.
[230] See above, n. 11.

esse quod habet res in sua natura. Tertium esse est suum esse in comparatione ad speciem, secundum quod relativa habent esse. Cum autem ratio istam naturam communem reperiri perceperit in multis individuis, tunc intellectus abstrahit similitudinem talis naturae communis; et tunc meretur nomen intentionis sive speciei. Et hoc est suum proprium esse in esse actuali.

30. Et per hoc respondendum est ad rationem quod Commentator intelligit sic, quod intellectus agens est causa ultimae actionis in quam potest universale, quia, etsi non esset intellectus, esset tamen communitas in rebus secundum actum existendi primo modo sumendo esse actuale **[O 92vb]** et secundum potentiam multiplicandi. Verbi gratia, si non esset sensus, non minus esset sensibile extra, sed tamen non esset in ultima actualitate sensibilis.

31. Ad aliud patet per distinctionem primam.

32. Ad aliud dicendum quod Aristoteles non intendit quod fiat demonstratio de linea quae est in anima, sed in veritate lineae protractae, quae non potest[a] omnino secundum rectum[b] protrahi, subest aliqua linea quae in veritate recta est, et haec est illa quae intelligitur per lineam descriptam. Unde, cum dicitur quod materia mathematicorum est intelligibilis etc.[c], huiusmodi propositiones sunt sic intelligendae quod mathematica secundum rem et esse sunt in materia naturali, sed considerantur[d] prout praecedunt[e] materiam naturalem sive transmutationem.

33. Ad aliud dicendum quod corrumpere aliquid est dupliciter: aut ratione qualitatum activarum et passivarum, et quod corrumpit sic aliud est vere individuum et vere res et natura in eodem, et sic non corrumpit excellens sensibile sensum, aut ratione improportionis, et sic corrumpit sensibile sensum. Et hoc quia organum sensus sub quadam proportione media se habet respectu sensibilium; ideo potest excellens sensibile corrumpere illam proportionem [f]oculo manente in debita quantitate.

34. Ad ultimum dicendum quod simulacra manent in organo et non in speculo, sed non quia potius recipiantur in organo quam in speculo, sed quia organum est speculum animatum, et ideo nobilioris compositionis. Et haec est una causa quare magis manent[g]. Et etiam in organo est anima praesens, quae potest convertere se super speciem post absentiam sensibilis, in speculo vere non; et ideo in organo manent simulacra, in speculo non.

[a] potest *scripsimus*; possunt O
[b] rectum *corr. ex* rationem (*lectio incerta*) O
[c] etc. *scripsimus*; et O
[d] considerantur] non *add. s.l.* O
[e] praecedunt *scripsimus cum* T; procedunt in O; excedunt C
[f] oculo ... quantitate *scripsimus cum* TC (qualitate *pro* quantitate C); conclusio manente in debita quantitate *scrips. sed del.* O
[g] manent *scripsimus*; manet O

the kind of being that belongs to relatives. Once reason has perceived that a particular common nature is found in many individuals, then the intellect abstracts a likeness of this common nature, and then a universal deserves the name of intention or of species. And this is the proper being of a universal in its actual being.

30. This provides the response to the argument that the Commentator means that the agent intellect is the cause of the last action which a universal can perform, because, even if there were no intellect, there would still be community in things in accordance with the act of existing, taking actual being in the first sense, and in accordance with the power of multiplication. For example, if there were no senses, there would nevertheless be something sensible in extra-mental reality, yet it would not possess the last actuality of a sensible thing.

31. The reply to the other argument[231] is clear from the first distinction.[232]

32. To the other argument[233] it must be said that Aristotle does not mean that a demonstration is about a line existing in the soul, but it is true that there is a line that is truly straight and which underlies the line which is prolonged and which cannot be fully prolonged in a rectilinear way, and this is the line understood through the line described. Thus, when it is said that the matter of mathematical things is intelligible and the like, similar propositions must be understood in the sense that mathematical things in reality and being exist in natural matter, but they are considered as preceding natural matter or change.

33. To the other argument[234] it must be said that to cause the corruption of something can happen in two ways: either (i) in virtue of active and passive qualities, and what causes the corruption of another thing in this way is truly an individual and truly a thing and a nature in that thing; the excess of something sensible does not cause the corruption of the senses in this way; or (ii) because of a lack of proportion, and in this way something sensible causes the corruption of the senses. And the reason for this is that the sense organ has an intermediate proportion to sensible things; therefore, the excess of a sensible can cause the corruption of that proportion, even when an eye remains in its appropriate quantity.

34. To the last argument[235] it must be said that sense images remain in an organ and not in a mirror, this is not, however, because they are more powerfully received in an organ than in a mirror, but because an organ is an animate mirror, and therefore of a nobler composition. This is one reason why sense images persist for a longer time <in an organ>. Another is that the soul, which can turn itself towards a species after the absence of a sensible thing, is present in an organ, but it is not truly present in a mirror; therefore, sense images remain in an organ and not in a mirror.

[231] See above, n. 12.
[232] See above, nn. 28–29.
[233] See above, n. 13.
[234] See above, n. 14.
[235] See above, n. 15.

<Quaestio 36>
Quaeritur utrum pars et totum sint idem.

1. Et videtur quod sic: circumscripta quantitate huius medietatis et illius et aliis accidentibus eorundem, manet eadem substantia materiae primae indivisibilis, una et simplex in illis; ergo una pars est id idem substantialiter quod alia pars et solum differunt penes accidentia.

2. Item, definitio animalis convenit cuilibet[a] parti animalis; ergo quaelibet pars[b] est animal.

3. Sed dicitur quod illa definitio convenit eis sic ut ista pars est animata anima communi et non propria sive anima totius et non propria; et propterea dicitur quod definitio [c]non inest eis immediate, sed mediante toto.

4. Sed contra: anima[d] absolute facit animatum et anima sensitiva facit animal et quaelibet pars dicitur animata ab anima; ergo similiter quaelibet[e] pars[f] erit animal[g] anima sensitiva[h].

5. Item, illud est ens magis quod magis est perfectum et quod magis de forma habet; ens et unum convertuntur; ergo quod simpliciter magis habet de forma et magis est[i] perfectum est magis unum. Sed unitas rei consistit in unitate et convenientia partium cum toto; ergo, ubi maior[j] est perfectio, ibi maior est unitas partium [k]cum toto. Sed maior est perfectio in animali quam in terra; sed tanta est perfectio in terra quod quaelibet pars terrae est terra; ergo multo fortius tanta erit identitas in animali quod quaelibet pars animalis erit animal.[l] <...>

6. Item, quaelibet pars substantiae est substantia; ergo quaelibet pars animalis est animal, quia eadem est proportio inter substantiam et suas partes quae est inter animal et suas partes.

7. Item, sequitur convertibiliter: ‘pars animalis, ergo animalis pars’ in se et in respectu. Verbi gratia: ‘animalis pars est minor toto; ergo pars animalis est minor toto’. Ergo eadem habent supposita. Sed supposita huius, ‘animalis pars’, sunt ista, hoc animal et hoc animal. Quod patet in sillogizando sic: ‘cuiuslibet animalis pars est minor toto; sed hoc est animal; ergo huius

[a] cuilibet *corr. s.l. ex* alicui O
[b] pars] animalis *add. s.l.* O
[c] non inest *suppl. mg.* O
[d] anima *suppl. mg.* O
[e] quaelibet *suppl. s.l.* O
[f] pars] sensitiva *add. sed del.* O
[g] animal] ab *add. s.l.* O
[h] sensitiva] totius *add. sed del.* O
[i] est *suppl. s.l.* O
[j] maior *corr. ex* magis O
[k] cum toto *suppl. s.l.* O
[l] animal *hic aliqua deesse videntur, fort. per hom. omissa; cf. n. 13 et TC n. 6*

<Question 36>
The question is whether a part and a whole are the same.

1. Arguments for the affirmative: once quantity and all the other accidents of this half and that half are removed, the same substance of prime matter remains in them as indivisible, one, and simple; therefore, one part is essentially the same as another part, and they only differ in their accidents.

2. Furthermore, the definition of animal applies to every part of an animal; therefore, every part is an animal.

3. But it is said that that definition applies to the parts in such a way that this part is animate in virtue of a common soul and not of a proper soul or in virtue of the soul of the whole and not of a proper soul; this is why it is said that the definition does not inhere in the parts immediately but through the whole.

4. Against this: the soul makes something animate absolutely and the sensitive soul makes something an animal, and every part is said to be animate in virtue of the soul; therefore, every part will similarly be an animal in virtue of the sensitive soul.

5. Furthermore, a thing that is more perfect and has more of a form is more of a being; but being and one are convertible; therefore, what has more of a form absolutely speaking and is more perfect is more one. But the unity of a thing consists in the unity and the agreement of its parts with the whole; therefore, where there is greater perfection there is greater unity of the parts with the whole. But there is a greater perfection in an animal than in the earth; but there is so much perfection in the earth that every part of earth is earth; therefore, there is a much stronger reason to hold that there is so much identity in an animal that every part of an animal will be an animal. <...>

6. Furthermore, every part of a substance is a substance; therefore, every part of an animal is an animal, because there is the same relation between a substance and its parts as between an animal and its parts.

7. Furthermore, the consequence 'part of an animal; therefore, an animal's part' is convertible both in itself and relatively. For example: 'an animal's part is less than the whole; therefore, the part of an animal is less than the whole'. Therefore, the antecedent and the consequent have the same supposits. But

animalis etc.'. Ergo similiter hoc animal et illud animal erunt supposita huius, 'pars animalis'. Contingit ergo praedicare unum de altero, sic dicendo: 'animal est pars animalis' sive 'pars animalis est animal'.

8. Item, nulla violentia producit speciem in esse nec rem naturalem; sed per decisionem, quae est violentia, inducitur anima sensitiva in parte animalis, ut patet in decisione anulosi[a]; ergo illa sensitiva ante decisionem infuit parti; ergo ante decisionem fuerunt partes in toto actu animalia. Et ita pars animalis est animal.

9. Ad quaestionem dicendum secundum magistros nostros quod pars nec est eadem toti nec diversa[b].

10. Ad primam rationem in oppositum dicendum quod, si circumscribamus omnem formam a parte et toto, cedunt in unum secundum materiam, quia materia unius non erit actu distinctum[c] a materia alterius. Materia tamen[d], cum confunditur in unum, est in potentia ut secundum partes distinguatur et extendatur, ita quod una pars materiae fiat una pars continui. Et sic patet quod non solum est diversitas penes accidentalia.

11. Ad secundam rationem dicendum[e] sicut dictum est.

12. Ad illud in oppositum dicendum quod maior perfectio facit ad minorem denominationem et minor perfectio ad maiorem denominationem. Et hoc dico in partibus respectu totius, sicut patet, quia substantia est minus perfecta quam animal, et ideo possum dicere quod partes substantiae sunt substantiae. Et similiter de corpore, minus tamen proprie quam de substantia. Possum enim dicere quod quaelibet pars corporis est corpus. Tertia autem forma est magis perfecta, ut animal; ideo nulla sua pars, nec per se nec in compositione, potest dici animal, loquendo de animali perfecto. Loquendo tamen de corpore imperfecto, ut de corpore[f] anulosi, quando pars dividitur, possumus dicere quod est animal, licet incomplete. In ultima forma, quae maxime est completa, ut in homine, possum dicere quod nulla sua pars, nec

[a] anulosi *corr. ex* anguli si O
[b] diversa] a toto *add. s.l.* O
[c] distinctum *lectio incerta* O
[d] tamen *suppl. s.l.* O
[e] dicendum *suppl. s.l.* O
[f] corpore] animalis *add. s.l.* O

the supposits of 'animal's part' are this animal and that animal, which is made clear by the following syllogism: 'every animal's part is less than the whole; but this is an animal; therefore, this animal's part <is less than the whole>'. Therefore, this animal and that animal will similarly be the supposits of 'part of an animal'; therefore, it is possible to predicate one of the other by saying that an animal is a part of an animal or a part of an animal is an animal.

8. Furthermore, no violence brings a species or a natural thing into being; but the sensitive soul is induced in a part of an animal by separation, which is a kind of violence, as is clear in the separation of an annular animal; therefore, that sensitive soul inhered in a part before being separated; therefore, the parts in the whole were animals in act before being separated, and so a part of an animal is an animal.

9. In accordance with our masters it must be replied to this question that a part is neither the same as nor different from a whole.[236]

10. To the first argument to the contrary[237] it must be said that, if we remove every form from a part and from a whole, they give place to one thing with respect to matter, because the matter of the part will not be distinct in act from the matter of the whole. But when matter is merged into one, it is in potency to being distinguished and extended in accordance with <its> parts, in such a way that one part of matter will become one part of a continuum. This makes it clear that diversity does not only occur with respect to accidental determinations.

11. To the second argument[238] it must be replied as has already been replied.[239]

12. To the argument to the contrary[240] it must be said that a greater perfection contributes to a minor denomination and a minor perfection to a greater denomination. I say this about <the naming of> parts in relation to the whole, as is clear. Thus, because substance is less perfect than animal, it follows that I can say that the parts of a substance are substances. The same applies to the case of a body, but less properly than in the case of a substance; for I can say that every part of a body is a body. There is a third kind of form, however, which is more perfect, such as that of an animal; therefore, none of the parts of an animal can be said to be an animal, either per se or in composition, when one speaks of a perfect animal. But if we speak of an imperfect body, for example of the body of an annular animal, then, when a part is divided, we can say that that part is an animal, although incompletely. In the

[236] See, for example, Roger Bacon, *Supra libros octo Physicorum*, 24.29–25.2; Adam of Buckfield, *In Phys.*, MS Venezia, Bibl. Naz. Marc., Lat. VI, 1, fol. 56rb: "In veritate tamen pars et totum nec sunt simpliciter idem nec simpliciter diversa; non enim sunt simpliciter idem, cum non possit dici hoc esse hoc, nec simpliciter diversa, cum sint sub uno actu existendi.".

[237] See above, n. 1.

[238] See above, n. 2.

[239] See above, n. 3.

[240] See above, n. 4.

complete nec incomplete, potest dici homo. In animali tamen, quia talis forma non est ita completa, ideo ibi potest dici aliquantulum et in quibusdam animalibus quod partes eorum sunt animal, et hoc incomplete. Unde non tenet, quia animatum simpliciter minus perficit quam animatum anima sensitiva.

13. Ista responsione dicendum est ad omnes rationes subsequentes et ad rationem quae fit a proportione.

14. Ad rationem logicalem dicendum quod, licet inter se convertantur, non tamen convertuntur quando additur distributio, quia non facit idem. In una enim subicitur li 'pars animalis' et in alia li 'animal' solum, et ita non habent eadem supposita.

15. Et **[O 93ra]** si quaeratur quare non idem subicitur utrobique, dicendum quod illud quod exigit exigit ratione qualitatis, illud quod exigitur exigitur[a] ratione substantiae. Illud autem quod exigit ratione qualitatis magis est[b] dependens quoad hoc quam illud quod exigitur ratione substantiae. Et ideo li 'pars', cum sua qualitas specificatur per li 'animalis', nullo modo potest distribui sine illo cum dicitur sic: 'quaelibet pars animalis est minor suo toto'. Li 'animalis', quia non exigitur ratione qualitatis, sed solum ratione substantiae, ideo non est ita dependens; et ideo potest distribui sine illo a quo exigitur, cum sic dicitur: 'cuiuslibet animalis pars est minor suo toto'. Et omni eodem modo est in hoc sophismate: 'cuiuslibet hominis asinus currit'.

16. Ad ultimam rationem dicendum quod violentia de se non producit speciem, potest tamen esse excitans, sicut violentia percussionis ferri[c] super lapidem excitat ignem, qui fuit in potentia in lapide. Unde, quia in qualibet parte animalis anulosi est anima in potentia, violentia potest conferre ad actualem productionem, non sicut perfectum agens, sed sicut excitans tantum.

[a] exigitur *suppl. s.l.* O
[b] est *suppl. s.l.* O
[c] ferri *corr. s.l. ex* ferei (*lectio incerta*) O

last kind of form, which is maximally complete, namely <that found> in man, none of its parts can be said to be a man, either completely or incompletely. But since the form of an animal is not so complete, we can then say to some extent and about some animals that their parts are an animal, and this incompletely. Thus, the argument does not hold because being animate absolutely speaking gives less perfection than being animate through a sensitive soul.

13. This is the reply to be given to all the arguments that followed[241] and to the argument based on the identity of the relations <between a substance and its parts and between an animal and its parts>.[242]

14. To the logical argument[243] it must be said that, although <'part of an animal' and 'an animal's part'> are convertible, they are not, however, convertible when distribution is added, because the distribution does not produce the same proposition; for in one proposition the subject is 'part of an animal' and in the other it is only 'animal', and thus the two expressions do not have the same supposits.

15. And if it is asked why it is the case that the subject is not the same in both propositions, it must be said that what requires <a subject> requires it because of a quality, <and> <a subject> that is required, is required because of a substance, but what requires <a subject> because of a quality is more dependent in this respect than <the subject that> is required because of a substance. Therefore, 'part', because its quality is specified by 'of an animal', cannot be distributed without that qualifying phrase when the <universal> proposition 'every part of an animal is less than its whole' is expressed. The expression 'animal's', given that <its subject> is not required because of a quality, but because of a substance, is not dependent in that way. Therefore, <the subject> can be distributed without <the qualifying phrase> that requires it, as happens when the <universal> proposition 'every animal's part is less than its whole' is expressed. And the case of the sophism 'every man's ass runs' is absolutely the same.

16. To the last argument[244] it must be said that violence by itself does not produce a species, but it can be a stimulating factor, as when the violence of the striking of iron on a stone stimulates fire, which was in potency in that stone. Thus, since the soul exists in potency in every part of an annular animal, violence can contribute to its actual production, not as a perfect agent, but only as something that stimulates.

[241] See above, n. 5. However, the plural 'arguments' seems to indicate that an argument is missing in version O; see Appendix, Book I, q. 36, n. 6.

[242] See above, n. 6.

[243] See above, n. 7.

[244] See above, n. 8.

17. Ratio autem quare[a] in animalibus anulosi corporis plus quam in aliis post divisionem est in qualibet parte anima et quaelibet pars vivit et sentit est huiusmodi, secundum quosdam: animalia anulosi corporis non habent cor, sed habent aliquid proportionale cordi; et illud proportionale cordi est oblongae figurae et in[b] divisione salvatur proportionale cordi[c] in debita figura. Et in propria figura[d] in ipso viget anima; unde, si dividatur anguilla, manet semper illud proportionale cordi sub eadem figura, et ex fecunditate suae formae potest anima multiplicare se et una anima potest fieri plures animae. Nec impedit[e] quod est cordi proportionale, quia semper post divisionem est eiusdem figurae. In aliis autem animalibus est cor circularis figurae, et quacumque parte divisa a figura circulari perit figura; et ita in his non potest esse aliqua lesio cordis praeter[f] corruptionem animae.

18. Contra: lupus aquaticus habet cor circularis figurae et si removeatur[g] cor, per magnum tempus movetur corpus et etiam cor vivit per magnum tempus et movetur et, si pungatur, sentit acutum. Unde super septimum huius disputat Avicenna contra Averroem[h], qui dixit quod principium vitae fuit in capite et non in corde. Et secundum Commentatorem, ratio sua fuit quia, remoto corde quorundam animalium, adhuc moventur et vivunt, quod non contingeret si in corde esset principium vitae. Et ita principium vitae est in capite. Et dicit Avicenna contra ipsum quod ipse vidit arietem post decisionem capitis movere se, quod non contingeret si esset in capite.

19. Alii dicunt sic quod, cum quaedam animalia sint per putrefactionem generata et quaedam per propagationem, et in his quae generantur per propagationem ordinavit natura ut anima ex fecunditate suae formae posset multiplicare se per decisionem seminis eius in quo est, ideo in his quae per putrefactionem generantur ordinavit natura ut anima posset ex fecunditate suae formae se ipsam multiplicare per[i] divisionem eius in quo est.

20. Contra: si hoc[j] esset verum, tunc lupus aquaticus, cum generetur per propagationem, non posset moveri[k] in partibus, cum partes essent divisae a suo toto. Et iterum, si hoc esset verum, tunc illa proprietas conveniret omni[l] eo quod generatur per putrefactionem, quod scimus esse falsum.

[a] quare *scripsimus cum* TC; qualiter O
[b] in *suppl. s.l.* O
[c] cordi] oblongae figurae *add. sed del.* O
[d] figura] et *add. s.l.* O
[e] impedit] eius quod est *add. sed del.* O
[f] praeter *scripsimus cum* T; post O; propter C
[g] removeatur *scripsimus cum* TC; moveatur O
[h] Averroem] Averroys O
[i] per *suppl. s.l.* O
[j] hoc *suppl. s.l.* O
[k] moveri *scripsimus*; movere O
[l] omni *scripsimus cum* TC; cum O

17. According to some people, the reason why it is in animals with an annular body more than in others that the soul remains in every part after their division and that every part is alive and senses is the following: animals with an annular body do not have a heart, but have something analogous to a heart; the thing analogous to the heart has an elongated shape, and when the animals are divided it retains its appropriate shape, and when it has its appropriate shape the soul has power over it. Thus, if an eel is divided, the thing analogous to the heart always retains the same shape, and because of the fertility of its form the soul can multiply itself and one soul can become many souls. Nor does the thing analogous to the heart hinder this process, because it always retains the same shape after a division. But in other animals the heart has a circular shape, and when any part is divided from a circular shape, that shape perishes, so that in these animals their heart cannot be damaged without their soul being corrupted.

18. Against this: the heart of a wolf fish has a circular shape and if its heart is removed, its body goes on moving for a long time and its heart also remains alive and goes on moving for a long time, and if its heart is pierced, it perceives the piercing. Thus, in his comment on *Physics* VII Avicenna argues against Averroes,[245] who said that the principle of life was in the head and not in the heart. According to the Commentator, Avicenna's argument was that, when the heart of some animals is removed, these animals go on moving and remain alive, which would not happen if the principle of life were in the heart; hence, the principle of life is in the head. Avicenna says against Averroes that he saw a ram that moved after its head had been cut off, which would not happen if the principle of life were in the head.

19. Other people say that, given that some animals are generated by putrefaction and others by reproduction, and that in those generated by reproduction nature ordered that their soul in virtue of the fertility of its form could multiply itself through the separation of the semen of the animal in which it exists, it follows that, in the animals generated by putrefaction nature ordered that their soul in virtue of the fertility of its form could multiply itself through the division of the animal in which it exists.

20. Against this: if this were true, then since a wolf fish is generated by reproduction, its parts could not move, when they had been divided from their whole. Also, if this were true, then that property would also belong to everything generated by putrefaction, which we know to be false.

[245] The correct reference is to Averroes who argues against Galen in Aver. *In Phys*. VII, t.c. 4, fol. 309ra–b.

21. Ideo dicendum, ut mihi videtur, quod duplex est causa huius: una a[a] parte animae, scilicet quod in huiusmodi animalibus magis naturalis est inhaerentia[b] cum corpore; et a parte corporis est dicta convenientia partium cum toto et convenientia partium ad invicem, sicut videmus in plantis[c] quod, propter naturalem inhaerentiam animae cum corpore et convenientia partium cum toto et inter se, <cum[d]> dividatur ramus, si plantetur, iterum crescit.

1*. Quaesito de convenientia partium ad totum, quaeritur de convenientia partium inter se. Et primo quaeritur de identitate partium substantiae et secundo de convenientia partium quarum una est substantia, reliqua accidens. Circa primum quaeritur primo de partibus continuis, secundo de partibus contiguis [e]quarum una est corruptibilis, reliqua incorruptibilis, ut de convenientia inter ignem et concavum orbis lunae. Circa secundum, primo an substantia et accidens sint idem, secundo utrum species[f] praeexigat speciem[g] ad sui multiplicationem vel simul [[h]cum specie] multiplicetur species formae substantialis cum multiplicatione formae accidentalis.

<Quaestio 37>
Circa primum an omnes partes eiusdem continui sint eaedem in substantia.

1. Et quod sic videtur: forma dividit et distinguit; ergo forma numerat; sed partes in eodem continuo non habent formas diversas; ergo non numerantur, quia forma facit numerationem.

2. Item, quaecumque numerantur sunt diversa supposita; sed partes in continuo non sunt diversa supposita, sed unum, quia totum est suppositum; unde et partes non numerantur.

3. Item, quaecumque numerantur numerantur sub aliquo communi; sed partes in continuo non numerantur in aliquo communi, quia non in substantia nec in corpore; ergo etc.

[a] a *suppl. s.l.* O
[b] inhaerentia] animae *add. s.l.* O
[c] plantis *scripsimus cum* TC; *lectio incerta* O
[d] cum *supplevimus*
[e] quarum … incorruptibilis *transp. post* multiplicationem (rum *pro* quarum) *sed corr. mg.* O (quarum una est corruptibilis *transp. post* multiplicationem *sed corr. mg.* [et alia *pro* reliqua] T; quarum una est corruptibilis et alia incorruptibilis *transp. post* multiplicationem C)
[f] species] accidentis *add. s.l.* O
[g] speciem] substantiae *add. s.l.* O
[h] cum specie *delevimus*

21. Therefore, in my view, it must be said that there are two causes of the fact < that it is in animals with an annular body more than in others that the soul remains in every part after their division and that every part is alive and senses>: one on the side of the soul, that is, that in animals of this kind the inherence of the soul in the body is more natural; the other on the side of the body, that is, the aforesaid agreement of the parts with the whole and the agreement of one part with another. In the same way, we see in plants that, because of the natural inherence of the soul in the body and of the agreement of the parts with the whole and one with another, when a branch is divided, then if it is planted, it grows again.

1*. After the question about the agreement of the parts with the whole,[246] questions about the agreement of the parts one with another should now be raised. The questions about the identity of the parts of a substance come first, and secondly those about the agreement of parts when one of them is a substance and the other an accident. As to the first topic, the question about continuous parts comes first,[247] and secondly that about contiguous parts one of which is corruptible and the other incorruptible, for example, about the agreement between fire and the concavity of the lunar orb.[248] As to the second topic, the first question is whether a substance and an accident are the same,[249] and the second question is whether a species is a prerequisite for the multiplication of a species or whether the species of a substantial form is simultaneously multiplied with the multiplication of an accidental form.[250]

<Question 37>
As to the first topic it is asked whether the parts of the same continuum are the same in substance.

1. Arguments for the affirmative: a form divides and distinguishes; therefore, a form produces a number; but the parts of the same continuum do not have different forms; therefore, they are not numbered, because a form produces a number.

2. Furthermore, all the things that are numbered are different supposits; but the parts of a continuum are not different supposits, but one supposit, because the whole is the supposit; hence, the parts are not numbered.

3. Furthermore, all the things that are numbered are numbered under something common; but the parts of a continuum are not numbered under something common, because they are neither <numbered> under substance nor under body; therefore, <the parts of a continuum are not numbered>.

[246] That is, q. 36.
[247] See below, q. 37.
[248] See below, q. 38.
[249] See below, q. 39.
[250] See below, q. 40.

4. Solutio: numerare partes alicuius continui, hoc est dupliciter: aut sub forma accidentali aut sub forma substantiali. Si primo modo, hoc est dupliciter: aut sub forma accidentali absoluta aut sub forma accidentali comparata. Si ultimo modo, hoc dupliciter: aut recipiunt suam comparationem ex sua per se existentia et separatione aut recipiunt suam comparationem ex inhaerentia sua in aliquo toto. Et solum hoc ultimo modo numerantur **[O 93rb]** partes in continuo. Cum accidente[a] comparato[b] ex per se existentia possunt numerari, sed tamen non proprie. Nullo vero aliorum modorum contingit ipsas numerari, quia non contingit dicere quod sint duae substantiae vel duo corpora nec etiam contingit dicere quod sint duo colorata proprie nec etiam contingit dicere quod sint duo tota, sed solum contingit dicere quod sunt duae partes, et pars est semper sub inhaerentia.

<Quaestio 38>

Quaeritur de partibus diversorum totorum, ut corruptibili et incorruptibili, utrum conveniant in aliqua natura.

1. Et videtur quod sic: quaecumque simul sunt non solum tactu convenientia, conveniunt in natura et essentia formae; sed pars ignis corruptibilis et pars orbis lunae incorruptibilis simul sunt, quia habent ultima simul et contiguantur et non solo tactu convenientia. Probatio: plus conveniunt illa pars ignis et illa pars orbis quam aliqua pars terrae, [c]si esset ibi, et illa pars orbis; sed si terra esset ibi, convenirent tactu[d]; ergo illa plus convenient quam tactu; ergo etc.

2. Item, plus differunt corruptibile corporeum et incorruptibile incorporeum quam corruptibile corporeum et incorruptibile corporeum; sed primae conveniunt in aliqua una essentia. Probatio: anima rationalis, si non esset de genere substantiae, tunc non esset perfectio hominis; relinquitur ergo quod substantia de ea praedicatur. Ergo multo fortius et secundae simul erunt et habebunt aliquam naturam communem.

3. Item, dicit Augustinus quod una et eadem est materia in corruptibilibus et in incorruptibilibus, quia dicit quod Deus[e] creavit omnia simul, hoc est materiam primam omnium, tam [f]corruptibilium quam incorruptibilium. Supponit igitur quod materia prima est eadem omnium. Et quorum materia est una, et genus est unum; ergo etc.

[a] accidente] tamen *add. s.l.* O
[b] comparato] quae est *add. sed del.* O
[c] si esset ibi *post* orbis *transp.* (l. 21) *sed corr.* O
[d] tactu *corr. ex* tacta O
[e] Deus *suppl. s.l.* O
[f] corruptibilium quam incorruptibilium *scripsimus cum* TC; generabilium quam corruptibilium O

4. Solution. The numbering of the parts of a continuum can be understood in two ways: either (i) under an accidental form or (ii) under a substantial form. Case (i) is twofold: either (ia) under an absolute accidental form or (ib) under a comparative accidental form. Case (ib) is twofold: either (ib') the parts receive their comparison from their per se existence and separation or (ib'') they receive their comparison from their inherence in the whole. The parts of a continuum are numbered only in way (ib''). They can also be numbered (ib') from their per se existence together with a compared accident, but this is not in any proper sense <of numbering>. In no other way is it possible for the parts to be numbered, because it is not possible to say that they are two substances or two bodies nor is it possible to say that they are two coloured things in any proper sense or that they are two wholes, but it is only possible to say that they are two parts, and a part always exists through inherence <in the whole>.

<Question 38>
The question is whether the parts of different wholes, such as a corruptible whole and an incorruptible whole, agree in a nature.

1. Arguments for the affirmative: all the things that are in the same place and do not only agree in being in contact agree in the nature and essence of a form; but a part of fire, which is corruptible, and a part of the lunar orb, which is incorruptible, are in the same place, because their extremities are in the same place, and are contiguous, and they do not only agree in being in contact. Proof: that part of fire and that part of the orb agree more than would a part of earth, if that part of earth were there, and that part of the orb; but if some earth were there, that earth and that part of the lunar orb would agree in being in contact; therefore, a part of fire and a part of the lunar orb agree in more than in being in contact; therefore, <they agree in the nature and essence of a form>.

2. Furthermore, what is corruptible and corporeal and what is incorruptible and incorporeal differ more than what is corruptible and corporeal and what is incorruptible and corporeal; but the first pair of things agree in a single essence. Proof: if the rational soul did not belong to the genus of substance, then it would not be the perfection of man; therefore, it follows that substance is predicated of the rational soul. Therefore, there is a much stronger reason to say that the second pair of things will also exist together and will have a common nature.

3. Furthermore, Augustine says that one and the same matter exists in corruptible things and in incorruptible things, because he says that God created all things simultaneously,[251] that is, the prime matter of all things, both corruptible and incorruptible. He assumes, therefore, that prime matter is the same in all things. And things that share one matter also share one genus; therefore, <corruptible and incorruptible things share one genus>.

[251] See, e.g., Aug. *De genesi ad litt.* V.5, VIII.20, CSEL 28.1, 146.6–15, 258.25–259.5.

4. Item, conveniunt univoce in quantitate, quia eadem est demonstratio ibi de quantitate; ergo similiter convenient in substantia cuius est quantitas.

5. Item, habent eandem mensuram.

6. Item, non est comparatio nisi sub univoco; sed dicit Aristoteles quod quidam aestimabant quod sol non est nisi pedalis quantitatis, cum tamen secundum veritatem sit maior tota terra. Cum ergo sic comparantur ad invicem, erit aliquid univocum illis.

7. Ad oppositum: corruptibile et incorruptibile plus differunt quam secundum genus.

8. Solutio. Duplex est hic opinio: una quae dicit quod conveniunt in genere generalissimo et non solum in hoc, sed etiam[a] in corpore; alia quae dicit quod[b] nulla species est in genere substantiae nisi educatur de potentia generis, et quod nihil educitur de potentia generis nisi per actionem naturalem, unde, quia corpora caelestia non sic educuntur, ideo ponunt quidam quod non sunt in genere substantiae.

9. Sustinendo quod in nulla natura et essentia conveniunt, dicendum ad primam rationem, primo ad maiorem, quod ignem in sua sphaera et orbem lunae non solo tactu convenire, hoc est dupliciter: aut quia conveniunt in aliqua natura analogice participata ab illis, quae res sive natura non est[c] pars compositi alicuius nec praedicabilis de aliquo, quae res est virtus caelestis diffusa per universum. Et secundum hoc non contingit concludere quod, si[d] non solo tactu conveniunt, quod conveniant in aliqua natura et essentia. Aut quia conveniunt in aliqua forma quae communiter est praedicabilis de utroque et etiam quae est altera pars eorum. Et sic est minor falsa sustinendo opinionem[e].

10. Vel dicendum quod in nullo conveniunt. Sed quia ignis est corpus levissimum, ideo appetit esse super aerem. Et quia orbis lunae est faex omnium orbium sicut terra est faex elementorum, ideo appetit infimum locum corporis caelestis. Et ideo possumus dicere quod in nullo plus quam in tactu conveniunt, praeterquam in solo situ.

[a] etiam] solum *add. sed del.* O
[b] quod *scripsimus*; quia O
[c] est *suppl. s.l.* O
[d] si *scripsimus*; hoc *(lectio incerta)* O
[e] opinionem *scripsimus*; oppositum O

4. Furthermore, they agree univocally in quantity, because the demonstration concerning quantity is the same in both cases; therefore, they will similarly agree in the substance to which the quantity belongs.

5. Furthermore, they have the same measure.

6. Furthermore, there is no comparison except under something univocal; but Aristotle says that some people thought that the quantity of the sun is only one foot, although the sun is in truth bigger than the whole earth.[252] Therefore, since the sun and the earth are compared with each other in this way, there will be something univocal belonging to them.

7. To the contrary: the corruptible and the incorruptible differ in more than genus.

8. Solution. There are two opinions. One says that the corruptible and the incorruptible agree in the most general genus and not only in this, but also in the genus of body. The other says that no species is in the genus of substance unless it is educed from the potency of that genus, and that nothing is educed from the potency of that genus except through a natural action; therefore, since celestial bodies are not educed in this way, some people assume that they are not in the genus of substance.

9. If we accept the view that the corruptible and the incorruptible do not agree in any nature or essence, in reply to the first argument[253] it must first be said that the claim in the major premise that fire in its sphere and the lunar orb do not agree only in being in contact can be taken in two ways: either (i) because they agree in a nature that they participate analogically, and that this thing or nature is neither a part of a composite nor predicable of anything, but is a celestial power diffused throughout the universe. And if the claim is interpreted in this way it is not possible to conclude that, if fire and the lunar orb do not only agree in being in contact, they agree in a nature and essence. Or (ii) because they agree in a form which is a common predicable of both and is also is a part of them. And in this case the minor premise is false according to this opinion.

10. Alternatively it may be said that the corruptible and the incorruptible agree in nothing. But fire desires to be above air because it is the lightest body, and the lunar orb desires the lowest place proper to a celestial body because it is the most impure of all the orbs just as earth is the most impure of all elements. Therefore we can say that they agree in nothing more than being in contact, except for location alone.

[252] Arist. *De insomniis* 2, 460b18–20.

[253] See above, n. 1.

<Quaestio 39>

Quaeritur[a] [b]utrum unum et idem sit substantia et accidens.

<Quaestio 39.1>

1. Et videtur quod sic, quia lux est substantia. [c]Probatio: quicquid est [d]perfectivum nobilis substantiae est substantia; [e]sed lux est huiusmodi, secundum Commentatorem, quia perficit stellam.

2. Item, lux corporis caelestis est causa istorum inferiorum; et quod est causa substantiae est substantia; ergo etc.

3. Et lux est accidens. Probatio, quia est sensibilis, et omne tale in quantum tale est accidens, quia sensus est qualitatis[f].

4. Item, corpora caelestia, communicando sua lumina cum istis inferioribus, non communicant suam substantiam.

5. Item, quod est formale in luce, ut splendor, est accidens.

6. Item, lux est perfectio coloris, et color est accidens; ergo lux est accidens.

7. Ad[g] idem[h] super septimum *Metaphysicae*.

8. Item, in minus potest lux quam accidens sive species accidentis; ergo multo fortius lux est accidens. Quod possit in minus videtur sic. Si domus sit valde opaca et tamen[i] ingrediatur sol per forum[j] aliquod[k] protendens se contra parietem in domo illa, constat quod non illuminabit nisi partem sibi obiectam. Unde in aliis partibus domus[l] adhuc est tenebra. Sed si oculus esset in tenebra, videret coloratum, quod non contingeret nisi [m]species eius

[a] Quaeritur] circa secundum *praem.* T; circa secundum primo C
[b] utrum … accidens] an substantia et accidens sint idem TC
[c] Probatio: quicquid] et lux est accidens ergo idem est substantia et accidens. Quod sit substantia videtur, quia quod TC
[d] perfectivum nobilis] perfectio TC
[e] sed … splendor (n. 5)] lux est perfectio substantiae, ut stellae, quia ut (ut *om.* C) dicit Commentator super Caelum et mundum quod stella non est nisi maior congregatio (aggregatio *pro* congregatio C) lucis; ergo lux est substantia. Item lumina corporum supracaelestium causant vegetativam et sensitivam in his inferioribus; sed accidens non est causa substantiae; ergo lux (lux *om.* C) non (non *s.l.* T) est accidens. Quod autem lux sit accidens videtur sic: omne sensibile est accidens; lux est sensibile; ergo lux est accidens. Item, corpus caeleste multiplicat suum lumen hic inferius et non multiplicans (*corr. in* multiplicat T) suam substantiam, quia absurdum est dicere hoc; ergo lux TC
[f] qualitatis *scripsimus*; qualitas O
[g] Ad] illud *add.* C
[h] idem] dicendum *add.* C
[i] tamen] licet cum hoc *scrips. sed del.* T; et cum hoc C
[j] forum] foramen TC
[k] aliquod] et *add.* O
[l] domus] tenebrae O
[m] species eius *om.* O

<Question 39>

The question is whether one and the same thing is a substance and an accident.

<Question 39.1>

1. Arguments for the affirmative: light is a substance. Proof: everything that gives perfection to a noble substance is a substance; but according to the Commentator, light is something of this kind, because it gives perfection to a star.[254]

2. Furthermore, the light of a celestial body is a cause of the things here below; and what is a cause of a substance is a substance; therefore, <light is a substance>.

3. Light is also an accident. Proof: light is a sensible thing, and everything of this kind in so far as it is of this kind is an accident, given that the senses are of a quality.

4. Furthermore, in communicating their lights to the bodies here below, the celestial bodies do not communicate their substance to them.

5. Furthermore, what is formal in light, namely brightness, is an accident.

6. Furthermore, light is the perfection of colour, and colour is an accident; therefore, light is an accident.

7. Arguments for the same conclusion are given in the commentary on *Metaphysics* VII.[255]

8. Furthermore, light has less power than an accident or the species of an accident; therefore, there is a much stronger reason to hold that light is an accident. That light has less power is shown as follows: if a house is very dark and a ray of sunlight enters through a hole and extends itself to a wall in that house, it is apparent that the ray of sunlight will only illuminate the part opposite to it, so that the other parts of the house will still be in darkness. But if an eye were in darkness, it would see a coloured body, which would not happen if its species were not multiplied in the darkness, where the species of

[254] Aver. *In De caelo* II, t.c. 42, ed. Carmody, II, 355.198–231.

[255] This topic is not discussed in Aspall's Questions on the *Metaphysics* preserved in MS C.

multiplicaretur[a] in tenebra [b]ubi species lucis non multiplicatur. Si ergo species coloris sive color potest multiplicare se ubi lux non potest, tunc potentior est accidens aliquod[c] quam lux. [d]Cum ergo magis potens quam lux est accidens, tunc lux non erit substantia, sed potius accidens. **[O 93va]**

<Quaestio 40>

Quaeritur[e] an accidens, cum[f] multiplicat speciem suam in medio vel ad sensum, simul cum specie accidentis[g] commultiplicetur[h] aliquid substantiae deferentis illud accidens.

<Quaestio 40.1>

1. Et[i] quod non videtur: formae luminares corporis caelestis multiplicant se in diaphano; et ipsa lux est accidens, quia omne sensibile est accidens; sed[j] lux in diaphano est sensibile; ergo lux illa est accidens.

2. Quod autem nihil formae substantialis commultiplicetur[k] videtur sic: dicit Aristoteles in libro *De generatione* quod non exterius se faciunt quaecumque nec sunt contraria nec ex contrariis; [l]sed forma substantialis [m]corporis caelestis nec est ex contrariis nec est ei aliquid contrarium; ergo forma substantialis eiusdem non facit se extra se.

3. Item, minus transmutabiles [n]sunt species quam formae earundem specierum. Et hoc quia propter materiam est transmutabilitas; quod ergo magis unitur materiae magis est transmutabile. Sed[o] forma magis unitur materiae[p] quam sua species; ergo species minus transmutabilis est. Sed forma substantialis caeli non in tantum transmutatur quod faciat se extra se per multiplicationem; ergo nec species eius similiter.

[a] multiplicaretur *corr. s.l. ex* multiplicatur T; *lectio incerta* C
[b] ubi … multiplicatur] et etiam species lucis TC
[c] aliquod *om.* TC
[d] Cum … accidens] ergo lux non est substantia TC
[e] Quaeritur] consequenter *praem.* TC
[f] cum *om.* TC *sed suppl. s.l.* T
[g] accidentis *om.* TC
[h] commultiplicetur] multiplicet TC
[i] Et *om.* TC
[j] sed *om.* TC
[k] commultiplicetur] multiplicetur T; multi^{tur} C
[l] sed … eiusdem] ergo forma substantialis eiusdem cum (cum *s.l.*) non sit huiusmodi C
[m] corporis ... contrarium] corporis caelestis (corporis caelestis *mg.*) nec est (*corr. mg. in* habet) aliquid contrarium nec est (est *s.l.*) ex contrariis T
[n] sunt species *suppl. mg.* T
[o] Sed] quare cum C
[p] materiae *suppl. s.l.* OT

light is not multiplied. Therefore, if the species of a colour or a colour can multiply itself where light cannot, then an accident is more powerful than light. Therefore, since an accident is more powerful than light, light will not be a substance, but rather an accident.

<Question 40>

When an accident multiplies its species in the medium or as far as the senses, the question is whether something belonging to the substance that carries that accident is multiplied together with the species of that accident.

<Question 40.1>

1. Arguments for the negative: the illuminating forms of a celestial body multiply themselves in a transparent medium; and light itself is an accident, because every sensible thing is an accident; but light in a transparent medium is a sensible thing; therefore, that light is an accident.

2. That nothing of a substantial form is multiplied together <with the species of an accident> is shown as follows: in *On Generation* Aristotle says that no things that are not contraries or from contraries extend outside themselves;[256] but the substantial form of a celestial body is neither from contraries nor is there something contrary to it; therefore, the substantial form of a celestial body does not extend outside itself.

3. Furthermore, species are less changeable than the forms corresponding to those same species. The reason for this is that being changeable is due to matter; therefore, what is more joined to matter is more changeable; but a form is more joined to matter than is its species; therefore, its species is less changeable. But the substantial form of the heavens does not change in such a way as to extend itself outside itself through multiplication; therefore, neither does its species.

[256] Arist. *De gen. et corr.* I.7, 323b28–29.

4. Item, quae magis videntur multiplicare speciem ad sensum quam forma substantialis caeli non multiplicant[a]; [b]ergo nec forma substantialis caeli[c]. Numerus enim et magnitudo et huiusmodi sensibilia[d] non multiplicant speciem suam in sensum, quia, si multiplicarent, essent activa in sensum; sed non sunt activa in sensum. [e]Probatio huius assumpti: oculus caret omni colore ut recipiat omnem colorem indifferenter; si ergo oculus deberet speciem magnitudinis recipere, necesse esset quod similiter[f] careret omni magnitudine. Sed hoc falsum; ergo primum.

5. Ad oppositum: nobilitas[g] formae est se multiplicare; [h]ergo quanto forma est nobilior, tanto magis multiplicabit se; sed forma substantialis [i]est nobilior quam accidentalis.

6. Item, sicut[j] res ad rem, sic[k] species ad speciem; sed [l]accidens praeexigit substantiam; ergo et[m] species accidentis speciem[n] substantiae.

7. Item, nisi [o]multiplicaretur species substantiae ad sensum, [p]tunc non[q] cognosceretur substantia, quia omnis cognitio[r] nostra ortum habet a sensu.

8. Item, [s]totus ignis agit in totum aerem, per Aristotelem, libro *De generatione*; sed hoc non posset esse nisi forma substantialis emitteret suam speciem in profundum patientis.

9. Item[t], minus nobile non excitat magis nobile; sed forma accidentalis minus nobilis est quam forma substantialis; immissio ergo formae accidentalis

[a] multiplicant] speciem suam in sensum, quia, si multiplicarent *add. sed del.* O
[b] ergo … caeli] speciem ergo etc. minor patet T
[c] caeli] antecedens manifestatur sic *add.* C
[d] sensibilia] communia *add.* TC
[e] Probatio huius assumpti] ergo non faciunt speciem in sensum. Item (ad idem C) TC
[f] similiter *om.* TC
[g] nobilitas] ex nobilitate TC
[h] ergo … se *om.* TC
[i] est ... accidentalis] est multo nobilior (multo nobilior est C) omni forma accidentali; ergo forma substantialis multotius (multo fortius C) se ipsam multiplicat quam forma accidentalis TC
[j] sicut] se habet *add.* TC
[k] sic] se habet *add.* T
[l] accidens praeexigit substantiam] res accidentis non est sine re substantiae TC
[m] et *om.* TC
[n] speciem] non est sine specie TC
[o] multiplicaretur species substantiae] species substantiae esset delata TC
[p] tunc non cognosceretur] numquam intelligeretur TC
[q] non *suppl. s.l.* O
[r] cognitio] scientia TC
[s] totus … patientis] Aristoteles in (libro *add.* C) De generatione dicit quod in generatione conversiva non transmutatur materia in materiam (naturam T) nec forma in formam sed totum in totum, sed hoc non contingeret nisi speciem totius (totius *mg.* T) immitteret; species ergo aggregati immittitur (immitteret... immittitur] aggregati immitteretur C), et ita forma substantialis simul cum forma (forma *om.* C) accidentali speciem suam immittit TC
[t] Item] ad idem *add.* T

4. Furthermore, things that seem more likely than the substantial form of the heavens to multiply their species as far as the senses do not in fact multiply their species; therefore, neither does the substantial form of the heavens; for a number and a magnitude and sensible things of this kind do not multiply their species as far as the senses, because, if they multiplied them, they would act on the senses; but they do not act on the senses. Proof of this assumption: an eye lacks every colour so that it can receive every colour indifferently; therefore, if an eye had to receive the species of a magnitude, it would be necessary for it also to lack every magnitude. But this <consequent> is false; therefore, the antecedent <is false>.

5. To the contrary: the nobility of a form consists in multiplying itself; therefore, the nobler a form is the more it will multiply itself; but a substantial form is nobler than an accidental form.

6. Furthermore, just as a thing is related to a thing, so a species is related to a species; but an accident presupposes a substance; therefore, the species of an accident also presupposes the species of a substance.

7. Furthermore, a substance would not be known unless the species of that substance were multiplied as far as the senses, because all our cognition originates from the senses.

8. Furthermore, the whole fire acts on the whole air, according to Aristotle in *On Generation*;[257] but this could not happen unless a substantial form introduced its species into the depths of a patient.

9. Furthermore, the less noble does not stimulate the more noble; but an accidental form is less noble than a substantial form; therefore, the introduction

[257] Arist. *De gen. et corr.* I.9, 326b31–33.

ignis in aerem non excitat ad productionem alterius formae[a] substantialis. Sed constat quod altera forma[b] substantialis inducitur in materia aeris; ergo ignis non solum immittit speciem formae accidentalis, sed simul cum hoc speciem formae substantialis.

<Quaestio 39.2>

1. Solutio[c]. Ad primum problema dicendum quod substantia et accidens non sunt idem.

2. Quidam tamen dicunt quod substantia et accidens possunt esse idem. Sed illud est substantia uno respectu et alio respectu accidens, sicut dicunt quidam quod forma prima est accidens in se; [d]in quantum tamen perficit hanc[e] materiam sive ponit compositum in specie, sic est substantia.

3. Sed illud [f]improbatum est super septimum *Metaphysicae*.

4. [g]Argumentis in oppositum dicunt quidam quod lux in medio nec est substantia nec accidens. In suo[h] fonte, ut in corpore solis vel in alio corpore luminoso, in veritate[i] est accidens, licet[j] tamen propter defectum nominum formarum substantialium dicimus quod lux est forma substantialis corporis luminosi, sicut dicit Commentator quod stella non est nisi maior aggregatio lucis. Unde huic consentit Commentator alibi dicens quod, quia formae substantiales elementorum[k] sunt ignotae nec sunt nomina illis imposita, utimur gravi et levi pro formis substantialibus elementorum. Ponunt ergo quod lux in medio nec est substantia nec accidens.

5. Contra: nulla species denominat illud in quo est nomine sui principalis[l]. Non enim dicimus quod medium est coloratum, licet species coloris in medio multiplicetur; possumus tamen bene dicere quod medium est immutatum[m]; [n][o]et sic <color[p]> non denominat medium nomine sui principalis. Sed lux[q]

[a] formae] scilicet *add. s.l.* O
[b] forma] scilicet *add. s.l.* O
[c] Solutio *om.* TC
[d] in quantum *corr. mg. ex* numquam T
[e] hanc *om.* TC
[f] improbatum est] improbatur C
[g] Argumentis … quidam] alii dicunt contrariun C
[h] suo] tamen *add. s.l.* O
[i] veritate] tamen *add. sed del.* T *add. mg.* C
[j] licet] sed T
[k] elementorum] ponunt ergo quod lux in medio nec est substantia nec accidens *add. sed del.* T
[l] principalis] principii O
[m] immutatum *sc.* a specie coloris; luce *add. s.l.* O
[n] et ... illuminatum *om.* O;
[o] et... principalis *om.* C
[p] color *supplevimus*
[q] lux] non *add. sed del.* T

of an accidental form of fire into the air does not stimulate the production of the other substantial form. But it is apparent that the other substantial form is induced into the matter of air; therefore, fire does not only introduce the species of an accidental form, but it also <introduces> the species of a substantial form together with it.

<Question 39.2>

1. Solution. To the first problem it must be said that a substance and an accident are not the same thing.

2. Some people, however, say that a substance and an accident can be the same thing, but that thing is a substance in one respect and an accident in another, just as some people say that the first form <that comes to be in matter> is an accident in itself, but it is a substance in so far as it gives perfection to this matter or locates a composite in its species.

3. But this position has been disproved in the commentary on *Metaphysics* VII.[258]

4. To the arguments to the contrary some people reply that light in the medium is neither a substance nor an accident. In its source, for example in the body of the sun or in another luminous body, light is truly an accident, although because we are short of names for substantial forms we say that light is the substantial form of a luminous body, just as the Commentator says that a star is nothing other than a greater aggregate of light.[259] Thus, the Commentator agrees with this position when he says elsewhere that since the substantial forms of the elements are not known nor are there names imposed on them, we use the names 'heavy' and 'light' to refer to the substantial forms of the elements.[260] Therefore, these people posit that light in the medium is neither a substance nor an accident.[261]

5. Against this: no species denominates the thing in which it exists from its primitive name; for we do not say that the medium is coloured, although the species of a colour is multiplied in the medium; but we can indeed say that the medium is changed, and so a colour does not denominate the medium from

[258] This topic is not discussed in Aspall's Questions on the *Metaphysics* preserved in MS C.

[259] A passage from Peter of Auvergne's commentary on the *De caelo* suggests that this claim possibly originates from Averroes' idea that a star is the thickest part of its orb (Aver. *In De caelo* II, t.c. 42, ed. Carmody, II, 350.61–351.86; *Auctoritates Aristotelis*, 165, n. 72); see Peter of Auvergne, *In De caelo*, 238.76–78: "Dicit enim Averroes quod stella nihil aliud est quam densior pars sui orbis, quia lumen magis aggregat et terminat.".

[260] Aver. *In Metaph.* VIII, t.c. 5, fols. 212vb–213rb. In this passage, however, Averroes mentions the qualities 'subtle' and 'thick', not the qualities 'light' and 'heavy'.

[261] Versions of the view according to which the light existing in the medium is neither a substance nor a real accident, but an intentional being are reported, for example, by Bonaventure and Aquinas in their commentaries on the *Sentences*; see Bonaventure, *In Sent.* II, dist. 13, a. 3, q. 2, 330; Thomas Aquinas, *In Sent.* II, dist. 13, q. 1, a. 3, 334.

denominat medium nomine sui principalis, quia dicimus quod medium est illuminatum; ergo non sola species lucis in medio multiplicatur.

6. Item, nihil videtur per solam speciem nisi illud quod est directe obiectum visui, quia non videmus coloratum situm[a] a latere nostro neque retro nos, sed solum coloratum [b]quod est nobis directe obiectum. Arguo[c] tunc sic: nihil multiplicans se[d] per solam speciem videtur a latere; sed lux videtur a latere, [e]ut si imaginetur domus, si aliquis sit in domo[f], bene videbit[g] lucem quae causabitur a latere; ergo etc.

7. Item, speciei non est species; si ergo lux esset in medio ut sola species, tunc non faceret speciem in alio. Sed si teneatur speculum[h] contra obiectum aliquod, erit species obiecti in speculo et etiam species medii illuminati, quia in speculo apparebit simul cum specie obiecti totum medium illuminatum. Et si hoc, lux in medio non est species.

8. Item, a nostris magistris datur talis distinctio[i]: formarum quaedam multiplicat se[j] et extendit, ut forma carnis secundum speciem, quia caro secundum speciem est[k] illud radicale principium quod sumpsimus a primis parentibus nostris. [l]Sed nisi illa caro esset ex fecunditate suae formae sui ipsius multiplicativa, aliquando[m] consumpta esset et etiam nisi extenderet se[n], non esset simul cum carne secundum materiam in qualibet parte corporis. Et ista multiplicatio et extensio non est [o]ut multiplicatio speciei, sed rei. Alia est forma quae extendit se[p] et non multiplicat, ut forma elementaris. Et haec tantum[q] secundum quod res est extendit se[r]. Et tertio modo est forma quae multiplicat se[s] et non extendit [t]secundum ipsos, ut forma lucis. Cum ergo omnes aliae partes dentur **[O 93vb]** secundum quod res sunt, ergo lux multiplicabit[u] se [v]tantum secundum quod res est.

[a] situm] si sit C
[b] quod est *om.* C
[c] Arguo] videmus (nos *add.* C) TC
[d] se *suppl. s.l.* O; *ante* multiplicans T
[e] ut ... latere *scrips. sed del.* O
[f] domo] medio TC *sed* vel domo *add. s.l.* T
[g] videbit] videbat O
[h] speculum] in alio *add. sed del.* O
[i] distinctio] quod *add.* TC
[j] se *om.* OTC *sed suppl. s.l.* OT
[k] est *om.* O *sed suppl. s.l.* scilicet O
[l] Sed nisi] si nisi *scrips. sed corr. in* si non *(lectio incerta)* O
[m] aliquando] olim TC
[n] se *om.* OTC *sed suppl. s.l.* O
[o] ut multiplicatio *suppl. mg.* T
[p] se *om.* OTC *sed suppl. s.l.* O
[q] tantum *lectio incerta* O
[r] se *om.* OTC *sed suppl. s.l.* O
[s] se *om.* OTC *sed suppl. s.l.* O
[t] secundum ipsos *corr. mg. ex* secundum speciem lucis T; *om.* C
[u] multiplicabit] multiplicat TC
[v] tantum secundum quod] ut C

its primitive name. But light denominates the medium from its primitive name, because we say that the medium is illuminated; therefore, it is not only the species of light that is multiplied in the medium.

6. Furthermore, nothing other than what is directly opposite to sight is seen only through its species, for we do not see a coloured thing situated at our side or behind us, but only a coloured thing which is directly opposite to us. Then, I argue as follows: nothing which multiplies itself only through its species is seen from the side; but light is seen from the side, for example, if we imagine a house and someone in the house, he will indeed see a light which is caused from the side; therefore, <light does not only multiply itself through its species>.

7. Furthermore, there is no species of a species; therefore, if light were in the medium only as a species, it would not produce a species in another thing. But if a mirror is held in front of an object, the species of that object and the species of the illuminated medium will both exist in the mirror, because the whole illuminated medium will appear in the mirror together with the species of that object. If this is the case, then light in the medium is not a species.

8. Furthermore, our masters make the following distinctions: (i) one kind of form multiplies and extends itself, as does the form of flesh in accordance with its species, because flesh in accordance with its species is that original principle which we took from our first parents. But if that flesh were not capable of multiplying itself in virtue of the fertility of its form, it would at some time be consumed; also, if it did not extend itself, it would not exist in every part of the body together with the flesh in accordance with its matter. And this kind of multiplication and extension is not the multiplication of a species, but of a thing. (ii) Another kind of form extends itself and does not multiply itself, as does an elementary form. This kind of form extends itself in so far as it is a thing. (iii) A third kind of form according to these people multiplies itself but does not extend itself, as does the form of light. Since all the other members of this distinction are about forms in so far as they are things, therefore, light will multiply itself only in so far as it is a thing.[262]

[262] This threefold distinction is mentioned by Robert Kilwardby in his commentary on the *Sentences*; see Robert Kilwardby, *In Sent.* II, q. 69, 194.60–69.

9. Quidam bene concedunt quod lux non multiplicatur[a] in medio secundum quod est sola intentio. Unde quidam dicunt sic, quod lux in medio est accidens propter hoc quod est quoddam visibile, et omne tale est in tertia specie qualitatis. Et cum hoc ponunt quod nihil substantiae caeli multiplicatur hic inferius.

10. Sed quod aliquid substantiale ipsius caeli commultiplicetur[b], videtur sic: omne accidens habet subiectum aliquod a quo dependet; sed lux non dependet a medio; aliquid[c] ergo commultiplicatur[d] cum luce, [e]quae quidem accidens est, a quo dependet lux et quod est subiectum proprium ipsius[f] lucis. Probatio minoris: quod dependet ab alio ut a subiecto non numeratur in eodem; sed lux numeratur in aere sive in medio. Probatio[g]: si inter duo corpora luminosa ponatur aliquod corpus opacum, apparebunt ibi duae tenebrae sive duae privationes luminis, [h]quod non contingeret nisi prius in illa parte essent duo lumina numerata. [i]Numerantur ergo duo lumina in eadem parte medii. Et si hoc, lux non dependet a medio. Et [j]si hoc, [k]non est in illo ut in subiecto. Et hoc adhuc a quibusdam conceditur.

11. Aliter dicitur sic quod est substantia lucis et splendor. Unde lux secundum quod splendor, sic est visibile. Et sic est qualitas sive accidens. [l]Secundum vero substantiam lucis, substantia. [m]Est etiam multiplicatio substantiae lucis ubi non est lux splendens, quia per virtutes luminares generantur mineralia in ventre terrae, et tamen non est ibi aliquis splendor lucis. Et etiam de nocte fiunt generationes mediantibus virtutibus luminaribus, quae sunt agentia universalia.

12. Contra: si substantia lucis multiplicatur hic inferius, aut ergo est illa[n] substantia corporea aut incorporea. Si sit substantia incorporea, tunc est intelligentia vel aliquod huiusmodi separatum. Hoc falsum est[o]; ergo primum. Si corporea, cum sit simul in loco cum aere sive medio, tunc duo corpora [p]erunt in eodem loco.

[a] multiplicatur] lux *add. sed del.* O
[b] commultiplicetur] multiplicatur T; multiplicetur C
[c] aliquid] aliud O
[d] commultiplicatur] multiplicatur TC *sed corr.* T
[e] quae ... est *scrips. sed del.* O; *om.* C
[f] ipsius *om.* TC
[g] Probatio] ergo etc. Probatio minoris huius C
[h] quod ... numerata *suppl. mg.* T
[i] Numerantur ... medii] numerantur et lumina *scrips. sed del.* O
[j] si hoc] ita T
[k] non ... illo] est in alio C
[l] Secundum ... lucis] secundum autem subiectum lucis est TC
[m] Est etiam] et TC
[n] illa *om.* TC
[o] est *om.* TC
[p] erunt ... loco] in eodem loco erunt et hoc falsum C

9. Some people do correctly concede that light is not multiplied in the medium in so far as it is only an intention. Thus, some people say that light in the medium is an accident in virtue of the fact that it is something visible, and everything of this kind is in the third species of quality. In addition to this they assume that nothing belonging to the substance of the heavens is multiplied here below.[263]

10. But that something belonging to the substance of the heavens themselves is multiplied together <with light> is shown as follows: every accident has a subject on which it depends; but light does not depend on the medium; therefore, something on which light, which is an accident, depends and which is the proper subject of light itself will be multiplied together with light. Proof of the minor premise: what depends on something else as a subject is not numbered in that thing; but light is numbered in the air or in the medium. Proof: if a dark body is located between two luminous bodies, two instances of darkness or two privations of light will appear there, which would not happen if there had not first been two numbered lights in that part <of the medium>. Therefore, two lights are numbered in the same part of the medium. And if this is the case, light does not depend on the medium, and therefore light is not in the medium as its subject. This is also conceded by some people.

11. Another view speaks of the substance of light and its brightness in this way. Thus, light in so far as it is brightness is visible, and in this way it is a quality or an accident. But light in so far as it is the substance of light is a substance. The multiplication of the substance of light also occurs where light is not bright, because minerals come to be in virtue of illuminating powers in the depths of the earth, although there is no brightness of light there. It is also the case that generation occurs at night in virtue of illuminating powers, which are universal agents.[264]

12. Against this: if the substance of light is multiplied here below, then either that substance is corporeal or incorporeal. If it is an incorporeal substance, then it is an intelligence or something separate of that kind. But this <consequent> is false; therefore, the antecedent <is false>. If it is a corporeal substance, then, since it is in the same place as the air or the medium, two bodies will be in the same place <which is absurd>.

[263] This solution is supported, for example, by Aquinas; see Thomas Aquinas, *In Sent.* II, dist. 13, q. 1, a. 3, 334–335.

[264] A version of this theory is supported by Bonaventure; see Bonaventure, *In Sent.* II, dist. 13, a. 3, q. 2, 330–331.

13. [a]Ad[b] argumentum[c] probans quod lux in medio nullo modo substantia est, dicendum quod lux non est incorporeum, sed corporeum, quia est habitus procedens a corpore et est habitus medii nec dependet a medio. Nec etiam potest lux multiplicari nisi in medio[d] diaphano, quia Aristoteles in secundo *De anima* improbat opinionem Empedoclis ponentis vacuum esse, quia Empedocles posuit quod propter vacuum quod est inter nos et caelum, si in caelo esset formica, videremus[e] formicam. Et improbat Aristoteles hoc dicens quod, si vacuum esset, nihil videretur omnino, quia non esset aliquid illuminatum. Sic, secundum Aristotelem, patet quod lux non[f] multiplicatur in vacuo.

14. [g]Dicitur igitur sic, quod quaedam corpora sunt absolutae penetrationis, [h]sicut lux, et sic lux potest esse simul loco cum [i]aere sive medio[j]. Et quaedam sunt corpora absolutae expulsionis; et huiusmodi [k]nullo modo compatiuntur se in eodem loco, cuiusmodi sunt lapides, ligna etc.[l] Quaedam autem sunt corpora medio modo se habentia, scilicet quae nec sunt absolutae expulsionis nec absolutae penetrationis, [m]sicut vapor aqueus: [n]vapor ille non est absolutae penetrationis, quia[o] <...>, quando vapores ascendunt a fundo aquae, non expellunt simpliciter; et propter hoc[p] adgeneratur maior quantitas aquae, [q]scilicet quia vapor recipit resistentiam ab aqua. Et quia[r] iste modus loquendi non est multum naturalis, [s]sed potius theologicus, ideo omittatur.

15. Aliter dicitur sic, et forte melius, [t]secundum mentem Aristotelis, quod non est substantia absolute, sed est forma substantialis medii, ut aeris. Sed[u]

[a] Ad … vacuo *scrips. sed ante* ad secundum problema (q. 40.2, n. 1) *transp. mg.* T
[b] Ad] illud *add. s.l.* O
[c] argumentum] ultimum *add. s.l.* O
[d] medio] ut *add. s.l.* O
[e] videremus *corr. ex* videamus O
[f] non *om.* OC
[g] Dicitur igitur sic] ad hoc dicitur TC
[h] sicut … medio *om. sed suppl. mg. et s.l.* ...(?) autem simul possunt esse sic (?) lux C
[i] aere sive] eo in quo est lux et sic possunt omnia alia quae sunt absolutae penetrationis T
[j] medio] tunc duo corpora erunt in illa quae sunt absolutae penetrationis *add. sed del.* O
[k] nullo modo] non C
[l] etc.] et huiusmodi TC
[m] sicut … penetrationis *om. hom.* O
[n] vapor ille] cum aqua quia vaporabile T
[o] quia *hic aliqua deesse videntur fort. per hom. omissa*
[p] hoc] non *add.* O
[q] scilicet … aqua *scrips. sed del.* O
[r] quia] aqua re *add. sed del.* C
[s] sed potius theologicus *om.* TC
[t] secundum mentem Aristotelis] quod sententia Aristotelis est C
[u] Sed] haec *add.* O

13. To the argument that proves that light in the medium is in no way a substance[265] it must be said that light is not incorporeal, but corporeal, because it is a habit proceeding from a body and is a habit of the medium but does not depend on the medium. Nor can light be multiplied except in a transparent medium, given that in *On the Soul* II Aristotle refutes the opinion of Empedocles, who posited that a void exists, on the grounds that it is because of the void located between us and the heavens that we would see an ant if it were in the heavens. Aristotle rejects this view, saying that if a void existed, nothing would be seen at all, because nothing would be illuminated.[266] Thus, according to Aristotle, it is clear that light is not multiplied in a void.

14. Therefore, one view is that (i) some bodies are completely able to penetrate, such as light, and thus light can be in the same place as air or the medium. (ii) Other bodies are completely able to expel; and bodies of this kind are in no way compatible with one another in the same place, and stones, wood, and the like are of this kind. (iii) Other bodies are intermediate, that is, neither completely able to expel nor completely able to penetrate, such as the steam produced from water.[267] This kind of steam is not completely able to penetrate <water> because <...> when steam rises from the bottom of water it does not completely expel <that water>. And it is because steam encounters resistance from water that a greater quantity of water is generated. Since this way of speaking is not very natural but rather theological, let it therefore be omitted.

15. Another and perhaps better view, in accordance with the intention of Aristotle, is that <light> is not absolutely a substance, but it is the substantial

[265] See above, n. 12.
[266] Arist. *De anima* II.7, 419a15–21. However, the philosopher criticised by Aristotle is Democritus, not Empedocles.
[267] This view is ascribed to Augustine by Adam of Buckfield. Cf. Adam of Buckfield, *In De anima* II, MS Oxford, Bodl. Libr., Can. Misc. 322, fol. 31rb: "Secundum autem Augustinum, ut dicitur, est lux corpus. Sed dividit corpus tripliciter, quorum unum habet penetrationem et non expulsionem sive naturam penetrandi et non expellendi, aliud habet expulsionem et penetrationem, tertium expulsionem sed non penetrationem. Et primo modo dicitur lux corpus, secundo modo ignis et aer, tertio modo corpora solida, quae solum apud vulgum dicuntur corpora. Et primo modo et secundo non est inconveniens duo corpora esse in eodem loco, sed tertio modo. Hoc tamen non intendit Aristoteles.".

forma substantialis dicitur dupliciter, vel [a]quia sit forma quae est altera pars compositi, sine qua[b] non potest compositum manere, ut aeritas est forma substantialis aeris, quia pars essentialis et sine qua forma non potest aer in esse suo[c] conservari, et sic non est lux forma substantialis aeris. Aliter est forma substantialis [d]quae est[e] substantialiter ad habitum quendam essentialem medii, ad cuius habitus destructionem non sequitur destructio medii. Et [f]quod huic consentiat Aristoteles patet[g] in secundo *De anima*[h], capitulo de sensu in[i] communi, in illo passu: "Cuius quidem[j] est visus [k]hoc est visibile", [l]dicens quod "Lumen est actus lucidi secundum quod lucidum". Et paulo post: "Est autem tenebra privatio habitus[m]". Et sic, per Aristotelem, patet quod lumen est habitus medii ad cuius destructionem non destruitur medium.

16. Item, Commentator: "Lux non est nisi habitus medii".

17. [n]Inter illos textus praeterea[o], dicit Aristoteles quod neque est corpus neque fluxus a corpore. Et exponit Commentator 'fluxum a corpore' dicens quod potest intendere per illud corpora mathematica[p], quae fluunt a corpore naturali secundum considerationem, licet sint secundum esse in corpore.

<Quaestio 40.2>

1. Solutio[q]. Ad secundum problema dicendum quod non solum est multiplicatio speciei accidentis in medio, sed aliquo modo species formae substantialis commultiplicatur[r]. [s]Hoc tamen est primo et[t] per se, quod color multiplicet se. Quod autem forma substantialis multiplicet se, hoc est per se et

[a] quia sit *om.* C
[b] qua *scripsimus*; quo OTC
[c] suo] suum O
[d] quae est *corr. mg. in* quia confert T
[e] est *fortasse cum* T *mg.* confert *scribendum*
[f] quod huic consentiat] huic consentit TC
[g] patet *om.* OTC *sed suppl. s.l.* O
[h] anima] in *add.* TC
[i] in *om.* TC
[j] quidem] quid igitur O
[k] hoc est visibile *suppl. mg.* T
[l] dicens … habitus *suppl. mg. et* ex lucido *add. mg.* T
[m] habitus] autem ex lucido *add.* C
[n] Inter illos textus] inter illos *scrips. sed del.* O; inter textus *scrips. sed corr. s.l. in* post praedictos textus T
[o] praeterea *om.* TC
[p] mathematica] manifesta O; *corr. s.l. ex* methe[ca] T
[q] Solutio *om.* TC
[r] commultiplicatur] multiplicatur TC
[s] Hoc tamen est] cum hoc sit C
[t] et] non *add.s.l.* O

form of the medium, for example of air. But a substantial form is spoken of in two ways: either (i) it is said to be the form which is the other part of a composite, without which a composite cannot remain, in the way that air-ness is the substantial form of air, because it is an essential part <of air> and the form without which air cannot be preserved in its being. Light is not the substantial form of air in this sense. (ii) In another way a substantial form is said to be that form which is substantial with respect to an essential habit of the medium, which is such that the destruction of the medium does not follow from the destruction of that habit. That this view is in agreement with Aristotle is clear from *On the Soul* II, chapter about the common sense, in the passage "The visible is the thing of which there is sight",[268] where he says that "light is the act of the translucent in so far as it is translucent".[269] And shortly after <he says>: "but darkness is the privation of a habit".[270] Thus it is clear that, according to Aristotle, light is a habit of the medium such that the medium is not destroyed because of the destruction of that habit.

16. Furthermore, the Commentator <says>: "Light is nothing but a habit of the medium".[271]

17. Furthermore, in those texts Aristotle says that light is neither a body nor anything flowing from a body.[272] And the Commentator explains the expression 'something flowing from a body' by saying that Aristotle may be referring to mathematical bodies, which flow from a natural body in accordance with the way of considering it, although in accordance with their being they exist in a <natural> body.[273]

<Question 40.2>

1. Solution. To the second problem it must be said that not only the multiplication of the species of an accident <is found> in the medium, but the species of a substantial form is <also found as> in some way multiplied together <with the species of an accident>. What happens primarily and per se, however, is that a colour multiplies itself, whereas that a substantial form

[268] Actually, it is in the chapter on sight. Cf. Arist. *De anima* II.7, 418a26.
[269] Arist. *De anima* II.7, 418b9–10.
[270] Arist. *De anima* II.7, 418b18–19.
[271] Aver. *In De anima* II, t.c. 69, ed. Crawford, 236.32–33.
[272] Arist. *De anima* II.7, 418b14–15.
[273] Not found in Aver. *In De anima* II, t.c. 69 (on Arist. *De anima* II.7, 418b9–17). Interestingly, t.c. 69 ends *ex abrupto* before the explanation of this point.

non primo, propter [a]hoc quod sub multiplicatione formae accidentalis latet[b] multiplicatio formae substantialis.

2. Et huic videtur Aristoteles consentire cum dicit[c] quod lapis non est in[d] anima, sed species lapidis. Sed si solum color lapidis esset in[e] anima, non esset tota species lapidis in anima. Manifestum est[f] ergo, secundum Aristotelem, quod species formae substantialis commultiplicatur.

3. Item, de nullo iudicatur in intellectu nisi de illo quod prius fuit [g]in sensu; sed tam de substantia quam de qualitate obiecti fit iudicium in intellectu; ergo tam species unius quam alterius in sensum[h] multiplicatur.

4. [ij]Item, si solus color faceret se ad sensum, tunc, cum bruta non[k] iudicent nisi tantum de hoc quod in sensum[l] recipitur, tunc brutum, ut canis, non plus cognosceret dominum suum quam aliquod[m] aliud animal, dummodo essent eiusdem coloris, quod [n]scimus esse falsum; [o]ergo primum.

5. Item, si hoc est verum, quod sola species coloris recipitur in organo, tunc [p]agnus non plus fugeret propter praesentiam lupi quam propter praesentiam alterius animalis eiusdem coloris, quod scimus similiter[q] esse falsum. **[O 94ra]**

6. Dicendum ergo quod utrumque speciem multiplicat, tamen color est illud quod primo et[r] per se immutat sensum et[s] sub specie coloris commultiplicatur forma substantialis.

7. Unde [t]primae rationi dicendum quod forma substantialis ipsius caeli multiplicat speciem suam hic inferius.

[a] hoc quod sub] quod C
[b] latet] patet O
[c] dicit] dicat TC
[d] in *om.* O
[e] in *suppl. s.l.* O
[f] est *om.* TC
[g] in sensu *suppl. s.l.* O
[h] sensum] sensu TC
[i] Item ... falsum *suppl. mg*. T
[j] Item ... primum *transp. post* scimus similiter esse falsum (n. 5) C
[k] non] vident *add. sed del.* O
[l] sensum] sensu TC
[m] aliquod *om.* T *(mg.)* C
[n] scimus esse] experimur T *(mg.)*
[o] ergo primum *om.* T *(mg.)* C
[p] agnus non *corr. mg. ex* agens rerum T
[q] similiter *om.* TC
[r] et] non *add. s.l.* O
[s] et] quia *add. s.l.* O
[t] primae rationi *scrips. sed del.* O

multiplies itself happens per se but not primarily, because the multiplication of a substantial form is latent in the multiplication of an accidental form.

2. Aristotle seems to agree with this view when he says that a stone is not in the soul, but the species of a stone.[274] But if only the colour of a stone were in the soul, the whole species of a stone would not be in the soul. It is manifest, therefore, that, according to Aristotle, the species of a substantial form is multiplied together <with the species of an accident>.

3. Furthermore, the intellect only makes a judgment about a thing which was in the senses beforehand; but the intellect makes a judgment about both the substance and the quality of an object; therefore, both the species of a substance and of a quality are multiplied in the senses.

4. Furthermore, if only colour reached the senses, then, since brute animals only judge about what is received in the senses, a brute animal, say a dog, would not recognise his owner more than another animal if they were of the same colour, but we know this to be false; therefore, the antecedent <i.e., that only colour reaches the senses, is false>.

5. Furthermore, if it is true that only the species of a colour is received in the <sense> organ, then a lamb would no more run away because of the presence of a wolf than because of the presence of another animal of the same colour, which we likewise know to be false.

6. Therefore, it must be said that each of the two multiplies its species, but colour is that which primarily and per se modifies the senses and a substantial form is multiplied together with it under the species of a colour.

7. Thus, to the first argument[275] it must be said that the substantial form of the heavens themselves multiplies its species here below.

[274] Arist. *De anima* III.8, 431b29–432a1.

[275] See above, q. 40.1, n. 1.

8. Ad primum argumentum in contrarium dicendum quod, cum dicit Aristoteles quod [a]non exterius se faciunt [bc]quaecumque etc., haec propositio est intelligenda quod non exterius se faciunt ad generationem et corruptionem [d]quae [e]nec sunt contraria[f] etc.[g] Bene enim scimus quod corpora caelestia multiplicant suas virtutes hic inferius ad conservationem inferiorum et similiter[h] virtutes luminares corporum caelestium faciunt se inferius ad[i] immutationem, et tamen virtuti caelesti non est aliquid contrarium.

9. Secundo argumento probanti minorem dicendum per distinctionem huius quod est 'alligari materiae', quia, comparando [j]in gradu rei, quaedam forma magis est condicionibus materiae ligata, [k]et quaedam minus, sicut forma intelligentiae minus ligatur materiae, hoc est minus sequitur condiciones materiae, quam forma caeli, et illa minus[l] quam forma animalis[m], et illa minus quam forma mixti, et sic semper descendendo. Vel minus ligatur materiae non secundum gradum rei, sed unum est res et aliud intentio; et sic forma quae res est in subiecto[n] magis alligatur materiae quam sua species sive intentio. Sed hoc habito, distinguo[o] sic: quaedam est transmutatio quae est per contrarietatem, et est in deficiendo, quia contrarietas est causa corruptionis, et haec transmutatio est in elementis. Alia est transmutatio quae est in proficiendo, et hoc tantum [p]in praesentia agentis, [qr]sicut multiplicatio habitus realis in medio quae est lux; et multiplicatio fit ad habitum, et sic in proficiendo; et haec fit per praesentiam corporis luminosi. Vel in deficiendo, et hoc per absentiam agentis, quia absente corpore luminoso fit mutatio habitus in

[a] non exterius se *corr. mg. ex* exinsi[ne] T
[b] quaecumque ... corruptionem *om.* T
[c] quaecumque ... faciunt *om. hom. sed suppl. mg.* hoc intelligendum est quantum C
[d] quae nec *corr. s.l. ex* quin T
[e] nec sunt] non C
[f] contraria] nec ex contrariis, haec propositio est intelligenda quod non exterius se faciunt ad generationem et corruptionem quin (quin *corr. s.l. in* quae non) sunt contraria *add.* T; nec ex contrariis *add.* C
[g] etc. *corr. s.l. ex* nec etiam cum T
[h] similiter] etiam TC
[i] ad] sive extra *praem.* T
[j] in gradu *corr. ex* in gradum T; gradum C
[k] et quaedam minus *om.* O; quaedam minus C
[l] minus] ligatur *add. sed del.* O
[m] animalis *scripsimus*; elementaris OTC
[n] subiecto] substando O
[o] distinguo] transmutationem *add.* TC
[p] in praesentia] per praesentiam C
[q] sicut ... habitum *corr. mg. ex* et multiplicatio fit ad habitum (quia illa multiplicatio fit ad habitum quendam *pro* et multiplicatio fit ad habitum) T
[r] sicut ... agentis *om. hom sed suppl. mg.* C

8. To the first argument to the contrary[276] it must be said that, when Aristotle says that no things <that are not contraries or from contraries> extend outside themselves,[277] this proposition should be understood in the sense that things that are not contrary <or from contraries> do not extend outside themselves with regard to generation and corruption; for we do indeed know that celestial bodies multiply their powers here below for the preservation of the things here below, and similarly the illuminating powers of the celestial bodies extend themselves here below with regard to change, although there is nothing contrary to a celestial power.

9. To the second argument proving the minor premise[278] one must reply by drawing a distinction about what it is to be joined to matter, because (i) if we compare the degrees of things, one form is more fully joined to the conditions of matter and another less fully. For example, the form of an intelligence is less fully joined to matter —that is, it follows the conditions of matter less fully— than the form of the heavens, and the form of the heavens less fully than the form of an animal, and the form of an animal less fully than the form of a mixed body, and so on always proceeding downwards in this way. Or (ii) something is less fully joined to matter <than something else> not with regard to the degrees of things, but because one is a thing and the other is an intention; in this way the form which is a thing in a subject is more fully joined to matter than its species or intention. But given this, I make the following distinctions: (a) some kind of change occurs through contrariety, and this kind of change is a deterioration, because contrariety is a cause of corruption. This kind of change occurs in the elements. (b) Another kind of change is that which is an improvement, and this occurs only in the presence of an agent, as does the multiplication of a real habit, like light, in the medium. This kind of multiplication is directed towards a habit, and so it is an improvement, and it occurs because of the presence of a luminous body. Or (c) <another kind of change> is a deterioration, and this occurs because of the absence of an agent, given that when a luminous body is absent, the change of a habit into

[276] See above, q. 40.1, n. 2.
[277] Arist. *De gen. et corr.* I.7, 323b28–29.
[278] See above, q. 40.1, n. 3.

privationem[a], ut illuminationis in tenebra, quae mutatio est aliqualis transmutatio.

10. [b]Unde multiplicatio est duplex: vel realiter vel intentionaliter. Si realiter, hoc dupliciter, quia aut multiplicatur commune in plura supposita, et sic multiplicatur forma communis in sua supposita, aut est multiplicatio alicuius per aliquem suum habitum realem, et sic multiplicatur lux in medio. Aut est multiplicatio intentionalis, sicut speciei. Si autem loquamur de multiplicatione formae per supposita, sic quanto forma, quae res est, magis ligatur materiae, tanto magis[c] multiplicatur. Si autem loquamur de multiplicatione[d] substantialis habitus in alio, sic[e] in corporalibus forma minus ligata materiae magis se multiplicat. Magis enim se multiplicat forma caeli quam aliqua forma elementaris et formae stellarum magis quam formae aliarum partium orbis. Et[f] sic forma minus ligata materiae magis transmutatur transmutatione quae est multiplicatio sic dicta [g]in genere corporum. Item si loquamur de multiplicatione intentionali, tunc, cum corpora caelestia apertius videantur quam[h] alia corpora, sic adhuc inter formas corporales forma materiae magis ligata minus multiplicat suam speciem in sensum. Hoc viso patet responsio argumento.

11. Ad alia duo argumenta patet responsio super quaternum *De memoria et reminiscentia*, super illam quaestionem 'An magnitudo immittat[i] speciem in organo.'

<Quaestio 41>
Consequenter arguit Aristoteles sic: si omnia sunt unum, tunc idem est bonum et malum.

1. Quod autem non sit inconveniens idem esse bonum et malum probatio[j]: esse est per se entis; ergo esse mali est per se entis. Sed malum non est ens nisi a bono, quod est per se ens. Sed a quocumque est malum, ipsum est malum; bonum ergo est malum.

[a] privationem] illi *add. sed del.* O

[b] Unde … dupliciter] Unde multiplicatio dupliciter *scrips. sed corr. mg. in* Item nota quod multiplicatio est dupliciter vel realis vel intentionalis si realis adhuc dupliciter T; si multiplicatio dupliciter *scrips. sed corr. s.l. et mg. in* unde multiplicatio dupliciter aut realiter aut intentionaliter (?) si realiter C

[c] magis] ligatur materiae *add. sed del.* T

[d] multiplicatione] aliqualis *add. sed del.* O

[e] sic] sed O; licet C; *corr. s.l. in* cum T

[f] Et] (tunc *s.l.* T) in genere (talium *s.l.* T) corporum *add.* TC

[g] in genere corporum *om.* TC

[h] quam] aliqua *add.* O

[i] immittat] immutat O; immutet C

[j] probatio] videtur sic T; videtur C

its privation occurs, such as the change of light into darkness, and this change is a kind of transformation.

10. Thus, there are two kinds of multiplication: either (i) real or (ii) intentional. Case (i) is twofold, because either (ia) what is common is multiplied in many supposits, and in this way a common form is multiplied in its supposits, or (ib) it is the multiplication of something in virtue of a real habit it has, and in this way light is multiplied in the medium. Case (ii) is an intentional multiplication, such as that of a species. If we are speaking of (ia), the multiplication of a form in its supposits, then in this sense the more fully joined to matter is the form that is a thing, the more it is multiplied. If we speak of (ib), the multiplication of a substantial habit in another thing, then in this sense a form less fully joined to matter multiplies itself more in corporeal things; for the form of the heavens multiplies itself more than does an elementary form and the forms of the stars multiply themselves more than do the forms of the other parts of the orb. Thus in the genus of bodies a form less fully joined to matter is more changed by the change which is called a multiplication in sense (ib). Furthermore, if we are speaking of an intentional multiplication, then, since the celestial bodies are seen more openly than other bodies, then in this sense too among corporeal forms a form more fully joined to matter is less apt to multiply the species <it sends> towards the senses. Now that we have seen this, the reply to the argument is clear.

11. The reply to the other two arguments[279] is made clear in the Question 'Whether a magnitude introduces its species into the <sense> organ' which is found in the quire containing *On Memory*.[280]

<Question 41>
Next Aristotle argues as follows: if all things are one, then the same thing is <both> good and bad.[281]

1. Proof that it is not absurd for the same thing to be <both> good and bad: being belongs to a being per se; therefore, the being of something bad belongs to a being per se. But what is bad is a being only in virtue of what is good, which is a being per se. But everything from which what is bad derives is itself bad; therefore, what is good is bad.

[279] See above, q. 40.1, n. 4. This reply refers to the fourth argument for the negative and to the proof of the assumption on which that argument relies.
[280] Not found in the anonymous questions on *De memoria et reminiscentia*, which are possibly by Aspall, preserved in MS N.
[281] Arist. *Phys*. I.2, 185b19–22.

2. Item, quidam supposuerunt[a] quod est summe malum sicut est aliquod summe bonum, quos reprobat Aristoteles, dicens quod non est summe malum. [b]Tunc sic: omne malum est causatum; aut ergo a bono aut a malo. Si a malo, tunc eadem est quaestio, et ita aut erit procedere in infinitum aut tandem [c]dandum est quod aliquod malum est causatum a bono. Sed quod est causa mali est malum; ergo aliquod malum est bonum.

3. Item, Aristoteles in *Topicis* dicit quod non bona[d] sunt magis eligenda bonis, ut si non bona contineant in se unum valde bonum et [e]aliud[f] parvum malum et si bona contineant in se duo parva bona. Arguo[g] tunc sic: cum sic dicitur: 'non bona', cum sit pluralis numeri, geminat suum singulare; idem ergo est 'non bona'[h] [i]quod non[j] bonum et non bonum. Si ergo haec sunt magis eligenda bonis, haec erunt vel aequaliter bona vel [k]maius bonum. Et ita non bonum est bonum, et non bonum est malum; ergo etc.

4. Item, eadem sunt principia essendi et cognoscendi; sed habitus est principium cognoscendi privationem, quia dicit Aristoteles quod rectus[l] est iudex sui et obliqui; bonum ergo est principium cognoscendi malum. Sed quod est principium cognoscendi est principium essendi; bonum ergo est principium essendi malum. Et quod est causa[m] etc.

5. Item, Aristoteles in libro[n] *De generatione*: "Propter longe distare a primo sunt quaedam corruptibilia et quaedam incorruptibilia, et quaedam [o]bona et quaedam mala". Si ergo propter longe distare a primo sunt quaedam mala, cum ergo omnia aequaliter distent a primo, quia inter creatorem et creaturam est infinita distantia, et infinitum infinito non est [p]maius [q]neque minus, tunc omnia sunt aequaliter mala. Sed aliqua sunt bona; [r]ergo [s]aliqua sunt bona et mala. **[O 94rb]**

[a] supposuerunt] posuerunt TC
[b] Tunc sic] quia si sic *scrips. sed corr. s.l. in* cum C
[c] dandum est] standum est et dicendum C
[d] bona] bonis *add.* C
[e] aliud … se *om.* C
[f] aliud] ad O
[g] Arguo *om.* TC
[h] bona *om.* C
[i] quod non bonum *suppl. mg.* T
[j] non] est *add.* O
[k] maius bonum] unum erit unus *scrips. sed corr. mg. in* magis bonum T, sed sive sic sive sic semper habeo propositum quod utrumque est bonum *add. sed del.* T
[l] rectus] idem O
[m] causa] mali est malum ergo *add.* T
[n] libro *om.* TC
[o] bona et quaedam *om. hom. sed suppl. mg*. T
[p] maius neque *om.* C
[q] neque minus *om.* T
[a] ergo ... mala] bona ergo sunt mala C
[b] aliqua … mala] etc. *scrips. sed corr. mg. in* bona sunt mala T

2. Furthermore, some people assumed that there is something maximally bad just as there is something maximally good, and these people are refuted by Aristotle, who says that there is nothing maximally bad.[282] Then <it is argued> as follows: everything bad is caused; therefore, either it comes from something good or from something bad. If it comes from something bad, then the same question arises, and thus there will either be an infinite regress or at some point one must concede that there is something bad caused by something good. But what is a cause of something bad is bad; therefore, something bad is good.

3. Furthermore, in the *Topics* Aristotle says that things that are not good are more to be chosen than good things,[283] for example if the things that are not good contain in themselves one thing that is a great good and another thing that is a small bad thing, and if the things that are good contain in themselves two small good things. Then I argue as follows: when he says 'things that are not good', since this is plural, it duplicates its singular; therefore, 'things that are not good' signifies the same as 'a thing which is not good and a thing which is not good.' Therefore, if these things ought to be chosen rather than good things, these things will be either equally good or greater goods. And thus what is not good is good, and what is not good is bad; therefore, <the same thing is both good and bad >.

4. Furthermore, the principles of being and of knowing are the same; but habit is the principle of knowing a privation, because Aristotle says that the straight is the judge of itself and of the oblique;[284] therefore, what is good is the principle of knowing what is bad. But what is a principle of knowing is a principle of being; therefore, what is good is the principle of being for what is bad. And what is a cause <of what is bad is bad>.

5. Furthermore, in *On Generation* Aristotle says: it is because of being at a great distance from the first being that some things are corruptible and others are incorruptible, and some things are good and others are bad.[285] Therefore, if some things are bad because they are at a great distance from the first being, then since all things are equally distant from the first being, given that between the creator and a creature there is an infinite distance, and one infinite is neither greater nor smaller than another infinite, it follows that all things are equally bad. But some things are good; therefore, some things are <both> good and bad.

[282] This is probably a reference to Arist. *Eth. Nic.* IV.12, 1126a12–13. But see also, e.g., Aug. *De diversis quaestionibus octoginta tribus,* q. 6, CCSL 44A, 14.2–9.

[283] Arist. *Top.* III.2, 117a21–22.

[284] Arist. *De anima* I.5, 411a5.

[285] Arist. *De gen. et corr.* II.10, 336b27–34.

6. Item, quaecumque reperiuntur ut res simul in eodem indivisibili sunt unum; sed suppono quod virtus et vitium et huiusmodi non faciunt speciem in anima, sed sunt ibi ut res et qualitates animae; tunc, si aliquis sit virtuosus, virtus est qualitas animae ipsius. Et potest esse quod ille idem cognoscat vitium, et non cognoscitur per speciem. Est ergo malum sive vitium in ipso sicut[a] res et natura. Et ita malum et bonum sunt in eodem indivisibili, et ita sunt unum.

7. Dicendum quod quoddam est ens simpliciter in actu, ita quod nihil[b] admiscetur illi de potentia; et sic est tantum prima causa in actu, ita quod nullo modo in potentia, secundum Aristotelem. Et, secundum Aristotelem, non est malum nisi post potentiam. Cum ergo in primo nullo modo sit potentia, in primo nullo modo erit malum nec est[c] aliquo modo potentia ad non esse. Aliter est aliquod[d] ens in actu ita quod, cum [e]sit creatum ex nihilo, potentiam quantum[f] [g]est[h] de se habet ad non esse propter hoc[i] quod est ex nihilo. Verumtamen non habet potentiam inclinativam ad non esse, quia non est [j]unio formae cum materia mediante aliqua privatione sive contrarietate. Et huiusmodi sunt intelligentiae et omnia[k] caelestia. Et in his, quantum est[l] de se, est potentia ad malum, licet non sit ibi potentia inclinativa ad malum. Tertio modo est aliquid ens quod habet potentiam inclinativam ad non esse sive ad malum propter hoc quod est ex contrariis. Et tale ens sunt huiusmodi mixta elementaria [m]et elementa; unde ista inclinantur ad malum.

8. Sed duplex est malum, [n]scilicet malum incommodi et malum poenae. Si loquamur de malo incommodi[o], sic dicit Augustinus quod quoddam est malum incommodi quod solam privationem dicit, sicut caecitas in oculo solam privationem dicit. Aliud est malum incommodi quod non solam privationem dicit, sicut dicimus quod dolere de aliquo est malum incommodi, et tamen dolere aliquid positionis[p] dicit. Aliud est malum iniustitiae et[q] malum

[a] sicut] ut TC
[b] nihil *corr. s.l. ex* nec aliquid T
[c] est] in ipso *add. s.l.* O
[d] aliquod] aliquid TC
[e] sit creatum] creatur C
[f] quantum] quam TC *sed corr. s.l.* T
[g] est ... habet] de se habet est C
[h] est *suppl. s.l.* O
[i] hoc] illud T; id C
[j] unio formae] uniforme TC *sed corr. mg. in* in ipso unio formae T; *corr.* C
[k] omnia] intelligentia *add. sed del.* O; *corr. s.l. in* corpora T
[l] est *om.* OTC *sed suppl. s.l.* (*post* se T) OT
[m] et elementa *suppl. s.l.* T
[n] scilicet ... poenae] aut malum incommodi aut malum iniustitiae TC
[o] incommodi *om. sed* primo modo *suppl. s.l.* O
[p] positionis *corr. mg. ex* privatio est sibi T
[q] et] est *add.* T; quod est C

6. Furthermore, all the things that are found as things in the same place and in the same indivisible are one; but I assume that virtue and vice and the like do not produce a species in the soul, but that they are there as things and qualities of the soul. Then, if someone is virtuous, virtue is a quality of his soul, and it can happen that that same man knows vice, and vice is not known through a species. Therefore, what is bad, namely a vice, exists in that man as a thing and a nature. And thus what is bad and what is good are in the same indivisible, and so they are one.

7. It must be said that (i) one kind of being is absolutely in act, so that nothing of potency is mixed with it; and only the first cause is in act in this way, so that it is in no way in potency, according to Aristotle. And according to Aristotle, what is bad only exists subsequent to a potency.[286] Therefore, since there is in no way any potency in the first being, in no way can what is bad or what is in some way a potency to non-being exist in the first being. (ii) Another kind of being in act is such that, since it is created from nothing, it has a potency to non-being in so far as it depends on itself, because it comes from nothing. It does not, however, have a potency that inclines it towards non-being, because the union of its form with matter does not happen because of a privation or contrariety. The intelligences and all celestial bodies are of this kind. In these things, in so far as it depends on them, there is a potency to what is bad, although there is no potency there that inclines <them> to what is bad. (iii) A third kind of being has a potency that inclines <it> to non-being or to what is bad because of the fact that a being of this kind is composed of contraries. The bodies mixed from the elements and the elements <themselves> are beings of this kind; thus, these beings have an inclination towards what is bad.

8. But there are two kinds of bad: (i) <one kind of bad is> the bad of inconvenience and the bad of penalty. Speaking of the bad of inconvenience, Augustine says that (ia) one kind of bad of inconvenience only indicates a privation, just as blindness only indicates a privation in an eye. (ib) Another kind of bad of inconvenience does not only indicate a privation, as when we say that to feel pain in respect of something is a bad of inconvenience, and yet to feel pain indicates something positive. (ii) Another kind of bad is the bad

[286] Arist. *Metaph.* IX.9, 1051a18–19.

vitii sive peccati. Et illud malum nihil est nisi carentia quaedam sive privatio. Sed, secundum Aristotelem, [a]nullus actus est peccatum sive malum, sed circumstantiae actus, secundum Aristotelem, faciunt quod actus est malus. Et quod[b] vitium non sit aliquid per positionem ostendit Augustinus tali exemplo: si esset furiosus aliquis ligatus catena et contingeret quod ipse rumperet[c] catenam et[d] multa mala faceret, causa totius mali esset catenae ruptio. [e]Et tamen in veritate ruptio catenae nihil est. Unde sic est in proposito quod sensus debet succumbere rationi; si ergo ratio succumbat sensui, tunc defectus dominii rationis est causa mali [f]causati a dominio sensus; et tamen defectus dominii rationis[g] nihil positionis est, sed sola privatio. Et sic dicunt quod peccatum non est nisi privatio quaedam.

9. Secundum autem magistros nostros dicitur [h]sic, quod malum [i]potest considerari vel ratione illius a quo imponitur nomen eius, et sic nihil positionis dicit, sed solam carentiam boni, quia a sola carentia imponitur nomen; vel [j]potest considerari ratione eius cui imponitur [k]nomen, et sic malum non dicit solam carentiam, quia 'malum' imponitur a privatione boni in debente habere bonum, sicut 'caecum' imponitur privato visu et debenti habere visum. [l]Et ideo male dicitur lapis[m] caecus, quia lapis non est [n]aptus natus habere visum, et sic non dicit solam privationem.

10. Quaestioni ergo dicendum quod duplex est esse mali: vel esse suae purae privationis, et sic nullo modo est idem[o] esse boni et mali; aliud est esse mali quod est esse suae substentationis, scilicet secundum quod nominat privationem boni cum eo in quo natum est esse bonum, et sic potest esse idem esse boni et mali.

11. Et nota quod bonum non est substentativum mali in quantum ipsum bonum est, sed unde bonum deficiens, [p]et deficiens unde[q] ex nihilo.

[a] nullus actus *corr. ex* nullum accidens T
[b] quod] illud *add.* TC *sed del.* T
[c] rumperet] rumpet O
[d] et *suppl. s.l. (lectio incerta)* O
[e] Et tamen] nihil tamen unum *scrips. sed corr. in* tamen T
[f] causati … sensus *om.* C
[g] rationis] est causa mali *add. hom. sed corr.* C
[h] sic, quod malum] malum quia C
[i] potest considerari] dupliciter potest considerari vel T
[j] potest considerari *suppl. mg.* T
[k] nomen, et sic *suppl. mg.* C
[l] Et … visum *om. hom. sed corr. mg.* T
[m] lapis *suppl. s.l.* O
[n] aptus natus] aptum natum OT *(mg.)*
[o] idem *suppl. s.l.* O
[p] et deficiens *om. hom.* C
[q] unde] quia TC

of injustice and the bad of vice or sin.[287] And that kind of bad is a mere lack or privation. But according to Aristotle, no act is a sin or something bad <in itself>, but it is the circumstances of an act that make an act bad.[288] And Augustine shows that a vice is not something positive by the following example: if a wild man were tied to a chain and it came about that he broke the chain and committed many bad acts, the cause of the whole bad would be the breaking of the chain. In truth, however, the breaking of the chain is just nothing.[289] Accordingly, in the present case, <we know> that the senses should submit to reason; therefore, if reason submits to the senses, then the failure of reason's domination is the cause of the bad caused by the domination of the senses; the failure of reason's domination, however, is not something positive, but only a privation. In this way these people say that a sin is only a privation.

9. According to our masters, however, it is said that what is bad can be considered either (i) in accordance with the thing from which the name is imposed, and in this way what is bad does not express something positive, but only the lack of what is good, because the name is only imposed from the lack; or (ii) it can be considered in accordance with the thing on which the name is imposed, and in this way what is bad does not only express a lack, because 'bad' is imposed in virtue of the privation of what is good in something that should be good, just as 'blind' is imposed on a thing deprived of sight and that should have sight. Thus it is not appropriate to say that a stone is blind, because a stone is not apt by nature to have sight, and so 'blind' does not only express a privation.

10. To the question, therefore, it must be said that what is bad has two kinds of being: either (i) the being of its pure privation, and in this way the being of what is good and of what is bad are in no way the same; (ii) another kind of being of what is bad is the being of what is a subject for it, that is, in so far as the name 'bad' expresses the privation of what is good together with the thing in which what is good is apt by nature to exist, and in this way the being of what is good and of what is bad can be the same.

11. Note that what is good is not a subject for what is bad in so far as it is good, but in so far as it is a defective good and it is defective because it comes from nothing.

[287] The correct reference is Anselm, *De casu diaboli,* c. 26, ed. Schmitt, 274.8–15.

[288] Arist. *Eth. Nic.* III.2, 1111a3–19.

[289] Cf. Anselm, *De casu diaboli,* c. 26, ed. Schmitt, 274.16–24. Anselm's explanation is not based on the example of a man's breaking the chain, but on similar ones.

12. Primae ergo et secundae rationi dicendum quod bonum est causa mali, sed non unde bonum, sed unde deficiens, et deficiens[a] unde[b] ex nihilo.

13. Argumento *Topicorum* dicendum quod li 'non bona' potest negare multitudinem bonorum, et sic intendit Aristoteles, cum[c] vult quod, si unum sit magnum bonum et reliquum parvum malum, ista sunt[d] non bona, hoc est quae non sunt multa bona sunt magis [e]etc., sicut in proposito casu; et sic non procedit argumentum de geminatione[f] singularis. Vel potest li 'non bona' privare qualitatem [g]boni de duobus, et sic non intendit Aristoteles, et sic procedit geminatio[h] singularis.

14. Ad aliud, cum dicit quod eadem sunt principia essendi et cognoscendi, mihi videtur quod non est textus[i], sed[j] quod sunt proportionalia etc.[k] Et etiam dicit Aristoteles[l] quod secundum proportionem sunt eadem. [m]Vel si textus[n] sit, tunc [o]est propositio sic intelligenda, quod eadem sunt in his quae cognoscuntur[p] ratione propria, sicut eadem[q] principia[r] per quae cognoscitur[s] homo per[t] suam causalitatem [u]eadem sunt principia essendi hominem. In his autem[v] quae cognoscuntur non in ratione propria, sed mediante alio, ut mediante habitu[w], [x]sicut cognoscitur privatio, non sunt eadem principia[y].

15. Ad ultimum dicendum quod nec virtus nec vitium faciunt speciem in sensum sive intellectum, sed cognoscuntur per suos effectus, ut virtus ex frequenter bene operari et vitium ex male operari. Unde virtuosus potest cognoscere vitium, [z]etsi ipse non operetur, sicut patet prius.

[a] deficiens *scrips. sed del.* O
[b] unde *om.* O
[c] cum] quia TC
[d] sunt *suppl. s.l.* O
[e] etc. ... casu] etc. TC *sed del. et corr. mg. in* elicienda duobus bonis parvis T
[f] geminatione *corr. s.l. ex* singnatione C
[g] boni de] in C
[h] geminatio *post* singularis T; generatio O; *corr. s.l. ex* singnatio C
[i] textus] tixtus OT
[j] sed] si est textus scilicet (*corr. s.l.* si *in* sed) T
[k] etc.] principia essendi *praem.* TC
[l] Aristoteles *om.* TC
[m] Vel ... sit] si autem sit textus (tixtus T) TC
[n] textus] tixtus O
[o] est ... sunt] sic (sic *om.* C) propositio est vera TC
[p] cognoscuntur] in *add.* TC
[q] eadem] sunt *add.* TC
[r] principia *om.* OC *sed suppl. s.l.* O
[s] cognoscitur] cognoscit (*lectio incerta*) O
[t] per *om.* O
[u] eadem sunt] et T
[v] autem *om.* TC *sed suppl. s.l.* T
[w] habitu] quae sic cognoscuntur *add. (lectio incerta)* C
[x] sicut cognoscitur privatio *suppl. mg.* C
[y] principia] etc. *add.* TC
[z] etsi ipse] et ipse si *scrips. sed corr.* O

12. To the first and second arguments,[290] therefore, it must be said that what is good is a cause of what is bad, but not in so far as it is good, but in so far as it is defective and it is defective in so far as it comes from nothing.

13. To the argument from the *Topics*[291] it must be said that (i) the expression 'things that are not good' can negate a multitude of good things, and this is what Aristotle means, since he wants to say that, if one thing is a great good and another is a small bad, these things are not good things, that is, the things that are not many good things are more <to be chosen than the things that are good>, as happens in the present case. In this way the argument about the duplication of the singular does not succeed. Or (ii) the expression 'things that are not good' can deny the quality of good of two things, and this is not what Aristotle means, and in this way the duplication of the singular succeeds.

14. To the other argument,[292] when Aristotle says that the principles of being and of knowing are the same,[293] it seems to me that this is not what the text of Aristotle says, but the text says that <the principles of being and of knowing> are proportional. Aristotle also says that they are the same by analogy. Or if this is what the text says, then the proposition should be understood in the sense that these principles are the same in the things known through their proper *ratio*, for example the principles through which a man is known through his causality are the same as the principles of his being. But in the things not known through their proper *ratio*, but through something else, for example through a habit, in the way that a privation is known, those principles are not the same.

15. To the last argument[294] it must be said that neither virtue nor vice produces a species in the senses or in the intellect, but they are known through their effects, so that virtue is known from acting well frequently and vice from acting badly. Thus, a virtuous man can know vice even if he himself does not act, as was made clear before.

[290] See above, nn. 1–2.
[291] See above, n. 3.
[292] See above, n. 4.
[293] This is probably a reference to Arist. *Metaph*. II.1, 993b23–31.
[294] See above, n. 6.

1*. Quaeritur de principiis corporis mobilis. Et primo de materia, circa quam **[O 94va]** quaeruntur viginti quaestiones.

<Quaestio 42>
Prima est utrum materia prima sit.

<Quaestio 42.1>

1. Et videtur quod non, quia quanto plus habet aliquid de forma, tanto plus habet de esse; ergo quanto minime de forma, tanto minime de esse. Sed materia prima minime habet de forma, quia omnino caret forma; ergo etc.

2. Item, omne quod est aut est universale aut particulare; sed materia prima nec[a] universalis, quia universale est forma; quod ergo est sine forma, illud est sine universalitate; sed materia prima non est forma nec [b]habet formam; ergo non erit universalis. Nec particularis, quia universale praedicatur de particulari secundum identitatem; sed universale non sic potest praedicari de materia prima. Tum quia universale est forma; cum ergo materia prima et forma non sunt idem, nec erunt[c] idem universale et materia prima, et ita unum non potest praedicari de altero etc. Tum quia dicit Aristoteles in septimo[d] *Metaphysicae* quod [e]alia est natura materiae primae et cuiuscumque praedicabilis; ergo etc.

3. Item, omne quod est aut est hoc aliquid aut essentia. Sed materia prima non est hoc aliquid, quia sic esset in praedicamento, quod falsum est. Nec essentia. Probatio: omnis essentia aut est generalis aut specialis; sed prima materia neque est[f] essentia generalis neque essentia specialis; ergo nullo modo est essentia; ergo etc.

4. Item, ens et unum convertuntur; quod ergo non est unum non est ens. Sed materia prima non est unum, quia omne quod est unum est forma vel formam habens; sed materia prima non est forma nec formam habens, [g]quia solus actus sive sola forma dividit et distinguit; ergo materia prima non est unum; ergo nec ens.

[a] nec] non est N
[b] habet formam] formam habens N
[c] erunt] erit O
[d] septimo] secundo O
[e] alia ... cuiuscumque] natura materiae primae est alia ab essentia cuiuslibet N
[f] est *suppl. s.l.* O
[g] quia ... distinguit *fort. post* vel formam habens *legendum*

1*. Questions about the principles of a mobile body. And first about matter, about which there are twenty questions.

<Question 42>
The first question is whether prime matter exists.

<Question 42.1>

1. Arguments for the negative: the more form something has the more being it has; therefore, if something has a minimal amount of form, it has a minimal amount of being. But prime matter has a minimal amount of form, since it lacks every form; therefore, <prime matter has a minimal amount of being>.

2. Furthermore, everything that exists is either a universal or a particular; but prime matter is neither a universal, because a universal is a form, and so what is without form is without universality; but prime matter is neither a form nor has a form; therefore, it will not be a universal. Nor is prime matter a particular, because a universal is predicated of a particular by identity; but a universal cannot be predicated in this way of prime matter: both because (i) a universal is a form; therefore, since prime matter and a form are not the same, a universal and prime matter will not be the same either, and so one cannot be predicated of the other <by identity>; and (ii) because in *Metaphysics* VII Aristotle says that the natures of prime matter and of any predicable are not the same;[295] therefore, <a universal cannot be predicated of prime matter by identity>.

3. Furthermore, everything that exists is either a determinate thing or an essence. But prime matter is not a determinate thing, because if it were it would be in a category, which is false. Nor is prime matter an essence. Proof: every essence is either generic or specific; but prime matter is neither a generic essence nor a specific essence; therefore, it is in no way an essence; therefore, <prime matter does not exist>.

4. Furthermore, being and one are convertible; therefore, what is not one is not a being. But prime matter is not one, because everything that is one is a form or has a form; but prime matter neither is a form nor has a form, because only an act or only a form divides and distinguishes; therefore, prime matter is not one, and so it is not a being either.

[295] Arist. *Metaph.* VII.3, 1029a20–23.

5. Item, materia prima est incorruptibilis, ut probat Aristoteles in fine primi *Physicorum*. [a]Sed quod non sit incorruptibilis probatio, quia materia prima est aeterna, et aeternum non est pars rei[b] corruptibilis, secundum Commentatorem[c], et si[d] esset, esset pars rei corruptibilis; ergo etc.

6. Item, Commentator paulo post: de necessitate omne passivum componitur ex passivis; sed materia prima non est[e] passiva; ergo non potest esse pars rei passivae, et si esset, tale esset. Quare nullo modo est[f].

7. Si materia prima esset, tunc esset substantia. Sed quod non sit substantia probatio: dicit Aristoteles in secundo *Physicorum* quod substantia est ad aliquid; sed materia prima non est ad aliquid, quia, si esset ad aliquid, tunc non esset prima, quia simul tunc esset aliud cum ea[g]; ergo non est substantia; ergo etc.

8. Item, si materia prima esset, tunc esset principium existendi compositum.

9. Sed contra: ista tria se habent secundum ordinem: ens solum in potentia, ens in potentia ordinata ad actum, ens complete in[h] actu. Sed solum ens in potentia, ut materia prima, non est principium existendi ens[i] quod est sibi propinquius; ergo nec eius[j] quod est remotius a se. Sed[k] materia prima non est principium existendi formae, quia sic ex ea[l] fieret forma, et hoc est[m] falsum. Et forma propinquius se tenet materiae[n] quam [o]ens completum; quare etc.

10. Item, ista tria se habent secundum ordinem: pure nihil et ens in potentia et ens actu [p]pure et complete, quia ens in potentia est medium inter pure nihil et pure ens actu; ergo quae est proportio, ex quo se habent secundum ordinem, inter primum et secundum, talis erit proportio inter secundum et tertium. Sed pure nihil[q] non est principium existendi ens in potentia, quia infinita est distantia inter pure nihil et ens in potentia; ergo similiter nec ens in potentia erit principium existendi ens actu complete, ut compositum; ergo etc.

[a] Sed … quia *scrips. sed del.* O
[b] rei *ex. corr. s.l.* O
[c] Commentatorem] in secundo Metaphysicae *add.* N
[d] si] materia prima *add. mg.* N
[e] est] potest esse N
[f] est] corruptibilis *add.* ON *sed del.* O
[g] ea *scripsimus*; eo OTC
[h] in] ultimo *add.* N
[i] ens] eius N
[j] eius] ens N
[k] Sed *scripsimus*; quia ON
[l] ea] eo O
[m] est *suppl. s.l.* O
[n] materiae *om.* ON *sed suppl. s.l.* O
[o] ens completum] eius compositum N
[p] pure et complete] puro et completo N
[q] nihil *suppl. s.l.* O

5. Furthermore, prime matter is incorruptible, as Aristotle proves at the end of *Physics* I.[296] The following argument, however, proves that prime matter is not incorruptible: prime matter is eternal, and what is eternal is not a part of a corruptible thing, according to the Commentator;[297] but, if prime matter existed, it would be a part of a corruptible thing; therefore, <prime matter does not exist>.

6. Furthermore, the Commentator says shortly after: everything that is a subject for a passion is necessarily composed of things that are subjects for a passion;[298] but prime matter is not a thing that is a subject for a passion; therefore, it cannot be a part of a thing that is a subject for a passion; but if prime matter existed, it would be such a part. Hence, prime matter in no way exists.

7. If prime matter existed, it would be a substance. But the following argument proves that it is not a substance: in *Physics* II Aristotle says that a substance is a relation;[299] but prime matter is not a relation, because if it were a relation then it would not be first, because there would be something else simultaneous with it; therefore, it is not a substance; therefore, <prime matter does not exist>.

8. Furthermore, if prime matter existed, then it would be a principle of existence for the composite.

9. But the following arguments are against this: these three, a being only in potency, a being in a potency ordered to an act, <and> a being completely in act, are in an order; but a being that is only in potency, namely prime matter, is not a principle of existence for what is closer to itself; therefore, neither is it <such a principle> for what is more remote from itself. But prime matter is not a principle of existence for form, because if it were, form would come to be from prime matter, and this is false; but form is closer to matter than is a complete being; therefore, <prime matter is not a principle of existence for the composite>.

10. Furthermore, these three, what is absolutely nothing, what is in potency, and what is absolutely and completely in act, are in an order, because what is in potency is intermediate between what is absolutely nothing and what is absolutely in act; therefore, given this ordering, the relation between the second and the third is the same as the relation between the first and the second. But what is absolutely nothing is not a principle of existence for what is in potency, because there is an infinite distance between what is absolutely nothing and what is in potency; therefore, in the same way what is in potency will not be a principle of existence for what is completely in act, namely the composite; therefore, <prime matter is not a principle of existence for the composite>.

[296] Arist. *Phys.* I.9, 192a25–29.
[297] Aver. *In Metaph.* II, t.c. 7, fol. 32va.
[298] Aver. *In Metaph.* I, t.c. 38, fol. 22va.
[299] Arist. *Phys.* II.2, 194b8–9.

11. Item, dicit Aristoteles in septimo[a] *Metaphysicae* sic: “Dico autem materiam neque quale neque quantum neque aliquod[b] eorum per quae dividitur ens”; ergo materia prima non est ens.

<Quaestio 43>

[c]Quaeritur secundo, supposito quod materia prima[d] sit, utrum sit prima.

<Quaestio 43.1>

1. Et videtur quod non, quia continuum est divisibile in infinitum, et hoc ratione materiae; ergo materia est infinita. Sed in[e] infinitis non est sumere primum; ergo non erit aliqua materia prima.

2. Item, dicitur quod materia est infinita secundum potentias. Sed potentia proportionatur substantiae et sic substantia proportionatur potentiae; tunc, si sua potentia erit[f] infinita, et sua substantia erit infinita; et ulterius sicut prius.

3. Item, suppono quod materia prima[g] habeat partem [h]et partem in sua substantia, non tamen suppono quod habeat partem extra partem, cum[i] materia prima habeat[j] in se partes infinitas in potentia. Sed inter partes infinitas in potentia non est sumere primum; quare etc.

4. Item, si aliqua forma substantialis debeat exire in esse per generationem, licet ultima inductio sit subita, tamen illam[k] subitam inductionem praecedit successio[l], quia, si ex aere debeat fieri ignis, tunc successio est in ista [m]factione et in ista productione formae ignis, ut quando ignis exterior dessiccat aerem et rarefacit illam materiam aeris magis. In successivis autem non est sumere primum; ergo in ista inductione formae non est sumere primum. Sicut ergo non est sumere primum in gradibus formalibus, nec similiter erit sumere primum in gradibus materialibus, quia, sicut fit[n] transmutatio a parte formae, similiter est ibi transmutatio a parte materiae et nova materia inducta sicut et nova forma[o].

[a] septimo *scripsimus*; secundo ON
[b] aliquod] aliquid N
[c] Quaeritur ... est aliqua materia una (n. 9) *transp. post* creatura (q. 42.2, n. 14) N
[d] prima *om.* N
[e] in *suppl. s.l.* O
[f] erit] est N
[g] prima *om.* N
[h] et partem *om. hom.* N
[i] cum] tunc O
[j] habeat] habet O
[k] illam] istam N
[l] successio] et longa alteratio *add.* N
[m] factione ... formae] productione N
[n] fit] est N
[o] forma] ut videtur *add.* N

11. Furthermore, in *Metaphysics* VII Aristotle says: "By matter I mean <that which in itself> is neither a quality nor a quantity nor any one of the things through which being is divided";[300] therefore, prime matter is not a being.

<Question 43>

Granted that prime matter exists, the second question is whether it is first.

<Question 43.1>

1. Arguments for the negative: a continuum is divisible to infinity, and this in virtue of matter; therefore, matter is infinite. But it is not possible to find anything first in infinites; therefore, there will not be any prime matter.

2. Furthermore, it is said that matter is infinite with regard to its potencies. But a potency is proportional to a substance, and so a substance is proportional to a potency; then, if the potency of matter is infinite, its substance will also be infinite, and then one can argue as before.[301]

3. Furthermore, I assume that prime matter has many parts in its substance, but I do not assume that it has parts such that one is outside another, since prime matter has in itself infinitely many parts in potency. But it is not possible to find a first <part> among infinitely many parts in potency; therefore, <there is no prime matter>.

4. Furthermore, if a substantial form must come into being through generation, even though its last induction is instantaneous, there is a succession preceding that instantaneous induction, because if fire must come to be from air, there is a succession in this coming to be and in this production of the form of fire, as when an exterior fire makes air dry and makes the matter of air more rare. But it is not possible to find a first <thing> in successive things; therefore, in this induction of a form it is not possible to find a first <stage>. Therefore, just as it is not possible to find a first <stage> in formal stages, it will also not be possible to find a first <stage> in material stages, because, just as a change occurs on the side of a form, so a change will occur there on the side of matter, and just as a new form is induced, so a new matter is also induced.

[300] Arist. *Metaph.* VII.3, 1029a20–21.
[301] See above, n. 1.

5. Item, si aliquid indivisibile se habeat ita quod ex ipso fluat aliquid continuum, non est sumere primum post indivisibile[a] in illa fluxione; sed tota multitudo creaturarum facta est ex non esse quasi ex quodam indivisibili et omnes creaturae fluunt ex illo non esse indivisibili; ergo in creaturis non est sumere primum. Ergo non est sumere aliquam materiam primam.

6. Item, non est aliqua prima forma; ergo nec erit aliqua prima materia eadem ratione.

7. Item, non est una sola forma quae est principium, sed duae; ergo similiter nec erit una sola materia quae est principium, sed duae[b] eadem ratione. Probatio antecedentis: extrema contrarietatis in substantia sunt principia; sed extrema sunt duo et omne contrarium est forma; ergo ibi erunt duae formae quae sunt principia et non una sola.

8. Item, dicit Aristoteles in nono *Philosophiae primae* quod non est malum nisi post potentiam, et intelligit ibi de potentia passiva; sed potentia passiva debetur materiae; si[c] ergo non est potentia passiva aliqua prima, nec erit aliqua materia prima. Sed nulla **[O 94vb]** potentia passiva est prima, quia non est aliquod summe malum; ergo non est sumere primum in malo. Et si hoc, non erit sumere primum in potentia passiva.

9. Ad oppositum: quod sit aliqua materia una, et ita[d] prima, probatio: circumscripta aliqua substantia ab omni forma et ab omni actualitate, ita quod non sit[e] nisi quoddam ens in potentia, tunc illud ens aut est unum aut multa. Non multa, quia sic esset ibi distinctio et nulla potest esse ibi, quia solus actus etc. Ergo oportet quod sit[f] unum. Ergo est aliqua materia una[g].

<Quaestio 42.2>

1. [h]Solutio primae quaestionis et veritas problematis talis est, quod materia prima est ens in potentia et non in actu.

2. Primae rationi in oppositum dicendum quod maior [i]vera est sic: quanto plus habet aliquid de forma, tanto plus habet de esse actuali. Et sic vera est conclusio.

[a] indivisibile] ut *add.* O
[b] duae] ergo similiter nec erit una *add. hom. sed del.* N
[c] si *suppl. s.l. post* ergo O
[d] ita *corr. in* ita est (*lectio incerta*) O
[e] sit] ibi *add.* N
[f] sit *suppl. s.l.* O
[g] una] et ita prima *add.* N
[h] Solutio ... problematis] veritas huius quaestionis N
[i] vera est] intelligitur N

5. Furthermore, if an indivisible is such that a continuum flows from it, in this flowing it is not possible to find a first <thing> after that indivisible; but the whole multitude of creatures comes to be from non-being as if from an indivisible, and all creatures flow from that indivisible non-being; therefore, it is not possible to find a first <thing> in creatures. Therefore, it is not possible to find any prime matter.

6. Furthermore, there is no first form; therefore, for the same reason there will not be any prime matter.

7. Furthermore, there is not just one form that is a principle, but two; therefore, for the same reason there will not be just one matter that is a principle, but two. Proof of the antecedent: the extremes of a contrariety in a substance are principles; but the extremes are two and every contrary is a form; therefore, there will be two forms there that are principles and not only one.

8. Furthermore, in *First philosophy* IX Aristotle says that something bad only occurs after a potency,[302] and in that passage he means a passive potency; but a passive potency is owing to matter; therefore, if there is no first passive potency, there will be no prime matter either. But no passive potency is something first, because there is nothing maximally bad; therefore, it is not possible to find something that is first in what is bad. If this is the case, it will not be possible to find anything first in a passive potency.

9. To the contrary, proof that there is one matter and hence a prime matter: when every form and every actuality is removed from a substance, so that only a being in potency is left, then that being is either one or many. Not many, because if so there would be a distinction, but there can be no distinction there, because only an act <distinguishes>. Therefore, it is necessary that that being should be one. Therefore, there is one matter.

<Question 42.2>

1. The solution to the first question and the true answer to the problem is that prime matter is a being in potency and not in act.

2. To the first argument to the contrary[303] it must be said that the major premise is true in the following sense: the more form something has the more actual being it has. And in this way the conclusion is true.

[302] Arist. *Metaph.* IX.9, 1051a18–19.

[303] See above, q. 42.1, n. 1.

3. [a]Secundae rationi dicendum quod divisio data est de illis quae sunt in praedicamento ordinabilia; talia vero non sunt principia; unde de principiis non[b] procedit divisio.

4. Vel dicendum est quod materia prima est particulare. Sed duplex est particulare: unum quod fit per signationem formae[c]. Et hoc dupliciter: aut per signationem formae completae, et sic est particulare speciei, aut per signationem formae incompletae, et sic est particulare generis. Et haec est differentia inter particulare generis et particulare speciei. Aliud est per indistinctionem formae sive essentiae. Et sic materia prima est particulare, quia [d]substantia sive essentia materiae primae est essentia una. Et non est multa per essentias, est tamen multa per potentias.

5. [e]Tertiae rationi dicendum quod illa divisio procedit de his quae sunt ordinabilia in praedicamento, et ideo [f]de materia non procedit[g].

6. [h]Quartae rationi dicendum quod, sicut materia prima est ens, similiter est unum, quia, sicut est ens solum in potentia, sic est unum solum in potentia. Illa autem propositio, 'omne quod est unum est forma vel formam habens', intelligenda est de illo quod est unum in actu et non solum potentia, ut materia prima.

7. Vel sic: duplex est unitas: una quae fit per signationem formae, et sic materia prima non est unum; alia quae est indivisibilitas essentiae, et tali unitate est materia prima una[i].

8. [j]Quintae rationi dicendum quod aeternum est dupliciter: aut et secundum substantiam et secundum esse aut secundum substantiam aeternum et secundum esse [k]non aeternum, [l]sed corruptibile. Primo modo est intelligenda propositio Aristotelis, secundo vero modo procedit de materia, quia materia prima secundum suam substantiam est incorruptibilis, secundum esse tamen quod habet in illo vel isto[m] est corruptibilis.

9. [n]Sextae rationi sic: passivum est duplex: unum quod patitur contrarietate naturali a suo contrario abiciendo aliquid a substantia, et sic non patitur materia prima; aliud quod[o] est receptio, et sic patitur materia prima, quia in

[a] Secundae rationi dicendum] ad secundam dico N
[b] non *suppl. s.l.* O
[c] formae] in materia *add.* N
[d] substantia sive essentia] essentia subiecta O
[e] Tertiae rationi] ad tertium N
[f] de materia *om.* ON *sed suppl. s.l.* O
[g] procedit] concludit N
[h] Quartae rationi] ad quartum N
[i] una] unum; est enim unum per privationem formae et per positionem suae essentiae N
[j] Quintae rationi] ad quintum N
[k] non aeternum, sed *om.* N
[l] sed corruptibile *ex corr.* O
[m] isto] in illo O
[n] Sextae rationi sic] ad sextum dicendum quod N
[o] quod] quae O

3. To the second argument[304] it must be said that that division applies to those things that can be ordered in a category; but things of this kind are not principles; thus, that division does not apply to principles.

4. Alternatively, it must be said that prime matter is a particular. But there are two kinds of particular: (i) one that comes to be through the designation of a form. And this case is twofold: either (ia) through the designation of a complete form, and in this way the particular is a particular of a species, or (ib) through the designation of an incomplete form, and in this way the particular is a particular of a genus. This is the difference between a particular of a genus and a particular of a species. (ii) The other kind of particular comes to be through the lack of distinction by a form or essence. Prime matter is a particular of kind (ii), because the substance or essence of prime matter is one essence. Thus prime matter is not many because of <different> essences, but because of <different> potencies.

5. To the third argument[305] it must be said that that division applies to the things that can be ordered in a category, and therefore it does not apply to matter.

6. To the fourth argument[306] it must be said that, just as prime matter is a being so it is one, because just as it is a being only in potency so it is one only in potency. But the proposition 'everything that is one is a form or has a form' must be understood as being about a thing that is one in act and not only in potency, such as prime matter.

7. Another reply is that there are two kinds of unity: (i) one that comes to be through the designation of a form, and prime matter is not one in this way; (ii) the other is the indivisibility of an essence, and prime matter is one in virtue of this kind of unity.

8. To the fifth argument[307] it must be said that what is eternal is of two kinds: either (i) both in substance and in being or (ii) eternal in substance and non-eternal but corruptible in being. Aristotle's proposition must be understood as being about what is eternal in way (i), but way (ii) applies to matter, because prime matter is incorruptible in its substance, but it is corruptible with respect to the being which it has in this or that thing.

9. To the sixth argument[308] it must be said that there are two kinds of passive thing: (i) one is acted on by its contrary through natural contrariety, giving up something of its substance; and prime matter is not passive in this way. (ii) Another involves reception, and prime matter is passive in this way,

[304] See above, q. 42.1, n. 2.
[305] See above, q. 42.1, n. 3.
[306] See above, q. 42.1, n. 4.
[307] See above, q. 42.1, n. 5.
[308] See above, q. 42.1, n. 6.

recipiendo formam. Et iste[a] modus ultimus est extendendo nomen passivi communiter.

10. [b]Septimae rationi dicendum quod duplex est 'ad aliquid': una est quae accidens est, alia quae est substantia. Ad 'ad[c] aliquid' quae accidens est duae condiciones requiruntur[d]: una est quod habeat aliquid extra ipsum ad quod dicatur; secunda condicio[e], quod sequatur esse formae substantialis in materia. Ad 'ad aliquid' quae est substantia non requiritur nisi una condicio tantum, scilicet quod habeat aliquid extra ipsum ad quod dicatur. Unde materia prima sic est ad aliquid, quia[f] habet aliquid extra ipsum ad quod dicatur et non propter ultimam condicionem, quia hoc deficit in materia prima, scilicet quod sequatur esse formae substantialis.

11. [g]Octavae rationi dicendum quod materia prima est principium formae. Sed esse principium formae, hoc est dupliciter: aut ita quod ex eius substantia producatur forma, et sic materia prima non est principium formae; aut ita quod eius substantia subsistat formae, et sic materia prima est principium per se formae.

12. [h]Nonae rationi dicendum quod argumentum concluderet si esset aequalis distantia inter secundum et tertium et inter primum et secundum; sed non est aequalis distantia, quia inter pure nihil et ens in potentia est infinita distantia. Sic autem non est inter ens in potentia et ens actu; unde non est aequalis distantia etc.

13. [i]Decimae rationi dicendum quod Aristoteles intelligit quod materia prima[j] non sit ens quod dividitur per praedicamenta nec aliquod praedicamentorum [k]in se nec ordinabile in praedicamento, quia illud ens quod dividitur in decem praedicamenta est ens actu [l]et quodlibet praedicamentum, et sic quodlibet[m] in praedicamento est ens actu. Sed materia prima non est ens actu. Unde non procedit ratio.

14. Et notandum quod, licet pure nihil non sit causa entis in potentia, ut materiae primae, est tamen eiusdem[n] occasio, quia, nisi esset[o] pure nihil, numquam fieret[p] aliqua creatura.

[a] iste] ille O
[b] Septimae rationi] ad septimum N
[c] ad *om.* O
[d] requiruntur] quaeruntur O
[e] condicio] est *add.* N
[f] quia] quod N
[g] Octavae rationi] ad octavam rationem N
[h] Nonae rationi] ad nonum (*lectio incerta*) N
[i] Decimae rationi] ad decimam N
[j] prima *om.* N
[k] in se *om.* N
[l] et quodlibet praedicametum *sc.* est ens actu
[m] quodlibet *om.* N
[n] eiusdem] eius N
[o] esset] fuisset N
[p] fieret] esset N

because it is passive in receiving a form. This last way (ii) involves extending the term 'passive thing' to be taken in a broad sense.

10. To the seventh argument[309] it must be said that there are two kinds of things that are relations: (i) one is an accident, (ii) the other is a substance. (i) For the relation that is an accident two conditions are required: one is that it should have something outside itself to which it is said to be related; the second condition is that it should follow from the being of a substantial form in matter. (ii) For the relation that is a substance only one condition is required, namely, that it should have something outside itself to which it is said to be related. Thus, prime matter is relational in the second way, because it has something outside itself to which it is said to be related and not because of the last condition <of (i)>, that is, that it should follow from the being of a substantial form, given that this condition is not met in the case of prime matter.

11. To the eighth argument[310] it must be said that prime matter is a principle of a form. But something can be a principle of a form in two ways: either (i) in such a way that a form is produced from its substance, and prime matter is not a principle of a form in this way; or (ii) in such a way that its substance is a subject for a form, and in this way prime matter is a per se principle of a form.

12. To the ninth argument[311] it must be said that the argument would be conclusive if there were an equal distance between the second and the third items and between the first and the second items; but there is not an equal distance, because between what is absolutely nothing and what is in potency there is an infinite distance, but there is not an infinite distance between what is in potency and what is in act; thus, there is not an equal distance <between the second and the third items and between the first and the second items>.

13. To the tenth argument[312] it must be said that Aristotle means that prime matter is neither a being that is divided through the categories nor one of the categories in itself nor something that can be ordered in a category, because the being that is divided through the ten categories is being in act and every category <is a being in act>, and so everything belonging to a category is a being in act. But prime matter is not a being in act. Thus, the argument does not succeed.

14. It must be noted that, although what is absolutely nothing is not a cause of what is in potency, namely of prime matter, it is, however, its occasion, because if what is absolutely nothing did not exist, a creature would never come to be.

[309] See above, q. 42.1, n. 7.
[310] See above, q. 42.1, nn. 8–9.
[311] See above, q. 42.1, n. 10.
[312] See above, q. 42.1, n. 11.

<Quaestio 43.2>

1. [a]Solutio secundae quaestionis talis est et etiam veritas problematis, quod est aliqua materia prima.

2. Ad tres primas rationes in oppositum dicendum quod infinitatem esse in materia prima, hoc est dupliciter: aut ratione suae potentiae aut ratione suae substantiae. Si primo modo, sic est infinitas in materia prima, nec etiam erit devenire ad materiam primam.

3. Et dicunt[b] quidam hoc esse verum vocando infinitatem indeterminationem suae potentiae, quia materia prima est indeterminate[c] se habens respectu cuiuscumque formae.

4. Alii dicunt quod hoc verum est etiam vocando infinitatem indeterminationem suae potentiae et etiam vocando infinitatem numerum infinitum potentiarum, quia dicunt quod tot sunt formae[d] in materia prima in potentia quot sunt formae sive ideae[e] in motore primo sive in ipsa prima causa in actu. Et ipsi ponunt quod infinitae sunt formae et infinitae ideae in prima causa; unde ponunt similiter infinitas esse potentias in materia prima iuxta ideas infinitas in causa prima.

5. Si secundo modo, hoc dupliciter: aut loquendo de substantia materierum diversarum[f] quarum **[O 95ra]** quaelibet ordinata est ad aliam secundum prius et posterius in causalitate in componendo et[g] in resolvendo[h] formas completas in formas incompletiores, et sic non est infinitas in materia prima. Verbi gratia: accipiamus aliquod corpus vel aliquod animal, ut hominem, resolvamus primo ab eo humanitatem, quae est completa forma; tunc iam destruitur materia quae fuit necessitas in comparatione ad hanc formam completam. Postea resolvamus ab eo corporeitatem[i], et sic destruitur materia quae fuit necessitas in comparatione ad corporeitatem. Postea resolvamus ab eo substantialitatem; tunc destruitur materia quae fuit necessitas in comparatione ad hanc formam. Et tunc deventum est ad materiam primam, quia tunc non manet nisi quoddam ens in potentia, quod est materia prima. Ecce quod hoc modo resolvendo formam completam in formam incompletam est devenire ad materiam primam, sic procedendo secundum ordinem in causalitate.

[a] Solutio ... problematis] veritas huius quaestionis N
[b] dicunt *explicit ex abrupto* N
[c] indeterminate *ex corr.* O
[d] formae *scripsimus cum* TC; potentiae O
[e] ideae *scripsimus cum* TC; potentiae O
[f] diversarum] partium materiae *add. sed del.* O
[g] et *suppl. s.l. (lectio incerta)* O
[h] resolvendo *corr. s.l. ex* solvendo O
[i] corporeitatem] postea resolvamus ab eo substantialitatem *add. hom. sed del.* O

<Question 43.2>

1. The solution to the second question and the true answer to the problem as well is that there is a prime matter.

2. To the first three arguments to the contrary[313] it must be said that the claim that infinity exists in prime matter can be taken in two ways: either (i) because of its potency or (ii) because of its substance. In way (i), infinity exists in prime matter and it is not possible to arrive at prime matter.

3. Some people say that this is true if one calls 'infinity' the indetermination of the potency of prime matter, because prime matter is indeterminate in respect of every form.[314]

4. Other people say that this is true both if one calls 'infinity' the indetermination of the potency of prime matter and if one calls 'infinity' the infinite number of potencies <of prime matter>, because they say that there are as many forms that exist in potency in prime matter as there are forms or ideas that exist in act in the first mover or in the first cause itself. These people assume that there are infinitely many forms and infinitely many ideas in the first cause; thus, they similarly assume that there are infinitely many potencies in prime matter, which correspond to the infinitely many ideas in the first cause.

5. Case (ii) <that is, if we take that infinity exists in prime matter because of its substance> is twofold: (iia) In one way if we speak of the substance of different matters each one of which is ordered to another in accordance with the prior and the posterior in causality, by composing and resolving complete forms into more incomplete forms; in this way there is no infinity in prime matter. For example: let us take a body or an animal, for instance, a man, and let us first remove from it humanity, which is a complete form; then, the matter that was a necessity with respect to this complete form is already destroyed. Let us then remove from it corporeity, and in this way the matter that was a necessity with respect to corporeity is destroyed. Let us then remove from it substantiality; then the matter that was a necessity with respect to this form is destroyed. And then we arrive at prime matter, because then only a being in potency, which is prime matter, remains. Thus it is that by this way of resolving a complete form into an incomplete form, and by proceeding in accordance with the order of causality, it is possible to arrive at prime matter.

[313] See above, q. 43.1, nn. 1–3.

[314] The author of the commentary on *De generatione et corruptione* attributed to Richard Rufus holds a similar view; see Richard Rufus of Cornwall, *In De gen. et corr.*, 166.56–57.

6. Aut loquendo de substantia materierum diversarum quarum nulla ordinata est alii secundum prius et posterius in causalitate, nec in componendo nec in resolvendo, nec etiam secundum formas completiores et incompletiores, sic est sumere infinitatem in materia nec est devenire ad materiam primam. Verbi gratia: accipiamus aliquod corpus; corpus illud resolvamus in duas medietates et unam medietatem, cum sit corpus, adhuc in alias duas medietates. Ecce qualiter sic numquam erit devenire ad materiam primam quin contingat resolvere ulterius. Et sic procedit ratio, primo modo non.

7. Quartae rationi dicendum quod nos possumus considerare illam successionem dupliciter: vel respiciendo illam successionem tamquam aliquod continuum, et sic non contingit sumere primum gradum materiae illius successionis; vel respiciendo totam successionem in se secundum quod tota est, et sic est successio [a]illa finita, et sic contingit sumere primum in illa successione. Verbi gratia, si Socrates moveatur ab A spatio in B spatium, totus iste motus et tota ista successio, si sumatur in se et secundum se totam, sic finita est et est sumere primum huius successionis, ut A. Si vero sumatur ista successio secundum quod continuum est et secundum hoc respiciatur, tunc non est sumere primum. Verbi gratia: non potest dici quod A est primum huius successionis, quia, cum A sit quoddam continuum, potest dividi in infinitum. Et secundum hoc non est sumere primum.

8. Sextae[b] rationi dicendum quod non est nisi una sola forma, proprie sumendo formam, quia forma, secundum quod proprie sumitur et pro nobiliori extremo contrarietatis, sic non est nisi una sola forma. Prout tamen transmutatur ad nominandum vilius extremum contrarietatis, sic aliquo modo potest dici quod privatio[c] est forma. Et sic communiter loquitur Aristoteles in libro *Physicorum*, cum dicit quod privatio est forma.

9. Quintae[d] rationi dicendum quod haec est falsa, quod omnes creaturae successive fluunt ab ipso non esse, et quod non[e] est ibi fluxus sive successio proprie loquendo, quia dicit Augustinus quod Deus creavit omnia simul et subito sive in instanti[f]. Et si esset ibi fluxus, tunc nullo modo fierent in instanti, quia fluxus dicit successionem. Est tamen quidam modus loquendi dicendo sic quod creaturae[g] fluunt a non esse, sed tamen non proprie[h] loquuntur, sed est usus quidam sic dicendi.

[a] illa finita *inv. sed corr.* O
[b] Sextae *corr. s.l. ex* quintae
[c] privatio] et f *add. sed del.* O
[d] Quintae *corr. s.l. ex* sextae
[e] non *suppl. s.l.* O
[f] instanti *corr. s.l. ex* stanti O
[g] creaturae *iter. sed corr.* O
[h] proprie] illi *add. s.l.* O

6. (iib) In another way we speak of the substance of different matters none of which is ordered to another in accordance with the prior and the posterior in causality, whether in composition or in resolution, or even in accordance with more complete and more incomplete forms. In this way it is possible to apply infinity to matter and it is not possible to arrive at prime matter. For example: let us take a body. Let us resolve that body into two halves, and <then resolve> one half, since it is a body, once more into two halves. This is the way in which it will never be possible to arrive at prime matter in such a way that it is not possible to resolve further. The argument succeeds in way (iib), but not in way (iia).

7. To the fourth argument[315] it must be said that we can consider that succession in two ways: either (i) by taking that succession as a continuum, and in this way it is not possible to find a first stage of matter in that succession; or (ii) by taking the whole succession in itself in so far as it is a whole, and in this way that succession is finite, and in this way it is possible to find a first < stage> in that succession. For example, if Socrates moves from space A to space B, this whole motion and this whole succession, if it is taken in itself and as a whole, is finite and it is possible to find a first <space> in this succession, that is, A. But if this succession is taken in so far as it is a continuum and it is considered in this way, then it is not possible to find a first <space>. For example, it cannot be said that A is the first in this succession, because since A is a continuum, it can be divided to infinity. Hence, in way (i) it is not possible to find a first <space>.

8. To the sixth argument[316] it must be said that there is only one form, if one takes 'form' in a proper sense, because if 'form' is taken in a proper sense and for the nobler extreme of a contrariety, then there is only one form. But in so far as the term 'form' is modified so that it names the meaner extreme of a contrariety, it can be said that a privation is in some sense a form. Aristotle takes 'form' in this broad sense in the *Physics*, when he says that a privation is a form.[317]

9. To the fifth argument[318] it must be said that the proposition that all creatures successively flow from non-being is false and that there is no flowing or succession properly speaking there, because Augustine says that God created all things simultaneously and immediately or at an instant.[319] But if there were a flowing there, then things would in no way come to be at an instant, because flowing expresses a succession. There is, however, a way of speaking by which we say that creatures flow from non-being, but this is not properly speaking but merely a case of <common> usage.

[315] See above, q. 43.1, n. 4.
[316] See above, q. 43.1, nn. 6–7.
[317] Arist. *Phys.* II.1, 193b19–20.
[318] See above, q. 43.1, n. 5.
[319] See, e.g., Aug. *De genesi ad litt.* I.15, IV.33, CSEL 28.1, 21.7–22.14, 131.23–132.13.

10. Septimae rationi dicendum quod potentiam passivam deberi malo[a], hoc est dupliciter: aut ita quod potentia passiva sit eius occasio aut ita quod sit eius per se causa. Si secundo modo, sic est[b] falsa, quia, si esset eius per se causa, tunc ad extensionem mali sequeretur extensio potentiae passivae. Et hoc falsum est, quia per unicam potentiam passivam et non extensam possumus infinities male operari, [c]ita quod non contingat ad primum malum <devenire[d]> propter extensionem eius quod est male operari. Si primo modo, sic vera est, quia passivum sonat in defectum. Et defectus qui[e] per virtutes materiales fit est causa mali. Et secundum hoc non sequitur quod, licet sit eius occasio, quod, si[f] in uno non est sumere primum, nec similiter in reliquo[g].

<Quaestio 44>
Quaeritur tertio de materia prima an incepit esse.

1. Et quod non videtur quinque rationibus, quia, si inciperet, tunc haberet prius non esse. Quaeritur ergo an suum esse et non esse simul sint aut non. Si simul, tunc contradictoria simul. Si non, sed est distantia inter ea, tunc est ibi medium, quia inter quaelibet duo instantia cadit tempus medium, et ita contradictoria habent medium, quod est contra Aristotelem in primo *Posteriorum*.

2. Item, finis et principium sunt correlativa; cui ergo non inest unum, nec reliquum. Sed materia prima non habet finem; ergo non habet principium.

3. Item, dicit Aristoteles in fine primi *Physicorum* quod materia est ingenita; ergo non processit in esse.

4. Item, potentia est essentiale materiae; sed potentia sua non processit in esse nec incepit; ergo etc. Probatio minoris talis est: si sua potentia inceperit esse, tunc ante suum esse aut fuit in potentia aut non. Si fuit in potentia, tunc potentia praecedit potentiam, et hoc est inconveniens. Si non fuit in potentia, tunc numquam exiret in esse, et hoc est inconveniens; quare etc. Ista est ratio Gundissalini.

[a] malo *scripsimus*; materiae O
[b] est *suppl. s.l.* O
[c] ita … operari *scrips. sed del.* O
[d] devenire *supplevimus*
[e] qui *scripsimus*; quae O
[f] si *scripsimus*; licet O
[g] Q. 50 *hic transp.* C

10. To the seventh argument[320] it must be said that the claim that a passive potency is attributable to what is bad can be taken in two ways: either (i) in such a way that a passive potency is an occasion of what is bad or (ii) in such a way that it is a per se cause of it. In way (ii) the claim is false, because if a passive potency were a per se cause of what is bad, then the extension of a passive potency would follow from the extension of what is bad. But this is false, because in virtue of only one unextended passive potency we can act badly an infinite number of times, in such a way that we do not arrive at a first bad through the extension of bad actions. In way (i) the proposition is true, because what is passive is spoken of as a defect. And a defect that occurs because of material powers is a cause of what is bad. In this way, although a passive potency is an occasion of what is bad, it does not follow that, if it is not possible to find something that is first in one of the two senses, then it is similarly not possible in the other sense.

<Question 44>
The third question is whether prime matter began to exist.

1. There are five arguments for the negative: if matter began, it would first have non-being. It is therefore asked whether its being and its non-being are simultaneous or not. If they are simultaneous, then contradictories are simultaneous. If they are not, but there is a distance between them, then there is something intermediate between them, because an intermediate time intervenes between any two instants, and thus contradictories have something intermediate between them, which is against Aristotle in *Posterior Analytics* I.[321]

2. Furthermore, an end and a beginning are correlative; therefore, if one of them does not exist in something neither does the other; but prime matter does not have an end; therefore, it does not have a beginning.

3. Furthermore, at the end of *Physics* I Aristotle says that matter is non-generated;[322] therefore, it did not proceed into being.

4. Furthermore, potency is essential to matter; but the potency of matter did not proceed into being nor did it begin; therefore, <matter did not proceed into being nor did it begin>. Proof of the minor premise: if the potency of matter began to exist, then before its being it was either in potency or not. If it was in potency, then a potency precedes a potency, and this is absurd. If it was not in potency, then it would never have come to be, and this is absurd; therefore, <the potency of matter did not begin to exist>. This is the argument of Gundissalinus.[323]

[320] See above, q. 43.1, n. 8.
[321] Arist. *An. post.* I.2, 72a12–13.
[322] Arist. *Phys.* I.9, 192a29–31.
[323] Gundissalinus, *De processione mundi*, ed. Bülow, 33.14–34.15.

5. Item, si materia prima est in potentia, tunc habuit esse post non esse, et si hoc, exivit in esse per mutationem. Sed quod non possit exire in esse per mutationem probatio: ista mutatio est; ergo exivit in esse. Constat quod **[O 95rb]** non nisi per mutationem. Quaero tunc de illa mutatione eodem modo quo prius. Et sic erit processus in infinitum vel dicendum est quod non exivit in esse per mutationem; sed primum est impossibile; ergo secundum erit possibile; ergo etc.

6. Ad oppositum: causa prima per Aristotelem in *Metaphysica* incidit in triplici genere cause; sed nisi materia prima inciperet, non esset in triplici genere causae; ergo etc. Probatio minoris: si materia prima non inciperet, tunc non esset efficiens respectu materiae primae. Probatio consequentiae: duae sunt condiciones requisitae ad hoc quod aliquid sit efficiens: prima est quod agat, secunda est quod conservat rem effectam in esse. Sed prima condicio magis facit ad efficientiam quam secunda, quia res proprie dicitur efficiens ab agere sive facere, non autem a conservare, quia a conservatione non denominatur. Sed si materia prima non inciperet, non ageret tunc causa prima in materiam primam; ergo, si materia prima non inciperet, non esset efficiens respectu materiae primae; ergo etc.

7. Solutio. Veritas problematis talis est, secundum Aristotelem, quod incepit esse.

8. Ad primum in oppositum dicendum quod esse et non esse simul esse, hoc est dupliciter: aut simul ita quod in eodem instanti aut simul ita quod in uno instanti immediate post aliud. Si primo modo, sic non sunt contradictoria simul nec sic sunt esse et non esse simul. Si secundo modo, sic sunt contradictoria simul nec etiam sic habent medium.

9. Sed contra: dicit Aristoteles, sexto *Physicorum*, quod inter quaelibet duo puncta cadit linea media et etiam inter quaelibet duo instantia cadit tempus medium; ergo similiter inter esse et non esse [a]diversorum instantium cadet tempus medium.

[a] diversorum instantium *scripsimus*; diversarum instantiarum O

5. Furthermore, if prime matter is in potency, then it had being after non-being, and if this was the case, it came to be through a change. Proof that it cannot come to be through a change: this change exists; therefore, it came to be. It is manifest that this happens only through a change. Then I raise the same question as before about that change, and so there will be an infinite regress, or it must be said that that change did not come to be through a change; but the first case is impossible; therefore, the second case will be possible; therefore, <prime matter is not in potency>.

6. To the contrary: according to Aristotle in the *Metaphysics*, the first cause falls into three genera of cause;[324] but unless prime matter began, the first cause would not be in three genera of cause; therefore, <prime matter began>. Proof of the minor premise: if prime matter did not begin, then the first cause would not be an efficient cause of prime matter. Proof of the consequence: two conditions are required in order for something to be an efficient cause: the first is that it acts, the second is that it preserves the thing brought into being. But the first condition contributes more to efficiency than does the second, because a thing is properly said to be an efficient cause from acting or producing, but not from preserving, since it is not denominated from preserving. But if prime matter did not begin, then the first cause would not act on prime matter; therefore, if prime matter did not begin, the first cause would not be an efficient cause with respect to prime matter; therefore, <prime matter began>.[325]

7. Solution. According to Aristotle, the true answer to the problem is that prime matter began to exist.

8. To the first argument to the contrary[326] it must be said that the claim that being and non-being are simultaneous can be taken in two ways: either (i) simultaneous in such a way that they exist at the same instant or (ii) simultaneous in such a way that they exist at two instants one of which is immediately after the other. Contradictories are not simultaneous in way (i), nor are being and non-being simultaneous in this way, but contradictories are simultaneous in way (ii), and in this way they do not have something intermediate between them.

9. Against this reply: in *Physics* VI Aristotle says that between every two points there is an intermediate line and between every two instants there is an intermediate time;[327] therefore, there will similarly be an intermediate time between the being and the non-being of different instants.

[324] This claim is not found in Aristotle but it probably originates from Averroes' idea that, according to Aristotle, metaphysics inquires into the formal, the efficient and the final causes; see Aver. *In Phys.* I, t.c. 1, fol. 6ra.

[325] The author of the commentary on the *Physics* attributed to Richard Rufus makes a similar point in the discussion on the eternity of the world; see Richard Rufus of Cornwall, *In Phys.*, 222.136–223.151; see also below, book VIII, q. 5, n. 39.

[326] See above, n. 1.

[327] Arist. *Phys.* VI.1, 231b9–10.

10. Sed dicendum quod Aristoteles in libro *Physicorum* intendit si illa duo puncta fuerint eiusdem lineae et similiter si illa duo instantia fuerint eiusdem temporis; si vero illa duo puncta fuerint diversarum linearum, tunc non est necesse cadere lineam mediam. Similiter, si duo fuerint instantia eiusdem temporis, tunc necesse est cadere tempus medium; si vero fuerint duo instantia, et hoc diversorum, sicut est in proposito, quia ultimum instans sui non esse est instans aeternitatis et primum instans sui esse est instans temporis, tunc non necesse est cadere tempus medium.

11. Aliter dicendum est secundum quosdam qui ponunt gradus infinitos in instanti. Ipsi enim dicunt quod duo contradictoria esse in eodem gradu instantis simul, hoc est inconveniens; tamen duo contradictoria esse in diversis gradibus instantis non est inconveniens, quia prius per naturam est unum extremum contradictoriorum quam alterum. Unde secundum hoc dicunt esse possibile contradictoria esse simul, ita quod simul tempore, unum tamen prius per naturam et reliquum posterius per naturam. Et simul stant quod aliquid possit esse simul tempore cum alio, posterius tamen per naturam. Et sic est in proposito.

12. Secundae rationi dicendum quod quaedam sunt relativa quorum utrumque extremum [a]relative dicitur ad aliud, ut pater et filius, quaedam quorum unum solum dicitur relative. Et sic est in proposito, quia finis dicitur in comparatione ad principium et non econverso, secundum Augustinum dicentem quod in quibusdam est accipere principium ubi non finem. Verbi gratia in numeris, quia, cum numerus crescat in infinitum per appositionem, non est sumere finem; est tamen ibi sumere primum aliquo modo, quia primum numerum. Huius etiam causa talis est: prius non dependet a posteriori, sed posterius dependet a priori, secundum quod dicit logicus; unde, cum principium sit quasi prius fine et finis posterius, erit relatio inter finem et principium ita quod illud quod habet finem habuit principium, et non omne illud quod habuit principium habebit[b] finem. Verbi gratia: quod enim est homo est animal, non tamen omne quod est animal est homo. Et eodem modo est a parte ista.

[a] relative … aliud *corr. ex* dicitur ad aliud relative O
[b] habebit *scripsimus cum* TC; *corr. ex* habuit (*lectio incerta*) O

10. It must be said, however, that in the *Physics* Aristotle has in mind the case of two points belonging to the same line and similarly of two instants belonging to the same time; but if those two points belong to different lines, then it is not necessary for there to be an intermediate line between them. Similarly, if two instants belong to the same time, then it is necessary for there to be an intermediate time between them; but if there are two instants belonging to different things, as happens in the present case, because the last instant of the non-being of prime matter is the instant of eternity and the first instant of its being is an instant of time, then it is not necessary for there to be an intermediate time.

11. Another reply is that of those people who assume that there are infinitely many stages in an instant. They say that it is absurd for two contradictories to be in the same stage of an instant simultaneously; but for two contradictories to be in different stages of an instant is not absurd, because one of the two extremes of a contradiction is prior by nature to the other. Accordingly, they say that it is possible for contradictories to be simultaneous in such a way that they are simultaneous in time, but one is prior by nature and the other posterior by nature. And for something to be simultaneous in time with something else and for it to be posterior to it by nature are compatible. And this is what happens in the present case.

12. To the second argument[328] it must be said that some relatives are such that each extreme is said relatively to the other, such as a father and a son, other relatives are such that only one of them is said relatively to the other. And this is what happens in the present case, because an end is said in relation to a beginning and not the other way around, according to Augustine, who says that in some things it is possible to find a beginning but not an end, for example, in numbers, because, given that a number increases infinitely by addition, it is not possible to reach an end; but it is possible to find a first in some way, because it is possible to find a first number.[329] The reason for this is the following: what is prior does not depend on what is posterior, but what is posterior depends on what is prior, according to a logician; thus, since a beginning is, as it were, prior to an end, and an end is posterior, an end and a beginning will be related in such a way that a thing that has an end had a beginning, and not everything that had a beginning will have an end. For example: that which is a man is an animal, but it is not the case that everything that is an animal is a man. And the same happens in the present case.

[328] See above, n. 2.

[329] The idea that there can be a beginning without an end does not sound genuinely Augustinian. The example of number may be suggested by Aristotle's remark in *Phys.* III.7, 207b1–3 that in numbers there is a minimum but not a maximum.

13. Sed contra: accipiamus aliquam partem; illa pars habet finem. Et similiter est de omnibus partibus eadem ratione; ergo etc. Verbi gratia: totum esse materiae primae a suo principio nunc habet finem sive in illa[a] die, et etiam totum esse materiae primae quod habuit in Socrate, mortuo Socrate, habet finem in ipso.

14. Sed dicendum quod duplex est finis: unus qui est comparatus, et talem finem habet materia prima, quia si comparetur quoad suum esse praecedens, sic suum esse in illa[b] die sive illa dies est terminus sui esse praecedentis; et sic materia prima habet finem. Alius qui est absolutus; et est ille finis quem non sequitur aliquid. Et sic materia prima non habet finem, quia numquam ultimum sui esse sumere contingit quin remaneat aliud esse ulterius sumendum.

15. Tertiae rationi dicendum quod concludit bene quod non possit exire in esse per generationem; non tamen concludit quin possit exire per creationem.

16. Quartae rationi dicendum illam potentiam praecedere aliam potentiam est dupliciter: aut potentia passiva, et sic non praecedit eam alia potentia et sic procedit ratio; aut potentia efficientis, et talis potentia praecedit potentiam materiae primae. Nec est inconveniens quod talis potentia praecedat, dummodo scilicet non sit eadem potentia numero, ut scilicet passiva.

17. Quintae rationi dicendum quod non contingit procedere in infinitum, quia mutatio procedit in esse non alia mutatione a se ipsa; procedit tamen in esse per aliud agens, ut per agens extrinsecum, sed tamen via per quam exit in esse non est alia a se ipsa. [c]Unde solum illi qui per se sunt termini exeunt in esse alia via. Unde, cum sic dicitur: 'omne quod exit in esse exit vel motu vel mutatione', hoc est intelligendum de rebus sive formis in quas terminatur motus et non de viis in formam. **[O 95va]**

[a] illa *scripsimus*; illo O
[b] illa *scripsimus*; illo O
[c] Unde …via *scrips. sed del.* O

13. But against this: let us take a part; that part has an end. And for the same reason all the parts have an end; therefore, <the whole has an end>. For example: the whole being of prime matter starting from its beginning has an end now or in that day, and the whole being which prime matter had in Socrates, once Socrates is dead, also has an end in Socrates.

14. But it must be said that there are two kinds of end: (i) one is relative, and prime matter has this kind of end, because if prime matter is taken in relation to its preceding being, then its being in this day or that day is the terminus of its preceding being; in this way prime matter has an end. (ii) Another is absolute, and this is the kind of end to which nothing is subsequent; in this way prime matter does not have an end, because it is not possible to find a last stage of its being in such a way that no further being can be found.

15. To the third argument[330] it must be said that it correctly concludes that prime matter cannot come to be through generation, but it does not prove that prime matter cannot come to be through creation.

16. To the fourth argument[331] it must be said that the claim that another potency precedes that potency can be taken in two ways: either (i) <the potency that precedes> is a passive potency, and in this way another potency does not precede that potency, and the argument succeeds in this way; or (ii) <the potency that precedes is> the potency of an efficient cause, and this kind of potency precedes the potency of prime matter. And it is not absurd for this kind of potency to precede, as long as it is not numerically the same potency as the passive potency.

17. To the fifth argument[332] it must be said that there is no infinite regress, because a change comes into being not through a change distinct from itself, but through another agent, namely through an extrinsic agent; however, the way through which it comes to be is not distinct from itself. Thus, only the things that are per se termini come into being through a way distinct from them. Therefore, the claim 'everything that comes to be comes to be either through a motion or through a change' should be understood to be about the things or forms that are the termini of a motion, and not about the ways towards a form.

[330] See above, n. 3.
[331] See above, n. 4.
[332] See above, n. 5.

<Quaestio 45>
Quaeritur quarto, habito quod materia prima incepit esse, a quo incepit esse.

1. Et videtur quod non a causa prima sic: causa prima est uniformis, nullam habens innovationem nec transmutationem; ergo, si materia prima incepit esse a causa prima, cum causa prima fuerit ab aeterno, et ipsa erit ab aeterno, quia, si materia prima non esset ab aeterno et esset a Deo, tunc in eo fieret transmutatio, quod est inconveniens. Et similiter etiam inconveniens est materiam primam esse ab aeterno; quare etc.

2. Item, si prima materia esset a Deo, tunc exiret in esse per creationem, et ita crearetur semel in se[a], et tunc sequeretur ipsam[b] multotiens creari, quod est inconveniens. Probatio consequentiae: pono quod materia prima sit pars cuiuslibet compositi. Arguo tunc sic: omne quod creatur creatur secundum se totum; ergo Deus, in creando istam animam, creat materiam primam. Et semel creavit ipsam in se; ergo multotiens creat unum et idem, quod, cum sit inconveniens, relinquitur quod etc.

3. Item, materia prima est pars corporis mobilis; et pars est prior toto; ergo materia prima prius est corpore mobili. Et si hoc, prius est motu. Et si hoc, prius est tempore, quia tempus est passio motus et eius mensura. Et si hoc, est prius eis quae sunt in tempore; ergo prius est instanti temporis. Sed si materia prima habeat esse a prima causa, tunc exit in esse per creationem. Cum ergo per creationem habeat esse, suum esse[c] non est in instanti temporis; oportet ergo quod sit in instanti aeternitatis. Similiter ante esse habuit non esse; constat quod non in instanti temporis; ergo in instanti aeternitatis. Ergo, dato quod materia creatur, erit dare unum illorum: vel quod contradictoria simul sunt in instanti vel quod in instanti aeternitatis est prius et posterius, quorum utrumque est impossibile. Quod primum sit impossibile patet. Et quod secundum similiter patet per Boethium dicentem quod aeternitas est interminabilis vitae possessio tota simul, ita quod non prius et posterius; ergo etc.

4. Item, si Deus creavit materiam primam, tunc eam intellexit et scivit; sed in scientia Dei non cadit prius et posterius; ergo simul scivit suum esse et non esse, dato quod ipsam creavit. Sed quae scita sunt vera sunt; ergo, dato quod

[a] se] et iterum in alio *add. s.l.* O
[b] ipsam *scripsimus*; ipsum O
[c] esse *ex corr. (lectio incerta)* O

<Question 45>

Now that it has been established that prime matter began to exist, the fourth question is in virtue of what it began to exist.

1. Arguments that this is not in virtue of the first cause: the first cause is uniform, without any innovation or change; therefore, if prime matter began to exist in virtue of the first cause, since the first cause existed from eternity, prime matter will also exist from eternity, because, if prime matter did not exist from eternity and was brought into existence by God, then there would be a change in God, which is absurd. And similarly it is also absurd <to hold> that prime matter exists from eternity; hence, <prime matter is not brought into existence by the first cause>.

2. Furthermore, if prime matter were brought into existence by God, then it would come to be through creation, and so it would be created once in itself, and then it would follow that it is created many times, which is absurd. Proof of the consequence: I assume that prime matter is a part of every composite. Then I argue as follows: everything created is created in its totality; therefore, in creating this soul God created prime matter. And he created prime matter in itself once; therefore, God creates one and the same thing many times, which is absurd, so that we are left with the conclusion that <prime matter is not brought into existence by God>.

3. Furthermore, prime matter is part of a mobile body; but a part is prior to the whole; therefore, prime matter is prior to a mobile body. And if that is the case, it is prior to a motion. And if that is the case, it is prior to time, because time is a passion of a motion and its measure. And if that is the case, it is prior to the things that are in time; therefore, prime matter is prior to an instant of time. But if prime matter receives its being from the first cause, then it comes to be through creation. Therefore, since prime matter has being through creation, its being is not at an instant of time; it is necessary, therefore, for it to exist at an instant of eternity. In the same way, prime matter had non-being before being; it is apparent that its non-being is not at an instant of time; therefore, it is at an instant of eternity. Therefore, if one assumes that matter is created, one of the following two cases must be put forward: either (i) that contradictories exist simultaneously at an instant or (ii) that there is a prior and a posterior at the instant of eternity, each of which is impossible. That case (i) is impossible is clear. That case (ii) is similarly impossible is clear from Boethius's statement that eternity is the total simultaneous possession of an endless life, in such a way that there is no prior and posterior; therefore, <prime matter is not created>.

4. Furthermore, if God created prime matter, then he understood and knew it; but in God's knowledge there is no prior and posterior; therefore, God simultaneously knew the being of prime matter and its non-being, if he

materia creetur, contradictoria erunt simul vera, quod est inconveniens; ergo etc.

5. Item, oppositio fundatur super aliquam naturam. Cum ergo oppositio sit inter esse et non esse, ista oppositio fundatur[a] super aliquid, ergo suum non esse fundatur super aliquid a quo suum esse processit. Sed si crearetur a Deo, tunc fieret ex nihilo; quare etc.

6. Item, agens et actum proportionantur quoad nobilitatem et ignobilitatem; cum ergo prima causa sit agens nobilissimum, ei debetur actum nobilissimum. Sed materia prima inter omnia creata est vilissimum, quia plus accedit ad non esse et, quanto plus ad non esse, tanto plus[b] ad ignobilitatem; ergo non potest creare materiam primam.

7. Item, si causa prima materiam primam crearet, tunc ipsam intelligeret; sed non potest ipsam intelligere. Probatio: quod intelligit causa prima intelligit per ideas et rationes causales rerum; sed materia prima non habet ideam; ergo etc. Probatio minoris: proportionaliter se habent[c] species ad mentem causatam et ideae ad mentem incausatam; sed materia prima non facit speciem in mentem causatam; ergo nec ideam in mentem incausatam. Minor patet, quia omne quod facit speciem est visibile; sed materia prima non est visibile, sed solum imaginabile; quare etc.

8. Item, dicit Aristoteles in undecimo *Metaphysicae* quod quaedam sunt quae melius est causam primam nobilem non intelligere quam intelligere, ut turpia et vilia; sed inter omnia entia materia prima est minus nobile, quia plus accedit ad non esse; quare etc.

9. Ad oppositum: materia prima creatur; aut ergo a se aut ab alio. Non a se, quia nihil se ipsum componit nec ad esse producit, ut dicitur libro *De causis*; ergo ab alio. Sed quod a nulla creatura probatio: sequitur: 'creatura est; ergo materia prima est', quia, cum materia prima sit altera pars compositi, necessario sequitur et non convertitur; ergo materia prima est in plus et prius.

[a] fundatur *scripsimus*; fundetur O
[b] plus *suppl. s.l.* O
[c] habent *scripsimus cum* TC; habet O

created it. But things that are known are true; therefore, if matter is created, contradictories will be simultaneously true, which is absurd; therefore, <God did not create prime matter>.

5. Furthermore, opposition is founded on a nature. Therefore, since there is an opposition between being and non-being, this opposition is founded on something; therefore, the non-being of prime matter is founded on something from which its being proceeded. But if prime matter were created by God, then it would come to be from nothing; therefore, <prime matter is not created by God>.

6. Furthermore, an agent and what is produced by it are proportional with respect to nobility and baseness; therefore, since the first cause is the noblest agent, the noblest thing produced is attributable to it. But prime matter is the basest of all creatures, because it is the closest to non-being, and the closer something is to non-being the closer it is to baseness; therefore, God cannot create prime matter.

7. Furthermore, if the first cause created prime matter, then the first cause would understand prime matter; but the first cause cannot understand prime matter. Proof: the first cause understands what it understands through the ideas and the causal principles of things; but prime matter does not have an idea; therefore, <the first cause does not understand prime matter>. Proof of the minor premise: species are related to a caused mind as are ideas to an uncaused mind; but prime matter does not produce a species in a caused mind; therefore, neither does it produce an idea in an uncaused mind. The minor premise is clear because everything that produces a species is visible; but prime matter is not visible, but only imaginable; therefore, <prime matter does not produce a species>.

8. Furthermore, in *Metaphysics* XI Aristotle says that some things are such that it is better for the noble first cause not to understand them than to understand them,[333] for instance, disgraceful and base things; but prime matter is the least noble of all beings because it is the closest to non-being; hence, <it is better for the first cause not to understand prime matter>.

9. To the contrary: prime matter is created; therefore, either <it is created> by itself or by something else. It is not created by itself because "nothing is responsible for its own composition or brings itself into being", as is said in the book *On Causes*;[334] therefore, <it is created> by something else. Proof that <it is> not <created> by a creature: 'a creature exists; therefore, prime matter exists', follows because since prime matter is the other part of a composite, this consequence is necessary and not convertible; therefore, prime matter is more common than and prior to <any other creature>. But

[333] Arist. *Metaph.* XII.9, 1074b32–33.

[334] The correct reference is Nicolas of Amiens, *Ars catholicae fidei* I.1, prop. 3, ed. Dreyer, 80.11.

Sed nullum posterius producit prius ad esse; ergo etc. Oportet ergo quod causa prima.

10. Item, quod creatur ex nihilo non[a] producitur in esse ab alio nisi a causa prima; sed materia prima creatur ex nihilo; ergo etc. Probatio maioris: creatio fit ex nihilo et inter nihil et aliquid est infinita distantia; ergo oportet quod productio alicuius ex tali infinita distantia fiat per habentem potentiam infinitam. Et infinitam potentiam nulla creatura habet, nisi solum causa prima; quare etc. Probatio minoris: materia prima inter omnia entia maxime est non ens et propinquius non enti; ergo, cum nihil fuit ei propinquius ex quo potest produci quam non ens oportet ut producatur ex non ente; quare etc.

11. Solutio problematis et etiam veritas talis est, quod creatur a causa prima.

12. Ad primum in oppositum dicendum quod Deus est intransmutabilis quantum ad operationes quae terminantur in se ipsum, cuiusmodi est intelligere, non autem quantum ad operationes quae terminantur in aliud, ut in ista inferiora, cuiusmodi est creare.

13. Secundae rationi dicendum quod materiam primam multotiens creari diversimode non est inconveniens, ut semel substantialiter, multotiens accidentaliter. Et est accidentaliter creari quando multotiens creatur aliquid secundum esse diversa, ita scilicet quod creatio non terminetur ad illud per se, sed ad quoddam compositum. Substantialiter autem creari est quando creatur aliquid secundum suum unum esse, ita scilicet quod creatio terminetur ad illud per se secundum quod simplex **[O 95vb]** est et non secundum quod est pars compositi. Et sic materia prima substantialiter creatur isto ultimo modo et priori modo accidentaliter. Et hoc non est inconveniens. Hoc modo dicunt illi qui non ponunt partibilitatem in materia.

14. Aliter dicunt illi qui ponunt infinitatem partium in materia prima et quod ibi sint partes infinitae, ita quod pars sub parte et non pars extra partem. Dicunt enim quod materia prima multotiens creari potest secundum diversas partes, ita quod secundum unam sui partem creatur quando illud compositum creatur et secundum aliam quando aliud compositum creatur. Et dicunt hoc non esse inconveniens sic, ut[b] prius.

[a] non *suppl. mg.* O
[b] ut *scripsimus*; nec O

nothing posterior produces in being what is prior; therefore, <prime matter is not created by a creature>. Therefore, it is necessary that the first cause <creates the prime matter>.

10. Furthermore, what is created from nothing can only be brought into being by the first cause; but prime matter is created from nothing; therefore, <it is brought into being by the first cause>. Proof of the major premise: creation is from nothing, and between nothing and something there is an infinite distance; therefore, it is necessary for the production of something from such an infinite distance to come about through an infinite power. And no creature has infinite power but only the first cause does; hence, <only the first cause creates>. Proof of the minor premise: among all beings prime matter is maximally a non-being and closer to non-being; therefore, since nothing from which prime matter could be produced was closer to prime matter than non-being, it is necessary for prime matter to be produced from non-being; hence, <prime matter is produced from nothing>.

11. The solution to the problem and also the truth of the matter is that prime matter is created by the first cause.

12. To the first argument to the contrary[335] it must be said that God is unchangeable with regard to the operations that end in himself, such as understanding, but not with regard to the operations that end in something else, for example in the things here below, and creation is of this kind.

13. To the second argument[336] it must be said that it is not absurd for prime matter to be created at many times in different ways, that is, substantially at one time and accidentally at many other times. (i) Something is created accidentally when it is created at many times in accordance with different beings, that is, in such a way that the creation does not terminate in that thing per se but in some composite. (ii) Something is created substantially when it is created in accordance with its one being, that is, in such a way that the creation terminates in that thing per se in so far as it is simple and not in so far as it is a part of a composite. Prime matter is created substantially in way (ii), and accidentally in way (i). And this is not absurd. This is the reply of those people who do not attribute divisibility into parts to matter.

14. A different reply is given by those people who posit an infinity of parts in prime matter and hold that there are infinitely many parts <found> in prime matter in such a way that any one part is under another and not outside another. They say that prime matter can be created at many times with respect to different parts, so that in accordance with one of its parts it is created when one composite is created and in accordance with another part it is created when another composite is created. They say that in this way it is not absurd <that matter should be created several times> as in the preceding answer.

[335] See above, n. 1.
[336] See above, n. 2.

15. Tertiae rationi dicendum quod materia prima est prius instanti temporis per naturam et non sequitur propter hoc quod non sit in illo, quia simul stant quod sint simul secundum existentiam et tamen unum eorum prius per naturam. Verbi gratia: prius est aliquid animal quam homo et tamen simul secundum existentiam est animal et homo. Et similiter est in multis aliis exemplis.

16. Aliter potest responderi quod infiniti sunt gradus in instanti, sicut prius patuit.

17. Quartae rationi dicendum quod simul scivit Deus ipsa esse, non tamen scivit ipsa esse simul; unde est ibi compositio et divisio ex eo quod illa adverbialis determinatio 'simul' potest componi cum hoc verbo 'scivit' vel cum hoc verbo 'esse'. Si primo modo, sic vera est et sic non sequitur conclusio, quia in praemissa determinat li 'simul' le 'scire' et in conclusione li 'esse', cum dicit: 'ergo simul sunt vera'. Si secundo modo, sic falsa est et sic sequitur conclusio et similiter conclusio falsa.

18. Vel dicendum quod Deus non scivit ipsa simul, ita quod [non[a]] in eodem instanti scivit ipsa esse, sed simul ita quod in uno instanti unum et in alio instanti reliquum. Et sic non sequitur conclusio, quia non sequitur nisi in eodem esset praemissa vera.

19. Quintae rationi dicendum secundum quosdam quod esse et non esse materiae primae non sunt contradictoria, quia dicunt quod esse et non esse, quod esse nullum aliud esse praecessit sed solum non esse, non sunt contradictoria nec debent dici. Sed non esse quod subsequitur esse contradicit eidem, quod forte falsum est.

[a] non *delevimus*

15. To the third argument[337] it must be said that prime matter is prior to an instant of time by nature but it does not follow from this that prime matter does not exist at that instant, because that two things are simultaneous in existence and that one of them is prior by nature <to the other> are compatible claims. For example, something's being an animal is prior to its being a man; in existence, however, it is simultaneously an animal and a man. There are many other similar examples.

16. Another possible reply is that there are infinitely many stages in an instant, as was made clear before.[338]

17. To the fourth argument[339] it must be said that God knew simultaneously the being <of prime matter and the non-being of prime matter> but he did not know that these were simultaneous; thus, there is composition and division in that proposition because the adverbial determination 'simultaneously' can be composed with the verb 'knew' or with this verb 'to be'. In the first way, the proposition is true and the conclusion does not follow, because in the premise 'simultaneously' qualifies 'to know', and in the conclusion it qualifies 'to be', when he says 'therefore, they are simultaneously true'. In the second way, the proposition is false and the conclusion does follow, but it is also false.

18. Alternatively it should be said that God did not know them simultaneously in the sense that he knew that they existed at the same instant, but he knew them simultaneously in the sense that he knew that one of them <existed> at one instant, and he knew that the other <existed> at another instant. In this way the conclusion does not follow because the conclusion only follows if the premise is true for the same instant.

19. To the fifth argument[340] it must be said, according to some people, that the being and the non-being of prime matter are not contradictories, because they say that, when a being and a non-being are such that no other being but only a non-being preceded that being, that being and that non-being are not contradictories and they should not be said <to be contradictories>. <They also say that> a non-being that follows a being is contradictory with it,[341] which is perhaps false.

[337] See above, n. 3.
[338] See above, q. 44, n. 11.
[339] See above, n. 4.
[340] See above, n. 5.
[341] Albert the Great holds this view in his discussion of the origin of matter; see Albert the Great, *Summa de creaturis*, tr. 1, q. 2, art. 3, 326b–327a: "... Haec propositio est falsa: Omne cujus esse praecedit non esse, radicatur in aliquo. Nec valet probatio: Non enim oppositio contradictionis est inter esse et non esse in quibus non esse praecedit esse, sed potius inter ea in quibus esse praecedit non esse... Unde in oppositione contradictionis esse praecedit non esse, et non e converso.".

20. Aliter dicitur quod extremum negativum contradictionis non requirit aliquam naturam, sed solummodo extremum affirmativum.

21. Sextae rationi dicendum quod proportio in nobilitate inter agens et actum non tenet in voluntarie agentibus, cuiusmodi est prima causa, sed solum in naturaliter agentibus[a].

22. Vel dicendum quod creatio non concernit actum, sed solum distantiam inter rem actam et illud ex quo agitur. Et patet pertractatio.

23. Septimae rationi dicendum quod non sunt proportionalia, quia mens incausata non recipit ideas materiae in quantum materia nec formae in quantum forma nec compositi in quantum compositum, sed in quantum ens vel substantia et etiam in quantum causata sunt. Mens vero causata recipit species compositi in quantum compositum, et similiter de aliis.

24. Octavae rationi dicendum quod duplex est malum et similiter duplex vitium et similiter duplex turpe: unum quod est turpe naturae, ut est in monstruosis generationibus, quando puer sex digitos habet et huiusmodi, et de tali turpe non intendit ibi Aristoteles; aliud quod est turpe moris, ut esse luxuriosus et superbus et huiusmodi, et forte de tali vitio intendit ibi. Et sic non sequitur conclusio de materia prima.

25. Vel dicendum quod Aristoteles vocat ibi vilia quae non debet Deus intelligere omnia ea quae sunt per receptionem ab extra, quia, si prima causa intelligeret per receptionem ab extra, ita quod suus intellectus terminetur super receptas species, sequitur quod efficeretur vilis, quod est inconveniens. Et huic responsioni videtur Aristoteles consentire, quia intendit ibi probare quod Deus non potest intelligere <[b]per receptionem ab extra>, quia dicit quod in ipso relucent omnes rationes causales rerum et, quando intelligit se ipsum, intelligit omnia quae sunt, non ita quod fiat receptio ab extra, sed per hoc quod rationes omnium relucent in ipso.

[a] agentibus] Septimae rationi dicendum quod non sunt proportionalia, quia mens incausata *add. sed del.* O

[b] per ... extra *supplevimus*; alia a se *suppl. s.l.* O

20. Another reply is that the negative extreme of a contradiction does not require a nature, but only the positive extreme does.

21. To the sixth argument[342] it must be said that the relation of nobility between an agent and a thing produced by it does not hold for voluntary agents, such as the first cause, but only for natural agents.

22. Alternatively it must be said that what creation concerns is not the thing produced but only the distance between the thing produced and that from which it is produced. And the discussion is then clear.

23. To the seventh argument[343] it must be said that an uncaused and a caused mind are not proportional, because an uncaused mind does not receive the ideas of matter in so far as it is matter or of a form in so far as it is a form or of a composite in so far as it is a composite, but in so far as it is a being or a substance or in so far as these things are caused. But a caused mind receives the species of a composite in so far as it is a composite, and similarly for the other things.

24. To the eighth argument[344] it must be said that there are two kinds of bad and similarly two kinds of vice and two kinds of disgrace: (i) one is the disgrace of nature, for example in the generation of monsters, as when a child has six fingers and the like; and Aristotle does not have this kind of disgrace in mind here. (ii) Another is the disgrace of manners, such as being extravagant and proud and the like; Aristotle perhaps has this kind of vice in mind here, and in this sense the conclusion does not apply to prime matter.

25. Alternatively it should be said that in that passage Aristotle calls all those things —that is, the things that God ought not to understand— base which are understood through a reception from outside, because if the first cause understood through a reception from outside, in such a way that its intellect terminated at a received species, it would follow that the first cause would become base, which is absurd. Aristotle seems to assent to this reply, because in that passage[345] he intends to prove that God cannot understand through a reception from outside, because he says that all the causal principles of things shine in God, and when he understands himself, he understands all existing things, not in such a way that there is a reception from outside, but in virtue of the fact that the principles of all things shine in him.[346]

[342] See above, n. 6.
[343] See above, n. 7.
[344] See above, n. 8.
[345] Arist. *Metaph.* XII.9, 1074b21–34.
[346] The correct reference is Aver. *In Metaph.* XII, t.c. 51, fol. 337ra.

<Quaestio 46>

Quaeritur quinto, habito quod materia prima incepit esse, et hoc a causa prima, qualiter incepit esse, an scilicet per formam vel per se ipsam.

1. Et videtur quod per[a] formam sic: dicit Aristoteles in primo libro *Physicorum* quod materia cognoscitur secundum analogiam, hoc est per comparationem ad formam; sed ab eodem est sua entitas a quo est sua cognitio; ergo sua entitas est ab entitate formae; ergo, cum producuntur in esse, principalius producitur forma quam materia; ergo etc.

2. Item, quando aliquid dependet ab alio, habita cognitione de uno sufficienter, habetur cognitio de reliquo, quia cognito uno correlativo, cognoscitur et reliquum. Sed materia secundum substantiam suam dependet a forma; ergo etc. Probatio minoris: materia, quia imperfecta, appetit formam; ergo, quanto magis imperfecta, tanto magis appetet formam. Sed materia prima maxime est imperfecta, quia minus habet de esse; ergo maximam habebit dependentiam ad formam. Hoc autem est secundum substantiam, quare dependebit secundum substantiam.

3. Item, plus convenit materia cum forma quam habitus cum privatione, quia forma perficit materiam et est simul cum ea, habitus vero non perficit privationem nec etiam simul est cum ea. Sed in tantum convenit habitus cum privatione quod habitus est principium existendi eiusdem et etiam cognoscendi; ergo multo fortius in tantum conveniet materia cum forma quod etiam erit et principium existendi[b] et cognoscendi; quare etc.

4. Item, dicitur in septimo *Metaphysicae* quod forma est magis substantia quam materia, et in primo *Physicorum* quod forma est magis natura. Et sequitur tunc quod forma est nobilior. Et quod nobilius **[O 96ra]** prius; ergo forma prior est materia. Et priori debetur prior productio; quare etc.

[a] per] se ipsam *add. sed del.* O
[b] existendi] ipsam *add. s.l.* O

<Question 46>

Now that it has been established that prime matter began to exist, and this by an action of the first cause, the fifth question is in what way it began to exist, that is, whether in virtue of form or by itself.

1. Arguments that <it began to exist> in virtue of form: in *Physics* I Aristotle says that matter is known by analogy,[347] that is, in relation to form; but the entity of something comes from the same thing as does the cognition of it; therefore, the entity of matter comes from the entity of form; therefore, when matter and form are brought into being, the form is more principally produced than the matter; therefore, <prime matter begins to exist in virtue of form>.

2. Furthermore, when one thing depends on another, once the cognition of one of the two things has been sufficiently acquired, the cognition of the other is also acquired, because, once one of two correlatives is known, the other is also known. But matter in its substance depends on form; therefore, <once form is known, matter is also known>. Proof of the minor premise: since matter is imperfect, it desires form; therefore, the more imperfect matter is, the more it will desire form. But prime matter is maximally imperfect, because it has less being; therefore, it will have a maximal dependence on form. But the maximal dependence is that found with respect to substance; hence, matter will depend <on form> in its substance.

3. Furthermore, matter agrees with form more than habit agrees with privation, because form gives perfection to matter and is simultaneous with it, whereas habit does not give perfection to privation and is not simultaneous with it. But habit agrees with privation to the extent that habit is the principle of the existence of privation and also of the cognition of it; therefore, there is a much stronger reason to hold that matter will agree with form to the extent that form will be the principle of its existence and of the cognition of it; hence, <matter is produced in virtue of form>.

4. Furthermore, in *Metaphysics* VII it is said that a form is more of a substance than matter,[348] and in *Physics* I that a form is more of a nature than matter.[349] It then follows that form is nobler. And what is nobler is prior; therefore, form is prior to matter. And to what is prior a prior production is attributed; hence, <the production of form is prior to that of matter and so matter is produced in virtue of form>.

[347] Arist. *Phys*. I.7, 191a7–8.
[348] Arist. *Metaph*. VII.3, 1029a29–30.
[349] The correct reference is Arist. *Phys*. II.1, 193b6–8.

5. Item, Aristoteles determinat hanc quaestionem in nono, utrum actus sit[a] potentia prior vel econverso. Et dicit quod in eodem potentia est prior; loquendo tamen simpliciter et in diversis, prior est actus. Sed actus reducitur ad formam et potentia reducitur ad materiam; ergo forma simpliciter prior erit materia. Et priori debetur etc.

6. Ad[b] oppositum: prius est deferens quam delatum, prius recipiens quam receptum; sed materia est deferens et recipiens, forma vero est delata et recepta[c]. Et priori debetur prior productio; quare materia primo creatur.

7. Item, si forma esset prius materia, tunc illa forma esset indifferens a causa prima, quod est inconveniens; oportet ergo quod materia sit prior quae possit esse subiectum formae. Et priori debetur etc.

8. Item, materia est unum principium secundum rem; ergo aliquid entitatis habet quae non est entitas formae; ergo et aliquid cognitionis, quae cognitio est alia a cognitione formae; ergo per se produci potest in esse, non mediante forma.

9. Item, quod habet ideam in causa prima potest intelligi sine alio; sed materia prima habet ideam in prima causa; ergo etc. Probatio minoris: materia prima est aliquid ens; ergo aliquid veritatis habet. Et veritati unde creata[d] est respondet idea; quare etc. Et quod potest intelligi sine alio per se creari potest; quare etc.

10. Circa solutionem huius quaestionis duplex est opinio. Una talis: materiae primae per se debetur creatio et non formae, et hoc dixerunt propter secundam rationem ad oppositum. Et dicunt quod causa prima, in creando materiam primam, simul cum ea concreavit suas potentias et ex illis potentiis fecit Deus formam. Unde dicendum quod non potest vere dici et proprie quod forma creatur, quia vocando actionem Dei quae fit ex nihilo, sic creatur materia proprie secundum eos et non forma. Vocando tamen actionem Dei quae non fit ex nihilo, sic dicunt quod quodam modo potest dici quod forma creatur, sed non proprie.

11. Alia est opinio talis quae dicit quod creatio non debetur materiae per se, sed toti composito. Et ipsi dicunt quod creatio non terminatur ad materiam

[a] sit *suppl. s.l.* O
[b] Ad] item *praem. sed del.* O
[c] recepta] *illegibile add. sed del.* O
[d] creata *scripsimus*; causativa (*lectio incerta*) O; causatum T; creatum C

5. Furthermore, in *Metaphysics* IX Aristotle settles the question of whether act is prior to potency or the other way around.[350] He says that in one and the same thing potency is prior; but if we speak absolutely and about different things, act is prior. But act is reduced to form and potency is reduced to matter; therefore, form is absolutely prior to matter. And to what is prior <a prior production> is attributed.

6. To the contrary: what carries <something> is prior to what is carried, and what receives <is> prior to what is received; but matter is a carrier and is a receiver, whereas form is carried and received. And to what is prior a prior production is attributed; therefore, matter is created first.

7. Furthermore, if a form were prior to matter, then that form would not be different from the first cause, which is absurd; therefore, it is necessary for matter, which can be the subject of a form, to be prior. And to what is prior <a prior production> is attributed.

8. Furthermore, matter is one principle in reality; therefore, it has an entity distinct from the entity of form; therefore, it also gives rise to a cognition distinct from the cognition of form; therefore, it can be brought into being per se, and not by means of form.

9. Furthermore, what has an idea in the first cause can be understood without another thing; but prime matter has an idea in the first cause; therefore, <it can be understood without another thing>. Proof of the minor premise: prime matter is a being; therefore, it has some truth. And an idea corresponds to a truth in so far as it is created; hence, <prime matter has an idea in the first cause>. And what can be understood without another thing can be created per se; hence, <prime matter can be created per se>.

10. There are two opinions about the solution to this question. (i) One is that creation is per se ascribed to prime matter and not to form. These people maintained this because of the second argument to the contrary.[351] They also say that in creating prime matter, the first cause co-created its potencies simultaneously with it and that God produced form from those potencies. Thus, it must be said that it cannot truly and properly be said that form is created, because if we call <‘creation’> an action of God that starts from nothing, then according to these people it is matter and not form that is properly created. But if we call <‘creation’> an action of God that does not start from nothing, then they say that it can in some way be said that form is created, although not properly.

11. (ii) Another opinion says that creation is not ascribed to matter per se, but to the whole composite. These people say that creation does not terminate

[350] Arist. *Metaph.* IX.8, 1049b10–1051a3.
[351] See above, n. 7.

per se nec ad formam per se, sed ad totum compositum; et ideo dicunt totum proprie creari. Sed si quaeratur quod istorum componentium magis creatur, dicunt quod forma, et hoc quia, cum forma sit finis et terminus, magis creatur quam materia, quia magis ad ipsam terminatur creatio.

12. Sustinendo primam opinionem ad praesens, ad primam rationem in oppositum dicendum quod Aristoteles loquitur ibi de materia physica et[a] in comparatione ad intellectum nostrum et non in comparatione ad intellectum causae primae, quia[b] nos cognoscimus, per Aristotelem, posteriora magis quam priora, quia cognitio nostra a sensu incipit et terminatur ad priora et notiora naturae. Et ideo, quantum ad intellectum nostrum, licet non possumus cognoscere materiam primam nisi sub comparatione ad formam, materia[c] prima potest esse simpliciter prior. E contrario modo est in intellectu causae primae, quia ipse veras naturas rerum considerat et videt; ideo quoad suum intellectum materia prima est prior et potest intelligi sine comparatione ad formam. Et prima causa creat res[d] secundum quod sunt priores apud intellectum suum; et ideo primo creavit materiam.

13. Secundae rationi dicendum quod materia prima secundum suam substantiam non appetit formam.

14. Et dicendum[e] ad illud quod obicitur contra quod materia prima non habet dependentiam ad formam quia appetit eam. Et dicit Aristoteles in fine primi *Physicorum* quod materia non appetit formam nisi quia privata est et propter privationem, ubi dicit quod materia non est turpis per se nisi per privationem; ergo materia prima de se dependentiam ad formam non habet nisi per privationem. Hoc idem patet per Commentatorem dicentem quod de materia, prout consideratur sub forma, sic de ea determinat naturalis; si autem prout informis est, sic de ea considerat metaphysicus.

[a] et *scripsimus cum* TC; *om.* O
[b] quia] enim *add. sed del.* O
[c] materia] tamen *add. s.l.* O
[d] res *suppl. s.l.* O
[e] dicendum] quod *add. sed del.* O

in matter per se or in form per se but in the whole composite; therefore, they say that it is a whole that is properly created. But if it is asked which of these components is more properly created, they reply that it is form. The reason for this is that, since form is an end and a terminus, it is more properly created than matter, because creation more properly terminates in it.

12. If we accept the first opinion for the moment, it must be replied to the first argument to the contrary[352] that Aristotle is speaking there of physical matter and is also speaking in relation to our intellect and not in relation to the intellect of the first cause, because, according to Aristotle, we know posterior things better than prior things,[353] given that our cognition starts from the senses and terminates at things that are prior and better known to nature. Therefore, although our intellect cannot know prime matter except in relation to form, prime matter can be absolutely prior. The contrary happens in the intellect of the first cause, because it considers and sees the true natures of things; therefore, in relation to the intellect of the first cause, prime matter is prior and can be understood without relation to form. And the first cause creates things in so far as they are prior in its intellect; therefore, it created matter first.

13. To the second argument[354] it must be said that prime matter in its substance does not desire form.

14. To the objection against this[355] it must be said that it is not because it desires form that prime matter has any dependence on form. At the end of *Physics* I, where he says that matter is not disgraceful per se but only through privation, Aristotle says that matter does not desire form unless it has been deprived of form and because of privation;[356] therefore, prime matter does not by itself have any dependence on form unless through privation. The same is clear from the Commentator's statement that a natural philosopher establishes definite conclusions about matter in so far as it is considered under form; but if matter is considered in so far as it is without form, then it is a metaphysician who considers it.[357]

[352] See above, n. 1.

[353] Arist. *Phys*. I.1, 184a18–23.

[354] See above, n. 2.

[355] That is, the argument in favour of the claim that matter depends on form in n. 2 above.

[356] Arist. *Phys*. I.9, 192a22–25.

[357] Averroes does maintain that prime matter can be known only insofar as it is considered under a form (see Aver. *In Phys.* I, t.c. 69, fol. 40va-b) and he does contrast physical and metaphysical knowledge of matter (see Aver. *In Phys*. I, t.c. 83, fol. 47va, and *In Metaph*. VII, t.c. 9, fols. 159vb–160ra). However, his point in contrasting physics and metaphysics is that physics considers prime matter in its role of substrate of generation, whereas metaphysics considers prime matter in its role of constitutive principle of substance.

15. Vel supposito quod materia in sui substantia dependeat a forma, tunc dicendum quod ista dependentia non dicitur relatio quae est accidens, de qua relatione dicitur quod cognito uno etc., sed est quidam modus loquendi in substantialibus et non meretur nomen relationis quae est[a] accidens, sed nomen relationis quae substantia est, sicut dictum est superius.

16. Vel dicendum quod habet dependentiam ad formam, sed illud quod dicit[b] Aristoteles, "Cognito uno etc.", non intelligendum est sic quod ratio unius sit ratio alterius ita quod species unius sit species alterius, sed quod unum potest cognosci per reliquum. Et non sequitur propter hoc quin materia prima per se creetur, dummodo[c] ratio unius non[d] sit ratio alterius.

17. Tertiae rationi dicendum quod plus sibi[e] pertinere, hoc est dupliciter: aut in simul existendo aut in occasionando. Primo modo plus pertinent sibi[f] materia et forma, secundo modo non, sed sic[g] conveniunt habitus et privatio magis, quia privatio nihil entitatis habet nisi per respectum ad habitum sive[h] ad subiectum in quo est habitus. Primo modo plus conveniunt materia et forma, quoniam esse materiae actu dependet a forma, tamen suum esse potentia non.

18. Quartae rationi dicendum quod non sequitur 'nobilius; ergo prius', sed universaliter deficit. Verbi gratia: homo est nobilius animali et tamen animal prius, quia incompletum universaliter prius est completo; unde deficit illa consequentia. Et similiter in proposito, quia, licet forma sit nobilior, tamen materia est incompletior, et ita prior.

19. Quintae rationi, ad cognoscendum qualiter et quando actus praecedit potentiam et quando econverso, nota quod actus et potentia aut sunt in eodem supposito secundum substantiam et in eodem secundum numerum aut in diversis subiectis. Si secundo modo, hoc dupliciter: aut una[i] illarum substantiarum est causa alterius in ratione efficientiae et est sic ordinata ad aliam[j] ut sit effectivum eius, aut non, sed est ordinata ad ipsam[k] solum ut ipsam deferat et ut ipsam[l] recipiat. Si primo modo, hoc dupliciter: aut enim sunt illae diversae substantiae sic **[O 96rb]** ordinatae ad invicem secundum

[a] est *suppl. s.l.* O
[b] dicit *iter. sed corr.* O
[c] dummodo] quod *add. sed del.* O
[d] non *suppl. s.l.* O
[e] sibi *suppl. s.l.* O
[f] sibi *suppl. s.l.* O
[g] sic *iter. sed corr.* O
[h] sive *scripsimus*; sicut O
[i] una *scripsimus*; unum O
[j] aliam *scripsimus*; alium O
[k] ipsam *scripsimus*; ipsum O
[l] ipsam *scripsimus*; ipsum O

15. Alternatively, if we assume that matter in its substance depends on form, then it must be replied that this dependence does not express a relation that is an accident —of which kind of relation it is said that once one <of two correlatives> is known, <the other is also known>— but is just a mode of speaking about substantial things and does not deserve the name of a relation that is an accident, but <rather> the name of a relation that is a substance, as has been said above.[358]

16. Alternatively, it must be replied that matter depends on form, but the claim of Aristotle "once one <of two correlatives> is known, <the other is also known>"[359] must not be understood in the sense that the *ratio* of one of them is the *ratio* of the other in such a way that the species of one of them is the species of the other, but that one of them can be known through the other. And it does not follow because of this that matter is not created per se, as long as the *ratio* of one of them is not the *ratio* of the other.

17. To the third argument[360] it must be said that things can agree more in two ways: either (i) in existing simultaneously or (ii) in occasioning. Matter and form agree more in way (i), but not in way (ii), but habit and privation agree more in way (ii), because a privation has no entity except through a relation to a habit or to the subject in which a habit exists. Matter and form agree more in way (i) because the being of matter in act depends on form, but its being in potency does not.

18. To the fourth argument[361] it must be said that the consequence 'nobler; therefore, prior' does not hold, but it fails in all cases. For example: a man is nobler than an animal, but an animal is prior, because what is incomplete is universally prior to what is complete; thus, that consequence fails. The same happens in the present case, because, although form is nobler, matter, however, is more incomplete and hence it is prior.

19. To the fifth argument,[362] in order to know how and when an act precedes a potency and when it is the other way around, note that an act and a potency are either (i) in the same supposit in substance and in the same supposit in number or (ii) in different subjects. Case (ii) is twofold: either (iia) one of those substances is an efficient cause of the other and it is ordered to the other so that it is an efficient cause of it, or (iib) this is not so, but it is ordered to the other only because it carries the other and receives it. Case (iia) is twofold: either (iia') those different substances are ordered one to the other with regard to efficiency in a genus or (iia'') outside a genus. Case (iia'') is that

[358] See above q. 42.2, n. 10.
[359] Arist. *Cat*. 7, 8b14–15.
[360] See above, n. 3.
[361] See above, n. 4.
[362] See above, n. 5.

efficientiam in genere aut extra. Si secundo modo, sic est prima causa. Si primo modo, hoc est dupliciter: aut sunt in eodem genere, ut ille homo et iste homo et haec planta et illa planta, aut in diversis generibus, ut substantia et accidens. Si autem sint in eodem secundum numerum vel in diversis et sint ordinata tantum secundum recipientiam et delacionem et non secundum efficientiam, tunc semper potentia praecedit actum, quia in talibus potentia est recipiens et actus est receptus, et recipiens est prius recepto. Unde, quia materia est quasi recipiens et forma recepta, ideo materia prius forma. Et sic patet quod non procedit[a] ratio. Si vero actus et potentia sint in diversis subiectis et se habeant secundum efficientiam, sive sint in genere, et hoc sive in eodem genere sive in diversis generibus, [b]sive extra, semper tunc actus praecedit potentiam, quia omne agens praecedit illud quod debet produci ex eo quod est in potentia. Unde, si[c] fas est comparare primam causam ad creaturam iam factam, eius actus est prior, et universaliter[d] agens illo[e] quod debet produci in esse; et sic est si ex homine generetur homo vel ex substantia accidens vel si Deus faciat creaturam.

<Quaestio 47>
Quaeritur sexto quid sit, utrum scilicet materia prima sit substantia vel accidens.

1. Et videtur quod accidens sic: propter debilitatem esse dicitur accidens esse accidens; sed inter omnia entia materia prima minus habet de entitate et debilissimum esse habet; ergo maxime erit accidens.

2. Item, Commentator dicit quod omnia accidentia causantur a materia et praecipue quantitas; sed principiata sunt unigenea[f] cum principiis; quare etc.

3. Huic rationi respondent dupliciter quidam: uno modo sic, quod materia secundum sui substantiam est causa substantiarum, materia autem sub quodam esse accidentali causa est[g] accidentium. Primo modo necesse est principiata et principia esse unigenea[h], secundo modo non.

4. Sed contra hoc sic: illud esse superadditum aut est esse substantiale aut est esse accidentale. Si primo modo, habetur propositum. Si secundo modo, quaeritur tunc ulterius: aut est causatum a substantia aut ab esse. Si est causatum a substantia, habetur propositum. Si ab esse, quaeritur tunc

[a] procedit *scripsimus*; praecedit O
[b] sive extra *suppl. s.l.* O
[c] si *scripsimus cum* TC; sicut O; *corr. in* sicut T
[d] universaliter] ubi est *add. sed del.* O
[e] illo *scripsimus*; illud O (quam causata TC)
[f] unigenea *scripsimus*; unigenita O; unige[m] TC
[g] est *suppl. s.l.* O
[h] unigenea *ex corr.* O

of the first cause. Case (iia') is twofold: either (iia'1) they are in the same genus, such as that man and this man, and this plant and that plant, or (iia'2) they are in different genera, such as a substance and an accident. However, if they are in the same supposit in number or in different supposits and they are only ordered with regard to reception and carrying and not with regard to efficiency, then a potency always precedes an act, because in this case the potency is receiving and the act is received, and what receives is prior to what is received. Thus, since matter is receiving, as it were, and form is received, it follows that matter is prior to form. It is then clear that the argument does not succeed. But if an act and a potency are in different subjects and are related with regard to efficiency, or are in a genus, whether the same genus or different genera, or are outside a genus, then an act always precedes a potency, because every agent precedes what must be produced from what is in potency. Thus, if it is allowed to compare the first cause to a creature already produced, its act is prior, and an agent is universally prior to what must be brought into being; and this is what happens if a man is generated by a man or an accident by a substance or if God produces a creature.

<Question 47>
The sixth question is what prime matter is, that is, whether prime matter is a substance or an accident.

1. Arguments that it is an accident: an accident is said to be an accident because of the weakness of its being; but among all beings prime matter has the least entity and the weakest being; therefore, it will maximally be an accident.

2. Furthermore, the Commentator says that all accidents are caused by matter and principally quantity;[363] but the things derived from principles are homogeneous with their principles; hence, <matter is an accident>.

3. Some people reply to this argument in two ways: in one way that matter in its substance is a cause of substances, but matter under an accidental being is a cause of accidents. In the first way it is necessary for the things derived from principles to be homogeneous with their principles, in the second way this is not necessary.

4. But against this: that additional being is either (i) a substantial being or (ii) an accidental being. If (i), then we get the intended conclusion. If (ii), then the question arises again whether (iia) that being is caused by a substance or (iib) by a being. If (iia) holds, we get the intended conclusion. If (iib)

[363] On matter as the cause of all accidental determinations, see, e.g., Aver. *In Metaph.* XII, t.c. 25, fol. 309vb; on quantity as the most basic determination of matter, see, e.g., Aver. *In Metaph.* VII, t.c. 8, fol. 159ra–b; *De sub. orb.* 1, fols. 3vb–4rb.

ulterius: aut ab esse substantiali aut ab esse accidentali. Si primo modo, habetur propositum. Si secundo modo, tunc quaerendum ulterius ut prius. Et tandem erit processus in infinitum, quod est inconveniens; quare responsio nulla.

5. Aliter dicunt quidam sic, quod materia secundum esse diversum et diversum est causa accidentium et substantiarum.

6. Sed contra: ista esse aut sunt idem aut diversa. Si idem, tunc substantia et accidens idem, quia idem sub eodem esse non est causa diversorum. Si diversa[a], quia unum et idem per idem esse non est causa diversorum esse, oportet ergo quod hoc sit per alia esse diversa. Et sic[b] tandem erit[c] processus in infinitum, quod est inconveniens; quare responsio nulla.

7. Item, materia est[d] accidentis; ergo materia est accidens. Praemissa patet, quoniam materia est subiectum et causa accidentium. Consequentia patet per Aristotelem in septimo *Metaphysicae* sic arguendo: accidentia sunt entis, ergo sunt entia. A simili a parte ista.

8. Item, omne quod de se est extensum est accidens; sed materia prima de se est extensa, circumscriptis omnibus accidentibus; ergo etc. Probatio minoris: esse quantitatis causatur a materia; ergo tota extensio quantitatis causabitur ab extensione materiae. Sed prius manet dempto posteriori; ergo materia actualiter extensa est circumscripta quantitate.

9. Item, haec est falsa: 'materia est forma'; ergo haec erit falsa: 'materia est substantia', quia praedicatum dicit formam et propterea[e] genus generalissimum dicit formam et praedicat formam. Sed aut substantia aut accidens; et non substantia; erit ergo accidens.

10. Item, materia est causa substantiarum et substantiae fiunt ex ea; sed per Aristotelem, primo *Physicorum*, quidlibet non fit ex quolibet, sed oppositum ex opposito; ergo materia erit non substantia. Et vel substantia vel accidens; ergo accidens.

11. Ad oppositum: ex non substantiis non fit substantia; sed ex materia et forma fit substantia; ergo utrumque est substantia[f]; ergo materia erit substantia.

[a] diversa] et non causantur a materia per idem esse *add. s.l.* O
[b] sic *suppl. s.l.* O
[c] erit *suppl. s.l.* O
[d] est] causa *add. s.l.* O
[e] propterea *scripsimus*; praeterea O
[f] substantia] fiunt substantiae *add. sed del.* O

holds, then the question arises again whether (iib') it is caused by a substantial being or (iib'') by an accidental being. In case (iib'), we get the intended conclusion. In case (iib'') the same question arises, and there will be an infinite regress, which is absurd; hence, their reply is worth nothing.

5. Alternatively some people give a different reply, namely that matter with respect to diverse beings is a cause of accidents and of substances.

6. But against this: these beings are either the same or different. (i) If they are the same, then a substance and an accident are the same, because the same thing under the same being is not a cause of different things. (ii) If they are different, since what is one and the same thing in virtue of the same being is not a cause of different beings, it is therefore necessary that this should happen in virtue of other different beings. And then there will be an infinite regress, which is absurd; hence, their reply is worth nothing.

7. Furthermore, matter belongs to an accident; therefore, matter is an accident. The premise is clear because matter is a subject and a cause of accidents. The consequence is clear from Aristotle in *Metaphysics* VII, where he argues as follows: accidents belong to beings; therefore, they are beings.[364] The same applies in the present case.

8. Furthermore, everything that is extended by itself is an accident; but prime matter is extended by itself, once all accidents have been removed; therefore, <prime matter is an accident>. Proof of the minor premise: the being of a quantity is caused by matter; therefore, the whole extension of a quantity will be caused by the extension of matter. But what is prior remains, when what is posterior is removed; therefore, matter is actually extended, when quantity is removed.

9. Furthermore, the proposition 'matter is a form' is false; therefore, the proposition 'matter is a substance' will be false, because the predicate expresses a form and because of this the most general genus expresses a form and predicates a form. But matter is either a substance or an accident: and it is not a substance; therefore, it will be an accident.

10. Furthermore, matter is a cause of substances and substances come to be from it; but according to Aristotle in *Physics* I, it is not the case that anything comes to be from anything, but an opposite <comes> from its opposite;[365] therefore, matter will not be a substance. And it is either a substance or an accident; therefore, it is an accident.

11. To the contrary: a substance does not come to be from non-substances; but a substance comes to be from matter and a form; therefore, each of them is a substance; therefore, matter will be a substance.

[364] Arist. *Metaph.* VII.1, 1028a18–20.
[365] Arist. *Phys.* I.5, 188a30–b26.

12. Item, dicit Aristoteles in septimo *Metaphysicae* quod 'substantia' dividitur per materiam, formam et compositum; et divisum praedicatur de dividentibus; ergo materia erit substantia.

13. Solutio sive veritas problematis est quod materia prima est substantia.

14. Primae rationi in oppositum dicendum quod sufficiens causa accidentis non est habere esse debile, sed sua causa sufficiens est habere esse debile simul cum hoc quod causetur ex unione formae substantialis in materia. Et ultima condicio[a] deficit a parte materiae primae, quia praecessit omnem formam et omne compositum, accidentia vero subsequuntur compositum.

15. Vel dicendum quod habere esse debile, hoc est dupliciter: aut quoad modum essendi, et sic accidens habet nobilius esse materia, quia modus essendi accidentis est esse in actu et modus essendi materiae est esse in potentia passiva; et hoc modo materia magis habet esse debile. Aut quoad appropriationem esse, et sic materia et universaliter substantiae omnes nobilius esse habent accidentibus, quia accidens, cum suum esse dependeat totaliter ab alio, non appropriat sibi esse, quia esse in actu, quod quidem esse habent accidentia, est commune et nulli appropriatum. Materia vero et substantiae omnes, quae non dependent, habent esse sibi appropriatum.

16. Secundae rationi dicendum, sicut dictum fuit, quod materia secundum esse diversa est causa substantiarum et accidentium, videlicet sub esse substantiali est causa substantiarum et sub esse accidentali est causa accidentium.

17. Ad rationem in oppositum dicendum quod non contingit procedere in infinitum, immo erit sumere statum, quia unum et idem per idem esse potest esse causa diversorum secundum rationem et esse sive diversorum secundum esse solum. Dicendum est ergo quod eorum esse sunt diversa et tamen non erit procedere in infinitum.

18. Vel dicendum quod materia prima nec per idem esse nec per diversa [b]causa est illorum esse, sed se ipsa est causa illorum et non per aliud, ut patet alibi per exemplum de mutatione et similiter de creatione. **[O 96va]**

[a] condicio] sic *add. sed del.* O
[b] causa est *corr. in* est causa O

12. Furthermore, in *Metaphysics* VII Aristotle says that 'substance' is divided into matter, form, and the composite;[366] and what is divided is predicated of the things dividing it; therefore, matter will be a substance.

13. The solution of the problem or the truth of the matter is that prime matter is a substance.

14. To the first argument to the contrary[367] it must be said that the sufficient cause of an accident is not for it to have a weak being, but that its sufficient cause is for it to have a weak being and in addition to this to be caused by the union of a substantial form with matter. The last condition is not met by prime matter, because prime matter preceded every form and every composite, whereas accidents are subsequent to a composite.

15. Alternatively, it must be said that something can have a weak being in two ways: either (i) with respect to its mode of being, and in this way an accident has a nobler being than matter, because the mode of being of an accident is to be in act, whereas the mode of being of matter is to be in passive potency; and in this way matter has a weaker being. Or (ii) with respect to the appropriation of being, and in this way matter and all substances, taken universally, have a nobler being than accidents. This is because, since the being of an accident is totally dependent on something else, an accident does not appropriate being for itself, given that being in act —the kind of being that accidents have— is common and peculiar to nothing. But matter and all substances, which are not dependent, have a being which is their own.

16. To the second argument[368] it must be said, as was said beforehand,[369] that matter in accordance with different beings is a cause of substances and of accidents, that is, it is a cause of substances under a substantial being and a cause of accidents under an accidental being.

17. To the argument to the contrary[370] it must be said that there is no infinite regress; on the contrary, one comes to a halt, because one and the same thing through the same being can be the cause of things that are different conceptually and in being or <the cause> of things that are different in being alone. Therefore, it must be said that their beings are different, but there is no infinite regress.

18. Alternatively it should be said that prime matter is a cause of those beings neither through the same being nor through different beings, but it is a cause of those beings by itself and not through something else, as is made clear elsewhere through the examples of change and also of creation.

[366] Arist. *Metaph*. VII.3, 1029a1–3.
[367] See above, n. 1.
[368] See above, n. 2.
[369] See above, n. 5.
[370] See above, n. 6.

19. Tertiae rationi dicendum quod 'hoc esse huius' est dupliciter: aut enim dicitur hoc esse huius per intentionem nominis descendentis in ipso, per posterius tamen, et in talibus potest dici quod hoc[a] est hoc si est huius. Et sic est a parte accidentis, quia accidens per descensum quem habet ab ente accipit hoc nomen 'ens' aliquo modo per posterius, et ideo potest dici quod est ens. Aut etiam dicitur hoc esse huius non per intentionem nominis etc., sed per solam subiectionem et causalitatem, et tunc non potest dici quod hoc est hoc. Et sic est in proposito a parte materiae, quia materia prima nullum descensum habet ab accidente, et ideo nec per prius nec per posterius erit dicere quod materia est accidens. Sed quia est subiectum et similiter causa accidentium, propter nihil aliud dici potest quod materia est accidentis. Et sic patet in quibus debet ista consequentia tenere: 'hoc est huius; ergo hoc est hoc'.

20. Quartae rationi dicendum quod haec propositio, 'extensio quantitatis est ab extensione materiae', habet duos sensus: vel ita quod extensio quantitatis causatur a materia ita quod propter potentiam in materia ad extensionem consequitur quantitas et per adventum formae alicuius. Sic est propositio vera et sic non sequitur conclusio. Vel ita quod materia ante adventum quantitatis prius extendatur. Et sic est propositio falsa. Si dicatur primo modo, sic quod actualitas extensionis in quantitate sive quod sua actualis extensio producatur de potentia materiae, tunc est[b] vera, et quoad[c] hoc non procedit ratio. Sed si iste sit sensus: extensio quantitatis causatur a priori extensione materiae, sic est falsa et sequitur conclusio.

21. Quintae rationi patebit postea, quando quaeretur an aliquid praedicetur de materia.

22. Sextae rationi et ultimae dicendum quod materia est substantia et quod substantia fit ex materia, quia substantia composita. Concedendum est ergo quod non fit quodlibet ex quolibet, sed ex opposito. Et dicendum quod concludit verum cum sic arguit: 'ex materia fit substantia'; hoc est verum sic, scilicet substantia composita. Et cum concludit: 'ergo materia non est

[a] hoc] quod *add. sed del.* O
[b] est *suppl. s.l.* O
[c] quoad *lectio incerta* O

19. To the third argument[371] it must be said that the claim 'this belongs to this' can be understood in two ways: either (i) this is said to belong to this because the intention of the name applies to it but in a secondary sense, and for these things it can be said that this is this if this belongs to this. This is the case of accidents, because an accident receives the name 'being' in a secondary sense in virtue of the descent that an accident has from a being, and therefore it can be said that an accident is a being. Or (ii) this is said to belong to this not because the intention of the name < applies to it>, but only in virtue of its being a subject or a cause <of the thing to which it is said to belong>, and in this case it cannot be said that this is this. And this is what happens in the present case concerning matter, because prime matter does not descend from an accident, and therefore it will not be possible to say that matter is an accident either in a primary or in a secondary sense. But since matter is a subject and similarly a cause of accidents, there is nothing else that would allow one to say that matter belongs to an accident. And so it is clear in what cases the consequence 'this belongs to this; therefore, this is this' should hold.

20. To the fourth argument[372] it must be said that the proposition 'the extension of a quantity derives from the extension of matter' has two senses: (i) one is that the extension of a quantity is caused by matter in such a way that that quantity follows from a potency for extension in matter and through the supervening of a form. In this sense the proposition is true and the conclusion does not follow. (ii) The other sense is that matter is extended before a quantity supervenes. And in this sense the proposition is false. If the proposition is understood in sense (i) so that the actuality of an extension in a quantity or its actual extension is produced from a potency of matter, then the proposition is true, and in this case the argument does not succeed. But if the sense is that the extension of a quantity is caused by a prior extension of matter, then the proposition is false and the conclusion follows.

21. The reply to the fifth argument[373] will become clear later, when it will be asked whether something is predicated of matter.[374]

22. To the sixth and last argument[375] it must be said that matter is a substance and that a substance comes to be from matter, because a substance is composite. Therefore, it must be conceded that it is not the case that anything comes to be from anything, but rather from its opposite. And it must be said that the argument implies a true conclusion when it is argued: 'a substance comes to be from matter', because this is true of a composite

[371] See above, n. 7.
[372] See above, n. 8.
[373] See above, n. 9.
[374] See below, q. 48.
[375] See above, n. 10.

substantia', dicendum quod hoc verum est prout sequitur ex praemissa, quia non est substantia composita. Est tamen substantia simplex, et ita non sequitur quod nullo modo sit substantia.

23. Per iam dicta patet qualiter istae duae propositiones, 'ex non substantiis non fit substantia' et 'ex non substantiis fit substantia', per Aristotelem, qui[a] dicit: "Oppositum fit ex suo opposito", simul stant et etiam qualiter sunt intelligendae, quia ex nullo modo substantiis non fit substantia et ex non substantiis aliquo modo fit substantia. Et ita patet quod non est repugnantia inter dictas propositiones.

<Quaestio 48>

Quaeritur septimo, habito quod materia prima sit substantia, utrum sit substantia in actu vel non.

1. Et videtur quod in actu sic: materia est genus generalissimum; et genus sive generalissimum est actu substantia; ergo etc. Probatio maioris: arguit Aristoteles in quinto *Physicorum* sic: quorum genus idem, et materia est eadem et econverso; ergo materia et genus idem.

2. Item, illud est generalissimum quod est primum respectu omnium formarum subsequentium in illo genere; sed tale est materia prima; ergo materia prima est generalissimum. Et ulterius ut prius.

3. Item, circa genus generalissimum consistit contrarietas prima; sed circa primam materiam similiter consistit contrarietas prima; ergo materia est genus. Probatio maioris: genus est illud quod primo dividitur in contraria. Probatio minoris per Aristotelem primo libro *Physicorum* dicentem quod necesse est ponere tertium principium quod debet subsistere contrariis, quia non possunt subsistere sine tertio principio, quod est materia. Vult ergo quod materia est illud quod primo subsistit contrariis; ergo etc. Et ulterius tunc sicut prius.

[a] qui *suppl. s.l. (lectio incerta)* O

substance. And when the argument <further> concludes 'therefore, matter is not a substance', it must be said that this is true in so far as the conclusion follows from the premise, because matter is not a composite substance. Matter, however, is a simple substance, and so it does not follow that matter is in no way a substance.

23. Given what has already been said it is clear in what way the two propositions 'a substance does not come to be from non-substance' and 'a substance comes to be from non-substances' are consistent, according to Aristotle, who says "an opposite comes to be from its opposite",[376] and in what way they must be understood, because a substance does not come to be from things that are in no way substances, and a substance comes to be from things that are in some way not substances. It is then clear that there is no inconsistency between the two propositions above.

<Question 48>
Now that it has been established that prime matter is a substance, the seventh question is whether it is a substance in act or not.

1. Arguments that it is a substance in act: matter is the most general genus; and the most general genus is substance in act; therefore, <matter is substance in act>. Proof of the major premise: in *Physics* V Aristotle argues that the things that have the same genus also have the same matter and the other way around;[377] therefore, matter and genus are the same.

2. Furthermore, the most general genus is the first with respect to all the subsequent forms in that genus; prime matter is of that kind; therefore, prime matter is the most general genus. And then we get the same conclusion as before.[378]

3. Furthermore, the first contrariety exists in the most general genus; and the first contrariety also exists in prime matter; therefore, matter is the <most general> genus. Proof of the major premise: the <most general> genus is that which is primarily divided into contraries. Proof of the minor premise: in *Physics* I Aristotle says that it is necessary to assume a third principle that must be a subject for contraries,[379] because contraries cannot subsist without a third principle, which is matter. Therefore, Aristotle wants to hold that matter is the thing that is the first subject for contraries; therefore, <the first contrariety exists in matter>. And then again we get the same conclusion as before.[380]

[376] Arist. *Phys.* I.5, 188b25–26.
[377] The correct reference is Arist. *Metaph.* V.6, 1016a24–28.
[378] See above, n. 1.
[379] Arist. *Phys.* I.6, 189a34–b3.
[380] See above, n. 1.

4. Item, quanto aliquid magis est particulare, et magis habet de forma, et quanto aliquid communius, tanto minus habet de forma; quod est ergo communissimum minime habet de forma. Sed materia est illud quod minime habet de forma; materia ergo est illud quod est communissimum. Et tale est generalissimum; ergo etc.

5. Item, omnis forma actualis est a differentia actuali; sed materia prima caret omni differentia actuali et genus generalissimum [a]similiter. Probatio, quia genus dupliciter habet differentias; quasdam habet divisivas, et illae differentiae sunt in potentia in ipso; quasdam habet constitutivas, et illae sunt actualiter in ipso. Verbi gratia: differentia completiva[b] animalis est sensibile, et illam habet actualiter; differentiae suae divisivae sunt rationale, irrationale, et ista sunt potentia in ipso. Si ergo generalissimum haberet formam actualem, tunc haberet differentiam actualem, quae est completiva. Sed nullam talem potest habere generalissimum, quia omnis differentia completiva unius est differentia divisiva alterius. Verbi gratia: sensibile est differentia divisiva animati et sensibile est differentia completiva animalis. Si ergo haberet differentiam talem, tunc esset aliquod genus superius ipso quod divideretur per illam differentiam, et hoc est inconveniens. Et praeterea, genus omne est in potentia ad differentias. Materia ergo et genus sunt idem.

6. Item, 'substantia', per Aristotelem, septimo *Metaphysicae*, dividitur in materiam, formam et compositum; et divisum praedicatur de dividentibus; materia ergo est substantia. Et omne praedicatum dicit formam in materia; ergo materia est informata, et ita est actu.

7. Ad oppositum: Commentator super finem undecimi dicit quod omnes formae quae sunt in motore primo in actu sunt in potentia in materia prima; sed omnes formae sive rationes causales sunt actu in motore primo; ergo omnes formae sunt in potentia in materia prima, et ita nulla in actu.

8. Solutio sive veritas problematis talis est, quod materia prima in quantum materia nullam habet substantiam actu; materia tamen in quantum naturalis est est substantia in actu, licet incomplete, quia, si materia naturalis est, tunc corpus mobile est. Et si corpus mobile est, tunc motus est. Et si motus est, actio est. Et omnis actio est in actu. Unde materia naturalis semper est in actu, materia tamen simpliciter, in quantum materia, non est in actu.

[a] similiter. Probatio *inv. sed corr.* O
[b] completiva *scripsimus*; completive *scrips. et* vel constitutiva *add. s.l.* O

4. Furthermore, the more particular something is the more form it has, and the more common something is the less form it has; therefore, what is most common has a minimum amount of form. But matter is that which has a minimum amount of form; therefore, matter is that which is most common. And a most general genus is of this kind; therefore, <matter is the most general genus>.

5. Furthermore, every actual form derives from an actual difference; but prime matter lacks every actual difference, and so does the most general genus. Proof: a genus has differences of two kinds: some are divisive, and those differences exist in potency in it; others are constitutive, and those differences exist in act in it. For example, the completive difference of an animal is sensible, and an animal has that difference in act; its divisive differences are rational and irrational, and these exist in potency in it. Therefore, if the most general genus had an actual form, then it would have an actual difference, which is a completive difference. But the most general genus cannot have any such difference because every completive difference of one genus is a divisive difference of another genus. For example, being sensible is a divisive difference of something animate and being sensible is a completive difference of an animal. Therefore, if the most general genus had such a difference, then there would be a genus superior to it which would be divided through that difference, and this is absurd. Moreover, every genus is in potency to its differences. Therefore, matter and the <most general> genus are the same.

6. Furthermore, according to Aristotle in *Metaphysics* VII, 'substance' is divided into matter, form, and the composite;[381] and what is divided is predicated of the things dividing it; therefore, matter is a substance. And every predicate expresses a form in matter; therefore, matter is in-formed, and so it is in act.

7. To the contrary: in his commentary on the end of *Metaphysics* XI the Commentator says that all forms that exist in act in the first mover exist in potency in prime matter;[382] but all forms or causal principles exist in act in the first mover; therefore, all forms exist in potency in prime matter, and thus none of them is in act.

8. The solution of the problem or the truth of the matter is that prime matter in so far as it is matter has no substance in act; but matter in so far as it is natural is a substance in act, although incompletely, because, if natural matter exists, then a mobile body exists. And if a mobile body exists, then a motion exists. And if a motion exists, then an action exists. And every action is in act. Thus, natural matter is always in act, but matter taken absolutely in so far as it is matter is not in act.

[381] Arist. *Metaph.* VII.3, 1029a1–3.
[382] Aver. *In Metaph.* XII, t.c. 18, fol. 305va.

9. Ad primum in oppositum dicendum quod illa consequentia deficit, ‘hoc sequitur ad hoc et econverso; ergo sunt eadem’. Verbi gratia, instantia in terminis sic: quorumcumque homo est species, eorundem risibile est proprium et econverso; non tamen risibile et homo sunt eadem nec species et proprium.

10. Secundae rationi dicendum quod esse primum respectu subsequentium, hoc est dupliciter: aut primum ordinabile, et sic maior vera, aut primum, hoc est principium ordinabilis **[O 96vb]** ut generalissimi. Et hoc est dupliciter: aut primum nobilitate, et sic forma, aut primum natura, et sic materia. Hoc ultimo modo est minor vera. Unde aequivocatur ‘primum’ in argumento.

11. Tertiae rationi dicendum quod duplex est contrarietas: una in genere, alia extra genus. Secunda contrarietas sistit circa materiam, prima contrarietas circa generalissimum. Verbi gratia: sint tres substantiae et circumscribantur omnes aliae formae a substantia, scilicet humanitatem, animalitatem, carneitatem, corporalitatem, ita quod nulla forma remaneat nisi sola substantialitas; tunc illa forma quae sic praedicatur de istis tribus est genus generalissimum. Si velimus tunc procedere et facere plures formas, tunc dividendum est substantiam in duo contraria, ut in corporeitatem et in incorporeitatem; et haec sunt prima contraria circa generalissimum. Et postea, si velimus facere plures formas, procedendum est per divisionem illorum primorum contrariorum. Iam patet quae sunt contraria prima in genere et quae sunt contraria quibus subsistit genus generalissimum.

12. Quae autem sunt contraria extra genus sive quibus subsistit materia prima, sic apparet. Circumscribamus a digito meo omnes formas, tam substantiales quam accidentales, ita quod demamus ultimam formam in genere, quae est substantialitas; tunc remanet sola materia prima sive quiddam ens in potentia. Hic[a] est modus imaginandi materiam primam. Post vero circumscriptionem formarum accidentalium et etiam substantialium, ita quod demamus formam generis generalissimi, manet materia prima iam privata omni forma, et est omnino extra genus. Sua contraria tunc sunt ista, carentia formae generis generalissimi et forma generis generalissimi. Et ista sunt

[a] Hic *scripsimus*; his O

9. To the first argument to the contrary[383] it must be said that the consequence 'this follows from this and the other way around; therefore, they are the same' fails. Arguments about definitions provide the following counterexample: all the things which have man as a species, have being capable of laughing as a *proprium*, and the other way around; but being capable of laughing and man are not the same and neither are a species and a *proprium* the same.

10. To the second argument[384] it must be said that something can be first with respect to what comes after it in two ways: either (i) it is a first among what can be ordered, and in this way the major premise is true, or (ii) it is a first that is a principle of what can be ordered, for example, the most general genus. Case (ii) is twofold: either (iia) it is a first in nobility, and so it is a form, or (iib) it is a first by nature, and so it is matter. The minor premise is true in this last sense (iib). Thus the argument uses the term 'first' equivocally.

11. To the third argument[385] it must be said that there are two kinds of contrariety: (i) one is in a genus, and (ii) the other is outside a genus. The second kind of contrariety exists in matter, whereas the first kind of contrariety exists in the most general genus. For example, suppose that there are three substances and let all the other forms —that is, humanity, animality, fleshiness, corporeity— be separated from substance so that no form remains except substantiality alone; then the form that is predicated in this way of these three things is the most general genus. If we then want to continue and to produce many forms, we must divide substance into two contraries, such as corporeity and incorporeity; and these are the first contraries in the most general genus. And if we then want to produce many forms, it is necessary to proceed through the division of those first contraries. It is now clear what the first contraries in a genus are and to what contraries the most general genus is subject.

12. What the contraries outside a genus are, that is, those for which prime matter is a subject, becomes clear as follows. Let us remove from my finger all forms, both substantial and accidental, in such a way that we also remove the last form in the genus, which is substantiality; then only prime matter or a being in potency remains. This is a way of imagining prime matter. But after the removal of accidental and even of substantial forms, so that we also remove the form of the most general genus, prime matter already remains deprived of every form, and is absolutely outside a genus. Its contraries then are these: the lack of the form of the most general genus, and the form of the most general genus. And these contraries are outside a genus, and are the

[383] See above, n. 1.
[384] See above, n. 2.
[385] See above, n. 3.

contraria extra genus. Et ista contraria sunt principia naturae, quibus, ut[a] dicit Aristoteles, necesse est esse tertium principium, ut materiam, quod eis subsistat. Et ista est contrarietas prima in genere substantiae, ad quam omnes aliae reducuntur, ut dicit Aristoteles.

13. Iam patet quae contraria sunt principia et quae sunt contraria extra genus, quoniam contraria quae sunt circa materiam. Et etiam quae sunt contraria quae sunt in genere et quae nequaquam sunt principia, quoniam illa quae sunt circa genus generalissimum. Et ita patet quod est aequivocatio in argumento de eo quod est subsistere contrariis; diversimode enim[b] subsistit materia contrariis et genus generalissimum et etiam diversis contrariis subsistunt.

14. Quartae rationi dicendum quod[c] 'minime' dupliciter potest sumi: vel pro nihilo[d] vel pro aliquo quod tamen sit parvum valde [e]sive quod diminute est aliquid, secundum quod dicitur gallice: 'nent ou [f]trop peu'. Si primo modo sumatur, sic est haec vera, quod materia minime habet de forma, quia simpliciter nihil habet de forma. Si secundo modo, sic est haec vera, quod generalissimum minime habet de forma, quia generalissimum formam habet, licet incompletam. Et hoc modo sumitur 'minime' pro infimo in genere.

15. Quintae rationi dicendum quod, loquendo proprie de differentia, sicut loquitur Aristoteles in decimo *Metaphysicae* quando dat differentiam inter ista tria, non idem, diversum et differens, dicendo quod non idem est commune rebus in genere et extra genus, diversum vero est commune rebus diversorum generum, differens autem solum est commune rebus eiusdem generis —verbi gratia, non proprie dicitur quod Socrates differat ab albo, bene tamen dicitur quod Socrates est diversum ab albo et aliter quod non est idem—, loquendo ergo de differentia proprie, secundum quod est commune solum rebus eiusdem generis, est haec vera: 'omnis forma actualis est a differentia actuali', quia sic non est differentia nisi in illis quae sunt sub generalissimo, et sic in generalissimo non est proprie differentia actualis et tamen est ibi forma aliquo modo actualis. Si tamen loquamur communiter de differentia, maior est falsa, ut si sumamus differentiam communiter pro omni eo quod est causa differentiae et diversitatis. Et sic non est forma actualis a differentia actuali

[a] ut *lectio incerta* O
[b] enim *suppl. s.l.* O
[c] quod] omne (*lectio incerta*) *add. sed del.* O
[d] nihilo *scripsimus*; non O
[e] sive … gallice *corr. ex* secundum quod dicitur sive quod diminute est aliquid gallice O
[f] trop peu *scripsimus*; troppou O

principles of nature, about which Aristotle says that it is necessary for there to exist a third principle, namely matter, in which they subsist. This is the first contrariety in the genus of substance, to which all the other contrarieties are reduced, as Aristotle says.[386]

13. It is now clear which contraries are principles and which contraries are outside a genus, that is, the contraries that exist in matter. It is also clear which contraries are in a genus and are in no way principles, that is, those that exist in the most general genus. Thus, it is clear that to be a subject for contraries is taken equivocally in the argument; for matter is a subject for contraries in a different way from that in which the most general genus is, and moreover matter and the most general genus are subjects for different contraries.

14. To the fourth argument[387] it must be said that 'minimally' can be taken in two ways: either (i) to stand for nothing or (ii) to stand for something, which is however extremely small or which is something in a diminished way, in accordance with the French saying: 'néant ou trop peu' <'nothing or too little'>. In case (i) the proposition saying that matter has a minimum amount of form is true, because matter completely lacks form. In case (ii), the proposition saying that the most general genus has a minimum amount of form is true, because the most general genus has a form, although it is incomplete. And in this way 'minimum' is taken for the lowest in a genus.

15. To the fifth argument[388] it must be said that, if we speak of a difference in the proper sense, as Aristotle speaks in *Metaphysics* X when he distinguishes between these three items 'not the same', 'diverse', and 'different', saying that being not the same is common to the things in a genus and those outside a genus, being diverse is common to things of different genera, and being different is only common to things of the same genus[389] —for example, it is not appropriate to say that Socrates differs from a white person, while it is appropriate to say that Socrates is diverse from a white person, and also that he is not the same— it follows that, if we speak of a difference in the proper sense, in so far as it is common only to things of the same genus, the proposition 'every actual form comes from an actual difference' is true, because in this sense the difference only exists in the things that fall under the most general genus, and in this sense there is no actual difference in the most general genus <itself> in the proper sense, but there is a form there that is in some way actual. If we speak of difference in a broad sense, however, the major premise is false, as when we take difference in a broad sense for everything that is the cause of a difference and of a diversity. In this sense an actual form does not

[386] Arist. *Phys.* I.7, 190b17–191a14.
[387] See above, n. 4.
[388] See above, n. 5.
[389] Arist. *Metaph.* X.3, 1054b22–1055a2.

dividente genus, sed quia se ipsa faciat se differre ab alio. Ut si imaginaremur quod duo essent genera prima, unum differret ab alio et tamen non differentia actuali dividente genus, sed differentia communiter dicta, ut diversitate suae naturae, quia forma est quae se ipsa diversificatur ab alio. Unde dicendum quod maior intelligenda est in illis quae possunt esse et genus et species vel species tantum. Et ita non potest ratio procedere de materia nec de generalissimo.

16. Sextae rationi dicendum secundum quosdam quod, cum sic dicitur: 'materia est substantia', li 'substantia', secundum quod praedicatur de ea[a], secundum intellectum non est aliquid sumptum in rectitudine, sed in obliquitate, ut iste sit sensus: materia est deferens substantiam sive recipiens. Vel sic: materia est in potentia ad substantiam. Vel sic: materia est substantiae.

17. Sed contra: sicut est dicere scilicet 'pura potentia est potentia', similiter, ut videtur, erit dicere quod nuda substantia, cuiusmodi est materia prima, erit substantia, licet non habeat respectum ad formas, quod concedendum est.

18. Dicendum est ergo quod materia est substantia, sed non substantia quae est actu, sed potentia. Et dicendum quod divisio Aristotelis non est pure univoci nec pure aequivoci, sed analogici, et secundum diversam rationem dicitur de illis. Unde dicendum quod non dicitur de materia nisi secundum quod est in potentia. Et tunc, cum dicit quod omne praedicatum dicit formam, dicendum quod hoc intelligendum est si praedicatum, prout actu est, dicatur de subiecto et non solum secundum quod potentia est, sicut est in proposito.

19. Vel dicendum quod dupliciter possumus loqui de praedicatione: vel communiter vel proprie. Et est praedicatio communiter sumpta, et hoc affirmative, quando non plus requiritur nisi identitas praedicati ad subiectum in supposito solum, sicut est cum dicitur: 'haec materia est haec substantia'; ecce, hic non est convenientia nisi solum in supposito. Et in tali praedicatione non necesse est praedicatum dicere formam, sed solum suppositum. Et sic est cum dicitur: 'materia est substantia', quia significatur quod nuda substantia materiae[b] et sine omni forma est substantia. Et tunc ibi est solum convenientia in supposito, quia eidem supposito inest unum et reliquum, sicut homo et album conveniunt in supposito. <[c]Alia est praedicatio propria.> Et hoc in

[a] ea] materia *add. s.l.* O
[b] materiae *suppl. s.l.* O
[c] Alia … propria *supplevimus cum* TC; praedicatio autem propria *suppl. mg.* O

derive from an actual difference dividing a genus, but the actual form comes about because the form itself makes itself different from another. For example, if we were to imagine that there are two first genera, then one would differ from the other, but not in virtue of an actual difference dividing a genus, but in virtue of a difference taken in a broad sense, for example in virtue of a diversity of its nature, because a form is such that it is diversified from another by itself. Thus, it must be said that the major premise must be understood as applying to the case of things that can be both a genus and a species or only a species. And so the argument cannot succeed when applied to matter or the most general genus.

16. To the sixth argument[390] it must be said, according to some people, that in the claim 'matter is a substance', the term 'substance' in so far as it is predicated of matter is not to be understood as something taken in the nominative case but as something taken obliquely, so that the sense is either: matter is a bearer or receiver of substance; or this: matter is in potency to a substance; or this: matter belongs to a substance.

17. But against this: just as one can say 'a pure potency is a potency', it also seems that one can similarly say that a bare substance, such as prime matter, is a substance, even though it does not have a relation to forms. This must be conceded.

18. Therefore, it must be said that matter is a substance, although it is not a substance in act, but in potency. And it must be said that Aristotle's division is neither that of something purely univocal nor that of something purely equivocal, but of something analogical, and 'substance' is said of them in accordance with different *rationes*. Thus, it must be said that 'substance' is not said of matter except in so far as it is in potency. And then, when he says that every predicate expresses a form, it must be said that this is to be understood of the case in which a predicate is said of a subject in so far as it is in act and not only in so far as it is in potency, as happens in the present case.

19. Alternatively, it must be said that we can talk about a predication in two ways: either (i) in a broad sense or (ii) in a proper sense. (i) Predication is taken in a broad sense, and affirmatively, when nothing more is required than the identity of the predicate with the subject in a supposit, as happens in the predication 'this matter is this substance'. Here there is only an agreement in supposit. In this kind of predication it is not necessary for the predicate to express a form, but only a supposit. This is the case of the predication 'matter is a substance', because this signifies that the bare substance of matter without any form is a substance. In this case there is only an agreement in supposit, because the subject and the predicate inhere in the same supposit, as for example a man and a white thing agree in a supposit. (ii) <Alternatively,

[390] See above, n. 6.

affirmativis propositionibus est quando convenientia est in supposito et differentia in forma, sicut est quando praedicatur superius de inferiori; unde plus requiritur ibi quam sola convenientia in supposito, scilicet differentia in forma. Propter autem convenientiam in supposito fit veritas in propositione solum. Et ideo, licet homo et album sint res diversorum **[O 97ra]** generum et homo et asinus res eiusdem, quia tamen homo et album convenientiam habent in supposito et homo et asinus non, ideo haec praedicatio est vera: 'homo est album', haec autem falsa: 'homo est asinus'. Unde praedicatio propria est haec[a]: 'homo est animal', quia praedicatum et subiectum conveniunt in supposito, et ideo est ibi veritas; et quia animal dicit formam alterius animalis quam hominis, ideo est ibi differentia in forma. Et in tali praedicatione necesse est praedicatum dicere formam non gratia suppositi in quo conveniunt, sed gratia formae in qua differunt. Patet ergo iam in qua praedicatione necesse est praedicatum dicere formam et in qua non, et quod non necesse est [ut[b]] in ista: 'materia est substantia'.

20. Aliter adhuc dicunt quidam, scilicet quod nihil aliud est dictu, cum dicitur quod praedicatum dicit formam, quam si diceretur: 'praedicatum se habet ad modum formae', quia sicut forma est illud quod nobilius est et etiam illud quod ultimo advenit, similiter praedicatum est nobilius subiecto et ultimo loco advenit. Et illi dicunt quod non necesse est quod praedicatum dicat formam, nisi solum quod se habeat ad modum formae. Et similiter est cum dicitur: 'materia est substantia'. Et dicere quod praedicatum est forma et informans subiectum non est nisi modus loquendi quidam.

<Quaestio 49>
[c]Quaeritur octavo utrum materia prima possit intelligi ab intellectu possibili.

1. Et videtur quod non, quia dicit Aristoteles in tertio *De anima*: "Omnino sicut res sunt separabiles, sic et circa intellectum sunt", [d]hoc est intellectus noster [e]non intelligit nisi per abstractionem; quod ergo non potest abstrahi non potest intelligi. Sed materia prima non potest abstrahi; ergo etc. Probatio minoris: nihil abstrahitur a se ipso; sed non est aliud a quo posset materia abstrahi; ergo etc.

[a] haec *scripsimus*; hic O
[b] ut *delevimus*
[c] Quaeritur octavo] Consequenter quaeritur TC
[d] hoc est] tunc sic (arguo *add. mg.*) T; sed C
[e] non intelligit nisi] intelligit TC

predication can be taken in a proper sense>, and in affirmative propositions a proper predication is when there is an agreement in supposit and a difference in form, for example when a superior is predicated of its inferior; thus, not only an agreement in supposit but also a difference in form is required for proper predication. But the truth of a proposition only depends on an agreement in supposit. Therefore, while it is true that man and white thing belong to different genera, and that man and ass belong to the same genus, nonetheless, since man and white thing agree in supposit, whereas man and ass do not, it follows that the predication 'a man is a white thing' is true, but the predication 'a man is an ass' is false. Thus, 'man is an animal' is a proper predication because the predicate and the subject agree in supposit, and hence the proposition is true, and because 'animal' also expresses the form of an animal other than man, so that there is a difference in form. In this kind of predication it is necessary for the predicate to express a form not in virtue of the supposit in which the subject and the predicate agree, but in virtue of the form in which they differ. Therefore, it is clear in which kind of predication it is necessary for the predicate to express a form and in which kind it is not necessary, and it is clear that it is not necessary in the predication 'matter is a substance'.

20. Some people have still another view, namely that saying that a predicate expresses a form is to say nothing other than that a predicate has the mode of a form, because just as a form is what is nobler and is what supervenes last, a predicate is nobler than the subject and supervenes last. These people say that it is not necessary for the predicate to express a form except in the sense that it has the mode of a form. The same applies to the proposition 'matter is a substance'. Moreover, they maintain that to say that the predicate is a form and that it in-forms the subject is nothing other than a mode of speaking.

<Question 49>
The eighth question is whether prime matter can be understood by the possible intellect.

1. Arguments for the negative: in *On the Soul* III Aristotle says that just as things are separable, so they are in the intellect,[391] that is, our intellect does not understand except by abstraction; therefore, what cannot be abstracted cannot be understood. But prime matter cannot be abstracted; therefore, <it cannot be understood>. Proof of the minor premise: nothing is abstracted from itself; but there is nothing different from matter from which matter can be abstracted; therefore, <matter cannot be abstracted>.

[391] Arist. *De anima* III.8, 431b24–25.

2. Item, quanto aliquid minus habet de materia [a]et plus de forma, magis est intelligibile, et quanto magis de materia et minus de forma, tanto minus intelligibile; quod ergo est totaliter materia nullo modo est intelligibile[b].

3. Item, singulare est[c] dum sentitur, universale dum intelligitur, ut dicit Aristoteles; sed materia nec est universale nec singulare. Et hoc quia omne quod est universale vel singulare est forma vel habens formam; sed[d] materia prima nec est forma nec formam habens; ergo etc.

4. Item, Aristoteles, capitulo de phantasia: "Nos intelligimus cum volumus[e], sed non sentimus cum volumus[f]". Et causa huius est, [g]secundum ipsum, quia intelligibile semper est praesens intellectui sicut apprehensibile apprehendenti, [h]sensibile vero non est semper praesens sensui; sed materia prima non est[i] praesens intellectui sicut apprehensibile, quia, si esset, aut esset ibi ut res[j] aut ut species aut per effectum aliquem. Non ut res. Nullus enim dicit quod materia alicuius faciat se extra[k] ab eo cuius est materia ad intellectum ad hoc quod intelligatur. Nec per speciem, quia nihil facit speciem nisi forma vel formam habens. Nec per effectum, quia tunc efficeret materia aliquid, et sic coincideret cum forma, fine et efficiente, quod [l]negat Aristoteles.

5. Item, singulare non est intelligibile nisi in suo universali, et hoc quia forma ligatur materiae; si ergo ligatio cum materia impedit singulare ita quod non intelligitur, tunc quod totaliter est materia non intelligetur.

6. Item, Aristoteles dicit [m]in capitulo de phantasia: "Nihil intelligitur[n] nisi quod prius facit phantasma[o]"; sed materia prima phantasma non facit; ergo etc. Probatio minoris: solum illud facit phantasma[p] quod facit speciem in sensum: [q]sed [r]materia prima non facit speciem in sensum propter hoc quod[s] nec est [t]forma nec formam habens; ergo etc.

[a] et ... tanto] magis est intelligibile et quanto magis de materia TC
[b] intelligibile] quod totaliter est *add. sed del.* T
[c] est *om.* TC
[d] sed *om.* TC
[e] volumus] volimus OT *sed corr.* T
[f] volumus] volimus OT *sed corr.* T
[g] secundum ipsum *om.* TC
[h] sensibile ... est] et sic non est sensibile TC
[i] est] semper *add.* TC
[j] res] et natura *add. sed del.* T
[k] extra *om.* OTC *sed suppl. s.l.* O
[l] negat Aristoteles] est contra Aristotelem C
[m] in capitulo *om. sed suppl. s.l.* capitulo T
[n] intelligitur] intelligimus T; cadit in intellectu C
[o] phantasma *corr. ex* phantasiam OC
[p] phantasma] phantasiam OC *sed corr.* O
[q] sed ... sensum *om. sed suppl. mg.* talis non est materia T
[r] materia prima *suppl. s.l.* O
[s] quod] forma *add. sed del.* O
[t] forma nec formam] materia nec materiam C

2. Furthermore, the less matter and the more form something has, the more it is intelligible, and the more matter and the less form something has, the less it is intelligible; therefore, what is totally matter is in no way intelligible.

3. Furthermore, something is singular as long as it is sensed, and universal as long as it is understood, as Aristotle says;[392] but matter is neither universal nor singular. The reason for this is that everything that is either universal or singular is a form or has a form; but prime matter neither is a form nor has a form; therefore, <prime matter is neither universal nor singular>.

4. Furthermore, in the chapter on imagination Aristotle <says>: "We understand when we will",[393] but "we do not sense when we will".[394] According to Aristotle, the reason for this is that an intelligible thing is always present to the intellect in the way that something apprehensible <is present> to the thing that apprehends, whereas a sensible thing is not always present to the senses. But prime matter is not present to the intellect as something apprehensible, because, if this were the case, prime matter would be in the intellect either as a thing or as a species or through some effect. But prime matter is not in the intellect as a thing; for nobody says that the matter of a thing extends itself outside the thing whose matter it is up to the intellect in order to be understood. Nor is prime matter present through its species, because nothing produces a species except a form or something having a form. Nor is it present through an effect, because then matter would produce something, and so it would coincide with the form, the end and the efficient cause, which Aristotle denies.[395]

5. Furthermore, a singular is only intelligible in its universal, and this because it is joined to matter; therefore, if being joined to matter prevents a singular from being understood, then what is totally matter will not be understood.

6. Furthermore, in the chapter on imagination Aristotle says: "Nothing is understood other than what first produces a phantasm";[396] but prime matter does not produce a phantasm; therefore, <prime matter is not understood>. Proof of the minor premise: only the thing which produces a species in the senses produces a phantasm; but prime matter does not produce a species in the senses because it neither is a form nor has a form; therefore, <prime matter does not produce a phantasm>.

[392] Arist. *Phys.* I.5, 189a5–7; but see also above, q. 11, n. 4.
[393] This is probably a reference to Arist. *De anima* III.2, 427b17–20.
[394] This is probably a reference to Arist. *De anima* III.2, 428a8–9.
[395] Arist. *Phys.* II.7, 198a24–27.
[396] Arist. *De anima* III.7, 431a16–17.

7. Ad oppositum: intellecto posteriore intelligitur prius; ergo, intellecto composito, quod est posterius quam materia vel forma, quae ipsum componunt, intelligetur materia.

8. Item, dicit Aristoteles, [a]primo *Physicorum*, quod materia est scibilis per analogiam, et ita[b] est scibilis.

9. Item, ens in quantum ens cognoscitur, quia de ente in quantum ens est scientia, ut metaphysica. Sed intellecto aliquo, intelliguntur eius differentiae; et ens dividitur per actum et potentiam [c]prima divisione; ergo illud quod est in potentia intelligi potest. Sed actus reducitur ad formam et potentia ad materiam primam; ergo, cum potentia potest intelligi, et materia intelligetur.

10. Item, materia prima assimilatur maxime intellectui, quia intellectus est receptivus universalium, et dicit Aristoteles in decimo *Metaphysicae* quod materia prima recipit formas universales; ergo materia prima maxime proportionatur apprehensioni intellectus.

11. Item, ad hoc idem Aristoteles in tertio *De anima*: "Intellectus possibilis non est hoc antequam actu intelligat", hoc est intellectus non est aliquid in actu antequam perficiatur per intelligibile. Et ita materia non est aliquid in actu antequam perficiatur per formam; et ita materia maxime assimilatur intellectui; et ita maxime proportionatur apprehensioni intellectus.

12. [d]Solutio sive veritas problematis talis est, quod nos intelligimus materiam primam, quia aliter non esset scientia de ipsa, [e]et Aristoteles in primo *Metaphysicae* determinat de materia.

13. [f]Primae rationi dicendum quod est separatio duplex: una[g] secundum existentiam, ita quod intellectus separet aliquid ab alio ita quod faciat ipsum existere sine illo ad hoc quod intelligatur, et talis separatio est impossibilis; alia est separatio secundum modum considerandi, ita quod consideret illud quod est sub certis appendiciis[h] ad materiam et sub certis figurationibus non prout est[i] sub huiusmodi appendiciis[j]. Et istam abstractionem ultimam[k]

[a] primo Physicorum *om.* TC
[b] ita] sequitur quod *add.* TC
[c] prima … intelligetur] ergo potentia intelligitur; propter hoc ergo quod materia est in potentia tantum (tantum in potentia C) non impeditur quin materia (materia *om.* T) posset (possit C) intelligi TC
[d] Solutio … est] dicendum TC
[e] et … materia *om.* TC *sed suppl. mg.* (secundo *pro* primo Metaphysicae) C
[f] Primae rationi] ad primum argumentum TC
[g] una] aut TC
[h] appendiciis *lectio incerta* O; appraehendentiis *scrips. sed. corr. in* appendentiis T
[i] est] sunt OTC *sed corr. s.l.* T
[j] appendiciis] dependentis (*lectio incerta*) O; appraehendentiis *scrips. sed corr.* in appendentiis T
[k] ultimam *om.* TC

7. To the contrary: if what is posterior is understood, what is prior is understood; therefore, if a composite, which is posterior to the matter and form that are its components, is understood, matter will be understood.

8. Furthermore, in *Physics* I Aristotle says that matter is knowable by analogy,[397] and so it is knowable.

9. Furthermore, being in so far as it is being is known, since there is a science of being in so far as it is being, that is, metaphysics. But once something is understood, its differences are understood; and the first division of being is into act and potency; therefore, what is in potency can be understood. But act is reduced to form and potency to prime matter; therefore, since potency can be understood, matter will also be understood.

10. Furthermore, prime matter is maximally similar to the intellect, because the intellect is receptive of universals, and in *Metaphysics* X Aristotle says that prime matter receives universal forms;[398] therefore, prime matter is maximally proportional to the apprehension of the intellect.

11. Furthermore, in support of the same point there is Aristotle's claim in *On the Soul* III: "The possible intellect is not this something before it understands in act",[399] that is, the intellect is not something in act before it is perfected by an intelligible thing. And in the same way matter is not something in act before it is perfected by a form; thus matter is maximally similar to the intellect, and so it is maximally proportional to the apprehension of the intellect.

12. The solution of the problem or the truth of the matter is that we understand prime matter, because otherwise there would be no knowledge of it, whereas in *Metaphysics* I Aristotle draws definite conclusions about matter.[400]

13. To the first argument[401] it must be said that there are two kinds of separation: (i) one kind involves existence, so that the intellect would separate one thing from another in such a way that it would make that thing exist without the other in order to understand it; and this kind of separation is impossible. (ii) Another kind of separation is separation in accordance with the mode of considering, so that the intellect considers a thing that exists under certain appendages to matter and certain configurations but not in so far as it exists under such appendages. The intellect performs this second kind of abstraction in order to actually understand something, but it does not perform the first kind of abstraction. In this way it is possible to understand

[397] Arist. *Phys*. I.7, 191a7–12.

[398] The correct reference is Aver. *In Metaph.* I, t.c. 17, fol. 14vb.

[399] Arist. *De anima* III.4, 429a22–24, b30–31.

[400] This is probably a reference to Arist. *Metaph.* I.8, 989b6–14.

[401] See above, n. 1.

facit intellectus ad hoc quod actu intelligat [a]aliquid, primam vero non facit. Et sic contingit [b]intelligere materiam et abstrahere eam, quia materia est pars plurium individuorum et potest considerari non prout est pars [c]huius individui vel illius [d]determinate, sed secundum quod est natura in se considerata. Et [e]per talem abstractionem materia prima est intelligibilis.

14. Sed contra[f]: quod materia non possit sic abstrahi videtur, quia aliqua virtus animae componit hanc, scilicet quod materia prima est pars huius. Et non alia virtus quam intellectiva, tum quia nulla alia virtus componit, tum quia materia non cognoscitur nisi per decursum rationis a causato ad causam, quod non facit aliqua[g] alia virtus a virtute intellectiva.

15. [h]Quaeratur ratio huius.

16. [i]Secundae **[O 97rb]** rationi dicendum sicut prius, quod [j]non sequitur, si quanto aliquid plus habet de materia, tanto minus intelligibile, quod illud quod totum est materia nullo modo sit intelligibile, sed sequitur quod diminute est intelligibile, [k]quod verum est. Si autem concludatur sic: 'ergo minime est intelligibile', tunc conclusio distinguenda est [l]de li 'minime', sicut prius patuit.

17. [m]Tertiae rationi dicendum quod haec propositio: 'solum universale est [n]illud quod intelligitur', est per se falsa, quia summe singulare est summe intelligibile, sicut causa prima, et tamen non est[o] universale. Et ratio huius est propter hoc quod [p]tunc essent multi Dei, et hoc est inconveniens. Vel quia universale est principium ordinabilis in genere; cum ergo causa prima non sit in genere, non [q]est universale nec universale habens quod habet esse in rebus extra. Nec habet universale quod sit in anima, [r]quia universale quod est in

[a] aliquid … facit *om.* TC
[b] intelligere … eam] separare materiam TC
[c] huius individui vel] illius nec O
[d] determinate, sed *om.* TC *sed suppl. mg.* sed T
[e] per … est] sic est materia TC
[f] contra *om.* TC *sed. suppl. s.l.* C
[g] aliqua *om.* TC
[h] Quaeratur ratio huius *om.* TC *sed suppl. mg.* C; *suppl. mg.* quare intellectus poterit ipsam materiam abstrahere T
[i] Secundae rationi] ad secundum TC
[j] non ... sit] (licet *praem. s.l.* T) quanto aliquid magis habet de materia tanto minus (est *add. s.l.* T) intelligibile, non sequitur quod illud (quod *add.* T) totum est materia quod nullo modo est TC
[k] quod … patuit] quia minime est aequivocum ad non et ad diminute TC
[l] de li *lectio incerta* O
[m] Tertiae rationi] Ad tertium TC
[n] illud quod intelligitur] intelligibile TC *sed* illud quod *praem. mg.* C
[o] est] habet TC *sed corr. mg.* C
[p] tunc ... quia *om.* TC
[q] est ... habens] habet universale TC
[r] quia … anima est *om. sed corr. mg. et s.l.* T (est² *post* simplicius)

and to abstract matter, because matter is a part of many individuals and can be considered not in so far as it is determinately a part of this or that individual, but in so far as it is a nature considered in itself. Prime matter is intelligible in virtue of this kind of abstraction.

14. But against this: it seems that matter cannot be abstracted in this way because a power of the soul composes the proposition 'prime matter is a part of this thing'; and no power other than the intellectual power does this, both because no other power composes, and because matter is only known through a rational progression from effect to cause, which no power other than the intellectual power produces.

15. The reason for this should be investigated.

16. To the second argument[402] it must be said as before that from the premise 'the more matter something has, the less intelligible it is' it does not follow that what is totally matter is in no way intelligible, but it does follow that it is intelligible in a diminished way, and this is true. But if one draws the conclusion 'therefore, it is minimally intelligible', in this conclusion the senses of 'minimally' must be distinguished, as was made clear before.[403]

17. To the third argument[404] it must be said that the proposition 'only a universal is what is understood' is per se false, because what is maximally singular is maximally intelligible, namely the first cause, and yet it is not a universal. The reason for this is that, if the first cause were a universal, there would be many Gods, and this is absurd. Another reason is that a universal is the principle of something that can be ordered in a genus; therefore, since the first cause is not in a genus, it is not a universal and it does not have a universal

[402] See above, n. 2.
[403] See above, q. 48, n. 14.
[404] See above, n. 3.

anima est simplicius eo cuius est universale, et nihil est simplicius primo. Est etiam [a]illa propositio falsa de intellectu animae rationalis et intelligentiarum, quae intelligunt res sicut sunt et singularia secundum quod singularia. Unde haec propositio: 'solum universale est illud quod intelligitur', ad hoc quod sit vera debet condicionari[b] a parte intellectus et a parte intelligibilis. A parte intellectus sic, quod intelligatur de intellectu qui prius est potentia[c] postea in actu, ut de intellectu possibili. A parte intelligibilis sic debet condicionari, quod solum singularia[d] extra intelliguntur per suum universale.

18. Iterum, non omne singulare est sensibile, sed singulare in materia sensibili extra, quia intelligentiae et animae rationales[e] et huiusmodi non sentiuntur. [f]Et nisi praedicta propositio sumatur cum hac condicione solum, non sequitur conclusio, sed erit fallacia consequentis, sicut patet intuenti.

19. Aliter dicitur sic, quod materia prima aliquo modo est universale sive commune. Sed aliquid potest[g] [h]esse commune dupliciter: vel ita quod habeat realem communitatem alicuius formae per quam est multiplicativa sui in plura[i] supposita, sive actu multiplicatur [j]sive non, et hanc communitatem non habet materia prima. Alia est communitas secundum rationem, cui non respondet aliqua[k] realis natura multiplicabilis per plura; et haec est communitas solius intentionis. Sicut si dicam: 'infinitum in actu', est nomen commune, quia huic infinito in actu et illi, et tamen non est realis infinitas quae respondeat intentioni nominis nec propter defectum rei est minus nomen appellativum. Et isto modo est materia commune aliquod. Et hoc solum secundum intentionem nominis, quia[l] nomen appellativum. Et haec communitas sufficit ad hoc quod intelligatur.

20. Aliter dicitur sic, quod materiam habere communitatem, hoc[m] potest esse dupliciter: vel de se et essentialiter, et sic non habet communitatem, vel accidentaliter, ut sub esse formae [n]sive totius compositi, et sic habet[o] communitatem.

[a] illa ... rationalis] falsa (*illegibile et* vera *scrips. sed corr. in* falsa T) de intellectu (possibili *add. sed del.* T) primi TC
[b] condicionari] et *add.* TC
[c] potentia] in potentia et TC
[d] singularia] sensibilia *add.* TC
[e] rationales *ex corr. (?)* O; singulares TC
[f] Et ... intuenti *om.* TC
[g] potest *om.* TC
[h] esse commune] esse commune est T; est commune C
[i] plura] in potentia *add. sed del.* T
[j] sive non *om. sed suppl. mg.* sive non actu multiplicetur T; actu multiplicatur *add.* C
[k] aliqua *om.* TC
[l] quia] est *add. s. l.* O
[m] hoc *om.* TC
[n] sive totius compositi *om.* TC
[o] habet] aliquam *add.* TC

that has being in extra-mental things. Nor does it have a universal that exists in the soul, because a universal that exists in the soul is simpler than the thing of which it is a universal, and nothing is simpler than the first being. That proposition is also false about the intellect of the rational soul and of the intelligences, which understand things as they exist and singulars in so far as they are singulars. Accordingly, the proposition 'only a universal is what is understood' in order to be true must be qualified on the side of the intellect and on the side of the intelligible thing. On the side of the intellect, it must be qualified so that it is understood of the intellect that is first in potency and then in act, that is, of the possible intellect. On the side of the intelligible thing it must be qualified so that only extra-mental singulars are understood in virtue of their universal.

18. Furthermore, not every singular is sensible, but only a singular existing in external sensible matter, because intelligences and rational souls and the like are not sensed. And unless the proposition is taken with this qualification, the conclusion does not follow, but there will be a fallacy of the consequent, as is clear to anyone who examines the argument.

19. Another reply is that prime matter is in some way universal or common. But something can be common in two ways: (i) in such a way that it has the real community of a form in virtue of which a form can multiply itself in many supposits, regardless of whether it is multiplied in act or not; prime matter does not have this kind of community. (ii) Another kind of community is in accordance with reason, and there is no real nature which can be multiplied in many things that corresponds to it; this kind of community is only one of an intention. This is the situation when I say that 'infinite in act' is a common name because it would apply to this infinite in act and to that, even though there is no real infinity that corresponds to the intention of the name; and even though there is no such thing, the name is no less an appellative noun. And it is in this way that matter is something common, and this only happens in accordance with the intention of the name, because it is an appellative noun. This kind of community is sufficient for the intelligibility of matter.

20. Another reply is that the claim that matter has community can be understood in two ways: either (i) by itself and essentially, and in this way matter has no community, or (ii) accidentally, that is, under the being of a form or of a whole composite, and in this way matter has community.

21. [a]Quartae rationi dicendum quod materia prima est praesens[b] in nobis. Et hoc dupliciter, [c]quia et sicut pars compositionis [d]et sicut apprehensibile, [e]sed nec est apprehensibile per speciem nec ut res nec per effectum, sed solum per signa[f], [g]exempli gratia quia videmus quod in omni transmutatione est aliquid fixum permanens; unde in transmutatione quae est generatio[h], cum ibi transmutentur omnes formae, tam substantiales quam accidentales, necesse est aliquid manere fixum et deferens [i]omnes transmutationes, quod non est nisi ipsa materia. Aliud signum est quod bene scimus quod est status in causa materiali, resolvendo formam post formam, ut docet Aristoteles in secundo *Metaphysicae*. Aliud signum est quod non est forma extensa nisi in materia nec est forma delata nisi per materiam.

22. Aliter dicitur sic, quod materia non facit speciem de se, sed per accidens, ut sub esse formae—et non pono quod forma sola faciat speciem, sed totum compositum facit speciem—; et ita materia per accidens facit speciem.

23. Ad quintum dicendum quod alligatio formae cum materia non est tota causa quare singulare [j]non intelligitur, sed quia singulare est sub certis appendiciis[k] et figurationibus certis et dimensionibus certis; et huiusmodi solum respondent virtuti sensitivae, ut imaginativae et aliis virtutibus, et nullo modo intellectui, quia non intelligitur aliquid[l] sub his condicionibus, 'hic' et 'nunc', et appendiciis certis ad materiam.

24. Ad ultimum dicendum quod 'nihil intelligimus sine phantasmate', hoc est verum de rebus sensibilibus extra et de intellectu possibili, sicut prius patuit[m].

25. Vel sic: materia[n] aliquo modo facit phantasma, quia sub esse formae, ut prius patuit.

26. Quia ultimae rationes falso concludunt, licet non falsum, [o]ideo ad haec responde ut habetur postea.

[a] Quartae rationi] Ad quartum TC
[b] praesens *om.* TC
[c] quia et] vel C
[d] et sicut] et est (*sed corr. s.l. in* etiam) sicut T; vel sic C
[e] sed … apprehensibile *om. hom. sed mg. suppl.* T
[f] signa] signum TC
[g] exempli gratia *scripsimus*; efficientia O; *om.* TC
[h] generatio] in genere C
[i] omnes transmutationes, quod] et hoc C
[j] non intelligitur] non sentitur *scrips.* OC *sed corr. s.l.* O, *corr. mg.* C; sentitur *scrips. sed corr. mg.* T
[k] appendiciis] appendentiis O; *corr. mg. ex* apprehendentiis T; ad materiam *add.* C
[l] aliquid *om.* OTC *sed suppl. s.l.* O
[m] patuit] patet TC
[n] materia *om.* OC *sed suppl. s.l.* O
[o] ideo ... habetur] idem (idem *corr. s.l. in* ideo T; ad hoc idem *pro* idem ad hoc C) ad hoc respondebitur TC

21. To the fourth argument[405] it must be said that prime matter is present in us in two ways because it is present both as part of a composition and as apprehensible. Prime matter, however, is apprehensible neither through a species nor as a thing nor through its effect, but only through its signs, for example because we see that in every change there is something fixed and permanent. Hence, since in the change that is generation all forms, both substantial and accidental, change, it is necessary for something that is fixed and that bears all these changes to remain, and this is nothing other than matter itself. Another sign is that we certainly know that the resolution of one form after another stops at the material cause, as Aristotle teaches in *Metaphysics* II.[406] Another sign is that a form is not extended in anything other than matter and is not borne by anything other than matter.

22. Another reply is that matter does not produce a species by itself, but only accidentally, namely under the being of a form —and I do not assume that a form alone produces a species, but the whole composite produces a species—, and thus matter produces a species accidentally.[407]

23. To the fifth argument[408] it must be said that the joining of a form with matter is not the whole cause of a singular's not being understood, but the cause is that a singular exists under certain appendages, configurations, and dimensions; and things of this kind only relate to the sensitive powers, such as the imaginative power and the other powers, but in no way <do they relate to> the intellect, because something is not understood under the conditions 'here' and 'now', and under certain appendages of matter.

24. To the last argument[409] it must be said that the proposition 'we understand nothing without a phantasm' is true of extra-mental sensible things and of the possible intellect, as was made clear before.[410]

25. Another reply is that matter in some way produces a phantasm because <it does so> under the being of a form, as was made clear before.[411]

26. Since the last arguments[412] use a false deduction, although their conclusion is not false, you should reply to these as will be established later.[413]

[405] See above, n. 4.
[406] This is probably a reference to Arist. *Metaph.* I.8, 989b6–14; on the order of the first two books of the *Metaphysics* in the Arabic-Latin translation, see above, note on q. 9, n. 6.
[407] Bacon gives this reply. See Roger Bacon, *Supra libros octo Physicorum*, 56.11–22.
[408] See above, n. 5.
[409] See above, n. 6.
[410] See above, nn. 17–18.
[411] See above, n. 22.
[412] See above, n. 5–6.
[413] Not found.

<Quaestio 50>
[a]Quaeritur nono an materia prima sit una numero.

1. Et videtur quod non, quia receptum est in recipiente per modum recipientis et non per modum recepti; si ergo materia esset una numero, cum sit receptiva formarum, omnis forma in materia recepta esset una numero. Sed hoc est falsum, quia dicit Commentator super secundum[b] *Metaphysicae*, super illud [c]capitulum: "Quoniam autem in fundamento nihil est distinctum", quod materia prima primo recipit formas magis universales, sicut primo formam generalissimi, [d]secundo vero formas specialiores.

2. Item, unum et multa sunt [e]differentiae entis; et differentia posterior est eo cuius est differentia; ergo unum et multa sunt[f] posteriora ente. Sed materia est principium entis; et quod est posterius posteriore est posterius priore; materia ergo prima praecedit unum et multa. Et ita non est unum nec multa; et ita non est una numero.

3. Item, quod est unum numero est unum quia sub aliqua forma una numeratur; sed prima materia de se sub nulla una forma numeratur, quia non habet formam; materia ergo prima non est una numero.

4. Item, 'materia' est nomen commune [g]et 'prima' est nomen commune; et commune communi additum non facit aliquod signatum singulare; cum ergo sic dicitur: 'materia prima', non est singulare signatum, et ita non una numero.

5. Ad oppositum: [h]illud quod de se non est multiplicabile nec multiplicatum per plura supposita est unum **[O 97va]** numero. Sed materia prima de se nec est multiplicabilis nec actu[i] multiplicata per plura numero, quia quod multiplicatur multiplicatur ratione formae[j]; materia non habet formam; ergo etc.

6. [k]Solutio sive veritas problematis talis est, quod materiam esse unam numero, hoc[l] contingit dupliciter: vel essentialiter vel accidentaliter. Si accidentaliter, sic non est materia una numero, quia materia accidentaliter

[a] Quaeritur nono] consequenter quaeritur T; Quaeritur C
[b] secundum] primum C
[c] capitulum: "Quoniam autem *om.* T; quoniam C
[d] secundo vero] secundario TC
[e] differentiae entis *corr. s.l. ex* differentes T
[f] sunt] differentiae entis *add. sed del.* O
[g] et … commune *om. hom. sed suppl mg.* (*om.* nomen) C
[h] illud ... multiplicatum] sic quod est de se (non est *add. s.l.* C) multiplicabile non multiplicatum de se C
[i] actu *om.* TC
[j] formae] sed *add.* T; et *add.* C
[k] Solutio … est] Dicendum T; dicitur C
[l] hoc *om.* TC

<Question 50>

The ninth question is whether prime matter is one in number.

1. Arguments for the negative: what is received is in the recipient in the mode of the recipient and not in the mode of what is received; therefore, if matter were one in number, since it is receptive of forms, all the forms received in matter would be one in number. But this is false because in the chapter "Since in the foundation nothing is distinct" of his commentary on *Metaphysics* II the Commentator says that prime matter first receives more universal forms, so that it first receives the form of the most general genus, and secondly more special forms.[414]

2. Furthermore, being one and being many are differences of being; and a difference is posterior to the thing of which it is a difference; therefore, being one and being many are posterior to being. But matter is a principle of being; and what is posterior to what is posterior is posterior to what is prior; therefore, prime matter precedes being one and being many; therefore, it is neither one nor many, and thus it is not one in number.

3. Furthermore, what is one in number is one because it is numbered under some form; but prime matter is by itself numbered under no form, since it does not have a form; therefore, prime matter is not one in number.

4. Furthermore, 'matter' is a common name, and 'first' is a common name; but something common does not produce some determinate singular when added to something common; therefore, the expression 'prime matter' does not stand for a determinate singular, and so prime matter is not one in number.

5. To the contrary: what is not by itself multipliable or multiplied in many supposits is one in number; but prime matter is not by itself multipliable or multiplied in act in many things in number, because what is multiplied is multiplied because of a form, and matter does not have a form; therefore, <prime matter is one in number>.

6. The solution of the problem or the truth of the matter is that the claim that matter is one in number can be understood in two ways: either (i) essentially or (ii) accidentally. Matter is not one in number in way (ii) because matter is accidentally numbered under different forms; and thus, just as forms

[414] Aver. *In Metaph.* I, t.c. 17, fol. 14vb.

numeratur sub diversis formis; [a]et ita, sicut formae sunt[b] plures essentialiter, sic materia est [c]plura numero accidentaliter. Si essentialiter, hoc contingit dupliciter: vel quia[d] in sui essentia caret omni forma numerante et distinguente, et sic est materia prima una numero, secundum quod dicit Commentator quod materia prima est una numero per privationem omnium formarum. Vel una numero essentialiter [e]quia habet aliquam formam signatam essentialiter perficientem et distinguentem quae est quidditas individui; et ita non est materia una numero. Et ita[f] haec forma[g] facit unitatem essentialem, ad quam consequitur quod[h] sit in se indivisum et ab aliis divisum. Et ita habet comparationem ad alia ita quod potest ponere [i]in numerum cum aliis, et ita consequitur unitas accidentalis. Unde prius per naturam est unitas essentialis quam accidentalis, simul tamen tempore.

7. [j]Sed contra: per hoc quod dicis[k], quod materia est multa accidentaliter quia sub forma[l] multiplicata, videtur quod omnis forma sit accidentalis.

8. Dicendum quod dicere aliquid esse accidentale contingit dupliciter: vel quia accidentaliter inhaeret, et omnis forma accidentalis [m]sic est simpliciter accidens; vel quia illud non est de essentia eius cui inhaeret, et ita omnis forma est accidentalis materiae, essentialiter tamen inhaeret sicut perficiens suo perfectibili. Et ita est forma substantialis accidens materiae.

9. Item, materia prima est ita unum numero numerositate essentiae et essentialiter quod ad ipsam non consequitur unitas accidentalis, [n]et causa prima [o]est ita una numero essentialiter quod[p] ad ipsam non consequitur unitas accidentalis, quia cum aliqua creatura non numeratur; et ita materia prima est illud quod maxime assimilatur causae primae.

10. Dicendum quod materia prima est illud quod minime [q]causae primae assimilatur. Et hoc concludam ex argumento ad oppositum, quia prima causa est una numero essentialiter, et hoc in sua essentia formali, ita quod causa

[a] et ita, sicut] ita quod *scrips. sed corr. s.l. in* ita quod sicut T
[b] sunt *om.* OTC *sed suppl. s.l.* OT
[c] plura numero] plures C
[d] quia] ita quod TC
[e] quia habet] ita quod habeat T; quia habet quia *scrips. sed* quia[2] *del.* C
[f] ita] illa O; *om.* C
[g] forma] quae *add. s.l.* O
[h] quod] ipsum unum *add. s.l.* O
[i] in numerum *suppl. mg.* T
[j] Sed contra: per] super TC
[k] dicis] dicit TC
[l] forma] comparata *add. sed del.* O
[m] sic … accidentalis *om. hom. sed suppl. mg.* (est *pro* sic est) C
[n] et … accidentalis *scrips. sed del.* O
[o] est ita una *suppl. mg.* T
[p] quod] quia O
[q] causae primae *om. sed suppl. s.l.* ei O

are essentially many, so matter too is many in number accidentally. Way (i) can be understood in two ways: either (ia) in the sense that matter in its essence lacks every form that produces a number and a distinction, and in this way prime matter is one in number, in accordance with the Commentator's claim that prime matter is one in number because of the privation of all forms. [415] Or (ib) in the sense that matter has a determinate form, which gives it an essential perfection and distinction, and which is the quiddity of an individual; in this way matter is not one in number. This kind of <determinate> form is what produces an essential unity, and from that <essential unity> it follows that <a thing> is undivided in itself and divided from other things so that it can be taken in relation to other things in such a way that it can make a numerical multitude with them, and this is how accidental unity is produced. Thus, essential unity is prior by nature to accidental unity, but temporally simultaneous.

7. But against this: your claim that matter is accidentally many because it is multiplied under a form seems to imply that every form is accidental.

8. It must be said that the claim that something is accidental can be understood in two ways: either (i) because it inheres accidentally, and in this way every accidental form is absolutely an accident; or (ii) because it does not belong to the essence of the thing in which it inheres, and in this way every form is accidental to matter. However, it essentially inheres in matter as what perfects something perfectible, and in this way a substantial form is an accident of matter.

9. Furthermore, prime matter is one in number by the numerability of its essence and essentially in such a way that an accidental unity does not follow from it, and the first cause is essentially one in number in such a way that an accidental unity does not follow from it either, because it is not numbered together with any creature; therefore, prime matter is that which is maximally similar to the first cause.

10. It must be said that prime matter is that which is minimally similar to the first cause. I draw this conclusion from the argument to the contrary[416] because the first cause is one in number essentially, and this in its formal

[415] Aver. *In Metaph.* XII, t.c. 14, fols. 300vb–301ra.
[416] See above, n. 9.

prima est totaliter [a]forma et totaliter in actu cui non admiscetur aliquid de potentia, et propter sui nobilitatem non ponit in numerum cum aliqua creatura. [b]Materia prima opposito modo se habet, quia in sua essentia privatur omni forma et omni actualitate et propter sui[c] vilitatem non numeratur cum aliqua creatura alia. Et omnis creatura alia habet aliquid actualitatis, et ita materia prima est illud quod minime [d]convenit cum causa prima[e].

<Quaestio 51>
[f]Quaeritur decimo, dato quod [g]materia prima intelligitur[h], an materia prima intelligat.

1. Et videtur quod sic, quia dicit Commentator super [i]secundum *Metaphysicae* quod materia recipit primo formas magis universales; est ergo materia prima receptiva formarum universalium. Et intellectus est receptivus formarum universalium; si ergo intellectus [j]propter suam receptionem intelligit, et materia similiter.

2. Item, intelligentia se totam intelligit; ergo se tota [k]intelligit se. Ratio[l] huius consequentiae est quia in separatis[m] idem est quod intelligit et quod intelligitur. Sed[n] intelligentia intelligit materiam in ea. Probatio, quia nos intelligimus materiam in intelligentiis; ergo multo fortius intelligentia intelligit materiam in se ipsa. Si ergo idem est quod intelligit et quod intelligitur, tunc, cum intelligentia intelligat materiam primam in se ipsa, tunc materia prima de se erit intelligens.

3. Item, dicit Aristoteles in tertio *De anima* quod intellectus non est hoc antequam [o]actu intelligat sive recipiat species ab extra. Et sic[p] nec materia prima[q] est aliquid antequam recipiat formas ab extra; sicut ergo intellectus intelligit, sic et materia.

[a] forma et totaliter *om.* TC, *sed suppl. mg.* T; *suppl. s.l.* forma C
[b] Materia … creatura *iter. sed corr.* O
[c] sui *om.* TC *sed. suppl. s.l. post* vilitatem T
[d] convenit ... prima] assimilatur causae primae C
[e] Q. 52 *hic transp.* C
[f] Quaeritur decimo] Consequenter TC
[g] materia prima *om.* C
[h] intelligitur] intelligatur quaeritur TC
[i] secundum Metaphysicae] primum (*lectio incerta*) C
[j] propter suam receptionem *om.* C
[k] intelligit se] intelligit T; intelligitur C
[l] Ratio *suppl. s.l.* O
[m] separatis] a materia *add.* C
[n] Sed] ergo O
[o] actu intelligat sive *om.* TC
[p] sic] similiter TC
[q] prima *om.* TC

essence, in such a way that the first cause is totally a form and totally in an act with which nothing of potency is mixed, and it is because of its nobility that it is not numbered together with any creature. But the case of prime matter is the opposite to that of the first cause, because in its essence prime matter is deprived of every form and of every actuality and it is because of its baseness that prime matter is not numbered together with any other creature. Also, every other creature has some actuality, and thus prime matter is what is minimally in agreement with the first cause.

<Question 51>
Granted that prime matter is understood, the tenth question is whether prime matter understands.

1. Arguments for the affirmative: in his commentary on *Metaphysics* II the Commentator says that matter first receives more universal forms;[417] therefore, prime matter is receptive of universal forms. The intellect too is receptive of universal forms; therefore, if the intellect understands because of its reception, matter does as well.

2. Furthermore, an intelligence understands itself as a whole; therefore, it understands itself in virtue of itself as a whole. The reason for this consequence is that in separate <substances> what understands and what is understood are the same. But an intelligence understands matter in its own self. Proof: we understand matter in intelligences; therefore, there is a much stronger reason to hold that an intelligence understands matter in its own self. Therefore, if what understands and what is understood are the same, then, since an intelligence understands prime matter in its own self, prime matter will be intelligent by itself.

3. Furthermore, in *On the Soul* III Aristotle says that the intellect is not anything determinate until it understands in act or receives a species from outside.[418] In the same way, prime matter is not anything <determinate> before it receives forms from outside; therefore, just as the intellect understands, so does matter as well.

[417] Aver. *In Metaph.* I, t.c. 17, fol.14vb.
[418] Arist. *De anima* III.4, 429a22–24, b30–31.

4. Item, est reperire[a] compositum ex materia et forma quod intelligit et [b]est reperire[c] formam tantum quae intelligit, [d]sicut causam primam, quae tantum forma est; ergo erit reperire[e] [f]tantum materiam quae[g] intelliget[h]. Et ratio huius consequentiae est quia sic arguit Aristoteles in [i]undecimo *Metaphysicae* et octavo *Physicorum*: si est reperire movens et motum etc.

5. Ad oppositum: intelligere est agere; materia prima non agit[j]; ergo etc. Probatio maioris: intelligere non solum est recipere, sed post receptum iudicare; sed iudicare est agere; [k]intelligere ergo est agere. Probatio minoris principalis argumenti: si materia ageret, tunc coincideret cum forma et efficiente; sed hoc est contra Aristotelem dicentem quod tres causae coincidunt, sed materia nullo modo coincidit.

6. Item, intelligere est actio nobilissima, cum sit essentialis causae primae et etiam nobilissimis[l] creaturis, ut intelligentiis; sed enti vilissimo non debetur actio nobilissima; materia ergo prima non intelligit.

7. Item, Commentator dicit in secundo *Metaphysicae* quod intellectus est natus ad formas universales sicut materia prima ad formas particulares; et si hoc, materia prima solum nata est formis particularibus, et forma particularis non[m] proportionatur intellectui; materia ergo prima non intelligit.

8. [n]Solutio sive veritas problematis est quod materia prima non intelligit.

9. Primae rationi patebit solutio in proximo problemate.

10. Secundae rationi dicendum quod nullus intellectus causatus se toto intelligit, hoc est qualibet sui parte, sed solum causa prima se tota[o] intelligit. Unde intelligentia se tota intelligit, hoc est[p] [q]se perfecta, et non se tota, [r]hoc est qualibet parte sua. Et forte intelligentia tota sive[s] perfecta intelligit,

[a] reperire *corr. ex* recipere OC
[b] est reperire] reperit *scrips. sed del.* T
[c] reperire *corr. ex* recipere OC
[d] sicut causam *iter.* C
[e] reperire *corr. ex* recipere OC *et* tertium scilicet *add. s.l.* O
[f] tantum materiam] materiam solam C
[g] quae] non *add. sed del.* O
[h] intelliget] intelligit TC
[i] undecimo … Physicorum *scripsimus*; octavo Metaphysicae et octavo (et octavo *s.l.*) Physicorum O; octavo huius vel undecimo (vel undecimo *s.l.*) T; undecimo C
[j] agit] aliquid *add.* TC
[k] intelligere … agere] ergo et intelligere C
[l] nobilissimis *corr. ex* nobilissima *(lectio incerta)* O
[m] non] solum *add. sed del.* O
[n] Solutio … problematis] Dicendum TC
[o] tota] toto O
[p] est] verum *add.* TC
[q] se perfecta *inv. sed corr.* O; separata C
[r] hoc est] haec T
[s] sive] hoc est TC

4. Furthermore, it is possible to find a composite of matter and form that understands, and it is possible to find a form alone that understands, such as the first cause, which is a form alone; therefore, it will be possible to find matter alone that understands. The reason supporting this consequence is that in *Metaphysics* XI and in *Physics* VIII Aristotle argues in this way: if it is possible to find a mover and a moved thing <together and it is possible to find a moved thing alone, then it is possible to find a mover alone>.[419]

5. To the contrary: to understand is to act; prime matter does not act; therefore, <prime matter does not understand>. Proof of the major premise: to understand is not only to receive but also to judge what has been received; but to judge is to act; therefore, to understand is to act. Proof of the minor premise of the principal argument: if matter acted, then it would coincide with a form and with an efficient cause; but this is against Aristotle who says that three of the causes coincide, but that matter in no way coincides <with them>.[420]

6. Furthermore, to understand is the noblest action, since it is essential to the first cause and also to the noblest among creatures, namely the intelligences; but the noblest action is not attributed to the basest being; therefore, prime matter does not understand.

7. Furthermore, in *Metaphysics* II the Commentator says that the intellect has a natural inclination towards universal forms just as prime matter has towards particular forms;[421] and if this is the case, prime matter only has a natural inclination to particular forms, and a particular form is not proportional to the intellect; therefore, prime matter does not understand.

8. The solution of the problem or the truth of the matter is that prime matter does not understand.

9. The solution to the first argument[422] will be made clear in the next question.[423]

10. To the second argument[424] it must be said that no caused intellect understands in virtue of itself as a whole, that is, of each of its parts, but only the first cause understands in virtue of itself as a whole. Thus, an intelligence understands in virtue of itself as a whole in the sense that it understands in virtue of itself as perfect but not of itself as a whole, that is, of each of its parts. <However>, it is perhaps the case that an intelligence understands in virtue of itself as a whole or as perfect, and this in virtue of its formal part,

[419] Arist. *Metaph.* XII.7, 1072a21–26; *Phys.* VIII.5, 256b20–24.
[420] Arist. *Phys.* II.7, 198a24–27.
[421] The correct reference is Aver. *In De anima* III, t.c. 5, ed. Crawford, 387.27–388.37.
[422] See above, n. 1.
[423] See below, q. 52, nn. 12–13.
[424] See above, n. 2.

et hoc ratione partis formalis[a], [b]sicut dicimus quod homo intelligit ratione [c]partis formalis, ut intellectivae, et non ratione partis materialis. Unde, cum dicit quod in separatis non differt quod intelligit et quod intelligitur, hoc est verum de eo quod immediate est intellectum, quia immediate se totum intelligit et mediante toto intelligit suas partes. Unde, quia totum est illud quod immediate intelligitur, idem est quod intelligit et quod intelligitur. **[O 97vb]**

11. [d]Tertiae rationi dicendum quod haec propositio, 'intellectus non est aliquid in actu antequam recipiat species ab extra', habet unum sensum falsum et duos veros aliquo modo. Si autem is sit sensus: intellectus non est aliquid actu, hoc est: [e]nullam habet actualitatem antequam recipiat species ab extra, sic est propositio falsa, quia intellectiva est perfectio pueri[f] [g]ponens ipsum in specie antequam aliquid recipiat ab extra. Vel sic intelligatur[h] propositio, quod non habet aliquam formam intelligibilem in actu antequam recipiat species ab extra, et sic habet propositio veritatem. Vel sic [i]glossatur haec propositio: 'intellectus [j]est non hoc[k]', hoc est non operatur per instrumentum corporeum in quo sit media proportio respectu intelligibilium, sicut in instrumentis sensuum[l] est media proportio per quam potest quilibet sensus particularis super proprium sensibile sibi proportionatum, ita quod non super sensibile alterius sensus; unde intellectus est non hoc[m], hoc est non est determinatus[n] in instrumento corporeo per quod operetur; et propter hoc non potest sensus super omnia singularia, potest tamen intellectus super omnia universalia.

12. [o]Ultimae rationi dicendum quod regula Aristotelis[p] non est generalis, quia deficit quando unum plus[q] dependet [r]ab altero quam econverso, sicut in substantia et accidente; et ideo non valet: 'substantia et accidens sunt simul,

[a] formalis] formalem vel C
[b] sicut ... formalis *om. hom. sed suppl. mg.* T
[c] partis ... intellectivae *om. sed suppl. s.l.* formae C
[d] Tertiae rationi] Ad tertium TC
[e] nullam habet] non habet aliquam T
[f] pueri *corr. mg. in* hominis T; *scrips. sed del.* C
[g] ponens ... specie] ipsum in specie constituens C
[h] intelligatur] intelligitur TC
[i] glossatur haec propositio] glossetur C
[j] est non hoc *lectio incerta sed corr. mg. in* non est hoc T; non est etc. C
[k] hoc] homo (*lectio incerta*) O
[l] sensuum] sensus TC
[m] hoc] homo *(lectio incerta)* O
[n] determinatus] determinatum OTC *sed corr.* O
[o] Ultimae rationi] Ad ultimum TC
[p] Aristotelis] in undecimo *add.* TC *sed corr. mg. in* in octavo vel undecimo T
[q] plus *suppl. s.l.* O
[r] ab altero *om.* T

just as we say that a man understands in virtue of his formal part, namely in virtue of the intellectual part, and not of the material part. Thus, Aristotle's claim that in separate <substances> what understands and what is understood are not different[425] is true about what is immediately understood, because a separate substance immediately understands itself as a whole and by means of itself as a whole it understands its own parts. Thus, since the whole is what is immediately understood, what understands and what is understood are the same.

11. To the third argument[426] it must be said that the proposition 'the intellect is not something in act before it receives a species from outside' has one false sense and two senses that are true in some way. (i) If the sense is 'the intellect is not something in act, that is, it has no actuality, before it receives a species from outside', the proposition is false, because the intellectual soul is the perfection of a child, placing it in its species before it receives anything from outside. (ii) If the proposition is understood in the sense that the intellect does not have an intelligible form in act before it receives a species from outside, in this sense the proposition is true. (iii) If the proposition 'the intellect is not a determinate thing' is glossed as follows: that is, the intellect does not operate with a corporeal instrument in which there is an intermediate relation to intelligible things, in the way that in the instruments of the senses there is an intermediate relation in virtue of which a particular sense has a power over its proper sensible that is proportional to it, so that this power does not apply to the sensible of another sense. In this sense the intellect is not a determinate thing, that is, it is not determined with regard to a corporeal instrument through which it operates; and because of this <difference> a particular sense does not have power over all singulars, whereas the intellect does have power over all universals.

12. To the last argument[427] it must be said that Aristotle's rule is not general, because it fails when one of the two things depends on the other more than the other way around, as happens in the case of a substance and an

[425] Arist. *Metaph.* XII.9, 1075a3–5.
[426] See above, n. 3.
[427] See above, n. 4.

et substantia est per se; ergo accidens erit[a] per se', quia plus dependet accidens a substantia quam econverso. Et sic est hic: 'est reperire sensitivam et vegetativam simul[b] et vegetativam per se; ergo et sensitivam [c]per se sine vegetativa'. Non valet, quia sensitiva plus dependet a vegetativa quam econverso. Et sic est in proposito, quia[d] plus dependet materia [e]ad formam quam econverso.

13. Aliter deficit, [f]quia, cum dicit quod est reperire compositum intelligens, iam sumit aliquid in genere; 'et est reperire formam tantum quae intelligit', [g]bene iam sumit formam extra genus, ut causam primam. Et sic non est processus uniformis[h] [i]in proposito.

<Quaestio 52>

[j]Quaeritur undecimo utrum materia prima[k] recipiat formas universales [l]secundum quod universales.

1. Et videtur quod sic: dicit Commentator secundo[m] *Metaphysicae* quod[n] materia prima recipit primo [o]formas magis universales. Est ergo secundum Commentatorem materia prima receptiva formarum universalium.

2. Item, universale est unum in multis et de multis; et si hoc, de necessitate est in rebus. Et quod est in rebus recipitur in materia prima; forma ergo universalis secundum quod universalis recipitur in materia prima.

3. Item, dicit Commentator super undecimum *Metaphysicae* quod omnes formae quae sunt in actu in motore primo sunt in potentia in materia prima. Sed omnis forma est actu in motore primo; ergo et forma universalis. Et si hoc, forma universalis secundum quod universalis est in potentia in materia prima. [p]Sed quod est in potentia in materia prima potest ei actu inesse; potest ergo materia prima recipere formas universales secundum quod sunt[q] universales.

[a] erit] est TC
[b] simul *suppl. s.l.* O
[c] per ... vegetativa *om. sed suppl. mg.* per se C
[d] quia] quod O
[e] ad formam] a forma C
[f] quia, cum dicit *om.* C
[g] bene ... primam] tu iam sumis formam (iam sumitur forma C) extra genus TC
[h] uniformis *suppl. mg.* C
[i] in proposito *om.* TC
[j] Quaeritur undecimo] Consequenter quaeritur TC
[k] prima] quia prima *add.* C
[l] secundum quod] unde C
[m] secundo] super quintum TC *sed corr. s.l. in* super secundum T
[n] quod] intellectus (quod intellectus C) est natus ad formas universales sicut materia prima quia TC
[o] formas magis universales *corr. mg. ex* universales formas T; magis universales C
[p] Sed ... prima *om. hom.* C
[q] sunt *om.* TC

accident; therefore, the inference 'a substance and an accident exist together, and a substance exists by itself; therefore, an accident will exist by itself' is not valid, because an accident depends on a substance more than the other way around. The same observation applies to the inference 'it is possible to find a sensitive and a vegetative soul together, and it is possible to find a vegetative soul by itself; therefore, it is also possible to find a sensitive soul without a vegetative soul', which is not valid, because a sensitive soul depends on a vegetative soul more than the other way around. The same observation also applies to the present case, because matter depends more on form than the other way around.

13. Aristotle's rule also fails in another respect because when he says that it is possible to find a composite that understands, the remark presupposes something in a genus; and when he says 'and it is possible to find a form that understands alone', the remark presupposes a form outside a genus, such as the first cause. Hence there is no uniform process <of reasoning> in the present case.

<Question 52>
The eleventh question is whether prime matter receives universal forms in so far as they are universal.

1. Arguments for the affirmative: in *Metaphysics* II the Commentator says that prime matter first receives more universal forms.[428] Therefore, according to the Commentator, prime matter is receptive of universal forms.

2. Furthermore, a universal is one in many and of many; if this is the case, it is necessary for a universal to exist in things. And what exists in things is received in prime matter; therefore, a universal form in so far as it is universal is received in prime matter.

3. Furthermore, in his commentary on *Metaphysics* XI the Commentator says that all forms that are in act in the first mover are in potency in prime matter.[429] But every form is in act in the first mover; therefore, a universal form is as well. If this is the case, then a universal form in so far as it is universal is in potency in prime matter. But what is in potency in prime matter can inhere in it in act; therefore, prime matter can receive universal forms in so far as they are universal.

[428] Aver. *In Metaph.* I, t.c. 17, fol.14vb.
[429] Aver. *In Metaph.* XII, t.c. 18, fol. 305va.

4. Item, materia naturalis supponit materiam primam; sed aliquid[a] universale habet esse in materia naturali; ergo et[b] in materia prima. Probatio minoris: dicit Aristoteles in secundo *De anima* quod materia naturalis recipitur in definitione naturali[c]; sed omnes partes definitionis formae sunt communes et illud quod definitum est est universale, quia sola species definitur; materia ergo naturalis est universale; [d]ergo quod recipitur in illa erit universale.

5. Item, dicit Aristoteles in octavo *Metaphysicae* quod substantiae simplices non habent definitionem, sed omnis substantia definita est composita ex materia et forma; sed solum universale definitur; ergo universale habet materiam et formam. Et ita materia recipit formas universales.

6. Item, forma generis generalissimi non habet esse in anima, cum praecedat [e]per naturam et etiam cum nihil recipiatur in anima ut res; est ergo recepta in aliquo extra. Sed in communiori materia non potest recipi quam in materia prima; materia ergo prima recipit formas universales.

7. Item, Aristoteles dicit in libro[f] *Perihermeneias* quod rerum quaedam sunt universales, quaedam particulares, [g]et ita quaedam sunt composita universalia et quaedam particularia.

8. Ad oppositum: quorum condiciones repugnant, unum non est receptivum alterius; sed condiciones universalis et materiae primae repugnant; ergo etc. Probatio minoris: universale est universale per abstractionem, materia non est materia per abstractionem. [h]Et etiam materia prima est hic et nunc, universale nec est hic nec nunc, ut dicit Aristoteles in primo *Posteriorum*, ubi dicit quod nec in aliquo et[i] aliquo non [j]nec aliquando et aliquando non, sed etc.

9. Item, universale unde universale dicitur correlative ad singulare; sed materia prima non est singulare, quia omne singulare est compositum et materia non est compositum; [k]universale ergo per se non respicit [l]materiam primam.

[a] aliquid] aliquod TC
[b] et] est O
[c] naturali] naturalis TC
[d] ergo … universale *om. hom.* C
[e] per naturam] animam C
[f] libro] secundo T; primo C
[g] et … particularia *suppl. mg.* (composita *ante* particularia *add.*) T
[h] Et etiam] quia C
[i] et] nec in T; et in C
[j] nec … non *suppl. mg.* T
[k] universale … primam] ergo materia prima non recipit universale C
[l] materiam primam] singulare OT *sed corr. s.l.* O; *corr. mg. in* materiam; quare nec materia per se recipiet formam universalem T

4. Furthermore, natural matter presupposes prime matter; but something universal has being in natural matter; therefore, it also has being in prime matter. Proof of the minor premise: in *On the Soul* II Aristotle says that natural matter is included in a natural definition;[430] but all the parts of a definition are common forms and what is defined is a universal, because only a species is defined; therefore, natural matter is a universal, and so what is received in it will be a universal.

5. Furthermore, in *Metaphysics* VIII Aristotle says that simple substances do not have a definition,[431] but every substance that is defined is composed of matter and form; but only a universal is defined; therefore, a universal has matter and form. And thus matter receives universal forms.

6. Furthermore, the form of the most general genus does not have being in the soul, since it is prior by nature and because nothing is received in the soul as a thing; therefore, it is received in something outside the soul. But it cannot be received in any matter more common than prime matter; therefore, prime matter receives universal forms.

7. Furthermore, in *De Interpretatione* Aristotle says that some things are universal and others are particular,[432] and so some composites are universal and others are particular.

8. To the contrary: things with incompatible conditions are such that one is not receptive of the other; but the conditions of a universal and those of prime matter are incompatible; therefore, <prime matter is not receptive of a universal>. Proof of the minor premise: a universal is universal through abstraction, whereas matter is not matter through abstraction. Moreover, prime matter is here and now, whereas a universal is neither here nor now, as Aristotle says in *Posterior Analytics* I, where he says that a universal is not such that it exists in one thing, but not in another thing, or that it exists at one time, but not at another time, but <in all things and always>.[433]

9. Furthermore, a universal in so far as it is universal is spoken of in correlation with a singular; but prime matter is not a singular, because every singular is a composite and matter is not a composite; therefore, a universal is not per se related to prime matter.

[430] The correct reference is Arist. *De anima* I.1, 403a29–b12.
[431] Arist. *Metaph.* VIII.6, 1045a36–b7.
[432] Arist. *De int.* 7, 17a38–39.
[433] Arist. *An. post.* I.31, 87b31–33.

10. Item, Commentator dicit super secundum *Metaphysicae* quod, sicut se habet intellectus ad formas universales, sic se habet materia ad formas particulares; sed intellectus non recipit formas nisi universales; ergo materia non[a] nisi[b] particulares.

11. [c]Solutio sive veritas problematis talis est secundum quosdam qui dicunt quod materia recipit formas universales unde universales et quod universale secundum quod universale est compositum ex materia et forma.

12. Aliter [d]secundum alios et melius dicitur sic: universale unde universale non habet respectum ad materiam per se. Sed ad singulare per se habet respectum, quia, cum forma primo[e] advenit materiae, facit compositum; et tunc meretur nomen substantiae. Et posterius per [f]naturam est illa eadem [g]forma perfectio totius individui[h]. Et posterius per naturam non ita perficit unum individuum quin habeat aptitudinem ad perfectionem alterius; et sic est universale. Unde materia primo[i] recipit formas universales sub forma sua singulari, quia materia et forma primo faciunt hanc substantiam. Unde, cum universale habeat unum esse in anima et aliud esse habeat in sua natura et aliud esse habeat in re[j], secundum esse quod habet in anima nullo modo est in materia prima nec secundum esse suae naturae aliquo modo est in materia, quia, si non esset materia, esset universale in sua habitudine propria respectu sui superioris et respectu passionum inhaerentium ratione suae naturae. Secundum autem esse quod habet in rebus[k], sic habet esse aliquo modo in materia.

13. Sed dicere materiam primam [l]recipere formam universalem contingit dupliciter: aut formam particularem magis communem inter alias formas particulares, et sic materia prima recipit primo formas universales, quia prius est haec substantia quam hoc corpus et prius hoc corpus quam hoc corpus animatum, et sic descendendo a singulari communiori ad singulare minus commune; et [m]sic est propositio vera quae **[O 98ra]** dicit quod materia prima primo recipit formas universales. Vel in ratione universalis. Et hoc dupliciter: vel mediate vel immediate. Materia autem prima immediate non recipit formam universalem unde universalem, et hoc quia respectus universalis

[a] non] est *add. sed del.* T; recipit *add.* C
[b] nisi] formas *add.* TC
[c] Solutio … qui] Dicendum (dicitur C) quaestioni duobus modis (modis *om.* C). Quidam TC
[d] secundum ... sic] dicitur et melius quod TC
[e] primo *scripsimus*; prius OTC
[f] naturam ... per *om. hom. sed suppl. mg.* C (perfecta *pro* perfectio)
[g] forma perfectio *inv. sed corr.* O
[h] individui] et sic dicitur quidditas *add. mg.* O
[i] primo] prima TC *sed corr. s.l.* T
[j] re *corr. ex* se (*lectio dubia*) O
[k] rebus] sic habet esse quod habet in rebus *add. sed del.* T
[l] recipere formam universalem] accipere formas universales T
[m] sic est propositio] haec propositio sic est C

10. Furthermore, in his comment on *Metaphysics* II the Commentator says that just as the intellect is related to universal forms, so matter is related to particular forms;[434] but the intellect only receives universal forms; therefore, matter only receives particular forms.

11. The solution of the problem or the truth of the matter, according to some people, is that matter receives universal forms in so far as they are universal and that a universal in so far as it is universal is a composite of matter and a form.

12. Other people have the different and better view that a universal in so far as it is universal is not per se related to matter, but it is per se related to a singular, because, when a form first comes to be in matter, it produces a composite, and then it deserves the name of substance. At a later stage in the natural order, that same form is the perfection of the whole individual. And at a later stage in the natural order, a form does not give perfection to one individual unless it has an aptitude to be the perfection of another individual, and in this way it is a universal. Thus, matter first receives universal forms under its singular form, because matter and form first produce this <particular> substance. Hence, since a universal has one being in the soul and another being in its own nature and another being in things, in no way does it exist in prime matter in accordance with the being that it has in the soul. Nor does it exist in matter in some way or other in accordance with the being of its nature, because, if matter did not exist, the universal would exist in its proper habit with respect to its superior and with respect to those passions that inhere in it in virtue of its nature. But in accordance with the being that it has in things, a universal exists in matter in some way.

13. The claim that prime matter receives a universal form, however, can be understood in two ways: either (i) in the sense that prime matter receives a particular form which is more common than other particular forms; and in this way prime matter first receives universal forms, because it is this substance before it is this body and it is this body before it is this animate body, and so on, descending from a more common singular to a less common singular. In this way the proposition that says that prime matter first receives universal

[434] The correct reference is Aver. *In De anima* III, t.c. 5, ed. Crawford, 387.27–388.37.

[a]secundum quod universalis immediate terminatur ad suppositum. Si mediate, hoc contingit dupliciter: vel per[b] impressionem, et hoc est impossibile—nihil enim imprimitur materiae nisi quod est singulare—aut per relationem. Et sic materia prima recipit formas universales secundum quod universales solum secundum relationem, quia immediata relatio universalis [c]in quantum universale est ad singulare, et singulare est singulare a signatione formae in materia; si ergo universale immediate refertur ad singulare et singulare ad materiam, tunc universale mediate ad materiam habet relationem, quia mediante singulari.

14. Si autem quaeratur qualis debetur compositio definito, quidam dicunt quod, quia dicit Aristoteles quod partes definitionis formae sunt, ideo definitum primo significat formam, secundario autem aggregatum. Et dicunt quod illud aggregatum non est nisi respectus ad suppositum indefinite sumptum. Et hoc est: significat formam per respectum ad suppositum. [d]Primo tamen significat formam.

15. Aliter dicitur sic, quod species definitur, et illud est aggregatum quod definitur, hoc non est quia ipsa forma substantialis[e] sit composita, sed quia omnia contenta sub ipsa sive omnia in quibus salvatur [f]ipsa species sunt composita, [g]cuiusmodi sunt supposita sive res singulares speciei. Unde dicunt quod, sicut res ipsa [h]singularis est vere composita ex materia et forma, sic species est proportionaliter composita ex genere pro materia et differentia pro forma.

<Quaestio 53>

[i]Quaeritur duodecimo utrum materia possit per partem orationis significari.

1. Et quod non sic videtur: secundum Aristotelem, omne nomen[j] primo significat formam, secundario aggregatum; quod ergo nullo modo habet formam non poterit per partem orationis significari[k].

2. Item, omne nomen significat substantiam cum qualitate; sed materia secundum se non habet qualitatem; ergo [l]per nomen non significatur.

[a] secundum quod] unde C
[b] per *suppl. s.l.* O
[c] in quantum] secundum quod TC
[d] Primo … formam *om.* C
[e] substantialis] specialis TC *sed corr. mg.* C
[f] ipsa species *om.* C
[g] cuiusmodi sunt supposita *suppl. mg.* T
[h] singularis est] singularis (*lectio incerta*) et C
[i] Quaeritur duodecimo] Consequenter quaeritur TC
[j] nomen] quod est C
[k] significari] ergo etc. *add.* C
[l] per … significatur] etc. C

forms is true. Or (ii) in the sense that prime matter receives a universal form under the *ratio* of universal. And this can be understood in two ways: either (iia) mediately or (iib) immediately. Not (iib), because prime matter does not immediately receive a universal form in so far as it is universal, and the reason for this is that a universal in so far as it is universal is immediately related to its supposit. Case (iia) can also be understood in two ways: either (iia') through an impression, and this is impossible, for nothing is impressed on matter except what is singular, or (iia'') through a relation. Prime matter only receives universal forms in so far as they are universal in accordance with a relation, because the immediate relation of a universal in so far as it is universal is to a singular, and a singular is singular in virtue of the individualization of its form in matter; therefore, if a universal is immediately related to a singular and a singular <is related> to matter, then a universal is mediately related to matter, because it is related through a singular.

14. But if it is asked what kind of composition is to be ascribed to what is defined, some people maintain that, since Aristotle says that the parts of a definition are forms,[435] it follows that what is defined first signifies a form and secondarily an aggregate. And they say that that aggregate is nothing other than a relation to a supposit taken in an indefinite way, that is, what is defined signifies a form as related to a supposit, but it primarily signifies the form.

15. Another opinion is that a species is defined, and what is defined is an aggregate, not because a substantial form itself is composite, but because all the things falling under it or all the things in which that species is found are composite, namely the supposits or the singulars of a species. Thus, they say that, just as a singular thing itself is truly composed of matter and a form, so a species is analogically composed of a genus as matter and of a difference as a form.

<Question 53>
The twelfth question is whether matter can be signified by a part of speech.

1. Arguments for the negative: according to Aristotle, every name primarily signifies a form, and secondarily an aggregate;[436] therefore, what in no way has a form cannot be signified by a part of speech.

2. Furthermore, every name signifies a substance with a quality; but matter by itself does not have a quality; therefore, it is not signified by a name.

[435] Arist. *Metaph.* VII.10, 1035b33–34.
[436] Arist. *Metaph.* VIII.3, 1043a29–37; cf. Aver. *In Metaph.* VIII, t.c. 7, fol.215va–b.

3. Item, omne nomen imponitur a forma; quod ergo non habet formam, ei non imponitur nomen. [a]Sed materia prima non habet formam; ergo etc.

4. Item, omne nomen aut est commune aut proprium. Sed materiae non potest imponi nomen proprium, cum non sit singulare alicuius communis. [b]Et etiam nomen proprium potest dici de pluribus aequivoce[c], sicut 'Socrates' dicitur de pluribus aequivoce; sed 'materia' non potest dici de pluribus aequivoce. Nec[d] potest ei imponi nomen commune, cum omnis communitas a forma procedat[e] et materia non est forma.

5. Item, Aristoteles in primo *Perihermeneias* dicit quod voces sunt notae passionum quae sunt in anima; cum ergo materia non faciat passionem in anima, tunc materiae non imponetur vox. Probatio minoris: solum illud facit passionem in anima quod facit speciem in anima; [f]sed materia non facit speciem in anima, ut probatum est prius; non ergo[g] facit passionem in anima.

6. Ad oppositum: [h]omni ei quod potest intelligi, ei potest nomen imponi; sed materia prima potest intelligi, [i]sicut probatum est prius; ergo ei potest nomen imponi.

7. Item, si sic dicam: 'nuda substantia materiae primae', omnia ista nomina sunt [j]et nominant substantiam materiae primae; ergo nuda substantia materiae primae per nomen potest significari.

8. Item, quod significatur per circumlocutionem[k] potest significari per unum nomen, nisi sit propter defectum imponentis; sed materia prima[l] potest significari per circumlocutionem; ergo, quantum de se est, potest per unum nomen significari.

9. [m]Solutio sive veritas problematis secundum quosdam talis est, quia [n]nullum nomen potest meram substantiam significare, sed tantum pronomen, et materia prima[o] secundum se mera substantia[p] est et sine qualitate aliqua; ideo non potest per nomen significari, sed tantum per pronomen.

[a] Sed … etc. *om.* TC
[b] Et etiam] item si est C
[c] aequivoce] et etiam nomen proprium *add. sed del.* T
[d] Nec] non O
[e] procedat] procedit TC
[f] sed … anima *om. hom.* O
[g] ergo] materia *add. s.l.* O
[h] omni ei] in omni eo C
[i] sicut … prius *om.* T
[j] et … primae *suppl. mg.* C
[k] circumlocutionem] ergo quantum de se est *add. sed del.* O
[l] prima *om.* O
[m] Solutio … est] Quidam dicunt sic (sic *om.* C) quod TC
[n] nullum nomen] nomen *scrips. sed corr. mg. in* nomen non T
[p] substantia *om.* OTC *sed suppl. s.l.* T

3. Furthermore, every name is imposed from a form; therefore, a name is not imposed on what does not have a form. But prime matter does not have a form; therefore, <no name is imposed on matter>.

4. Furthermore, every name is either common or proper. But a proper name cannot be imposed on matter, because matter is not a singular instance of something common. Moreover, a proper name can be said of many things equivocally, for example 'Socrates' is said of many people equivocally; but 'matter' cannot be said of many things equivocally. Nor can a common name be imposed on it, because every community proceeds from a form and matter is not a form.

5. Furthermore, in *De Interpretatione* I Aristotle says that spoken words are signs of the passions that are in the soul;[437] therefore, since matter does not produce a passion in the soul, a spoken word will not be imposed on matter. Proof of the minor premise: only what produces a species in the soul produces a passion in the soul; but matter does not produce a species in the soul, as has been proved before;[438] therefore, matter does not produce a passion in the soul.

6. To the contrary: a name can be imposed on everything that can be understood; but prime matter can be understood, as has been proved before;[439] therefore, a name can be imposed on it.

7. Furthermore, if I say 'the bare substance of prime matter', all these words are names and they name the substance of prime matter; therefore, the bare substance of prime matter can be signified by a name.

8. Furthermore, what is signified by a circumlocution can be signified by one name, unless this is prevented by a deficiency in the person who imposes names; but prime matter can be signified by a circumlocution; therefore, it can, in itself, be signified by one name.

9. The solution of the problem or the truth of the matter, according to some people, is that since no name other than a pronoun can signify a mere substance, and prime matter is by itself a mere substance without any quality, it follows that prime matter cannot be signified by a noun, but only by a pronoun.

[437] Arist. *De int.* 1, 16a3–4.
[438] See above, q. 49, n. 4.
[439] See above, q. 49.

10. Aliter dicitur [a]secundum alios quod nos possumus sumere nomen communiter et proprie et communissime, sicut[b] dicimus quod 'substantia' dicitur de forma per prius, de composito secundario, tertio de materia, et etiam iuxta hoc quod dicit Aristoteles septimo[c] *Metaphysicae*, quod 'substantia' dicitur de forma et composito et, si ex consequenti dicamus quod dicatur de materia, quae est altera pars compositi, dicemus quod materia est substantia communissimo modo sumendo substantiam. Si ergo communissimo modo sumatur nomen, prout imponitur formae et composito et ex consequenti imponitur alteri parti compositi, ut materiae, sic possumus dicere quod nomen imponitur materiae. Et ita materia prima per nomen significatur communissimo modo sumendo nomen[d].

11. Aliter adhuc[e] dicitur [f]secundum alios quod materia prima potest per nomen proprie significari, sicut si[g] sic dicam: 'nuda substantia materiae primae'. Unde, sicut est hic si quaeratur: 'chimera, quae res[h] est?', nonne debet responderi quoniam nomen est? Et tamen nec est res extra sibi respondens pro substantia nec pro qualitate. Non ergo exigitur ad hoc quod nomen imponatur quod sit res extra [i]ratione substantiae et ratione qualitatis, [j]sed exigitur vel quod sit res extra vel quod sit intentio aliqua imaginata vel intellecta quae[k] sit[l] ratione qualitatis et etiam cum hoc quod[m] sit aliquo modo suppositum indefinite[n] sumptum imaginatum quod sit ratione substantiae. Si ergo talibus quae non sunt in rerum natura, sed tantum in imaginatione, imponantur proprie nomina, tunc multo fortius potest nomen imponi materiae, quae est aliquid in rerum natura; unde potest hoc nomen 'materia' imponi illi. Et 'materia' est nomen commune communitate rationis;

[a] secundum alios] sic TC
[b] sicut] et sic C
[c] septimo] in secundo (*lectio incerta* C) OC
[d] nomen] Aliter dicitur sic (sic *om. et* secundum alios *suppl s.l.* C) quod quia in omni eo quod est extra causam primam est reperire aliquam proprietatem quae est alia a substantia sua (sua sub. C) et (unde C) in materia prima est aliqua proprietas quae est alia a sua substantia (et in materia prima ... substantia *suppl. mg.* T), ut forte sua potentia passiva, et ita materiae primae potest nomen imponi, et est (*corr. s.l. ex* etiam T) sua substantia (in *add.* C) ratione substantiae nominis (nominis *om.* C) et illa proprietas est (est *om.* C) in ratione qualitatis et sufficit proprietas quaecumque ad qualitatem quae exigitur a grammatico *add.* TC
[e] adhuc *om.* TC
[f] secundum alios] sic TC
[g] si *om.* TC
[h] res *scripsimus*; pars OTC *sed* non *add. sed del.* O
[i] ratione ... qualitatis *om.* T
[j] sed exigitur *suppl. mg.* T
[k] quae] quod TC
[l] sit] loco *add. s.l.* O
[m] quod *om.* OTC *sed suppl. s.l.* OT *et* ibi *add. s.l.* O
[n] indefinite] infinite O

10. Another opinion is that we can take 'name' in a broad sense, in a proper sense, and in a broadest sense, just as we say that 'substance' is said primarily of a form, secondarily of a composite, and thirdly of matter, and just as we also say, in accordance with what Aristotle says in *Metaphysics* VII, that 'substance' is said of a form and of a composite,[440] and, if as a result we say that 'substance' is said of matter, which is the other part of a composite, we will be saying that matter is a substance in the broadest sense of that word. Therefore, if a name is taken in the broadest sense, in so far as it is imposed on a form and on a composite and, as a result, on the other part of a composite, namely matter, in this way we can say that a name is imposed on matter. Hence prime matter is signified by a name, taking the name in its broadest sense.

11. Another opinion is that prime matter can be properly signified by a name, for example when I say 'the bare substance of prime matter'. Thus, the situation here is like the situation when one asks 'what kind of thing is a chimera? ': shouldn't one reply by taking it that the word 'chimera' is a name? And yet there is no extra-mental thing, whether a substance or a quality, corresponding to it. Therefore, in order for a name to be imposed, it is not required that there should be an extra-mental thing having the *ratio* of a substance or of a quality, but it is required that there should be either an extra-mental thing or an intention, whether imagined or understood, which has the *ratio* of a quality and that together with this there should be some kind of imagined supposit, indefinitely taken, which has the *ratio* of a substance. Therefore, if names are properly imposed on things that do not exist in reality but only in imagination, then there is a much stronger reason to hold that a name can be imposed on matter, which is something in reality; thus, the name 'matter' can be imposed on matter. Moreover, 'matter' is a

[440] Arist. *Metaph.* VII.3, 1029a 2–3.

unde intentio imaginata communis est ei pro qualitate et suppositum imaginatum, si[a] fas est dicere quod materia est [b]suppositum, est[c] ei pro substantia. Unde, si plures essent materiae, adhuc esset materia communis omnibus illis, et hoc communitate suae intentionis.

<Quaestio 54>

[d]Quaeritur tertio decimo an materia prima[e] in sua substantia habeat aliquam partibilitatem aliam a partibilitate quantitatis.

1. Et quod materiae in sua substantia non debeatur partibilitas probatio: Aristoteles dicit in hoc[f] tertio, capitulo de infinito, quod, si esset[g] infinitum extra magnitudinem[h], illud infinitum esset indivisibile, innuens ibi[i] quod omnis divisibilitas est a quantitate; sed materia de se non est **[O 98rb]** quanta; ergo materiae de se non debetur divisio et, si hoc, nec partibilitas.

2. Item, Aristoteles dicit in eodem capitulo quod multitudo et magnitudo sunt causae divisibilitatis et partibilitatis. Vult [j]ergo ibi quod quantitas sit causa divisibilitatis. Sed materia non est quanta; ergo etc.

3. Item, demonstrato aliquo quanto, et dividatur per medietates duas[k]; circumscribamus ergo ab utraque [l]quantitatem et alias dispositiones accidentales; tunc illud quod remanet in utraque aut sunt unum aut duo. Si sunt unum, habeo propositum. Si sint duo, tunc una est in se una et altera pars est in se una; et ita in ipsis erit multitudo. Et si hoc, circumscripta omni quantitate, adhuc erit quantitas ut multitudo.

4. [m]Ad quaestionem dicunt quidam sic, quod habere partem et partem contingit dupliciter: vel partem sub parte, et sic materia prima[n] habet partem et partem; vel habere partem extra partem, et sic materia prima[o] non habet partem et partem.

[a] si] sicut OT
[b] suppositum, est ei *corr. mg. ex* ei suppositum T
[c] est *suppl. s.l.* O
[d] Quaeritur tertio decimo] consequenter quaeritur TC
[e] prima *om.* TC
[f] hoc] libro illo C; *fort.* libro *scribendum*
[g] esset *suppl. s.l.* O
[h] magnitudinem] imaginem O
[i] ibi] ibidem T; idem C
[j] ergo ibi] igitur C
[k] duas] suas TC *sed* duas *suppl. s.l. post* per T
[l] quantitatem et alias] quantitate omnes C
[m] Ad ... quidam] dicitur ad quaestionem T; ad quaestionem dicitur C
[n] prima *om.* TC
[o] prima *om.* TC

common name by community of *ratio*; thus, an imagined common intention has the *ratio* of a quality with respect to matter, and an imagined supposit, if it is legitimate to say that matter is a supposit, has the *ratio* of a substance with respect to it. Thus, if there were many matters, matter would still be common to all those matters, and this by the community of its intention.

<Question 54>

The thirteenth question is whether prime matter in its substance has a divisibility into parts distinct from the divisibility into parts of quantity.

1. Arguments for the negative: in *Physics* III in the chapter on the infinite, Aristotle says that if an infinite existed outside a magnitude, that infinite would be indivisible,[441] suggesting by this that every divisibility comes from a quantity; but matter by itself is not a quantity; therefore, division is not to be assigned to matter by itself, and if that is the case, neither is divisibility into parts.

2. Furthermore, in the same chapter Aristotle says that multitude and magnitude are the causes of divisibility and of divisibility into parts.[442] Therefore, he wants to maintain there that quantity is the cause of divisibility. But matter has no quantity; therefore, <matter does not have divisibility into parts>.

3. Furthermore, given a quantity, let it be divided into two halves. If we remove quantity and the other accidental dispositions from each half, then what remains of each half is either one or two. If <the remainder> is one, then I have the intended conclusion. If <the remainder> is two, then one half is one in itself and the other half is one in itself, and thus they will form a multitude. If that is the case, then when every quantity is removed, there will still be a quantity, that is, a multitude.

4. Some people reply to this question by saying that something can have distinct parts in two ways: either (i) it has parts such that one part is under another, and in this way prime matter has distinct parts; or (ii) it has parts such that one part is outside the other, and in this way prime matter does not have distinct parts.[443]

[441] Arist. *Phys*. III.5, 204a8–11.
[442] Arist. *Phys*. III.5, 204a11–12.
[443] Two versions of the theory according to which matter is divisible into parts that exist in the same place and are not extended are reported by Robert Kilwardby in the *De ortu scientiarum*; see Robert Kilwardby, *De ortu scientiarum*, c. 31, §§ 280–284, 102.7–103.32.

5. Contra hoc sic: illud quod est sub alio habet partem suppositam[a] illi[b]; sed ubi una pars supponitur[c] alii, ibi est una pars extra aliam partem; si ergo [d]in materia est pars sub parte, erit pars extra partem.

6. Item, [e]ad idem: si materia habet partem sub parte, aut ergo in[f] potentia aut in[g] actu. Si actu, tunc habet partem extra partem, quia quod actu[h] supponitur alteri est extra ipsum. Si potentia tantum, tunc, sicut est dicere quod[i] habet partem sub parte, quia in potentia solum, sic erit dicere quod materia prima habet partem extra partem, et hoc quia in potentia habet partem extra partem. Sed consequens est falsum; ergo et antecedens.

7. Item, ad principale: circumscripta omni forma tam substantiali quam accidentali, illud quod manet aut est unum[j] simpliciter, et [k]si hoc, habeo propositum. Si autem illud quod [l]manet habeat partem [m]et partem, tunc, cum solus actus dividat, materia prima sic divisa habebit formam. Quod falsum est; ergo primum.

8. Item, si materia prima habeat partem et partem, tunc erit una pars sub una forma particulari et alia pars sub alia. Et si hoc, tot erunt materiae primae quot sunt formae particulares. Et ita erunt quasi infinitae materiae primae, quod falsum est; ergo primum.

9. Item, tanta[n] est materia in minima parte mundi quanta est in maxima parte, quod non contingeret nisi eadem esset materia omnium. Et[o] materia omnium est impartibilis. Probatio antecedentis: ex uno pugillo terrae per conversivam generationem elementorum fiunt decem aquae et ex uno aquae decem aeris; et ita ex uno pugillo terrae [p]fiunt centum aeris, item ex uno pugillo aeris fiunt decem ignis. Et ita, a primo ad ultimum: ex uno pugillo terrae fiunt mille ignis. Et iterum, ex minimo ignis contingit per conversivam generationem fieri terram, et tunc iterum per conversivam generationem

[a] suppositam] suprapositam O
[b] illi] alii TC
[c] supponitur] superponitur O; *corr. ex* se ponitur. T
[d] in materia *om.* C
[e] ad idem *om.* T
[f] in *om.* TC
[g] in *om.* TC
[h] actu *om.* OTC *sed suppl. s.l.* O; *suppl. mg. ante* est T
[i] quod] materia *add.* TC
[j] unum] una TC *sed corr.* T
[k] si hoc] sic C
[l] manet habeat] remanet habet T
[m] et partem *om. hom.* C
[n] tanta *corr. s.l. ex* tam O; *suppl. mg.* T
[o] Et] ita *add. s.l.* T; quod *add.* C
[p] fiunt ... pugillo *om. sed suppl. mg.* fiunt centum aeris et ex uno pugillo T

5. Against this: the thing that is under another thing has a part placed under that thing; but if one part is placed under another, one part is outside the other; therefore, if there is one part under another in matter, there will also be one part outside another.

6. Moreover, against the same reply: if matter has one part under another, this happens either in potency or in act. (i) If in act, then matter has one part outside another, because what is under something <in act> is outside it. (ii) If only in potency, then, just as it is possible to say that matter has one part under another because this happens only in potency, so it will be possible to say that prime matter has one part outside another because it has one part outside another in potency. But the consequent is false; therefore, the antecedent is also false.

7. Furthermore, concerning the main question: when every form, both substantial and accidental, is removed, what remains is either absolutely one, and in this case I have the intended conclusion, or it has distinct parts, and then, since only an act divides, if prime matter is divided in this way, it will have a form; but this is false; therefore, the first premise is false too.

8. Furthermore, if prime matter has distinct parts, then one part will be under one particular form and another part under another. If this is the case, there will be as many prime matters as there are particular forms, and thus there will be almost infinitely many prime matters, which is false; therefore, the first premise is also false.

9. Furthermore, there is an equal amount of matter in a minimal part and in a maximal part of the world, which would not happen if the matter of all things were not the same. But the matter of all things is not divisible into parts. Proof of the antecedent: ten handfuls of water come to be from one handful of earth by the reciprocal generation of elements, and ten handfuls of air from one handful of water, and thus one hundred handfuls of air come to be from one handful of earth, and furthermore ten handfuls of fire come to be from one handful of air, and so by the argument from first to last, one thousand handfuls of fire come to be from one handful of earth. Moreover, earth can come to be from a minimal fire by reciprocal generation, and then

contingit ex illa [a]terra iterum facere millesies tantum ignis, ut prius patuit[b]. Et similiter[c] contingit demum ex uno pugillo terrae generare et replere totum mundum igne [d]vel aere vel aqua, et hoc semper[e] convertendo generationem. Et in isto argumento oportet supponere quod sit sumere minimum in elementis.

10. Item, si materia habeat[f] partem et partem, tunc erit verum dicere: 'haec pars materiae est materia' et 'illa pars materiae est materia'. Et si hoc, materia est communis[g] omnibus partibus contentis sub. Et omnis communitas a forma est; materia ergo habebit formam, quod est impossibile; ergo primum.

11. Item, quae differunt genere differunt specie et quae differunt specie differunt numero, et hoc quia semper ad maiorem diversitatem sequitur minor diversitas. Sed materia una numero [h]potest esse sub formis diversis genere et specie; ergo et[i] numero. [j]Et ita potest una materia numero subesse omnibus formis. Probatio minoris. Haec albedo et haec longitudo differunt genere et specie; ergo et numero. Sed una materia potest subesse utrisque[k]; potest ergo una materia subesse omnibus formis.

12. Item, si materia prima est secundum unam partem in uno individuo et secundum aliam partem in alio individuo et non tota[l] in quolibet individuo, tunc materia non est unum principiorum essentialium, et hoc quia principium est totum in quolibet principiato.

13. Item, si resolvantur omnes formae, erit status ad unam formam primam, ut ad formam generis generalissimi[m]. [n]Sed uni formae primae respondet una materia prima; ergo si resolvantur omnes formae, erit status ad unam materiam primam. Probatio maioris: si fiat resolutio omnium formarum usque ad formam generalissimi, tunc aut manet una forma generalissimi aut plures. Si una, habeo propositum. Si plures, ergo plura erunt generalissima. Sed hoc est falsum; ergo primum.

[a] terra iterum *om.* T
[b] patuit] factus fuit T
[c] similiter] sic TC; vel sic *add. s.l.* O
[d] vel aere] vel igne *scrips. sed del.* O
[e] semper *om.* TC
[f] habeat] habet TC
[g] communis] commune O
[h] potest esse sub] poterit (potest C) subesse TC
[i] et] poterit subesse diversis formis TC
[j] Et … numero *om. hom.* O
[k] utrisque] utrisque (*sed corr. in* utrique) istorum T; utrique illorum C
[l] tota] cum alio *add. sed del.* T
[m] generalissimi] generalissimum O
[n] Sed … generalissimi *om.* O

again one thousand times more of fire can be produced from that earth by reciprocal generation, as was made clear before. In the same way fire can then be generated from one handful of earth and one can fill the whole world with fire or water, and this by repeated reciprocal generations. In this argument it is necessary to assume that there is a minimum in the elements.

10. Furthermore, if matter has distinct parts, then it will be true to say 'this part of matter is matter' and 'that part of matter is matter'. If that is the case, then matter is common to all the parts contained in it. But every community comes from a form; therefore, matter will have a form, which is impossible; therefore, the first premise is impossible too.

11. Furthermore, things that differ in genus differ in species and things that differ in species differ in number, and the reason for this is that from a greater diversity a smaller diversity always follows. But matter as numerically one can be under forms that are different in genus and in species; therefore, it can also be under forms that are different in number, and thus matter as numerically one can be under all forms. Proof of the minor premise: this whiteness and this length differ in genus and species; therefore, they also differ in number; but one matter can be under both; therefore, one matter can be under all forms.

12. Furthermore, if prime matter with regard to one of its parts is in one individual and with regard to another part is in another individual so that the whole of matter is not in every individual, then matter is not one of the essential principles, and this is because a principle is as a whole in everything derived from it.

13. Furthermore, if all forms are resolved, this resolution will stop at one first form, that is, at the form of the most general genus. But to one first form there corresponds one prime matter; therefore, if all forms are resolved, this resolution will stop at one prime matter. Proof of the major premise: if the resolution of all forms reaches the form of the most general genus, then either one form of the most general genus remains or many <forms remain>. If one <form remains>, I have the intended conclusion. If many <forms remain>, then there will be many most general genera. But this is false; therefore, the first premise is also false.

14. Ad oppositum: est reperire plus et minus de materia [a]ubi non est reperire[b] plus et minus de quantitate; alia ergo est partibilitas materiae a partibilitate quantitatis. Probatio maioris sic: secundum Aristotelem, ex uno pugillo terrae fiunt decem[c] aquae; si ergo tantum esset [d]de materia in uno pugillo aquae quantum est in uno pugillo terrae, tunc ex uno pugillo aquae fieret [e]unus pugillus terrae, quod est contra Aristotelem dicentem quod decem pugilli aquae sunt minimum ex quo potest fieri unus pugillus terrae per conversivam generationem.

15. Item, Aristoteles dicit [f]in capitulo de tempore quod partibilitas mensurae est a partibilitate mensurati; sed quantitas est mensura substantiae materiae; si ergo quantitati debeatur partibilitas, hoc erit a partibilitate substantiae materiae. Sed quantitati[g] debetur partibilitas; ergo et materiae. Et si hoc, materia est partibilis.

16. Item, suppono quod in qualibet[h] anima[i] rationali est materia prima; tunc quaero: aut tota materia prima est in qualibet anima aut non. Si non, sed una pars sit in una anima et alia pars in alia anima, tunc habeo propositum. Si autem sit tota in qualibet[j], tunc, sicut multotiens creabitur[k] anima [l]alia et alia, sic multotiens creabitur[m] materia prima. Et sic hoc, unum et idem numero multotiens creabitur, quod est falsum; ergo primum.

17. Item[n], Aristoteles dicit[o] in[p] tertio *Caeli et mundi* quod subtile est parum in[q] partibus, grossum[r] est multum in partibus; sed non intendit quod subtile sit parum in partibus formae, quia in subtili maior est perfectio partium, nec etiam[s] est parum in partibus quantitatis, quia **[O 98va]**, quanto aliquid[t] subtilius et rarius, tanto maioris quantitatis est se ipso ingrossato et

[a] ubi … quantitate *om.* O
[b] reperire *om.* C
[c] decem] pugilli *add.* TC
[d] de materia *suppl. s.l.* C
[e] unus pugillus] unum pugillum TC *sed corr.* pugillum *in* pugillus T
[f] in … tempore *suppl. mg. post* Aristoteles T
[g] quantitati *corr. ex* quantitas non T
[h] qualibet] aliqua O; quolibet C
[i] anima *suppl. s.l.* TC
[j] qualibet] anima *add.* TC
[k] creabitur] creatur TC
[l] alia et alia *om.* T
[m] creabitur] creatur TC
[n] Item *om.* O
[o] dicit *transp. post* mundi T; *om.* C
[p] in] secundo vel *add.* O
[q] in *om.* O
[r] grossum] autem *add.* TC
[s] etiam *om.* TC
[t] aliquid] ad O; fuerit *add.* C

14. To the contrary: it is possible to find more and less matter where it is not possible to find more and less quantity; therefore, matter's divisibility into parts is distinct from a quantity's divisibility into parts. Proof of the major premise: according to Aristotle, ten handfuls of water come to be from one handful of earth;[444] therefore, if there were as much matter in one handful of water as there is in one handful of earth, then one handful of earth would be produced from one handful of water, which is against Aristotle, who says that ten handfuls of water are the minimum from which one handful of earth can be produced by reciprocal generation.

15. Furthermore, in the chapter on time Aristotle says that a measure's divisibility into parts comes from the divisibility into parts of the thing measured;[445] but quantity is the measure of the substance of matter; therefore, if divisibility into parts is ascribed to quantity, this divisibility will derive from the divisibility of the substance of matter. But the divisibility into parts is ascribed to quantity; therefore, it is also ascribed to matter. If so, matter is divisible into parts.

16. Furthermore, I assume that in every rational soul there is prime matter; then I ask whether the whole of prime matter is in every soul or not. If this is not the case, but one part of prime matter is in one soul and another part in another soul, then I have the intended conclusion. If the whole of prime matter is in every soul, then, just as many different souls are created, so prime matter will be created many times, and hence one and the same thing will be created many times, which is false; therefore, the first premise is also false.

17. Furthermore, in *On the Heavens* III Aristotle says that what is subtle is such that there is little of it in its parts, and what is thick is such that there is a lot of it in its parts.[446] But Aristotle does not mean that what is subtle is such that there is little of it in the parts of its form, because in what is subtle there is a greater perfection of the parts, nor does he mean that there is less of it in the parts of its quantity, because the more subtle and rare something is, the

[444] Arist. *De gen. et corr.* II.6, 333a20–22.
[445] Arist. *Phys.* IV.12, 220b23–28.
[446] Arist. *De caelo* III.5, 303b26–27.

condensato. Intendit ergo [a]de partibus materiae, ut subtile sit parum in partibus materiae, hoc est parum habet de materia in suis partibus, et grossum est multum in suis partibus, hoc est multum habet de materia in suis[b] partibus. Et si hoc, alia est partibilitas materiae a partibilitate quantitatis.

18. Item, demonstrato aliquo continuo diviso in medietates et sit una medietas A et alia medietas sit B. Tunc sic: non est una pars materiae primae sub A et sub B, quia nihil unum et idem numero est in diversis partibus secundum se totum, nisi sit anima vel naturam animae habens. Sed materia nec est anima nec naturam animae habens; ergo etc.

19. Item, signemus aliquod quantum sicut prius. Tunc quaero: aut una quantitas est alia ab alia quantitate aut non. Et constat quod sunt duae quantitates. Sed diversis accidentibus eiusdem speciei diversa respondent subiecta; aliud ergo est subiectum unius quantitatis et alterius. Sed quantitas non habet subiectum nisi substantiam sive materiam; alia ergo est partibilitas substantiae a partibilitate quantitatis.

20. Item, quantitas non est partibilis nisi quia in materia est; partibilitas ergo quantitatis dependet a materia. Et si hoc, materia est partibilis aliquo modo.

21. Item, [c]dicit Aristoteles quod omne habens partes propinque sitas est densum; si autem intelligat[d] de partibus quantitativis, cum omne continuum sive rarum habeat sic partes propinque sitas, tunc omne continuum esset densum, quod falsum est. Intelligit ergo sic, quod omne habens[e] partes materiae propinque sitas est densum. Et si hoc, materia habet partem et partem.

22. Item, Aristoteles dicit in septimo huius quod aequali[f] de aere et[g] aqua respondet multitudo aequivoce. Si intendat de partibus quantitativis, tunc est propositio falsa, quia sic est[h] aequale univoce in illis. Intendit ergo de partibus materiae. Et si hoc, partibilitas [i]in substantia materiae reperitur alia a partibilitate quae est in quantitate.

23. [j]Solutio sive rei veritas problematis secundum quosdam talis est, quia quidam dicunt quod materia prima habet partem sub parte secundum simultatem, hoc est habet partes simul inter se indistinctas, ita quod non habet

[a] de partibus *om.* C
[b] suis *corr. s.l. ex* habens T; *om.* C
[c] dicit Aristoteles quod] Aristoteles TC
[d] intelligat] Aristoteles *add.* TC
[e] habens] quod habet T
[f] aequali *corr. ex* aequaliter T; aequalitati C
[g] et] de *add.* TC
[h] est] esset O
[i] in substantia *suppl. mg.* T
[j] Solutio … quia *om.* TC

greater its quantity is than when it becomes thick and condensed. Therefore, Aristotle is referring to the parts of matter, and means that what is subtle is such that there is little of it in the parts of matter, that is, it has little matter in its parts, and what is thick is such that there is a lot of it in the parts of matter, that is, it has a lot of matter in its parts. If so, matter's divisibility into parts is distinct from quantity's divisibility into parts.

18. Furthermore, consider a continuum divided into halves, and let one half be A and the other half B. Then one can argue as follows: one and the same part of prime matter is not under A and under B, because nothing that is one and the same in number is in different parts as a whole, the only exception being the soul or something having the nature of a soul. But matter is neither a soul nor does it have the nature of a soul; therefore, <matter is not as one and the same in number in different parts as a whole>.

19. Furthermore, let us divide a quantity as before.[447] Then I ask whether one quantity is distinct from the other quantity or not. It is evident that they are two quantities. But different accidents of the same species have different subjects; therefore, the subject of one quantity is distinct from the subject of the other. But a quantity has only a substance or matter as subject; therefore, a substance's divisibility into parts is distinct from a quantity's divisibility into parts.

20. Furthermore, quantity is divisible into parts only because it exists in matter; therefore, quantity's divisibility into parts depends on matter. If that is the case, then matter is in some way divisible into parts.

21. Furthermore, Aristotle says that everything with parts located close one to another is dense;[448] but if he is referring to quantitative parts, then, since every continuum or rare thing has quantitative parts located close one to another, every continuum would be dense, which is false. Therefore, he means that everything with parts of matter located close one to another is dense. If that is the case, then matter has distinct parts.

22. Furthermore, in *Physics* VII Aristotle says that a multitude is found equivocally in an equal quantity of air and of water.[449] If he is referring to quantitative parts, then the proposition is false, because being equal is found univocally in them. Therefore, he is referring to the parts of matter. If that is the case, then the divisibility into parts found in the substance of matter is distinct from the divisibility into parts that exists in a quantity.

23. The solution of the problem or the truth of the matter, according to some people, is the following: some people say that prime matter has one part under another in accordance with their being in the same place, that is, it has

[447] See above, n. 18.
[448] Arist. *Cat.* 8, 10a20–21.
[449] Arist. *Phys.* VII.4, 248b13–16.

partes inter se extensas. Unde sic ponentes ponunt eandem simplicem essentiam materiae primae, [a]et non ita quod materia prima habeat partem sub parte secundum remotionem. [b]Et hoc modo concluderet ratio.

24. Aliter tamen[c] dicunt quidam quod materia[d] non habet partem et partem, sed quod tota materia prima indivisibilis est in quolibet. Nec ponunt quod per adventum formae extendatur materia, sed quod forma facit materiam se multiplicare.

25. Aliter dicitur, et mihi videtur quod melius, dicendo quod materia prima habet [e]partem et partem, nec est eadem pars materiae primae in duobus individuis sive in duabus partibus diversis alicuius continui. Unde ponunt quod, sicut est partibilitas in materia, ut in potentia, sic sunt infinitae partes in materia. Et ita est alia partibilitas materiae a partibilitate quantitatis, actualis tamen extensio materiae non est nisi ab extensione quantitatis. Unde isti ponunt quod nihil unum et idem numero potest reperiri in diversis individuis.

26. Primae et secundae rationi in contrarium dicendum quod Aristoteles, cum dicit quod, si infinitum esset etc., innuit quod non est divisio quae est cum actuali extensione nisi in quantitate. Et similiter, cum dicit quod magnitudo et multitudo etc., intendit quod sunt causae divisibilitatis[f] cum actuali extensione.

27. [g]Tertiae rationi dicendum quod alia et alia pars corporis substantiae est in diversis partibus eiusdem continui. Verumtamen, si circumscribatur quantitas in eis, manet pars et pars in potentia tantum et non sub actuali extensione; unde actualis extensio non est in materia nisi a quantitate.

28. Ad argumenta quae probant quod materia non habet partem sub parte dicendum, et primo primae[h], quod habere partem sub parte contingit dupliciter: aut secundum remotionem et distantiam aut secundum simultatem. Si secundum remotionem, ita scilicet quod una pars supponatur alii, sic bene [i]arguit, et sic non habet materia partem sub parte. Si secundum simultatem, ita quod omnes partes sunt simul, sicut vult Aristoteles quod fit condensatio et rarefactio nullo adveniente et nullo recedente, sed, quando fit condensatio, partes materiae magis cedunt in unum et, quando fit rarefactio, minus cedunt

[a] et ... ratio *om.* TC *sed suppl. mg.* (concludit *pro* concluderet) C
[b] Et ... ratio *scrips. sed del.* O
[c] tamen *om.* TC
[d] materia] prima *add.* TC
[e] partem et partem] partes T
[f] divisibilitatis] veritatis diversitatis *scrips. sed corr. mg. in* divisibilitatis et partibilitatis T; diversitatis C
[g] Tertiae rationi] ad tertium TC
[h] primae] dicendum *add. sed del.* O
[i] arguit, et] arguitur T

parts that are in the same place and not distinct one from another, in such a way that it does not have parts extended one with respect to another. Thus, those who hold this position assume that prime matter has the same simple essence, in such a way that prime matter does not have parts one under another and distant one from another. If this were the case, the argument would be conclusive.

24. Other people have another view, namely that matter does not have distinct parts, but that prime matter as indivisible exists in everything as a whole. Moreover, they do not assume that a form that comes to be in matter causes it to extend, but rather that a form causes matter to multiply itself.

25. Another opinion, which seems better to me, is that prime matter has distinct parts, and that the parts of prime matter existing in two individuals or in two distinct parts of a continuum are not the same. Thus, they assume that, just as there is a divisibility into parts in matter, that is, in potency, so there are infinitely many parts in matter, but the actual extension of matter only derives from the extension of a quantity. Accordingly, these people assume that nothing that is one and the same in number can be found in distinct individuals.

26. To the first and to the second arguments to the contrary[450] it must be said that when Aristotle says that if an infinite existed <outside a magnitude, that infinite would be indivisible>, he suggests that the division that is found together with an actual extension only exists in a quantity. Similarly, when he says that magnitude and multitude <are the causes of divisibility and of divisibility into parts>, he means that these are causes of the divisibility that is found together with an actual extension.

27. To the third argument[451] it must be said that different parts of a bodily substance are in different parts of the same continuum. But if quantity is removed from these parts, the parts remain distinct only in potency and not under an actual extension; thus, actual extension in matter only derives from quantity.

28. In reply to the arguments that prove that matter does not have parts that are one under another, to the first argument[452] it must be said that to have parts that are one under another comes about in two ways: either (i) these parts are far and distant one from another, or (ii) these parts are in the same place. In case (i), that is, when one part is placed under another, the argument is correct, and matter does not have parts that are one under another in this way. In case (ii), all the parts are in the same place in the way that Aristotle describes when he says that condensation and rarefaction occur with nothing

[450] See above, nn. 1–2.

[451] See above, n. 3.

[452] See above, n. 5.

partes in unum[a], sic sunt partes materiae simul, quia[b] simul cedunt in unum, et sic non procedit ratio.

29. [c]Secundae rationi dicendum quod illa propositio[d], 'materia prima habet partem sub parte', est aliquo modo vera actualiter, quia haec est actualiter et praesenter vera, 'materia prima habet partem sub parte'. Si autem sic intelligatur, scilicet quod materia prima habet actualiter partem sub parte, sic est falsa. Sed illa: 'materia prima habet partem extra partem', nec est uno sensu vera nec altero. Et sic patet ad [e]secundum contra illam responsionem.

30. [f]Quartae rationi dicendum ad principale quod 'solus actus dividit', hoc est[g] verum in his quae dividuntur actuali multiplicatione sive actuali extensione. Et sic non dividitur materia, nisi potentia solum, sicut habet [h]partes potentia.

31. [i]Quintae rationi dicendum sicut dictum est in principio quaestionum de principiis, super primum argumentum illius quaestionis: 'An sit aliqua materia omnino prima', quia ibi patet per distinctionem ad istud.

32. [j]Sextae rationi dicendum secundum quosdam quod non est sumere minimum.

33. Vel sic: cum ex uno pugillo terrae fiant[k] decem aquae, non potest iterum ex uno pugillo[l] aquae fieri unus terrae, quia, secundum Aristotelem, minimum aquae quae potest in generationem unius pugilli [m]terrae est decem pugilli aquae, et minimum aeris ex quo fiat unus pugillus terrae est centum pugilli aeris, et minimum ignis [n]mille. Et ita patet quod, cum **[O 98vb]** dicit quod ex minimo ignis potest fieri terra, [o]falsum dicit.

[a] unum] et *add.* TC *sed del.* T
[b] quia] eo quod T; quod C
[c] Secundae rationi] ad secundum TC
[d] propositio *om.* TC
[e] secundum ... responsionem] tertium TC
[f] Quartae ... principale] Ad quartum principale (quartam rationem rationem principalem C) dicendum TC
[g] est *suppl. s.l.* O
[h] partes potentia] potentiam O
[i] Quintae rationi] Ad quintum TC
[j] Sextae rationi] ad sextum TC
[k] fiant] fiunt TC
[l] pugillo *om.* TC
[m] terrae ... pugilli *om. hom. sed. suppl. mg.* terrae est decem C
[n] mille... ignis *om. hom. sed corr. mg. in* ex quo T
[o] falsum dicit *scrips. sed corr. mg. in* est mille ignis. Unde cum dicit quod ex minimo ignis potest fieri terra falsum dicit. Est enim determinatum a natura ex quanto ita quod non ex minori possit fieri terra vel aliud elementum T

coming in and nothing going out,[453] but that in condensation the parts of matter merge into one more fully, and in rarefaction they merge into one less fully. In this way, the parts of matter are in the same place because they merge into one through being in the same place. On this interpretation, the argument does not succeed.

29. To the second argument[454] it must be said that the proposition 'prime matter has parts that are one under another' is in some sense actually true, because the proposition 'prime matter has parts that are one under another' is actually and presently true, although if it is understood in the sense that prime matter actually has parts that are one under another, in this sense it is false. But the proposition 'prime matter has parts that are one outside another' is true neither in one sense nor in the other. And with this the reply to the second objection against that reply is clear.

30. To the fourth argument for the main conclusion[455] it must be said that the claim 'only an act divides' is true for the things divided by an actual multiplication or an actual extension. But matter is not divided in this way, but only in potency, in the way in which it has parts in potency.

31. To the fifth argument[456] it must be said as was said at the beginning of the questions about the principles, in connection with the first argument of the question 'Whether there is some matter which is absolutely first',[457] because the reply to the argument in the present question becomes clear with the distinction used for the argument in that question.

32. To the sixth argument[458] it must be said that, according to some people, it is not possible to find a minimum amount.

33. Another reply is the following: when ten handfuls of water come to be from one handful of earth, it is not possible for a handful of earth to come to be once more from one handful of water, because according to Aristotle the minimum amount of water that can produce one handful of earth is ten handfuls of water, and the minimum amount of air from which one handful of earth can come to be is one hundred handfuls of air, and the minimum amount of fire is one thousand handfuls.[459] Thus it is clear that the person in question says something false when he says that earth can come to be from a minimum amount of fire.

[453] Arist. *Phys.* IV.9, 217a26–33.
[454] See above, n. 6.
[455] See above, n. 7.
[456] See above, n. 8.
[457] See above, q. 43.2, nn. 2–6.
[458] See above, n. 9.
[459] Arist. *De gen. et corr.* II.6, 333a20–22.

34. [a]Septimae rationi[b] dicendum sicut dictum fuit in illa quaestione ubi[c] quaeritur an materia prima[d] sit substantia in actu super ultimum argumentum.

35. [e]Octavae rationi dicendum quod [fg]possibile est unam et eandem materiam primam esse sub diversis accidentibus [h]genere et [i]specie similiter diversis, sed eadem materia numero non potest esse sub diversis accidentibus numero eiusdem speciei. Et ratio huius est quia ex esse formae substantialis in materia causantur omnia accidentia, et diversa accidentia genere causantur a diversis esse genere; unde, quia una et eadem materia potest simul et semel esse sub diversis esse specie vel genere, ideo potest una et eadem[j] substantiae pars esse quanta et alba et huiusmodi. Sed quia una pars materiae non potest simul et semel esse sub diversis esse eiusdem speciei, sed[k] una pars substantiae est sub uno esse tantum respectu[l] albedinis, ideo non potest accidens eiusdem speciei numerari in eadem parte materiae.

36. Quod autem materia non habeat partem et partem videtur sic, quia, si pars sit in materia prima, illa ergo pars aut est[m] essentialis aut pars subiectiva aut pars integralis aut pars accidentalis. Sed quod non est pars accidentalis videtur sic, quia[n] in materia non est accidens; ergo non est pars accidentalis, sicut dicimus quod homo albus est pars accidentalis hominis. Nec pars integralis, cum pars integralis non [o]sit alicubi nisi ubi est actualis extensio. Nec est pars subiectiva, quia omnes partes subiectivae numerantur sub aliquo[p] communi ratione alicuius formae omnibus partibus convenientis; sed non est aliqua forma in materia. Nec est pars essentialis, quia tunc componeretur [q]materia ex essentiis, quod falsum est.

[a] Septimae rationi] Ad septimum TC
[b] rationi *scripsimus*; ratione O
[c] ubi] in qua TC
[d] prima *om.* TC
[e] Octavae rationi] ad octavum TC
[f] possibile est *suppl. s.l.* O
[g] possibile … sub] una et eadem materia potest subesse TC
[h] genere … accidentibus *om. hom. sed suppl. mg.* genere et diversis specie sed eadem materia numero non potest subesse diversis accidentibus T
[i] specie similiter diversis] diversis specie C
[j] eadem] species *add. sed del.* O
[k] sed] sicut TC *sed corr. s.l. in* quia T
[l] respectu *suppl. s.l.* O
[m] est] pars *add.* TC
[n] quia *om.* TC
[o] sit … extensio] respondeat nisi actualiter extenso TC
[p] aliquo] aliqua O
[q] materia ex essentiis *suppl. mg.* C

34. To the seventh argument[460] it must be replied as was replied to the last argument of the question that asks whether prime matter is a substance in act.[461]

35. To the eighth argument[462] it must be said that it is possible for one and the same prime matter to be under accidents that are different in genus and also in species, but numerically the same matter cannot be under numerically distinct accidents of the same species. The reason for this is that all accidents causally derive from the being of a substantial form in matter, and accidents different in genus causally derive from beings different in genus; thus, since one and the same matter can be at one and the same time under beings different in species or genus, it follows that one and the same part of a substance can be a quantity, white, and the like. But since one part of matter cannot at one and the same time be under different beings of the same species, but one part of a substance is under only one being with respect to whiteness, it follows that an accident of the same species cannot be numbered in the same part of matter.

36. That matter does not have distinct parts is shown as follows: if there is a part in prime matter, that part is either an essential part or a part that is a subject or an integral part or an accidental part. But that it is not an accidental part is shown by the fact that there is no accident in matter; therefore, there is not an accidental part either, in the way that we say that white man is an accidental part of man. Nor is it an integral part, since an integral part exists only where there is actual extension. Nor is it a part that is a subject, because all parts that are subjects are numbered under some common *ratio* of a form that belongs to all of them; but there is no form in matter. Nor is it an essential part because then matter would be composed of essences, which is false.

[460] See above, n. 10.
[461] See above, q. 48, nn. 18–20.
[462] See above, n. 11.

37. Item, dictum est quod materia habet partem [a]sub parte, hoc est[b] partem sub parte [c]secundum simultatem. Contra: si hoc est verum, tunc 'sub' dicit idem quod 'simul'; sed[d] eorum quae simul sunt non est maior ratio quare unum contineat alterum quam econverso. Et si hoc, aut quaelibet pars continet[e] aliam partem aut [f]nulla pars continet aliam. Sed [g]constat quod quaelibet pars non est continens respectu alterius partis; ergo nulla continet, et ita non habet partem et partem.

<Quaestio 55>

[h]Quaeritur quarto decimo an partibilitas materiae sit prior partibilitate quantitatis et causa eiusdem.

1. Et quod partibilitas [i]primo reperiatur in quantitate videtur sic: ubi actus et potentia sunt circa idem genus, actus simpliciter praecedit potentiam; sed in quantitate est partibilitas actualis et in materia est partibilitas secundum potentiam; partibilitas ergo quantitatis praecedit partibilitatem materiae.

2. Item, ibi prius est partibilitas ubi prius est infinitas; sed quantitati prius competit infinitas, cum dicat Aristoteles quod[j] infinitum quantitati congruat; ergo quantitati prius competit partibilitas.

3. Item, ubi maior est[k] materiae divisibilitas, ibi est minor divisibilitas quantitatis, ut in uno pugillo terrae est tantum de materia quantum in centum aeris et tamen in centum pugillis aeris est maior partibilitas quantitatis, [l]quia ibi est maior quantitas partium, et est ibi minor partibilitas materiae, quia minus est de materia in partibus [m]aeris quam in partibus terrae. Sicut ergo partibilitas quantitatis separatur a partibilitate materiae, ita scilicet quod non est causa illius, sic separabitur partibilitas materiae a partibilitate quantitatis, ita quod non est causa illius.

[a] sub parte *suppl. mg.* T
[b] est] habet *add.* TC
[c] secundum simultatem *suppl. mg.* C
[d] sed *suppl. s.l.* O
[e] continet] contineat TC *sed corr.* T
[f] nulla pars] nullum T; nulla C
[g] constat quod *om.* T
[h] Quaeritur quarto decimo] Consequenter quaeritur T; *om.* C
[i] primo reperiatur] reperiatur prius T
[j] quod] finitum et *add.* C
[k] est] diversitas *add. sed del.* O; *om.* TC *sed suppl. s.l. post* ubi T
[l] quia ... quantitas *om.* TC *sed suppl. mg.* T
[m] aeris ... partibus *om. hom.* C

37. Furthermore, it has been said that matter has parts that are one under another in accordance with their being in the same place. Against this: if this is true, then 'under' signifies the same as 'in the same place'; but things that are in the same place are such that there is no greater reason why one should contain the other rather than the other way around. If this is the case, then either every part contains another part or no part contains another. But it is apparent that it is not the case that every part is a container with respect to another part; therefore, no part contains <another part>, and thus matter does not have distinct parts.

<Question 55>
The fourteenth question is whether matter's divisibility into parts is prior to and a cause of quantity's divisibility into parts.

1. Arguments that divisibility into parts is first found in quantity: if an act and a potency are relative to the same genus, then the act absolutely precedes the potency; but in quantity there is a divisibility into parts in act and in matter there is a divisibility into parts in potency; therefore, quantity's divisibility into parts precedes matter's divisibility into parts.

2. Furthermore, divisibility into parts first exists where infinity first exists; but infinity first belongs to quantity, since Aristotle says that the infinite is appropriate to quantity;[463] therefore, divisibility into parts first belongs to quantity.

3. Furthermore, where there is more divisibility of matter there is less divisibility of a quantity, for example in one handful of earth there is as much matter as in one hundred handfuls of air, but in one hundred handfuls of air there is more divisibility into quantitative parts, because there is a greater quantity of parts, and there is less divisibility into material parts, because there is less matter in the parts of air than in the parts of earth. Therefore, just as quantity's divisibility into parts is separate from matter's divisibility into parts, in such a way that it is not the cause of matter's divisibility into parts, so matter's divisibility into parts will be separate from quantity's divisibility into parts in such a way that it is not the cause of quantity's divisibility into parts.

[463] Arist. *Phys.* I.2, 185a33–34.

4. Item, quia lux est causa coloris in corpore terminato, ideo est ita quod, ubi est maior incorporatio [abc]lucis, ibi est maior apparitio coloris, sicut maior incorporatio est causa albedinis et parva nigredinis. Si[d] ergo partibilitas materiae est causa partibilitatis quantitatis, [e]tunc, ubi maior est partibilitas materiae, erit et maior partibilitas quantitatis; sed hoc est falsum, ut prius patet; ergo [f]et primum.

5. Item, unum non est causa alterius [g]si ipsa duo fuerint sub oppositis dispositionibus. Sed materia habet partem sub parte, ut prius patuit, et quantum habet partem extra partem; et istae condiciones [h]sunt oppositae; ergo partibilitas materiae non est causa partibilitatis quantitatis nec econverso, et ita neutra prior alia.

6. Ad oppositum: quantitas, [i]hoc quod est, egreditur a substantia; et ita condicio quantitatis egreditur a condicione materiae; et ita partibilitas quantitatis est a partibilitate materiae.

7. Item, partibilitas mensurae est a partibilitate mensurati, ut dicit Aristoteles in capitulo de tempore; sed quantitas est mensura materiae; ergo partibilitas quantitatis est a partibilitate materiae.

8. Item, accidens duplicem recipit divisionem, scilicet aut secundum formam aut secundum subiectum, sicut color dividitur divisione formali in album, nigrum [j]et huiusmodi, divisione subiecti dividitur per divisionem eius in quo extenditur. Cum ergo quantitas sit accidens, tunc divisione formali dividetur in continuum et discretum et huiusmodi, et divisione subiecti similiter dividetur; sed materia est subiectum quantitatis; ergo[k] partibilitas quantitatis est a partibilitate materiae.

9. [l]Solutio sive rei veritas est quod alia est partibilitas materiae a partibilitate quantitatis, sicut prius patuit. Unde, si loquamur de partibilitate extendendo nomen partibilitatis ad illud quod habet partes simul, sic est partibilitas primo in materia, et hoc est potentia solum. Et haec partibilitas est causa partibilitatis quantitatis. Si autem loquamur proprie de divisione

[a] lucis … apparitio *om. sed suppl. mg.* (incorporatio *pro* apparitio) O
[b] lucis … incorporatio *om. hom.* T
[c] lucis … primum *corr. mg. in* lucis ibi est maior apparitio coloris quia magna incorporatio lucis est causa albedinis et parva nigredinis. Si ergo partibilitas materiae est causa partibilitatis quantitatis, tunc ubi est maior partibilitas materiae ibi erit maior partibilitas quantitatis; sed hoc est falsum, ut prius patuit; ergo primum T
[d] Si] sicut TC *sed corr.* C
[e] tunc … quantitatis *om. hom.* T
[f] et primum] etc. C
[g] si ipsa duo] si ipsa (*sed suppl. mg.* duo?) T; cum C
[h] sunt oppositae *suppl. s.l.* O
[i] hoc quod est *om.* TC
[j] et huiusmodi *om.* C
[k] ergo] et ita C
[l] Solutio … est] Dicendum TC

4. Furthermore, it is because light is the cause of colour in a limited body that where there is a greater embodiment of light there is a greater presence of colour, for example a greater embodiment <of light> is the cause of whiteness and a small embodiment <is the cause> of blackness. Therefore, if matter's divisibility into parts is the cause of quantity's divisibility into parts, then, where there is more divisibility into parts of matter there will also be more divisibility into the parts of quantity; but this is false, as was made clear before;[464] therefore, the first premise is also false.

5. Furthermore, one thing is not the cause of another if these two things are under opposite dispositions. But matter has parts that are one under another, as was made clear before,[465] and quantity has parts that are one outside another, and these conditions are opposite; therefore, matter's divisibility into parts is not the cause of quantity's divisibility into parts or the other way around, and thus neither of the two is prior to the other.

6. To the contrary: quantity in its essence comes from substance; thus a condition of quantity comes from a condition of matter; therefore, quantity's divisibility into parts comes from matter's divisibility.

7. Furthermore, measure's divisibility into parts comes from the divisibility into parts of the thing measured, as Aristotle says in the chapter on time;[466] but quantity is the measure of matter; therefore, quantity's divisibility into parts comes from matter's divisibility into parts.

8. Furthermore, an accident receives two kinds of division: in form or in subject, for example colour is divided by a formal division into white, black, and the like, and it is divided by a division in subject through the division of the thing in which it is extended. Therefore, since a quantity is an accident, it is divided by a formal division into the continuum, the discrete, and the like, and it is also divided by a division in subject; but matter is the subject of quantity; therefore, quantity's divisibility into parts comes from matter's divisibility into parts.

9. The solution of the problem or the truth of the matter is that matter's divisibility into parts is distinct from quantity's divisibility into parts, as was made clear before.[467] Thus, if we speak of divisibility into parts while extending the name of divisibility into parts to something with parts that exist in the same place, then divisibility into parts first exists in matter, but only in potency. And this kind of divisibility into parts is the cause of quantity's divisibility into parts. But if we speak in the proper sense of the division into parts one of which is outside

[464] See above, n. 3.
[465] See above, q. 54, n. 25, 28.
[466] Arist. *Phys.* IV.12, 220b23–28.
[467] See above, q. 54, n. 25.

quae habet partem extra partem, sic per prius est partibilitas in quantitate, quia substantia actu non extenditur nisi quia quanta.

10. [a]Primae rationi dicendum sicut patet prius in illa quaestione: 'An materia creatur ante formam'.

11. [b]Secundae rationi dicendum quod 'ubi prius[c] congruit etc.', hoc est verum de partibilitate quae fit cum extensione, et non de alia.

12. [d]Tertiae rationi dicendum quod, ubicumque[e] est maior partibilitas quantitatis, ibi est maior partibilitas materiae accidentaliter, scilicet per actualem extensionem; sed talis partibilitas accidentalis materiae causatur a partibilitate materiae primo modo **[O 99ra]** et simpliciter, quia materia per se et primo habet partem sub parte, hoc est omnes partes simul in potentia. Quod autem extendantur partes actualiter, hoc[f] est per accidens.

13. Vel dicendum quod non sequitur, quia substantia uniformiter est causa accidentis; et ideo non oportet quod partibilitas quantitatis possit separari a partibilitate substantiae, licet secundum naturam et essentiam contingat econverso.

14. [g]Istae quaestiones iam disputatae de materia disputantur de materia prout absolute et in se consideratur.

15. [h]Quartae rationi dicendum quod 'ubi est maior incorporatio' etc., hoc est verum, quia incorporatio lucis est causa efficiens coloris; et ita maior incorporatio maiorem facit coloris apparitionem. Sed materia est causa quantitatis non ut efficiens, sed ut subiectum; et ita non arguitur secundum idem [i]genus causae.

16. [j]Ultimae rationi dicendum quod oppositio condicionum non repugnat quin unum possit esse subiectum alterius[k], sicut materia privata et in potentia tantum potest esse subiectum formae, [l]quae[m] simpliciter est in actu et nullo modo privata.

[a] Primae rationi] ad primum argumentum TC
[b] Secundae rationi] ad secundum TC
[c] prius] plus TC
[d] Tertiae rationi] ad tertium TC
[e] ubicumque] ubi TC
[f] hoc] maiori extensione vel minori *praem.* TC
[g] Istae … consideratur *fort. post* modo privata (n. 16) *legendum*; *om.* TC
[h] Quartae rationi] ad quartum TC
[i] genus causae *corr. mg. ex* generatione T
[j] Ultimae rationi] ad ultimum TC
[k] alterius *om.* O
[l] quae simpliciter] quia simplex C
[m] quae] quod O

another, then in this way divisibility into parts first exists in quantity, because the only reason for a substance's being extended in act is that it is a quantity.

10. To the first argument[468] it must be replied in the way made clear earlier in the question 'Whether matter is created before form'.[469]

11. To the second argument[470] it must be said that the claim 'where <infinity> is first appropriate <there divisibility into parts first exists'> is true of the divisibility into parts that is accompanied by extension, and not of the other kind of divisibility into parts.

12. To the third argument[471] it must be said that, wherever there is a greater divisibility into parts of quantity there is accidentally a greater divisibility into parts of matter, that is, in virtue of an actual extension; but this kind of accidental divisibility into parts of matter is caused by matter's divisibility into parts primarily and absolutely, because matter per se and primarily has parts that are one under another, that is, all the parts are in the same place in potency. But that the parts <of matter> are actually extended is accidental.

13. Another reply is that the conclusion does not follow because a substance is uniformly the cause of an accident; therefore, it is not necessary for a quantity's divisibility into parts of a quantity to be separable from a substance's divisibility into parts, although in accordance with nature and essence the converse does occur.

14. The questions already raised about matter[472] concern matter considered absolutely and in itself.

15. To the fourth argument it must be said that the claim 'where there is a greater embodiment <of light there is a greater presence of colour>' is true because the embodiment of light is the efficient cause of colour, so that a greater embodiment produces a greater presence of colour. But matter is the cause of quantity not as its efficient cause but as its subject; therefore, the argument is not about the same genus of cause.

16. To the last argument it must be said that the opposition between the conditions of two things does not make it impossible for one thing to be the subject of the other, in the way that matter under a privation and only in potency can be the subject of a form which is absolutely in act and in no way under a privation.

[468] See above, n. 1.
[469] See above, q. 46, n. 19.
[470] See above, n. 2.
[471] See above, n. 3.
[472] That is, qq. 42–55.

1*. [a]Quaeritur quindecimo de materia in comparatione ad formam, [b]sexto decimo in comparatione ad compositum, [c]septimo decimo in comparatione ad potentias, duodevicesimo[d] in comparatione ad privationem, undevicesimo[e] de privatione in se, vicesimo[f] de forma in se.

<Quaestio 56>
Circa primum primo quaeritur utrum materia et forma sint essentiae simpliciter diversae an[g] sit aliqua natura communis illis.

1. Et quod sint eiusdem essentiae simpliciter, videtur sic: materia et forma plus conveniunt quam substantia et accidens; sed substantia et accidens in tantum conveniunt quod de ipsis dicitur quaedam[h] natura analoga et non pure aequivoca, ut ens; ergo materia et forma, cum plus conveniant, de ipsis dicetur natura aliqua simpliciter univoca. Et ita sunt eiusdem essentiae.

2. Item, omnis forma habet aliquam materiam proximam quae est necessitas respectu ipsius; [i]ergo forma omnino prima habet aliquam materiam quae est necessitas respectu sui. Sed non habet materiam nisi omnino primam; ergo ipsa est necessitas respectu formae primae. Sed materia quae est necessitas respectu formae est eiusdem essentiae cum forma; materia ergo prima et forma prima sunt eiusdem essentiae.

3. Item, in praedicationibus[j] essentialibus praedicatum est idem subiecto; sed[k] haec praedicatio est essentialis, 'materia est substantia', quia 'substantia' dividitur in materiam, formam et compositum; ergo praedicatum est idem subiecto. Et omne praedicatum praedicat formam; ergo forma et materia sunt idem in essentia.

4. Item, non est comparatio nisi in univoco; sed materia et forma comparantur in substantia, quia dicit Aristoteles in septimo[l] *Metaphysicae* quod forma est magis substantia quam materia; ergo substantia dicitur de illis univoce.

5. Item, actus et potentia non diversificant essentiam; sed materia est in potentia et forma in actu; ergo materia et forma non diversificant essentiam, et ita sunt eiusdem essentiae.

[a] Quaeritur quindecimo] consequenter quaeritur TC
[b] sexto decimo] secundo TC
[c] septimo decimo] tertio TC
[d] duodevicesimo] quarto TC
[e] undevicesimo] quinto TC
[f] vicesimo] sexto TC
[g] an] aut TC
[h] quaedam *om.* TC
[i] ergo … sui *om. sed suppl. mg.* ergo et forma omnino prima T
[j] praedicationibus *scripsimus*; praedicatis OTC
[k] sed] et sic C
[l] septimo *scripsimus* (*cf. q. 53, n. 10*); secundo OTC

1*. The fifteenth question[473] is about matter taken in relation to form, the sixteenth is about matter taken in relation to the composite,[474] the seventeenth is about matter taken in relation to its potencies,[475] the eighteenth is about matter taken in relation to privation,[476] the nineteenth is about privation in itself,[477] and the twentieth is about form in itself.[478]

<Question 56>

As to the first issue, the first question is whether matter and form are absolutely different essences or whether there is a nature common to them.

1. Arguments that they have the same essence absolutely: matter and form are more in agreement than are substance and accident; but substance and accident agree to the extent that an analogical and not purely equivocal nature, that is, being, is predicated of them; therefore, since matter and form agree more <than do substance and accident>, an absolutely univocal nature will be predicated of them, and so they have the same essence.

2. Furthermore, every form has some proximate matter that is a necessity for that form itself; therefore, a form that is absolutely first has some matter that is a necessity for it. But the only matter which such a form has is the matter that is absolutely first; therefore, prime matter is a necessity for the first form. But the matter that is a necessity for a form has the same essence as that form; therefore, prime matter and the first form have the same essence.

3. Furthermore, in essential predications the predicate is the same as the subject; but the predication 'matter is substance' is essential, because 'substance' is divided into matter, form, and the composite; therefore, the predicate is the same as the subject. And every predicate predicates a form; therefore, form and matter are the same in essence.

4. Furthermore, a comparison is only found in something univocal; but matter and form are compared with respect to substance, since in *Metaphysics* VII Aristotle says that form is more of a substance than is matter;[479] therefore, substance is said of matter and form univocally.

5. Furthermore, act and potency do not produce an essential difference; but matter is in potency and form is in act; therefore, matter and form do not produce an essential difference, and so they have the same essence.

[473] That is, qq. 56–57.
[474] That is, qq. 58–61.
[475] That is, qq. 62–65.
[476] That is, q. 72.
[477] That is, qq. 66–71.
[478] That is, qq. 73–77.
[479] Arist. *Metaph.* VII.3, 1029a29–30.

6. Ad oppositum sic[a]: nisi materia et forma essent diversae essentiae, non esset realis compositio substantiarum compositarum, quia tunc non essent [b]res diversae ex quibus fieret compositio substantiae.

7. Item, Aristoteles dicit in septimo *Metaphysicae* quod alia est essentia materiae [c]ab essentia cuiuslibet[d] praedicabilis.

8. Item, duo sunt principia secundum rem, scilicet materia et forma, et alia est res unius a re alterius, ut dicit Aristoteles.

9. Item, Aristoteles in secundo huius dividit naturam in materiam et formam; si ergo natura dividitur in haec, non est eadem natura in istis.

10. Item[e], [f]Aristoteles dicit quod forma, finis et efficiens [g]coincidunt, materia [h]autem non coincidit, hoc est forma et finis et efficiens possunt radicari in una natura sive essentia, materia autem cum illis in nulla natura radicatur.

11. [i]Solutio sive rei veritas problematis secundum quosdam talis est, quod materia et forma conveniunt in essentia generali, non autem in essentia speciali.

12. Contra: inter omnes essentias [j]generales non est aliqua essentia generalior quam essentia generis generalissimi; sed non conveniunt in essentia generis generalissimi, quia generalissimum non praedicatur nisi de contentis, et omnia contenta sub genere sunt singularia composita, et materia et forma non sunt composita; ergo non sunt sub generalissimo [k]etc., sed [l]sunt principia eorum quae sunt [m]sub generalissimo.

13. Item, materia et forma sunt ita prima quod nihil est[n] prius eis per naturam; sed si convenirent in aliqua natura communi, illa natura esset prior illis, quod est impossibile; ergo primum.

14. Aliter [o]dicunt quidam quod materia et forma nec sunt simpliciter idem nec simpliciter diversa. Et ratio [p]eorum est quia dicunt quod materia est ens secundum quid tantum propter suum diminutum esse; et ens dividitur in idem et diversum; cum ergo materia participat divisum tantum secundum

[a] sic *om.* TC
[b] res diversae] diversae substantiae C
[c] ab essentia] a natura C
[d] cuiuslibet] alicuius O
[e] Item *om.* O
[f] Aristoteles dicit *corr. s.l. ex* dicit (Aristoteles *post* dicit) T; *inv.* C
[g] coincidunt … efficiens *om. hom.* C
[h] autem non coincidit] quia non coincidunt *scrips. sed corr. mg. in* autem cum nulla illarum coincidit T
[i] Solutio … est] Aliqui sic dicunt (dicunt *suppl. s.l.*) T; dicitur sic (*lectio incerta*) C
[j] generales … in essentia *om.* C
[k] etc., sed] praedicata *scrips. sed del.* O; et etiam C
[l] sunt ... generalissimo *scrips. sed del.* O
[m] sub generalissimo] generalissima C
[n] est *om.* TC *sed suppl. mg. post* prius T
[o] dicunt quidam] dicitur sic TC
[p] eorum est] huius C

6. To the contrary: if matter and form did not have different essences, there would be no real composition in composite substances, because then there would be no distinct things from which the composition of a substance would result.

7. Furthermore, in *Metaphysics* VII Aristotle says that the essence of matter is distinct from the essence of any predicable.[480]

8. Furthermore, there are two principles in reality, that is, matter and form, and the thing corresponding to one of them is distinct from the thing corresponding to the other, as Aristotle says.[481]

9. Furthermore, in *Physics* II Aristotle divides nature into matter and form;[482] therefore, if nature is divided into these things, these things do not have the same nature.

10. Furthermore, Aristotle says that form, end, and efficient cause coincide,[483] but matter does not coincide <with them>, that is, form, end, and efficient cause can be rooted in one nature or essence, but matter is not rooted in any nature together with them.

11. The solution of the problem or the truth of the matter, according to some people, is that matter and form agree in a general essence, but not in a special essence.

12. Against this: among all general essences there is no essence that is more general than that of the most general genus; but matter and form do not agree in the essence of the most general genus, because the most general genus is only predicated of the things contained under it, and all the things contained under a genus are singular composites, whereas matter and form are not composites; therefore, they are not contained under the most general genus, but are principles of the things contained under the most general genus.

13. Furthermore, matter and form are first in such a way that nothing is prior to them by nature; but if they agreed in some common nature, that nature would be prior to them, which is impossible; therefore, the first premise is impossible.

14. Some people have a different view, namely that matter and form are neither absolutely the same nor absolutely different. Their explanation is that matter is only a being in a qualified way because of its diminished being; and being is divided into being the same and being different; therefore, since matter participates in what is divided only in a qualified way, it cannot

[480] Arist. *Metaph.* VII.3, 1029a20–23.
[481] This is probably a reference to Arist. *Phys.* I.7, 190b17–23.
[482] Arist. *Phys.* II.1, 193a28–31.
[483] Arist. *Phys.* II.7, 198a24–27.

quid, non potest participare dividentia simpliciter. Et ita materia nec[a] est simpliciter eadem[b] alicui nec [c]simpliciter diversa.

15. Aliter dicitur [d]secundum quosdam et melius quod nihil unum et idem est commune materiae et formae; verumtamen inter materiam et formam mutua est inclinatio, ita quod ex ipsis fit essentialiter unum. Et ita magis conveniunt quam substantia et accidens, [e]eo quod mutuo inclinantur ad invicem. Accidens autem inclinatur ad substantiam, sed non econverso; et ideo ex ipsis non fit simpliciter unum nisi unum subiecto tantum.

16. Et per hoc patet responsio primae rationi, quia non plus conveniunt quia[f] conveniant in aliquo praedicabili, sed quia materia se tota est in potentia, forma autem se tota est in actu, mutuo dependent ab[g] invicem.

17. [h]Secundae rationi dicendum quod omnis forma habet materiam quae est necessitas; unde materia prima non est[i] necessitas respectu alicuius formae, sed materia cum quibusdam potentiis diminutis est necessitas respectu formae primae quae meretur nomen formae; et haec materia est idem in essentia cum forma.

18. [j]Tertiae rationi dicendum sicut patet prius in illa quaestione: [k]'An materia sit substantia in actu', ubi dicitur qualiter praedicatum dicit[l] formam.

19. [m]Quartae rationi dicendum quod est comparatio dupliciter: aut est comparatio in natura aliqua, et sic non comparantur, aut est comparatio in proprietate. [n]Unde materia et forma comparantur in proprietate, sicut dicimus quod materia magis **[O 99rb]** substat [o]quam forma, et hoc quia proprium est materiae substare, forma autem magis subsistit quam materia, cum formae sit subsistere fluxibilitati materiae et eam terminare. Unde materia substat formis [p]et individuum substat formis, sed aequivoce, quia

[a] nec] non TC
[b] eadem] idem TC
[c] simpliciter diversa] diversum simpliciter T; simpliciter diversum C
[d] secundum quosdam *om.* TC
[e] eo quod] scilicet quod O; quia C
[f] quia] quamvis TC
[g] ab *om.* TC
[h] Secundae rationi] ad secundum TC
[i] est] materia *add.* OT
[j] Tertiae … prius] ad tertium patet TC
[k] An ... actu] utrum materia (materia *om.* C) sit generalissimum TC
[l] dicit] praedicat TC
[m] Quartae rationi] ad quartum TC
[n] Unde ... proprietate *iter. sed corr.* O; *om. hom.* C
[o] quam … substare *iter. sed corr.* T
[p] et … formis *om. hom. sed suppl. mg.* T

participate absolutely in the things that divide. Therefore, matter is neither absolutely the same as something nor absolutely different.

15. Some people have a different and better view, namely that nothing that is one and the same is common to matter and form; there is, however, a reciprocal inclination between matter and form so that something that is essentially one comes to be from them. Thus, matter and form agree more than do substance and accident, because they have a reciprocal inclination one towards the other, whereas an accident has an inclination towards a substance, but not the other way around; therefore, something that is absolutely one comes to be from a substance and an accident only in subject.

16. With this the reply to the first argument[484] is clear; for matter and form do not agree more because they agree in a predicable, but because they are dependent on one another, since matter as a whole is in potency and form as a whole is in act.

17. To the second argument[485] it must be said that every form has some matter that is a necessity <for it>; thus, prime matter is not a necessity for form, but matter together with some diminished potencies is a necessity for the first form that deserves the name of form; and this matter has the same essence as that form.

18. To the third argument[486] it must be replied in the way that was made clear in the question 'Whether matter is a substance in act', where it was explained how a predicate expresses a form.[487]

19. To the fourth argument[488] it must be said that there are two kinds of comparison: either (i) comparison with respect to a nature, and matter and form are not compared in this way or (ii) comparison with respect to a property. Matter and form are compared with respect to a property, as when we say that matter underlies <something> more <properly> than does form, and this because to underlie <something> is proper to matter, whereas form supports <something> more <properly> than does matter, because to support the flowing of matter and to terminate matter is proper to form. Thus, both matter and an individual underlie forms, but in an equivocal way, because an

[484] See above, n. 1.
[485] See above, n. 2.
[486] See above, n. 3.
[487] See above, q. 48, nn. 18–20.
[488] See above, n. 4.

individuum substat formis praedicabilibus, ut communibus, materia autem substat formis [a]constituentibus[b] compositum.

20. [c]Ultimae rationi dicendum quod, quando potentia fit actus, tunc non diversificant essentiam, [d]ut potentia activa et forma non diversificant essentiam; quando autem potentia fit sub forma ut tantum formam deferens, tunc diversificant essentiam, sicut contingit de materia et forma.

<Quaestio 57>

[e]Quia in secunda positione[f] ponitur quod materia est ens secundum quid, ideo[g] quaeritur utrum hoc sit verum.

1. Et videtur quod sic: materia est medium [h]inter simpliciter nihil et aliquid simpliciter; et quod est medium inter ista est illud quod maxime accedit ad non esse, [i]et ita est illud quod est maxime diminutum. Sed propter diminutum esse alicuius dicitur illud[j] ens secundum quid; [k]cum ergo materia in sua substantia sit maxime diminuta, maxime erit ens secundum quid.

2. Item[l], illud[m] dicitur ens secundum quid quod patitur[n] contradictionem sui simpliciter; [o]sed materia prima est ens ita quod patitur contradictionem sui simpliciter. Ens enim quod dicitur de materia compatitur secum quod non est simpliciter ens; materia ergo est ens secundum quid tantum.

3. Item, nos dicimus quod animal mortuum est animal secundum quid et non simpliciter, et tamen est in potentia ut ex ipso generetur animal, ut per putrefactionem; sed materia est ita diminute ens quod ex ipsa in sui substantia non potest aliquid [p]in esse produci; materia ergo multo fortius dicetur ens secundum quid.

[a] constituentibus compositum *corr. mg. ex* quaestionibus compositis T
[b] constituentibus] continentibus *scrips. sed* vel complentibus *add. mg.* O
[c] Ultimae rationi] ad ultimum TC
[d] ut … essentiam *om. hom. sed suppl. mg.* (quia *pro* ut) T
[e] Quia ... ponitur] consequenter quaeritur de (*corr. s.l. ex* in T) secunda positione TC
[f] positione] ratione O; scilicet *add. s.l.* T
[g] ideo *om.* TC
[h] inter simpliciter] inter T; *inv.* C
[i] et ita *om.* C
[j] illud] esse *add. s.l.* O
[k] cum ... erit] ergo materia in sua substantia est maxime C
[l] Item] materia prima *add. sed del.* O
[m] illud *om.* TC *sed suppl. mg.* T
[n] patitur *corr. in* compatitur OT
[o] sed … simpliciter *om. hom. sed suppl. mg.* T
[p] in esse *ex corr. s.l.* T

individual underlies predicable forms, namely common forms, whereas matter underlies the forms that constitute a composite.

20. To the last argument[489] it must be said that, when a potency becomes an act, then the potency and the act do not produce an essential difference, for example, an active potency and a form do not produce an essential difference; but when a potency comes to be under a form, as something that only carries that form, then they do produce an essential difference, as happens in the case of matter and form.

<Question 57>
Since the second position[490] assumes that matter is a being in a qualified way, the question then is whether this is true.

1. Arguments for the positive: matter is something intermediate between what is absolutely nothing and what is absolutely something; and what is intermediate between these things is what is maximally close to non-being, and so it is what is maximally diminished. But something is said to be a being in a qualified way because of its diminished being; therefore, since matter in its substance is maximally diminished, it will maximally be a being in a qualified way.

2. Furthermore, what is absolutely compatible with its own contradiction is said to be a being in a qualified way; but prime matter is a being such that it is absolutely compatible with its own contradiction; for the being said of matter is such that what is absolutely a non-being is compatible with it; therefore, matter is a being only in a qualified way.

3. Furthermore, we say that a dead animal is an animal in a qualified way and not absolutely, even though a dead animal is in potency towards an animal being generated from itself, that is, by putrefaction; but matter is a diminished being in such a way that something cannot be brought into being from it in its own substance; therefore, there is a much stronger reason to hold that matter should be said to be a being in a qualified way.

[489] See above, n. 5.
[490] See above, q. 56, n. 14.

4. Ad oppositum: nullum divisum dividitur in suum secundum quid tamquam in alterum dividentium, ut homo non dividitur in[a] hominem verum et hominem pictum; sed ens dividitur in ens actu et ens potentia; [b]ergo ens potentia non est secundum quid contentum sub ente. Et ita, cum materia sit ens potentia[c], non erit ens secundum quid.

5. [d]Item, nihil quod est secundum quid sub superiore est simpliciter sub inferiore; sed materia est simpliciter sub substantia, quia substantia dividitur in materiam, formam et compositum; ergo materia non est[e] secundum quid sub superiore, ut sub ente, et ita non est ens secundum quid.

6. Item, nihil quod est secundum quid sub superiore constituit realiter aliquod inferius; sed materia [f]est altera pars constituens compositum, quod est simpliciter sub ente; [g]ergo materia non est secundum quid sub ente, et ita non est[h] ens secundum quid.

7. [i]Solutio sive rei veritas problematis est quod materia[j] uno modo est simpliciter ens et alio modo non, quia 'simpliciter' potest privare determinationem. Et sic possumus dicere quod omne quod est, sive sit in potentia sive in actu, est isto modo ens simpliciter. Et sic dividitur ens in ens actu et ens potentia. Et sic recipiunt dividentia praedicationem divisi. Et[k] isto modo est materia ens simpliciter. Unde, si 'secundum quid' privet tale 'simpliciter', sic non est materia ens secundum quid. Alio modo dicitur 'ens simpliciter' ita quod li 'simpliciter' non solum privat determinationem sive condicionem determinationis, sed ponat condicionem completionis. Sic non est materia aliquid ens simpliciter, quia non est aliquid completum; unde, si li[l] 'secundum quid' privet condicionem completionis, sic est materia ens secundum quid, quia non est aliquid completum

8. Unde sumendo 'ens simpliciter[m]' primo modo procedunt omnes rationes probantes quod materia est ens simpliciter.

[a] in *suppl. s.l.* O
[b] ergo ... ente *corr. mg. ex* non est secundum quid contentum sub homine T
[c] potentia *om.* OTC *sed suppl. s.l.* O; *suppl. mg.* in potentia T
[d] Item ... ens secundum quid *om. hom. sed suppl. mg.* T
[e] est *suppl. s.l.* O
[f] est ... constituens] constituit realiter aliquod inferius ut C
[g] ergo ... ente *om. hom.* TC *sed suppl. mg.* ergo non est materia secundum quid sub ente T
[h] est *om.* OC *sed suppl. s.l.* O
[i] Solutio ... est] dicendum TC
[j] materia *om.* TC *sed suppl. s.l.* T
[k] Et] in *add.* O
[l] li *om.* TC
[m] simpliciter *om.* OTC *sed suppl. s.l.* O; *mg.* T

4. To the contrary: nothing that is divided is divided into itself as qualified as if into one of the two things dividing it. For example man is not divided into true man and painted man; but being is divided into being in act and being in potency; therefore, being in potency is not contained under being in a qualified way. Thus, since matter is a being in potency, it will not be a being in a qualified way.

5. Furthermore, nothing that falls in a qualified way under what is superior falls absolutely under what is inferior; but matter falls absolutely under substance, because substance is divided into matter, form, and the composite; therefore, matter does not fall in a qualified way under what is superior, such as being, and thus it is not a being in a qualified way.

6. Furthermore, nothing that falls in a qualified way under what is superior really constitutes something inferior; but matter is one of the two parts that constitute the composite, which falls absolutely under being; therefore, matter does not fall in a qualified way under being, and thus it is not a being in a qualified way.

7. The solution of the problem or the truth of the matter is that matter is in one way absolutely a being and in another way it is not, because (i) 'absolutely' can express the privation of a determination. In accordance with (i), we can say that everything that is, whether it is in potency or in act, is absolutely a being, and that that being is divided into being in act and being in potency, and that the things that divide receive the predication of what is divided. In this sense, matter is absolutely a being. Thus, if 'in a qualified way' is a privation of 'absolutely' understood as in (i), then matter is not a being in a qualified way. (ii) In another way we speak of 'being absolutely' in such a way that 'absolutely' does not only express the privation of a determination or of the condition of a determination, but it also posits the condition of completion. In accordance with (ii), matter is not absolutely a being, because it is not something complete; thus, if 'in a qualified way' expresses the privation of the condition of completion, then matter is a being in a qualified way, because it is not something complete.

8. Thus, the arguments that prove that matter is absolutely a being[491] succeed by taking 'being absolutely' in way (i).

[491] See above, nn. 4–6.

9. [a]Rationibus in oppositum dicendum, et[b] primo primae[c], quod, primo modo sumendo 'ens simpliciter', materia non est medium inter nihil et ens simpliciter, quia est ens simpliciter, sumendo 'ens' ut prius et in sua communitate, prout dividitur in ens actu et ens potentia.

10. [d]Secundae rationi dicendum quod non patitur secum contradictionem entis simpliciter, secundum quod conceditur quod est ens simpliciter[e], quia ens sic simpliciter dividitur etc.

11. [f]Ultimae rationi dicendum quod animal mortuum est animal secundum quid, quia non recipit praedicationem veri animalis nisi secundum similitudinem tantum; [g]est tamen in potentia ut ex ipso fiat animal per putrefactionem, sed non animal eiusdem speciei, sed animal alterius speciei. Sed materia non est in potentia ut fiat ens simpliciter, primo modo sumendo 'ens simpliciter', quia sic est semper ens simpliciter, communiter[h] sumendo 'ens', prout dividitur in ens[i] actu et ens potentia.

<Quaestio 58>

Quaeritur[j] de comparatione materiae ad compositum[k], utrum scilicet[l] materia in sua substantia sit principium compositi.

1. Et quod materia in sua substantia sit principium naturae videtur sic: materia et forma sunt principia per hoc quod unum est in potentia, ut materia, et [m]alterum in actu, ut forma. Sed materia, quanto magis absolvitur a forma, tanto magis est in potentia; sed in sua substantia maxime absolvitur [n]a forma; ergo in sua substantia maxime est in potentia, et ita in sua substantia maxime est principium.

2. Item, prius per naturam advenit forma materiae et facit compositum quam ei adveniat aliqua privatio quae sit principium transmutationis. Et si hoc est verum, tunc materia prius per naturam est principium compositi quam super ipsam addatur aliqua potentia sive aliqua forma incompleta; et ita in sua substantia [o]ante omnem privationem erit principium naturae.

[a] Rationibus in oppositum] responsionibus *scrips. sed corr. mg. in* rationibus T
[b] et *om.* TC
[c] primae] ad primum C
[d] Secundae rationi] ad secundum TC
[e] simpliciter] simpliciter ens *add.* C
[f] Ultimae rationi] ad ultimum TC
[g] est tamen *corr. s.l. ex* quia est T
[h] communiter] scilicet *add. s.l.* O
[i] ens *suppl. s.l.* O
[j] Quaeritur] consequenter *praem.* TC
[k] compositum] et ita *add. sed corr. s.l. in* et primo T; et etiam *add.* C
[l] scilicet *om.* TC
[m] alterum in] reliquum C
[n] a forma *suppl. s.l.* O
[o] ante omnem privationem] sine omni privatione C

9. To the arguments to the contrary, and first to the first argument[492] it must be said that, taking 'being absolutely' in way (i), matter is not something intermediate between nothing and a being absolutely, because it is a being absolutely, taking 'being' as before and in its commonness, in so far as it is divided into being in act and being in potency.

10. To the second argument[493] it must be said that matter is not absolutely compatible with the contradiction of being, in so far as it is conceded that it is a being absolutely, because being absolutely taken in this way is divided <into being in act and being in potency>.

11. To the last argument[494] it must be said that a dead animal is an animal in a qualified way because it only receives the predicate that applies to a genuine animal through similarity; however, it is in potency towards an animal being generated from itself by putrefaction, although not an animal of the same species, but an animal of another species. Matter, however, is not in potency to become a being absolutely, taking 'being absolutely' in way (i), because matter is always a being absolutely, taking 'being' in a broad sense, in so far as 'being' is divided into being in act and being in potency.

<Question 58>

The question is about the comparison of matter to the composite, that is, whether matter in its substance is a principle of the composite.

1. Arguments that matter in its substance is a principle of nature: matter and form are principles in virtue of the fact that one is in potency, that is, matter, and the other in act, that is, form. But the more matter is detached from form, the more it is in potency; but matter in its substance is maximally detached from form; therefore, matter in its substance is maximally in potency, and therefore matter in its substance is maximally a principle.

2. Furthermore, by nature, form comes to be in matter and produces the composite prior to the coming to be in matter of privation, which is the principle of change. If this is true, then matter's being a principle of the composite is prior by nature to the addition of a potency or an incomplete form to it; therefore, matter will be a principle of nature in its substance before every privation.

[492] See above, n. 1.
[493] See above, n. 2.
[494] See above, n. 3.

3. Item, corpora caelestia sunt composita ex materia [a]et forma, et tamen non est ibi aliqua privatio.

4. Item, materia dicitur esse principium per hoc quod inclinatur ad formam; [b]et non inclinatur ad formam nisi quia imperfecta. Sed in sua substantia est maxime imperfecta; ergo in sua substantia erit maxime principium.

5. Ad oppositum: Aristoteles in secundo huius dicit: "Quoniam[c] materia est illud [d]ex quo fit aliud cum insit", hoc est materia est illud ex quo fit forma; sed ex substantia materiae non fit forma, quia non transmutatur[e] ab uno in aliud; materia ergo in sua substantia, hoc est sine potentiis diminutis, non est principium compositi.

6. Item, Aristoteles dicit in primo **[O 99va]** *Physicorum* quod unum subiectum sufficit ad patiendum, sicut materia; sed materia in sua substantia non patitur, quia nulla substantia simplex patitur; ad hoc ergo quod materia patiatur necesse est [f]quod ei superaddantur [g]aliquae potentiae incompletae mediantibus quibus patiatur.

7. Item, illud est principium materiale quod est subiectum in transmutatione naturali; [h]sed materia nuda non est subiectum in transmutatione naturali, quia oportet symbolum manere, et ita non manet nuda substantia materiae. [i]Et si hoc, nuda substantia materiae non est principium.

8. [j]Solutio sive rei veritas quaestionis est per distinctionem, scilicet quod aliquid esse principium substantiae contingit dupliciter: vel principium substantiae in quantum substantia est, et sic [k]prima materia in sua substantia est principium compositi; aut principium substantiae [l]secundum quod substantia transmutabilis est, et sic non est materia in sua substantia principium, sed materia cum suis potentiis diminutis est principium talis substantiae, scilicet[m] transmutabilis. Et hoc quia non est transmutatio in materia nisi quia privationes et appetitus contrarii sunt in materia.

9. Et hoc viso, faciliter respondetur argumentis[n] [o]in contrarium.

[a] et forma *iter. sed corr.* T
[b] et ... formam *om. hom. sed suppl. mg.* (sed *pro* et) T
[c] Quoniam] quod TC
[d] ex ... illud *om. hom. sed suppl. mg.* T
[e] transmutatur] fit transmutatio *add. s.l.* O
[f] quod ei superaddantur] ei addere semper (*corr. s.l. in* superaddere) T; semper quod ei addantur C
[g] aliquae potentiae incompletae] aliquas potentias incompletas T
[h] sed ... naturali *om. hom. sed suppl. mg.* (naturali *om.*) T
[i] Et ... materiae *om. hom. sed suppl. s.l.* ergo T
[j] Solutio ... est] dicendum quaestioni scilicet (quod *pro* scilicet C) TC
[k] prima ... est] potest (substantia *add. sed del.* T) materia in sua substantia esse TC
[l] secundum quod] in quantum T
[m] scilicet] sicut *scrips. sed corr. mg. in* ut T; sic C
[n] argumentis] ad argumenta C
[o] in contrarium *om.* TC

3. Furthermore, the celestial bodies are composed of matter and of form, but there is no privation in them.

4. Furthermore, matter is said to be a principle in virtue of the fact that it has an inclination towards form; and matter only has an inclination towards form because it is imperfect. But matter in its substance is maximally imperfect; therefore, matter in its substance will maximally be a principle.

5. To the contrary: in *Physics* II Aristotle says "Since matter is that from which something else comes to be by inhering in it",[495] that is, matter is that from which a form comes to be; but a form does not come to be from the substance of matter, because the substance of matter is not changed from being one thing to being another; therefore, matter in its substance, that is, without diminished potencies, is not a principle of the composite.

6. Furthermore, in *Physics* I Aristotle says that it is sufficient <to posit> just one passive thing,[496] namely matter; but matter in its substance is not a passive thing, because no simple substance is a passive thing; therefore, in order for matter to be a passive thing it is necessary that some incomplete potencies, by means of which matter is a passive thing, should be added to it.

7. Furthermore, the material principle is the thing that is the subject in a natural change; but bare matter is not the subject in a natural change, because it is necessary for the quality in which the generated and the corrupted thing agree to remain, and thus the bare substance of matter does not remain. If that is the case, the bare substance of matter is not a principle.

8. The solution of the question or the truth of the matter is reached by distinguishing two ways in which something can be a principle of substance: either (i) it can be a principle of substance in so far as it is substance, and in this way prime matter in its substance is a principle of the composite; or (ii) it can be a principle of substance in so far as substance is subject to change, and in this way matter in its substance is not a principle, but matter with its diminished potencies is a principle of this kind of substance, that is, of substance subject to change. The reason for this is that there is only change in matter because privations and contrary appetites exist in matter.

9. With this it is easy to reply to the arguments to the contrary.[497]

[495] The correct reference is Arist. *Phys*. I.9, 192a31–32.
[496] Arist. *Phys*. I.6, 189b18–19.
[497] See above, nn. 1–4.

<Quaestio 59>

Quaeritur[a] an eadem sit essentia primae materiae in his inferioribus corruptibilibus et superioribus incorruptibilibus.

1. Et quod non videtur: dicit Aristoteles in undecimo[b] *Metaphysicae* quod substantia corruptibilis et aeterna non habent commune principium, et ita nec materiam.

2. Item, Aristoteles dicit in septimo, [c]in capitulo de eo quod generatur, quod[d] materia est illud quo[e] res potest esse et non esse; cum[f] aeterna non possunt non esse, tunc in aeternis aut non est materia aut non univoce cum his inferioribus.

3. Diceres sic: quod materia est[g] qua res[h] etc., hoc non est intelligendum de materia in sua substantia, sed de materia sub forma transmutabili.

4. Contra: materia non est qua res potest esse et non esse nisi propter suam imperfectionem; cum ergo materia in sua substantia sit multo magis imperfecta quam secundum quod est sub forma aliqua, [i]sive transmutabili sive intransmutabili, tunc materia in sua substantia magis erit [j]illud quo res potest esse et non esse quam materia sub forma transmutabili aliqua.

5. Item, in primo *Generationum* dicit Aristoteles[k] quod formae debetur actio, passio vero alterius potentiae est[l], ut materiae, et similiter scribitur conversa huius; sed in supracaelestibus nulla est passio; ergo in supracaelestibus nulla est materia.

6. Item, quorum genus idem, et materia eadem; quae ergo non sunt in eodem genere non[m] habent materiam eandem. Sed corruptibile et incorruptibile non conveniunt in genere, cum dicat Aristoteles in fine decimi quod corruptibile et incorruptibile plus differunt quam secundum genus; et ita non conveniunt in genere.

[a] Quaeritur] Consequenter quaeritur TC
[b] undecimo] tertio C
[c] in capitulo] Metaphysicae C
[d] quod] quia O
[e] quo *corr. s.l. in* per quod O; qua TC
[f] cum] igitur *add. s.l.* O
[g] est] res *add. sed del.* O
[h] res] potest esse *add.* C
[i] sive ... intransmutabili] sive transmutabili *sed del.* sive *et suppl. mg.* aut intransmutabili T; sive intransmutabili C
[j] illud quo *corr. s.l. in* illud per quod O; qua T; illud qua C
[k] Aristoteles *om.* TC
[l] est *om.* OTC *sed suppl. s.l.* O; *suppl. s.l. ante* potentiae T
[m] non *suppl. s.l.* O

<Question 59>

The question is whether the essence of prime matter is the same in the corruptible things here below and in the incorruptible things above.

1. Arguments for the negative: in *Metaphysics* XI Aristotle says that a corruptible substance and an eternal substance do not have a common principle,[498] and so they do not have matter in common either.

2. Furthermore, in *Metaphysics* VII in the chapter about what comes to be, Aristotle says that matter is that by which a thing can be and not be;[499] since eternal things cannot not be, then in eternal things either there is no matter or there is matter but not univocally with the things here below.

3. You might say that the claim that matter is that by which a thing <can be and not be> should not be understood of matter in its substance, but of matter under a changeable form.

4. Against this: matter is that by which a thing can be and not be only because of its imperfection; therefore, since matter in its substance is much more imperfect than matter in so far as it is under a form, whether changeable or unchangeable, then matter in its substance will be more properly that by which a thing can be and not be than will matter under some changeable form.

5. Furthermore, in *On Generation* I Aristotle says that action is ascribed to form, whereas passion belongs to another potency,[500] that is, to matter, and he also says the converse of this <that is, that matter belongs to passion>; but there is no passion in celestial things; therefore, there is no matter in celestial things.

6. Furthermore, things of the same genus have the same matter; therefore, things that are not of the same genus do not have the same matter. But what is corruptible and what is incorruptible do not have a common genus, since at the end of *Metaphysics* X Aristotle says that what is corruptible and what is incorruptible differ in more than in genus,[501] and thus they do not have a common genus.

[498] Arist. *Metaph.* XII.2, 1069b24–26.

[499] Arist. *Metaph.* VII.7, 1032a20–22.

[500] The correct reference is Arist. *De gen. et corr.* II.9, 335b29–31.

[501] This is probably a reference to Arist. *Metaph.* X.10, 1058b26–1059a10, although in these passages Aristotle maintains that what is corruptible and what is incorruptible differ in genus instead of in more than genus.

7. Item, Commentator dicit super tertium[a] *Metaphysicae* quod 'corpus' dicitur aequivoce de [b]corruptibili et incorruptibili.

8. Item, Commentator dicit super [c]principium secundi *Metaphysicae* quod 'natura' dicitur aequivoce de supercaelestibus et inferioribus. Sed natura non est nisi materia vel forma. Sed si forma diversificatur, cum [d]ad diversificationem formae diversificetur[e] materia, tunc materia diversificabitur. Si autem natura quae est materia dicatur aequivoce, tunc habeo propositum.

9. Item, nihil est in genere quod non educitur de potentia materiae, quia, sicut in ratione differentiae educuntur de potentia generis, sic res respondens differentiae [f]debet educi de potentia materiae; sed nullum incorruptibile educitur de potentia materiae; ergo incorruptibile non est in genere.

10. Ad oppositum: Aristoteles dicit in principio *Metaphysicae* quod necesse est imaginari[g] materiam [h]in re mota; sed incorruptibilia, ut [i]corpora caelestia, moventur; ergo habent materiam. Sed quod habeant materiam univoce cum his inferioribus videtur, quia, si passio est univoca, et subiectum est univocum. Sed motus est univoce in his et in illis, ut videtur, quia motus caeli est mensura motuum inferiorum, quia, si quaeratur quantus[j] est motus [k]vel quantum durat motus, dicitur quoniam per unum diem vel per duos dies. Sed mensura est unigeneum cum mensurato, ergo in utroque est motus univoce. Et praeterea[l], inferiora et superiora comparantur in motu, quia superiora velocius moventur quam inferiora, et comparatio est in univoco[m]. Item, tempus est univoce mensura motuum superiorum et inferiorum.

11. Item ad principale, dicit Augustinus quod prima causa creavit primordialem materiam, et dicit quod illa[n] materia est eadem in omnibus quae sunt.

12. Item, accidentia reperiuntur univoce hic et ibi[o]; ergo et materia. Est enim quantitas univoce in superioribus et inferioribus, cum una sit scientia de quantitate hic et ibi.

[a] tertium (*lectio incerta* T) OTC *sed add s.l.* et similiter super decimum in fine (*lectio incerta*) O
[b] corruptibili et *corr. s.l. in* corpore corruptibili et corpore O
[c] principium secundi Metaphysicae] secundum Metaphysicae in principio TC
[d] ad diversificationem] adversitatem (*lectio incerta*) T; ad diversitatem C
[e] diversificetur] diversificatur TC *et transp. post* materia T
[f] debet educi] educitur C
[g] imaginari *corr. ex* imaginare O
[h] in re mota *corr. ex* in remota OT; non remota C
[i] corpora caelestia] supracaelestia TC
[j] quantus] qualiter TC *sed corr. s.l.* T
[k] vel … motus *om. hom. sed suppl. mg.* vel quantum durat T
[l] praeterea] etiam TC
[m] univoco] univocis ergo etc. C
[n] illa] una O
[o] ibi] et hoc *add.* C

7. Furthermore, in his commentary on *Metaphysics* III the Commentator says that 'body' is said equivocally of a corruptible body and of an incorruptible body.[502]

8. Furthermore, in his commentary on the beginning of *Metaphysics* II the Commentator says that 'nature' is said equivocally of the celestial things and of the things here below.[503] But a nature is nothing other than matter or form. But if form is diversified, then since matter is diversified following the diversification of form, matter will be diversified. But if the nature that is matter is said equivocally, then I have the intended conclusion.

9. Furthermore, nothing that is not educed from the potency of matter is in a genus, because just as at the rational level differences are educed from the potency of a genus, so the thing corresponding to a difference must be educed from the potency of matter; but no incorruptible thing is educed from the potency of matter; therefore, an incorruptible thing is not in a genus.

10. To the contrary: at the beginning of the *Metaphysics* Aristotle says that it is necessary to imagine matter in a thing subject to motion;[504] but incorruptible things, such as the celestial bodies, are subject to motion; therefore, they have matter. And that they have matter univocally with the things here below is shown because if a passion is univocal, its subject is also univocal. But it seems that motion exists univocally in these two kinds of thing, because the motion of the heavens is the measure of the motions here below, given that if one asks how much a motion is or how long a motion lasts, the reply is that it lasts for one day or for two days. But a measure is homogeneous with the thing measured; therefore, motion exists univocally in both kinds of thing. Moreover, the things here below and those above are compared with respect to motion, because the things above move faster than the things here below, and a comparison is with respect to something univocal. Furthermore, time is univocally the measure of the motions above and of those here below.

11. Furthermore, as to the main claim, Augustine says that the first cause created primordial matter, and he says that that matter is the same in all beings.[505]

12. Furthermore, accidents are found univocally both here and there; therefore, matter is as well; for quantity exists univocally in the things above and in the things here below, since the science dealing with quantity both here and there is one science.

[502] The correct reference is Aver. *In Metaph.* X, t.c. 26, fol. 276va.

[503] Aver. *In Metaph.* II, t.c. 16, fol. 36ra.

[504] Arist. *Metaph.* II.2, 994b25–26.

[505] This is probably a reference to Aug. *De genesi ad litt.* V.5, VIII.20, CSEL 28.1, 146.6–15, 258.25–259.5 or to Ps. Aug. *De mirabilibus sacrae scripturae* I.1, c.1, PL 35, 2151.

13. Item, Aristoteles in secundo *De anima* dicit quod diaphaneitas non est aequivoce in superioribus et[a] inferioribus; et si hoc, materia est eadem.

14. Item, ignis est sub globo lunari ut in sphaera sive in loco suo proprio, et plus convenit cum sphaera lunae quam solo contactu; et non videtur quod plus conveniat nisi quia in materia; ergo etc. Quod autem plus conveniat videtur, quia, si ibi esset terra, solo tactu conveniret; et ignis plus convenit; igitur plus quam solo tactu.

15. Item[b], Aristoteles dicit in undecimo *Metaphysicae*[c] quod tot sunt formae in potentia in materia prima quot sunt rationes causales [d]actu in mente divina sive primi; sed rationes causales tam formarum supercaelestium quam inferiorum sunt in actu in motore primo[e]; ergo tam formae superiorum quam inferiorum sunt in potentia in materia prima. Et ita in supracaelestibus est materia prima.

16. Item, quorum est idem genus, et eadem materia. Sed corruptibile et incorruptibile sunt in eodem genere, ut videtur, quia dividentia sunt in diviso; sed substantia dividitur in corporeum et in[f] incorporeum; ergo incorporeum est in eodem[g] genere cum corporeo, et ita descendendo.

17. Sed diceres sic **[O 99vb]**, scilicet quod dividitur per corporeum et incorporeum, sed sumitur ibi[h] 'incorporeum' pro incorporeo materiali, sicut est sensitiva vel vegetativa.

18. Contra: quod anima rationalis, quod est incorporeum incorruptibile, sit in eodem genere cum corporeo[i] corruptibili videtur sic, quia quod facit simpliciter unum cum aliquo est in eodem genere cum eodem; unde, quia albedo non est simpliciter in eodem genere cum corpore, non facit simpliciter unum cum corpore. [j]Sed anima rationalis facit simpliciter unum [k]cum corpore; ergo simpliciter[l] sunt in eodem genere. [m]Et etiam perfectio et illud cuius est perfectio sunt in eodem genere; sed anima rationalis est ultima perfectio hominis; ergo anima rationalis est in eodem genere cum corpore.

[a] et] in *add.* TC
[b] Item *om.* O
[c] Metaphysicae *om.* TC
[d] actu ... sive] in actu in mente TC
[e] primo] ergo tam formae superiorum quam inferiorum sunt in actu in motore primo *add.* O
[f] in *om.* TC *sed suppl. s.l.* T
[g] eodem *om.* OTC *sed suppl. s.l.* O
[h] ibi] hic TC
[i] corporeo] corpore TC
[j] Sed … corpore *om. hom. sed suppl. mg.* T
[k] cum corpore *om.* C
[l] simpliciter *scrips. sed del.* O
[m] Et etiam] item C

13. Furthermore, in *On the Soul* II Aristotle says that transparency does not exist equivocally in the things above and in those here below;[506] if that is the case, then matter is the same <in the things above and in those here below>.

14. Furthermore, fire is under the lunar globe as in its proper sphere or place, and it agrees with the lunar globe by more than contact alone, and this happens, it seems, only because it agrees in matter; therefore, <the matter of fire is the same as that of the lunar globe>. That fire agrees <with the lunar globe> by more <than contact alone> is shown because if earth were under the lunar globe, it would agree with the lunar globe by contact alone; and fire agrees <with the lunar globe> more <than does earth>; therefore, it agrees <with it> by more than contact alone.

15. Furthermore, in *Metaphysics* XI Aristotle says that as many forms are in potency in prime matter as there are causal principles in act in the divine mind or in the mind of the first being;[507] but the causal principles of the forms both of celestial things and of the things here below are in act in the first mover; therefore, the forms both of the things above and of the things here below are in potency in prime matter. Thus, there is prime matter in celestial things.

16. Furthermore, things that have the same genus have the same matter. But what is corruptible and what is incorruptible are in the same genus, it seems, because the things that divide are in the divided thing; but substance is divided into corporeal and incorporeal; therefore, what is incorporeal is in the same genus as what is corporeal, and so on descending <to the further subdivisions of the genus of substance>.

17. But you might say that in the claim that substance is divided into corporeal and incorporeal, 'incorporeal' is taken for what is material and incorporeal, such as the sensitive or the vegetative soul.

18. Against this: that the rational soul, which is incorruptible and incorporeal, is in the same genus as what is corruptible and corporeal is shown as follows: what produces something that is absolutely one with a thing is in the same genus as that thing; thus, since whiteness is not absolutely in the same genus as body, it does not produce something that is absolutely one with a body. But the rational soul produces something that is absolutely one with a body; therefore, they are absolutely in the same genus. Moreover, a perfection and the thing of which it is a perfection are in the same genus; but the rational soul is the final perfection of man; therefore, the rational soul is in the same genus as body.

[506] Arist. *De anima* II.7, 418b7–9.

[507] The correct reference is Aver. *In Metaph.* XII, t.c. 18, fol. 305va.

19. Diceres quod anima rationalis secundum suam substantiam non est perfectio substantiae, et sic est anima incorruptibilis, et sic non est in genere, sed secundum aliquod suum esse est perfectio substantiae, et secundum illud esse est corruptibile, et secundum illud esse ponitur in eodem praedicamento cum eo cuius est perfectio.

20. [a]Contra: probatio quod anima rationalis in sua substantia et non tantum secundum esse sit perfectio hominis, quia homo est vere res in praedicamento substantiae; ergo sua ultima perfectio erit vere res in eodem praedicamento. Et erit sua ultima perfectio prout res est; [b]et si hoc[c], prout res est, ponetur in eodem praedicamento cum corruptibili[d].

21. Item, illud esse aut est corruptibile aut incorruptibile. Si incorruptibile, habeo propositum, quia secundum illud esse est in eodem genere; ergo corruptibile [e]et incorruptibile erunt in eodem genere. [f]Si autem illud esse sit corruptibile, sed nullum esse corruptibile et accidentale ponit incorruptibile in genere, ergo illud esse nullo modo ponet animam rationalem in genere.

22. [g]Solutio sive rei veritas secundum quosdam huius problematis est quod corruptibilium et incorruptibilium non est eadem materia. Et ipsi[h] dicunt quod duae sunt materiae primae differentes per essentias et est una materia corruptibilium et alia incorruptibilium.

23. Et si obiciatur eis quoniam[i] actus dividit, dicunt quod hoc non est verum in his quae differunt in suis essentiis. Et etiam dicunt quod, si supracaelestia denudarentur[j] ab omnibus formis [k]et potentiis usque ad [l]materiam primam illorum et corruptibilia inferiora similiter denudarentur ab omnibus suis formis usque ad eorum materiam primam, adhuc materia superiorum sive incorruptibilium [m]esset multo nobilior quam materia inferiorum corruptibilium.

[a] Contra: probatio] contra videtur T; item videtur C
[b] et … est *om. hom.* C
[c] hoc] anima *add. s.l.* O
[d] corruptibili] corpore T; corpore *praem. s.l.* O
[e] et incorruptibile erunt] erit C
[f] Si … corruptibile *om. sed suppl. mg.* si corruptibile T
[g] Solutio … est] Quidam (hic *add.* C) dicunt TC
[h] ipsi *om.* TC
[i] quoniam] quod TC
[j] denudarentur] denudantur TC
[k] et potentiis *om.* C
[l] materiam … ad *om. hom.* C
[m] esset multo] esse multo adhuc C

19. You might say that the rational soul in its substance is not the perfection of a substance —and in this way the <rational> soul is incorruptible and is not in a genus— but it is the perfection of a substance in accordance with one aspect of its being, and in accordance with that existence it is corruptible and is placed in the same category as the thing of which it is the perfection.

20. Against this: a proof that the rational soul is the perfection of man in its substance and not only in accordance with its existence is that a man is truly a thing in the category of substance; therefore, his ultimate perfection will truly be a thing in the same category, and that thing will be his ultimate perfection in so far as it is a thing; if that is the case, the rational soul in so far as it is a thing will be placed in the same category as what is corruptible.

21. Furthermore, that being is either corruptible or incorruptible. If it is incorruptible, I have the intended conclusion, because the rational soul is in the same genus in accordance with that being; therefore, what is corruptible and what is incorruptible will be in the same genus. If that being is corruptible, then, since no corruptible and accidental being places what is incorruptible in a genus, it follows that that being will in no way place the rational soul in a genus.

22. The solution of the problem or the truth of the matter, according to some people, is that corruptible and incorruptible things do not have the same matter. They say that there are two kinds of prime matter that are different in essence, one of them belonging to corruptible things and the other to incorruptible things.

23. If it is objected to these people that an act is what divides, they say that this is not true of the things that differ in their essences. They also say that, if the celestial bodies were stripped of all forms and potencies up to their prime matter, and likewise corruptible things here below were stripped of all their forms until prime matter was reached, the matter of the things above, that is, of incorruptible things, would be much nobler than the matter of the corruptible things here below.

24. Aliter dicitur sic, quod eadem in essentia est materia prima omnium, scilicet tam corruptibilium quam incorruptibilium. Et illud magis concordat cum philosophia nostra et [a]etiam cum Augustino quam prima opinio. [b]Tamen magister Willelmus[c] Lupus fuit primae opinionis.

25. Et quia [d]cum illa quaestione in disputando coniungitur alia [e]quaestio, scilicet utrum corruptibile et incorruptibile sint in eodem genere[f], dicendum [g]secundum quosdam quod [h]nullum vere incorruptibile est in eodem praedicamento cum corruptibili. Et hoc quia nihil est in praedicamento sive in genere [i]quod non educitur de potentia materiae; et nullum incorruptibile est de potentia materiae eductum[j].

26. Aliter dicitur et melius, scilicet quod corruptibile et incorruptibile sunt simpliciter in eodem genere, quia substantia dividitur penes corporeum et incorporeum, et[k] incorporeum dividitur, quia quoddam est unibile, ut anima, quoddam[l] non unibile, sicut[m] intelligentia. Si[n] unibile, quoddam[o] est apprehendens, sicut sensitiva, quoddam[p] non apprehendens, sicut[q] vegetativa. Divisio autem substantiae corporeae patet per Porphyrium.

27. Unde argumentis [r]in contrarium dicendum, et primo auctoritatibus.

[a] etiam cum Augustino] est melior C
[b] Tamen … opinionis *om.* TC
[c] Willelmus] W. O
[d] cum illa] in ista T
[e] quaestio, scilicet *suppl. mg.* T
[f] genere] cui *add. s.l. (lectio incerta)* O
[g] secundum quosdam *om.* C
[h] nullum vere *corr. mg. ex* hoc (*lectio incerta*) T
[i] quod non] nisi quod C
[j] eductum] et (et *s.l.*) si sic obiciatur: rationale et (incorruptibile *add. et del.*) irrationale sunt in eodem genere; ergo corruptibile et incorruptibile, dicerent quod non sequitur. Et causa huius est quia corruptibile dicit primam potentiam ad transmutationem [in] intrinsecam rei in genere; cum ergo incorruptibile privat illam (illam *s.l.*) primam potentiam, non potest manere sub eodem genere; rationale autem non dicit primam potentiam (non potest ... potentiam *mg.*) et ideo irrationale, etsi privet potentiam importatam per rationale (ponitur tamen *add. et del.*), ponit tamen potentiam priorem, et ideo sunt in eodem genere. Sed qualiter ergo est una scientia, ut naturalis, de corruptibili et incorruptibili? Dicunt quod naturalis ponit terminum (*corr. mg. ex* tantum) suae considerationis ad magnitudinem sensibilem, in qua conveniunt corruptibile et incorruptibile, et ideo sunt in eodem genere apud ipsum, licet in suis essentiis sint diversorum generum *add.* T
[k] et *suppl. mg.* O
[l] quoddam *corr. ex* quod O
[m] sicut] ut TC
[n] Si] sed non C
[o] quoddam] quaedam O
[p] quoddam] quaedam O; et *praem.* C
[q] sicut] ut TC
[r] in … auctoritatibus *om.* TC

24. Another opinion is that the prime matter of all things, that is, both corruptible and incorruptible, is the same in essence. This opinion is more in accordance with our philosophy, and also with Augustine, than is the first opinion. However, master W. Lupus[508] supported the first opinion.

25. Since the discussion of this question is connected with another question, that is, whether what is corruptible and what is incorruptible are in the same genus, then, according to some people, it must be said that nothing truly incorruptible is in the same category as what is corruptible. The reason for this is that nothing that is not educed from a potency of matter is in a category or in a genus; and no incorruptible thing is educed from a potency of matter.

26. Another and better view is that what is corruptible and what is incorruptible are absolutely in the same genus, because substance is divided into corporeal and incorporeal, and incorporeal substance is divided because one kind, such as the soul, can be united <to something corporeal>, and another kind, such as the intelligences, cannot be united to it. One kind of substance that can be united <to something corporeal>, for instance, the sensitive soul, is able to apprehend, and another kind, for instance, the vegetative soul, is not able to apprehend. The division of corporeal substance is clear from Porphyry.[509]

27. Now, we have to reply to the arguments to the contrary and first to those taken from authoritative texts.

[508] William Lupus (le Lou), M.A., dead after 1253; cf. Emden 1958, 1178–1179; Glorieux 1971, 398.

[509] Porph. *Isag.*, AL I.6–7, 9.18–10.13.

28. [a]Primae rationi dicendum[b] quod, cum dicit Aristoteles quod non est principium commune, ipse intendit de principio proximo et non de principio remoto.

29. [c]Secundae rationi dicendum[d] quod 'materia est qua res potest esse et non esse', hoc non est intelligendum de materia secundum quod materia, [e]sed de materia [f]secundum quod privata est. Unde imperfectio materiae non est causa absoluta qua res potest esse etc., sed materia unde privata est, hoc est[g]: materia cum fuerit sub forma una habente in se contrarietatem, in materia est potentia diminuta respectu alterius formae, et ratione illius potentiae diminutae appetit illam formam; unde materia ratione illius appetitus est illud [h]quo res etc.

30. [i]Tertiae rationi dicendum quod haec propositio: 'materiae debetur passio', habet duos sensus. Vel sic: materia de se patitur, et iste sensus [j]est fallens[k]. Vel sic, [l]quod materia, prout est sub forma contraria vel contrarietatem habente, patitur, et sic materia patitur, hoc est materia est subiectum deferens actionem et passionem quae fit inter illa contraria. Vel sic: 'materiae debetur passio', hoc est intelligendum de materia quae recipit contrarium respectu alterius formae.

31. [m]Quartae rationi dicendum quod Aristoteles, cum dicit quod [n]corruptibile et incorruptibile etc., intendit de incorruptibili quod nec potest recipere non esse nec habet potentiam inclinativam ad non esse, cuiusmodi est prima causa.

32. Vel sic: 'corruptibile et incorruptibile plus differunt quam secundum genus', hoc est quam secundum genus subiectum; et non intendit de genere logico praedicabili.

33. Vel sic: '[o]plus differunt quam secundum genus', hoc est: [p]plus differunt quam aliqua alia quae sunt sub genere. Unde in fine decimi distinguit Aristoteles diversos modos contrariorum et consequenter dicit quod

[a] Primae rationi] circa primum T; circa primam quaestionem C
[b] dicendum] et primo auctoritatibus et primo ad primam dicendum *add.* C
[c] Secundae rationi] ad secundum TC
[d] dicendum] cum dicit C
[e] sed … quod *suppl. mg.* T
[f] secundum quod *suppl. mg.* C
[g] est] in *add. s.l.* O
[h] quo res *corr. in* per quod res O; qua res TC
[i] Tertiae rationi] Ad tertium TC
[j] est fallens] falsus est TC
[k] fallens] vel falsus *add. s.l.* O
[l] quod … et sic *om. hom.* T *sed suppl. mg. et add.* est verum quod T
[m] Quartae rationi] ad quartum TC
[n] corruptibile et incorruptibile] incorruptibile T; corruptibile C
[o] plus differunt *om.* T
[p] plus differunt *om.* T

28. To the first argument[510] it must be said that, when Aristotle says that there is no common principle, he is referring to a proximate principle and not to a remote principle.

29. To the second argument[511] it must be said that the claim ‘matter is that by which a thing can be and not be’ is not to be understood of matter in so far as it is matter, but of matter in so far as it is under a privation. Thus, the imperfection of matter is not an absolute cause of the fact that a thing can be <and not be>, but matter in so far as it is under a privation <is such a cause>, that is, when matter is under one form that has contrariety in itself, there is a diminished potency with respect to another form in matter, and matter has an appetite for that form because of that diminished potency; thus, matter is that by which a thing <can be and not be> because of that appetite.

30. To the third argument[512] it must be said that the proposition ‘passion is ascribed to matter’ has two senses. Either (i) matter is by itself subject to passion, and this sense is fallacious; or (ii) matter in so far as it is under a contrary form or a form having a contrariety is subject to passion, and in this way matter is subject to passion, that is, matter is the subject that carries the action and the passion that occur between those contraries. Alternatively, the claim ‘passion is ascribed to matter’ must be understood of the matter that receives a contrary with respect to another form.

31. To the fourth argument[513] it must be said that when Aristotle says that what is corruptible and what is incorruptible <differ in more than in genus>, he is referring to an incorruptible thing that can neither receive non-being nor has a potency that inclines towards non-being, and the first cause is of this kind.

32. Another reply is that the claim ‘what is corruptible and what is incorruptible differ in more than in genus’ refers to the genus of a real subject and not to a logical and predicable genus.

33. Another reply is that the claim ‘they differ in more than in genus’ means that they differ in more than do the other things that are under a genus. Thus, at the end of *Metaphysics* X Aristotle distinguishes different modes of contraries[514] and he then says that what is corruptible and what is incorruptible

[510] See above, n. 1.
[511] See above, n. 2.
[512] See above, n. 5.
[513] See above, n. 6.
[514] Arist. *Metaph.* X.10, 1058b36–1059a8.

corruptibile et incorruptibile sunt diversa genere, hoc est: [a]diverso modo se habentia respectu generis quam aliqua alia contraria. [b]Unde esse diversa genere, hoc[c] est dupliciter: vel ita quod genus sit diversum respectu ipsorum, et sic non sunt corruptibile et incorruptibile etc.; vel ita quod diversimode se habeant respectu generis quam aliqua alia contraria, et sic est propositio Aristotelis vera. **[O 100ra]**

34. [d]Quintae rationi dicendum quod intendit de aequivocatione naturali sive [e]metaphysice loquendo, sicut dicit quod 'genus' dicitur aequivoce de suis speciebus; logice autem loquendo dicit quod dicitur univoce, quia univoca est praedicatio et logicus considerat modos praedicandi.

35. [f]Sextae rationi dicendum quod est loqui de natura vel secundum communes condiciones naturae, cuiusmodi sunt movere et huiusmodi, et sic dicitur 'natura' univoce de utroque; vel secundum speciales condiciones naturae, cuiusmodi sunt esse[g] transmutabile et esse intransmutabile[h], sic dicitur 'natura' aequivoce.

36. [i]Septimae rationi dicendum quod, ad hoc quod aliquid sit in genere[j] sicut differentia generis vel sicut species in illo genere vel oportet quod educatur[k] de potentia generis et[l] res quae [m]ei respondet educatur de potentia materiae, sicut est de speciebus irrationalibus sive eorum differentiis, vel oportet quod aliquid quod exit de potentia materiae sit in potentia et inclinetur ad illud ut ad suam perfectionem et etiam quod [n]ipsum inclinetur ad alterum ut ad suum perfectibile, ita quod propter mutuam inclinationem ipsorum ex ipsis fiat vere unum. Et sic dicimus quod rationale est differentia generis et in potentia generis isto ultimo modo et res ei respondens illo modo est in potentia materiae.

37. Aliter dicitur super [o]octavum huius, in illa quaestione: 'An intelligentiae sint [p]substantiae in genere'.

[a] diverso modo] e diverso C
[b] Unde ... contraria *iter. hom. sed corr.* T
[c] hoc *om.* TC
[d] Quintae rationi] ad quintum TC
[e] metaphysice loquendo] metaphysica C
[f] Sextae rationi] ad sextum TC
[g] esse *om.* TC *sed suppl. s.l.* T
[h] intransmutabile] et *add.* TC
[i] Septimae rationi] ad septimum TC
[j] genere] oportet quod sit in genere *add.* TC *sed del.* T
[k] educatur *corr. ex* praedicatur O
[l] et] ut TC *sed corr. s.l.* T
[m] ei respondet *suppl. mg.* T
[n] ipsum ... alterum] alterum inclinetur ad ipsum TC
[o] octavum huius] septimum TC
[p] substantiae in genere] in genere substantiae (*lectio incerta* T) TC *sed corr. mg.* T

are different in genus,[515] that is, they are related to a genus differently from other contraries. Accordingly, things can be different in genus in two ways: either (i) in such a way that there is a different genus for each of them, and this is not the case of what is corruptible and of what is incorruptible; or (ii) in such a way that they are related to a genus differently from other contraries, and in this way the proposition of Aristotle is true.

34. To the fifth argument[516] it must be said that the Commentator is referring to natural equivocity or equivocity metaphysically speaking when he says that a genus is said equivocally of its species; but he says that a genus is said univocally logically speaking, because its predication is univocal and a logician considers modes of predication.

35. To the sixth argument[517] it must be said that we can speak of a nature either (i) in accordance with the common conditions of a nature, such as to move and the like, and in this way nature is said univocally of both; or (ii) in accordance with the special conditions of a nature, such as being changeable and being unchangeable, and in this way nature is said equivocally.

36. To the seventh argument[518] it must be said that in order for something to be in a genus as a difference or as a species of that genus either (i) it is necessary for it to be educed from a potency of that genus and for the thing corresponding to it to be educed from a potency of matter, as happens with the species of irrational things or with their differences, or (ii) it is necessary for something that comes from a potency of matter to be in potency and to have an inclination towards that thing as its perfection, and moreover for that thing to have an inclination towards the other thing as its perfectible, in such a way that because of their reciprocal inclination, something truly one will come to be from them. So we say that being rational is a difference of a genus and is in a potency of a genus in this last way and that the thing corresponding to <the difference> 'being rational' is in a potency of matter in the same way.

37. Another reply is given in the commentary on *Physics* VIII in the question 'Whether the intelligences are substances in a genus'.[519]

[515] Arist. *Metaph.* X.10, 1059a9–10.
[516] See above, n. 7.
[517] See above, n. 8.
[518] See above, n. 9.
[519] See below, Book VIII, q. 39.

<Quaestio 60>
Quaeritur[a] utrum materia prima secundum suam substantiam sit subiectum generationis.

1. Et quod sic videtur, quia dicit Aristoteles [b]in libro *De generatione*[c] quod yle est subiectum generationis, et exponitur de materia prima.

2. Item, solum illud est subiectum generationis quod manet in tota transmutatione unius elementi in alterum; sed sola substantia materiae primae manet. Probatio: generetur ignis ex aere. Tunc illud quod prius fuit individuum aeris fit individuum [d]alterius[e] speciei, [f]ut ignis; tunc aut est unum individuum numero post transmutationem et ante aut diversa. Si idem, tunc erit idem individuum numero sub diversis speciebus, sicut sub aere et igne, [g]et hoc est impossibile; ergo primum. Si sunt[h] diversa individua diversarum specierum, sed in quocumque numeratur[i] species, in eodem numeratur genus, ergo quaecumque sunt diversarum specierum, in ipsis numeratur genus sic: si iste homo et iste asinus [j]sint sub diversis speciebus, tunc sunt duae substantiae et duo corpora, et ita numerantur omnia in illis, ita quod nihil praeter materiam manet idem numero in illis. Si ergo eadem materia post transmutationem et ante sit sub diversis speciebus, tunc[k] nulla forma manet eadem numero in utroque. [l]Et si hoc, tunc manet sola[m] substantia materiae primae in transmutatione, et ita est illud subiectum.

3. Item, Aristoteles dicit in secundo *Generationum*[n] quod subiectum generationis non est aliquid in actu sicut subiectum loci mutationis; sed nihil praeter materiam primam est quin sit aliquo modo[o] actu; ergo nihil praeter materiam primam est subiectum generationis.

4. Item, in primo *Generationum* dicit Aristoteles quod, si subiectum generationis est aliquid actu, tunc non differt generatio ab alteratione, quia alteratio est transmutatio in qualitatibus subiecto manente in actu per definitionem alterationis.

[a] Quaeritur] consequenter *praem.* TC
[b] in libro *om.* TC
[c] generatione] et corruptione *add.* T; etc. *add.* C
[d] alterius speciei, ut *om.* C
[e] alterius *suppl. s.l. post* speciei O
[f] ut ignis *om.* OT *sed suppl. s.l.* O; *mg.* T
[g] et ... impossibile] possibile *scrips. sed corr. s.l. in* quod est impossibile O; hoc est impossibile C
[h] sunt] sint T; sint duo C
[i] numeratur] numerantur O
[j] sint sub] sunt sub T; sunt in C
[k] tunc] cum TC
[l] Et ... sola] tantum manet T; et si hoc tantum manet C
[m] sola *suppl. s.l.* O
[n] Generationum] de generatione C
[o] modo] in *add.* TC

<Question 60>
The question is whether prime matter in its substance is the subject of generation.

1. Arguments for the affirmative: in *On Generation* Aristotle says that *yle* is the subject of generation,[520] and this claim is understood to be about prime matter.

2. Furthermore, the subject of generation is only the thing that remains throughout the entire change of one element into another; but only the substance of prime matter remains. Proof: suppose that fire is generated from air. That thing which was at first an individual of air becomes an individual of another species, namely of fire. Then either there is one and the same individual in number before and after that change or there are different individuals. If it is the same individual, then the same individual in number will fall under different species, namely under air and fire, which is impossible; therefore, the first premise is impossible. Alternatively, they are different individuals of different species, but wherever a species is numbered a genus is also numbered, therefore, in all things belonging to different species the genera are numbered as follows: if this man and this ass fall under different species, then they are two substances and two bodies, and in this way all genera are numbered in them, in such a way that nothing in them other than matter remains the same in number. Therefore, if the same matter falls under different species before and after a change, then no form remains the same in number in both <the corrupted and the generated thing>. If that is the case, then only the substance of prime matter remains in a change, and thus it is the subject of the change.

3. Furthermore, in *On Generation* II Aristotle says that the subject of generation is not something in act like the subject of locomotion;[521] but nothing other than prime matter is such that it is not in some way in act; therefore, nothing other than prime matter is the subject of generation.

4. Furthermore, in *On Generation* I Aristotle says that, if the subject of generation is something in act, then generation does not differ from alteration, because alteration is by definition a change in qualities while the subject remains in act.[522]

[520] Arist. *De gen. et corr*. I.4, 320a2–3.
[521] On the substrate of substantial change as something potential, see Arist. *De gen. et corr.* II.4, 329a24–b3 (see also *De gen. et corr.* I.3, 317b18–25); on the substrate of locomotion as something complete, and therefore something in actuality, see Arist. *Phys*. VIII.7, 260b30–33.
[522] Arist. *De gen. et corr*. I.1, 314b1–4 .

5. Ad oppositum: illud est subiectum ex quo generatur aliud; sed ex contrario generatur contrarium, et quod est contrarium est aliquid in actu; ergo aliquid in actu est subiectum generationis.

6. Item, non solum materia manet in transmutatione elementorum, sed et[a] qualitates aliquae, quia dicit Aristoteles quod in transmutatione elementorum facilior est transmutatio ubi symbolum manet, hoc est passio communis.

7. Item, quantitas manet et corporeitas, et hoc quia quod corrumpitur corrumpitur a contrario; sed nec corporeitati nec quantitati est aliquid contrarium; ergo neutrum corrumpitur.

8. Item, in tota transmutatione est actio et passio naturalis; sed sola substantia materiae primae non patitur; ergo sola [b]substantia [c]primae materiae non manet in transmutatione.

9. Item, in tota transmutatione est actio et passio per contactum; sed materia in sua substantia non est tangibilis.

10. Quidam dicunt sic, quod materia in sua substantia non est subiectum generationis, sed si debeat transmutari aer in ignem, quicquid est[d] commune utrique manet, sicut substantialitas et corporeitas et symbolum manet in utroque [e]non transmutatum, nec corrumpitur [f]aliquid nisi contrarietas in ipsis.

11. Sed mihi videtur quod nihil manet idem numero in illis nisi materia, quando individuum [g]aeris fit individuum ignis.

12. Aliter dicitur sic, quod materia in sua substantia non manet, sed materia cum quibusdam potentiis diminutis manet et incompletis. Et dicitur[h] quod cum materia nihil manet quod meretur nomen actus sive formae.

13. Unde primae rationi dicendum quod duplex est subiectum: vel subiectum accidentaliter vel subiectum essentialiter. Subiectum autem essentialiter est illud quod manet et sustinet totam transmutationem, et tale est ipsum subiectum cum quibusdam potentiis diminutis. Et sic est subiectum solum in potentia. Subiectum autem accidentaliter est illud quod primo recipit actionem[i], in quod primo agit[j] transmutans. Et huiusmodi est contrarium et aliquod[k] ens actu.

[a] et *om.* TC
[b] substantia primae materiae] materia prima TC
[c] primae materiae *ex corr.* O
[d] est] in generatione quicquid est *add. sed del.* T
[e] non transmutatum] nec transmutatur T
[f] aliquid nisi contrarietas *om.* C
[g] aeris fit individuum *om. hom.* T *sed post* ignis *suppl mg.* fit ex aere T
[h] dicitur] dicunt TC
[i] actionem] et *add. s.l.* T; et *add.* C
[j] agit *corr. mg. ex* quid est subiectum transmutationis T
[k] aliquod] aliquid TC

5. To the contrary: the subject is that from which another thing is generated; but a contrary is generated from a contrary, and what is contrary is something in act; therefore, something in act is the subject of a generation.

6. Furthermore, not only matter but also some qualities remain in a change of elements, since Aristotle says that in a change of elements the change is easier where the quality in which the elements agree,[523] that is, a common passion, remains.

7. Furthermore, quantity and corporeity remain, because what is corrupted is corrupted by its contrary; but neither corporeity nor quantity have a contrary; therefore, neither of them is corrupted.

8. Furthermore, in an entire change there are a natural action and passion; but it is not only the substance of prime matter that is subject to passion; therefore, it is not only the substance of prime matter that fails to remain in a change.

9. Furthermore, in an entire change there are an action and a passion through contact; but matter in its substance is not tangible.

10. Some people say that matter in its substance is not the subject of generation, but if air has to be transformed into fire, whatever is common to both, such as substantiality and corporeity, remains, and the quality in which they agree remains unchanged in both, and nothing apart from contrariety is corrupted in them.[524]

11. But it seems to me that when an individual of air becomes an individual of fire, nothing other than matter remains numerically the same in those things.

12. Another view is that matter in its substance does not remain but matter with some diminished and incomplete potencies does remain.[525] It is also said that nothing deserving the name of act or form remains along with that matter.

13. Thus, to the first argument[526] it must be said that there are two kinds of subject: either (i) a subject in an accidental way or (ii) a subject in an essential way. (ii) A subject in an essential way is that which remains and sustains the entire change, and the subject itself, as having some diminished potencies, is of this kind. In this way, the subject is only in potency. (i) A subject in an accidental way is that which first receives an action, and is that on which the agent of a change first acts; and a contrary and some sort of being in act are subjects of this kind.

[523] Arist. *De gen. et corr.* II.4, 331a23–25.

[524] On this theory within the English commentary tradition of the *Physics*, see Donati, 2002b, 227–230.

[525] The author of the commentary on *De generatione et corruptione* attributed to Richard Rufus holds a version of the theory according to which the substrate of generation is matter considered in so far as it includes its (active) potencies; see Richard Rufus of Cornwall, *In De gen. et corr.*, 114.171–177.

[526] See above, n. 5.

14. [a]Secundae rationi dicendum quod nulla qualitas manet eadem numero, sed eadem specie; unde non manet symbolum idem numero, sed idem specie.

15. [b]Tertiae rationi dicendum quod nihil corrumpitur corruptione propria nisi quod est contrarium vel habet contrarium; sed aliquid potest corrumpi corruptione[c] aliena sive per accidens[d], sicut forma substantialis corrumpitur corruptione qualitatum contrariarum quae sunt in composito[e]. Unde, si ex aere generetur ignis, caliditas ignis[f] non corrumpit caliditatem [g]in aere, quia non contrariatur eidem, sed corruptione sui subiecti corrumpitur sua caliditas[h].

16. [i]Aliis duabus rationibus dicendum quod materia cum suis **[O 100rb]** potentiis diminutis potest pati et etiam per contactum potest pati. Unde materia cum talibus potentiis [j]est subiectum manens in transmutatione.

<Quaestio 61>
Quaeritur[k] utrum materia in sua substantia possit pati.

1. Et quod sic videtur, quia, quanto aliquid plus habet de materia, tanto magis patitur, et quanto [l]minus, minus; quod ergo est totaliter materia, [m]totaliter patitur et secundum se totum.

2. Item[n], quod totaliter est forma totaliter agit, sicut prima causa; ergo per oppositum quod totaliter est materia totaliter patitur. Et tale est materia in sua substantia.

3. Item, Aristoteles[o] in libro [p]*De generatione* [q]dicit quod materiae debetur pati, agere vero alterius potentiae[r] est.

[a] Secundae rationi] ad secundum TC
[b] Tertiae rationi] ad tertium TC
[c] corruptione] qualitatum *add. sed del.* O
[d] accidens] sed etc. *add. s.l. (lectio incerta)* O
[e] composito *scripsimus*; ipso OTC
[f] ignis] in igne TC
[g] in aere] aeris T
[h] caliditas] etc. *add. s.l.* O
[i] Aliis duabus rationibus] ad alia duo TC
[j] est ... transmutatione *om.* OTC *sed suppl. mg.* O
[k] Quaeritur] consequenter *praem.* TC
[l] minus, minus *scripsimus*; magis magis OTC
[m] totaliter … totum] secundum se totum patitur (et *s.l. add.*) totaliter patitur et tale est materia in sua substantia T; secundum se totum patitur C
[n] Item] item in libro generationum *praem. sed del.* T
[o] Aristoteles *om.* TC
[p] De generatione] generationum T
[q] dicit quod *om.* TC
[r] potentiae] scilicet formae *add. s.l.* O

14. To the second argument[527] it must be said that no quality remains the same in number, but only the same in species; thus, the quality in which the generated and the corrupted thing agree does not remain the same in number, but only the same in species.

15. To the third argument[528] it must be said that nothing is corrupted by its own corruption except what is a contrary or has a contrary; but something can be corrupted by the corruption of something else or accidentally, in the way that a substantial form is corrupted by the corruption of the contrary qualities that exist in the composite. Thus, if fire is generated from air, the heat of fire does not cause a corruption of the heat in air, because it is not contrary to it, but the heat of air is corrupted by the corruption of its subject.

16. To the other two arguments[529] it must be said that matter with its diminished potencies can be subject to passion and can also be subject to passion through contact. Thus, matter as having such potencies is the subject that remains in a change.

<Question 61>

The question is whether matter in its substance can be subject to passion.

1. Arguments for the affirmative: the more matter something has the more it is subject to passion, and the less matter it has, the less it is subject to passion; therefore, what is entirely matter, is entirely and as a whole subject to passion.

2. Furthermore, what is entirely form is entirely active, as is the first cause; therefore, by opposition, what is entirely matter is entirely subject to passion. Matter in its substance is of this kind.

3. Furthermore, in *On Generation* Aristotle says that being subject to passion is ascribed to matter, whereas acting belongs to another power.[530]

[527] See above, n. 6.
[528] See above, n. 7.
[529] See above, nn. 8–9.
[530] Arist. *De gen. et corr.* II.9, 335b29–31.

4. Ad oppositum: quod naturaliter patitur patitur ratione alicuius contrarietatis; sed materiae in sua substantia [a]non est aliquid contrarium; ergo etc.

5. Item, quod patitur patitur per contactum; sed materia prima nec tangit nec tangitur; ergo etc.

6. [b]Solutio sive rei veritas problematis talis est per distinctionem passionis, quod duplex est passio: vel passio quae est receptio vel passio quae est per contrarietatem naturalem; unde materia patitur passione quae est receptio et non passione quae fit per contrarietatem. Et isto modo solvit Aristoteles in tertio[c] *De anima* quaestionem quae quaerit utrum intellectus sit passibilis.

7. [d]Primae rationi dicendum quod haec propositio: 'quanto aliquid plus habet de materia' etc.[e], est falsa si intelligatur quod materia de se sit illud quod patitur. Si autem sic intelligatur, quanto plus habet de materia, hoc est quanto forma magis alligatur materiae, tanto plus patitur, sic[f] est vera. Unde materia non est causa passionis nisi sicut occasio; unde quanto forma magis materialis, magis patitur, et quanto minus materialis, et[g] minus patitur.

8. [h]Secundae rationi dicendum quod, sicut illud quod est totaliter forma[i] agit, sic materia[j] patitur, quia quod totaliter est forma, ut prima causa, totaliter agit communicatione et largiendo suam bonitatem, et ita materia totaliter patitur in recipiendo. Et hoc verum est.

9. Vel sic: non uniformiter arguitur, quia illud quod totaliter [k]est forma, ut causa prima, est extra [l]omne genus, et illud quod est totaliter materia, est in genere. Et etiam[m] non est naturale agens[n].

10. Ad tertium patet[o].

1*. Quaeritur[p] de potentiis materiae. Et primo utrum in materia sint multae potentiae aeque primae aut una simpliciter prima; secundo, utrum sit eadem potentia respectu formarum substantialium et accidentalium; tertio,

[a] non est aliquid] nihil est C
[b] Solutio … passionis] dicendum TC
[c] tertio] secundo TC
[d] Primae rationi] Ad primum TC
[e] etc.] dicendum quod *add.* TC
[f] sic *om.* TC
[g] et *om.* TC
[h] Secundae rationi] ad secundum TC
[i] forma] totaliter *add. s.l.* O
[j] materia] prima *add.* TC *sed del.* T
[k] est forma] agit C
[l] omne genus, et] et omne *scrips. sed ante* et *suppl. s.l.* genus T; genus et omne C
[m] etiam] causa prima *add. s.l.* O
[n] agens] sicut etc. *add. s.l.* O
[o] patet] per solutionem secundi *add. mg.* T; ex praedictis *add.* C
[p] Quaeritur] omnes quaestiones hucusque disputatae de materia disputantur de ea in se considerata *praem.* O; consequenter *praem.* TC

4. To the contrary: what is naturally subject to passion is subject to passion because of some contrariety; but there is nothing contrary to matter in its substance; therefore, <matter in its substance is not subject to passion>.

5. Furthermore, what is subject to passion is subject to passion through contact; but prime matter neither touches nor is touched; therefore, <prime matter is not subject to passion>.

6. The solution of the problem or the truth of the matter is achieved by distinguishing two kinds of passion: (i) the passion that is a reception and (ii) the passion that comes from natural contrariety. Matter is subject to the passion that is a reception and not to the passion that occurs through contrariety. In *On the Soul* III Aristotle gives this solution to the question of whether the intellect is passive.[531]

7. To the first argument[532] it must be said that the proposition 'the more matter something has <the more it is subject to passion>' is false if it is understood in the sense that matter in itself is the thing subject to passion. But if it is understood in the sense that the more matter something has, that is, the more form is joined to matter, the more it is subject to passion, then it is true. Thus, matter is a cause of passion only as an occasion; therefore, the more material a form is, the more it is subject to passion, and the less material it is, the less it is subject to passion.

8. To the second argument[533] it must be said that the way in which matter is subject to passion corresponds to the way in which a thing that is entirely form acts, because what is entirely form, such as the first cause, acts entirely by communicating and bestowing its goodness, and thus matter is entirely subject to passion in receiving. And this is true.

9. Another reply is that it is not argued uniformly, because what is entirely form, such as the first cause, is outside every genus, and what is entirely matter is in a genus. Moreover, what is entirely form is not a natural agent.

10. The reply to the third argument[534] is clear.

1*. Questions about the potencies of matter. The first question is whether in matter there are many potencies that are equally first or one potency that is absolutely first;[535] the second question is whether potency with respect to substantial and accidental forms is the same;[536] the third question is whether

[531] Arist. *De anima* III.4, 429a13–b9.
[532] See above, n. 1.
[533] See above, n. 2.
[534] See above, n. 3.
[535] See below, q. 62.
[536] See below, q. 63.

utrum potentia sit substantialis materiae vel accidentalis; quarto, utrum potentia activa sit forma aliquo modo.

<Quaestio 62>

1. De primo sic. Primo videtur quod plures sint potentiae in materia sic: si in materia esset[a] tantum una potentia, et potentia et actus sunt unum per essentiam, et forma et actus idem, tunc potentia erit idem[b] in essentia cum forma. Si ergo tantum una esset potentia in materia, tunc non possent[c] esse multae formae in materia. Consequens est[d] falsum; ergo et antecedens.

2. Item, sunt multae formae primae. [e]Probatio: ut decem praedicamenta. Sed cuilibet formae primae respondet una potentia prima; et si hoc, tunc decem erunt potentiae primae [f]in materia prima.

3. Ad oppositum: est ordo in formis ita quod quaedam sunt priores et quaedam posteriores; et formae respondet potentia; cum ergo una [g]est forma[h] prima[i], una erit potentia prima.

4. Item, Aristoteles dicit in quinto[j] huius quod ubi non est primum, ibi non est prius, ergo[k], ubi est prius, ibi est primum; sed in potentiis est prius; ergo in potentiis erit aliquid simpliciter primum.

5. Dicendum quod est loqui [l]de potentia dupliciter: aut de potentia eiusdem rationis aut de potentia diversarum rationum. Si de potentia eiusdem rationis, sic non est nisi una potentia prima, quia, si loquamur de potentia passiva tantum, tunc, sicut eadem potentia receptiva [m]dicimus quod doleum vacuum recipit diversos licores, sic dicimus quod, cum materia denudatur omnibus formis per intellectum, una potentia passiva sive receptiva recipit omnes formas. Et sic est una tantum. Si loquamur de potentia activa tantum, cum[n] omnes potentiae activae sint[o] eiusdem rationis, sic est una potentia prima. Et sic est ordo in potentiis sicut est ordo in formis; unde illa potentia

[a] esset] non esset nisi C
[b] idem *suppl. s.l.* O
[c] possent] possunt TC
[d] est *om.* TC
[e] Probatio: ut] quia TC
[f] in materia prima *om.* C
[g] est forma prima] sit forma prima T; forma prima sit C
[h] forma *corr. ex* potentia O
[i] prima] potentia *add.* O
[j] quinto] septimo (*lectio incerta*) T; quarto C
[k] ergo *suppl. s.l.* O
[l] de potentia *suppl. s.l.* T
[m] dicimus quod *scrips. sed del. et suppl. s.l. post* sicut O
[n] cum] non *add. s.l.* O
[o] sint] sunt O

potency is substantial in matter or accidental;[537] the fourth question is whether the active potency is in some way a form.[538]

<Question 62>

1. As to the first question, first the arguments that there are many potencies in matter: if there were only one potency in matter, since potency and act are one in essence, and form and act are the same, then potency would be the same in essence as form. Therefore, if there were only one potency in matter, then there could not be many forms in matter. The consequent is false; therefore, the antecedent is also false.

2. Furthermore, there are many first forms, as the fact that there are ten categories proves. But to every first form there corresponds one first potency; and if this is the case, there will be ten first potencies in prime matter.

3. To the contrary: there is an order in forms so that some are prior and some are posterior; and a potency corresponds to a form; therefore, since there is one first form, there will be one first potency.

4. Furthermore, in *Physics* V Aristotle says that where there is no first there is no prior; [539] therefore, where there is a prior there is a first; but in potencies there is a prior; therefore, in potencies there will be something absolutely first.

5. It must be said that it is possible to speak of a potency in two ways: either (i) of a potency of the same kind or (ii) of potencies of different kinds. In case (i) there is only one first potency, because, (ia) if we speak only of a passive potency, then, just as we say that an empty vessel receives different liquids in virtue of the same receptive potency, so we say that, when matter is stripped of all forms through the intellect, it receives all forms in virtue of one passive or receptive potency. In this way then there is only one potency. (ib) If we speak only of an active potency, then, since all active potencies are of the same kind, then there is only one first potency. In this way, just as there is an order in forms, so there is an order in potencies; thus, the potency relative to

[537] See below, q. 64.
[538] See below, q. 65.
[539] The correct reference is Arist. *Metaph.* V.11, 1018b9–14.

quae est respectu formae primae est potentia prima. Et[a] sicut formarum quaedam sunt priores et quaedam posteriores, [b]sic est in potentiis quod quaedam sunt priores et quaedam posteriores, et hoc per suos respectus ad formas.

6. Si autem loquamur de potentiis diversarum rationum, tunc sunt duae potentiae primae, sicut potentia[c] activa et passiva. Et dicuntur istae potentiae potentiae diversarum rationum, et hoc quia diversa est ratio potentiae activae et passivae. Et sic sunt duae potentiae primae.

<Quaestio 63>
Quaeritur consequenter utrum eadem sit potentia respectu formarum substantialium et accidentalium.

1. Et quod non videtur sic: non est idem esse in potentia ignem et in potentia calidum. [d]Sicut enim differunt ignis, quae est forma substantialis, et calidum, quae est forma accidentalis, sic differunt potentiae respectu eorum. Sed ipsa differunt; ergo et eorum potentiae.

2. Item, actui respondet potentia, et forma et actus idem; ergo formae respondet potentia. Sed forma substantialis et accidentalis differunt; ergo et earum potentiae.

3. Ad oppositum: si esset potentia [e]in materia respectu formarum accidentalium sicut respectu formarum substantialium, tunc, sicut ex materia et forma fit vere unum, sic fieret ex materia et accidente vere unum. Et ratio huius est quia materia habet potentiam [f]per quam inclinatur ad formam[g], et forma inclinatur ad materiam de se; ideo propter ipsam mutuam inclinationem ex ipsis fit vere unum. Si ergo in materia esset **[O 100va]** potentia aliqua respectu formarum accidentalium per quam inclinaretur ad accidens, et accidens de se dependet ad materiam ut ad suum subiectum, tunc propter mutuam dependentiam inter ipsa ex ipsis fieret vere unum, quod scimus esse falsum.

4. [h]Solutio sive rei veritas problematis secundum quosdam talis est, quod in materia prima non est potentia activa ad formam accidentalem, sed est potentia[i] tantum ad formam substantialem.

[a] Et *om.* TC *sed suppl. s.l.* T
[b] sic … posteriores *om. hom.* TC *sed suppl. mg.* T
[c] potentia *om.* TC
[d] Sicut enim] sicut ergo T; sic C
[e] in materia *om.* C
[f] per quam] qua T
[g] formam] et forma inclinatur ad formam *add. sed del.* T
[h] Solutio … est] dicitur sic T; dicendum C
[i] potentia] activa *add.* C

the first form is the first potency. Moreover, just as forms are such that some are prior and some are posterior, so potencies are such that some are prior and some are posterior, and this is because of their relations to forms.

6. (ii) If we speak of potencies of different kinds, then there are two first potencies, that is, an active potency and a passive potency. These potencies are said to be potencies of different kinds, because the *ratio* of an active potency is different from that of a passive potency. In this way then there are two first potencies.

<Question 63>
The next question is whether potency relative to substantial and accidental forms is the same.

1. Arguments for the negative: being fire in potency is not the same as being hot in potency; for just as fire, which is a substantial form, and something's being hot, given that heat is an accidental form, differ, so the potencies relative to them also differ. But fire and something's being hot differ; therefore, their potencies also differ.

2. Furthermore, a potency corresponds to an act, and form and act are the same; therefore, a potency corresponds to a form. But a substantial form and an accidental form differ; therefore, their potencies also differ.

3. To the contrary: if there were a potency relating to accidental forms in matter just as there is a potency relating to substantial forms, then just as something that is truly one comes to be from matter and a <substantial> form, so something that is truly one would come to be from matter and an accident. The reason for this is that matter has a potency in virtue of which it has an inclination towards form, and form by itself has an inclination towards matter; therefore, because of their reciprocal inclination something that is truly one comes to be from them. Therefore, if there were in matter a potency relating to accidental forms in virtue of which matter had an inclination towards an accident, and, given that an accident in itself depends on matter as its subject, then in virtue of their reciprocal dependence something that is truly one would come to be from them, which we know to be false.

4. The solution of the problem or the truth of the matter, according to some people, is that in prime matter there is no active potency <that inclines> towards an accidental form, but only a potency <that inclines> towards a substantial form.

5. [a]Secundum alios veritas problematis talis est, quod una est potentia respectu formarum substantialium et alia respectu formarum accidentalium, et sicut formae sunt diversae, sic et potentiae diversificantur.

6. [b]Secundum alios adhuc potius dicitur sic, quod, sicut forma una et eadem est causa substantiae et accidentis, sed secundum suam substantiam est causa substantiae et secundum suum esse est causa accidentis, sic est de potentia. Unde Aristoteles dicit in decimo quod proportionaliter se habent principia substantiae et accidentis, et alibi quod eadem sunt principia substantiae et accidentis. Et secundum magistros nostros[c] exponitur sic, quod eadem sunt principia secundum rem, diversa tamen secundum rationem; unde illud idem quod est secundum rem principium substantiae per esse suum est principium accidentis. Consimiliter[d] ergo est de potentia, quod illa eadem potentia quae per suam substantiam est causa formae substantialis per suum[e] esse est causa accidentis, sicut eadem potentia quae est ad igneitatem per esse suum potest esse ad caliditatem in eadem materia, dummodo recipiat variationem per esse aliquod.

7. Ad argumenta patet.

<Quaestio 64>
Quaeritur[f] utrum potentia sit essentialis materiae aut accidentalis.

1. Et quod non sit essentialis videtur sic: dicit Commentator [g]super hunc primum librum quod, inducta forma, corrumpitur potentia et manet materia in composito completo; sed si[h] esset essentialis materiae, materia manente, non corrumperetur potentia; potentia ergo materiae est accidentalis.

2. Item, potentia est ad aliquid, materia vero non; ergo potentia materiae non est essentialis sed accidentalis.

3. Ad oppositum: Aristoteles in hoc primo dicit quod materia non est scibilis nisi per analogiam, hoc est per relationem ad formam; sed sua relatio ad formam non est nisi sua potentia; ergo materia scitur per suam potentiam. Sed nihil scitur nisi per suam quidditatem; potentia ergo est quidditas materiae, et si hoc, potentia[i] est eidem essentialis.

[a] Secundum … est] aliter ponitur sic TC
[b] Secundum … sic] aliter dicitur sic (sic *om.* C) et melius TC
[c] nostros *om.* TC
[d] Consimiliter] similiter TC
[e] suum *om.* TC *sed suppl. mg.* T
[f] Quaeritur] consequenter *praem.* TC
[g] super … corrumpitur] quod inter diversatorum (*lectio incerta*) formas privatur C
[h] si] potentia *add. s.l.* O
[i] potentia *om.* TC

5. According to other people the true answer to the problem is that there is one potency relating to substantial forms and another potency relating to accidental forms, and just as these forms are different, so the potencies are also different.

6. According to other people it should rather be said that, just as one and the same form is the cause of a substance and of an accident, but in virtue of its substance it is the cause of a substance and in virtue of its being it is the cause of an accident, so it is in the case of potency. Thus, in *Metaphysics* X Aristotle says that the principles of a substance and of an accident are proportionally related,[540] and he says elsewhere that the principles of a substance and of an accident are the same.[541] According to our masters, this must be explained in the sense that principles are the same in reality, but they are different in *ratio*; thus, the same thing that is in reality the principle of a substance is also the principle of an accident in virtue of its being. Therefore, the same happens in the case of potency, that is, the same potency that in virtue of its substance is the cause of a substantial form is the cause of an accident in virtue of its being, just as the same potency that inclines towards being fire can be a potency <that inclines> towards heat in the same matter in virtue of its being, provided that it is diversified through some being or other.

7. The reply to the arguments[542] is clear.

<Question 64>
The question is whether potency is essential to matter or accidental.

1. Arguments that it is not essential: in his commentary on *Physics* I the Commentator says that, once form has been induced, potency is corrupted and matter remains in the complete composite; but if potency were essential to matter, then since matter remains, the potency would not be corrupted; therefore, potency is accidental to matter.[543]

2. Furthermore, potency <inclines> towards something, whereas matter does not; therefore, potency is not essential but accidental to matter.

3. To the contrary: in *Physics* I Aristotle says that matter is only knowable by analogy,[544] that is, through its relation to form; but matter's relation to form is nothing other than its potency; therefore, matter is known in virtue of its potency. But nothing is known except in virtue of its quiddity; therefore,

[540] The correct reference is Arist. *Metaph.* XII.4, 1070b16–21; XII.5, 1071a29–34.
[541] This is probably a reference to Arist. *Metaph.* XII.5, 1071a34–36.
[542] See above, nn. 1–3.
[543] Aver. *In Phys.* I, t.c. 70, fol. 41rb.
[544] Arist. *Phys.* I.7, 191a7–8.

4. [a]Solutio sive rei veritas problematis talis est, quod duplex est potentia, scilicet activa et passiva. Unde, si quaereretur utrum potentia esset substantia vel[b] accidens, deberet dici quod neque[c] substantia neque accidens, et hoc quia nomen potentiae est nomen virtutis et non est nomen entis vel rei naturalis. Dicendum ergo quod potentia passiva est materiae essentialis, quia non est possibile intelligere materiam praeter illam potentiam receptivam sive passivam. Si[d] loquamur de potentia activa, tunc est dicere quod uno modo est essentialis[e], alio modo non, sicut dicimus quod anima rationalis est essentialis corpori non quia essentia corporis non possit intelligi sine anima, sed quia anima essentialiter inclinatur ad corpus ut ad illud quod perficit et cum quo facit simpliciter unum. Unde consimiliter est ex hac parte, quia dicere potentiam activam essentialem materiae contingit dupliciter: vel quia essentialiter inclinatur et advenit et perficit ipsam, ita quod ex ipsis fiat vere unum, et sic est potentia activa essentialis materiae; vel quia essentia materiae non intelligatur sine illa potentia, et sic non est potentia activa essentialis materiae.

<Quaestio 65>

Quaeritur consequenter utrum potentia activa sit forma aliquo modo.

1. Et videtur quod sic, quia formae est agere sive agere est a forma; sed ipsa potentia activa agit; ergo etc.

2. Ad oppositum: materia per suam potentiam est principium naturale[f]; sed quod est principium materiale[g] est in potentia ad omnes formas, et ita est in potentia respectu formae generis generalissimi. Sed si illa potentia esset forma aliquo modo, tunc mediante illa non esset in potentia ad formam, [h]et hoc quia ipsa esset forma.

3. [i]Solutio sive rei veritas problematis talis est, quod ista potentia activa quae adicitur materiae[j] ad hoc quod recipiat formam est aliquid incompletum quod non meretur aliquo modo nomen formae in actu; est tamen aliquid de essentia formae.

4. [k]Contra: si potentia sit[l] aliquid quod est idem in essentia cum forma, et in materia est potentia activa ad omnes formas, cum illa potentia sit idem

[a] Solutio … est] dicendum TC
[b] vel] aut TC
[c] neque] est *add.* TC
[d] Si *om.* O
[e] essentialis] et *add.* TC
[f] naturale] et materiale constat *add. s.l.* O
[g] materiale *an* naturale *cum* TC *scribendum?*
[h] et … forma *om.* C
[i] Solutio … est] dicendum TC
[j] materiae] me O
[k] Contra … forma (n. 6) *post* extra consimilis speciei (n. 7) TC
[l] sit] fundatur super TC

potency is the quiddity of matter, and if that is the case, potency is essential to matter.

4. The solution of the problem or the truth of the matter is that there are two kinds of potency, that is, active and passive. Thus, if it were asked whether potency is a substance or an accident, it should be replied that it is neither a substance nor an accident, and this is because the name of a potency is the name of a power and not that of a being or of a natural thing. Therefore, it must be said that the passive potency is essential to matter because it is not possible to understand matter apart from that receptive or passive potency. If we speak of the active potency, then we can say that in one way this kind of potency is essential to matter, and in another way it is not, just as we say that the rational soul is essential to the body not because the essence of the body cannot be understood without the soul, but because the soul has an essential inclination towards the body as the thing to which it gives perfection and together with which it produces something that is absolutely one. The present case is similar, because to say that the active potency is essential to matter can be understood in two ways: either (i) because the active potency has an essential inclination towards matter and comes to be in matter and gives it perfection, in such a way that something that is truly one comes to be from them, and in this way the active potency is essential to matter; or (ii) because the essence of matter is not understood without that potency, and in this way the active potency is not essential to matter.

<Question 65>

The next question is whether the active potency is in some way a form.

1. Arguments for the affirmative: acting belongs to a form or acting derives from a form; but the active potency acts; therefore, <the active potency is a form>.

2. To the contrary: matter is a natural principle in virtue of its potency; but what is a material principle is in potency to all forms, and so it is in potency to the form of the most general genus. But if that potency were a form in some way, then matter would not be in potency to form by means of that potency, and this is because that potency itself would be a form.

3. The solution of the problem or the truth of the matter is that the kind of active potency that is added to matter in order for matter to receive a form is something incomplete which in some way does not deserve the name 'form in act'; nonetheless it is something belonging to the essence of a form.

4. Against this: if a potency is something that is the same in essence as a form, and if there is an active potency to all forms in matter, then since that potency is the same in essence as a form, nature would bring nothing new into

[a]in essentia cum forma, tunc natura nihil novum produceret in esse, sed tantum manifestaret quod in materia esset occultum. Et sic non differret [b]forma naturalis a forma incisionis.

5. Item, si dicatur quod materia in sua substantia non est principium, sed sub illa potentia diminuta, tunc, cum illa potentia sit idem in essentia cum forma, compositum ex materia et illa potentia erit [c]idem in essentia cum composito ex materia et forma. Et si hoc non sunt[d] duo principia realiter distincta et essentialiter[e].

6. Ad hoc[f] quaere in *Metaphysica*[g], in illa quaestione: 'Utrum potentia activa sit idem in essentia cum forma'.

7. Unde primae rationi [h]in contrarium dicendum quod duobus modis est actio: vel de se non circumscribendo[i] virtutes caelestes[j], et sic soli formae in actu debetur actio, et sic non agit potentia activa; vel actio per adiutorium[k] alicuius consimilis speciei, et sic non est actio formae in actu, sed formae diminutae. Et sic agit potentia activa, quia tantum agit quando excitatur per agens extra consimilis speciei.

1*. Quaeritur[l] [m]de secundo principio naturae, scilicet de privatione, [n]circa quam quaeruntur sex quaestiones.

<Quaestio 66>
Primo quaeritur utrum [o]privatio dicat solam negationem.

1. Et primo videtur quod non, quia illud quod dicit puram negationem non determinat subiectum[p] susceptibile sui; sed privatio naturalis determinat sibi subiectum, sicut materiam, cum dicat Aristoteles quod necesse est tertium

[a] in essentia *om.* TC
[b] forma naturalis *om.* OTC *sed suppl. s.l.* O
[c] idem in essentia] in esse idem *scrips. sed corr. s.l. in* in essentia idem T
[d] sunt] materia et forma *add. s.l.* O
[e] essentialiter] sicut nec materia et illa potentia *add. s.l.* O
[f] hoc] haec T; *lectio incerta* C
[g] *Metaphysica*] super textum *add.* O
[h] in contrarium *om.* C
[i] circumscribendo] tamen *add. s.l.* O
[j] caelestes *om.* TC
[k] adiutorium] aliquid incorporeum O
[l] Quaeritur] consequenter *praem.* TC
[m] de … scilicet *om. hom.* TC
[n] circa … quaeritur] et primo TC
[o] privatio dicat] dicit T
[p] subiectum] sibi sicut naturam *add. sed del.* T

being, but it would only make manifest what is hidden in matter. Thus, a natural form would not be different from a form produced by sculpting.

5. Furthermore, if it is said that matter is not a principle in its substance, but <as falling> under that diminished potency, then, since that potency is the same in essence as a form, then the composite of matter and of that potency will be the same in essence as the composite of matter and form. If this is the case, matter and form are not two principles that are distinct in reality and essentially.

6. Investigate this topic in the commentary on the *Metaphysics* in the question 'Whether an active potency is the same in essence as a form'.[545]

7. Thus to the first argument to the contrary[546] it must be said that an action can occur in two ways: either (i) by itself, albeit without removing the celestial powers, and this kind of action is only ascribed to a form in act, and an active potency does not act with this kind of action; or (ii) an action occurs through the assistance of something of a similar species, and this kind of action does not belong to a form in act, but to a diminished form. An active potency acts in this way because it only acts when it is stimulated by an external agent of a similar species.

1*. Questions about the second principle of nature, namely privation, about which six questions are raised.[547]

<Question 66>
The first question is whether privation expresses only negation.

1. Arguments for the negative: what expresses pure negation does not determine a subject capable of receiving it; but natural privation determines a subject for itself, namely matter, since Aristotle says that it is necessary for there to be a third principle, that is, a subject for these things, namely form

[545] See Aspall, *Qq. in Metaph.* II, MS C, fol. 68va–b: "De potentia activa... an sit eiusdem essentiae cum forma".
[546] See above, n. 1.
[547] See below, qq. 66–71.

principium esse, **[O 100vb]** scilicet subiectum his, ut formae et suae privationi. Et si hoc, privatio determinat sibi [a]subiectum, sicut materiam.

2. Item, differentia est inter contradictionem et privationem, ut[b] si sic dicam: 'videns, [c]non videns' et 'videns et caecus'; sed[d] utrobique est li 'videns' idem; ergo tota differentia est inter li[e] 'non videns' et li[f] 'caecum'. Sed 'non videns' nullo modo ponit aliquid; ergo 'caecum' aliquo modo ponit aliquid sive est aliquid.

3. Item, Aristoteles dividit privationem in [g]nono, dicens quod quaedam privatio est pura negatio, et est illa quae privat actum et potentiam, alia est privatio quae non est pura negatio et haec privat actum, non tamen potentiam ad actum. [h]Sed privatio quae est principium per accidens non privat potentiam et actum, sed actum tantum, et hoc quia, si utrumque privaret, tunc mediante ipsa materia non appeteret formam; non ergo privat utrumque. Et si hoc, privatio quae est tertium principium non dicit puram negationem.

4. Item, materia appetit formam per privationem et non per puram negationem, [i]quia, si per puram negationem appeteret, cum in materia rei generabilis et corruptibilis sit pura negatio formae caelestis et econverso[j], utrumque appeteret alterum, et ita inter ipsa esset mutuo[k] conversiva generatio, quod falsum est; ergo primum.

5. Item, Aristoteles dicit in primo huius quod materia appetit[l] formam sicut femina virum. Sed femina non propter puram negationem appetit virum, scilicet quia[m] non sit vir, sed tantum propter imperfectionem; sed materia appetit formam propter privationem; privatio ergo non nihil est [n]sed aliquid.

6. Ad oppositum: Aristoteles dicit quod materia est non ens per accidens, quia per privationem, et privatio est non ens per se, quia non per aliud; maior ergo debetur non entitas privationi quam materiae. Sed materia est ita [o]non ens quod non ens actu; ergo privatio erit ita non ens quod neque ens actu neque potentia.

[a] subiectum, sicut] primum ut C
[b] ut *ex corr.* (*lectio incerta*) O; sed TC
[c] non videns *om. hom.* C
[d] sed *om.* TC *sed suppl. s.l.* T
[e] li *om.* TC
[f] li *om.* TC
[g] nono, dicens quod *corr. s.l. et mg. ex* continuo T
[h] Sed … actum *om. hom. sed post* actum tantum *suppl. mg.* sed privatio quae est tertium principiorum non privat utrumque T
[i] quia … negationem *om. hom.* C
[j] econverso] tunc *add.* TC
[k] mutuo] mutua TC
[l] appetit] substantiam *add. sed del.* O
[m] quia] quod ita C
[n] sed aliquid *om.* TC
[o] non ens *om.* C

and its privation.[548] If this is the case, privation determines a subject for itself, namely matter.

2. Furthermore, there is a difference between contradiction and privation, for example when I say 'what sees' and 'what does not see', and 'what sees' and 'what is blind'. In both cases 'what sees' is the same; therefore, the whole difference is between 'what does not see' and 'what is blind'. But 'does not see' in no way puts forward anything <positive>; therefore, 'is blind' does in some way put forward something <positive>, that is, <being blind> is something.

3. Furthermore, in *Metaphysics* IX Aristotle divides privations, saying that one kind of privation is a pure negation, and it is that which takes away act and potency, <and> another kind of privation is that which is not a pure negation, and this takes away the act, but not the potency to the act.[549] But the privation that is a principle accidentally does not take away <both> potency and act, but only act. The reason for this is that, if it took both away, then by means of it matter would not have an appetite for form; therefore, it does not take both away. If this is the case, the privation that is the third principle does not express pure negation.

4. Furthermore, matter has an appetite for form in virtue of privation and not in virtue of pure negation, because, if it had an appetite for form in virtue of pure negation, then, since in the matter of a generable and corruptible thing there is a pure negation of a celestial form and the other way around, each of these two kinds of thing would have an appetite for the other. As a result, there would be a reciprocal generation between them; but this is false; therefore, the first premise is false.

5. Furthermore, in *Physics* I Aristotle says that matter has an appetite for form in the way that a female has an appetite for a male.[550] But a female does not have an appetite for a male in virtue of a pure negation, that is, because she is not a man, but only because of her imperfection; but matter has an appetite for form in virtue of privation. Therefore, privation is not nothing but something.

6. To the contrary: Aristotle says that matter is a non-being accidentally because this comes about in virtue of privation, and privation is a non-being per se, because it is not a non-being in virtue of something else;[551] therefore, a greater non-entity is ascribed to privation than to matter. But matter is a non-being in the sense that it is not a being in act; therefore, privation will be a non-being in the sense that it is neither a being in act nor in potency.

[548] Arist. *Phys.* I.7, 191a4–5, 16–17.

[549] The correct reference is Arist. *Metaph.* X.4, 1055b3–6; see also Arist. *Metaph.* V.22, 1022b22–27.

[550] Arist. *Phys.* I.9, 192a22–23.

[551] Arist. *Phys.* I.9, 192a3–5.

7. Item, privationes posteriores dependent a privatione prima; sed privationes posteriores non solum privant actum, sed etiam[a] potentiam ad actum; [b]ergo privatio prima non solum privabit actum, sed etiam potentiam ad actum. Et si hoc, non differt a pura negatione.

8. [c]Solutio sive rei veritas problematis est talis, quod privatio ponit aliquid, quia dicit idem quod privationem[d] formae cum aptitudine ad eandem; unde illud quod est in ipsa[e] positionis est aptitudo ad formam. Unde duplex est significatum privationis sive dupliciter est considerare privationem: vel ratione eius a quo imponitur vel ratione eius cui imponitur. Si autem consideretur ratione eius a quo imponitur, sic solam carentiam dicit et defectum; et secundum illud solum non est principium naturae. Si consideretur ratione eius cui imponitur, sic nominat illud quod caret habitu, habet tamen aptitudinem ad habitum; et sic est privatio[f] principium, quia [g]nominat carentiam formae cum aptitudine ad eandem[h], sicut 'caecum' non solum nominat carentiam[i], sed[j] illud[k] quod aptum natum est habere visum.

9. Primae rationi in contrarium dicendum quod, eo modo quo materia est non ens per accidens, quia non est ens actu, [l]et hoc per privationem, eo modo privatio est non ens per se, quia privatio per se, hoc est non per aliud, est non ens actu. Nec in plus potest sua conclusio; [m]unde non concludit.

<Quaestio 67>

Habito[n] quod privatio aliquid ponat, quaeritur [o]nunc secundo quid ponat.

1. Et videtur quod non ponat materiam, et hoc quia materia non contrariatur alicui, et privatio contrariatur[p] formae. Non formam, [q]et hoc quia[r] contrariatur formae. Non compositum, quia est principium compositi

[a] etiam *om.* TC
[b] ergo … ad actum *om. hom.* C
[c] Solutio … talis] dicendum TC
[d] privationem] vel carentiam *add. s.l.* O; privatio C
[e] ipsa *scripsimus*; ipso OTC *sed corr.* T; ipso (ipso *scrips. sed del.*) privatione est (est *scrips. sed del.*) loco *add. mg.* T
[f] privatio *om.* TC *sed suppl. mg.* T
[g] nominat carentiam formae] carentiam dicit C
[h] eandem] illam TC
[i] carentiam] visus *add. s.l.* O
[j] sed] et *add. s.l.* O
[k] illud *om.* TC
[l] et hoc per *corr. s.l. ex* erit hoc propter T; sed habet C
[m] unde non concludit *om.* TC *sed* ad illud quaere *add.* T
[n] Habito] consequenter dato TC
[o] nunc secundo *om.* TC
[p] contrariatur] scilicet *add. s.l.* O
[q] et hoc *om.* C
[r] quia] privatio *add. s.l.* O

7. Furthermore, posterior privations depend on the first privation; but posterior privations are not only privations of the act, but also of the potency to the act; therefore, the first privation will not only be the privation of the act, but also of the potency to the act. If that is the case, the first privation does not differ from pure negation.

8. The solution of the problem or the truth of the matter is that privation puts forward something <positive>, because it expresses the privation of form together with an aptitude to <receive> form; thus, the positive element in privation is the aptitude to <receive> form. Accordingly, there are two significates of the term 'privation' or two ways in which it is possible to consider privation: either (i) with regard to that from which the name 'privation' is imposed or (ii) with regard to that on which the name 'privation' is imposed. In case (i) privation only expresses lack and defect; and when considered only in that respect privation is not a principle of nature. In case (ii) privation names what lacks a habit but has an aptitude to a habit; and in this way privation is a principle, because it names the lack of form together with the aptitude to <receive> form, just as 'blind' does not only name the lack <of sight>, but also the natural aptitude to have sight.

9. To the first argument to the contrary[552] it must be said that privation is a non-being per se in a way that corresponds to the way in which matter is a non-being accidentally, because matter is not a being in act, and this is in virtue of privation, whereas privation per se, that is, not in virtue of something else, is a non-being in act. And the conclusion of this argument does not imply anything more; thus, the argument is not conclusive.

<Question 67>
Now that it has been established that privation involves the postulation of something, the second question is what it postulates.

1. It seems that privation involves the postulation of matter because matter is not contrary to anything, whereas privation is contrary to form. Nor does privation involve the postulation of form, because it is contrary to form. Nor does it involve the postulation of the composite, because it is a principle

[552] See above, n. 6.

et prius composito. Nec etiam ponit quidditatem compositi, et hoc quia quidditas se habet per modum positionis, et privatio non. Et etiam quidditas est perfectio individui, et privatio non sic. Nec etiam ponit accidens, et hoc quia est principium substantiae, et nullum accidens est principium substantiae. Et etiam privatio praecedit accidens, quia privatio praecedit esse formae substantialis in materia, cum sit principium substantiae[a], et illud esse est causa accidentis, et quod est prius priore est prius posteriore.

2. [b]Ad oppositum: quod[c] ponat aliquid in genere videtur sic[d], quia in[e] quocumque genere est contrarietas prima, in eodem[f] sunt extrema contrarietatis; sed prima contrarietas est in genere substantiae; ergo et extrema. Cum ergo privatio sit alterum extremum contrarietatis primae, privatio ponet aliquid quod est in genere substantiae.

3. Item, in quocumque genere est unum contrariorum, et reliquum[g]; sed forma, quae est alterum extremum contrarietatis, est in genere substantiae[h]; ergo privatio secundum aliquid sui erit in eodem genere.

4. [i]Solutio sive rei veritas [j]problematis est talis, quod privatio ponit aliquid quod est in genere substantiae, non tamen sicut aliquid ordinatum in genere, sed sicut aliquo modo principium ordinabilis in genere, sicut est de unitate, quae est pura entis indivisio[k]: non est sub quantitate discreta sicut ordinabile in ipsa[l], sed sicut principium ordinabilium illius generis. Unde mihi videtur quod res respondens privationi non est aliquid[m] aliud quam potentia activa diminuta, et hoc quia potentia activa diminuta[n] nominat carentiam formae cum aptitudine ad eandem[o].

[a] substantiae *om.* TC *sed suppl. mg.* T
[b] Ad oppositum *om.* TC
[c] quod] autem *add.* TC
[d] sic *om.* TC
[e] in *suppl. s.l.* O
[f] eodem] genere *add.* TC
[g] reliquum] alterum TC
[h] substantiae *om.* TC *sed suppl. mg.* T
[i] Solutio … quod] ad quod privatio *scrips. sed corr. s.l. in* dicendum quod T; dicendum quod C
[j] problematis est talis *corr. ex* est talis problematis O
[k] indivisio] quod ipsa scilicet *add. s.l.* O; enim *add. s.l.* T
[l] ipsa *scripsimus*; ipso OT; *lectio incerta* C
[m] aliquid *om.* TC
[n] diminuta *om.* TC
[o] eandem] et similiter privatio *add.* O; et patet ad argumenta *add. mg.* T

of the composite and prior to the composite. Nor does it involve the postulation of the quiddity of the composite, because quiddity is something positive, whereas privation is not. Moreover, quiddity is the perfection of an individual, and privation is not. Nor does it involve the postulation of an accident either, because it is a principle of substance, and no accident is a principle of substance. Moreover, privation precedes an accident, because privation precedes the being of substantial form in matter, since it is a principle of substance, and that <substantial> being is the cause of an accident, and what is prior to what is prior is prior to what is posterior.

2. To the contrary: it seems that privation involves the postulation of something belonging to a genus, because in whatever genus the first contrariety is found, the extremes of that contrariety are also found; but the first contrariety is in the genus of substance; therefore, its extremes are as well. Therefore, since privation is the other extreme of the first contrariety, privation involves the postulation of something that belongs to the genus of substance.

3. Furthermore, in whatever genus one of two contraries is found, the other is also found; but form, which is one of the two extremes of the contrariety, is in the genus of substance; therefore, privation in accordance with something belonging to it will be in the same genus.

4. The solution of the problem or the truth of the matter is that privation involves the postulation of something that belongs to the genus of substance, not as something ordered in that genus, but rather as in some way a principle of what can be ordered in that genus, just as unity, which is the pure indivision of a being, is not contained under discrete quantity as something that can be ordered in it, but rather as a principle of the things that can be ordered in that genus. Thus, it seems to me that the thing corresponding to privation is not anything other than a diminished active potency, and this is because a diminished active potency expresses the lack of a form together with an aptitude to <receive> that form.

<Quaestio 68>

[a]Quaeritur nunc tertio an sit multitudo privationum in materia.

1. Et quod non videtur, quia, ubi sunt plura, ibi est distinctio inter illa; sed in materia nulla est distinctio; ergo nulla multitudo, et ita non multitudo privationum[b].

2. Ad oppositum: multiplex est indigentia materiae, et hoc non nisi per privationem; multipliciter ergo **[O 101ra]** sunt privationes in materia, et ita non est ibi tantum una privatio.

3. [c]Solutio sive rei veritas est problematis quod est considerare privationem dupliciter: vel in se et absolute vel in comparatione. Si in se consideretur, sic non sunt plures privationes. Si in comparatione, hoc contingit dupliciter: vel in comparatione ad materiam, et sic adhuc est tantum unica privatio, vel in comparatione ad formam, et tunc, sicut diversificantur formae et sunt plures, sic diversificantur privationes et sunt plures.

4. Mihi[d] tamen videtur quod, sicut est una potentia prima sive una forma prima, sic est una privatio prima et consimiliter[e] diversificatur privatio sicut et potentia.

<Quaestio 69>

[f]Quaeritur quarto an privatio sit principium per accidens.

1. Et quod non videtur sic: privatio non solum dicit carentiam, sed inclinationem[g] ad formam; ergo [h]res aliqua subest mediante qua est principium. Aut ergo materia aut forma aut aliquid materiae aut aliquid formae. Si aliquid materiae, ergo est principium reale sicut materia. Si formae, adhuc erit principium reale sicut forma. Et ita [i]nullo modo est[j] principium per accidens.

2. [k]Ad oppositum: solum illud est principium per se quod manet in composito; [l]sed privatio non manet in composito; ergo privatio non est principium per se.

[a] Quaeritur nunc tertio] consequenter quaeritur TC
[b] privationum] potentiarum OTC *sed* seu privationum *add. mg.* O; *corr. mg. in* privationum T
[c] Solutio ... problematis] Dicendum TC
[d] Mihi] quibusdam C
[e] consimiliter] similiter TC
[f] Quaeritur quarto] consequenter quaeritur TC
[g] inclinationem] vel aptitudinem *add. s.l.* O
[h] res aliqua subest] est res aliqua substantiae C
[i] nullo modo] non TC
[j] est *om. sed suppl. s.l.* erit O
[k] Ad oppositum] contra TC
[l] sed ... composito *om. hom.* OT *sed suppl. mg.* T

<Question 68>
The third question is whether there are many privations in matter.

1. Argument for the negative: where there are many things there is a distinction between them; but in matter there is no distinction; therefore, there is no multitude, and thus there are not many privations.

2. To the contrary: there are many things lacking in matter, and this is only because of privation; therefore, there are many privations in matter, and thus there is not only one privation in it.

3. The solution of the problem or the truth of the matter is that it is possible to consider privation in two ways: either (i) in itself and absolutely or (ii) in a relative way. In case (i), there are not many privations. Case (ii) is twofold: either (iia) in relation to matter, and in this case there is still only one privation, or (iib) in relation to form, and then, just as forms are different and many, so privations are different and many.

4. It seems to me, however, that just as there is only one first potency or first form, so there is only one first privation, and privation is diversified in the same way as potency is.

<Question 69>
The fourth question is whether privation is a principle accidentally.

1. Argument for the negative: privation expresses not only a lack, but also an inclination towards form; therefore, there is a thing belonging to substance by means of which privation is a principle. Therefore, this is either matter or form or something belonging to matter or something belonging to form. If it is something belonging to matter, then it is a real principle just as matter is. If it is something belonging to form, it will still be a real principle just as form is. Thus, privation is in no way a principle accidentally.

2. To the contrary: only what remains in the composite is a principle per se; but privation does not remain in the composite; therefore, privation is not a principle per se.

3. [a]Solutio sive rei veritas problematis est sicut dicit Aristoteles, quod privatio est principium per accidens, quia non manet. Unde privatio dicit carentiam[b] cum aptitudine quadam[c]. Unde in inductione formae destruitur privatio solum in ratione carentiae, et ratione eius est principium per accidens.

4. Et per hoc patet responsio argumento.

<Quaestio 70>

[d]Quaeritur quinto, dato quod sit principium per accidens, ad quod principium [e]per se habet reduci.

1. Et[f] quod non ad formam videtur sic: nullum per accidens reducitur ad aliquid ut ad suum per se quod est destructivum sui; sed, secundum Aristotelem, forma destruit privationem; ergo privatio non reducitur ad formam ut ad suum per se.

2. Item, nullum per accidens est prius suo per se neque[g] exsecutione neque nobilitate; sed privatio est prior[h] exsecutione quam forma; ergo privatio non reducitur ad formam ut ad suum per se.

3. Ad oppositum: forma est habitus respectu privationis. Sed habitus est principium cognoscendi privationem[i], et eadem sunt principia [j]essendi et cognoscendi; ergo habitus est principium essendi privationem[k], et ita privatio reducitur ad habitum sive formam ut ad suum per se.

4. Item, res quae subest privationi plus pertinet formae, secundum quod prius dictum est, quam materiae; ergo potius habet reduci ad formam quam ad materiam.

5. [l]Solutio sive rei veritas problematis talis est secundum quosdam, quod reducitur ad materiam et ad formam, diversimode tamen[m], quia privatio potest considerari vel secundum quod privatio est, et sic se tenet a parte eius ex quo, cum dicat Aristoteles quod ex quo fit aliud est dupliciter, vel ex quo ut [n]materia vel ex quo ut privatio; et sic considerando reducitur ad materiam. Vel potest considerari privatio sub esse completiori vel minus completo, et sic habet reduci ad formam ut ad suum per se.

[a] Solutio … est] dicendum TC
[b] carentiam] formae *add. s.l.* O
[c] quadam] ad ipsam *add. s.l. (lectio incerta)* O
[d] Quaeritur ... accidens] consequenter dato quod sit principium per accidens quaeritur TC
[e] per se *om.* C
[f] Et *om.* TC
[g] neque] nec TC
[h] prior] vatione (*sic*) *add. sed del.* O
[i] privationem] privationis C
[j] essendi et cognoscendi] cognoscendi (privationem *add. sed del.* T) et essendi TC
[k] privationem] privationis T
[l] Solutio … quosdam] quidam dicunt TC
[m] tamen *om.* TC
[n] materia … considerari *om.* C

3. The solution of the problem or the truth of the matter is, as Aristotle says, that privation is a principle accidentally, because it does not remain.[553] Thus, privation expresses a lack together with an aptitude, and in the induction of form privation is destroyed only in so far as it is a lack, and because of this it is a principle accidentally.

4. With this the reply to the argument[554] is clear.

<Question 70>
Granted that privation is a principle accidentally, the fifth question is to what principle per se it must be reduced.

1. Arguments that it must not be reduced to form: nothing accidental is reduced to something as its per se <principle> if that thing is destructive of it; but according to Aristotle, form destroys privation;[555] therefore, privation is not reduced to form as its per se <principle>.

2. Furthermore, nothing that is by accident is prior to its per se <principle> either in production or in nobility; but privation is prior in production to form; therefore, privation is not reduced to form as its per se <principle>.

3. To the contrary: form is a habit with respect to privation. But habit is the principle for knowing privation; and the principles of being and of knowing are the same; therefore, habit is the principle for the being of privation, and so privation is reduced to habit or to form as its per se <principle>.

4. Furthermore, the thing that is the subject of privation has more to do with form than with matter, in accordance with what has been said before;[556] therefore, privation should be reduced to form rather than to matter.

5. The solution of the problem or the truth of the matter, according to some people, is that privation is reduced to matter and form, but in different ways. Privation can be considered either (i) in so far as it is privation, and in this way it concerns the thing from which, because Aristotle says that the thing from which another thing comes to be is of two kinds, either in the way that matter is the thing from which or in the way that privation is the thing from which.[557] It follows that when privation is considered in this way it is reduced to matter. Or (ii) privation can be considered under a more complete or a less complete being, and in this way it must be reduced to form as its per se <principle>.

[553] Arist. *Phys.* I.7, 190b26–27; cf. Aver. *In Phys.* I, t.c. 66, fol. 39va.
[554] See above, n. 1.
[555] Arist. *Phys.* I.9, 192a21–22.
[556] See above, q. 65, n. 3.
[557] Arist. *Phys.* I.7, 189b34–190a31, 190b12–17.

6. Aliter dicitur sic, quod reducitur ad materiam et formam, sed aliter quam prius, quia materia, secundum quod est principium naturale, habet possibilitatem ad formam et etiam[a] actum incompletum. Unde, si respiciamus privationem secundum quod nominat actum incompletum, sic reducitur privatio ad formam, quia actus incompletus est eiusdem essentiae cum forma. Si autem respiciamus privationem prout nominat possibilitatem ad formam, cum possibilitas se teneat a parte materiae, sic reducitur ad materiam.

7. [b]Primae rationi dicendum quod privatio corrumpitur per inductionem formae, et hoc solum ratione suae carentiae, non ratione aptitudinis; unde non simpliciter destruitur, sed secundum [c]aliquid sui.

8. [d]Secundae rationi dicendum quod, sicut dicimus quod non semper reducitur per accidens ad aliquod per se quod sit sui generis, sed ad [e]aliquid quod sit alterius generis, sicut dicit Aristoteles quod casus et fortuna, quae sunt causae per accidens, reducuntur ad naturam et intellectum sicut ad per se et tamen non sunt in eodem genere cum ipsis, similiter non est inconveniens reducere aliquod per accidens ad aliquod per se quod est posterius illo. Unde, licet privatio sit [f]prior forma secundum exsecutionem, est tamen forma prior privatione secundum nobilitatem et naturae intentionem. Unde non[g] est inconveniens privationem reduci ad formam [h]ut ad suum per se secundum naturae intentionem.

<Quaestio 71>

[ij]Quaeritur sexto utrum in supracaelestibus sit aliqua privatio.

1. Et quod non videtur: dicit Commentator super finem primi quod privatio non est nisi in corruptibili; sed supracaelestia sunt incorruptibilia; ergo in ipsis non est privatio.

2. Item, privatio est principium fieri et non existentiae rei; cum ergo in supracaelestibus non sit fieri sive transmutatio aliqua, in ipsis non erit privatio.

3. [k]Ad oppositum: privatio admiscetur cum substantia materiae; sed in supracaelestibus[l] est materia; ergo et privatio.

[a] etiam] ad *add. s.l.* T; ad *add.* C
[b] Primae rationi] ad primum TC
[c] aliquid sui] quid C
[d] Secundae rationi] ad secundum TC
[e] aliquid quod sit] aliquod TC
[f] prior … forma *om. hom. sed suppl. mg.* O
[g] non *om.* TC *sed suppl. s.l.* T
[h] ut … intentionem *om. sed suppl. s.l.* ut ad suum per se T; ad suum per se C
[i] Quaeritur … privationis aut (n. 4) *suppl. mg.* O
[j] Quaeritur sexto] consequenter quaeritur TC
[k] Ad oppositum] contra TC
[l] supracaelestibus] caelestibus TC

6. Another opinion is that privation is reduced to matter and form, but in a way different from that presented above, because matter in so far as it is a natural principle has a potentiality for form and also for an incomplete act. Thus, if we consider privation in so far as it names an incomplete act, in this way privation is reduced to form, because an incomplete act is of the same essence as a form. But if we consider privation in so far as it names a potentiality for form, then since potentiality concerns matter, in this way privation is reduced to matter.

7. To the first argument[558] it must be said that privation is corrupted by the induction of form, and this occurs only in virtue of its lack, not in virtue of its aptitude; thus, it is not absolutely destroyed, but something belonging to it is destroyed.

8. To the second argument[559] it must be said that, just as we say that what is accidental is not always reduced to something existing per se that belongs to its genus, but to something that belongs to another genus—and just as Aristotle says that chance and fortune, which are causes accidentally, are reduced to nature and to the intellect as their per se principles,[560] even though they are not in the same genus as them— similarly, it is not absurd to reduce something accidental to something existing per se that is posterior to it. Thus, although privation is prior to form in production, form, however, is prior to privation in nobility and in the intention of nature. Thus, it is not absurd to reduce privation to form as its per se <principle> in accordance with the intention of nature.

<Question 71>
The sixth question is whether there is privation in celestial things.

1. Arguments for the negative: in his commentary on the end of *Physics* I the Commentator says that privation only exists in corruptible things; but the celestial bodies are incorruptible; therefore, there is no privation in them.[561]

2. Furthermore, privation is a principle of the becoming and not of the existence of a thing; therefore, since there is no becoming or change in celestial things, there will be no privation in them.

3. To the contrary: privation is mixed with the substance of matter; but there is matter in celestial things; therefore, there is privation in them as well.

[558] See above, n. 1.
[559] See above, n. 2.
[560] Arist. *Phys*. II.5, 196b21–24.
[561] Aver. *In Phys*. I, t.c. 79, fol. 45rb.

4. [a]Solutio sive rei veritas problematis secundum quosdam talis est, quod est loqui de privatione dupliciter: aut de privatione in ratione privationis aut de re quae subest[b]. Si loquamur de privatione in ratione privationis, ut in ratione carentiae cum inclinatione ad formam, sic non est in supracaelestibus, et hoc quia ibi est carentia multarum formarum, non tamen est ibi inclinatio ad illas formas, quia sua forma ita complet materiam suam[c] quod non derelinquitur appetitus ad aliam. Si autem loquamur de re privationis[d], quae est eadem cum re formae, sic est dicere aliquo modo quod ibi[e] sit privatio[f].

5. Aliter dicitur sic, quod non est ibi res privationis, quia res privationis est forma incompleta sive potentia aliqua mediante qua inclinatur et appetit formam[g]. Unde dicunt quod forma caeli, nulla potentia diminuta praecedente, unit se suae materiae et[h] eam **[O 101rb]** perficit nullum derelinquendo appetitum ad formam contrariam. Unde non ponunt[i] in supracaelestibus nisi principia existentiae rei, scilicet materiam et formam.

<Quaestio 72>
[j]Quaeritur septimo utrum materia per suam privationem appetat[k] formam.

1. Et videtur quod non, quia privatio contrariatur formae; si ergo materia appeteret formam, et hoc per suam privationem, tunc privatio per se formam appeteret. Et si hoc, contrarium appeteret suum contrarium, quod est falsum; [l]ergo primum.

2. [m]Solutio sive rei veritas problematis talis est uno modo, quod privatio duo dicit, scilicet carentiam formae ratione privationis et aptitudinem ad eandem ratione rei quae subest. Unde, si consideremus privationem secundum quod tantum dicit carentiam et in quantum est in ratione privationis, sic contrariatur formae; et sic non appetit formam nec mediante ipsa solum sic se habente appetit materia formam. Si autem consideremus privationem ratione rei quae subest, sic [n]non contrariatur formae, quia res ipsa quae subest est

[a] Solutio … est] dicitur TC
[b] subest] ei *add. s.l.* O
[c] suam *om.* OTC *sed suppl. s.l.* O; *suppl. s.l. ante* materiam T
[d] privationis *om.* TC *sed suppl. mg.* T
[e] ibi *om.* OT *sed. suppl. s.l.* O
[f] privatio] in supracaelestibus *add.* T
[g] formam] incompletam *add.* O
[h] et *corr. s.l. ex* sed O
[i] ponunt *lectio dubia* O
[j] Quaeritur septimo] consequenter quaeritur TC
[k] appetat] suam *add.* TC
[l] ergo primum *om.* C
[m] Solutio … modo] dicitur TC
[n] non … subest *scrips.* (quae quidem res *pro* quia res ipsa) *sed del.* T

4. The solution of the problem or the truth of the matter, according to some people, is that we can speak of privation in two ways: either (i) of privation in so far as it is privation or (ii) of the thing subject to privation. Privation understood in way (i), that is, in so far as it is a lack together with an inclination towards form, does not exist in celestial things, because in these things there is a lack of many forms, but there is no inclination towards those forms, given that the form of a celestial thing completes matter in such a way that it does not leave an appetite for another form. In way (ii), that is, if we speak of the thing belonging to privation, which is the same as the thing belonging to form, it is possible to say that there is in some way privation in celestial things.

5. Another opinion is that in celestial things the thing belonging to privation is not found because the thing belonging to privation is an incomplete form or a potency by means of which something has an inclination and an appetite for form. Thus, some people say that the form of the heavens, without any preceding diminished potency, joins itself to matter and gives perfection to matter without leaving any appetite for a contrary form. Thus, they only posit the principles of the existence of a thing, that is, matter and form, in the celestial bodies.

<Question 72>
The seventh question is whether matter has an appetite for form in virtue of its privation.

1. Arguments for the negative: privation is contrary to form; therefore, if matter had an appetite for form in virtue of its privation, then privation would per se have an appetite for form. If that were the case, a contrary would have an appetite for its contrary, which is false; therefore, the first premise is false.

2. One view about the solution of the problem or the truth of the matter is that 'privation' expresses two things, that is, (i) the lack of a form by reason of privation and (ii) an aptitude for that form by reason of the thing that is subject to privation. Thus, if we consider privation in so far as it only expresses a lack and in so far as it is privation, then in this way it is contrary to form; and in this way it does not involve an appetite for form nor does matter have an appetite for form by means of privation that is only considered in this way. But if we consider privation by reason of the thing that is its subject, then in this way it is not contrary to form, because the thing that is its subject is

aliquid incompletum quod est idem in essentia cum forma; et sic appetit formam [a]et ratione eius illo modo appetit materia formam.

3. Aliter dicitur sic, quod, cum dicit: "si materia appetit formam per privationem", non sequitur quod privatio magis appetat, quia privatio non est aliquid ordinatum ad formam, sed solum est principium ordinandi aliud, ut materiam, ad receptionem formae. Si autem esset aliquid[b] ordinatum ad formam, tunc bene sequeretur, [c]aliter autem non.

1*. [d]Quaeritur de tertio principio naturae, scilicet de forma, circa quam quaeruntur quattuor quaestiones.

<Quaestio 73>
Primo [e]quaeritur quae sit illa forma quae est alterum principiorum.

1. Et videtur quod causa prima sit illa forma [f]quae est principium sive forma prima quae est alterum extremum contrarietatis sic: illa forma quae est principium est illa forma ad quam resolvit naturalis; sed physicus sive naturalis resolvit usque ad[g] primum motorem ut ad formam primam; ergo primus motor, qui est causa prima, est forma prima.

2. Item, illud quod est minimum in genere substantiae est forma prima et alterum principiorum; sed dicit Commentator super decimum[h] quod motor primus est minimum[i] in genere substantiae; ergo motor primus est forma prima. Sed motor primus est causa prima; ergo prima causa est forma prima.

3. Item, illa forma quae perficit rem substantiae et non est principiata est forma prima [j]et principium; sed causa prima est talis forma; ergo etc. Probatio minoris: [k]primo quod [l]non est principiata, cum sit omnino prima. Et quod haec[m] perficiat rem substantiae videtur, quia forma magis ligata materiae minus perficit et minus ligata magis perficit, sicut dicimus quod substantialitas propter suam imperfectionem sive incompletionem aequaliter perficit, et totum et partes, ita quod, sicut praedicatur de toto, sic et adeo

[a] et … formam *om. hom.* C
[b] aliquid *om.* TC
[c] aliter autem non *om. sed suppl. mg.* et aliter non C
[d] Quaeritur … quaestiones] consequenter quaeritur de forma et TC
[e] quaeritur quae sit] de C
[f] quae ... est illa] prima quae (quae *suppl. s.l.*) est alterum extremum contrarietatis illa (illa *scrips. sed del.*) est T; quia forma prima quae est alterum extremum contrarietatis est illa C
[g] ad *om.* O
[h] decimum] primum T; nonum C
[i] minimum] minimus O
[j] et … forma *om. sed suppl. mg.* sed prima causa est talis forma T
[k] primo quod] prima forma T
[l] non est *om.* C
[m] haec *om.* TC

something incomplete, which is the same in essence as form; and in this way privation does have an appetite for form and matter does have an appetite for form in virtue of privation considered in this way.

3. Another opinion is that, when Aristotle says "if matter has an appetite for form in virtue of privation",[562] it does not follow that privation has more appetite, because privation is not something ordered towards form, but it is only a principle of ordering something else, such as matter, towards the reception of form. But if privation were something ordered towards form, then the conclusion would certainly follow but not otherwise.

1*. Questions about the third principle of nature, that is, about form, about which four questions are raised.[563]

<Question 73>
The first question is what is the form that is the other principle.

1. Arguments that the first cause is the form that is a principle, namely, the first form that is the other extreme of the <first> contrariety: the form that is a principle is that at which a natural philosopher arrives by the method of resolution; but by this method a physical or natural philosopher arrives at the first mover as if at the first form; therefore, the first mover, which is the first cause, is the first form.

2. Furthermore, what is minimum in the genus of substance is the first form and the other principle; but in his commentary on *Metaphysics* X the Commentator says that the first mover is the minimum in the genus of substance;[564] therefore, the first mover is the first form. But the first mover is the first cause; therefore, the first cause is the first form.

3. Furthermore, the form that gives perfection to a thing belonging to the category of substance and that is not derived from a principle is the first form and a principle; but the first cause is a form of this kind; therefore, <the first cause is the first form and a principle>. Proof of the minor premise: in the first place, that the first cause is not derived from a principle <is proved> because it is absolutely first. And that the first cause gives perfection to a thing belonging to the category of substance is shown because a form that is more joined to matter gives less perfection and a form that is less joined to matter

[562] Arist. *Phys.* I. 9, 192a22–25
[563] That is, qq. 73–77
[564] Aver. *In Metaph.* X, t.c. 7, fol. 257ra

essentialiter praedicatur de qualibet parte. Unde, quia vegetativa incompletius perficit quam sensitiva, possum dicere quod totum vegetat et quaelibet pars in toto aliquo modo vegetat. Et de sensitiva est ita quod totum dicitur animal propter sensitivam totius et nulla pars dicitur animal propter sensitivam quae est in ipsa[a], sed aliquo modo dicitur sentiens. Item, rationalis perficit magis quam sensitiva, et rationalis ita denominat totum quod nullam partem, et hoc [b]propter hoc quod ita complete perficit. Et[c] ista forma inter omnes est minime ligata materiae. Cum ergo causa prima sit forma quae nullo modo est ligata materiae, [d]causa prima erit forma[e] maxime perficiens.

4. Item, illa materia quae est [f]principium est totaliter materia et totaliter in potentia; ergo illa forma quae [g]est principium primum est totaliter[h] forma et totaliter in actu. Sed tantum causa prima est talis forma[i].

5. Item, Augustinus dicit quod, sicut anima est ubique tota in minore mundo, [j]id est in homine, sic causa prima in maiori[k]. Sed anima est ita in minori mundo quod est pars eius; ergo sic erit ex alia parte; et ita causa prima erit pars compositi.

6. Ad oppositum: illa forma quae est alterum principiorum contrariatur privationi; sed causa prima nulli contrariatur; ergo etc.

7. Item, forma quae est alterum principiorum exigit aliquid sibi subici; sed causa prima non [l]exigit aliquid sibi subici, quia hoc esset absurdum dicere[m].

8. Item, Aristoteles dicit quod materia et forma manent partes compositi; sed prima causa nullius est pars; ergo etc.

9. Item, principia sunt incompleta respectu principiatorum; sed prima causa est completissimum; [n]ergo ipsa non est principium sive alterum principiorum.

[a] ipsa] ipso T
[b] propter ... perficit] propter hoc quod propter hoc (propter ... hoc *scrips. sed del.*) quia incomplete (*corr. in* magis complete) perficitur T
[c] Et] quia T; quod C
[d] causa prima *om.* C
[e] forma] prima *add.* TC
[f] principium est *om. sed suppl. s.l.* principium erit C
[g] est principium primum] principium est C
[h] totaliter] in actu *add. sed del.* C
[i] forma] ergo etc. *add.* C
[j] id est in] ut T; ut in C
[k] maiori] maiore TC
[l] exigit ... subici] est *scrips. sed del.* T; *om.* C
[m] dicere] ergo etc. *add.* C
[n] ergo ipsa] et ita TC

gives more perfection, just as we say that substantiality because of its imperfection or incompletion gives perfection in the same degree to the whole and to the parts, in such a way that it is as essentially predicated of every part as it is predicated of the whole. Thus, since the vegetative soul gives a less complete perfection than does the sensitive soul, I can say both that the whole has a vegetative life and that every part in the whole in some way has a vegetative life. In the case of the sensitive soul, the whole is said to be an animal in virtue of the sensitive soul of the whole, but no part is said to be an animal in virtue of the sensitive soul which is in that part, although it is in some way said to be capable of sense perception. Furthermore, the rational soul gives more perfection than does the sensitive soul, and the rational soul denominates the whole in such a way that it denominates no part, and this is in virtue of the fact that it gives perfection in a complete way. And this form is of all these forms the least joined to matter. Therefore, since the first cause is the form which is in no way joined to matter, the first cause will be the form that gives maximal perfection.

4. Furthermore, the matter that is a principle is entirely matter and entirely in potency; therefore, the form that is a first principle is entirely form and entirely in act; but only the first cause is a form of this kind.

5. Furthermore, Augustine says that just as soul is everywhere as a whole in the minor world, that is, in man, so is the first cause <everywhere as a whole> in the major world.[565] But soul is in the minor world in such a way that it is a part of it; therefore, the same will apply to the first cause, so that the first cause will be part of the composite.

6. To the contrary: the form that is the other principle is contrary to privation; but the first cause is contrary to nothing; therefore, <the first cause is not the form that is a principle>.

7. Furthermore, the form that is the other principle requires something that is its subject; but the first cause does not require anything as its subject, because it would be absurd to say this.

8. Furthermore, Aristotle says that matter and form remain as parts of the composite;[566] but the first cause is a part of nothing; therefore, <the first cause is not the form that is a principle>.

9. Furthermore, principles are incomplete as compared to the things derived from them; but the first cause is most complete; therefore, it is not one or other of <the two> principles.

[565] Aug. *Epistulae* 166.4, CSEL 44, 550.10–553.4; Ps. Aug. *De spiritu et anima,* c. 18, PL 40, 793–794.

[566] Arist. *Phys*. I.7, 190b17–29.

10. [a]Solutio sive rei veritas problematis secundum quosdam talis est, quod prima causa est principium de quo [b]scribit Aristoteles in primo *Physicorum*.

11. Aliter dicitur sic, quod est loqui de contrarietate[c] communiter, ut[d] comprehendamus habitum et privationem; et tunc est dicere quod causa prima est alterum extremum, sicut extremum habitus. Si autem [e]sumatur contrarietas proprie, sic non est aliquo modo extremum contrarietatis.

12. [f]Mihi tamen videtur quod causa prima nullo modo est principium neque alterum extremum contrarietatis, et loquor de principio quod[g] est pars rei. Unde[h] duplex est principium formale, scilicet vel principium formale conservans, non tamen est pars rei; et tale principium est causa prima, et tale principium praecedit omnem naturam; et de tali non loquitur Aristoteles in primo *Physicorum*. Alio modo est principium formale quod est [i]pars rei permanens in eius compositione; et de tali loquitur Aristoteles **[O 101va]** in primo huius et tale non est causa prima.

13. Ad primum argumentum dicendum quod resolvere ad motorem primum contingit dupliciter: vel ut ad [j]principium motivum tantum vel ut ad principium formale. Si[k] ad principium[l] motivum, hoc contingit dupliciter: vel[m] ad motivum appropriatum, et sic non resolvit Aristoteles in octavo ad motorem primum; vel[n] ad principium motivum [o]non appropriatum, et sic resolvit Aristoteles ad principium motivum. Si autem loquamur de principio formali, hoc dupliciter contingit: vel quod[p] est pars rei, et sic non resolvit; vel quod[q] non est pars rei, sed conservans; sic resolvit ad causam primam ut ad principium formale.

[a] Solutio … est] dicunt quidam TC
[b] scribit Aristoteles] loquitur Aristoteles T; loquitur C
[c] contrarietate] dupliciter communiter et proprie *add.* C
[d] ut] loquendo prout C
[e] sumatur contrarietas] sumamus contrarietatem C
[f] Mihi tamen] quia tamen mihi *scrips. sed* mihi *del.* C
[g] quod] quem O
[h] Unde] dico quod C
[i] pars … compositione] permanens in eius compositione et pars rei C
[j] principium … ad *om. hom. sed suppl. mg.* T
[k] Si] ut *add. s.l.* O
[l] principium *om.* TC
[m] vel] ut O
[n] vel] ut *add. s.l.* O
[o] non … motivum *om. hom. sed suppl. mg.* T (motorem primum *pro* principium motivum)
[p] quod] quae TC
[q] quod] quae OTC *sed corr.* O

10. The solution of the problem or the truth of the matter, according to some people, is that the first cause is the principle about which Aristotle writes in *Physics* I.[567]

11. Another opinion is that if we speak of contrariety in a broad sense, so that we include both habit and privation, then it is possible to say that the first cause is the other extreme, that is, the extreme which is a habit. But if we take contrariety in a proper sense, then the first cause is not in any way an extreme of a contrariety.

12. It seems to me, however, that the first cause is in no way a principle nor is it the other extreme of the <first> contrariety, and I am speaking of the principle that is a part of a thing. Thus, there are two kinds of formal principle: either (i) a preserving formal principle, which however is not a part of a thing; such a principle is the first cause, and it precedes every nature; and Aristotle is not speaking of this principle in *Physics* I. (ii) Another kind of formal principle is that which is a part of a thing and remains in its composition; and Aristotle is speaking of this kind of principle in *Physics* I, and the first cause is not a principle of this kind.

13. To the first argument[568] it must be said that it is possible to arrive at the first mover by the method of resolution in two ways: either (i) as if it is only a moving principle or (ii) as if it is a formal principle. Case (i) is twofold: either (ia) as if it is an appropriate mover, and in this way Aristotle does not arrive at the first mover by the method of resolution in *Physics* VIII; or (ib) as if it is a moving principle which is not appropriate, and taken in this way Aristotle does arrive at a moving principle by the method of resolution. Case (ii) is twofold: either (iia) as if it is a principle which is a part of a thing, and Aristotle does not arrive at this kind of principle by the method of resolution, or (iib) as if it is a principle which is not a part of a thing, but that preserves a thing. By the method of resolution, Aristotle arrives at the first cause as a formal principle taken in way (iib).

[567] Arist. *Phys*. I.7–9.
[568] See above, n. 1.

<Quaestio 74>

[a]Quaeritur secundo utrum forma primo adveniens materiae sit alterum principiorum.

1. Et videtur primo quod sic, quia illa forma est principium quae[b] primo exigit materiam sibi subici; sed talis est forma quae primo advenit materiae; ergo etc.

2. Item, Aristoteles dicit in hoc primo quod necesse est principia manere in esse compositi; forma ergo [c]quae primo advenit et manet in esse compositi melius debet esse forma prima. Et ita forma primo adveniens erit principium.

3. Item, illa forma est prima forma quae opposita est primae privationi; sed talis est forma primo adveniens, quia dicit Commentator super primum quod, inducta forma, [d]corrumpitur privatio; ergo inducta prima forma, corrumpitur privatio prima. Sed non corrumpitur nisi quia forma est ipsi contraria; ergo forma primo adveniens materiae est contraria privationi, et ita ipsa est principium.

4. Ad oppositum: Aristoteles dicit quod necesse est principia non esse ex aliis neque ex se invicem, sed alia ex ipsis; sed omnis forma quae est in materia educitur de potentia materiae; ergo nulla forma [e]in naturali materia est prima forma. Sed forma primo adveniens est in materia naturali; ergo non est primum principium.

5. Item, Aristoteles dicit quod oportet principia manere et nec generari nec corrumpi; sed omnis forma rei naturalis [f]est generabilis et corruptibilis; ergo nulla forma rei naturalis est[g] principium.

6. Item, Aristoteles dicit quod materia est ingenita et incorruptibilis. Et si hoc, alterum principiorum erit huiusmodi. Sed omnis forma [h]in re naturali est corruptibilis; ergo nulla est principium primum formale.

7. [i]Solutio sive rei veritas problematis talis est, quod forma quae primo advenit materiae est alterum principiorum. Et haec est forma generis generalissimi, non unde communis est et praedicabilis, sed unde res est et

[a] Quaeritur secundo] consequenter quaeritur TC
[b] quae *scripsimus*; quod OTC
[c] quae … manet] primo adveniens et manens C
[d] corrumpitur … forma *om. hom.* TC *sed suppl. mg.* T
[e] in naturali materia] naturalis T; naturaliter C
[f] est … naturalis *om. hom. sed suppl. mg.* (rei *om. mg.*) T
[g] est] generabilis *add. sed del.* O
[h] in re naturali] rei naturalis T
[i] Solutio … est] dicendum TC

<Question 74>
The second question is whether the form that first comes to be in matter is the other principle.

1. Arguments for the affirmative: the form that first requires matter to be its subject is a principle; but the form that first comes to be in matter is of this kind; therefore, <the form that first comes to be in matter is the other principle>.

2. Furthermore, in *Physics* I Aristotle says that it is necessary for the principles to remain in the being of the composite;[569] therefore, the form that first comes to be in matter and remains in the being of the composite is a better candidate for being the first form. Thus the form that first comes to be in matter will be a principle.

3. Furthermore, the form opposite to the first privation is the first form; but the form that first comes to be in matter is of this kind, because in his commentary on *Physics* 1 the Commentator says that, once a form is induced, its privation is corrupted; [570] therefore, once the first form has been induced, the first privation is corrupted. But the first privation is only corrupted because that form is contrary to it; therefore, the form that first comes to be in matter is contrary to privation, and so it is a principle.

4. To the contrary: Aristotle says that it is necessary that the principles should not derive from other things or from one another, but that other things should derive from them;[571] but every form existing in matter is educed from a potency of matter; therefore, no form existing in natural matter is the first form. But the form that first comes to be in matter exists in natural matter; therefore, it is not the first principle.

5. Furthermore, Aristotle says that it is necessary that the principles should remain and should be neither generated nor corrupted;[572] but every form of a natural thing is generable and corruptible; therefore, no form of a natural thing is a principle.

6. Furthermore, Aristotle says that matter is ingenerable and incorruptible.[573] If so, the other principle will be of this kind too. But every form of a natural thing is corruptible; therefore, no such form is the first formal principle.

7. The solution of the problem or the truth of the matter is that the form that first comes to be in matter is the other principle. And this is the form of the most general genus, not in so far as it is common and predicable, but in so

[569] Arist. *Phys.* I.7, 190b17–23.
[570] Aver. *In Phys.* I, t.c. 60, fol. 36ra.
[571] Arist. *Phys.* I.5, 188a27–28.
[572] Arist. *Phys.* I.6, 189a19–20.
[573] Arist. *Phys.* I.9, 192a27–29.

altera pars individui ipsius generis. Sed de hac forma contingit loqui dupliciter: vel in sua essentia, non prout determinatur in aliquo individuo, et sic est ingenita et incorruptibilis; aut est loqui de illa secundum esse quod habet in individuo signato. Et tunc contingit dupliciter loqui de eius generatione et corruptione: vel essentialiter vel accidentaliter. Si essentialiter, sic adhuc est ingenerabilis et incorruptibilis. Si per accidens, ut per generationem et corruptionem totius, sic est generabilis et corruptibilis.

8. Primae[a] rationi dicendum[b] sic, quod, cum dicat Aristoteles quod principia non fiunt ex se invicem, hoc verum est per se loquendo, per accidens tamen bene contingit.

9. Aliter dicitur sic, quod Aristoteles, cum[c] dicit quod principia non fiunt ex se invicem, dat rationem huius, dicens: quoniam sunt contraria; unde forma non fit ex privatione secundum quod privatio contrariatur eidem.

10. Vel sic: principia non fiunt ex se invicem [d]neque ex aliis sicut compositum fit ex aliis.

<Quaestio 75>

[e]Quaeritur tertio utrum forma prima sit multa secundum supposita aut sit tantum una secundum suppositum.

1. Quod sit multa videtur: principium formale respondet principiato per se; ergo diversitati principiatorum respondet diversitas principii formalis. Sed principiata sunt plura secundum suppositum; ergo et principium erit multa[f] secundum supposita.

2. Item, principium formale est in isto et in illo, et constat quod nihil idem est [g]in utroque; ergo nec principium formale.

3. Ad oppositum: omne suppositum aut est suppositum generis aut speciei; sed principium formale nec est tale nec tale, quia quod est suppositum est compositum, et tale non est forma prima.

4. Item, omne suppositum est subiectum universali; sed primum principium non est subiectum universali, cum sit prius quolibet universali.

[a] Primae] ergo *add.* TC
[b] dicendum *corr. in* dicitur T; dicitur C
[c] cum] quando TC
[d] neque ex aliis] etc. C
[e] Quaeritur tertio] consequenter quaeritur T
[f] multa] multum O
[g] in utroque] utrobique TC

far as it is a thing and the other part of an individual of that genus. But we can speak of this form in two ways: either (i) in its essence, <and> not in so far as it is determined in an individual; and in this way it is ingenerable and incorruptible; or (ii) in accordance with the being that it has in a given individual. In case (ii) we can speak of its generation and corruption in two ways: either (iia) essentially or (iib) accidentally. That form is ingenerable and incorruptible in way (iia), but it is generable and corruptible in way (iib), that is, accidentally, namely through the generation and the corruption of the whole <individual>.

8. To the first argument[574] it must be said that when Aristotle says that the principles do not come to be from one another, this is true speaking of them per se, but this is certainly possible accidentally.

9. Another reply is that when Aristotle says that the principles do not come to be from one another, the explanation he gives for this claim is that they are contraries; thus, form does not come to be from privation in so far as privation is contrary to form.

10. Another reply is the following: the principles do not come to be from one another nor < do they come> from other things in the way in which the composite comes to be from other things.

<Question 75>
The third question is whether the first form is many in relation to its supposits or whether it is only one form in relation to one supposit.

1. Arguments that it is many: a formal principle per se corresponds to a thing derived from that principle; therefore, the diversity of a formal principle corresponds to the diversity of the things derived from it. But the things derived from a principle are many in relation to their supposits; therefore, that principle will also be many in relation to its supposits.

2. Furthermore, a formal principle exists in this thing and in that thing, and it is apparent that nothing that is the same exists in both things; therefore, neither does the formal principle.

3. To the contrary: every supposit is either a supposit of a genus or of a species; but the formal principle is neither this kind nor that kind of supposit, because what is a supposit is composite, and the first form is not of this kind.

4. Furthermore, every supposit is the subject of a universal; but the first principle is not the subject of a universal, since it is prior to every universal.

[574] See above, n. 4.

5. Item, Commentator dicit quod materia prima est una numero in omnibus; et si hoc, principium formale erit unum numero in omnibus.

6. [a]Solutio sive rei veritas problematis talis est, quod est considerare principium formale dupliciter: vel secundum suam essentiam, et sic non est multa secundum suppositum neque una, quia sic non concernit suppositum; vel secundum suum esse in supposito. Et tunc contingit considerare eius individuationem dupliciter: vel per se vel per accidens. Si per se, sic non est plura secundum supposita. Si per accidens, sic est multa secundum supposita, quia, sicut est quod accidit superiori quod signetur in hoc individuo et hoc individuum essentialiter continetur sub ipso, sic accidit formae quod sit multa secundum supposita.

1*. [b]Quaeritur quarto de contrarietate in substantia.

<Quaestio 76>
Et primo utrum formae substantiali sit aliquid contrarium sive formae primae.

1. Et quia multi textus[c] sunt ad utramque partem, dicendum quod duplex est contrarietas, scilicet vel secundum formam et subiectum, et de tali contrarietate dicit Aristoteles quod prima contrarietas est in loco sursum et deorsum; et est ibi contrarietas secundum formam, quia potentia loci sursum contrariatur potentiae loci deorsum, et secundum subiectum, quia in contentis his locis est aliquo modo contrarietas, sicut dicimus quod ignis contrariatur terrae. Aut est contrarietas secundum formam tantum. Et hoc contingit[d] vel respectu subiecti constituti in esse specifico, et sic non est contrarietas nisi in [e]formis accidentalibus, aut respectu entis in potentia. Et hoc vel respectu alicuius [f]ut entis **[O 101vb]** in potentia, et sic sunt differentiae contrariae respectu generis, aut respectu alicuius vere entis in potentia, et sic sunt forma et privatio contraria respectu materiae. Et tantum secundo modo negat Aristoteles simpliciter contrarietatem esse in substantia.

[a] Solutio … est] dicendum TC
[b] Quaeritur ... contrarietate] consequenter quaeritur de contrarietate quae est TC
[c] textus] tixtus OT
[d] contingit] dupliciter *add.* TC
[e] formis accidentalibus *corr. mg. ex* forma actuali T
[f] ut … alicuius *om. hom.* O

5. Furthermore, the Commentator says that prime matter is one in number in all things;[575] if that is the case, the formal principle will be one in number in all things.

6. The solution of the problem or the truth of the matter is that the formal principle can be considered in two ways: either (i) in its essence, and in this way it is not many in relation to its supposits nor is it one, because it does not relate to a supposit in this way; or (ii) in accordance with its being in a supposit. In case (ii) there are two ways of considering its individuation: either (iia) per se or (iib) accidentally. In case (iia), it is not many in relation to its supposits. In case (iib), it is many in relation to its supposits because just as it is accidental to what is superior that it should be individualized in this individual and that this individual should be essentially contained under it, so it is accidental to a form to be many in relation to its supposits.

1*. The fourth question[576] is about contrariety in substance.

<Question 76>
The first question is whether there is something contrary to substantial form or to the first form.

1. Since there are many texts supporting both sides of this question, it must be said that there are two kinds of contrariety: either (i) in form and subject. Speaking of this kind of contrariety, Aristotle says that the first contrariety is in the downward and upward places.[577] There is both a contrariety in form in these places, because the power of the upward place is contrary to the power of the downward place, and a contrariety in subject, because the things contained by those places are in some way contrary, just as we say that fire is contrary to earth. Or (ii) a contrariety in form alone. This kind of contrariety is either (iia) with respect to a subject constituted in its species, and this kind of contrariety only exists in accidental forms, or (iib) with respect to a being in potency. Case (iib) is twofold: either (iib') with respect to something thought of as a being in potency, and in this way the differences are contrary with respect to the genus or (iib'') with respect to a genuine being in potency, and in this way form and privation are contrary with respect to matter. It is only in case (iia) that Aristotle absolutely denies that there is contrariety in a substance.[578]

[575] See, e.g. Aver. *In Metaph.* XII, t.c. 14, fols. 300vb–301ra.
[576] See below, qq. 76–77.
[577] Arist. *Cat.* 6, 6a12–13.
[578] Arist. *Cat.* 3, 3b24–25.

2. Iterum[a], [b]suscipere magis et minus contingit dupliciter: vel per maiorem et minorem participationem sui contrarii, et sic tantum forma accidentalis recipit magis et minus, vel suscipere magis[c] et minus penes actum et potentiam, et sic suscipit forma substantialis magis[d] et minus, quia una forma potest esse magis [e]in actu quam alia. In hoc tamen conveniunt intensio et remissio formae substantialis et accidentalis, quod neutra fit nisi secundum quod forma est in materia. Forma tamen substantialis, [f]quia substantialis potest considerari in se praeter hoc quod est actus materiae, et illo modo nullo modo recipit magis[g] [h]et minus; sed forma accidentalis non potest considerari praeter hoc quod est in materia sive in subiecto et a subiecto causatur[i].

<Quaestio 77>

Habito quod potentia activa sit idem in essentia cum forma et quod potentia activa non movet de se nisi excitetur ab agente extra, propter hoc quaeritur quid immittit agens in patiens.

1. Et primo videtur quod nihil immittat, quia, si aliquid immittat, aut [j]immittit substantiam aut accidens. Si substantiam, aut ergo materiam aut

[a] Iterum] item est *scrips. sed corr. mg. in* item aliquid T; item C
[b] suscipere ... minus, vel *om. hom.* TC *sed suppl. mg.* T
[c] magis] maius O
[d] magis] maius O
[e] in actu *om. sed suppl. s.l.* actualis O
[f] quia substantialis *suppl. mg.* T
[g] magis] maius O
[h] et minus *om.* C
[i] causatur] quod substantia suscipiat (recipiat *pro* suscipiat C) magis et minus videtur sic: illud gratia cuius nomen suscipit (recipit *pro* suscipit C) comparationem recipit magis et minus; sed hoc est (in *add.* C) ratione substantiae, ut (sic *pro* ut C) videtur sic: abstractum non recipit comparationem sed (per *scrips. sed corr. s.l.* T) solum concretum, et si hoc, concretum solum recipit comparationem ratione illa qua diversificatur ab abstracto; sed non diversificatur nisi ratione subiecti quod cum qualitate significat; ergo concretum ratione subiecti comparatur. Si dicatur quod concretum diversificatur non quia significat subiectum sed quia significat cum (sub *pro* cum C) inclinatione ad subiectum, contra: omne accidens in quantum accidens significat cum (sub *pro* cum C) inclinatione ad subiectum, et ita abstractum posset comparari. Item, si in (in *om.* C) ratione inclinationis ad subiectum (substantiam *pro* subiectum C) cum omne concretum aequaliter inclinet se ad subiectum, tunc omne concretum compararetur (comparetur *pro* compararetur C). Item, omne concretum comparatur ratione principalis significati quia ab intensione vel remissione significati imponitur vox significans secundum (secundum *om.* C) intensius et remissius. Si (sed *pro* si C) concretum principaliter significat substantiam, ut videtur, sic concretum est posterius abstracto, et hoc quia significatum (significat *pro* significatum C) suum est (esse *pro* est C) posterius. Sed non est posterius quoad accidens; ergo erit posterius quoad subiectum. Et illud ratione cuius est posterius est eius principale significatum vel in principali (*corr. ex* principale T) significato (vel in principali significato *om.* C), et ita ratione subiecti comparabitur (habebunt comparationem *pro* comparabitur C) *add.* TC
[j] immittit substantiam] immittat substantiam T; est substantia C

2. Furthermore, something can take on the more and the less in two ways: either (i) by more or less participation in its contrary, and in this way only an accidental form receives the more and the less, or (ii) with respect to act and potency, and in this way a substantial form receives the more and the less, because one <substantial> form can be more in act than another. The intension and remission of a substantial form and that of an accidental form, however, agree in that each of them occurs only in so far as the form exists in matter. But a substantial form, since it is substantial, can be considered in itself apart from its being an act of matter, and so considered, in no way does it receive the more and the less; but an accidental form cannot be considered apart from the fact that it exists in matter or in a subject and is caused by a subject.

<Question 77>
Since it has been established that active potency is the same in essence with form[579] and that active potency does not cause motion by itself unless it is stimulated by an external agent,[580] the question now is what the agent introduces into a patient.

1. First the argument for the negative: if an agent introduces something, it introduces either (i) a substance or (ii) an accident. In case (i), it introduces either matter or a form or a composite. It does not introduce only matter,

[579] See above, q. 65, n. 3.
[580] See above, q. 65, n. 7.

formam aut compositum. Non materiam tantum, quia materia non est sine forma. Nec formam tantum, quia forma non est sine materia. Nec compositum, quia cum continue immittat, si semper continue immitteret compositum, tunc tandem consumeretur agens per suam immissionem, quod falsum est. Non accidens, quia accidens non est sine [a]suo subiecto; si ergo accidens immitteretur, necesse esset [b]subiectum simul immitti.

2. Item, si aliquid immittatur, aut est generatum aut creatum. Non creatum, cum sit ab agente naturali, [c]et etiam cum non sit factum ex nihilo. Nec est generatum, quia omne quod generatur exigit sibi materiam pro subiecto; si ergo illud quod immittitur esset generatum et in profundum patientis delatum, cum tale non sit sine materia, tunc secum deferret aliquid materiae agentis, et ita demum per immissionem consumeretur.

3. Si dicatur quod illud quod immittitur est sola species et nec est simpliciter substantia nec simpliciter accidens neque[d] ens simpliciter, sed aliquid entis, contra: omnis actio naturalis fit ratione contrarietatis; sed illud immissum agit naturaliter in patiens; ergo illud immissum est contrarium. Sed sola species nulli contrariatur; ergo sola species non est illud quod immittitur. Quod autem sola species nulli contrariatur videtur, quia contraria non compatiuntur se simul in eodem; sed species albi[e] et nigri [f]et contrariorum aliorum[g] compatiuntur se in eadem parte medii et etiam in eadem parte organi, ut in pupilla; ergo species contrariorum non sunt contrariae. Et si hoc, species secundum quod talis nulli contrariatur.

4. Item, [h]distinctio est quod quaedam forma multiplicat se et non extendit[i], ut forma lucis, et quaedam multiplicat[j] et extendit, ut forma carnis secundum speciem, et quaedam extendit et non multiplicat, sicut forma elementi. Si ergo haec distinctio sit bona, tunc forma elementaris[k] non potest facere se extra se per multiplicationem; et ita non potest multiplicare speciem in materia aliena, et ita nihil immittere.

5. [l]Solutio sive rei veritas problematis talis est, quod agens aliquid immittit in patiens, et quod non solum immittit in superficie, sed in profundo. Et hoc quia movens proximum et immediatum est simul cum moto; unde,

[a] suo subiecto] substantia C
[b] subiectum simul immitti] immittere simul (simul *s.l.*) subiectum C
[c] et ... nihilo *om.* C
[d] neque] nec TC
[e] albi *corr. ex* albo O
[f] et contrariorum aliorum *om. sed suppl. mg.* et quorumcumque contrariorum T
[g] aliorum *suppl. s.l.* O
[h] distinctio est quod] illa est distinctio magistralis C
[i] extendit] se *add.* C
[j] multiplicat] se *add. s.l.* O
[k] elementaris] elementorum TC
[l] Solutio ... est] dicendum TC

because matter does not exist without form. Nor does it introduce only a form, because form does not exist without matter. Nor does it introduce a composite, because since an agent introduces something continuously, if it introduced a composite continuously, then the agent would eventually be consumed through its introduction, which is false. (ii) Nor does an agent introduce an accident, because an accident does not exist without its subject; therefore, if an accident were introduced, it would also be necessary for its subject to be introduced at the same time.

2. Furthermore, if something is introduced, this is either (i) generated or (ii) created. Not (ii), because it is produced by a natural agent, and because it is not produced from nothing. Not (i) either, because everything generated requires matter as its subject. Therefore, if what is introduced were generated and carried into the depths of a patient, then, since something of this kind does not exist without matter, it would carry with itself something of the matter of the agent, and thus the agent would eventually be consumed through its introduction.

3. If it is said that what is introduced is only a species and is neither absolutely a substance nor absolutely an accident nor absolutely a being, but something of a being, one can argue against this: every natural action occurs because of a contrariety; but the thing introduced acts naturally on the patient; therefore, the thing introduced is a contrary. But a species alone is contrary to nothing; therefore, a species alone is not the thing introduced. That a species alone is contrary to nothing is shown because contraries are not simultaneously compatible in the same thing; but the species of white and of black and of the other contraries are compatible in the same part of the medium and also in the same part of an organ, for example in the pupil; therefore, the species of the contraries are not contrary. If so, a species as such is contrary to nothing.

4. Furthermore, there is the following distinction: one kind of form, such as the form of light, multiplies itself but does not extend itself, and another kind of form, such as the form of flesh in accordance with its species, multiplies and extends itself, and another kind of form, such as the form of an element, extends itself but does not multiply itself. Therefore, if this distinction is correct, then an elementary form cannot go outside itself through multiplication, and thus it cannot multiply its species in an extraneous matter, and so it can introduce nothing.

5. The solution of the problem or the truth of the matter is that an agent introduces something into a patient, and that it introduces something not

cum illud immissum non solum moveat[a] in superficie, sed etiam in profundo, necesse est ponere quod aliquid immittatur in profundo patientis. Et[b] illud immissum non est res aliqua, sed est species sola. [c]Et forte, cum ignis immittit speciem in aere, non solum immittit speciem accidentis, ut suae caliditatis, sed speciem[d] totius aggregati[e], ut prius patuit, quando quaesitum fuit utrum cum specie accidentis commultiplicetur[f] species [g]substantiae. Unde illa species quam facit agens extra [h]se diversos habet effectus secundum diversitatem recipientium, quia, quando recipitur in sensum[i], recipitur ut immutans, quando autem recipitur in suo contrario vel in suo proprio passivo, recipitur ut excitans potentiam [j]ibi repertam quae est sibi similis.

6. Unde argumento in contrarium dicendum quod haec propositio 'omne agens naturale agit per contrarietatem', hoc est intelligendum de illo quod proprie agit et est res et natura in se. Si autem intelligatur de agente tantum excitante aliud ad agendum, sic est propositio falsa. Et sic agit species, quia non facit nisi excitat potentiam activam ad agendum.

7. Aliter dicitur[k] sic, quod species aliquando recipitur [l]ubi consequitur effectum speciei tantum, ut ubi est immutans, scilicet[m] in organo et medio, et sic non habet contrarium. Aliquando recipitur in aliquo ut aliquid educatur in actu quod prius fuit in potentia et ut corrumpatur forma eius [n]actualis in quo[o] recipitur, et sic aliquo modo habet species ipsa contrarium. Et hoc patet si apponatur manus igni. Primo immutabitur tactus a specie ignis, et tunc consequitur effectum speciei unde species est, et sic non reperitur ibi contrarietatem. Verumtamen, per tantum tempus poterit manus teneri contra ignem quod corrumpetur sua forma substantialis et inducetur forma ignis. Et hoc quia illa species excitabit potentiam ad ignem quae fuit in manu, et ita[p]

[a] moveat] vel immittatur *add. s.l.* O
[b] Et *suppl. s.l.* O
[c] Et ... accidentis] et forte tamen ignis immittat speciem in aere accidentis *scrips. sed corr. s.l. et mg. in* unde forte ponerem quod cum ignis debeat agere in aerem non solum immittit speciem accidentis in aerem T
[d] speciem] species OC
[e] aggregati] immittitur *add. mg.* O
[f] commultiplicetur] commultiplicatur TC
[g] substantiae ... species *om. hom. sed suppl. s.l.* (et notandum quod *pro* unde illa) O
[h] se diversos] se TC *sed corr. mg. in* diversos T
[i] sensum] sensu T
[j] ibi repertam] sibi separatam *scrips. sed del.* O
[k] dicitur] est dicere C
[l] ubi consequitur *corr. s.l. ex* nisi consequitur T; ut consequatur C
[m] scilicet] ut T; et C
[n] actualis *corr. mg. ex* accidentalis T; activa vel C
[o] quo *scripsimus*; quod O; quam TC *sed corr. s.l. in* quod T
[p] ita *corr. s.l. ex* ibi T; ibi *add.* C

only onto the surface of the patient but also into its depths.[581] The reason for this is that the proximate and immediate mover is in the same place as the thing moved; thus, since the thing that is introduced moves not only on the surface but also in the depths <of the patient>, it is necessary to assume that something is introduced into the depths of the patient. What it is that is introduced is not a thing, but only a species. It is perhaps the case that when fire introduces its species into air, it does not only introduce the species of an accident, for example of its heat, but the species of the whole aggregate, as was made clear before when it was asked whether the species of a substance is multiplied together with the species of an accident.[582] Hence, the species which an agent produces outside itself has different effects in accordance with the diversity of the things receiving it, because, when it is received in the senses, it is received as something that modifies the senses, but when it is received in its contrary or in the passive thing appropriate to it it is received as something that stimulates a potency similar to itself which exists there.

6. Thus, to the argument to the contrary[583] it must be said that the proposition 'every natural agent acts by contrariety' must be understood as being about an agent that acts properly and is a thing and a nature in itself. But if it is understood as being about an agent that only stimulates something else to act, then the proposition is false. A species acts in this way, because it does nothing apart from stimulating an active potency to act.

7. Another reply is that sometimes a species is received in something in which only the effect of that species follows from that species, as happens in the thing in which it produces a change, that is, in an organ and in the medium, and in this case the species does not have a contrary. But at other times a species is received in something in order to bring into act something that was first in potency and in order to corrupt the actual form of the thing in which it is received, and in this case the species itself has a kind of contrary. This is clear if a hand is put in contact with fire. First of all the sense of touch is changed by the species of fire, and then the effect of that species in so far as it is a species follows, and in this case there is no contrariety there. However, a hand can be kept in contact with fire for such a long time that its substantial form is corrupted and the form of fire is induced. This happens because that species stimulates the potency to fire that existed in the hand, and in this case there is a kind of contrariety there. The first reply, however, is better.

[581] The author of the commentary on the *De generatione et corruptione* attributed to Richard Rufus holds a version of this theory; see Richard Rufus of Cornwall, *In De gen. et corr.*, 151.790–153.836.

[582] See above, q. 40.2, n. 1.

[583] See above, nn. 1–3.

reperiet[a] ibi aliquo modo contrarietatem. Sed[b] prima responsio melior est[c].

8. [d]Secundae rationi dicendum quod forma elementaris de sua natura particulari non potest speciem suam multiplicare, **[O 102ra]** tamen mediante natura universali, ut luce corporis caelestis, luce dico non unde splendens est, sed secundum suam [e]substantiam sive secundum substantiam lucis. Unde, sicut dicimus quod color, qui[f] est proprium obiectum visus, non multiplicat se nisi mediante luce splendente, sic dicimus quod sonus[g], odor et huiusmodi non multiplicant se nisi mediante virtute lucis caelestis, licet non unde splendens est, quia in ventre terrae dicimus quod virtutes luminares caeli sunt causae generationis mineralium[h], et tamen ibi non est lux splendens, sed substantia lucis, quae undique in universo diffusa est.

9. Si dicatur quod species immissa est idem in essentia cum forma inducenda, et potentia activa similiter est eadem cum forma inducenda, tunc, cum inducitur forma, utrumque est[i] in ratione perfectionis; et ita duae sunt perfectiones eiusdem [j]speciei in eodem.

10. [k]Ad hoc respondetur sic, quod duae perfectiones eiusdem speciei non sunt in eodem, sed quando duae sic veniunt, fit perfectio intensior sive maior, sicut dicimus quod[l], si caro adveniat formae carnis, ut contingit in nutritione et augmentatione, non faciunt duas perfectiones, sed unam maiorem. Et etiam impossibile est plura accidentia sive formas eiusdem speciei numerari in eadem parte materiae.

11. Sed contra: lux est forma medii, et tamen lux numeratur in eadem parte medii, quia, si ponatur corpus opacum inter duo corpora luminosa, constat quod faciet [m]duas umbras, et ita duas privationes, quod non contingeret nisi lux esset numerata in eadem parte medii.

12. Dicitur ad hoc sic, quod quaedam sunt formae [n]quae sunt in materia ut a qua dependent et sunt aliquo modo perfectivae eius in quo[o] sunt; et

[a] reperiet] reperiret TC *sed corr. s.l.* T
[b] Sed *om.* TC
[c] est *om.* TC *sed suppl. s.l. ante* melior T
[d] Secundae rationi dicendum] secundum videtur C
[e] substantiam … lucis] lucis substantiam T; substantiam lucis C
[f] qui] lucis quod C
[g] sonus] et *add.* TC
[h] mineralium *corr. s.l. ex* universalium O
[i] est] ibi *add.* TC
[j] speciei in eodem *scrips. sed del.* T; in eodem O; speciei C
[k] Ad hoc respondetur] dicitur ad hoc TC
[l] quod] et *add.* TC
[m] duas … ita *om. hom.* T
[n] quae sunt *iter. sed corr.* T
[o] quo] qua O

8. To the second argument[584] it must be said that an elementary form can multiply its species not through its particular nature, but by means of a universal nature, such as that of the light of a celestial body. <Here> I mean light not in so far as it is bright, but through its substance or through the substance of light. Thus, just as we say that colour, which is the proper object of sight, does not multiply itself except by means of a bright light, so we say that sound, smell and the like do not multiply themselves except by means of the power of a celestial light, although not in so far as it is bright, but because we say that the illuminating powers of the heavens are the causes of the generation of minerals in the depths of earth where there is no bright light but only the substance of light, which is diffused throughout the universe.

9. If someone says that the species introduced is the same in essence with the form to be induced, and that the active potency is also the same as the form to be induced; and then, when the form is induced, both the species and the active potency will have the role of a perfection, and so there will be two perfections of the same species in the same thing.

10. To this conclusion it is replied that two perfections of the same species do not exist in the same thing; but when two perfections of this kind come about, they produce a more intense or greater perfection, just as we say that, if flesh supervenes on the form of flesh, as happens in nutrition and growth, they do not produce two perfections, but one greater perfection. It is also impossible for many accidents or forms of the same species to be numbered in the same part of matter.

11. But against this it is argued: light is the form of the medium, and yet light is numbered in the same part of the medium, because, if a dark body is placed between two luminous bodies, it is evident that it produces two shadows, and thus <it produces> two privations, which would not happen if light were not numbered in the same part of the medium.

12. To this argument it is replied that (i) some forms are such that they exist in matter as in a thing on which they depend and are in some way a perfection of the thing in which they exist; and the forms of this kind are not numbered in the same part of the thing in which they exist. (ii) Another kind

[584] See above, n. 4.

huiusmodi non numerantur in eadem parte eius in quo sunt. Alia est forma quae est in alio tantum ut in suo susceptibili, et non dependet ab illo; et talis forma est lux in medio, et talis forma bene potest numerari in eadem parte medii. Et etiam species in medio sive in organo sive in anima similiter se habet.

13. [a]Tot et tanta de principiis sint determinata.

<Quaestio 78>

[b]Quaeritur de hoc quod dicit consequenter, quod est sumere carnem minimam, an scilicet[c] sit sumere minimum in re naturali.

1. Et videtur primo quod sic, quia Aristoteles dicit in secundo *De anima* quod omnium natura constantium determinata est ratio magnitudinis et augmenti. Et si hoc, sicut determinatur in maximo, sic debet determinari in minimo.

2. Item, Aristoteles dicit[d]: "In unoquoque determinatum est[e] quantum debet esse secundum magnum et secundum parvum"; sicut ergo [f]est maximum, et sic et minimum.

3. Item, fieri cuiuslibet rei naturalis est in tempore determinato secundum longum et breve; ergo sic erit ex parte continui agentis quod erit minimum determinatum in quo minore non potest agere. Et si hoc, cum proprium sit naturalis agere, erit quodlibet naturale determinatum in minimo et in maximo.

4. Ad oppositum: sumatur caro quae dicitur minima; illa caro quanta est; ergo divisibilis in infinitum. Aut ergo divideretur[g] in naturalia aut in mathematica. Non in mathematica, quia sic essent mathematica principia naturae, sicut posuit Plato. Si in naturalia, sed non in alia quam in carnes, dividetur[h] ergo in carnes. Et si hoc, erit sic procedere in infinitum[i]. [j]Vel ad minus sequitur[k] quod, si non dividitur in mathematica, quod dividitur in naturalia, et ita in naturalibus non est minimum.

[a] Tot … determinata *om.* TC
[b] Quaeritur ... minimam] consequenter quaeritur (quaeritur *om.* C) super hoc (hoc *om.* C) quod dicit quod est sumere carnem minimam quaeritur TC
[c] scilicet *om.* TC
[d] dicit *om.* TC
[e] est] in *add.* O
[f] est … minimum] etc. C
[g] divideretur] dividitur T; dividetur C
[h] dividetur] dividitur TC
[i] infinitum] quod est falsum *add. sed del.* O
[j] Vel ... naturalia] minus sequitur quod, si non dividitur in mathematica, quod dividitur in naturalia *scrips. sed del.* O
[k] sequitur] sequetur T; *om.* C

of form is such that it exists in another thing as in a thing that only receives it and does not depend on it; light in the medium is a form of this kind, and a form of this kind can be numbered in the same part of the medium. A species in the medium or in an organ or in the soul also is of this kind.

13. A great number of important conclusions have been established about the principles.

<Question 78>

The question is about what Aristotle says next, namely that it is possible to take a minimal <amount of> flesh:[585] that is, the question is whether it is possible to find a minimum in a natural thing.

1. First the arguments for the affirmative: in *On the Soul* II Aristotle says that for all things that exist by nature there is a determinate ratio of their size and growth.[586] If so, just as the maximum of a natural thing is determined, so its minimum must be determined.

2. Furthermore, Aristotle says: "For each thing it is determined what its amount will be in relation to being large and being small"[587]; therefore, just as there is a maximum, so there is also a minimum.

3. Furthermore, the becoming of every natural thing occurs in a time determined with respect to length and brevity; therefore, the same will apply where a continuous agent is concerned, namely that there will be a determinate minimum such that that agent cannot act on something less than it. If this is the case, since acting is proper to a natural agent, every natural thing will be determined with respect to its minimum and its maximum.

4. To the contrary: take some flesh that is said to be minimal in amount; that flesh has a quantity; therefore, it is divisible to infinity. Therefore, either it is divisible into natural things or into mathematical things. It is not divisible into mathematical things, because in that case mathematical things would be principles of nature, just as Plato assumed. If it is divisible into natural things, but not into things other than flesh, it follows that it is divided into flesh. If so, there will be an infinite regress. Alternatively, if that flesh is not divided into mathematical things, it would at least follow that it is divided into natural things, so that there is no minimum in natural things.

[585] Arist. *Phys*. I.4, 187b27–188a2. Since they arise from a passage in *Phys*. I.4, qq. 78–79 are misplaced at the end of the questions on Book I, and in particular after the questions on *Phys*. I.7–9. That these two questions are not in their original location is proved by the fact that in the title of q. 78 the passage of the *Physics* giving rise to them (I.4, 187b27–188a2) is said to follow those giving rise to the preceding questions (I.7–9). Note, however, that a reference to the present location of qq. 78–79 at the end of the questions on Book I is found in Book VI (q. 11, n. 1*) and in Aspall's questions on *De caelo* (see Donati, 2013, 241, note 79).

[586] Arist. *De anima* II.4, 416a16–17.

[587] Arist. *Phys*. I.4, 187b20–21.

5. Item, natura carnis est natura homogenea; et si hoc, habet partes eiusdem naturae [a]cum toto[b]. Cum caro quae dicitur minima sit quanta[c], si dividatur, dividetur in carnes.

6. Item, sumatur caro minima; constat quod est caro composita sive mixta. Sed quae est proportio totorum componentium ad totam carnem, eadem[d] proportio partium componentium ad partem carnis. Sed talis est proportio totius ad totum quod facit totam carnem; ergo talis erit proportio partis ad partem quod[e] facit[f] eam esse partem carnis. Et ita erit[g] procedere in infinitum.

7. Item, sumatur minimum carnis in aliquo corpore animato; haec pars potest sentire, quia est animata. Si ergo sumatur pars minor, constat quod est animata[h], quia anima est ubique; ergo potest sentire. Et si hoc, non est sumere tam partem parvam quin potest[i] sentire. Et quod potest sentire potest sentiri; ergo non est sumere tam parvum quin possit sentiri. Et si[j] hoc, non est minimum.

8. Item, sumatur caro minima; convertit alimentum in sui naturam. Sed sicut caro est [k]divisibilis in infinitum, sic alimentum; sicut ergo totum convertit[l] totum, et pars partem. Et si hoc, sicut totum ad totum, et pars ad partem. Et ita non est minimum in re naturali.

<Quaestio 79>
Iuxta hoc quaeritur an sit sumere minimum grave.

1. Et videtur quod sit minimum actione, quia[m] Aristoteles dicit in septimo huius quod, si aliqua virtus potest movere aliquod corpus in aliquo tempore, medietas virtutis movebit medietatem corporis in medietate temporis. Et adicitur[n] quod ad veritatem huius propositionis exigitur quod, sicut fit divisio in mobili et [o]etiam in virtute movente, sic fiat medii[p] rarefactio. Sed

[a] cum toto] in toto et in parte C
[b] toto] igitur *add. s.l.* O
[c] quanta] et *add.* TC
[d] eadem] est *add.* TC
[e] quod] partes componentes *add. s.l.* O
[f] facit] facient O; faciunt T
[g] erit] est TC
[h] animata *corr. in* animatum O; animatum C
[i] potest] possit TC
[j] si *om.* O
[k] divisibilis in infinitum *corr. mg. ex* divisibilis *(lectio incerta)* T
[l] convertit] ad partem *add. sed del.* O
[m] quia *om.* TC *sed suppl. s.l.* T
[n] adicitur] dicitur TC
[o] etiam in] in T; *om.* C
[p] medii *suppl. s.l.* O

5. Furthermore, the nature of flesh is a homogeneous nature, and if this is the case, it has parts of the same nature as the whole. Since the flesh that is said to be minimal <in amount> is a quantity, if it is divided, it is divided into flesh.

6. Furthermore, take a minimal <amount of> flesh; it is evident that that minimal <amount of> flesh is composite or mixed. But the relation of the components of the parts to the parts of flesh is the same as the relation of the whole components to the whole <piece of> flesh. But the relation of the whole <components> to the whole <piece of flesh> is such that the whole <components> produce the whole <piece of> flesh; therefore, the relation of a part <of a component> to a part <of the flesh> is that the part <of the component> produces a part of the flesh, so that there will be an infinite regress.

7. Furthermore, take a minimum <amount of> of flesh in an animate body; this part can have sense perception because it is animate. If a smaller part is taken, it is evident that this part is animate, because the soul is everywhere; therefore, it can have sense perception. If so, it is not possible to take a part so small that it cannot have sense perception. And what can have sense perception can be perceived; therefore, it is not possible to take something so small that it cannot be perceived. If so, there is no minimum.

8. Furthermore, take a minimal <amount of> flesh; it converts food into its own nature. But just as flesh is divisible to infinity, so is food; therefore, just as the whole converts the whole, so a part converts a part. If this is the case, then just as the whole is to the whole, so a part is to a part. Thus, there is no minimum in a natural thing.

<Question 79>
In connection with this the question is whether it is possible to find a minimal heavy body.

1. Arguments that there is a minimum <required for performing> an action: in *Physics* VII Aristotle says that, if a power can move a body in some time, half of that power will move half of that body in half of the time.[588] And it is added that for the truth of this proposition it is required that just as the mobile thing and also the moving power are divided, so the medium is rarefied.

[588] Arist. *Phys.* VII.5, 250a4–7. In this passage, however, Aristotle says that half of the power will move half of the body (over the same space) in the same time, and not in half of the time. The formulation of Aristotle's rule given here by Aspall is found in Aver. *In Phys.* VII, t.c. 36, fol. 335va.

constat quod medium est determinatum in raritate et densitate, ita quod aer potest ita fieri rarus **[O 102rb]** quod, si fiat rarior, non manet in sua specie, vel ita densus quod, si fiat densior, non manet in specie aeris, sed transmutatur in aquam, [a]et ita de aliis mediis. Si ergo divisioni mobilis respondeat medii rarefactio ad hoc quod mobile possit in suam operationem sive in suum motum, cum ergo contingat stare in medii rarefactione, erit stare in divisione corporis, ita quod, si minus sumatur, non poterit in suam operationem. Et si hoc, erit necessario invenire minimum actione.

2. Item, nisi esset resistentia in medio, non fieret aliquis motus sive successio. Tunc sic: Aristoteles dicit in capitulo de infinito quod in tantum posset[b] aliquid crescere super aliud quod non reciperet resistentiam in illo, sed[c] statim corrumperet illud; ergo in tantum poterit aliquid decrescere quod non resisteret suo patienti. Si ergo medium sit passivum motus gravis, tunc in tantum poterit grave decrescere quod non poterit in [d]actionem quae debetur ei in medio; et ita erit minimum grave actione.

3. Ad oppositum: Aristoteles dicit quod omne grave gravius, et hoc quia est gravius sua parte; totum ergo grave et pars sua comparantur in gravitate. Et si hoc, gravitas est univoce in illis. Sed in toto est grave actione; ergo et[e] in sua parte erit grave actione; et ita non est ibi[f] minimum actione.

4. Item, Aristoteles in septimo huius dicit quod, si aliqua virtus potest super aliquod totum, medietas virtutis potest super medietatem illius totius. Sed si esset minimum actione, esset tunc [g]proportio Aristotelis falsa. Videtur ergo per Aristotelem quod non sit minimum actione.

5. Item, omne grave, secundum quod grave est, inclinatur ad suum locum. Si ergo sit minimum actione, est grave. Et si hoc, partes eius sunt graves. Et si hoc, inclinantur naturaliter deorsum sive ad suum locum. Sed inclinatio naturalis est causa motus naturalis; ergo naturaliter moventur[h] deorsum. Et si hoc, non est minimum grave.

[a] et … mediis . *corr. mg. in* vel ignem per nimiam rarefactionem O
[b] posset] potest T; possit C
[c] sed] et ita TC
[d] actionem quae] actione quod C
[e] et *om.* TC
[f] ibi *om.* TC
[g] proportio Aristotelis] propositio T; propositio Aristotelis C
[h] moventur] movetur OTC *sed corr. mg.* T

But it is evident that the medium is determinate in rarity and density in such a way that air can become so rare that, if it becomes rarer, it does not remain in its species, and it can become so dense that, if it becomes denser, it does not remain in the species of air, but is changed into water; and the same applies to the other media. Therefore, if in order for a mobile thing to be able to perform its operation or its motion the rarefaction of the medium must correspond to the division of the mobile thing, and since the rarefaction of the medium comes to an end, so the division of a body will also come to an end, in such a way that, if something even smaller were reached, it would not be able to perform its operation. If this is the case, it will be necessary to find a minimum <required for performing> an action.

2. Furthermore, if the medium offered no resistance, motion or succession would not take place. Then it can be argued as follows: in the chapter on the infinite Aristotle says that a thing can increase so much with respect to another thing that it would not meet any resistance from that other thing, but would immediately corrupt that thing;[589] therefore, something can decrease so much that it would offer no resistance to its patient. Therefore, if the medium can be subject to the motion of a heavy body, then a heavy body can decrease so much that it will not be able to perform the action ascribed to it in the medium; and thus there will be a minimal heavy body that can perform an action.

3. To the contrary: Aristotle says that every heavy body is heavier, [590] and this is because it is heavier than its part; therefore, the whole heavy body and its part can be compared with respect to heaviness. If this is the case, then heaviness exists univocally in them. But the whole body is heavy with respect to performing an action; therefore, its part will also be heavy with respect to performing an action; and thus there will be no minimum heaviness for performing an action in that body.

4. Furthermore, in *Physics* VII Aristotle says if a power is effective on a whole, half of the power is effective on half of that whole.[591] But if there were a minimum <power required> for an action, then the relation assumed by Aristotle would be false. Therefore, it seems that according to Aristotle there is no minimum <power required> for an action.

5. Furthermore, every heavy body in so far as it is heavy has an inclination towards its place. Therefore, if there is a minimum <required> for <the performance of> an action, this is a heavy body. If so, its parts are heavy, and thus they have a natural inclination downwards or towards their place. But a natural inclination is the cause of a natural motion; therefore, the parts move naturally downwards. If so, there is no minimal heavy body.

[589] Arist. *Phys*. III.5, 204b14–19.
[590] Arist. *De caelo* III.1, 299a31–b1.
[591] Arist. *Phys*. VII.5, 250a4–7.

6. Item, si sumatur minimum grave, si dividatur, non dividitur in materiam tantum vel formam tantum; dividitur ergo in compositum. Sed omnis forma in materia naturali habet operationem[a] naturalem; forma ergo partis in sua materia habet actionem naturalem, et ita non est minimum actione.

7. Item, sit minimum grave actione et sit deorsum. Si dividatur, partes eius[b] non moventur[c] sursum; quiescunt ergo deorsum. Sed eadem natura qua grave quiescit in suo loco movetur ad suum locum, [d]et hoc si sit extra illum. Cum ergo[e] non sit tam parvum grave quin, si sit in suo loco, ibi quiescat, ergo non est tam parvum grave quin, si sit extra suum locum, movebitur ad suum locum.

8. [f]Solutio sive rei veritas problematis secundum quosdam talis est, quod in corpore naturali nec est sumere minimum actione neque virtute. Et glossant propositionem[g] secundi *De anima*, cum dicit quod "Omnium natura constantium", hoc est a natura animae. Et aliter non intendit [h]secundum eos, quia statim post dicit quod, si igni apponerentur ligna in infinitum, ignis cresceret in infinitum. Non ergo intendit nisi de illis quae [i]constant a natura animae.

9. Aliter dicitur quod est minimum dupliciter: vel actione vel virtute[j]. Minimum virtute non est possibile reperire, minimum autem actione est possibile. Et istam distinctionem dicit Aristoteles in *De sensu et sensato*, ibi: "Obiciet autem", ubi dicit quod passio sensibilis est duobus modis, scilicet aut actione aut virtute. Et est passio divisibilis in infinitum virtute secundum divisionem subiecti, quia non est aliqua tam parva pars passionis sensibilis quin virtute et natura sua sit sensibilis. Actione vero non est passio divisibilis in infinitum. Est enim quantitas determinata in qua ita quod non in minori potest sensibile movere sensum.

[a] operationem] comparationem TC *sed corr.* T
[b] eius *om.* TC *sed suppl. mg.* T
[c] moventur] moveretur O
[d] et hoc *om.* TC
[e] ergo *om.* OTC *sed suppl. s.l.* OT
[f] Solutio … est] quidam dicunt TC
[g] propositionem] propositiones TC
[h] secundum eos *om.* TC
[i] constant … est minimum (n. 9) *suppl. mg.* T
[j] virtute] ad virtutem *add. sed del.* O

6. Furthermore, take a minimal heavy body. If it is divided, it is not divided into matter alone or into form alone; therefore, it is divided into a composite. But every form existing in natural matter has a natural operation; therefore, the form of a part in its matter has a natural action, and thus there is no minimum <required> for <the performance of> an action.

7. Furthermore, suppose that there is a minimal heavy body required for <the performance of> an action and that it is in the downward place. If it is divided, its parts do not move upwards; therefore, they are at rest in the downward place. But a heavy body when it is outside its place moves towards its place in virtue of the same nature in virtue of which it is at rest in its place. Therefore, since there is no heavy body so small that it is not at rest in its place, when it is in its place, therefore there is no heavy body so small that it will not move towards its place, when it is outside its place.

8. The solution of the problem or the truth of the matter according to some people is that in a natural body it is not possible to find a minimum <required> for <the performance of> an action or for a power. These people gloss Aristotle's proposition in *On the Soul* II when he says that "Of all things that exist by nature"[592] by saying that it refers to the nature of the soul. According to them, Aristotle does not intend the proposition to mean something else, because immediately afterwards he says that if wood were added to wood to infinity, fire would increase to infinity.[593] Therefore, Aristotle is only referring to things that exist in virtue of the nature of the soul.[594]

9. Another opinion is that there are two kinds of minimum: either for action or for power, and that it is not possible to find a minimum for a power, but it is possible to find a minimum for an action. Aristotle presents this distinction in *On Sense*, in the passage "One may object", where he says that a passion is sensible in two ways, that is, either with respect to action or with respect to power.[595] And a passion is divisible to infinity with respect to power in accordance with the division of its subject, because there is no part of a sensible passion so small that it is not sensible with respect to its power and its nature. But a passion is not divisible to infinity with respect to an action; for there is a determinate quantity such that a sensible thing of a quantity less than that quantity cannot move the senses.

[592] Arist. *De anima* II.4, 416a16–17.
[593] Arist. *De anima* II.4, 416a15–16. Note that in Aristotle's text this passage precedes instead of following the passage mentioned in the previous quotation.
[594] The author of the commentary on the *De generatione et corruptione* attributed to Richard Rufus and Roger Bacon hold a similar position; see Richard Rufus of Cornwall, *In De gen. et corr.*, 133.455–465; Roger Bacon, *Supra libros octo Physicorum*, 33.4–15.
[595] Arist. *De sensu* 6, 446a4–7.

10. Aliter dicitur sic, quod est loqui de re naturali dupliciter: vel in quantum res est et informata vel in quantum naturalis est. Si loquamur [a]in quantum res est et informata, sic non est sumere minimum, quia non est tam parvum sumere quin sit informatum[b] sive quin sua forma perficiat suam materiam. Si autem loquamur de ea prout naturalis est, sic est sumere minimum, et hoc[c] quia natura unde natura[d] ordinata est ad motum et[e] actionem; unde, quia est sumere minimum quod potest ad motum sive actionem suae[f] speciei deputatam, dicimus quod illo modo est sumere minimum, sicut patet in specie[g] gravis[h].

11. [i]Aliter dicitur sic, quod est sumere minimum. Et[j] si dividatur[k], erit tandem devenire ad minimum. Et[l] si ulterius dividatur, cedet in medium divisionis et non in extrema divisionis.

12. Aliter dicitur sic et melius, ut mihi videtur, quod est loqui[m] de re naturali dupliciter. Aut enim[n] de illa re quae, si dividatur, partes manent aequivoce partes, sicut est de carne. Et si loquamur de divisione actuali huiusmodi rerum, sic est sumere minimum, quia totum est minimum quia quaelibet pars carnis, cum est in toto et animata ab ipsa[o] anima totius, sentit; si autem dividatur a toto, nec sentit nec sentire potest. Si autem in huiusmodi loquamur de divisione mentali, dicendum sicut postea dicetur. Si autem loquamur de toto cuius partes divisae sunt partes univoce, tunc est loqui de minimo dupliciter: vel de minimo virtute vel actione. Si **[O 102va]** autem loquamur de minimo virtute, sic non est possibile sumere minimum. Si[p] loquamur de minimo actione, tunc contingit loqui dupliciter de actione: vel de actione quae est perficere suam materiam, et haec est actio suae speciei deputata, et sic non est sumere minimum; aut de actione [q]in extrinsecus. Sed agere in extrinsecus contingit dupliciter: vel ratione contrarietatis, ut aliud convertat in sui naturam, vel ratione convenientiae cum loco, ut acquirat sibi

[a] in quantum *om.* C
[b] informatum] informata OT
[c] hoc *om.* OTC *sed suppl. s.l.* O
[d] natura *suppl. s.l.* O
[e] et] ad *add.* TC
[f] suae *ex corr.* O
[g] specie *corr. mg. in* genere T
[h] gravis] generis OT *sed corr. mg.* T
[i] Aliter ... extrema divisionis *scrips. sed del.* O
[j] Et] quia T; quod C
[k] dividatur] dicatur O
[l] Et] quia T; quod C
[m] loqui *suppl. s.l.* O
[n] enim *om.* TC
[o] ipsa *om.* TC
[p] Si] autem *add.* TC
[q] in ... extrinsecus] intrinsecus sed agere intrinsecus C

10. Another opinion is that it is possible to speak of a natural thing in two ways: either (i) in so far as it is a thing and has a form or (ii) in so far as it is natural. In case (i) it is not possible to find a minimum because it is not possible to find a thing so small that it does not have a form or that its form does not give perfection to its matter. In case (ii), it is possible to find a minimum, and this is because nature in so far as it is nature is ordered to motion and to action; thus, since it is possible to find the minimum thing that can perform a motion or an action proper to its species, we say that it is possible to find a minimum in that sense, as is clear where the species of a heavy body is concerned.

11. Another opinion is that it is possible to find a minimum. And if that minimum is divided, we will eventually arrive at a minimum. If it is further divided, it falls into the middle of the division and not into the extremes of the division.

12. Another and better opinion, in my view, is that it is possible to speak of a natural thing in two ways. Either (i) of a thing such that, if it is divided, its parts remain parts only equivocally, as happens in the case of flesh. (ia) If we speak of the actual division of a thing of this kind, then it is possible to find a minimum, because the whole is a minimum since every part of flesh has sense perception when it is in the whole and is animated by the soul itself of the whole, but if it is divided from the whole, it does not have sense perception nor can it have it. (ib) If we speak of the mental division of a thing of this kind, then the reply will be the one given later. (ii) If we speak of a whole whose parts once divided are <still> univocally parts, then we can speak of a minimum in two ways: either (iia) of the minimum <required> for a power or (iib) <of the minimum required> for an action. In case (iia) it is not possible to find a minimum. Case (iib) is twofold: either (iib') we are speaking of an action that consists in giving perfection to its matter, and this is an action proper to its species, and in this case it is not possible to find a minimum; or (iib'') we are speaking of an action affecting something external. But something can act on something external to it either in virtue of a contrariety, in order to convert the other thing into its own nature, or in virtue of the

locum. Sed sive sic sive sic, est loqui dupliciter de illa actione: vel de actione inclinante et consequente effectum, et sic est minimum, quia potest dividi in tam parva quod, si in minora dividatur, neutra pars poterit aliud convertere actu in sui naturam neque movere de se actualiter ad suum locum. Vel de actione inclinante, non tamen effectum consequente, et sic non est sumere minimum, quia non est dividere aliquid in tam parva quin semper in partibus sit inclinatio naturalis ad suum locum et etiam ut agat [a]in aliud ut aliud convertat, licet non possit consequi effectum.

[a] in ... ut aliud] ut aliud *scrips. sed corr. s.l. in* in aliud ut C

suitability of a place, in order to acquire a place for itself. In both cases we can speak of an action in two ways: either (iib''1) we can speak of an action that inclines towards and achieves its effect; and in this way there is a minimum, because a thing can be divided into something so small that, if it is divided into even smaller parts, none of the parts will be able to actually convert another thing into its own nature or to actually move by itself towards its place. Or (iib''2) we can speak of an action that inclines towards but does not achieve its effect; and in this way it is not possible to find a minimum, because it is not possible to divide a thing into a part so small that the part does not have a natural inclination to move towards its place and also to act on another thing in order to convert it, although that part cannot achieve its effect.

LIBER II

1*. [a]*Eorum quae sunt aliqua quidem sunt natura.*

2*. [b]Quaeritur de natura primo. Et hic primo quaeritur utrum natura sit principium; secundo, dato quod sit principium, quaeritur utrum[c] materia vel forma; tertio, [d]dato quod sit forma, utrum sit forma actu vel [e]non, sed potentia.

<Quaestio 1>

1. [f]Quod autem natura non sit principium videtur: principium unde principium est in potentia est; [g]natura autem non est aliquid in potentia, quia, si sic, aut est in potentia passiva aut in potentia activa. Si in[h] activa, cum, secundum Aristotelem in quinto *Metaphysicae*, potentia activa est principium transmutandi aliud secundum quod aliud, tunc natura esset principium transmutandi aliud secundum quod aliud. Sed hoc est falsum, cum dicat Aristoteles quod natura est principium motus eius in quo est et non alterius. Si passiva, cum, secundum Aristotelem in eodem passu [i]in quinto *Metaphysicae*, potentia passiva est principium transmutandi ab alio, tunc esset natura [j]principium transmutandi ab alio, quod falsum est, quia est principium[k] eius in quo est etc.

2. Item, si natura est principium, aut ergo materia aut forma aut privatio; sed non est aliquid istorum, sicut postea patebit.

3. Ad oppositum: natura est principium motus et quietis eius etc., secundum Aristotelem; ergo, secundum Aristotelem, natura est principium.

4. Item, Aristoteles in secundo huius definit naturam et postea dividit eam in materiam et formam; sed constat quod eandem naturam quam definit dividit; natura ergo ibi definita aut est materia aut forma, [l]et omne tale est principium; quare etc.

[a] Eorum ... natura *om.* T; Eorum autem quae sunt C
[b] Quaeritur ... quaeritur] <S>uper secundum Physicorum quaeritur de natura et primo T; propterea (*lectio incerta*) quaeritur C
[c] utrum] sit *add.* TC
[d] dato quod] si TC
[e] non, sed *om.* TC
[f] Quod autem] Circa primum quod C
[g] natura ... potentia *om. hom. sed suppl. mg.* T
[h] in] potentia *add.* TC
[i] in quinto] quinti TC
[j] principium ... alio] huiusmodi C
[k] principium] motus *add.* TC
[l] et ... etc. *suppl. mg.* T (ergo *pro* quare)

BOOK II

1*. *Of things that exist, some exist by nature.*[1]

2*. Questions about nature are raised first. The first question is whether nature is a principle; granted that nature is a principle, the second question is whether nature is matter or form; granted that nature is form, the third question is whether nature is a form in act, or not in act, but in potency.

<Question 1>

1. Arguments that nature is not a principle: a principle in so far as it is a principle is in potency; but nature is not something in potency; for if nature is in potency, it is either (i) in passive potency or (ii) in active potency. In case (ii), since according to Aristotle in *Metaphysics* V, an active potency is a principle of changing another thing in so far as it is other,[2] then nature would be a principle of changing another thing in so far as it is other. But this is false, since Aristotle says that a nature is the principle of motion of the thing in which the nature exists and not of another thing.[3] In case (i), since according to Aristotle in the same passage of *Metaphysics* V, a passive potency is a principle of being changed by another thing,[4] then nature would be a principle of being changed by another thing, which is false, because a nature is a principle of the thing in which it exists <and not of another thing>.

2. Furthermore, if nature is a principle, then it is either matter or form or privation; but nature is none of these, as will be clear below.[5]

3. To the contrary: according to Aristotle, a nature is the principle of motion and rest of the thing <in which it exists>;[6] therefore, according to Aristotle, nature is a principle.

4. Furthermore, in *Physics* II Aristotle first gives the definition of nature and then divides nature into matter and form;[7] but it is apparent that he divides the same nature that he defines; therefore, the nature defined there is either matter or form; and everything of this kind is a principle; hence, <nature is a principle>.

[1] Arist. *Phys.* II.1, 192b8–9..
[2] Arist. *Metaph.* V.12, 1019a15–23.
[3] Arist. *Phys.* II.1, 192b20–22.
[4] Arist. *Metaph.* V.12, 1019a15–23.
[5] See below, q. 2, nn. 1–2; q. 3, nn. 1–6..
[6] Arist. *Phys.* II.1, 192b20–23.
[7] Arist. *Phys.* II.1, 192b8–193a9, 193a9–b6.

<Quaestio 2>
Consequenter, dato quod sit principium, quaeritur an sit materia.

1. Quod autem non sit materia videtur sic: natura est principium motus et quietis non cuiuscumque, sed motuum[a] determinatorum a natura; sed materia est aequaliter principium determinatorum et non determinatorum; ergo natura non est materia.

2. Item, per eandem naturam per quam fit motus naturalis ad quietem naturalem et quies naturalis in suo loco, per illud idem appetit res[b] esse in suo loco cum est extra; sed materia nec appetit moveri ad locum nec quiescere de se; ergo natura non est materia.

3. Ad oppositum: Aristoteles primo definit naturam et consequenter dividit eam in materiam et formam, quod non faceret nisi materia esset natura.

4. Item, super illum passum dicit Commentator quod materia omnium prima est natura omnium.

<Quaestio 3>
Quaeritur consequenter an sit forma.

1. Et primo videtur quod non: transmutatio naturalis est propter formam ut propter finem, et ita transmutatio praecedit formam; sed natura praecedit transmutationem; si ergo forma est posterius transmutatione et natura prius transmutatione, tunc natura non est forma.

2. Item, si in transmutatione naturali natura est forma, aut est forma agentis[c] aut forma passi. Quod non sit forma agentis videtur, quia illa est principium [d]transmutandi aliud secundum quod aliud, et natura est principium motus [e]et quietis eius in quo est et non alterius. Nec est forma passi, quia, si esset, tunc aut esset forma reperta in passo in principio transmutationis aut forma quae debet induci post transmutationem. Non forma reperta in passo, quia natura operatur et nititur ad corruptionem illius et inductionem alterius, et nihil operatur ad sui ipsius corruptionem; ergo natura non est illa forma.

[a] motuum *transp. post* principium TC
[b] res *suppl. s.l.* O; *om.* TC
[c] agentis *ex corr.* O
[d] transmutandi ... principium *om. hom.* C
[e] et quietis *om.* TC

<Question 2>
Granted that nature is a principle, the next question is whether nature is matter.

1. Arguments that nature is not matter: nature is a principle of motion and rest, not of any motions but only of those determined by nature; but matter is equally a principle of motions determined by nature and those not so determined; therefore, nature is not matter.

2. Furthermore, the nature in virtue of which a thing moves naturally towards its natural rest and is at rest naturally in its own place is the same principle in virtue of which that thing has the appetite for being in its own place when it is outside it; but matter by itself has neither the appetite for motion towards a place nor for rest in a place; therefore, nature is not matter.

3. To the contrary: Aristotle first gives the definition of nature and then divides it into matter and form;[8] but he would not proceed in this way unless matter were a nature.

4. Furthermore, in his comment on that text the Commentator says that the prime matter of all things is the nature of all things.[9]

<Question 3>
The next question is whether nature is form.

1. First the arguments for the negative: natural change is prior to form because form is the end for the sake of which change occurs; but nature is prior to change; therefore, since form is posterior to change and nature is prior to it, nature is not form.

2. Furthermore, if in natural change a nature is a form, it is either (i) the form of the agent or (ii) the form of the patient. Nature, however, is not (i) because that form is a principle of changing another thing in so far as it is other, whereas a nature is the principle of motion and rest of the thing in which it exists and not of another thing. Nature is not (ii) either, because if it were, it would either be (iia) the form existing in the patient at the beginning of the change or (iib) the form to be induced in the patient at the end of the change. But a nature is not (iia), because a nature acts and strives for the corruption of that form and for the induction of another form, and nothing acts for the corruption of itself; therefore, a nature is not that form.

[8] Arist. *Phys.* II.1, 192b8–193a9, 193a9–b6.
[9] Aver. *In Phys.* II, t.c. 9, fol. 51va.

3. Item, [a]ad idem: natura est principium transmutationis inducens[b] formam contrariam formae inventae, [c]et ita operatur ad corruptionem formae inventae [d]et, si hoc, non fuit illa.

4. Et etiam natura manet in tota transmutatione, et forma prius reperta non manet in tota transmutatione; ergo etc.

5. Et praeterea[e], illa forma[f] in tota transmutatione magis debilitatur et magis, et natura magis vigoratur, quia motus naturalis est fortior in fine quam in principio.

6. Non forma quae debet induci, quia illa est finis transmutationis, et natura est principium transmutationis; ergo natura non[g] est illa forma.

7. Ad oppositum: Aristoteles dividit naturam in materiam et formam, et ita natura est forma.

8. Et etiam ibidem dicit quod forma est magis digna dici natura quam materia.

<Quaestio 4>
Consequenter, dato quod[h] sit forma, quaeritur utrum sit forma actu aut forma potentia.

1. Quod sit forma in potentia videtur: principium unde principium est in potentia; si ergo natura sit principium et sit forma, erit forma in potentia et non in actu.

2. Item, forma in actu est perficiens et complens et quiescens; sed principium sive natura est transmutans et in fieri **[O 102vb]** respectu completi et semper appetit esse sub forma completa; natura ergo non debet dici forma in actu.

3. Ad oppositum: omne movens secundum quod tale est in actu; sed natura movet; ergo natura est aliquid in actu; si ergo sit forma, est forma in actu.

4. Item, Aristoteles in hoc secundo probat formam esse naturam. [i]Et arguit sic: unumquodque naturale magis est in sua natura quando est in actu quam quando est in potentia; sed forma est in[j] actu et materia in potentia;

[a] ad idem *om.* C
[b] inducens *lectio incerta* (*corr. ex* producens ?) O
[c] et ... inventae *om. hom.* C
[d] et ... fuit *iter.* C
[e] praeterea] etiam TC
[f] forma] *illegibile add.* O
[g] non *suppl. mg.* O; *om.* C
[h] quod] natura *add. s.l.* O
[i] Et arguit sic *om.* C
[j] in *om.* O

3. Furthermore, to the same conclusion: a nature is a principle of change, a principle that induces a form contrary to the form initially existing in the patient, so that a nature acts for the corruption of that initial form, and, if so, then a nature is not that form.

4. Furthermore, a nature remains throughout the whole change; but the form initially existing in the patient does not remain throughout the whole change; therefore, <a nature is not that form>.

5. Furthermore, the initial form becomes weaker and weaker throughout the whole change; but the nature becomes stronger, since the natural motion is stronger at the end than at the beginning.

6. A nature is not (iib) the form to be induced, because that form is the end of the change, whereas a nature is the principle of the change; therefore, a nature is not that form.

7. To the contrary: Aristotle divides nature into matter and form,[10] so that nature is form.

8. Also, Aristotle says there that form is more worthy of being called nature than matter.[11]

<Question 4>
Granted that nature is form, the next question is whether nature is a form in act or a form in potency.

1. Arguments that a nature is a form in potency: a principle in so far as it is a principle is in potency; therefore, if a nature is both a principle and a form, it will be a form in potency and not in act.

2. Furthermore, a form in act gives perfection and completion and is at rest; but a principle or nature is changing and in becoming with respect to something complete, and always has an appetite for being under the complete form; therefore, a nature should not be said to be the form in act.

3. To the contrary: every mover in so far as it is a mover is in act; but a nature is a mover; therefore, a nature is something in act; therefore, if a nature is a form, it is a form in act.

4. Furthermore, in *Physics* II Aristotle proves that a form is a nature, arguing as follows: every natural thing is in its nature more properly when it is in act than when it is in potency; but a form is in act and matter is in

[10] Arist. *Phys.* II.1, 193a9–b6.
[11] Arist. *Phys.* II.1, 193b3–8.

ergo forma magis digna est dici natura quam materia; ergo, secundum Aristotelem, forma in actu magis est natura quam forma in potentia.

5. Item, forma quae solum est in potentia non meretur nomen formae; sed una pars naturae meretur nomen formae, quia Aristoteles dividit naturam in materiam et formam; ergo natura est forma in actu.

6. Quidam dicunt sic quod natura nec est materia nec forma quae sunt partes compositi, sed virtus caelestis sive natura lucis caelestis quae omnia disponit in suis naturis et est causa ordinationis universi, secundum quod dicit Aristoteles in libro[a] *De generatione* quod elementa deterius agunt etc.

7. Contra istud sic: Aristoteles dividit naturam[b] in materiam et formam quae sunt partes compositi.

8. Item, dicit Aristoteles quod unumquodque magis est in sua natura cum est in actu quam cum est in potentia; intendit ergo quod illud sit natura quod prius est in potentia et postea in actu; et huiusmodi sunt partes compositi.

9. Aliter dicitur sic quod natura nec est materia nec forma, sed est quoddam[c] aggregatum ex materia et forma[d]. Unde debemus imaginari quod omnes formae speciales et distinguentes unum ab alio circumscriberentur, ita quod non maneret[e] nisi forma prima sive[f] substantialitas. Tunc illud quod manet est homogeneum et est principium omnium et est quoddam aggregatum confusum. Et est alicubi sub maiori densitate, ut prope centrum, et alicubi sub maiori raritate, ut prope circumferentiam, et recipit distinctionem secundum diversitatem formarum advenientium. Et est aliquo modo actu, quia incomplete, et aliquo modo potentia, quia non est sub actu completo. Et illud aggregatum est natura.

10. Contra hoc [g]est quod[h] prius, et etiam ratio Aristotelis quae vult quod forma quae dat complementum est natura.

11. Et etiam si quaeratur de terra, cum sit sursum et descendat, constat quod per naturam descendit. Et non natura praedicta, quia tunc quidlibet descenderet, cum sit reperire illam naturam in quolibet; descendit ergo sua forma naturali, ut forma gravis, et ita sua forma naturalis est sua natura. Et ita de igne sursum, si sit extra locum suum.

[a] libro *om.* TC
[b] naturam *suppl. mg.* O
[c] quoddam *ex corr.* O
[d] forma] Et imaginatur sic istud (sic istud *suppl. s.l.* T; istud *om.* C): in toto universo diffusa est materia (natura C) quia (quae C) nihil est in universo (diffusa *add. sed del.* T) quin sit compositum ex materia et forma *add.* TC
[e] maneret *ex corr.* O
[f] sive] sicut O
[g] est quod] argutum est C
[h] quod *ex corr.* O

potency; therefore, form is more worthy of being called nature than matter.[12] Therefore, according to Aristotle, a form in act is more properly a nature than a form in potency.

5. Furthermore, a form that is only in potency does not deserve the name of form; but one part of a nature deserves the name of form, because Aristotle divides nature into matter and form;[13] therefore, a nature is a form in act.

6. Some people say that a nature is not the matter or the form that are the parts of a composite, but a celestial power or the nature of the celestial light, which arranges all things in their natures and is the cause of the order of the universe, in accordance with Aristotle's statement in *On Generation* that the elements are weaker in their actions <than the tools or instruments>.[14]

7. Against this: Aristotle divides a nature into the matter and the form that are the parts of the composite.[15]

8. Furthermore, Aristotle says that everything is in its nature more properly when it is in act than when it is in potency;[16] therefore, Aristotle means that a nature is that thing which is first in potency and then in act; and the parts of a composite are of this kind.

9. Another view is that a nature is neither matter nor form but an aggregate of matter and form. Thus, we have to imagine that all the special forms distinguishing different kinds of thing are removed, so that only the first form, or the substantiality, is left. What is left then is something homogeneous, and is the principle of all things and a confused aggregate. This aggregate has more density somewhere, as in the proximity of the centre, and more rarity elsewhere, as in the proximity of the circumference, and it varies in accordance with the diversity of the forms supervening on it. This aggregate is in a way in act, that is, incompletely, and in a way in potency, because it is not under a complete act. And that aggregate is a nature.

10. Against this are the objections above,[17] and also the argument of Aristotle which contends that the form that gives completion is a nature.[18]

11. Furthermore, if one asks about the earth, when it is up and then is moved downwards, it is apparent that it is moved downwards by nature. But not by nature as posited above; for all things should then be moved downwards, since that nature is found in all things. Therefore, the earth is moved downwards in virtue of its natural form, that is, the form of a heavy thing, so that its natural form is its nature. The same applies to the upward motion of the fire if it exists outside its place.

[12] Arist. *Phys.* II.1, 193b6–8.
[13] Arist. *Phys.* II.1, 193a28–31.
[14] Arist. *De gen. et corr.* II.9, 336a11–12.
[15] See above, note 13.
[16] See above, note 12.
[17] See above, nn. 3–5.
[18] Arist. *Phys.* II.1, 193b6–8.

12. Aliter dicitur[a], et melius, quod natura, cum sit principium, sonat in potentiam. Et quia forma est magis natura quam materia, ideo natura sonat in potentiam activam. Sed potentia activa aut fundatur super aliquid completum in actu aut super formam in potentia et diminutam. Si super formam in potentia et diminutam, tunc, cum ipsa non possit agere de se, sed excitata et mota a specie agentis immissa, cum fuerit species immissa, tunc[b] forma diminuta [c]ibi reperta cum illa specie fiunt[d] unum agens. Et totum illud aggregatum est natura in generatione naturali et principium transmutationis eius in quo est non per accidens sed per se, et in transmutatione est natura, in fine autem transmutationis est perficiens sive forma.

13. Si autem fundetur[e] super formam in actu, aut ordinatur ad agendum in aliud aut ad disponendum illud in quo est. Si autem ordinetur ad agendum in aliud, sic meretur nomen efficientis et non nomen naturae, quia sic est eius ut efficientis et non ut natura. Et sic definitur potentia activa in quinto *Metaphysicae*[f], ubi dicitur quod est principium transmutandi aliud etc. Si autem ordinetur ad disponendum illud in quo est, aut ergo disponit ipsum ad motum localem aut disponit compositum sive materiam suam. Si disponat ad locum, sic est sua forma in actu sua natura, quia forma actualis est natura in motu locali, quia ibi est principium motus. Si autem disponat suam materiam sive compositum, sic est forma quae est alterum principiorum natura rei. Et sic intendit Aristoteles et sic dividit naturam in materiam et formam.

1*. [g]Quaeritur de definitione in se.

<Quaestio 5>
[h]Et primo quaeritur utrum natura sit principium motus et quietis.

1. Et primo in communi videtur quod natura non sit principium motus sic: sicut prius patet, natura rei est sua forma vel aliquid quod fit[i] forma in actu; sed forma nata est perficere suam materiam et eam[j] complere; ergo natura similiter, cum idem sint. Sed quod est natum perficere materiam est

[a] dicitur *ex corr.* O
[b] tunc] cum *add.* OC (*s.l.* O)
[c] ibi reperta *om. sed suppl. s.l.* simul O
[d] fiunt] fuerit; *lac.* C
[e] fundetur *corr. ex* fundatur O
[f] Metaphysicae] et similiter in nono *add. s.l.* O
[g] Quaeritur ... se *om.* TC
[h] Et primo] consequenter TC
[i] fit] sit TC
[j] eam] ipsam TC

12. Another and better view is that since nature is a principle, then it is spoken of as in potency. And since a form is more a nature than is matter, then nature is spoken of as in an active potency. But an active potency is either founded (i) on something complete and in act or (ii) on a form in potency and diminished. In case (ii), since that form cannot act by itself, but only when it is stimulated and moved by the introduced species of the agent, then when that species is introduced, the diminished form existing there together with that species becomes one agent. And it is the whole aggregate that is the nature in natural generation and the principle of change of the thing in which it exists not accidentally but per se. This aggregate is the nature during the change whereas at the end of the change this aggregate is what gives perfection, that is, is the form.

13. In case (i), then it is ordered either (ia) to act on another thing or (ib) to confer some disposition on the thing in which it exists. In case (ia) it deserves the name of efficient cause and not that of nature, because it belongs to that thing as to an efficient cause and not as its nature. Aristotle defines active potency understood in this way in *Metaphysics* V where he says that an active potency is a principle of changing another thing <in so far as it is other>.[19] In case (ib), the active potency either (ib') confers a disposition to locomotion on the thing in which it exists or (ib'') a disposition on the composite or on its matter. In case (ib'), the form in act of the composite is its nature, since the form in act is the nature in locomotion, given that it is the principle of that motion. In case (ib''), the form that is one of the two principles of the composite is the nature of the thing. Aristotle refers to form understood in this way when he divides nature into matter and form.[20]

1*. Questions about the definition as such.

<Question 5>
The first question is whether nature is a principle of motion and rest.

1. First the general arguments that nature is not a principle of motion: as is clear from what precedes,[21] the nature of a thing is its form or something that becomes its form in act; but form is apt by nature to confer perfection and completion on its matter, and therefore nature too, since form and nature are the same. But what is apt by nature to confer perfection on matter is apt

[19] Arist. *Metaph.* V.12, 1019a15–16.
[20] Arist. *Phys.* II.1, 193a28–31.
[21] See above, q. 4, nn. 12–13.

natum ipsam quietare; [a]ergo natura nata est quietare illud in quo est et, si hoc, non est principium motus in eodem.

2. Item, motus facit distare id[b] in quo est, quia, quando aliquid movetur, respectu alicuius distantis movetur. Sed contra appetitum naturalem est distare; motus ergo non [c]est appetitus a natura et, si hoc, natura non est principium motus.

3. Quod autem natura non sit principium quietis videtur: locus facit rem quietare, quia unumquodque naturaliter quiescens quiescit propter convenientiam quam habet cum suo loco; sed locus non est forma, et solum forma est natura; ergo natura non est principium quietis.

4. Item, motus et quies sunt contraria; sed natura est principium motus; ergo non erit principium sui contrarii, scilicet quietis.

5. [d]Quod autem natura non sit principium in motu generationis videtur[e] sic: secundum Aristotelem, generatio et corruptio sunt mutationes subitae; sed natura non est principium alicuius nisi eius cuius motus est in tempore; natura ergo non **[O 103ra]** est principium motus [f]in generatione.

6. Quod autem non sit principium motus in augmento videtur: natura non est principium motus alicuius nisi illius cuius terminus motus est naturalis; sed terminus augmenti non est naturalis, quia terminus eius est magnitudo, quae est res mathematica; ergo natura non est principium motus augmenti.

7. Quod autem non sit principium motus decrementi sive corruptionis[g], et ita nec est principium suorum contrariorum, videtur sic: dicit Aristoteles in libro *Caeli et mundi* quod omnis debilitatio est extra naturam; sed corruptio et diminutio sunt debilitationes; ergo utrumque est extra naturam et, si hoc, natura non est principium ipsorum.

8. Item, omne quod est a natura appetitur a natura; sed corruptio sive debilitatio non appetitur a natura; ergo natura non est principium ipsorum. Probatio minoris: natura appetit esse [h]et se ipsam salvare; sed corruptio operatur ad non esse; ergo corruptio sive diminutio non appetitur a natura.

9. Contra: Aristoteles dicit in quinto huius quod quaedam quies est naturalis et quaedam violenta. Quies enim naturalis opponitur motui violento [i]et

[a] ergo ... quietare *om. hom. sed suppl. s.l.* et O
[b] id] illud TC
[c] est appetitus *lectio incerta* O
[d] Quod autem] consequenter videtur quod TC
[e] videtur *om.* TC
[f] in generatione ... motus (n.6) *om. hom. sed suppl. mg.* in motu generationis. Nec est principium motus T
[g] corruptionis *ex corr.* O
[h] et se *suppl. s.l. et* in se *add.* O
[i] et ... violentae *om.* O

by nature to confer rest on it; therefore, the nature of a thing is apt by nature to make that thing at rest, and so it is not a principle of motion in that thing.

2. Furthermore, motion makes the thing in which it exists distant from itself, since the motion of a thing occurs with respect to something distant from it; but being distant is contrary to the natural appetite; therefore, nature does not have an appetite for motion, and if so, nature is not a principle of motion.

3. Arguments that nature is not a principle of rest: place is what makes a thing at rest, since anything that is naturally at rest is at rest because of its agreement with its place; but place is not form, and only form is nature; therefore, nature is not a principle of rest.

4. Furthermore, motion and rest are contraries; therefore, since nature is a principle of motion, it will not be the principle of its contrary, namely, of rest.

5. Argument that nature is not a principle of the motion of generation: according to Aristotle, generation and corruption are sudden changes;[22] but nature is only the principle of a thing whose motion is in time; therefore, nature is not a principle of motion in the case of generation.

6. Argument that nature is not a principle of the motion of increase: nature is only the principle of a motion whose terminus is natural; but the terminus of increase is not natural, since its terminus is a magnitude, which is a mathematical and not a natural thing; therefore, nature is not a principle of the motion of increase.

7. Argument that nature is not a principle of decrease or corruption, and so is not a principle of the changes contrary to these: in *On the Heavens*, Aristotle says that every weakening is outside nature;[23] but corruption and decrease are a weakening; therefore, they are both outside nature, and if so, nature is not a principle for them.

8. Furthermore, nature has an appetite for all that which is by nature; but nature does not have an appetite for corruption or weakening; therefore, nature is not a principle for them. Proof of the minor premise: nature has an appetite for being and for preserving itself; but corruption is directed towards non-being; therefore, nature does not have an appetite for corruption or diminution.

9. Against this: in *Physics* V Aristotle says that some rest is natural and some is violent. Natural rest is opposite to violent motion, and natural

[22] The claim that generation and corruption are sudden changes is not explicitly found in Aristotle but is an interpretation of Aristotle's view that a substantial change is not a motion in the strict sense in *Phys.* V.1, 225a20–2, 225b11. Cf. Hamesse, 1974, 152. This interpretation is strongly supported by Avicenna. Cf. Avic. *Liber Primus Naturalium*, tr. 2, c. 3, ed. Van Riet-Janssens-Allard, 187.14–188.31.

[23] Arist. *De caelo* II.6, 288b14.

motus naturalis quieti violentae. Et etiam generatio naturalis opponitur corruptioni violentae [a]et econverso; ergo corruptio naturalis opponitur generationi violentae. Et ita, secundum Aristotelem, aliqua generatio est naturalis et aliqua corruptio similiter.

10. Item, Aristoteles in hoc secundo dicit quod generatio est naturalis quia est via in naturam; sed generatio est via in formam in actu et non in formam in potentia; forma ergo in actu est natura, secundum Aristotelem, sicut prius probatur. Et [b]ita vult Aristoteles quod motus generationis est motus naturalis.

11. Item, Aristoteles dicit in quinto huius[c] quod, sicut quies est naturalis et appetitur a natura, sic est mors naturalis et a natura appetitur, et propter hoc per tantum tempus potest homo durare quod mors erit ei suavis et dulcis; et ita corruptio appetitur a natura et ita naturalis est.

12. Solutio[d]. Dicendum ad primum quod natura est principium motus et quietis, sicut determinat Aristoteles. Unde, sicut prius dictum est, illud quod est natura in generatione naturali est aggregatum ex potentia activa cum specie agentis immissa; unde illud quod est ibi natura est prius movens et postea fit perfectio rei.

13. Unde dicendum primae rationi quod natura, praeter hoc quod est perfectio rei[e] sive forma rei, potest considerari ut est principium ordinandi aliud ad formam, et unum non circumscribit aliud. Et hoc patet[f], quia illud aggregatum ex potentia activa et specie immissa est natura in generatione[g], hoc est in motu generationis, et est principium ordinandi materiam[h] ad formam. Et illa eadem, completa transmutatione, fit forma perfectiva rei generatae et quietans rem in esse specifico[i].

14. Ad aliam rationem patebit.

15. Argumentis quae probant quod natura non est causa quietis dicendum primo primae quod locus non est solum causa quietis alicuius in loco, sed[j] et[k] forma naturalis est causa. Et hoc patet, quia per idem movetur aliquid ad

[a] et econverso *om.* C
[b] ita vult *ex corr.* O
[c] huius] versus finem *add. s.l.* O
[d] Solutio *om.* TC
[e] rei *ex corr.* O
[f] patet *om.* TC
[g] generatione *ex corr.* O
[h] materiam] naturam *scrips. sed corr. s.l. in* aliud T
[i] specifico] Ad aliam rationem dicendum quod sicut motus facit distare sic natura rei quia utrumque facit distare mobile actio (actio *pro* a termino) a quo, neutrum autem facit essentiam sive naturam rei generatae *add. sed del.* T
[j] sed *corr. ex* et O
[k] et *suppl. s.l.* O; in *add. sed del.* T; *om.* C

motion is opposite to violent rest. Moreover, natural generation is opposite to violent corruption and the other way around; therefore, natural corruption is opposite to violent generation.[24] Hence, according to Aristotle, some generation is natural, and similarly some corruption is natural.

10. Furthermore, in *Physics* II Aristotle says that generation is natural because it is a way towards a nature;[25] but generation is a way towards a form in act and not towards a form in potency; therefore, according to Aristotle, a form in act is a nature, as has been proved above.[26] Hence, Aristotle wants the motion of generation to be a natural motion.

11. Furthermore, in *Physics* V Aristotle says that just as rest is natural and nature has an appetite for it, so death is similarly natural and nature has an appetite for it[27]—this is why a man can live so long that death will be pleasant and sweet to him. Hence, nature has an appetite for corruption, and thus corruption is natural.

12. Solution. To the first <question about the definition> it must be said that nature is a principle of motion and rest, as Aristotle establishes.[28] Accordingly, as has been said above,[29] the thing that is the nature in natural generation is the aggregate of the active potency and of the introduced species of the agent, so that the thing that is the nature in natural generation is first a mover and then becomes the perfection of the thing.

13. Thus, in reply to the first argument[30] it must be said that in addition to being the perfection or the form of the thing, a nature can also be considered as a principle of ordering another thing to the form, and one of these properties does not exclude the other. This is clear because the aggregate of the active potency and of the introduced species is the nature in generation, that is, in the motion of generation, and is also the principle of ordering the matter to the form. And that same thing at the end of the change becomes the form that confers perfection and rest in its specific being on the thing generated.

14. The reply to the other argument[31] will be made clear.[32]

15. As to the arguments showing that nature is not a cause of rest, to the first[33] it must be said that not only place but also the natural form of a thing

[24] Arist. *Phys*. V.6, 230a18–b21.
[25] Arist. *Phys*. II.1, 193b12–13.
[26] See above, q. 4, nn. 3–5.
[27] Arist. *Phys*. V.6, 230a26–28.
[28] Arist. *Phys*. II.1, 192b20–22.
[29] See above, q. 4, n. 12.
[30] See above, n. 1.
[31] See above, n. 2.
[32] This could be a reference to Book III, q. 1, n. 17.
[33] See above, n. 3

locum per quod quiescit in loco et econverso; sed si grave sit sursum, sua forma naturali[a] movetur deorsum; et si hoc, sua forma naturali quiescit deorsum, cum fuerit ibi.

16. Ad secundum dicendum quod quies non dicit privationem simpliciter, sed quo[b] ad nos, sicut dicit Aristoteles in quinto[c] *Metaphysicae* quod unum est privatio multitudinis [d]secundum sensum, et hoc quia multitudo magis est sensibilis quam unum, unum[e] tamen simpliciter dicit positionem. [f]Et ita est quod quies dicit privationem motus secundum sensum, quia motus est magis sensibile quam quies, quies tamen simpliciter dicit positionem sive habitum.

17. Ad tertium dicendum quod motus et quies non sunt contraria quocumque[g] modo, sed motus naturalis contrariatur quieti violentae et quies naturalis contrariatur motui violento. Et sic, secundum quod sunt contraria, non est natura principium utriusque, quia natura non est principium violenti. Motus autem naturalis non contrariatur quieti naturali, quia eadem natura qua aliquid movetur ad suum locum naturalem, eadem natura quiescit ibidem. Et isto modo est natura principium utriusque, quia in talibus est natura primo principium motus et postea fit illud idem terminus motus.

18. Ad argumentum quod probat quod natura[h] non est principium motus in generatione dicendum quod generatio potest nominare idem quod ultimam mutationem, [i]et sic non est motus, vel[j] potest nominare ultimam mutationem cum tota transmutatione praecedente quae est ad mutationem, et sic generatio est motus. Et sic est natura principium sui motus.

19. Ad argumentum quod probat quod non est principium motus[k] augmenti dicendum quod est considerare[l] quantum unde quantum est, et sic mathematicum est, et sic non est natura principium motus in ipsum, vel est[m] considerare quantum secundum quod naturale est, et sic est natura principium motus in ipsum[n].

[a] naturali *om.* TC
[b] quo] quoquo modo O
[c] quinto] *recte* decimo
[d] secundum sensum *om.* C
[e] unum *om.* TC *sed suppl. mg.* non T
[f] Et ... dicit *iter.* C
[g] quocumque *corr. ex* quoquo T; quoquo C
[h] natura *om.* TC
[i] et ... mutationem *om. hom. sed* vel potest nominare ultimam mutationem et sic non est motus *suppl. mg. post* sui motus T
[j] vel *ex corr.* O
[k] motus *om.* TC
[l] considerare *om.* OC
[m] est *suppl. s.l.* O
[n] ipsum *scripsimus*; quantis (*lectio incerta, corr. ex* ipsis (?)) O; ipsis TC

is a cause of its rest in a place. The reason for this is that in virtue of the same principle a thing moves towards a place and is at rest in it, and the other way around; but a heavy body placed up moves downwards in virtue of its natural form; therefore, it is in virtue of its natural form that it is at rest in the downward place, when it gets there.

16. To the second argument[34] it must be said that rest is not a privation absolutely speaking but in comparison to us, just as in *Metaphysics* V Aristotle says that the one is a privation of a multitude in accordance with the senses, because a multitude is more sensible than the one, but the one is a position absolutely speaking.[35] Similarly, rest is a privation of motion in accordance with the senses, because motion is more sensible than rest, but rest is a position or a habit absolutely speaking.

17. To the third argument[36] it must be said that motion and rest are not contraries in every way, but natural motion is contrary to violent rest and natural rest is contrary to violent motion. Thus, in so far as they are contraries, nature is not a principle of both, because nature is not a principle of what is violent. On the other hand, natural motion is not contrary to natural rest, since it is in virtue of the same nature that a thing moves towards its natural place and is at rest there. And so nature is a principle of both motion and rest, because in these cases nature is first a principle of the motion and then becomes the terminus of that motion.

18. To the argument showing that nature is not a principle of the motion in generation[37] it must be said that generation can be taken either (i) as the same as the last change, and in this way generation is not a motion; or (ii) as the last change together with the whole preceding change directed to it, and in this way generation is a motion and nature is a principle of the motion in generation.

19. To the argument showing that nature is not a principle of the motion of increase[38] it must be said that a quantity can be considered either (i) in so far as it is a quantity, and in this way it is mathematical, and nature is not a principle of the motion towards it; or (ii) in so far as it is natural, and in this way nature is a principle of the motion towards it.

[34] The second argument is missing.
[35] The correct reference is Arist. *Metaph.* X.3, 1054a26–29.
[36] See above, n. 4.
[37] See above, n. 5.
[38] See above, n. 6.

20. Ad argumenta quae probant quod natura non est principium corruptionis sive diminutionis dicendum quod est loqui de natura rei corruptibilis dupliciter: aut de natura principiorum componentium[a] inclinantium se ad [b]esse nova aut de natura principii complentis et perficientis compositum. Si primo modo, hoc contingit dupliciter: aut de natura principiorum componentium, [c]quae corpora sunt, nitentium et inclinantium se ad sua loca; et sic dicimus quod in mixtis est corruptio naturalis[d], quia elementa quae sunt de eius compositione naturalem habent inclinationem suis locis et naturaliter nituntur ad dissolutionem mixti. Aut de natura alicuius quod est in composito, ut alicuius potentiae activae, inclinantis se naturaliter suae perfectioni, quae est forma alia quam <[e]ea quae> insit; et sic est corruptio naturalis in simplicibus, quia **[O 103rb]** potentia activa quae est in aere ad ignem est causa corruptionis naturalis aeris et inductionis formae ignis. Si autem loquamur de natura principii complentis et perficientis compositum, sic [f]est corruptio cuiuslibet contra suam naturam, quia haec natura appetit salvare se ipsam, et sic sumendo naturam, est corruptio ab extra, hoc est ab alia natura quam sit natura complens et perficiens compositum. Et sic patet responsio[g] primae rationi.

21. Ad secundam etiam patet, quia ista natura perficiens appetit esse, et ita est[h] corruptio[i] contra talem naturam. Principia intrinseca nituntur ad sua loca, et sic naturaliter nituntur [j]ad corruptionem. Et ab ista natura appetitur[k] corruptio mixti.

[a] componentium] quae corrumpuntur sunt intendentium *add. sed del. et* et *add.* T; quae corpora sunt moventium et *add.* C

[b] esse ... ad *om. hom.* TC

[c] quae corpora sunt *scrips. sed del.* O

[d] naturalis] quia corpora simplicia quae sunt ibi solum in potentia naturaliter appetunt esse in suis actualitatibus et etiam *add.* T

[e] ea quae *supplevimus*

[f] est ... sic *om. hom.* C

[g] responsio *om.* TC

[h] est *lectio incerta* O

[i] corruptio *om.* OTC *sed suppl. s.l.* T

[j] ad corruptionem] corruptioni O; *corr. ex* ad sua loca corruptionem T

[k] appetitur] accidit O

20. To the arguments showing that nature is not a principle of corruption or decrease,[39] it must be said that we can talk about the nature of a corruptible thing in two ways: either (i) as the nature of the composing principles that have an inclination towards new beings; or (ii) as the nature of the principle that confers completion and perfection on the composite. In case (i), we talk either (ia) about the nature of the composing principles that are bodies and have a tendency and inclination towards their places; and in this way we say that corruption is natural in mixed bodies, because the elements composing them have a natural inclination towards their places and naturally strive for the dissolution of the mixed. Or (ib) about the nature of something in the composite, such as an active potency, having a natural inclination towards its perfection, which is a form distinct from the form that exists in the composite; and in this way corruption is natural in simple bodies, because the active potency towards fire that exists in the air is the cause of the natural corruption of the air and of the induction of the form of fire. In case (ii), the corruption of any thing is against its nature, since this nature has an appetite for preserving itself; and taking nature in this way corruption comes from outside, that is, from a nature different from that which confers completion and perfection on the composite. And thus the reply to the first argument[40] is clear.

21. The reply to the second argument[41] is also clear, because the nature that confers perfection has an appetite for its own being, and so the corruption <of the mixed body> is against such a nature. The intrinsic principles strive for their places, and hence they naturally strive for the corruption <of the mixed body>. And it is this nature <of the intrinsic principles> that has an appetite for the corruption of the mixed body.

[39] See above, nn. 7–8.
[40] See above, n. 7.
[41] See above, n. 8.

<Quaestio 6>
Quaeritur[a] an natura sit principium motus in supracaelestibus et consequenter utrum sit principium quietis ibidem.

1. Quod autem natura non sit principium motus videtur sic: [b]in motis a natura motor non appropriat sibi magis unam partem quam aliam; sed[c] motus caeli appropriatur ad unam partem tantum, quia suus motus semper incipit ab oriente.

2. Item, omnis motor voluntarius movens aliquid cum inclinatione propria ipsius mobilis, si apponetur plus mobilis, aequaliter vel melius movebit idem; sed dicit Aristoteles quod, si stella una apponatur ipsi caelo, movebit cum labore et poena; ergo motor caeli non movet[d] caelum cum sua naturali inclinatione, et [e]si hoc, motus caeli non est a naturali inclinatione, et ita non a natura.

3. Item, omnis motus naturalis fit ut aliquid de novo acquiratur; sed caelo nihil novum acquiritur; ergo caelum naturaliter non movetur.

4. Item, Aristoteles in tertio huius dicit quod motus est actus imperfecti; sed corpus caeli nullo modo est imperfectum; ergo corpus caeli non movetur.

5. Ad oppositum: Aristoteles dicit in [f]libro *Caeli et mundi* quod motus naturalis sequitur figuram naturalem; sed figura naturalis caeli[g] est figura circularis; ergo motus eius naturalis est motus circularis.

6. Item, Aristoteles in principio *Caeli et mundi* dividit motum naturalem sic: quidam est ad[h] medium et quidam est[i] a medio et quidam circa medium; sed motus circa medium est motus circularis; motus ergo circularis est motus naturalis.

<Quaestio 7>

1. Quod autem natura sit principium quietis in caelo videtur: pono quod sit aliquod corpus caeleste non motum, sicut in quo est locus Dei et spirituum. Et bene scimus quod illud quod est infimum, ut[j] terra, est immobile. Sed si extrema sunt immobilia et quiescunt, tunc media inter haec aut quiescunt aut saltem quiescere possunt.

[a] Quaeritur] consequenter *praem.* TC
[b] in motis] motus *ex corr.* O
[c] sed *om.* O
[d] movet *corr. ex* movebit T; movebit C
[e] si ... et *om. hom.* TC
[f] libro ... mundi] de caelo et mundo TC
[g] caeli *suppl. s.l.* OT; *om.* C
[h] ad *suppl. s.l.* O
[i] est *om.* TC
[j] ut *om.* O

<Question 6>

The question here is whether nature is a principle of motion in the celestial bodies, and the next is whether it is a principle of rest there.

1. Arguments that nature is not a principle of motion: in things moved by nature the mover does not make one part of a thing more its own than another; but the motion of the heavens is proper to one part only, since their motion always starts from the east.

2. Furthermore, every voluntary mover that moves a mobile thing in accordance with its proper inclination is such that if the mobile thing is increased, it will move it equally well or better; but Aristotle says that, if a star is added to the heavens, they will move with effort and pain;[42] therefore, the mover of the heavens does not move them in accordance with their natural inclination, and if so, the motion of the heavens does not result from a natural inclination; therefore, their motion is not by nature.

3. Furthermore, every natural motion occurs for the sake of the acquisition of something new; but the heavens acquire nothing new; therefore, the heavens do not move naturally.

4. Furthermore, in *Physics* III Aristotle says that motion is the act of the imperfect;[43] but the body of the heavens is in no way imperfect; therefore, the body of the heavens does not move.

5. To the contrary: in *On the Heavens* Aristotle says that natural motion is that which follows from a natural shape;[44] but the natural shape of the heavens is circular; therefore, the natural motion of the heavens is circular motion.

6. Furthermore, at the beginning of *On the Heavens* Aristotle divides natural motion as follows: some is towards the centre, some is from the centre, and some is around the centre;[45] but motion around the centre is circular motion; therefore, circular motion is a natural motion.

<Question 7>

1. Arguments that nature is a principle of rest in the heavens: I assume that there is a celestial body not subject to motion, such as the one which is the place of God and of the spirits. Also, it is well known that the lowest body, that is, the earth, is immobile. But if these two extremes are immobile and at rest, then the intermediates between them are also at rest or at least can be at rest.

[42] Arist. *De caelo* II.12, 293a10–11. Cf. the Latin-Arabic translation of Aristotle's *De caelo* in Aver. *In De caelo* II, t.c. 71, ed. Carmody, II, 407.1–3.
[43] Arist. *Phys.* III.2, 201b31–33.
[44] Arist. *De caelo* I.2, 268b15–20.
[45] Arist. *De caelo* I.2, 268b22–24.

2. Item, cessante causa finali, cessat illud quod est ad finem; sed motus caeli propter generationem inferiorum ordinatur; et pono ex fide quod deficiet generatio; ergo tunc cessabit motus caeli.

3. Item, motus caeli est voluntarius, secundum quod dicit Commentator super secundum *De anima* quod motor caeli habet duas virtutes, et est una illarum virtus concupiscibilis, et ita est movens voluntarie; sed ita est in inferioribus moventibus voluntarie quod non imponitur eis necessitas movendi; ergo sic erit in superiore movente quod movebit ita quod non imponatur ei necessitas movendi. Et si hoc, poterit quiescere a suo motu.

4. Item, secundum Aristotelem, natura est principium motus et quietis.

5. Item, Commentator dicit quod non est[a] potentia in caelo nisi ut quiescat.

6. Ad oppositum: Aristoteles in primo *De anima*, ubi ponit opinionem Platonis, dicit quod illud quod movetur circulariter[b] non potest fieri quiescens; sed caelum movetur circulariter[c]; ergo etc.[d]

7. Dicunt quidam quod caelum movetur naturali motu, ut a figura circulari. Et ista natura non potest perpetuare motum caeli, sed potest facere unam revolutionem ex sua forma naturali; unde ad perpetuandum suum motum exigitur intelligentia movens et motum suum disponens.

8. Contra: ista forma manet semper non transmutata aliquo modo[e]; est ergo forma semper uniformiter se habens. Si ergo de sua natura potest in unam revolutionem, eadem ratione poterit in aliam, et ita sua forma naturalis poterit suum motum continuare[f] et, si hoc, non exigitur intelligentia ad continuationem motus caeli.

9. Solutio[g]. Dicendum, ut mihi videtur, quod caelum movetur a forma sua naturali sive corporali, et semper manente sua forma corporali in ea dispositione in qua nunc est poterit motum suum continuare. Unde ratio quare forma corporalis caeli[h] est suus motor secundum circulum est duplex. Una est: forma corporalis caeli est inter formas corporales nobilissima, et ideo est forma suae materiae maxime perfectiva. Sed forma corporalis magis perficiens magis extendit suam materiam, [i]et hoc ut aequaliter omnes partes perficiat; cum igitur forma corporalis caeli sit forma nobilissima, suam

[a] est *suppl. s.l.* O
[b] circulariter *scripsimus*; naturaliter OTC
[c] circulariter *scripsimus*; naturaliter OTC
[d] etc.] Item non est quies alicuius nisi cum fuerit aliquid de novo acquisitum (aliquid de novo acquiritur *pro* fuerit...acquisitum C); sed in caelo nihil acquiritur de novo; ergo etc. *add.* TC
[e] modo] *illegibile add.* O; sed *add.* C
[f] continuare *ex corr.* O
[g] Solutio *om.* TC
[h] caeli] inter omnes formas corporales nobilissima *add. sed* omnes *del.* C
[i] et ... materiam *om. hom.* C

2. Furthermore, when the final cause ceases to exist, that which is ordered to the final cause also does; but the motion of the heavens is ordered to the generation of the things here below; and I take it by faith that this generation will cease to exist; therefore, the motion of the heavens will also cease to exist then.

3. Furthermore, the motion of the heavens is voluntary, as is shown by the Commentator's statement in his commentary on *On the Soul* II that the mover of the heavens has two powers, one of which is an appetitive power,[46] and thus it is a voluntary mover; but it is not necessary for inferior voluntary movers to cause motion; therefore, the superior mover will also cause motion without this being necessary for it. And if so, the heavens can stop their motion.

4. Furthermore, according to Aristotle, nature is a principle of motion and rest.[47]

5. Furthermore, the Commentator says that the only potency there is in the heavens is the potency to be at rest.[48]

6. To the contrary: in *On the Soul* I in presenting Plato's opinion Aristotle says that what moves with a circular motion cannot come to be at rest;[49] but the heavens move with a circular motion; therefore, <they cannot come to be at rest>.

7. Some people say that the heavens are subject to natural motion, that is, that deriving from their circular shape. This nature, however, cannot perpetuate the movement of the heavens but can only produce a single revolution in virtue of its natural form; hence, an intelligence that moves the heavens and gives them the disposition to their motion is required to perpetuate their motion.

8. Against this: this form always remains unchanged in any way; hence, it is a form that always remains in the same disposition. Therefore, if it can cause one revolution by its nature, it can also cause any other revolution, so that the natural form of the heavens can perpetuate their motion, and if so, an intelligence is not required to perpetuate the celestial motion.

9. Solution. It must be said, in my view, that the heavens are moved by their natural or corporeal form, and that since this form always remains in that disposition in which it is now, the heavens can perpetuate their motion. Hence, there are two reasons why the corporeal form of the heavens is the cause of their circular motion. (i) One is that the corporeal form of the heavens is the noblest among the corporeal forms, and therefore it is a form

[46] Aver. *In De anima* II, t.c. 32, ed. Crawford, 178.22–25.

[47] Arist. *Phys.* II.1, 192b20–22.

[48] Aver. *In Metaph.* XII, t.c. 41, fol. 324va.

[49] Arist. *De anima* I.3, 407a19–23

materiam inter omnes formas corporales magis extendit. Cum ergo figura circularis sit figura [a]extensioni aptissima, tunc forma corporalis caeli sub figura[b] circulari suam materiam extendit et eam perficit, et ita figura caeli est figura circularis. Sed inter formas corporales ita est quod, quanto forma sit nobilior, tanto maiorem facit in partibus [c]suae materiae[d] unigeneitatem, et ita **[O 103va]** in partibus caeli est maxime unigeneitas. Et ita, propter unigeneitatem et convenientiam appetit quaelibet pars esse simul in[e] loco cum alia parte. Sed hoc non est possibile propter partium extensionem; ideo appetitum suum consequuntur in quantum possunt, et ita quaelibet pars succedit alii in loco. Unde appetitus partium ut simul sint est causa motus circularis totius caeli.

10. Alia causa est ista. Forma lucis est forma nobilissima sita in caelo ut caelum per suas virtutes luminares conservet haec inferiora et causet generationes in haec inferiora. Unde, si caelum staret, stellae sitae in caelo et corpora luminosa sita in caelo diversas non causarent generationes in his inferioribus nec omnes partes terrae illuminarent. Ut ergo caelum aequaliter conservet haec inferiora et diversas causet generationes et lucem suam in omnem partem [f]inferiorem communicet, appetit sua forma naturalis ipsum[g] circulariter movere.

11. Si autem velimus ponere intelligentiam circulariter moventem, tunc duplex potest esse ratio. Una est quia intelligentia appetit assimilari ipsi primo sive suo creatori, qui[h] est ubique, [i]et etiam animae, quae est ubique in suo[j] minori mundo, ut in corpore. Et hoc non potest facere de se, quia non potest esse in qualibet parte simul; ideo facit quamlibet partem loco alterius, et hoc non potest facere nisi per motum circularem.

12. Aliter dicitur sic et melius quod caelum propter sui nobilitatem exigit aliquod nobile movens se sive regulans suum motum, et etiam quia caelum causat generationes hic inferius et aequaliter est causa generationis formae corporalis et formae[k] spiritualis, ut vegetativae et sensitivae. Licet ergo forma corporalis caeli possit super corporalia, non tamen potest super spiritum vegetalem et sensibilem, quia sunt nobiliora omni forma corporali; exigitur

[a] extensioni aptissima] amplissima *ex corr.* O
[b] figura] forma O
[c] suae materiae *om.* T
[d] materiae *ex corr.* O
[e] in *om.* TC
[f] inferiorem communicet *om.* C
[g] ipsum *scripsimus*; ipsam OT; *om.* C
[h] qui] quae OC
[i] et ... ubique *om.* O; *scrips. sed del.* T
[j] suo *om.* O
[k] formae] specialis *add. sed del.* O

that gives perfection to its matter to the maximum degree. But a corporeal form that gives more perfection to its matter also gives it more extension, and does this in order to give perfection to all the parts of its matter to an equal degree; therefore, since the corporeal form of the heavens is the noblest, it is the one among all the corporeal forms that gives maximal extension to its matter. And, since a circular shape is that which is most apt by nature to be extended, then the corporeal form of the heavens gives extension and perfection to its matter under a circular shape, and thus the shape of the heavens is a circular shape. Moreover, corporeal forms are such that the nobler a form is, the greater homogeneity it produces in the parts of its matter, and so there is maximal homogeneity in the parts of the heavens. And because of this homogeneity and agreement each part has the appetite for being in the same place as another part. This, however, is not possible because these parts are extended; therefore, these parts follow their appetite as much as possible, in that each part successively takes the place of another part. Hence, a cause of the circular motion of the whole heavens is the appetite of the parts for being in the same place.

10. (ii) The other reason is that the form of light is the noblest form situated in the heavens so that the heavens in virtue of their illuminating powers can preserve the things here below and cause their generations. Accordingly, if the heavens were at rest, the stars and the other luminous bodies situated in the heavens would not cause the different generations in the things here below nor would they illuminate all the parts of the earth. Therefore, the natural form of the heavens has an appetite for moving the heavens with a circular motion in order that the heavens equally preserve the things here below, and produce different generations, and communicate their light to every part here below.

11. There can be two reasons, however, for positing an intelligence as a cause of the circular motion of the heavens. (i) One is that an intelligence has an appetite for assimilating itself to the first being, that is, its creator, which is everywhere, and also to the soul, which is everywhere in its minor world, that is, in the body. But it cannot achieve this itself because it cannot be in every part at the same time; therefore, it makes every part to be in the place of another part, and it can do this only by the circular motion.

12. (ii) Another and better reason is that the heavens because of their nobility require something noble that produces and regulates their motion, and also that the heavens cause the generations here below and they are equally the cause of the generation of both corporeal and spiritual forms, such as the vegetative and the sensitive souls. Thus, although the corporeal form of the heavens can produce corporeal beings, it cannot, however, produce vegetative and sensible spiritual beings, because these are nobler

ergo spiritus in caelo sive ad motum caeli, ad hoc quod caelum [a]suo motu possit super productionem spiritus.

13. [b]Ad illud quaere super principium *Metaphysicae*, super quaestionem illam 'An caelum movetur etc.'

14. Una ratio ad probandum quod potest quiescere est talis: natura est principium motus et quietis. Et fuit penultima.

15. Et dicunt quidam huic sic quod in caelo est motus continuus, et semper manente caelo in illa dispositione in qua nunc est movebitur. Et tamen dicunt quod in caelo est potentia ad quietem. Nec [c]dicunt quod est illa potentia frustra, licet numquam consequatur effectum, quia, quando aliquid ordinatur ad duo, si unum consequatur, licet alterum non consequatur, non dicimus[d] ipsum[e] esse frustra, [f]sed quando aliquid ordinatur ad unum tantum et non consequatur illud, tunc dicimus quod est frustra.

16. Contra: in his inferioribus ita est quod motus est propter quietem. Et quando unum est propter aliud et consequatur unum tantum, tunc dicimus [g]quod est frustra etc.[h] Si enim grave moveretur semper[i] deorsum et numquam quiesceret deorsum, diceremus quod frustra haberet potentiam quiescendi deorsum. Ergo sic erit in superioribus quod, si semper moveantur et numquam quietem consequantur, potentia quiescendi in ipsis erit frustra.

17. Si dicatur[j] quod non est ibi motus propter quietem, tunc sic: si neutrum propter alterum et habet potentiam ad utrumque, tunc, si consequatur unum et non alterum, erit una potentia frustra, sicut, si neutrum[k] consequeretur, esset totum frustra.

18. Propter hoc dicitur sic quod eadem natura quae est causa motus partium caeli est causa quietis totius. Unde, quia partes moventur circa centrum, quod est immobile, quiescit totum circa centrum immobile, et ita eadem natura est principium motus et quietis in caelo.

[a] suo motu *om.* C
[b] Ad ... etc. *om.* TC
[c] dicunt quod *om. sed* ut dicunt *post* frustra *suppl.* T
[d] dicimus] potentiam *add. s.l.* O
[e] ipsum] ipsam OC
[f] sed ... frustra *om. hom.* C
[g] quod est *om. sed* illud *suppl. s.l.* O; *om.* C
[h] etc.] est aliqua potentia C
[i] semper *ex corr.* O
[j] dicatur] dicas TC
[k] neutrum] neutram O

than every corporeal form; therefore, a spiritual being is required in the heavens or for the motion of the heavens so that the heavens by their motion can produce spiritual beings.

13. On this topic see the question 'Whether the heavens are moved <by their natural form>' at the beginning of the *Metaphysics*.[50]

14. The penultimate argument showing that the heavens can be at rest is that nature is a principle of motion and rest.[51]

15. In reply to this argument some people say that the heavens move with a continuous motion, and if the heavens remain in their present disposition, they will always move. They maintain, however, that in the heavens there is a potency to rest and that that potency is not in vain, despite the fact that it never achieves its effect. The reason for this is that when something has a potency ordered to two things, if it achieves one of the two but does not achieve the other, we do not say that that it is in vain, whereas when something is ordered to only one thing and does not achieve that thing, then we say that that it is in vain.

16. Against this: in the bodies here below motion occurs for the sake of rest. And when one of two things is for the sake of the other and only one of them is achieved, then we do say that it is in vain. For example, if a heavy body were moved downwards for ever and were never at rest in the downward place, we would say that it is in vain that it has the potency of being at rest in the downward place. Similarly, therefore, in the superior bodies, if they are always in motion and they never come to be at rest, their potency to be at rest will be in vain.

17. It might be said that in superior bodies motion does not occur for the sake of rest. Against this: if neither of two things is for the sake of the other and there is a potency to each of them, then if one of the two and not the other is achieved, one of the two potencies is in vain, in the same way that if neither of the two things were achieved, the whole potency would be in vain.

18. Because of this it is said that the same nature is the cause of the motion of the parts of the heavens and of the rest of the heavens as a whole. Hence, because the parts of the heavens move around the centre, which is immobile, the heavens as a whole are at rest around the immobile centre, and thus the same nature is the principle of motion and of rest in the heavens.

[50] This question is not found in Aspall's Questions on the *Metaphysics* preserved in MS C. Cf. Donati, 2013a, 316–317.

[51] See above, n. 4.

<Quaestio 8>

Habito[a] quod caelum movetur naturaliter, ita quod de sua natura non possit[b] quiescere secundum partes et quiescit secundum totum, ita quod secundum[c] totum non movetur, quaeritur utrum sit possibile[d] quod caelum vel aliquid aliud moveatur secundum totum et quiescat secundum partes.

1. Quod autem sit possibile videtur: privativa nata sunt fieri circa idem, et hoc vel circa idem subiectum vel circa idem specie vel circa idem genere. Si ergo in caelo sit quies secundum totum et motus secundum partes, aut ergo in caelo aut in aliquo sui generis erit motus secundum totum et quies secundum partes.

2. Item, quando aliqua dispositio inest alicui gratia suae extensionis et gratia cuiuslibet partis, tunc inerit toti, sicut patet de albedine in suo subiecto; sed motus inest caelo gratia suae extensionis et suarum partium; ergo et toti inerit.

3. Item, sicut aliquid est in loco, sic quiescit; sed secundum Commentatorem in quarto, [e]caelum non est in loco nisi per suum centrum; ergo per suum centrum quiescit et, si hoc, non quiescit per se totum[f].

4. Dicendum quod non est possibile ut[g] aliquid moveatur secundum totum et quiescat secundum partes[h], sed via versa est possibile, quia eadem natura qua caelum movetur secundum partes quiescit secundum totum, ut prius patet.

5. Unde primae rationi dicendum quod aliquando sunt privativa talia quod unum extremum privat actum et potentiam respectu alterius et non econverso, sicut caecitas privat potentiam videndi et actum similiter, et hoc in

[a] Habito *ex corr.* O
[b] possit] potest TC
[c] secundum *om.* O
[d] possibile] contrarium scilicet *add. s.l.* O
[e] caelum ... totum *corr. mg. ex* totum caelum quiescit (quiescit *del.*) per accidens quia centrum per se; sed quando aliqua dispositio inest alicui per accidens potest privari a tali dispositione; quare totum caelum potest privari a quiete et ita totum moveri T (suum (1)-(2) *om. et* ergo caelum *pro* et si hoc T)
[f] totum] Ad hoc dicendum quod haec propositio Commentatoris 'caelum est in loco per accidens quia per centrum per se' et (et *del.*) absolute loquendo falsa est quia adeo per se est caelum in loco ita quod causans suum (suum *corr. s.l. ex* super) locum sicut centrum est in loco ita quod causatum est a (a *corr. s.l. ex* in) loco. Potest tamen sic verificari quod caelum est per accidens in loco eo quod non participat condiciones locati per se quae sunt indigere loco et quod locus sit aliud a locato etc. Vel sic: continuitas essentialiter inest loco; quia ergo continuitas spatii a centro causatur potest dici quod centrum primo et per se est in loco, circumferentia vero non eo quod continuitas loci eius causatur a centro et ita (et ita *suppl. s.l.*) per accidens aliquo modo est in loco. Vel sic: caelum non est in loco nisi per suum centrum quiescit et si hoc non quiescit per se totum *add.* T
[g] ut] quod TC
[h] partes] partem O

<Question 8>

Now that it has been established that the heavens move naturally so that they cannot by their nature be at rest with respect to their parts, and that they are at rest as a whole so that they do not move as a whole, the question is whether it is possible for the heavens or some other thing to move as a whole and to be at rest with respect to their parts.

1. Arguments for the affirmative: privative things are apt by nature to come about in the same thing, and this happens either in the same subject or in the same species or in the same genus. Therefore, if in the heavens there is rest with respect to the whole and motion with respect to their parts, then either in the heavens themselves or in something of their genus there will be motion with respect to the whole and rest with respect to their parts.

2. Furthermore, when a disposition inheres in a thing in virtue of its extension and in virtue of each of its parts, it will also inhere in the whole, as is clear in the case of whiteness in its subject; but motion inheres in the heavens in virtue of their extension and of their parts; therefore, it will also inhere in the whole.

3. Furthermore, just as a thing is in a place so it is at rest; but according to the Commentator in *Physics* IV, the heavens are in a place only in virtue of their centre;[52] therefore, they are at rest in virtue of their centre, and if so, they are not at rest in virtue of themselves as a whole.

4. It must said that it is not possible that something should move as a whole and be at rest with respect to its parts, but the other way around is possible, because it is the same nature in virtue of which the heavens move with respect to their parts and are at rest as a whole, as was made clear before.

5. Accordingly, to the first argument[53] it must be replied that sometimes privative things are such that one of the two extremes is a privation of both the act and the potency with respect to the other and not the other way around, just as blindness is a privation of both the power and the act of seeing, and this with regard to its proper determinate subject, that is, the eye.

[52] Aver. *In Phys.* IV, t.c. 43, fol. 142va.
[53] See above, n. 1.

comparatione ad subiectum proprium determinatum[a], **[O 103vb]** ut in comparatione ad oculum. Et talia nata sunt inesse eidem[b]. Alia sunt privativa quae, in comparatione ad aliquod subiectum, utrumque privat actum et potentiam respectu alterius et in comparatione ad aliud non, ut patet de raro et denso. In his autem inferioribus quod est rarum nunc potest fieri densum et econverso, in superioribus autem [c]quod est rarum non potest fieri[d] densum nec econverso, secundum quod dicit Commentator quod rarum et densum aequivoce sunt in inferioribus et supracaelestibus. Unde ita est de his privativis, motus et quies. In inferioribus autem quod potest moveri, et[e] quiescere, et, quod potest moveri secundum partes, et secundum totum, et tamen[f] non in omnibus, ut forte in terra. In caelo autem non est ita, quia ibi semper unum privative oppositorum[g] privat actum et potentiam respectu alterius. Unde, quia ibi est quies secundum totum, non potest esse motus secundum totum, et quia in partibus est motus, non potest ibi esse quies.

6. Ad secundum argumentum dicendum quod quaedam insunt alicui absolute, et de talibus est propositio vera. Si autem insint per comparationem ad aliud, non necesse est quod sit vera. Unde motus non inest caelo absolute, sed per comparationem ad spatium; ideo est quod non est inconveniens quod partes alicuius transmutant locum sive spatium, licet totum non transmutet spatium vel locum.

<Quaestio 9>

Quaeritur consequenter utrum natura sit principium quietis in terra secundum totalitatem terrae.

1. Quod autem tota quiescat violenter videtur, quia cuius quaelibet pars quiescit violenter, et totum quiescit violenter. Sed quaelibet pars terrae quiescit violenter, [h]ut videtur, quia violenter quiescit alicubi quod, amoto prohibente, ab illo loco recedit; sed quaelibet pars terrae est huiusmodi, quia, remota parte inferiore, descendit pars superior, et ita est de qualibet parte terrae; ergo quaelibet pars[i] terrae quiescit violenter.

[a] determinatum] determinatur TC
[b] eidem *corr. s.l. ex* idem O
[c] quod ... Commentator quod *om.* C
[d] fieri *ex corr.* O
[e] et] potest TC
[f] tamen *ex corr.* O
[g] oppositorum] oppositum O
[h] ut ... quiescit violenter] quia amota inferiore descendit superior; ergo etc. C
[i] pars *suppl. s.l.* O

And privative things of this kind are apt by nature to inhere in the same thing. Other privative things, however, are such that, with regard to a given subject, each of the two is a privation of the act and the potency with respect to the other, but not with regard to another subject, as is clear in the case of the rare and the dense. In the region here below a thing that is now rare can become dense and the other way around, whereas in the superior region a thing that is now rare cannot become dense or the other way around, in agreement with the claim of the Commentator that the rare and the dense are equivocally in things here below and in celestial things.[54] The same applies to these privative things, motion and rest. In the region here below a thing that can move can also be at rest, and a thing that can move with respect to its parts can also move as a whole, but this is not the case with all things, such as perhaps the earth. But in the heavens the case is different because there one of the two opposites by privation is always a privation of both act and potency with respect to the other. Therefore, since in the heavens as a whole there is rest, there cannot be motion of the whole there, and since there is motion in the parts, there cannot be rest there.

6. To the second argument[55] it must be said that some things inhere in something in an absolute way, and of these things the proposition is true. But in the case of things inhering with respect to something else, it is not necessary that the proposition be true. Hence, motion does not inhere in the heavens in an absolute way, but with respect to space; this is why it is not absurd that the parts of a thing should change place or space, while the whole does not change space or place.

<Question 9>
The next question is whether nature is a principle of rest of the earth in its totality.

1. Arguments that the whole earth is at rest by violence: a thing such that each of its parts is at rest by violence is as a whole at rest by violence. But each part of the earth is at rest by violence, as it seems, because a thing such that it recedes from a place when a hindrance is removed is at rest by violence in that place; but each part of the earth is of that kind, because, when an inferior part is removed, the superior part moves downwards, and this is so for every part of the earth; therefore, every part of the earth is at rest by violence.

[54] Aver. *De sub. orb.* 2, fol. 7va.
[55] See above, n. 2.

2. Item, si terra quiescit[a] naturaliter in medio, [b]quia medium mundi est medium terrae et quia medium terrae est medium mundi, et hoc propter hoc quod partes terrae propter suam gravitatem moventur et inclinantur et nituntur ad medium, ut ad centrum, tunc, si in una parte elevaretur[c] civitas vel mons vel castrum, quantum illa pars plus ponderaret quam prius, in tantum descenderet tota terra quousque adaequarentur partes circumstantes terram, et ita esset tota terra aliquo modo mobilis.

3. Item, causa prima est [d]simpliciter immobilis et intransmutabilis; quanto ergo aliquid est propinquius causae primae, tanto minus transmutabile et, quanto remotius, tanto magis[e] transmutabile. Sed terra est illud [f]quod [g]est maxime remotum ab illo; ergo terra est illud quod est maxime mobile et transmutabile, et ita moveri potest secundum locum.

4. Item, Aristoteles [h]dicit in undecimo quod res ordinantur dupliciter, quia unus est ordo rerum ad invicem et alius est ordo rerum ad causam primam, et [i]dicit quod primus ordo non est nisi propter secundum. Si ergo omnia ordinantur ad primum, quae plus approximantur plus participant condicionem[j] ipsius et quae minus minus. Sed causa prima est omnino immobilis; quod ergo remotissimum est[k] ab [l]ipso est maxime mobile. Et terra est tale.

5. Ad oppositum: [m]dicit Aristoteles in secundo[n] *Caeli et mundi* quod corpus circulare sempiterni motus movetur circa aliquod fixum et permanens, ut circa terram.

[a] quiescit] quiescat TC

[b] quia ... mobilis] ita quod medium terrae sit medium mundi, si una pars terrae elevaretur in terra vel civitas vel castrum ubi prius non fuit (non erit circa medium ubi prius fuit *add. sed del.*) quatenus ergo medium terrae fiat medium mundi necesse esset totam moveri; ergo tota terra est mobilis C

[c] elevaretur] construeretur (*lectio incerta*) *add. s.l.* O

[d] simpliciter immobilis et *om.* C

[e] magis] maius O

[f] quod … illud *om. hom. sed suppl. mg.* quod est remotius a prima causa; ergo terra est illud T

[g] est ... transmutabile] remotissimum est a causa prima; ergo terra maxime transmutabile C

[h] dicit ... invicem] in undecimo: unus est ordo rerum inter se C

[i] dicit quod *om.* C

[j] condicionem] condicione O; condiciones C

[k] est *iter.* O

[l] ipso ... tale] ipsa ut terra est maxime mobile C

[m] dicit ... terram] in Caelo et Mundo: corpus circulare quia est sempiterni motus movetur circa aliquod quod est fixum sempiterna fixione, quod est terra C

[n] secundo *ex. corr.* O; *om.* TC

2. Furthermore, if the earth is at rest by nature in the middle, because the middle of the world is the middle of the earth and because the middle of the earth is the middle of the world, and this happens because in virtue of their heaviness the parts of the earth move and have an inclination and strive towards the middle, that is, towards the centre, then, if a town or a mountain or fortifications were raised in a part of the earth, the heavier that part is than before, the more the whole earth would move downwards until the parts surrounding the earth were levelled, and thus the whole earth would be mobile in some way.

3. Furthermore, the first cause is absolutely immobile and unchangeable; therefore, the closer something is to the first cause, the less changeable it is, and the farther something is from it, the more changeable it is. But the earth is the thing most distant from the first cause; therefore, the earth is the thing that is maximally mobile and changeable, and thus it can move with respect to place.

4. Furthermore, in *Metaphysics* XI Aristotle says that things are ordered in two ways, because one order is that of things one to the other, and the other order is that of things to the first cause; and he says that the first order only exists for the sake of the second order.[56] Therefore, if all things are ordered to the first cause, those things that are closer to the first cause participate in the condition of the first cause more fully and those that are less close participate less fully. But the first cause is absolutely immobile; therefore, what is most distant from the first cause is maximally mobile. And the earth is such a thing.

5. To the contrary: in *On the Heavens* II Aristotle says that the circular body endowed with eternal motion moves around something fixed and permanent, namely, around the earth.[57]

[56] Arist. *Metaph.* XII.10, 1075a11–15.
[57] Arist. *De caelo* II.3, 286a13–21.

6. Item, [a]dicit Aristoteles in tertio[b] *Caeli et mundi* quod terra movetur ad centrum non quia centrum terrae, sed quia est medium mundi sive centrum mundi. Et hoc apparet quia, si tota terra esset super globum lunarem et hic esset gleba terrae, gleba non ascenderet usque ad terram, sed descenderet usque ad medium mundi, ita quod demum fieret medium glebae medium mundi. Et si hoc, cum medium terrae sit medium mundi, terra naturaliter quiesceret.

7. Solutio[c]. Dicendum quod terra secundum se totam quiescit naturaliter. Unde[d] motus quandoque ordinatur ad quietem eius cuius est, ut patet in motu gravium et levium, [e]quia moventur ad locum ut quiescant in loco, quandoque autem non ordinatur motus ad quietem eius cuius est, sicut partes caeli moventur, et hoc ad quietem totius, et hoc quia eadem natura qua partes caeli moventur secundum circulum circa terram, eadem natura quiescit totum. [f]Et ita est de terra quod, quia partes caeli moventur circulariter circa terram, necesse est terram esse fixam et immobilem sicut centrum respectu circumferentiae. Unde motus partium caeli ordinatur ad quietem terrae[g] sicut ad quietem totius[h]; [i]et ita motus earum non est ad quietem earum, sed ad quietem aliorum. Unde tota terra quiescit in suo loco naturali et partes similiter in eodem loco. Nec est spatium continens terram absolute locus terrae, sed spatium per respectum ad centrum, quia, si spatium per se esset locus terrae, tunc, si spatium esset sine terra et esset hic gleba, non plus moveretur[j] ad unam partem quam ad aliam, quod est falsum. Non ergo spatium absolute est eius **[O 104ra]** locus, sed per respectum ad centrum.

[a] dicit ... quiesceret] in tertio eiusdem: terra movetur ad centrum non quia est medium terrae, sed quia est medium mundi, quia si tota terra esset supra glebam (*pro* globum) lunarem et gleba esset in medio, non ascenderet, immo tota terra descenderet, ita quod demum medium totius fieret medium mundi, quod non contingeret nisi terra naturaliter quiesceret in medio C

[b] tertio] *recte* secundo

[c] Solutio *om.* TC

[d] Unde] dico ergo quod C

[e] quia ... terram] quae moventur ad loca sua ut ibi quiescant, quandoque autem non ordinatur motus ad quietem eius cuius est, sicut est in motu partium caeli, moventur enim ad quietem totius, et hoc quia eadem natura qua partes caeli moventur circulariter C

[f] Et ita] eodem modo C

[g] terrae] similiter *add. s.l.* O

[h] totius] caeli *add.* C

[i] et ... centrum] Unde tota terra quiescit naturaliter in suo loco naturali. Nec est dicere quod spatium continens terram sit eius locus absolute, sed per respectum ad centrum, quia, si sic, tunc, si tota terra amoveretur a suo loco et gleba esset in aere, non plus moveret ad unam partem spatii quam ad aliam, quod falsum est C

[j] moveretur] moveret OT

6. Furthermore, in *On the Heavens* III Aristotle says that the earth moves towards the centre not because it is the centre of the earth but because it is the middle or centre of the world.[58] And this is evident because if the whole earth were above the globe of the moon and a lump of earth were here, that lump would not move up to the earth, but it would move down to the middle of the world, in such a way that the middle of that lump of earth would at last become the middle of the world. If this is the case, since the middle of the earth is the middle of the world, then the earth would be at rest by nature.

7. Solution. It must be said that the earth as a whole is at rest by nature. Accordingly, in some cases motion is ordered to the rest of the thing to which it belongs, as is clear in the motion of heavy and light bodies, because they move towards a place in order to be at rest in that place; but in other cases motion is not ordered to the rest of the thing to which it belongs, as for instance the parts of the heavens move and their motion is ordered to the rest of the whole, and this happens because by the same nature by which the parts of the heavens move in a circle around the earth, the heavens as a whole are at rest. And so in the case of the earth, since the parts of the heavens move in a circle around the earth, it is necessary for the earth to be fixed and immobile in the way that the centre is in relation to the circumference. Accordingly, the motion of the parts of the heavens is ordered to the rest of the earth as it is to the rest of the whole, and so their motion is not ordered to their own rest but to the rest of other things. Thus, the whole earth is at rest in its natural place and similarly its parts are at rest in the same place. Also, the place of the earth is not the space containing the earth absolutely speaking but that space in relation to the centre, because if that space by itself were the place of the earth, then, if that space were without the earth and a lump of earth were here, that lump would no more move towards one part than towards another, which is false. Therefore, it is not the space <containing the earth> absolutely speaking that is the place of the earth, but that space in relation to the centre.

[58] The correct reference is Arist. *De caelo* II.14, 296b7–18.

8. [a]Unde primae rationi dicendum[b] quod, si aliquid sit extra suum locum et, remoto[c] prohibente[d], descendit[e], tunc quievit ibi violenter. Si[f] autem sit in suo loco [g]super partes consimilis speciei, licet descendat per remotionem partis inferioris, non quievit ibi violenter, quia, quando removetur pars inferior, non descendit pars superior quia ibi quievit violenter, sed, quia grave est, descendit quatenus plus approximetur[h] centro et ut continuetur[i] suo toti et ut levius non subsideat ei. Nec movetur de loco ad locum, sed movetur in loco, et hoc quia idem est locus totius et partis quantum ad salutem, non autem quantum ad continentiam, quia totum occupat maiorem locum quam sua pars.

9. [j]In hac responsione ponitur quod spatium terrae per respectum ad centrum est locus terrae et quod spatium determinatum est.

10. Contra: si imaginaretur per impossibile quod terra cresceret usque ad [k]globum lunarem, adhuc quiesceret naturaliter in suo loco sicut nunc facit [l]et, si hoc, non est determinatum spatium quod est locus terrae. Ubicumque ergo ponatur gleba, movebitur ut[m] in suo loco et ad suum totum.

11. Item, si poneremus terram decrescere quousque fieret in tertio vel[n] quarto minor quam nunc est, quiesceret sicut nunc. Et si esset gleba in illa distantia in qua est modo extimum terrae et ibi quiesceret, quiesceret violenter. Et ita tunc non esset locus terrae qui[o] nunc est [p]et, si hoc, loco terrae non determinatur spatium aliquod[q].

[a] Unde ... dicendum] Dico ergo primae rationi C
[b] dicendum *suppl. s.l.* O
[c] remoto] amoto C
[d] prohibente] tunc *add. sed del.* O
[e] descendit] moveatur ad suum locum C
[f] Si *ex corr.* O
[g] super ... superior] si amoveatur pars prohibens motum consimilis speciei non sequitur quod quievit violenter licet moveatur ulterius. Si enim amoveatur pars inferior terrae, tunc pars superior descendit non C
[h] approximetur *ex corr.* O
[i] continuetur] continetur O; *corr. ex* continetur T
[j] In ... ponitur] haec responsio ponit C
[k] globum lunarem] orbem lunae C
[l] et ... terrae] spatium ergo quod est locus terrae non est determinatum C
[m] ut *om.* TC
[n] vel] in *add.* TC
[o] qui] quod O; *corr. ex* quod T
[p] et ... aliquod] ergo non est spatium aliquod determinatum quod sit locus terrae C
[q] aliquod] Notandum quod pars terrae antequam separatur a toto naturaliter quiescit quia est de continuitate totius, sed cum separatur, non amplius est pars sed quoddam totum in se, et tunc secundum quosdam reperta (reperta *corr. ex* repertuatur) in pristino loco numquam quiesceret naturaliter, quia enim est quoddam totum in se et grave non quiesceret naturaliter antequam medium eius esset medium mundi. Unde post separationem partis quae fuit sub ea movetur deorsum, non quia prius quievit violenter sed per naturam de novo sibi acquisitam movetur, scilicet per naturam qua totum est grave. Per hoc patet ad secundum. Non enim quiescit tota terra naturaliter quia partes inclinantur ad centrum sed quia partes naturaliter quiescunt in toto

8. Accordingly, to the first argument[59] it must be said that, if something is outside its place, and it moves downwards when a hindrance is removed, then it was at rest there by violence. But if it is in its place and is above parts of the same species, although it moves downwards when an inferior part is removed, it was not at rest there by violence, because when the inferior part is removed, the superior part moves downwards not because it was at rest there by violence, but because, being heavy, it moves downwards to get closer to the centre and to be continuous with its whole and to prevent something lighter than it from resting below it. Nor does that part move from a place to a place, but it moves in a place, and this is because the place of the whole and of the part is the same with respect to safety, but not with respect to containment, since the whole occupies a greater place than its part.

9. This reply assumes that the space of the earth in relation to the centre is the place of the earth and that that space is determined.

10. Against this: imagine the impossible situation in which the earth expands as far as the globe of the moon. Even then the earth would by nature be at rest in its place as it is now, and if so, the space that is the place of the earth is not determined. Therefore, wherever a lump of earth is placed, it will move there as in its place and towards its whole.

11. Furthermore, if we assume that the earth decreases so that it becomes one third or one fourth smaller than it is now, it would be at rest then as it is now. And if a lump of earth were at that distant place where the border of the earth is now and were at rest there, it would be at rest by violence. In that situation, the place of the earth would not be its present place and if so, there is no space determined by the place of the earth.

[59] See above, n. 1.

12. [a]Ad idem super finem libri *De vegetabilibus*.

<Quaestio 10>
[b]Habito qualiter natura est principium motus in caelestibus et quietis in terra, quaeritur de motu animatorum. Et quia[c] moventur ab anima, quaeritur[d] utrum anima sit natura.

1. Et [e]quod anima sit natura videtur sic: Aristoteles in secundo huius dicit quod quaedam sunt natura et quaedam[f] propter alias causas; natura ut animalia[g], plantae [h]et huiusmodi; sed non est animal nisi per animam; anima ergo est natura.

2. Item, Commentator dicit quod animalia moventur naturaliter; sed non moventur nisi ab anima; [i]anima ergo est natura[j].

3. Ad oppositum: duo sunt principia operationum: natura et voluntas; sed voluntas reducitur ad animam, secundum Aristotelem, et non reducitur ad naturam; anima ergo non est natura.

4. Item, ubi Aristoteles dicit quod quaedam sunt propter alias causas[k], dicit Commentator quod ea quae sunt a voluntate [l]non sunt a natura; sed quae sunt a voluntate sunt ab anima; ergo anima non est natura.

5. Item, ea quae sunt a natura non possunt ad opposita, ea quae sunt ab anima possunt ad opposita, secundum quod dicit Aristoteles in fine *Perihermeneias* quod potentia rationalis valet ad opposita, et ita anima non est natura.

---

[e]t est totum unum continuum. Ad tertium dicendum et ad quartum quod quanto aliquid propinquius sit primae causae tanto magis assimilatur ipsi, verum est quoad quasdam condiciones, ut quoad nobilitatem naturae et intransmutabilitatem in sua natura, non tamen (tamen *suppl. mg.*) quoad omnes condiciones nec etiam quoad loci mutationem. Movetur enim caelum ad quietem terrae et inferiorum *add.* T

[a] Ad ... vegetabilibus *om.* TC

[b] Habito ... quaeritur] unde habito quod natura sit principium motus in superioribus et quietis in terra, quaero C

[c] quia] huiusmodi *add.* C

[d] quaeritur] quaero C

[e] quod ... quod] videtur quod sic: Aristoteles in isto secundo C

[f] quaedam] sunt *add.* TC

[g] animalia] et *add.* TC

[h] et huiusmodi *om.* C

[i] anima ... natura] ergo etc. C

[j] natura] Item, augmentatio et diminutio sunt motus naturales; sunt ergo a motore naturali; sed non sunt (sunt *om.* T) nisi ab anima; anima (anima *om.* C; *suppl. s.l.* T) ergo est natura (etc. *pro* est natura C) *add.* TC

[k] causas] in hoc secundo ibi *add.* C

[l] non ... voluntate *om. hom.* C

12. For the same point see the end of *On Plants*.[60]

<Question 10>

Now that it has been established in what way nature is a principle of motion in the celestial bodies and of rest in the earth, it is then asked about the motion of animate things. And since animate things are moved by the soul, the question is whether the soul is their nature.

1. Arguments for the affirmative: in *Physics* II Aristotle says that some things exist by nature and some things exist from other causes; things that exist by nature are animals, plants and the like;[61] but an animal is such only in virtue of the soul; therefore, the soul is its nature.

2. Furthermore, the Commentator says that animals are moved by nature;[62] but they are only moved by the soul; therefore, a soul is a nature.

3. To the contrary: there are two principles of action: nature and will; but a will is reduced to a soul, according to Aristotle,[63] and it is not reduced to a nature; therefore, a soul is not a nature.

4. Furthermore, in commenting on Aristotle's statement that some things are from other causes,[64] the Commentator says that the things that are from the will are not from nature;[65] but the things that are from the will are from the soul; therefore, a soul is not a nature.

5. Furthermore, the things that are from nature do not have a power for opposites, whereas the things that are from the soul have a power for opposites, according to Aristotle's statement at the end of *De Interpretatione* that a rational power is a power for opposites;[66] hence, a soul is not a nature.

[60] This is probably a reference to a commentary on *De Plantis* by Aspall, which has not yet been identified.

[61] Arist. *Phys*. II.1, 192b8–10.

[62] Aver. *In Phys*. VIII, t.c. 27, fol. 365ra–b.

[63] Arist. *De anima* III.9, 432b5–6.

[64] Arist. *Phys.* II.1, 192b8–9.

[65] Aver. *In Phys.* II, t.c. 1, fol. 48ra.

[66] Arist. *De int.* 13, 23a1–2

6. Quidam dicunt quod anima non est natura, quia nullus motus naturalis est ab anima, quia nec motus augmentationis[a] neque[b] generationis neque nutritionis [c]est ab anima, sed huiusmodi motus sunt a triplici calore[d], [e]quia ad talem motum concurrit triplex[f] calor[g]: animalis et [h]calor vitalis et calor caelestis. Termini autem motus[i] sunt ab anima, secundum quod dicit Aristoteles quod in unoquoque [j]natura constantium determinatur ratio magnitudinis et augmenti. Et intendit de his quae sunt per naturam [k]ab anima. Quod apparet, quia statim post[l] dicit quod, si ligna apponerentur igni in infinitum, ignis cresceret[m] in infinitum. Unde animalia dicuntur transmutabilia[n] ab anima non quia motus[o] qui sunt in animalibus sint[p] ab anima[q], sed quia termini motus in animalibus sunt [r]ab anima.

7. Aliter dicitur [s]sic, sicut dicit Aristoteles, quod quaedam sunt a natura et quaedam propter alias causas; a natura sicut animalia et plantae. [t]Vult ergo Aristoteles quod animalia sint[u] a natura. Sed non sunt animalia nisi ab anima; ergo aliquo modo est anima natura in animalibus. Unde natura potest dici duobus modis: aut communiter [v]et secundum proportionem aut [w]simpliciter et proprie. Si autem sumatur natura simpliciter et proprie, sic est principium motus eius in quo est nec distat huiusmodi motor essentialiter a moto nisi secundum quid tantum, sicut patet in motu gravis et levis, quia motor in[x] huiusmodi est sua forma, scilicet vel forma gravis vel forma levis, quae non

[a] augmentationis *ex corr.* O
[b] neque] nec TC
[c] est ... motus] sed C
[d] calore *corr. ex* colore O
[e] quia ... caelestis] in tali animali et virtute caelesti C
[f] triplex *corr. ex* triples O
[g] calor] scilicet calor *add.* T
[h] calor vitalis et *om.* O
[i] motus] istorum motuum C
[j] natura constantium *suppl. s.l.* O; *om.* TC
[k] ab anima] animae TC
[l] post] consequenter (*post* dicit) C; *recte* prius
[m] cresceret] crescet O
[n] transmutabilia *ex corr.* O; *om.* T; animalia C
[o] motus] naturales *add. mg.* O
[p] sint] sunt TC
[q] anima] ut a natura *add. s.l.* O
[r] ab anima *corr. ex* a natura OT
[s] sic, sicut dicit] quod cum dicat C
[t] Vult ergo Aristoteles] plane vult C
[u] sint] sunt TC
[v] et ... proportionem *om.* T
[w] simpliciter et *om.* T
[x] in *ex corr.* O

6. Some people say that a soul is not a nature because no natural motion comes from the soul; for, neither the motion of increase nor that of generation nor that of nutrition come from the soul, but these motions come from three kinds of heat, since three kinds of heat are involved in them, that is, animal heat, vital heat, and celestial heat. The termini of these motions, however, are from the soul, according to Aristotle's statement that in all the things that exist by nature the ratio of magnitude and increase is determined.[67] And he is talking about the things that come naturally from the soul. This is clear because he says immediately after, that if wood were added to a fire without any limit, the fire would increase without any limit.[68] Thus, animals are said to be transmutable by the soul, not because the motions that are in the animals are from the soul, but because the termini of the motions in animals are from the soul.

7. Another view is that, as Aristotle says, some things exist from nature, such as animals and plants, and some exist from other causes.[69] Therefore, Aristotle wants to say that animals exist from nature. But animals are animals only from the soul; therefore, the soul is in some way the nature in animals. Thus, nature can be said in two ways: either (i) in a broad sense and by analogy or (ii) absolutely and properly speaking. In case (ii), nature is a principle of motion of the thing in which it exists, and such a mover is not essentially distant from the moved but only in a qualified way, as is clear in the motion of heavy and light things, because in these things the mover is their form, that is, the form of a heavy thing or the form of a light thing, which do not differ

[67] Arist. *De anima* II.4, 416a16–17.
[68] Arist. *De anima* II.4, 416a15–16.
[69] See above, n. 1.

differunt secundum substantiam [a]ab illo moto quod est vel ipsum grave vel ipsum leve, sed secundum quid tantum. Et sic non est anima[b] natura. Si autem loquamur de natura communiter et secundum proportionem, sic non meretur proprie nomen naturae, sed nomen motoris proprie. Sic anima[c] est animati[d] natura, quia est motor corporis. Et [e]in huiusmodi differt movens a moto essentialiter, quia anima distat essentialiter a corpore, quod est illud quod est motum.

8. [f]Unde anima potest comparari[g] ad suas operationes dupliciter, quia ad quasdam ordinatur ut natura, ad quasdam autem[h] ordinatur ut voluntas. Si [i]autem ordinetur ut natura, ut patet in augmentatione, aut ordinatur ad motum regendum aut ad terminum motus sive finem efficiendum. Unde per se non operatur motum nec ordinatur ad motum, sed per accidens, quia non ordinatur nisi ut regat qualitates in suo motu quibus attribuitur motus per se ut[j] efficienti. Ad terminum autem[k] efficiendum ordinatur anima per se, secundum quod dicit Aristoteles quod omnium natura constantium, hoc est a natura animae existentium, determinata est ratio magnitudinis et augmenti. Et ita anima per se est causa terminationis **[O 104rb]** motus in augmento. Si autem [l]ordinetur ut voluntas, aut ordinatur ad motum aut ad terminum motus. Et sive sic sive sic, [m]est anima efficiens utriusque, sicut anima mea[n] mota per praesentiam alicuius speciei est motor in motu meo pro voluntate sua et terminat motum similiter pro voluntate sua.

[a] ab ... leve] a gravi et levi C
[b] anima] a *add.* O
[c] anima *suppl. mg.* O
[d] animati *om.* TC
[e] in ... motum] ita huiusmodi differunt movens et motum essentialiter, scilicet anima et corpus C
[f] Unde anima] natura ergo C
[g] comparari *corr. s.l. ex* considerari O; considerari C
[h] autem *om.* TC
[i] autem ... accidens] primo modo, sic est in augmentatione, tunc aut ordinatur ad motum regendum aut ad terminum motus efficiendum, ad motum regendum sive ad movendum ordinatur accidentaliter C
[j] ut *corr. ex* vel O; vel T
[k] autem] motus *add.* C
[l] ordinetur ut voluntas] secundo modo ordinetur scilicet ut voluntas, tunc C
[m] est ... similiter] attribuitur efficientia ipsi animae. Anima enim mea mota per praesentiam alicuius speciei movet corpus meum pro voluntate sua et similiter terminat motum istum C
[n] mea *corr. ex* me O

in substance from the moved, which is either the heavy thing itself or the light thing itself, but only in a qualified way. The soul is not a nature in this way. In case (i), however, nature does not properly deserve the name of nature, but the name of mover. A soul is the nature of an animate thing in this way, because it is the mover of the body. And in this kind of thing the mover is essentially different from the moved, because the soul is essentially distinct from the body, which is the moved.

8. Accordingly, the soul can be compared to its actions in two ways, because (i) to some of them it is ordered as nature, (ii) to others it is ordered as will. In case (i), as is clear in the case of increase, the soul is ordered either to direct the motion or to producing the terminus or end of the motion. Accordingly, the soul neither produces the motion nor is ordered to the motion per se but accidentally, since it is ordered to the motion only in order to direct the qualities in their motion and the qualities per se are the efficient cause of motion. The soul, however, is per se ordered to producing the terminus of the motion, according to Aristotle's statement that for all the things that exist by nature, that is, that exist by the nature of the soul, the ratio of magnitude and increase is determined.[70] The soul, then, is the per se cause of the limitation of motion in the case of increase. In case (ii), the soul is either ordered to the motion or to the terminus of the motion. In either way, the soul is the efficient cause of both <the motion and its terminus>. Thus when my soul is moved because of the presence of a species, it is the mover of my motion by its will and it similarly terminates that motion by its will.

[70] See above, n. 6.

<Quaestio 11>

[a]Habito quae est natura quae est principium motus caeli et principium quietis terrae et principium motus in animalibus, quaeritur quae natura est motor gravis et levis. Et quaeritur de gravi utrum motus eius sit naturalis.

1. Et [b]quod non videtur: omne quod movetur naturaliter ex se movetur[c]; sed grave non movetur ex se, quia dicit Aristoteles in octavo huius quod moveri ex se est proprium animatorum; sed grave secundum quod grave non est animatum; ergo secundum quod tale non movetur ex se.

2. Item, in secundo[d] *De anima* dicit Aristoteles quod in moto[e] ex se movens et motum distant essentialiter; [f]sed in gravi non differt movens a moto essentialiter; ergo etc. Probatio minoris: in gravi in actu non sunt nisi duo, scilicet materia et forma; aut ergo forma est movens et materia motum aut forma est movens et totum compositum est motum. Si [g]autem forma sit movens et totum compositum sit motum, tunc adhuc non differunt movens et motum essentialiter, quia forma non differt essentialiter a composito. Si [h]autem forma sit movens et materia sit motum, sed[i] contra: materia non est mobilis secundum locum, secundum quod dicit Aristoteles quod mobile secundum locum est corpus perfectum in actu. Et iterum, nihil movetur motu locali nisi quod appetit locum; sed materia non plus appetit unum locum quam alium; materia ergo non movetur localiter.

3. Si dicatur quod totum compositum est motum et virtus caelestis est movens sive motor, contra: secundum Aristotelem, motus simpliciter est a forma simpliciter; ergo motus contrarii sunt a formis contrariis; si ergo grave moveatur deorsum et leve sursum, isti motus erunt a formis contrariis. Sed virtuti caelesti non est aliquid contrarium nec in se includit contrarietatem; virtus ergo caelestis non potest esse motor in motu gravis.

4. Si dicatur quod virtus caelestis per se non est motor gravis deorsum et levis sursum, sed prout incorporatur gravi movet deorsum, et prout incorporatur levi movet sursum, et secundum quod diversis mobilibus incorporatur [j]diversos habet motus, contra: si hoc est verum, tunc virtus

[a] Habito … naturalis] Habito qualiter natura est principium motus et quietis in terra et etiam motus animatorum quaeritur de motu gravium et levium utrum sit a natura C

[b] quod non videtur] videtur quod non quia probatio quod motus gravis sit violentus et non naturalis quia C

[c] movetur *ex corr.* O

[d] secundo] *recte* primo

[e] moto] motis TC

[f] sed ... essentialiter *om. hom.* C

[g] autem ... motum] secundo modo C

[h] autem ... motum] primo modo C

[i] sed *om.* TC

[j] diversos habet motus] est causa motus diversorum C

<Question 11>

Now that it has been established which nature is a principle of the motion of the heavens, and a principle of the rest of the earth, and a principle of motion in animals, the question is which nature is the mover of heavy and light things. And the question is whether the motion of a heavy thing is natural.

1. Arguments for the negative: everything that is subject to natural motion is moved by itself; but a heavy thing is not moved by itself, since in *Physics* VIII Aristotle says that to be moved by itself is proper to animate things;[71] but a heavy thing qua heavy is not animate; therefore, a heavy thing qua heavy is not moved by itself.

2. Furthermore, in *On the Soul* II Aristotle says that in a thing moved by itself the mover and the moved are essentially distant;[72] but in a heavy thing the mover does not essentially differ from the moved; therefore, <a heavy thing is not moved by itself>. Proof of the minor premise: there are only two things in a heavy thing in act, namely, matter and form; therefore, either the form is the mover and the matter is the moved or the form is the mover and the whole composite is the moved. If the form were the mover and the whole composite were the moved, then in this case the mover and the moved would not be essentially different, because the form does not essentially differ from the composite. However, if the form were the mover and the matter were the moved, one could argue against this: the matter is not mobile with respect to place, according to Aristotle's statement that the mobile with respect to place is a body perfect in act.[73] Also, nothing moves with locomotion unless it has the appetite for a place; but matter has no more appetite for one place than for another; therefore, matter does not move with locomotion.

3. If it is said that the whole composite is moved and the celestial power is the mover, one can argue against this: according to Aristotle, motion taken absolutely derives from form taken absolutely;[74] therefore, contrary motions derive from contrary forms; thus, if a heavy thing moves downwards and a light thing moves upwards, these motions derive from contrary forms. But neither is there anything contrary to the celestial power nor does this power contain in itself a contrariety; therefore, the celestial power cannot be the mover in the motion of a heavy thing.

4. If it is said that the celestial power is not by itself the mover of a heavy thing downwards and of a light thing upwards, but that in so far as it is embodied in a heavy thing it produces downward motion and in so far as it is embodied in a light thing it produces upward motion, and in so far as it is embodied in

[71] Arist. *Phys.* VIII.4, 255a5–7.

[72] The correct reference is Arist. *De anima* I.4, 408b33–409a3.

[73] Arist. *Phys.* VIII.7, 260b30–33.

[74] Arist. *Phys.* VIII.4, 255b5–17.

caelestis cum forma gravis est motor gravis et, si hoc, aggregatum ex virtute et forma est motor et materia est [a]illud quod est motum. [b]Et tunc sicut prius obiciatur.

5. Solutio[c]. Dicendum quod est loqui de gravi dupliciter: aut de gravi in potentia essentiali aut accidentali. Si de gravi in potentia essentiali, [d]tunc est quaerere de eius motore dupliciter, scilicet aut de[e] motore ut efficiente aut de motore ut natura. Si loquamur de motore ut efficiente, sic est [f]dicere cum Aristotele quod agens extra est motor ut efficiens. Si autem loquamur de motore ut natura, sic est dicere quod potentia activa cum specie agentis extra est motor ut natura. [g]Unde hic posset quaeri: cum ignis sit in potentia essentiali terra sive grave, si terra sive grave immittat speciem suam in ipsum ita quod excitet potentiam activam quae est in igne ad terram, quae excitata promovet se ad actum, utrum, sicut successive acquirit formam, sic successive acquirat locum sibi, [h]quod non est hic disputare.

6. Si autem loquamur de gravi in potentia accidentali [i]et sic extra suum locum, respondet Commentator sic, quod motor ut efficiens est removens[j] prohibens; unde, remoto prohibente, movetur grave ex se per accidens, [k]quia movet aerem et aer motus movet grave, sicut est de nauta[l] in navi quod per se movet navem et navis mota movet nautam per accidens. Et hoc est exemplum Commentatoris. [m]Unde ad hoc quod dicit Aristoteles quod movens et motum distant essentialiter respondet[n] quod hoc est verum in his quae moventur ex se per se, cuiusmodi sunt animata[o]. Sed grave non [p]ita movetur, sed ex se movetur

[a] illud quod est *om.* T
[b] Et ... obiciatur] hoc autem prius improbatum est C
[c] Solutio *om.* TC
[d] tunc ... agentis extra] aut quaeritur de eius motore ut efficiente aut de eius motore ut natura. Si de motore ut efficiente, sic est dicere quod agens extra est motor ut efficiens. Si de motore ut natura, sic est dicere quod potentia activa cum specie immissa C
[e] de *om.* O
[f] dicere cum *corr. ex* dicit O
[g] Unde ... quae] et iuxta hoc accidit quaestio cum ignis sit terra sive grave in potentia accidentali, si terra immittat speciem in ignem quae immissa excitat potentiam (immissam *add. sed del.*) repertam in igne quatenus C
[h] quod ... disputare] sed istud non est hic disputandum C
[i] et ... sic] dicendum sicut dicit Commentator C
[j] removens] movens et O
[k] quia ... nautam] quia grave movet aerem et aer motus movet ipsum grave, sicut nauta in navi per se movet navem et mota navi movetur nauta C
[l] nauta *ex corr.* O
[m] Unde ... Aristoteles] Et ad illud Aristotelis C
[n] respondet] supplet (*lectio incerta*) Commentator dicendo C
[o] animata *ex corr.* O
[p] ita movetur] movetur ex se illo modo C

different mobile things it produces different motions, one can argue against this: if this is true, then the celestial power together with the form of a heavy thing is the mover of a heavy thing, and, if so, the aggregate of the power and the form is the mover and the matter is the moved. But this runs into the objection raised before.[75]

5. Solution. It must be said that there are two ways of talking about a heavy thing: either (i) about a heavy thing in essential potency or (ii) about a heavy thing in accidental potency. In case (i), there are two ways of asking for its mover: either (ia) for the mover as efficient cause or (ib) for the mover as nature. In case (ia), it must be said, following Aristotle, that the external agent is the mover as efficient cause. In case (ib), it must be said that the active power together with the species of the external agent is the mover as nature. Thus, the following question might arise here: since fire is earth or a heavy thing in essential potency, if earth or a heavy thing introduces its species into fire in such a way that it stimulates the active potency to earth that exists in fire, and this power, once stimulated, progresses towards the act, is it the case that just as fire acquires the form successively so too it acquires its place successively? But this question should not be discussed here.

6. In case (ii), that is, when a heavy thing is in accidental potency and thus outside its place, the Commentator replies that the mover as efficient cause is what removes the hindrance; thus, once the hindrance has been removed, a heavy thing is moved by itself accidentally, because it moves the air and the air that is moved moves the heavy thing, as happens with a sailor in a ship: he moves the ship per se and the ship that is moved moves the sailor accidentally. This is the Commentator's example.[76] Accordingly, to Aristotle's statement that the mover and the moved are essentially distant, the Commentator replies

[75] See above, n. 2.
[76] Aver. *In Phys.* VIII, t.c. 32, fol. 371va.

per accidens, quia grave[a] movet[b] medium per se[c], et medium motum movet grave per accidens.

7. Contra illud sic: remoto prohibente, constat quod lapis prius movet aerem quam aer moveat ipsum; quaero: in illo priori quid est motor eius? Et constat quod non aer.

8. Dicendum quod in motu gravis in potentia accidentali est totum mobile motum. Et motor universalis est [d]natura caelestis sive virtus caelestis, motor autem[e] particularis est[f] forma gravis rememorata a virtute caelesti, et[g] est motor ut natura in tali motu. [h]Nec differunt movens et motum essentialiter, hoc est movens non differt complete a moto. Nec exigitur hoc nisi in illis quae simpliciter ex se et per se moventur, cuiusmodi sunt animata, secundum Aristotelem. Unde non valet: 'non movetur simpliciter ex se et per se; ergo non movetur naturaliter'. Sed magis debet sequi oppositum, quia illud quod movetur ratione alicuius quod non est distans ab eo sed est idem quod sua forma [i]vel forma ipsa magis movetur naturaliter quam illud quod movetur per aliquid quod est distans ab eo essentialiter. Unde forma substantialis alicuius est causa cuiuslibet actionis eius. Sed actio est dupliciter: vel perficere suam materiam, et hoc facit forma substantialis tantum; alia est actio alicuius in extrinsecus. Sed hoc contingit dupliciter: vel ut aliud convertat in sui naturam et ipsum corrumpat, et sic agit mediantibus his qualitatibus, calidum, frigidum; aut ut acquirat sibi salutem[j] [k]sive locum, **[O 104va]** et sic agit per formam substantialem mediantibus his qualitatibus, gravitas et levitas.

<Quaestio 12>

Propter hoc quaeritur utrum forma substantialis [l]mediantibus accidentibus, ut gravitate et levitate, possit acquirere sibi locum.

1. Et videtur quod non [m]sic: Aristoteles dicit quod quantum[n] habet aliquid de loco, tantum habet de forma, et hoc quia locus est forma locati; et forma non acquiritur mediante accidente aliquo.

[a] grave] per se *add.* O
[b] movet] per se *add.* T
[c] se] et ita movetur ex se *add.* OT
[d] natura caelestis sive *om.* C
[e] autem *corr. ex* tunc O
[f] est *ex corr.* O
[g] et] quod O
[h] Nec differunt] non differt TC
[i] vel forma ipsa *om.* C
[j] salutem] saltem O
[k] sive locum *om.* C
[l] mediantibus accidentibus, ut] mediante C
[m] sic ... quod] quia C
[n] quantum *ex corr.* O

that this is true in the things that are moved by themselves per se, such as animate things.[77] However, a heavy thing is not moved in this way, but it is moved by itself accidentally, because a heavy thing moves the medium per se, and the medium that is moved moves a heavy thing accidentally.

7. Against this: when the hindrance is removed, it is apparent that the stone moves the air before the air moves the stone. Then I ask: in that prior time what is the mover of the stone? And it is apparent that its mover is not the air.

8. It must be said that in the motion of a heavy thing in accidental potency the whole mobile thing is the moved. And the universal mover is the celestial nature or celestial power, whereas the particular mover is the form of a heavy thing recalled by the celestial power, and it is the mover as nature in such motion. Also, the mover and the moved do not differ essentially, that is, the mover does not completely differ from the moved. Nor is this required except in those things that are moved by themselves absolutely and per se, as are animate things, according to Aristotle.[78] Accordingly, the inference 'it is not absolutely moved by itself and per se; therefore, it is not moved naturally' is not valid. Rather, the opposite conclusion should follow, because the thing that is moved in virtue of something that is not distant from it but which is the same as its form or is the form itself is moved more naturally than the thing that is moved in virtue of something essentially distant from it. Thus, the substantial form of a thing is the cause of all of its actions. But there are two kinds of such actions: (i) one is to confer perfection on its matter, and this belongs to the substantial form alone; (ii) the other action is applied to something extrinsic. And (ii) happens in two ways: either (iia) in order to convert the extrinsic thing into its own nature and so to corrupt it, and in this case the thing acts by means of these qualities: hot, cold; or (iib) in order to acquire its safety or place, and in this case the thing acts through the substantial form by means of these qualities: heaviness and lightness.

<Question 12>
Because of this the question arises whether the substantial form of a thing can make it acquire its place by means of accidents, such as heaviness and lightness.

1. Arguments for the negative: Aristotle says that however much something has of its place, it has that much of its form,[79] and this is because place is the form of the located body; but the form is not acquired by means of an accident.

[77] Aver. *In De anima* I, t.c. 69, ed. Crawford, 94.22–95.41.
[78] See above, n. 1.
[79] Arist. *De caelo* IV.3, 310a33–b1.

2. Item, Commentator[a] dicit, capitulo de loco, quod locus est perfectio locati; et[b] perfectio non acquiritur nisi per substantiale aliquod.

3. Contra: locus accidit locato, quod patet, quia locatum potest esse in specie sua completum sine loco continente.

4. Solutio[c]. Dicendum quod locus potest dupliciter considerari: aut solum prout [d]superficies est, et ita prout est quantitas, et sic est accidens locato; vel potest considerari non solum prout est ultimum sive superficies, sed simul cum hoc secundum[e] quod salvans est et influens [f]naturam sive virtutem aliquam in locatum ad salutem locati, et sic est aliquo modo essentiale. Sed aliquid potest esse essentiale alicui dupliciter: vel quia ponat ipsum in specie, et sic [g]de essentia ipsius est altera pars compositi, et isto modo non est locus essentialis locato; vel quia ipsum ordinat ad salutem[h] [i]et ipsum quietat et salvat, et sic est locus essentialis locato. Et quod sic est essentiale alicui potest simpliciter esse eius accidens. Et ita est locus accidens locato.

5. Quod autem motor gravis sit accidentalis videtur: grave in potentia essentiali habet motorem essentialem; ergo grave in potentia accidentali habebit motorem accidentalem.

6. Dicendum quod habet motorem accidentalem. Et dicitur motor accidentalis quia movet ipsum sub esse accidentali, [j]et hoc quia accidit ei quod sit extra suum locum proprium vel quia acquirit sibi aliquod esse accidentale, sicut aliquod 'ubi', quod non est de sua essentia. Vel dicitur accidentalis quia, cum est extra suum locum, accidit ei quod moveatur inferius et sit in illo fieri, quia accidit quod sit ibi medium aliquo modo resistens, quod est causa motus, quia, si non esset medium, subito faceret se deorsum. Unde motus deorsum est ei accidentale, quiescere tamen deorsum est ei essentiale.

1*. [k][l]In capitulo de natura consequenter dicit Aristoteles quod forma artificialis est accidens. Propter hoc quaeritur de forma artificiali primo[m]

[a] Commentator] Aristoteles T; *om.* C
[b] et] sed TC
[c] Solutio *om.* TC
[d] superficies ... superficies] continens est et sic est accidens locato aut secundum quod non est solum continens sive sola superficies C
[e] secundum *corr. ex* quod O
[f] naturam ... aliquam] virtutem C
[g] de ... est *om.* C
[h] salutem] saltem OC
[i] et ... salvat *om.* C
[j] et ... essentia] accidit enim gravi quod sit extra locum proprium; vel potest dici motor eius accidentalis quia acquirit sibi aliquod esse accidentale, sicut aliquod ubi C
[k] In ... natura *om.* T
[l] In ... quaeritur] Consequenter quia Aristoteles dicit in hoc capitulo quod forma artificialis est accidens quaero C
[m] primo *om.* O

2. Furthermore, in the chapter on place the Commentator says that place is the perfection of a located body;[80] but a perfection is acquired only in virtue of something substantial.

3. Against this: place is an accident of a located body, which is clear because a located body can be complete in its species without a containing place.

4. Solution. It must be said that place can be considered in two ways: either (i) only in so far as it is a surface, and thus in so far as it is a quantity, and in this way it is an accident of a located body; or (ii) it can be considered not only in so far as it is a boundary or a surface, but also in so far as it is something that saves and pours a nature or power into a located body for its safety, and in this way place is in some way essential <to a located body>. But something can be essential to a thing in two ways: either (a) because it places that thing in its species, and thus the other part of the composite belongs to the essence of that thing, and in this way place is not essential to a located body; or (b) because it orders that thing to its safety and makes it to be at rest and saves it, and in this way place is essential to a located body. And what is essential to something as in (b) can absolutely speaking be one of its accidents. And so place is accidental to a located body.

5. It seems to be the case that the mover of a heavy thing is accidental: a heavy thing in essential potency has an essential mover; therefore, a heavy thing in accidental potency will have an accidental mover.

6. It must be said that a heavy thing has an accidental mover. It is said to be an accidental mover because it moves a heavy thing under accidental being, and this is either because it is accidental to the heavy thing that it is outside its proper place or because it makes the heavy thing acquire an accidental being, that is, a 'where', which does not belong to its essence. Alternatively, it is said to be accidental because, when the heavy thing is outside its place, it is accidental to it that it moves downwards and that it undergoes that process. This is because it is accidental that there is a medium there that is in some way resistant, and which is the cause of the motion, for, if there were no such medium, the heavy thing would bring itself down immediately. Thus, the downward motion is accidental to a heavy thing, whereas to be at rest in the downward place is essential to it.

1*. In the chapter on nature Aristotle next says that an artificial form is an accident.[81] Because of this there are the following questions about an artificial

[80] Aver. *In Phys.* IV, t.c. 4, fol. 122vb.
[81] Arist. *Phys.* II.1, 193a11–21 (cf. Aver. *In Phys.* II, t.c. 7, fol. 51ra).

utrum sit substantia vel accidens, secundo in quo praedicamento [a]sit forma artificialis, tertio utrum formae artificiales habeant universale, quarto utrum vera forma artificialis sit in anima.

<Quaestio 13>

1. [b]Circa primum sic. Quod forma artificialis sit substantia videtur: Aristoteles dicit quod ex forma artificiali [c]et sua materia fit unum proportionaliter sicut ex forma naturali et sua materia; sed ex forma naturali et sua[d] materia fit[e] vere unum per essentiam; ergo similiter ex forma artificiali et sua materia fiet vere unum per essentiam. Sed ex materia et accidente non fit tale unum; forma ergo artificialis non est accidens sed substantia.

2. Item, per[f] Aristotelem, ars imitatur naturam in quantum potest; sed natura principaliter producit formas substantiales et ex incidenti formas accidentales; ergo [g]sic erit de arte quod principaliter producit etc.

3. Item, Aristoteles dicit quod ars producit domum eodem modo sicut natura eam produceret; sed si natura eam produceret, produceret [h]eam ut formam substantialem et non ut accidens; [i]ergo ars similiter producit eam ut substantiam et non ut accidens.

4. Item, Aristoteles dicit in quarto *Meteororum* quod natura et ars possunt super idem, sicut super metallum [j]potest utrumque; sed forma metallica est [k]substantialis sive substantia; ergo ars poterit super substantiam.

5. Item, nobilitati agentis respondet nobilitas acti; sed anima est nobilior agens quam natura; cum ergo natura possit super formam substantialem, tunc anima poterit super adeo nobile ad minus. Sed nullum accidens est adeo nobile sicut substantia, [l]sed substantia est adeo nobile; anima ergo poterit super substantiam in producendo. Sed est artificiale agens; ergo etc.

6. Item, omne quod est per appositionem respectu alterius est nobilius eo; sed artificiale se habet per appositionem respectu formae substantialis; ergo artificiale est nobilius quam forma substantialis. Sed nullum accidens est nobilius substantia; forma ergo artificialis non est accidens.

[a] sit forma artificialis] debet poni C
[b] Circa ... artificialis] Circa primum quod C
[c] et sua materia *suppl. s.l.* O
[d] sua *suppl. s.l.* O
[e] fit] fiet TC
[f] per] secundum TC
[g] sic ... etc.] ars similiter C
[h] eam ut *suppl. s.l.* O; ut (*s.l.*) T; tamquam C
[i] ergo ... accidens *om. hom.* T
[j] potest utrumque *om.* C
[k] substantialis ... substantiam] forma substantialis; ergo ars potest producere formam substantialem C
[l] sed ... nobilius substantia (n.6) *om. hom.* C

form: first, whether it is a substance or an accident; second, to what category an artificial form belongs; third, whether artificial forms have a universal; fourth, whether a true artificial form is in the soul.

<Question 13>

1. As to the first question, there are arguments that the artificial form is a substance: Aristotle says that from an artificial form and its matter something that is one comes to be by analogy with what comes to be from a natural form and its matter;[82] but from a natural form and its matter something that is really one in essence comes to be; therefore, in the same way, from an artificial form and its matter something that is really one in essence comes to be. However, something that is one in this way does not come to be from matter and an accident; therefore, an artificial form is not an accident but a substance.

2. Furthermore, according to Aristotle, art imitates nature in so far as it can;[83] but nature produces substantial forms principally and accidental forms incidentally; therefore, the same will apply to art, namely that it principally produces <substantial forms and incidentally accidental forms>.

3. Furthermore, Aristotle says that art produces a house in the same way that nature would produce it;[84] but if nature produced a house, it would produce it as a substantial form and not as an accident; therefore, art similarly produces it as a substance and not as an accident.

4. Furthermore, in *Meteorology* IV Aristotle says that nature and art can produce the same thing, for instance, both can produce metal;[85] but the form of metal is substantial or a substance; therefore, art can produce a substance.

5. Furthermore, the nobility of a thing produced corresponds to the nobility of the agent; but the soul is a nobler agent than nature; therefore, since nature can produce a substantial form, then the soul too can produce something that is at least as noble. But no accident is as noble as a substance, but a substance is as noble <as a substantial form>; therefore, the soul can produce a substance. But the soul is an artificial agent; therefore, <art can produce a substance>.

6. Furthermore, everything that is by addition with respect to another thing is nobler than that other thing; but an artificial thing is by addition with respect to a substantial form; therefore, an artificial thing is nobler than a substantial form. But no accident is nobler than a substance; therefore, an artificial form is not an accident.

[82] Arist. *Metaph.* VII.8, 1033b11–19.
[83] Arist. *Phys.* II.2, 194a21–23; II.8, 199a15–17.
[84] Arist. *Phys.* II.8, 199a12–13.
[85] Arist. *Meteor.* IV.12, 390b11–14.

7. Item, aliquod est[a] agens voluntarium quod potest super productionem formae substantialis, sicut causa prima et, si hoc, tunc voluntas non impedit quin aliquid[b] [c]possit prosequi effectum quod est substantia. Sed omne artificialiter agens est voluntarie agens; ergo in artificialiter agentibus non impedit voluntas quin possit super substantiam in producendo.

8. Ad oppositum: dicit Aristoteles quod formae artificiales sunt per appositionem formae substantiali; sed formae substantiali non apponitur nisi accidens; forma ergo artificialis est accidens.

9. Item, quod [d]appositum alii et remotum ab eodem non minus[e] manet in sua specie est accidens eidem[f]; sed ita se habet forma artificialis respectu[g] formae naturalis[h].

10. Probato quod forma artificialis [i]sit substantia [j]et quod non, probatio quod non sit accidens sic: omne accidens egreditur ex principiis subiecti; sed forma artificialis non egre **[O 104vb]** ditur ex principiis subiecti; ergo etc.

11. Item, dicit Commentator super principium undecimi quod omnia accidentia causantur a materia et praecipue quantitas; sed forma artificialis non causatur a materia; ergo etc.

12. Item, Aristoteles in tertio[k] *Metaphysicae* quaerit utrum sint eadem principia substantiae et accidentis. Et [l]secundum magistros solvitur sic: eadem sunt secundum rem, non autem eadem sunt secundum rationem. Sed quod est principium substantiae non est principium formae artificialis; ergo etc.

13. Solutio[m]. Dicendum quod forma artificialis in comparatione ad illa quae perficit, ut in comparatione ad artificiale[n], est essentialis, in se tamen et respectu substantiae est accidens. [o]Sed aliquid esse accidens contingit dupliciter: aut quia accidentaliter egreditur de principiis subiecti, et huiusmodi est proprie accidens, aut[p] quia accidentaliter apponitur subiecto vel[q]

[a] est *suppl. s.l.* O
[b] aliquid] aliud OT
[c] possit ... substantia] agens voluntarium possit super substantiam C
[d] appositum ... manet] potest apponi alicui et ab eo removeri illo cui apponitur non minus manente C
[e] minus] plus OT
[f] eidem *corr. s.l. ex* idem O
[g] respectu *suppl. s.l.* O; *om.* T
[h] naturalis] ergo etc. *add.* TC
[i] sit ... accidens] non sit accidens videtur C
[j] et quod non *fortasse delendum; scrips. sed del.* T
[k] tertio] *recte* undecimo
[l] secundum ... sic] dicunt magistri quod C
[m] Solutio *om.* TC
[n] artificiale *corr. s.l. in* artificiatum O
[o] Sed ... accidens] et hoc C
[p] aut *om.* O
[q] vel] quia *add. ex corr.* O

7. Furthermore, there is a voluntary agent that can produce a substantial form, that is, the first cause, and if this is the case, then the will does not prevent anything from producing a substance as its effect. But every artificial agent is a voluntary agent; therefore, in artificial agents the will does not prevent anything from producing a substance.

8. To the contrary: Aristotle says that artificial forms exist by addition to a substantial form;[86] but only an accident is added to a substantial form; therefore, an artificial form is an accident.

9. Furthermore, something whose addition to a thing and removal from it nonetheless allows that thing to remain in its species is an accident of that thing; but an artificial form is of this kind with respect to a natural form.

10. Now that proofs that an artificial form is a substance and that it is not a substance have been considered, it is now proved that it is not an accident: every accident emerges from the principles of its subject; but an artificial form does not emerge from the principles of its subject; therefore, <an artificial form is not an accident>.

11. Furthermore, in his comment on the beginning of *Metaphysics* XI the Commentator says that all accidents are caused by matter and principally quantity;[87] but an artificial form is not caused by matter; therefore, <an artificial form is not an accident>.

12. Furthermore, in *Metaphysics* III Aristotle asks whether the principles of a substance and those of an accident are the same.[88] According to the masters, the solution to this question is that these principles are the same in reality but are not the same in *ratio*. But what is the principle of a substance is not a principle of the artificial form; therefore, <an artificial form is not an accident>.

13. Solution. It must be said that an artificial form is essential with respect to the thing on which it confers perfection, that is, to the artificial thing; taken in itself, however, and in relation to a substance it is an accident. But something can be an accident in two ways: either (i) because it accidentally emerges from the principles of the subject, and this is properly an accident, or (ii) because it is accidentally added to a subject or removed from it, and this is not

[86] Arist. *Phys*. II.1, 193a12–17.
[87] The reference is to Aver. *In Metaph*. XII, t.c. 25, fol. 309vb, but in this passage Averroes only says that all accidents are caused by matter and not also that this is principally true of quantity. In other texts, however, Averroes does maintain that quantity is the most basic determination of matter. See, e.g., Aver. *In Metaph*. VII, t.c. 8, fol. 159ra-b; *De sub. orb*. 1, fols. 4vb–5rb.
[88] The correct reference is Arist. *Metaph*. XII.4, 1070a31–35.

removetur[a] ab eodem[b], et huiusmodi non est adeo proprie accidens sicut accidens primo modo dictum. Et isto modo dicitur forma artificialis accidens.

14. Unde omnes rationes quae probant quod non est accidens [cd]probant quod non est accidens primo modo.

15. Ad primam rationem [e]quae [f]probat eam esse substantiam dicendum quod non sequitur quod sit unum per essentiam, sed quod sit unum subiecto[g] tantum.

16. Ad illud argumentum[h] de nobilitate actionis animae dicendum quod, quando agens animatum agit in aliud, [i]aut agit in ipsum aliquid immittendo aut non. Si [j]agat aliquid immittendo, aut ergo immittit speciem, et tunc agit per contrarietatem, aut immittit aliquid quod est pars sui, et sic fit generatio per propagationem. Et utroque istorum modorum potest super productionem[k] substantiae, et sic sequitur proportio. [l]Aut agit non immittendo, sed aliquid ab extrinseco apponendo vel aliquid removendo, et tunc non producit substantiam, sed accidens, et sic producitur[m] artificiale.

17. Ad argumentum de actione primae causae dicendum quod quoddam est voluntarium agens quod agit per suam substantiam et non differt sua actio a sua[n] substantia. Et tale agens potest super productionem substantiae, et tale est causa prima. Quoddam est agens quod non agit per suam substantiam, sed est sua actio alia a substantia sua, et tale impossibile est quod per solam voluntatem producat substantiam.

18. Possunt tamen homines provocare malignos spiritus qui cognoscunt constellationes et sciunt virtutes rerum, ita quod possunt talia simul unire et in tali constellatione situare quod excitabitur potentia activa quae est in ipsis respectu alicuius formae ita quod fiet in actu. Et ita[o] artificium potest adiuvare ad productionem formae substantialis, [p]ut patuit in virgis pharaonis, ex quibus factae fuerunt serpentes.

[a] removetur *corr. ex* movetur O; removeatur T
[b] eodem] eo aliquid OT *sed* aliquid *del.* T
[c] probant ... accidens *om. hom sed* procedunt *suppl. mg.* T
[d] probant ... modo] primo modo procedunt C
[e] quae ... substantiam] ad hoc quod sit substantia T
[f] probat ... dicendum] videtur probare quod sit substantia dico C
[g] subiecto *corr. ex* substantia O
[h] argumentum] quod est *add.* O (est *s.l.*)
[i] aut ... ipsum *om. sed* aut agit *suppl. mg.* O
[j] agat aliquid immittendo] primo modo C
[k] productionem] eductionem TC
[l] Aut agit non] si autem agit nihil C
[m] producitur] producit TC
[n] sua *suppl. s.l.* O
[o] ita] ibi (*lectio incerta*) O
[p] ut ... serpentes] non tamen producit ipsam absolute sic fuit de magis (?) pharaonum qui faciebat serpentes ex virgis suis C

as properly an accident as is an accident of the first sort. And an artificial form is an accident of the second sort.

14. Accordingly, all the arguments proving that an artificial form is not an accident[89] prove that it is not an accident of the first sort.

15. To the first argument that proves that an artificial form is a substance,[90] it must be replied that it does not follow that it is one in essence, but that it is one only in subject.

16. To the argument about the nobility of the action of the soul[91] it must be said that, when an animated agent acts on another thing, either (i) it acts on it by introducing something or (ii) not. In case (i), it either introduces a species, and in this case it acts by contrariety, or it introduces something that is its part, and in this case there is generation by reproduction. And an agent can produce a substance in both these ways, and thus the analogy holds. In case (ii) it acts by adding something from outside or by removing something, and then it does not produce a substance, but an accident. An artificial thing is produced in the second way.

17. To the argument about the action of the first cause[92] it must be said that one sort of agent is a voluntary agent that acts in virtue of its substance and is such that its action does not differ from its substance. And such an agent can produce a substance, and the first cause is such an agent. Another agent is such that it does not act in virtue of its substance, but its action is distinct from its substance, and it is impossible for this agent to produce a substance in virtue of its will alone.

18. Men, however, can provoke malicious spirits who know the constellations and the powers of things, in such a way that the spirits can join these things together and situate them in a constellation with the result that the active power that exists in them with respect to a form will be stimulated and that the form will become in act. Thus an artefact can be of help in the production of the substantial form, as was made clear by the rods of the Pharaoh, from which serpents were produced.

[89] See above, nn. 10–12.

[90] See above, n. 1.

[91] See above, n. 5.

[92] See above, n. 7.

<Quaestio 14>
Quaeritur consequenter in quo praedicamento est forma artificialis[a], quia de aggregato ex forma et subiecto non est quaestio, quia non est in aliquo uno praedicamento.

1. Quod autem forma artificialis non sit in aliquo praedicamento videtur. Non est quantitas. Probatio, quia neque continua neque discreta, ut patet. Neque qualitas, quia[b] qualitas est secundum quam quales dicimur; sed non dicimur quales ab artificio. Non enim dicimur domificati ab artificio domus. Neque est positio, ut videtur, quia positio non nominat ipsum ordinatum, sed ipsam ordinationem. Non enim dicimus quod sedens sit in praedicamento positionis, sed[c] ipsa sessio tantum. Sed artificialia nominant ipsum ordinatum et non ipsam ordinationem; ergo artificialia non sunt in praedicamento positionis.

2. Potest dici quod non proprie sunt in praedicamento sicut non sunt proprie accidentia, sed, sicut communiter sunt accidentia, sicut[d] patuit, sic sunt in praedicamento. Unde artificiale[e] potest dupliciter facere suam denominationem, quia aut denominat suum proprium subiectum, sicut dicimus quod pannus fit supertunica; et sic est artificiale in praedicamento positionis, quia forma supertunicae situat quoquo modo partes panni ad hoc quod fiat supertunica. Aut potest denominare nos, sicut dicimus quod nos sumus supertunicati; et sic sunt in prima specie qualitatis.

<Quaestio 15>
Quaeritur consequenter utrum artificialia habeant universalia.

1. Et quod non videtur: universalia sunt perpetua, et artificialia non sunt perpetua; ergo artificialia non habent universalia.

2. Item, si artificialia habent[f] universalia, tunc universalia eorum erunt in anima, secundum quod dicit Aristoteles: “[g]Singulare dum sentitur et universale dum intelligitur”. Sed illud quod est in anima de artificiali non est universale. Probatio: [h]solum singulare est principium operationis; sed illud quod est in anima est principium operationis artificialis; ergo illud quod est in anima non est universale.

[a] artificialis] substantialis O
[b] quia *om. sed* enim *suppl. s.l. post* qualitas O
[c] sed] si O
[d] sicut] ut prius TC
[e] artificiale] accidentale TC
[f] habent] habeant TC
[g] Singulare ... et *om.* C
[h] solum ... artificialis *om.* O

<Question 14>
The next question is about the category that an artificial form belongs to, because there is no question about the aggregate of the form and the subject, given that it is not in a single category.

1. Arguments that an artificial form is not in a category. It is not a quantity, because it is clear that it is neither a continuous quantity nor a discrete quantity. It is not a quality, because a quality is a thing in accordance with which we are said to be such; but we are not said to be such in virtue of an artefact—for instance, we are not said to be housed by the artefact of the house. It is not a position, it seems, because a position does not designate the thing ordered but the ordering itself—for instance, we do not say that what is sitting is in the category of position, but only the sitting itself. But artificial things designate the thing ordered and not the ordering itself; therefore, artificial things are not in the category of position.

2. It can be said that artificial forms are not properly in a category as they are not properly accidents, but, given that they are accidents in a broad sense, as was made clear,[93] they are in a category. Thus, an artificial thing can denominate in two ways: either (i) it denominates its proper subject, as when we say that a piece of cloth becomes a surcoat; in this way the artificial thing is in the category of position, because the form of the surcoat situates the parts of the cloth in some way in order for the surcoat to come to be. Or (ii) it can denominate us, as when we say that we are surcoated; in this way artificial things are in the first species of quality.

<Question 15>
The next question is whether artificial things have universals.

1. Arguments for the negative: universals are perpetual, and artificial things are not perpetual; therefore, artificial things do not have universals.

2. Furthermore, if artificial things have universals, then their universals are in the soul, according to Aristotle's statement "Something is singular as long as it is sensed and universal as long as it is understood".[94] But the aspect of the artificial thing that exists in the soul is not a universal. Proof of this: only a singular is a principle of action; but the aspect of the artificial thing that exists in the soul is the principle of an artificial action; therefore, the aspect that exists in the soul is not a universal.

[93] See above, q. 13, n. 13.
[94] Arist. *Phys*. I.5, 189a5–7.

3. [a]Ad oppositum: secundum Aristotelem, domus et haec domus differunt sicut universale et particulare.

4. Item, signum non additur nisi termino communi; sed bene dicitur 'omnis domus'.

5. Item, secundum Aristotelem, 'sol' est commune de se, licet non habeat nisi unum appellatum; sed 'domus' habet plura [b]appellata; ergo multo fortius est commune.

6. Solutio[c]. Dicendum quod habent universalia et sunt eorum universalia perpetua, et hoc perpetuitate suae naturae sive perpetuitate habitudinis. [de]Et patebit ad illud super septimum, in illa quaestione 'Utrum universali secundum quod perpetuum respondet [f]singulare aliquod'. **[O 105ra]**

<Quaestio 16>
Quaeritur[g] utrum vera forma artificialis sit in anima.

1. Et quod sic videtur: agens naturale habet eandem formam specie quam habet actum; ergo similiter erit de agente artificiali.

2. Item, si vera forma artificialis non esset in anima, tunc non plus de artificiali esset in anima quam de naturali; sed per illud quod habet de naturali non[h] potest super productionem naturalis; [i]ergo per illud quod habet artificialis non potest super artificiale[j].

3. Item, [k]nihil denominatur a forma aliqua nisi sit in ipso[l] ut res; sed anima denominatur ab arte, quia dicitur artifex; ergo forma artificialis est in anima ut res est.

4. Item, [m]non est species sola in anima nisi recipiatur ab extra; sed contingit aliquando quod artifex facit artificium aliquod cuius similitudinem numquam vidit; ergo illius artificii non habet solam speciem. Et constat quod aliquid habet; habet ergo veram formam.

5. Item, plus est cognoscere et efficere quam cognoscere tantum. Sed artifex per illud quod est in anima cognoscit artificiatum et ipsum ad esse

[a] Ad oppositum] Contra TC
[b] appellata ... commune *om.* C
[c] Solutio *om.* TC
[d] Et ... aliquod *om.* T
[e] Et ... quaestione] sicut patet super septimum (?) Metaphysicae ubi quaeritur C
[f] singulare aliquod] universaliter singulare C
[g] Quaeritur] consequenter *praem.* TC
[h] non *suppl. s.l.* O
[i] ergo ... artificiale *om.* C
[j] artificiale] productionem artificialis T
[k] nihil ... res] nulla forma denominat nisi ubi est ut res C
[l] ipso] ipsa OT
[m] non ... nisi] nec est in anima ut species sola nisi quod C

3. To the contrary: according to Aristotle, house and this house differ as universal and particular.[95]

4. Furthermore, a sign is only added to a common term; but it is correct to say 'every house'.

5. Furthermore, according to Aristotle, the term 'sun' is common by itself, although it designates only one thing;[96] but the term 'house' designates many things; therefore, there is a much stronger reason to say that it is a common term.

6. Solution. It must be said that artificial things have universals, and their universals are perpetual either by their nature or by habit. This will be made clear in the comment on *Metaphysics* VII in the question 'Whether something singular corresponds to a universal in so far as it is perpetual'.[97]

<Question 16>

The question is whether a true artificial form is in the soul.

1. Arguments for the affirmative: a natural agent has specifically the same form as what is produced by it; therefore, the same applies to an artificial agent.

2. Furthermore, if a true artificial form were not in the soul, then no more of the artificial thing than of a natural thing would be in the soul; but the soul cannot produce a natural thing in virtue of what it has of the natural thing; therefore, the soul cannot produce an artificial thing in virtue of what it has of the artificial thing.

3. Furthermore, nothing is denominated by a form unless that form exists in it as a thing; but the soul is denominated by art, because it is said to be an artisan; therefore, an artificial form exists in the soul as a thing.

4. Furthermore, a species alone is not in the soul unless it is received from outside; but it sometimes happens that an artisan produces an artefact whose likeness he never saw; therefore, he does not have only the species of that artefact. And it is apparent that he has something of it; therefore, he has the true form.

5. Furthermore, knowing and producing is more than simply knowing. But an artisan knows the artefact in virtue of that aspect of it which is in the

[95] Arist. *Metaph.* VII.17, 1041b4–7.

[96] Arist. *De caelo* I.9, 278a23–b7.

[97] This question is not found in Aspall's Questions on the *Metaphysics* preserved in MS C.

producit; sed sola species solam facit cognitionem; ergo artifex plus habet artificiati quam solam speciem.

6. Item, Aristoteles dicit in septimo *Metaphysicae* quod naturalia fiunt ex sibi convenientibus in materia et forma, et artificialia ex sibi convenientibus in forma; sed quod producit naturale habet materiam et formam eiusdem speciei prout res est[a]; ergo producens artificiale habet formam prout res est.

7. Ad oppositum: formae artificiales non sunt formae ut res nisi in propriis subiectis; sed anima non est subiectum formae artificialis; ergo etc.

8. Solutio[b]. Dicendum quod vera forma artificiati non est in anima, sed solum species eius. Causa autem quare Aristoteles dicit quod forma artificiati est in anima, [c]et hoc innuit in multis locis, est ista: ad hoc quod anima sive artifex producat artificiatum necesse est quod habeat cognitionem cuiuslibet eius partis et totius, et ita per speciem quam habet artifex de artificiato potest cognoscere totum et partes. Per speciem autem rei naturalis vix concipimus naturale, quia exigitur lux abstrahens et medium et debita condicio organi. Unde tot exiguntur ad receptionem speciei rei naturalis quod vix cognoscimus ipsum. Possumus autem nobis fingere species artificiatorum.

9. Alia causa est ista: species artificiati et rei naturalis opposito modo se habent ad animam, quia species rei naturalis ab extra facit se per medium ad animam et non est principium educendi naturale in esse. Species autem rei artificialis est causa rei extra, quia anima iuxta speciem quae in ipsa est est causa efficiens artificiatum. Unde, quia species artificiati quae est in anima est causa efficiens artificiati extra et species rei naturalis non, ideo dicit Aristoteles [d]omnes suas auctoritates per quas videtur quod vera[e] forma artificiati[f] sit in anima.

10. [g]Secundae rationi dicendum quod, si li 'plus' dicat plus habere quia rem vel veram formam habere, sic est propositio falsa. Vel 'plus habet', hoc est 'verius habet speciem', quia de sua potentia est prosequi ut actu producat; sic est propositio vera.

[a] est] sunt TC
[b] Solutio *om.* TC
[c] et ... locis *om.* C
[d] omnes ... videtur] in pluribus locis C
[e] vera *om.* TC
[f] artificiati *om.* T; artificialis *suppl. s.l.* O
[g] Secundae ... vera] Ad secundam dico quod plus habere potest esse dupliciter: vel ita quod li 'plus' dicat rem, et sic est falsa. Vel 'plus', id est 'verius', et sic est vera. Hoc modo plus de artificiali est (in artificiali est *add. sed del.*) in anima (est + *illegibile* essentiae artificiati extra *add. sed del.*) et hoc est anima verius habet speciem artificiati quia de sua potentia est ut actu producat artificiatum per illam speciem C

soul and he brings it into being; but the species alone only produces a cognition; therefore, an artisan has more of the artificial thing than the species alone.

6. Furthermore, in *Metaphysics* VII Aristotle says that natural things come to be from things that are similar to them in matter and in form, and artificial things <come to be> from things that are similar to them in form;[98] but what produces a natural thing has the matter and form of the same species in so far as that form is a thing; therefore, what produces an artificial thing has the form in so far as that form is a thing.

7. To the contrary: artificial forms are not forms as things except in their proper subjects; but the soul is not the subject of an artificial form; therefore, <a true artificial form is not in the soul>.

8. Solution. It must be said that the true form of the artefact is not in the soul but only its species. The reason why Aristotle says that the form of the artefact is in the soul, and suggests this in many places,[99] is that in order for the soul or the artisan to produce the artefact it is necessary for it to have the cognition of each of its parts and of the whole, and thus in virtue of the species of the artefact that he has the artisan can know the whole and the parts. In virtue of the species of a natural thing, however, we barely conceive the natural thing, because the abstracting light and the medium and the appropriate condition of the organ are required. Accordingly, so many things are required for the reception of the species of a natural thing that we barely know it. We can, however, fashion for ourselves the species of artefacts.

9. Another reason <for what Aristotle said> is that the species of an artefact and of a natural thing have opposite relations to the soul, because the species of a natural thing reaches the soul from outside through the medium and is not a principle for bringing the natural thing into being. The species of an artificial thing, however, is the cause of the external thing, because the soul is the efficient cause of the artefact in virtue of the species that exists within it. Thus, it is because the species of an artefact existing in the soul is the efficient cause of an external artefact and the species of a natural thing is not, that Aristotle presents all his authoritative texts that seem to prove that the true form of an artefact is in the soul.[100]

10. To the second argument[101] it must be said that, if the word ‘more’ means to have more in the sense of having the thing or the true form, then the proposition is false. Alternatively, <one can say that> ‘it has more’ is equivalent to ‘it has the species more truly’, because it is in the soul’s power to produce it in act, and in this case the proposition is true.

[98] Arist. *Metaph.* VII.9, 1034a21–25.

[99] Arist. *Metaph.* VII.7, 1032a32-b1. Cf. also *Metaph.* XII.3, 1070a13–17, 30.

[100] See above, n. 8.

[101] See above, n. 2.

11. Ad tertium dicendum quod aliquid denominare aliud contingit dupliciter: aut sicut qualitas denominat suum proprium subiectum, et sic ars[a] non denominat[b] animam, et ita denominat [c]sola forma ut res est; aut [d]sicut efficiens sive condicione efficientiae[e], et sic denominat ars animam, [f]et sic non denominat [g]vera forma.

12. Ad alia patebit [h]super septimum *Metaphysicae*.

1*. Consequenter determinat Aristoteles de causis. Et primo de causis per se, secundo de causis per accidens. De[i] causis per se dicit primo quod quattuor sunt causae, et dat tres proprietates. Prima est quod multae possunt esse causae eiusdem causati. Secunda est quod idem potest esse causa contrariorum. Tertia est quod aliqua possunt esse causa sibi invicem.

2*. Cum Aristoteles determinet de causis per se et de causis per accidens, primo quaeritur de causis per se, circa quas quaeruntur quattuor quaestiones. Prima quaestio est de numero earum[j], secunda autem quaestio et tertia et quarta sunt circa tres proprietates ipsarum quas assignat Aristoteles.

<Quaestio 17>
Quaeritur ergo primo an quattuor sint causae sicut narrat Aristoteles.

1. Et videtur quod non, quia principium et causa idem; sed non sunt nisi tria principia, per Aristotelem; ergo nec nisi tres causae.

[a] ars *om.* TC; *suppl. mg.* O
[b] denominat] forma artificiati *add.* C
[c] sola forma] solam formam O; forma C
[d] sicut efficiens sive *om.* C
[e] efficientiae] efficientem O; essentiae C
[f] et ... forma *om.* T
[g] vera forma *om.* C
[h] super septimum Metaphysicae] alibi T
[i] De *ex corr.* O
[j] earum *scripsimus*; eorum O

11. To the third argument[102] it must be said that something can denominate another thing in two ways: either (i) as a quality denominates its proper subject, and in this way art does not denominate the soul, and only a form in so far as it is a thing denominates in this way; or (ii) as an efficient cause, that is, by the condition of efficiency, and in this way art denominates the soul, and a true form does not denominate in this way.

12. The replies to the other arguments[103] will be made clear in the comment on *Metaphysics* VII.[104]

1*. Next Aristotle establishes definite conclusions about causes,[105] firstly about per se causes,[106] secondly about accidental causes.[107] About per se causes he first says that there are four causes,[108] and he gives three of their properties. The first property is that there can be many causes of the same effect.[109] The second is that the same thing can be the cause of contrary things.[110] The third is that some things can be the cause of each other.[111]

2*. Given that Aristotle establishes definite conclusions about per se causes and about accidental causes, the questions about the per se causes are raised first, and there are four questions. The first question is about their number; the second, third, and fourth questions are about the three properties of these causes that Aristotle posits.

<Question 17>
The first question, therefore, is whether there are four causes, as Aristotle says.

1. Arguments for the negative: principle and cause are the same; but there are only three principles, according to Aristotle;[112] therefore, there are only three causes.

[102] See above, n. 3.
[103] See above, nn. 1, 4–6.
[104] The topic of q. 16, whether a true artificial form is in the soul, is also discussed in a question from Aspall's commentary on the *Metaphysics* preserved in MS C, fols. 108va–109ra ("Videtur enim ipsum (*sc.* Aristotelem) velle quod vera (videtur *ms.*) forma sit in anima, de quo quaeritur utrum hoc sit verum"). However, the arguments referred to here (nn. 1, 4–6) are not found in the question in the *Metaphysics* commentary.
[105] Arist. *Phys.* II.3–7.
[106] Arist. *Phys.* II.3.
[107] Arist. *Phys.* II.4–6.
[108] Arist. *Phys.* II.3, 194b23–195a3.
[109] Arist. *Phys.* II.3, 195a3–8.
[110] Arist. *Phys.* II.3, 195a11–14.
[111] Arist. *Phys.* II.3, 195a9–11.
[112] Arist. *Phys.* I.6–7.

2. Item, scire rem est scire suam causalitatem, et scire rem est scire suam essentiam; causalitas ergo et essentia idem. Sed unius rei tantum est una essentia; ergo et tantum una causa.

3. Dicendum cum Aristotele quod quattuor sunt causae.

4. Ad primum in oppositum dicendum quod, licet principium et causa sint idem secundum rem, tamen secundum quod sumuntur sumuntur a diversis, quia principium nominat illud quod est in composito vel ut res, et hoc propter materiam[a] et formam, vel ut defectus cum aptitudine quadam, et hoc propter privationem. **[O 105rb]** Et ideo, quia principium nominat illud quod est in re[b], in re autem non sunt nisi tria, videlicet materia et forma et privatio, ideo non sunt nisi tantum tria principia. Causa vero dicitur quia attribuit causalitatem rei. Sed attribuere causalitatem rei, hoc est dupliciter: aut rei prout est in fieri aut rei prout est in esse. Si rei prout est in esse, hoc est dupliciter: aut prout est in esse potentiali aut prout est in esse actuali. Primo modo est causa materialis, secundo modo causa formalis. Si rei prout est in fieri, cum fieri sit moveri, hoc est dupliciter. Aut enim res est movens et mota, et sic causa efficiens, secundum quod dicit Aristoteles quod non est movens nisi moveatur prius, et intelligit de movente quod est efficiens; aut est res movens non mota, et sic est causa finalis.

5. Patet iam ergo quod principium et causa sumuntur a diversis, quia principium sumitur primo, quia ab ipsa re, causa vero sumitur secundario, quia post ipsam rem sic constitutam.

6. Ad secundum dicendum quod causalitas et essentia non sunt simpliciter idem nec potest hoc per rationem praedictam concludi, sed sunt idem quodam modo, videlicet sicut illud quod constituit et illud quod constituitur, quia causa est quasi constituens essentiam. Essentia enim sive definitio integrat in se omnes causas. Non ergo sequitur quod, si tantum sit una essentia, quod propter hoc tantum sit unica causa unius rei, quia constituentia possunt esse multa et ipsum constitutum unum solum.

7. Vel sic: duplex est causa, una quae est sufficiens sive aggregans in se omnes causas, de qua causa dicitur quod, destructa causa, destruitur effectus. Non enim, destructa materia rei, destruitur res, nec etiam forma per se

[a] materiam *scripsimus*; naturam O
[b] re *ex corr.* O

2. Furthermore, to know a thing is to know its causality, and to know a thing is to know its essence; therefore, causality and essence are the same. But one thing has only one essence; therefore, it also has only one cause.

3. Following Aristotle it must be said that there are four causes.

4. To the first argument to the contrary[113] it must be said that, while a principle and a cause are the same in reality, however, in so far as they are taken from something, they are taken from different things, because principle designates what exists in the composite, either as a thing, and this because of matter and form, or as a defect with some aptitude, and this because of privation. Thus, since principle designates what exists in a thing and in a thing there are only three items, namely, matter, form, and privation, there are therefore only three principles. On the other hand, something is said to be a cause because it ascribes causality to a thing. But to ascribe causality to a thing happens in two ways: either (i) in so far as the thing is in becoming or (ii) in so far as it is in being. Case (ii) also happens in two ways: either (iia) in so far as it is in potential being or (iib) in so far as it is in actual being. In case (iia), the cause is the material cause, in case (iib) the formal cause. As to case (i), since becoming is a motion, this happens in two ways: either (ia) the thing is both a mover and moved, and this is the efficient cause, according to Aristotle's statement that something is not a mover unless it is moved beforehand, and he is speaking of the mover as an efficient cause;[114] or (ib) the thing is a mover but not moved, and this is a final cause.

5. It is already clear, therefore, that principle and cause are taken from different things: principle is taken primarily, because it is taken from the thing itself, whereas cause is taken secondarily, because it is taken after the thing itself has been constituted.

6. To the second argument[115] it must be said that causality and essence are not absolutely the same nor can this claim be proved by the argument presented above, but they are in some way the same, that is, in the way that that which constitutes is the same as that which is constituted by it, because a cause is, as it were, a constituent of an essence. Indeed, an essence or a definition includes in itself all the causes. Therefore, even if there is only one essence, it does not follow that just because of this there is only one cause of one thing, because the constituents can be many while the thing constituted is only one.

7. Another reply is the following: causes are of two kinds. (i) One is that which is sufficient, that is, it includes in itself all the causes. The claim 'once the cause is destroyed, the effect is destroyed' refers to this cause; for, it is not the case that, once matter is destroyed, a thing is destroyed, nor <is it

[113] See above, n. 1.
[114] This is probably a reference to Arist. *Phys*. II.7, 198a35–b4.
[115] See above, n. 2.

destructa nec fine nec efficiente; sed si omnes istae destruantur, non amplius manet res. Et haec causa est eadem cum essentia. Et sic loquendo de causa non est nisi unica causa unius rei. Alia est causa quae est communiter dicta de quacumque causa per se. Et sic loquendo de causa non sunt idem.

<Quaestio 18>
Quaeritur de prima proprietate quam assignat Aristoteles, quae est ista, quod plures possunt esse causae eiusdem causati.

1. Et videtur quod Aristoteles dicat falsum, quia causa est ad cuius esse sequitur aliud[a]; ergo <multae[b]> causae sunt ad quarum esse sequuntur multa esse; quare etc.

2. Item, idem[c] in quantum idem semper natum est facere idem; ergo idem effectus in quantum idem natus est fieri ab uno et eodem, et ita non a multis; quare etc.

3. Ad quaestionem dicendum cum Aristotele quod multae possunt esse causae eiusdem. Sed intelligendum est quod hoc est verum secundum quod illud habet in se esse diversa. Si vero non haberet nisi unum esse, quod quidem esset indivisibile, nec haberet nisi unam solam causam. Sed iam omnis res habet in se esse diversa, quia materia facit unum esse, forma vero facit aliud.

4. Unde dicendum quod causae quae perficiunt aliquod compositum dupliciter possunt considerari: vel in quantum concurrunt ad constitutionem illius rei vel prout in se considerantur. Si primo modo, sic istae quattuor causae unum faciunt esse, id est, rei existentiam, quia <nec[d]> sine materia nec sine forma nec sine motore nec sine fine posset fieri rei existentia. Si secundo modo, sic habent diversa esse et faciunt esse diversa secundum quod dicitur a materia subsistentia, quia materia est subsistens et substentans formam, a forma existentia, quia perficit et dat esse rei, a motore motus, a fine bonum.

5. Patet ergo iam qualiter secundum esse diversa possunt esse plures causae eiusdem.

6. Et per hoc patet responsio ad argumenta inspicienti.

[a] aliud *corr. mg. ex* esse in quantum idem O
[b] multae *supplevimus*
[c] idem] et eodem *add. sed del.* O
[d] nec *supplevimus*

destroyed> once the form per se is destroyed, or the end or the efficient cause, but if all these causes are destroyed, then the thing no longer persists. This cause is the same as essence. And if we talk about cause in this way, there is only one cause of one thing. (ii) Another kind of cause is that which is said in common of every per se cause. And if we talk about cause in this way < cause and essence> are not the same.

<Question 18>
This question is about the first property that Aristotle gives, which is that there can be many causes of the same caused thing.[116]

1. Arguments for the negative: a cause is a thing from whose being something else follows; therefore, many causes are those from whose being many beings follow; therefore, <there cannot be many causes of the same thing>.

2. Furthermore, the same thing in so far as it is the same is always apt by nature to produce the same; therefore, the same effect in so far as it is the same is apt by nature to be produced by one and the same thing, and thus not by many things; therefore, <there cannot be many causes of the same thing>.

3. To this question it must be said, following Aristotle, that there can be many causes of the same thing. It must be understood, however, that this is true in so far as that thing has in itself different beings. If it had only one being and this being were indivisible, it would have only one cause. Now, every thing has in itself different beings, because matter produces one being and form another being.

4. Accordingly, it must be said that the causes that confer perfection on a composite can be considered in two ways: either (i) in so far as they contribute to the constitution of that thing or (ii) in themselves. In case (i), these four causes produce one being, that is, the existence of the thing, because without the matter or without the form or without the mover or without the end the existence of the thing cannot come about. In case (ii), these causes have different beings and produce different beings so that subsistence is said to derive from matter, since matter is subsistent and supports the form, existence is said to derive from the form, since the form confers perfection and being on the thing, motion is said to derive from the mover, and the good from the end.

5. It is now clear in what way there can be many causes of the same thing with respect to different beings.

6. On this basis the reply to the arguments[117] is clear to those who consider it.

[116] See above, q. 17, n. 1*.
[117] See above, nn. 1–2.

<Quaestio 19>
Quaeritur de secunda proprietate, quae est quod idem potest esse causa contrariorum.

1. Contra: idem in quantum idem natum est facere idem, et idem non est sibi contrarium; ergo etc.

2. Istam rationem solvit Aristoteles dicens quod idem secundum esse diversa est causa contrariorum. Et exemplificat: verbi gratia, gubernator navis per sui praesentiam est causa salvationis, per sui absentiam est causa non salvationis.

3. Sed contra, quod per absentiam alicuius formae non possit haberi aliquis effectus per positionem probatio: inter omnes formas corporales quae procedunt a corpore lux est forma nobilissima; ergo ipsa est potentior. Sed lux per sui absentiam nullum effectum per positionem sive qui res est causat. Probatio: lux per sui absentiam non causat nisi tenebram, et tenebra nihil rei est. Ergo multo fortius nec aliqua alia forma.

4. Dicendum cum Aristotele quod per sui praesentiam et per sui absentiam est aliquid unum et idem causa contrariorum. Sed intelligendum est quod per sui praesentiam est causa ut causa, per sui absentiam causa ut occasio, non ut causa.

5. Ad argumentum in oppositum dicendum quod duplex est forma: una quae egreditur et educitur de potentiis materiae, et tales formae sunt substantialitas, corporeitas etc.; alia quae non educitur de potentiis materiae, sed est adveniens ab extra, et istam ponimus nos esse lucem[a]. Et ponimus istam formam primo advenire materiae ante aliquam aliam formam quae educitur vel simul advenire cum substantialitate, quae est prima forma educta de potentiis materiae. Unde ista forma quae est caliditas educitur de potentiis **[O 105va]** materiae et habet potentiam duplicem in materia: unam respectu sui ipsius, aliam respectu sui contrarii, licet radicaliter fuerit una et eadem potentia, quia una est potentia communis contrariorum. Quia ergo lux non egreditur de potentiis[b] materiae, sed est adveniens ab extra, ideo non est aliqua potentia in materia respectu lucis et respectu sui contrarii; et ideo illud quod derelinquitur per sui absentiam non est res aliqua, sed privatio. Patet ergo iam causa quare contrarium calidi ponit aliquid, contrarium autem lucis nihil ponit.

[a] lucem *scripsimus*; lucere O
[b] potentiis *scripsimus*; principiis O

<Question 19>
This question is about the second property, that is, that the same thing can be the cause of contrary things.[118]

1. Against this: the same thing in so far as it is the same is apt by nature to produce the same, and the same thing is not contrary to itself; therefore, <the same thing cannot be the cause of contrary things>.

2. Aristotle's solution to this argument is that the same thing in accordance with different beings is the cause of contrary things. And he gives the example of the pilot of the ship, who by his presence is the cause of the safety of the ship and by his absence is the cause of its lack of safety.[119]

3. Against this one can argue that the absence of a form cannot be responsible for a positive effect. Proof: light is the noblest among the corporeal forms that proceed from a body; therefore, light is more powerful <than the others>. But light by its absence does not cause a positive effect or an effect that is a thing. Proof: light by its absence only causes darkness, and darkness is nothing real. Therefore, there is a much stronger reason to hold that no other form <can cause a positive effect>.

4. Following Aristotle it must be said that one and the same thing by its presence and by its absence is the cause of contrary things.[120] It must be understood, however, that by its presence it is a cause taken as cause, whereas by its absence it is a cause taken as the occasion and not as the cause.

5. To the argument to the contrary it must be said that there are two kinds of form: (i) one kind is that which comes out of and is educed from the potencies of matter; and forms of this kind are substantiality, corporeality, and the like. (ii) Another kind is that which is not educed from the potencies of matter but comes from outside; we posit that light is a form of this kind. We also posit that this form inheres in matter either prior to any other form that is educed <from the potencies of matter> or simultaneously with substantiality, which is the first form educed from the potencies of matter. Accordingly, the form of heat is educed from the potencies of matter and has two potencies in matter: one with respect to itself and the other with respect to its contrary, although these originally were one and the same potency, because contraries have one potency common to them. Therefore, since light is not educed from the potencies of matter, but comes from outside, there is no potency in matter with respect to light and with respect to its contrary; hence, what is left by its absence is not a thing, but a privation. It is now clear why it is the case that the contrary of the hot is something positive, whereas the contrary of light is nothing positive.

[118] See above, q. 17, n. 1*.
[119] Arist. *Phys.* II.3, 195a12–14.
[120] See above, n. 2.

<Quaestio 20>
Quaeritur de tertia proprietate, quae est quod aliqua possunt esse causa sibi invicem.

1. Et exemplificat Aristoteles dicens “Verbi gratia: finis est causa efficientis et econverso”.

2. Contra: quicquid est causa causae est causa causati; ergo finis erit causa sui ipsius, quod est inconveniens per librum *De causis*: “Nihil se ipsum constituit nec ad esse producit”.

3. Solutio. Finis est causa efficientis per suam similitudinem et per suam speciem. Et sic non est inconveniens quod idem sit causa sui ipsius, quia species sive similitudo finis est causa finis ut[a] res est; unde idem non eodem modo potest esse causa sui ipsius.

4. Sed contra: movens unde movens efficiens est; sed si finis moveret efficientem, ut dictum est, tunc finis est efficiens, et sic efficientis esset efficiens, quod est inconveniens.

5. Item, eius quod nec est nec fuit nec est similitudo quae eius sit; ponis ergo falsum quando ponis similitudinem rei fiendae movere animam[b] efficientis.

6. Solutio. Dicendum est, sicut dictum est, quod species solum non est causa movens animam efficientis, sed unde comparata ad bonum. Unde species dupliciter potest considerari: vel prout species tantum, et sic solum est causa cognoscendi rem operandam; vel prout species ista est comparata, et hoc ad bonum simpliciter vel ad bonum apparens, et sic species non solum est causa cognoscendi, sed etiam operandi.

7. Ad primum in oppositum dicendum quod ‘movens’ dicitur dupliciter: uno modo in operando tantum, alio modo in excitando tantum. Si primo modo, sic est efficiens, verbi gratia carpentarius. Et tale movens est illud de quo loquitur Aristoteles, cum dicit quod movens non movet nisi prius moveatur. Et tale movens est proprie efficiens quia efficiens non est movens

[a] ut *ex corr.* O
[b] animam *lectio incerta* O

<Question 20>
This question is about the third property, which is that some things can be the cause of each other.[121]

1. Aristotle gives an example of this when he says that the end is the cause of the efficient cause and the other way around.[122]

2. Against this: whatever is the cause of the cause is the cause of the thing caused; therefore, the end will be the cause of itself, which is absurd according to the book *On Causes*: "Nothing is responsible for its own constitution or brings itself into being".[123]

3. Solution. The end is the cause of the efficient cause by its likeness and by its species. And it is not absurd that something is the cause of itself in this way, because the species or likeness of the end is the cause of the end as a thing; accordingly, the same thing can be cause of itself but not in the same way.

4. Against this: a mover in so far as it is a mover is an efficient cause; but if the end moved the efficient cause, as has been said,[124] then the end is the efficient cause, and so there will be an efficient cause of an efficient cause, which is absurd.

5. Furthermore, there is no likeness of something that does not exist and did not exist; therefore, you make a false assumption when you posit that the likeness of a future thing moves the soul of the efficient cause.

6. Solution. It must be said, as has already been said,[125] that a species alone is not the cause that moves the soul of the efficient cause, but a species in so far as it is compared to the good. Thus, a species can be considered in two ways: either (i) in so far as it is only a species, and in this way it is only the cause of knowing the thing to be produced; or (ii) in so far as the species is compared either to the absolute good or to the apparent good, and in this way a species is not only the cause of knowing but also of producing the thing.

7. To the first argument to the contrary[126] it must be said that 'mover' is said in two ways: (i) only with respect to acting, and (ii) only with respect to stimulating. In case (i) a mover is an efficient cause, for instance, the carpenter. And Aristotle refers to this mover when he says that a mover does not move unless it is moved beforehand.[127] This mover is an efficient cause properly speaking, because an efficient cause is not a mover before it is stimulated and

[121] See above, q. 17, n. 1*.
[122] Arist. *Phys.* II.3, 195a9–11.
[123] The correct reference is Nicolas of Amiens, *Ars catholicae fidei* I.1, prop. 3, ed. Dreyer, 80.12.
[124] See above, n. 3.
[125] See above, n. 3.
[126] See above, n. 4.
[127] See above, q. 17, n. 4.

antequam excitetur et moveatur per similitudinem et speciem rei fiendae. Unde tale movens proprie est efficiens. Et isto modo sumendi movens nullo modo est dicere quod finis movet. Si secundo modo, sic non est efficiens, sed finis. Verbi gratia, video poma, cogito quod vellem comedere poma; iam est hic aliquid excitans et movens, ut species pomi, non tamen est hic aliquid movens ut operans. Possum enim excitari a specie pomi, non tamen comedere poma. Unde tale movens non meretur nomen efficientis. Unde non tenet illud argumentum: 'movet; ergo est efficiens', nisi sic dicatur: 'movet ut operans; ergo est efficiens'. Unde sciendum quod finis est movens non mobile.

8. Ad secundum dicendum quod habere speciem penes animam est dupliciter: vel a re productam[a] et a re multiplicatam[b], et talis species est posterior eo cuius est species; vel secundum discursum rationis de re fienda coniecturando, et talis species praecedit rem fiendam et potest esse de eo quod non est, secundum quod dicit Aristoteles in septimo[c] *Metaphysicae* quod domus est a domo et sanitas a sanitate. De prima specie et de prima similitudine procedit argumentum, de secunda vero non, sicut patet.

1*. [d]Quaeritur de casu et fortuna.

<Quaestio 21>
Et quaeritur primo quid sit et quid significatur per hoc nomen 'casus'.

1. Et sumamus hoc nomen 'casus' etiam pro fortuna.

2. Et cum ibi non sint nisi duo, scilicet causa et eventus[e], videtur primo quod significet eventum, quia dicit Boethius quod casus est inopinatae rei eventus; cum ergo eventus cadat a parte ipsius effectus, videtur quod 'casus' significet [f]effectum sive eventum.

3. Item, 'casus' dicitur a cadendo, quod[g] vero est ab eventu cadit et non quod est a causa; immo a causa cadit aliud; ergo etc.

4. Item, non sunt in exemplo Aristotelis de casu nisi tria, scilicet ire ad forum, [h]obviare debitori[i], accipere pecuniam. Sed ire ad forum non est casus, quia hoc est a [j]suo proposito et intentum ab agente; oportet ergo quod casus

[a] productam *scripsimus*; producente O
[b] multiplicatam *scripsimus*; multiplicativa O
[c] septimo *ex corr.* O
[d] Quaeritur ... nomen] De casu (casu *corr. ex* determinatis (?)) et fortuna etc. Capitulum de casu et fortuna. Quaeratur ergo in principio quid sit C
[e] eventus *corr. ex* eventum O
[f] effectum sive *om.* C
[g] quod *corr. ex* quia O
[h] obviare ... forum] et hoc C
[i] debitori *scripsimus cum* C; creditori OT
[j] suo proposito] proposito euntis C

moved by the likeness and the species of the thing to be produced. Accordingly, this mover is an efficient cause properly speaking. Also, if mover is taken as in (i), there is no way of saying that the end moves. In case (ii), a mover is not an efficient cause but an end. For example, I see an apple and I think that I want to eat that apple. Now, in this case there is something that stimulates and moves, that is, the species of the apple, but there is not anything that moves as acting; for I can be stimulated by the species of the apple, but I can refrain from eating the apple. Accordingly, this mover does not deserve the name of efficient cause. Thus, that argument: 'it moves; therefore, it is an efficient cause' does not hold, unless it is formulated as follows: 'it moves as acting; therefore, it is an efficient cause'. Thus, the end is a non-mobile mover.

8. To the second argument[128] it must be said that to have a species in the soul comes about in two ways: either (i) as produced by the thing and multiplied by the thing, and this species is posterior to the thing of which it is the species; or (ii) in accordance with a rational process by conjecturing about the thing to be produced, and this species precedes the thing to be produced and can be of a thing that does not exist, in accordance with Aristotle's statement in *Metaphysics* VII that a house comes from a house and health comes from health.[129] The argument holds for a species and a likeness in case (i), but not for that in case (ii), as is clear.

1*. Questions about chance and fortune.

<Question 21>

The first question is what chance is and what the term 'chance' signifies.

1. And we take the term 'chance' to refer to fortune as well.

2. And since in this context there are only two things, namely, a cause and an event, we first consider <the arguments> that 'chance' signifies an event. Because Boethius says that chance is the event of an unexpected thing,[130] then, since an event falls on the side of the effect, it seems that 'chance' signifies an effect or event.

3. Furthermore, the word 'chance' comes from the word for falling; but what falls is from the event and not from the cause; rather, something else falls from the cause; therefore, <'chance' signifies an event>.

4. Furthermore, in Aristotle's example there are only three things, namely, going to the market, meeting the debtor, and getting the money.[131] But going to the market is not chance, because this is deliberated and intended by the

[128] See above, n. 5.
[129] Arist. *Metaph.* VII.7, 1032b11–12.
[130] Boethius, *De cons.* V.1.18, CCSL 94, 90.47–49.
[131] Arist. *Phys.* II.4, 196a3–4; II.5, 196b33–36.

sit aliquod[a] aliorum, scilicet [b]secundo modo vel tertio. Sed sive sic sive sic, utroque modo est eventus; quare etc.

5. Item, si casus sit causa per accidens, [c]ut vult Aristoteles, tunc album erit[d] casus respectu aedificationis [e]et musicum similiter, et ita dicemus quod in omni actione sit casus.

6. Ad oppositum: Aristoteles dicit quod[f] est causa[g] per accidens.

7. Item, eventus est a casu; tunc, si eventus esset casus, [h]sequeretur quod casus esset a casu, quod est inconveniens.

8. Item, casus est propter aliquid[i] et similiter causa; casus ergo cadit a[j] parte causae et non a[k] parte effectus; quare etc.

9. Quidam dicunt quod casus est aequivocum ad causam et ad eventum, et Aristoteles [l]definit et accipit eum[m] in uno sensu, **[O 105vb]** Boethius vero in alio sensu.

10. Sed[n] contra: tunc album et musicum et huiusmodi essent casus, quia sunt causa per accidens respectu aedificationis.

11. Et[o] contra hoc etiam sunt quaedam aliae rationes [p]priores, ut patet.

12. Quidam dicunt aliter quod 'casus' habet duplex significatum[q]: primum et secundarium. Primum significatum est illud a quo imponitur, et hoc est accidentalitas, et hoc cadit a parte eventus. Secundum significatum est illud cui imponitur, et hoc est causalitas, quia hoc est a parte naturae[r] agentis.

13. Sed[s] contra: Aristoteles dicit quod non est causa nisi secundum accidens, et ita accidentalitas cadit a parte causae, et ita utroque modo est accidentalitas. Nihil ergo dicunt.

[a] aliquod *ex corr.* O
[b] secundo ... tertio] vel obviare debitori vel accipere pecuniam C
[c] ut vult] sicut dicit C
[d] erit] esset TC
[e] et musicum similiter *om.* T
[f] quod] casus *add.* TC
[g] causa *ex corr.* O
[h] sequeretur quod casus *om. hom.* C
[i] aliquid] aliud TC
[j] a] ex TC
[k] a] ex TC
[l] definit et] definiens T; definiens casum C
[m] eum] eam O; *corr. ex* eam T
[n] Sed *om.* TC
[o] Et *om.* TC
[p] priores, ut patet] prius positae C
[q] significatum] scilicet *add.* TC
[r] naturae *om.* TC
[s] Sed *om.* TC

agent; therefore, it is necessary that chance is one of the other things, namely, the second or the third. But in either case it is an event; therefore, <'chance' signifies an event>.

5. Furthermore, if chance is an accidental cause, as Aristotle wishes,[132] then a white person and similarly a musician will be chance with respect to a building, and so we will say that there is chance in every action.

6. To the contrary: Aristotle says that chance is an accidental cause.[133]

7. Furthermore, an event is by chance; thus, if an event were chance, it would follow that chance would be by chance, which is absurd.

8. Furthermore, chance is for the sake of something and so is a cause; therefore, chance falls on the side of the cause and not on the side of the effect; therefore, <chance signifies a cause>.

9. Some people say that 'chance' is equivocal and applies to both the cause and the event, and Aristotle defines and takes 'chance' in one sense, and Boethius in the other sense.

10. Against this: then a white person and a musician and things of this sort would be chance, because they are accidental causes with respect to a building.

11. Against this view there are also some other prior arguments,[134] as is clear.

12. Some people have another view according to which 'chance' has two significates: a primary and a secondary. The primary significate is that from which the name is imposed, and this is accidentality, and falls on the side of the event. The secondary significate is that on which the name is imposed, and this is causality, because this is on the side of the nature of the agent.

13. Against this: Aristotle says that chance is a cause only accidentally,[135] and thus accidentality falls on the side of the cause, and thus in both ways there is accidentality. Therefore, these people say nothing.

[132] Arist. *Phys*. II.5, 196b21–29.
[133] See above, n. 5.
[134] See above, nn. 2–4.
[135] See above, n. 6.

14. Item, sicut accidentale est quod talis effectus fiat a tali causa, similiter accidentale est quod ista causa producat talem effectum, et ita accidentalitas cadit a parte causae.

15. Solutio[a]. Casus non est causa nec eventus, sed habitudo huius causae ad hunc effectum sive eventum ad quem non ordinatur. Et ista habitudo dupliciter potest considerari: vel prout incipit a causa et terminatur in causatum [b]sive eventum, et sic dicitur ista habitudo[c] causalitas accidentalis; vel prout incipit a causato sive eventu et consideratur[d] postea respectu causae, et sic dicitur ista habitudo effectus accidentalis sive eventus. Sed habitudo considerata prout incipit a causa et terminatur in causatum est prior quam habitudo econverso modo considerata. Et ideo Aristoteles, respiciens et considerans casum secundum suam principalem significationem et primariam, dicit ipsum esse causam per accidens. Boethius vero[e], considerans casum secundo modo, dixit ipsum esse eventum. Unde non est inter eos[f] contrarietas.

16. [g]Et sic patet responsio ad primum argumentum.

17. Ad secundum [h]per illud idem patet responsio.

18. Ad tertium dicendum quod casus [i]nullum illorum trium est per se, sed est habitudo quaedam media inter causam[j] et eventum, sicut praedictum est.

19. Ad quartum dicendum quod duplex est causa per accidens. Aut enim dicitur causa per accidens quia est accidens alicuius quod per se est causa illius, et [k]sic album et nigrum et huiusmodi dicuntur causae per accidens. Alio modo dicitur causa per accidens non quia est accidens alicuius quod est per se causa, sed quia [l]hoc quod ordinatur ad aliquid accidit quod producat alium[m] effectum, cuiusmodi[n] est accipere pecuniam a debitore, qui effectus nec intentus est a recipiente nec a solvente, vel aliquem talem. Et hoc modo dicuntur casus et fortuna esse causae per accidens. Et sic patet aequivocatio [o]in argumento. [p]Unde sciendum quod in comparatione ad causam potest dici casus et econverso similiter.

[a] Solutio] dicendum (dico C) ut mihi videtur quod TC
[b] sive eventum *om.* TC
[c] habitudo] effectus sive eventus accidentalis *add.* C
[d] consideratur *corr. ex* consideretur O; consideretur T
[e] vero *om.* TC
[f] eos *ex corr.* O
[g] Et ... idem patet responsio (n.17)] Sic ad primum argumentum et ad secundum C
[h] per ... responsio *om.* T
[i] nullum ... est] non est aliquid illorum trium TC
[j] causam] casum TC
[k] sic ... nigrum et *om.* C
[l] hoc ... producat] ad (*corr. mg. in* quam) hoc (quod *add. s.l.*) per se ordinatur aliud aliquid accidit quod producit O; accidit quod illud quod ordinatur ad aliquid producat aliud sive C
[m] alium] talem O; aliud C
[n] cuiusmodi] qui *ex corr.* O; cuius C
[o] in ... similiter *om.* T
[p] Unde ... similiter *om.* C

14. Furthermore, just as it is accidental that this effect comes from this cause, so likewise it is accidental that this cause produces this effect, and thus accidentality falls on the side of the cause.

15. Solution. Chance is neither the cause nor the event, but a relation of this cause to this effect or event to which this cause is not ordered. And this relation can be considered in two ways: either (i) in so far as it begins from the cause and ends in the caused thing or event, and in this way this relation is said to be accidental causality; or (ii) in so far as it begins from the caused thing or event and is later considered with respect to the cause, and in this way this relation is said to be an accidental effect or event. But the relation considered as in (i) is prior to the relation considered as in (ii). And this is why Aristotle, who looks at and considers chance with respect to its principal and primary signification, says that chance is an accidental cause.[136] Boethius, on the contrary, who considers chance as in (ii), said that chance is an event.[137] Accordingly, there is no contrast between Aristotle and Boethius.

16. And with this the reply to the first argument[138] is clear.

17. On the same basis the reply to the second argument[139] is clear too.

18. To the third argument[140] it must be said that chance is per se none of those three things, but it is an intermediate relation between a cause and an event, as has been said.[141]

19. To the fourth argument[142] it must be said that there are two kinds of accidental cause. Either (i) something is said to be an accidental cause <of an effect> because it is an accident of something that is the per se cause of that effect; and in this way white, black, and the like are said to be accidental causes. Or (ii) something is said to be an accidental cause not because it is an accident of something that is the per se cause, but because it is accidental that the thing ordered to something <such as going to the market> produces another effect, such as getting money from a debtor, which effect is intended neither by the man receiving the money nor by the man settling his debts, or some such effect. And in this way chance and fortune are said to be accidental causes. The equivocation in the argument is then clear. Accordingly, it must be known that chance can be said with respect to the cause and likewise the other way around.

[136] See above, n. 6.
[137] See above, n. 2.
[138] See above, n. 2.
[139] See above, n. 3.
[140] See above, n. 4.
[141] See above, n. 15.
[142] See above, n. 5.

<Quaestio 22>
Quaeritur an aliquid sit a casu.

1. Et videtur quod non, quia omne causatum quod est habet relationem ad causam primam, per Aristotelem in undecimo dicentem quod duplex ordo rerum est, scilicet unus in se et alius respectu causae primae. Sed relatio causati ad causam primam est sua essentia; ergo essentiale est quod consideretur prout est causatum a causa prima. Omne ergo quod est in mundo est aliter quam a casu; quare nihil a casu.

2. Item, casus, cum sit causa, per Aristotelem, aut est causata aut incausata. Non incausata, quia talis causa est causa prima. Oportet ergo quod sit causata. Sed hoc potest esse dupliciter: aut est causata[a] a causa per se aut a causa per accidens. Non a causa per accidens, quia talis est casus, et tunc erit processus in infinitum; oportet ergo quod a causa per se, et ita habet causam necessariam determinatam in aliquo subiecto; quare non est a casu.

3. Item, casus est causa immediata eventus casualis[b]. Sed talis causa necessario producit suum effectum; et quod est a necessaria productione non est a casu; quare. Probatio maioris: si non est causa immediata, ergo erit causa aliqua media[c] inter ipsum et eventum. Illa causa aut est per se aut per accidens. [d]Si per accidens, tunc erit processus in infinitum, quod est inconveniens; ergo erit causa per se. Et ita a causa immediata, et ita necessario producet illud quod producit et non a casu sive contingenter.

4. Item, casus reducitur ad causam per se; ergo et eventus reducitur ad illam causam quae est per se. Sed si reducitur ad causam per se, ergo est a causa per se, et nullum tale producitur in esse casualiter.

5. Item, omne quod est aut est substantia aut accidens. Si eventus rei[e] sit substantia, tunc non est a casu, quia substantia habet causas necessarias et determinatas praeexistentes et praecedentes suum esse. Si accidens, aut ergo per se, et tunc habetur propositum, sicut patet. Aut per accidens; sed omne tale habet esse per se in aliquo subiecto; quare nullo modo, ut videtur, est aliquid a casu.

[a] causata *scripsimus*; causatum O
[b] casualis *ex corr.* O
[c] media *ex corr*. O
[d] Si per accidens *om. hom. sed suppl. s.l.* O
[e] rei *ex corr.* O

<Question 22>
The question is whether something is by chance.

1. Arguments for the negative: every caused thing that exists has a relation to the first cause, according to Aristotle's statement in *Metaphysics* XI that there are two orders of things: one taken by itself and the other taken in relation to the first cause.[143] But the relation of a caused thing to the first cause is its essence; therefore, it is essential to it that it should be considered in so far as it is caused by the first cause. Therefore, everything in the world is in some other way than by chance, so that nothing is by chance.

2. Furthermore, since according to Aristotle, chance is a cause, it is either a caused cause or an uncaused cause. It is not an uncaused cause because such a cause is the first cause. It is necessary, then, for chance to be a caused cause. But this can happen in two ways: either it is caused by a per se cause or by an accidental cause. It is not caused by an accidental cause because chance is also such a cause, and then there will be an infinite regress. It is necessary, therefore, that it should be caused by a per se cause, so that it has a necessary definite cause in some subject; therefore, it is not by chance.

3. Furthermore, chance is the immediate cause of a casual event. But such a cause necessarily produces its effect; and what is produced by necessity is not by chance; therefore, <nothing is by chance>. Proof of the major premise: if chance is not the immediate cause, then there will be a cause intermediate between it and the event. That intermediate cause is either per se or accidental. If it is accidental, then there will be an infinite regress, which is absurd; therefore, it will be a per se cause. And thus <it comes> from an immediate cause, and thus this cause necessarily produces what it produces and not by chance or contingently.

4. Furthermore, chance is reduced to a per se cause; therefore, the event is also reduced to that per se cause. But if chance is reduced to a per se cause, then it derives from a per se cause, and nothing of that kind is brought into being by chance.

5. Furthermore, everything that exists is either a substance or an accident. If the event of a thing is a substance, then it is not by chance, because a substance has necessary and determined causes that are pre-existing and precede its being. If it is an accident, then either it is a per se accident, and then we get the intended conclusion, as is clear. Or it is an accidental accident; but everything of that kind has per se being in some subject; therefore, there is no way, as it seems, in which something is by chance.

[143] Arist. *Metaph.* XII.10, 1075a11–15.

6. Item, Plato dicit quod nihil est cuius ortum non praecessit legitima causa[a]; sed omne tale cuius ortum praecessit legitima causa est a causa per se; quare etc.

7. Item, Boethius: "Casus est inopinatae rei eventus ex confluentibus causis etc." Et dicit postea sic: "Concurrere autem et confluere dico ex divina providentia, quae omnia disponit in tempore et omnia in loco ordinat". Sed quod disponitur a divina providentia necessario eveniet; quare nihil erit a casu.

8. Item, per exemplum Aristotelis de casu patet quod nihil est a casu, quia Socratem esse sub casu lapidis, lapide tendente deorsum **[O 106ra]** cum tanto impetu, inducit fractionem capitis necessario.

9. Item, quicquid evenit scit Deus[b] evenire; et sua scientia est necessaria; ergo hanc rem evenire est necessarium[c]. Probatio consequentiae: dicit Averroes[d] in undecimo *Philosophiae primae* quod scientia humana et scientia divina differunt in hoc quod scientia divina est causa rerum, scientia humana non. Arguo tunc sic: sua scientia est causa rerum; et sua scientia[e] est necessaria; ergo res necessario[f] evenient[g]; quare nihil a casu.

10. Quidam dicunt sic quod ea quae accidunt in universo possunt dupliciter considerari: vel in comparatione ad suas causas universales vel in comparatione ad suas causas particulares. Si primo modo, sic nihil est a casu, quia causa universalis, ut causa prima, omnia disponit et ordinat suis temporibus atque locis. Si secundo modo, sic multa sunt a casu.

11. Sed contra: licet ab initio sciverit prima causa illud evenire, quia tamen non intendit de ista re, ideo non necessario fieret[h], sed a casu. Et praeterea multa sunt utroque modo necessaria; quare etc.

12. Solutio. Dicendum quod multa sunt a casu et quod, licet prima causa praesciverit[i] ea et ea disposuerit, tamen non tamquam[j] intendebat de eis; et ideo scientia Dei non aufert quin aliqua possunt esse casualia.

[a] causa] *illegibile add. s.l.* O
[b] Deus *suppl. s.l.* O
[c] necessarium *scripsimus*; necessario *ex corr.* O
[d] Averroes *scripsimus cum* TC (Commentator); Aristoteles O
[e] scientia *iter. sed corr.* O
[f] necessario *ex corr.* O
[g] evenient *scripsimus*; venient O
[h] fieret *lectio incerta* O
[i] praesciverit *scripsimus*; praescivit O
[j] tamquam *fortasse delendum*

6. Furthermore, Plato says that there is nothing such that a legitimate cause of it did not precede its coming into being;[144] but everything such that a legitimate cause preceded its coming into being is from a per se cause; therefore, <nothing is by chance>.

7. Furthermore, Boethius <says>: "Chance is the event of an unexpected thing deriving from causes coming together <among things done for some purpose>".[145] And he later says: "I mean that they concur and come together by divine providence, which orders all things in time and in place."[146] But what is ordered by divine providence will necessarily happen; therefore, nothing will be by chance.

8. Furthermore, the example of chance given by Aristotle[147] shows that nothing is by chance, because the fact that Socrates is under the falling stone, when the stone moves downwards with such an impetus, necessarily induces the fracture of his head.

9. Furthermore, whatever thing happens God knows that it happens, and his knowledge is necessary; therefore, it is necessary that this thing happens. Proof of the consequence: in *Metaphysics* XI Averroes says that the difference between human knowledge and divine knowledge is that divine knowledge is the cause of things, whereas human knowledge is not.[148] I then argue as follows: God's knowledge is the cause of things, and his knowledge is necessary; therefore, things will necessarily happen; therefore, nothing is by chance.

10. Some people say that the things that happen in the universe can be considered in two ways: either (i) in comparison to their universal causes or (ii) in comparison to their particular causes. In case (i) nothing is by chance, because a universal cause, that is, the first cause, arranges and orders all things in their times and places. In case (ii) many things are by chance.[149]

11. But against this: although the first cause knew from the beginning that that thing would happen, since, however, it does not intend that thing, therefore that thing did not happen necessarily but by chance. Also, many things are necessary in both ways; therefore, <they are not by chance in either way>.

12. Solution. It must be said that many things are by chance and that, although the first cause had foreknowledge of them and arranged them, this is not, however, as though it intended them; therefore, God's knowledge does not rule out the possibility that some things are by chance.

[144] Plato, *Tim.* 28A, ed. Waszink, 20.20–22.
[145] Boethius, *De cons.* V.1.18, CSSL 94, 90.47–49.
[146] Ibid., V.1.19, CSSL 94, 90.49–52.
[147] Arist. *Phys.* II.6, 197b30–32.
[148] Aver. *In Metaph.* XII, t.c. 51, fol. 337ra.
[149] A similar view is found in Richard Rufus of Cornwall, *In Phys.*, 125.302–126.314, and Roger Bacon, *Supra libros octo Physicorum*, 124.24–125.33.

13. Ad primum in oppositum dicendum quod omnes res, quotquot sunt in mundo, essentialiter ordinantur ad causam primam, sed quaedam[a] ordinantur ut intentae, ut quae habent causam necessariam sive per se, quaedam ut non[b] intentae, ut quae habent causam per accidens sive[c] casualia. Et ordinatio earum[d] quae ordinantur ad Deum ut intentae[e] est essentialis et essentia earum[f], ordinatio vero secundo modo est accidentalis.

14. Ad secundum dicendum quod reducitur ad causam per se. Sed reduci ad causam per se, hoc est dupliciter: aut per se respectu eius aut respectu alterius. Unde tenet illud argumentum: 'reducitur ad aliquod [g]per se respectu eius et ordinatur ab ipso, ergo habet condiciones et causas determinatas in illo subiecto', et aliter non. Et sic non est in casu et fortuna, quia casus et fortuna non reducuntur ad aliquod per se respectu eorum, sed respectu alterius. Unde in proposito deficit.

15. Et per hoc patet responsio ad omnes rationes sequentes usque ad auctoritates.

16. Ad primam auctoritatem dicendum quod non concludit. Non enim sequitur: 'nihil est cuius ortum non praecessit legitima causa, ut causa prima; ergo nihil est a casu' nisi illa causa legitima esset respectu illius; sed iam non est ita.

17. Ad secundam auctoritatem ipsius Boethii dicendum quod Deus ordinavit omnia in suis locis et temporibus, sicut prius dictum est, sed quaedam ut intenta[h] et quaedam ut non intenta[i].

18. Ad[j] rationem sequentem dicendum quod, licet necessario inducant, tamen, quia non est a proposito, quia tu non intendebas quod lapis caderet et frangeret caput tuum nec etiam lapis intendebat hoc—unde, si tu intendisses hoc et lapis similiter, certe tunc fractio capitis non esset a casu, sed a causa per se et a proposito—, quia ergo neutrum eorum intendebat fractionem capitis, ideo, licet necessario inducant, dicimus tamen quod est a casu, et hoc propter rationem praedictam.

[a] quaedam *corr. ex* quodam O
[b] non *suppl. s.l.* O
[c] sive] causalia *add. sed del.* O
[d] earum *scripsimus*; eorum O
[e] intentae *ex corr.* O
[f] earum *scripsimus*; eorum O
[g] per se *suppl. s.l.* O
[h] intenta *ex corr.* O
[i] intenta *ex corr.* O
[j] Ad] *illegibile add. s.l.* O

13. To the first argument to the contrary[150] it must be said that all things, however many there are in the world, are essentially ordered to the first cause, but some are ordered as intended, as those that have a per se or essential cause, others as not intended, as are those that have an accidental cause or are by chance. And the ordering of those things that are ordered to God as intended is essential and is their essence, whereas ordering in the second way is accidental.

14. To the second argument[151] it must be said that chance is reduced to a per se cause. But something can be reduced to a per se cause in two ways: either per se relative to itself or per se relative to something else. Hence the following argument holds: 'a thing is reduced to something per se relative to itself and is ordered by it, therefore that thing has determined conditions and causes in that subject'; but otherwise it does not apply. Thus it does not apply to chance and fortune because chance and fortune are not reduced to something per se relative to themselves, but relative to something else. Hence in the present case the argument fails.

15. And this makes clear the reply to all the arguments that followed up to the authoritative texts.[152]

16. To the first authoritative text[153] it must be said that it is not conclusive; for this does not follow: 'there is nothing such that a legitimate cause, such as the first cause, did not precede its coming into being; therefore, nothing is by chance' unless that legitimate cause were relative to that thing. But this is not the case here.

17. To the second authoritative text,[154] that of Boethius, it must be said that God ordered all things in their places and times, as has been said before, but some things <were ordered> as intended and others as not intended.

18. To the argument that followed[155] it must be said that, although those things necessarily induce the event, since, however, the event is not intended, because you did not intend that the stone should fall and fracture your head nor did the stone intend this—accordingly, if you had intended this and so had the stone, then certainly the fracture of the head would not have been by chance, but by a per se cause and intended—therefore, since neither of the two intended the fracture of the head, although they necessarily induce it, we say that the fracture is by chance, and this because of the aforesaid reason.

[150] See above, n. 1.
[151] See above, n. 2.
[152] See above, nn. 3–5.
[153] See above, n. 6.
[154] See above, n. 7.
[155] See above, n. 8.

19. Ad argumentum ultimum dicunt quidam quod istud argumentum non tenet: 'Deus scit hoc esse; ergo hoc est', sed est fallacia aequivocationis, quia in prima propositione contrahitur esse per li 'scire', in conclusione stat simpliciter sine omni determinatione.

20. Quidam dicunt aliter quod duplex est necessitas: a natura et a voluntate. Cum necessitate a natura bene tenet tale argumentum, cum necessitate voluntaria non, ut in proposito.

21. Quidam dicunt aliter et, secundum Boethium, quod duplex est necessitas, scilicet necessitas absoluta et necessitas condicionata sive respectiva[a]. Et est exemplum Boethii ut Socratem scitum currere est necessarium. De prima necessitate tenet istud argumentum, de secunda vero non. Et sic est in proposito.

22. Sed haec responsio parum differt a secunda responsione.

23. Quidam dicunt aliter sic quod, quantum est a parte scientiae Dei, est antecedens necessarium, quantum tamen est a parte huius effectus de[b] evenire, in quem effectum connotatur scientia Dei, est antecedens non necessarium.

<Quaestio 23>

Consequenter quaeritur utrum futurum aliquod sit a casu. Et hoc est quaerere de veritate propositionum de futuro.

1. Et quia quidam ponunt quod omnes sunt falsae, videtur primo quod omne quod eveniet necessario eveniet sic. Dicit Aristoteles in libro *Perihermeneias* quod, si A erit cras, est verum dicere quoniam A erit. [c]Et qua ratione [d]hodie respectu cras, et in millesimo die ante et, eadem ratione, in millesimo anno erat[e] verum dicere quoniam erit. Et ita in omni tempore fuit verum dicere quoniam erit; quare semper et ab aeterno fuit verum dicere quoniam erit. Et quod est perpetuum sive aeternum est necessarium; si ergo erit, necessario erit.

[a] respectiva *ex corr.* O

[b] de *ex corr.* O

[c] Et ... necessario erit] et non differt in millesimo anno quoniam A erit et in proximo ante ipsum eveniunt (eveniunt *corr. s. l. in* eventum). Si ergo non differt dicere in millesimo anno quoniam erit et in proximo, tunc non differt in aliquo alio tempore dicere illud et ita ab aeterno fuit verum dicere quoniam erit, et quod est perpetuum sive aeternum est necessarium. Si ergo erit, necessario erit. Item, si aliquid modo est, semper verum fuit (fuit *corr. mg. ex* est) dicere quoniam erit et ita ab aeterno, et si hoc, et haec fuit necessaria 'hoc erit' T; et non differt dicere in mille annis sicut in anno proximo quando erit et ita non differt in aliquo tempore dicere illud et ita ab aeterno fuit verum quoniam erit, et quod est perpetuum; sed quod est aeternum sive perpetuum est necessarium. Si ergo erit, necessario erit. Item, si aliquid modo est, semper fuit verum dicere quoniam erit et ita ab aeterno, et si hoc, haec fuit necessaria 'hoc erit' C

[d] hodie respectu *suppl. s.l.* O

[e] erat *corr. s.l. ex* erit O

19. To the last argument[156] some people say that the following argument does not hold: 'God knows that this is: therefore, this is', but there is a fallacy of equivocation, because in the first proposition 'to be' is limited by 'to know', whereas in the conclusion 'to be' is taken absolutely without any determination.

20. Some people have a different view, namely, that necessity is of two kinds: either by nature or by the will. The argument does hold for necessity by nature, but it does not for voluntary necessity, as in the present case.

21. Some people following Boethius have still another view, namely, that necessity is of two kinds: absolute necessity and conditioned or relative necessity. And Boethius's example is that it is necessary that Socrates, being known <to run>, should run.[157] The argument holds for necessity of the first kind, but not for necessity of the second kind, which is the one in the present case.

22. But there is little difference between this reply and the second.

23. Some people have still another view, namely, that the antecedent is necessary on the side of God's knowledge, but it is not necessary on the side of the occurrence of the effect, in which effect God's knowledge is connoted.

<Question 23>
The next question is whether something future is by chance. And this question is about the truth of propositions about the future.

1. And since some people posit that all these propositions are false, we first present the arguments showing that everything that will happen will necessarily happen. In *De Interpretatione* Aristotle says that if A will be tomorrow, it is true to say that A will be.[158] But for the same reason that it is true today relative to tomorrow, it was also true to say 'A will be' a thousand days ago or a thousand years ago. Thus, in every time it was true to say that A will be; therefore, always and from eternity it was true to say that A will be. And what is perpetual or eternal is necessary; therefore, if A will be, it will necessarily be.

[156] See above, n. 9.

[157] This is a reference to Boethius, *De cons*. V.6.27, CCSL 94, 104.91–94, although Boethius's example there is "ut si aliquem ambulare scias, eum ambulare necesse est", i.e. "if you know that someone walks, it is necessary that he walk.".

[158] Arist. *De int*. 9, 18b9–11.

2. Item, sequitur: 'A erit; ergo necessario erit', quia[a], si non sequitur, detur oppositum, scilicet 'non necessario erit'. Et sequitur: 'ergo possibile est non fore[b]'. Ponatur in esse sic: 'hoc non erit'. Haec [c]est contradictoria primae, [d]et ita ex opposito consequentis [e]etc. Quare consequentia prima est bona.

3. Item, si hoc erit, Deus[f] scit hoc evenire, [g]et quod Deus scit evenire necessario evenit[h]; futura ergo quae evenient necessario evenient.

4. Oppositum patet per Aristotelem in libro *Perihermeias*. **[O 106rb]**

5. Solutio[i]. Dicendum quod non omne quod evenit necessario evenit, quia aliqua necessario evenient, aliqua autem contingenter.

6. [j]Argumentis autem primis dicendum quod non valet: 'ab aeterno fuit haec vera, 'A erit'; ergo est necessaria', [k]sed coexigitur quod in toto illo tempore non potuit non fuisse vera[l]. Unde de ipso[m] casuali ita est quod[n], licet ab aeterno fuit haec vera, 'A erit', quia tamen in toto[o] tempore potuit non fuisse, ideo non sequitur quod necessario erit.

7. Contra: in perpetuis non differt[p] esse a[q] posse. Et si hoc, non stant ista duo simul, scilicet 'ab aeterno fuit haec vera, 'A erit'', et 'ab aeterno potuit non fuisse vera'.

8. Dicendum quod dicere veritatem propositionis esse perpetuam potest esse dupliciter: vel quia fundatur supra perpetuam rei existentiam, sicut haec 'Deus est' et 'homo est animal' et 'mundus est aeternus', [r]secundum philosophos; et cum tali perpetuitate non stat posse non esse; aut quia fundatur super perpetuum rei eventum in futuro, et cum hoc stat posse non esse per totum tempus suae perpetuitatis.

9. Ad rationem de praescientia Dei dicunt quidam, distinguendo scientiam, quod duplex est scientia Dei, scilicet scientia simplicis notitiae vel scientia

[a] quia *om.* TC
[b] fore] et *add.* TC
[c] est contradictoria] repugnant T; repugnat C
[d] et ... bona *om.* C
[e] etc. ... bona] infertur oppositum antecedentis T
[f] Deus *corr. ex* licet (*lectio incerta*) O
[g] et ... evenire *om. hom.* T
[h] evenit] eveniet TC
[i] Solutio *om.* TC
[j] Argumentis ... dicendum] Ad argumentum dicitur C
[k] sed coexigitur] immo exigitur ad eius veritatem C
[l] vera *ex corr.* O
[m] ipso *om.* TC
[n] quod *iter. sed corr.* O
[o] toto] illo *add.* TC
[p] differt] dicitur O
[q] a] et O
[r] secundum philosophos *om.* C

2. Furthermore, this follows 'A will be; therefore, it will necessarily be' because, if this does not follow, let the opposite <of the consequent> be conceded, that is, 'it will not necessarily be'. And this follows: 'therefore, it is possible that it will not be'. Let that be posited assertorically like this: 'this will not be'. This contradicts the first proposition, and thus from the opposite of the consequent <the opposite of the antecedent follows>. Therefore, the first consequence is valid.

3. Furthermore, if this thing will be, then God knows that this thing happens, and what God knows as happening necessarily happens; therefore, future things that will happen will necessarily happen.

4. The contrary of this is clear from Aristotle in *De Interpretatione*.[159]

5. Solution. It must be said that not everything that happens happens necessarily, because some things will happen necessarily, but others contingently.

6. To the first series of arguments[160] it must be said that the following consequence is not valid: 'the proposition 'A will be' was true from eternity; therefore, it is necessary', but it is also required that in all that time it was impossible for it not to be true. Thus, for a casual thing it is the case that, although the proposition 'A will be' was true from eternity, because, however, in all that time it was possible for A not to be, it does not follow that A will necessarily be.

7. Against this: in perpetual things being is not different from being possible. And if so, then the following two propositions are not compatible: "A will be' was true from eternity', and 'it was possible for it not to be true from eternity'.

8. It must be said that to say that the truth of a proposition is perpetual can be verified in two ways: either (i) because its truth is founded on the perpetual existence of the thing, as in the case of the propositions 'God exists', 'a man is an animal', and 'the world is eternal', according to the philosophers; and the possibility of its not being true is not compatible with this kind of perpetuity. Or (ii) because its truth is founded on the perpetual occurrence of the thing in the future; and the possibility of its not being true for the whole time of its perpetuity is compatible with this kind of perpetuity.

9. To the argument about God's foreknowledge[161] some people reply by drawing a distinction about knowledge, saying that God's knowledge is of two kinds: either (i) the knowledge of simple cognition or (ii) the knowledge

[159] Arist. *De int*. 9, 19a6–22.
[160] See above, nn. 1–2.
[161] See above, n. 3.

approbationis. Et [a]hoc dupliciter, quia aut approbat aliquid posse esse aut aliquid esse. Primo modo sciendi non sequitur 'scit hoc evenire; ergo necessario eveniet', quia sic scit casualia et omnia alia. Nec secundo modo sequitur, sed tantum tertio modo. Sed quia haec distinctio non [b]competit scientiae[c] nostrae, [d]sed theologiae[e], omittatur.

10. Aliter dicitur sic, quod bene sequitur 'Deus scit[f] hoc evenire; ergo eveniet'. Et [g]cum dicit 'antecedens est necessarium', verum est, sed non valet 'ergo consequens', quia aliquando cadit scientia Dei super res quae ab ipso non sunt intentae, et tunc non sequitur[h], quia [i]non fuit antecedens ut scientiae[j] Dei intendentis. Scit enim aliquod futurum contingens, non tamen illud intendit. Si autem scientia cadat super rem aliquam futuram ita quod illam rem intendat, bene sequitur 'scit hoc evenire; ergo eveniet', et est antecedens necessarium et consequens similiter.

11. Aliter dicitur sic[k]: 'scit hoc fore', hic sunt duo, scilicet scire Dei et effectus connotatus in creaturis per li 'hoc fore'. Unde[l] quantum [m]est de scientia Dei inclusum in antecedente, est necessitas[n], quantum tamen ad effectum connotatum in creaturis, potest esse contingentia. Et in consequente includitur solum ista contingentia[o]; ideo ex necessitate antecedentis non infertur necessitas consequentis.

[a] hoc dupliciter] scientia approbationis est duplex C
[b] competit ... omittatur] multum consonat philosophiae nostrae ideo eam omitto C
[c] scientiae *suppl. s.l.* O
[d] sed ... omittatur] omitto T
[e] theologiae] *illegibile add. s.l.* O
[f] scit *suppl. s.l.* O
[g] cum ... verum est] antecedens est necessarium C
[h] sequitur] sicut hic (*lectio incerta*) *add. s.l.* O
[i] non ... similiter] et si antecedens fuerit praescitum a Deo, non tamen fuit intentum aliquando aut cadit eius scientia super rem aliquam futuram ita quod illam intendat et tunc bene sequitur 'scit hoc evenire; ergo illud eveniet', et sicut antecedens est necessarium ita et consequens C
[j] scientiae *suppl. mg.* O
[k] sic] quod cum dico Deus C
[l] Unde] in *add. s.l.* O; *om.* T
[m] est ... inclusum] ad scientiam Dei inclusam TC
[n] necessitas] in *add.* O; *corr. mg. ex* necessarius T
[o] contingentia] necessitas OT

of approval. Case (ii) is twofold, because either (iia) God approves the possibility that something is or (iib) <God approves> that something is. In case (i), it does not follow 'God knows that this happens; therefore, this will necessarily happen', because God knows casual things and all other things in this way. Nor does it follow in case (iia), but it only follows in case (iib). But since this distinction does not belong to our science, but to theology, let it be omitted.

10. Another view is that the consequence 'God knows that this happens; therefore, this will happen' does hold. And when it is said that the antecedent is necessary, they say that this is true, but 'therefore, the consequent is necessary' does not hold, because God's knowledge sometimes falls on things that are not intended by him, and in that case the consequent does not follow, because the antecedent was not about God's knowledge as intentional; for God knows some future contingent, but he does not intend it. If God's knowledge, however, is about a future thing in such a way that God intends that thing, the consequence 'God knows that this happens; therefore, this will happen' does hold, and both the antecedent and the consequent are necessary.

11. Another view is that the claim 'God knows that this will be' involves two things, namely, God's knowledge and the effect connoted in creatures by the phrase 'this will be'. Accordingly, with respect to God's knowledge that is included in the antecedent there is necessity, but with respect to the effect connoted in creatures there can be contingency. And in this consequent only such contingency is included; therefore, the necessity of the consequent does not follow from the necessity of the antecedent.

<Quaestio 24>

Consequenter, quia quidam ponunt quod huiusmodi propositiones neque sunt verae neque falsae, et hoc quia, sicut res se habent ad esse, sic ad veritatem, [a]per Aristotelem, eo[b] quod res est vel non est etc., cum igitur res contingens futura se habeat indifferenter ad esse et ad non esse, sic et veritas propositionis.

1. Contra: omnis oratio enuntiativa et indicativa, [c]secundum Boethium, est vera vel falsa; sed haec 'A erit' est oratio indicativa [d]sive enuntiativa; ergo etc.

2. Item, secundum istam positionem, ista 'A erit' neque est vera neque falsa. Et si hoc, propositio asserens ipsam esse veram est falsa. Sed ipsa[e] se ipsam asserit esse veram, quia, si A erit, A fore[f] est verum. Et si hoc, haec erit[g] falsa 'A erit'.

3. Item, omnis propositio enuntians aliter quam est in re est falsa; sed sic dicto[h] 'A erit', enuntiatur aliter quam est, quia enuntiatur[i] evenire quod contingenter eveniet.

4. Item, sit A contingens; ut prius sequitur 'A erit; ergo necessario erit'. Consequens falsum; ergo antecedens. [j]Consequentia patet per[k] oppositum conclusionis,[l] [m]sicut prius patuit.

5. [n]Item, 'A erit' aut habet veritatem permutabilem aut non. Si non, ergo necessario eveniet, quod falsum est. Si permutabilem, permutetur ergo; sic ergo ab initio fuit haec vera 'A non erit'. Et prius fuit vera ab initio 'A erit'; ergo simul erit et non erit, [o]et contradictoria simul vera, quod falsum est; ergo primum.

[a] per ... etc. *om.* TC
[b] eo *lectio incerta* O
[c] secundum Boethium *om.* TC
[d] sive enuntiativa *om.* C
[e] ipsa *suppl. s.l.* O; *om.* TC
[f] fore *corr. s.l. ex* esse O; esse C
[g] erit] est TC
[h] dicto *ex corr.* O
[i] enuntiatur] enuntiat TC
[j] Consequentia patet] probatio consequentiae patet prius T; probatio consequentiae C
[k] per *ex corr.* O
[l] conclusionis] sumendo *add.* O
[m] sicut prius patuit *om.* T; sic non necessario erit et sequitur 'ergo possibile est non esse' et si hoc ponatur in esse sic 'A non erit'; sed haec est opposita primae; quare etc. C
[n] Item ... primum *om.* C
[o] et ... vera *om.* T

<Question 24>

The next question is about the opinion of those who posit that such propositions are neither true nor false, and that this is so because as things are related to being so they are related to truth, in accordance with Aristotle's statement: <a proposition is true or false> in virtue of the fact that a thing is or is not;[162] therefore, since a future contingent thing is related to being and non-being indifferently, so it is also for the truth of the proposition.

1. Against this: every assertive and declarative utterance is either true or false, according to Boethius;[163] but this proposition 'A will be' is a declarative or assertive utterance; therefore, <it is either true or false>.

2. Furthermore, according to this position, the proposition 'A will be' is neither true nor false. If so, the proposition asserting that it is true is false. But this proposition asserts that it itself is true, because if A will be, it is true that A will be. And if so, then the proposition 'A will be' will be false.

3. Furthermore, every proposition asserting something different from how it is in reality is false; but then the proposition 'A will be' asserts something different from how it is in reality, because it is asserted that the thing that will happen contingently will happen.

4. Furthermore, suppose that A is contingent. Then, as before, the following consequence holds 'A will be; therefore, it will necessarily be'. But the consequent is false; therefore the antecedent is too. The consequence is made obvious by taking the opposite of the conclusion, as was made clear before.[164]

5. Furthermore, 'A will be' either (i) has a changeable truth or (ii) it does not. If (ii), therefore it will necessarily happen, which is false. If (i), let then its truth be changed; therefore, in this case the proposition 'A will not be' was true from the beginning. And earlier the proposition 'A will be' was true from the beginning; therefore, A both will be and won't be at the same time, and contradictories will be true at the same time, which is false; therefore, the first premise (i) <is false>.

[162] Arist. *Metaph.* II.1, 993b30–31.

[163] Boethius, *In De int.* (*editio secunda*) II, PL 64, 454.

[164] See above, q. 23, n. 2.

6. Contra: secundum Aristotelem, differens est veritas in illis de futuro et de [a]praesenti et de praeterito. Et dicit quod in illis de futuro est veritas quae non est necessitas, et ita, secundum Aristotelem, in illis est veritas.

7. Item, si haec est falsa, 'A erit', et hoc quia determinate enuntiat evenire quod indeterminate[b] eventurum est, tunc erit haec falsa, 'A non erit', quia determinate enuntiat non evenire quod indeterminate non[c] eveniet, et ita erunt haec simul falsae[d], 'A erit', 'A non erit', quae sunt contradictoria.

9. Solutio[e]. Dicunt quidam quod ista non sunt contradictoria, '[f]A erit', 'A non erit', sed ista, 'A erit', 'A non necessario erit'.

9. Contra: sicut se habet esse ad necesse esse, sic non esse ad non necesse esse; ergo permutatim, sicut se habet necesse esse ad non necesse esse, sic esse ad non esse. Sed ei quod est necesse esse solum contradicit non necesse esse [g]et non aliud; ergo ei quod est esse solum debet contradicere non esse [h]et non aliud.

10. Item, si idem est dicere 'A erit' et 'A necessario erit', tunc convertuntur; ergo multo fortius convertuntur pro praesenti sic: 'est; ergo necessario est'. Sed hoc[i] falsum, quia sic nulla propositio esset[j] de inesse ut nunc, [k]sed simpliciter, [l]quod est contra Aristotelem **[O 106va]** libro *Priorum*.

11. [m]Solutio. Dicendum quod, si propositio de futuro enuntiet sicut res postea evenit, tunc est vera, si non, falsa[n].

12. [o]Hoc tamen apparet contra Boethium.

13. Sed forte intendit secundum opinionem aliorum.

14. Vel forte aequivocat 'determinate', quia enuntiare determinate vel determinate asserere contingit dupliciter: vel ita quod determinetur a natura

[a] praesenti et de *om. hom.* T
[b] indeterminate *corr. s.l. ex* interminate O
[c] non *om.* OC
[d] falsae] falsa OT
[e] Solutio *om.* TC
[f] A ... non erit *om.* C
[g] et non aliud *om.* C
[h] et non aliud *om.* C
[i] hoc] est *add.* TC
[j] esset *suppl. s.l.* O
[k] sed ... Priorum *om.* C
[l] quod ... Priorum *om.* T
[m] Solutio. Dicendum] mihi videtur TC
[n] falsa *ex corr.* O; non TC
[o] Hoc ... Boethium] Huic tamen videtur Boethius contradicere C

6. Against this: according to Aristotle, the truth of a proposition is not the same for propositions about the future, those about the present, and those about the past.[165] And Aristotle says that the truth of propositions about the future is not necessary,[166] and thus according to Aristotle there is truth in those propositions.

7. Furthermore, if the proposition 'A will be' is false and this because it determinately asserts that what is going to happen indeterminately does happen, then the proposition 'A will not be' will also be false, because it determinately asserts that what will not happen indeterminately does not happen, and thus the contradictory propositions 'A will be' and 'A will not be' will be simultaneously false.

8. Solution. Some people say that the propositions 'A will be' and 'A will not be' are not contradictories, but rather the propositions 'A will be' and 'A will not necessarily be'.

9. Against this: as being is related to necessary being, so non-being <is related to> to non-necessary being; therefore, reciprocally, as necessary being is related to non-necessary being, so being is related to non-being. But only non-necessary being and nothing else is in contradiction with necessary being; therefore non-being and nothing else must be in contradiction with being.

10. Furthermore, if to say 'A will be' and to say 'A will necessarily be' are the same, these propositions are convertible; therefore, there is a much stronger reason to say that they are convertible in the present tense in this way: 'something is; therefore, it necessarily is'. But this is false because if so, no proposition would be assertoric as of now, but simply, which is against Aristotle in the *Prior Analytics*.[167]

11. Solution. It must be said that if what a proposition about the future asserts accords with what happens afterwards, then it is true; if not, then it is false.

12. This, however, seems to be against Boethius.[168]

13. But perhaps Boethius refers to the opinion of others.

14. Or perhaps Boethius uses 'determinately' equivocally, because to assert determinately or to claim determinately happens in two ways: either (i) because the assertion is that things are determined by nature to be in this way

[165] Arist. *De int.* 9, 18a28–33.

[166] Arist. *De int.* 9, 19a27–32.

[167] Arist. *An. pr.* I.2, 25a1–2.

[168] This is probably a reference to Boethius, *In De int.* (*editio secunda*) III, PL 64, 498: "Non enim Aristoteles hoc dicit quod utraeque <i.e., 'hoc erit', 'hoc non erit'> nec verae nec falsae sint, sed quod una quidem ipsarum quaelibet aut vera aut falsa est. Non tamen quemadmodum in praeteritiis definite, nec quemadmodum in praesentibus, sed enuntiativarum vocum duplicem quodammodo esse naturam, quarum quaedam essent in quibus non modum verum et falsum inveniretur, sed in quibus etiam unam esse definite veram, alteram definite falsam, in aliis vero una quidem vera, altera falsa, sed indefinite et commutabiliter".

ad sic esse et non aliter et, si ita determinate enuntiet quod indeterminate evenit, falso enuntiat. Si autem sumatur 'determinate[a]', hoc est secundum veritatem eveniens, sic potest determinate enuntiari et compatitur secum accidentalitatem rei eventus.

15. Ad argumenta quae procedunt per oppositum conclusionis dicendum quod non sufficit quod propositio[b] de inesse repugnet, sed exigitur quod simul cum hoc ante positionem in esse repugnet.

<Quaestio 25>

Consequenter quaeritur de fortuitis, quia Augustinus ponit quod non sunt aliqua fortuita in comparatione ad causam primam.

1. Dicit enim Augustinus, quinto *De civitate Dei* in principio: "Quidam ponunt omnia fortuita, dicentes fortuita esse quae nullas causas habent aut ex nullo rationali ordine veniunt".

2. Sed contra eosdem[c] dicit Augustinus: "Eas causas quae fortuitae [d]sunt sive dicuntur, unde fortuna nomen accepit, non esse dicimus nullas, sed latentes easque tribuimus vel Dei veri vel quorumlibet spirituum voluntati[e]".

3. Item, de eisdem[f] Augustinus, libro primo, capitulo[g] primo: "Illa verba sunt quae religio nulla prohibet: 'forte', 'forsan', 'fortassis', 'fortasse', 'fortuitu', quod tamen totum ad divinam revocandum est providentiam. Etenim quae vulgo fortuna nominatur occulto quodam ordine regitur [h]nihilque aliud in [i]rebus casum[j] vocamus nisi cuius ratio et causa secreta est. Verumtamen penitet me sic[k] illam[l] nominasse fortunam cum videam homines habere[m] in pessima consuetudine, ut ubi debet dici 'hoc Deus voluit', dicatur 'hoc voluit fortuna'".

4. Sic ergo videtur, [n]secundum Augustinum, quod in comparatione ad causam primam non sunt aliqua casualia, sed omnia necessitate quadam evenientia; ponit ergo primam causam praescire omnia futura et in omnibus necessitatem imponere sui eventus.

[a] determinate] determinatum O
[b] propositio *scripsimus*; potentia O; positio T; posita C
[c] eosdem *ex corr.* O
[d] sunt sive *om.* TC
[e] voluntati *lectio incerta* O
[f] eisdem] eodem dicit TC
[g] capitulo] solutio (*lectio incerta*) O
[h] nihilque aliud] nihil quia T; nec aliud C
[i] rebus ... pessima *om.* O
[j] casum] causas T
[k] sic] si C
[l] illam] illa T
[m] habere] facere C
[n] secundum Augustinum] ex his verbis Augustini C

and not in another way, and if a proposition asserts determinately in this way what happens indeterminately then it is false. But (ii) if 'determinately' is taken in the sense of being asserted in accordance with the truth, then a proposition can assert determinately and yet be compatible with the accidentality of the thing's occurrence.

15. To the arguments that proceed from the opposite of the conclusion,[169] it must be said that it is not sufficient that the assertoric proposition should be inconsistent, but in addition to this it is also required that there should be an inconsistency before the positing of the assertoric proposition occurs.

<Question 25>

The next question about fortuitous things arises from Augustine's position that there are no fortuitous things in relation to the first cause.[170]

1. At the beginning of *On the City of God* V Augustine says: "Some people posit that all things are fortuitous, and they call fortuitous those things that have no causes or come about without any rational order".[171]

2. But against these people Augustine says: "We do not say that those causes that are or are said to be fortuitous (from which fortune takes its name) are not causes but that they are latent and we ascribe them to the will of the true God or of some spirits or other".[172]

3. Furthermore, in <*Retractions*> book I, chapter 1, Augustine says about these same people: "The following are words <connected with 'fortuitous'> that no religion prohibits, that is, *forte, forsan, fortassis, fortasse, fortuitu*. Nonetheless, everything must be brought back to divine providence, for a thing that people commonly call 'fortune' is regulated by some hidden order, and we only call it 'chance' when its reason and cause are hidden. Indeed, it pains me when it is called 'fortune', for I see that people have the worst habit of saying 'Fortune wanted this', when they ought to say 'God wanted this'".[173]

4. It seems, therefore, that according to Augustine there are no things that happen by chance in relation to the first cause, but all things happen by some necessity. Accordingly, he posits that the first cause has foreknowledge of all future things and imposes on all things the necessity of their occurrence.

[169] It is not clear which arguments are referred to here (one could be that at q. 23, n. 2).
[170] See below, n. 3.
[171] Aug. *De civ. Dei* V.1, CCSL 47, 128.2–4.
[172] Aug. *De civ. Dei* V.9, CCSL 47, 139.109–112.
[173] Aug. *Retract.* I.1, CCSL 57, 7.15–23.

5. Sed quia omnino huic non consentio, quia mihi videtur quod aliqua sunt a casu omni respectu, quia, si contingat duos homines obviare sibi invicem et unus occidat alterum vel aliud malum ei inferat, et etiam si contingat aliquem facere adulterium vel peccatum aliquod, et etiam si contingat aliquem ire sub casu alicuius lapidis ita quod lapis [a]illum occidat vel aliud malum ei inferat, non videtur mihi quod huiusmodi sunt intenta a causa prima sive quod habeant aliquam necessitatem [b]in comparatione ad causam primam. Sed si contingant, bene scio quod causa prima scit ipsa evenire, sed quod[c] scientia causae primae inferat necessitatem in eventu ipsorum, [d]hoc non videtur mihi.

6. Unde mihi videtur sic esse distinguendum: praescientia primi respectu futurorum est dupliciter, scilicet aut simplex praecognitio sine intentione ordinante, ut cognitio permissiva[e], et talis praecognitio non infert necessitatem in eventu aliquarum rerum, et sic scit casualia evenire; aut est praecognitio cum voluntate intendente et ordinante. Sed hoc potest esse dupliciter, quia aut potest esse intendens et ordinans sub condicione aut absolute. Si sub condicione, adhuc non infert necessitatem eventus, sicut audimus quod prima causa aliquando ordinavit subversionem civitatum [f]et tamen non [g]fecit subversionem earum, et hoc quia non ordinavit[h] nisi sub hac condicione: ‘[i]nisi emendarent se et peniterent’. Et quia penituerunt, retraxit manum suam a subversione civitatis. Si autem sit intendens et ordinans absolute, tunc inducit necessitatem in eventu rerum. [j]Et sic non praescit casualia[k].

7. [l]Quidam posuerunt omnia fatalia, sed quidam dixerunt fatum Dei voluntatem omnia immutabili[m] quodam ordine regentem. Et sic dicitur, ut dicit Augustinus, ‘fatum’ a fando, id est a[n] loquendo, quia verbum in quo disponit omnia et ordinat est immutabile. [o]Sed quia hoc nomen ‘fatum’ apud

[a] illum ... inferat] findat caput suum C
[b] in ... necessitatem *om. hom.* T
[c] quod *corr. ex* quia O
[d] hoc ... mihi] non intelligo C
[e] permissiva] parvissima O
[f] et ... earum *om.* T
[g] fecit subversionem earum] subvertebantur C
[h] ordinavit *corr. ex* ordinat O; earum subversionem *add.* C
[i] nisi ... et] si non C
[j] Et sic] sed isto modo C
[k] casualia] sive fortuita *add.* C
[l] Quidam ... sed] Consequenter gratia praedictorum quaerendum est de fato. Nota ergo in principio quod C
[m] immutabili] immutabilia O
[n] a *om.* TC
[o] Sed... linguam *om.* C

5. But since I do not agree with this at all, because it seems to me that some things are by chance in every respect, because if it happens that two men meet and one kills the other or inflicts something painful on him, or if it happens that someone commits adultery or some other sin, or again if it comes about that someone happens to go under a falling stone so that the stone kills him or inflicts something painful on him, it does not seem to me that these things are intended by the first cause or that they have some necessity in relation to the first cause. If these things happen, however, I do know that the first cause knows that these things happen, but it does not seem to me that the knowledge of the first cause causes their occurrence to be necessary.

6. Accordingly, it seems to me that one should distinguish two kinds of foreknowledge that the first cause has with respect to the future: either (i) a simple precognition without an ordering intention, that is, permissive cognition, and such foreknowledge does not imply any necessity in the occurrence of some things, and in this way God knows that casual things happen; or (ii) a precognition accompanied by an ordering and intending will. And this can happen in two ways: the will can be intending and ordering either (iia) conditionally or (iib) absolutely. In case (iia) as well, the precognition does not imply the necessity of the event, and so we hear that the first cause sometimes ordered the destruction of towns and yet did not actually cause their destruction, and this because God ordered it only under the following condition: 'unless they set themselves right and repented'. And since they repented, God withdrew his hand from the destruction of the town. In case (iib), <the precognition> induces necessity in the occurrence of the things; and God does not have foreknowledge of casual things in this way.

7. Some people posited that all things are ordained by fate, but some said that fate is the will of God that regulates all things with an unchangeable order. In this way, as Augustine said,[174] 'fate' is said to derive from 'to assent',

[174] Aug. *De civ. Dei* V.9, CCSL 47, 138.85–86.

eunucos[a] habet abusionem[b], dicit Augustinus: “Tene sententiam, sed corrige linguam”.

8. Alii dicunt fatum ordinem causarum quendam[c] quo omnia contingunt praeter Dei et hominis voluntatem cuiusdam ordinis necessitate, qui ordo incipit a stellis.

9. Primo modo, secundum Augustinum, fatum est aliquid, secundo modo fatum nihil est.

10. Sic ergo patet quod, proprie sumendo fatum, fatalia sunt quae necessariam causam habent sui eventus. Sed habere necessitatem sui eventus contingit dupliciter: vel a prima causa, secundum quod dicit Augustinus *De trinitate*, libro tertio, capitulo 4: “Voluntas Dei[d] prima et summa causa est omnium specierum atque motionum. Nihil enim fit quod non de intelligibili et interiori aula summi imperatoris egreditur”. Alio modo habent aliqua[e] necessitatem sui eventus, sicut prius patuit, ex constellationibus et motibus caelestium et huiusmodi, ita quod duo homines in tali constellatione possent obviare sibi ita[f] quod unus interficiet alterum.

<Quaestio 26>
Et quia dupliciter eveniunt[g] res in esse, scilicet et actione naturali et actione voluntaria, propter hoc primo quaeritur utrum in naturalibus sint aliqua sic fatalia, secundo in voluntariis.

1. Quod autem in naturalibus sint omnia isto modo fatalia videtur: si aliquid in naturalibus evenit in parte minori, **[O 106vb]** de quo minus videtur, constat quod causam habet. Et illa causa causam [h]habet, et sic est decurrere a causa usque ad causam quousque deveniamus ad motus caelestium sive ad eorum effectus, et ita demum erit concludere quoniam virtutes caelestes sunt causae necessariae eventus illius rei, ita [i]quod, quando est, habet causam necessariam sui eventus. Et si hoc, secundum Augustinum, est fatale.

2. Item, in omnibus eis quae habent ordinem ad invicem, si ibi sit ponere primum, erit vere dicere quod illud[j] est causa omnium consequentium, [k]sicut in generatione hominum fuit primus homo et ille fuit causa consequentium;

[a] eunucos] eunicos O
[b] abusionem] alvisionem O
[c] quendam] quarundam *ex corr.* O
[d] Dei *om.* O
[e] aliqua] aliam OT
[f] ita *om.* TC
[g] eveniunt *corr. ex* evenirent O; exeunt TC
[h] habet … quousque] et sic C
[i] quod, quando est *om.* C
[j] illud] ibi O
[k] sicut ... consequentium *om. hom.* C

that is, from 'to speak', because the word by which God disposes and orders all things is unchangeable. But since the name 'fate' is used loosely by the eunuchs, Augustine says: "Keep the meaning, but correct the language".[175]

8. Other people say that fate is an order of causes in virtue of which all things happen over and above the divine and human will by the necessity of an order, an order which comes from the stars.

9. According to Augustine, in the first way[176] fate is something, but in the second way[177] it is nothing.

10. It is clear, therefore, that, taking fate in the proper sense, things ordained by fate are those that have a necessary cause of their occurrence. But a thing can have the necessity of its occurrence in two ways: either (i) from the first cause, according to Augustine's statement in *On the Trinity*, book III, chapter 4: "The will of God is the first and supreme cause of all species and motions; for nothing happens that does not emerge from the intelligible and interior court of the supreme emperor".[178] Or (ii) as was made clear before,[179] from the constellations and the motions of the celestial bodies and the like, in such a way that two men in that constellation can meet with the result that one kills the other.

<Question 26>

And since things come to be in two ways, namely, by natural action and by voluntary action, then the first question is whether there are things ordained by fate among natural things, and the second question is whether the same applies to voluntary things.

1. Arguments that all natural things are ordained by fate: if among natural things something happens only rarely, it is evident that this event does have a cause, even though this is less obvious. And that cause has a cause, and so we pass from cause to cause until we arrive at the motions of the celestial bodies and their effects, and we will eventually reach the conclusion that the celestial powers are the necessary causes of the occurrence of that thing, so that, when that thing occurs, there is a necessary cause of its occurrence. And if this is the case, then according to Augustine,[180] that thing is ordained by fate.

2. Furthermore, in all things that are ordered one to the other, if one can posit a first in the series, then it will be true to say that that thing is the cause of all the things that come after it — for instance, in the generation of men

[175] Aug. *De civ. Dei* V.1, CCSL 47, 128.7–9.
[176] See above, n. 7.
[177] See above, n. 8.
[178] Aug. *De trin.* III.4, CCSL 50, 136.23–26.
[179] See above, n. 8.
[180] See above, q. 25, n. 7.

ergo ita erit in generatione motuum. Cum ergo sit ponere primum motum, erit causa subsequentium, et ita omnium generationum inferiorum.

3. Item, in materia naturali ita est quod in ipsa[a] sunt omnes formae in potentia sive rationes causales ad omnes formas. Et omnes aequaliter se habent respectu agentis et agens, ut corpus caeleste, per suas virtutes luminares est agens uniformiter; si ergo necessario producit unum, et aliud[b], et ita omnia. [c]Si autem loquamur de materia non disposita, illud idem est argumentum.

4. Si autem dicatur quod materia magis disposita est ad unam formam quam ad aliam, tunc arguatur[d] de illa dispositione qualiter se habet materia respectu illius dispositionis. Et ita in naturalibus sunt omnia isto modo fatalia.

5. Ad oppositum: si pluviae et siccitates et tempestates essent isto modo fatalia, tunc frustra fierent sacrificia et huiusmodi quae modo fiunt.

6. Item, prima causa est causa omnium eorum quae sunt, et hoc ut voluntas. Agit enim condicione electionis[e] et non necessitatis condicione. [f]Si ergo causa prima solum agit condicione electionis, tunc non imponit necessitatem eventui rerum.

7. Item, facilius est destruere quam construere; si ergo causa prima voluntate sua condidit omnia, tunc adeo faciliter potuit ea non fecisse antequam fecit; ergo non fuit necessitas in actione illius.

8. Item, Augustinus, libro tertio[g]: "Omnia sunt in electione vertibilis, non tamen mutabilis". Et dicitur quod scientia Dei est aliquo modo vertibilis et non tamen mutabilis, et hoc quia, si sciat aliquid evenire, dum tamen non intendat ipsum evenire, si non eveniat, aliquo modo vertitur scientia, non tamen mutatur. Si autem intenderet [h]ipsum[i] evenire et tunc non eveniret, mutabilis esset [j]sua scientia.

<Quaestio 27>
Quaeritur utrum res voluntariae sint fatalia ita quod corpora caelestia inducant necessitatem in illis.

1. Et primo videtur quod[k] res voluntariae habeant necessitatem a corporibus supracaelestibus sic: anima, eo quod coniuncta est corpori, patitur,

[a] ipsa *corr. ex* ipso O
[b] aliud *ex corr.* O
[c] Si ... argumentum] eadem ratione; illud argumentum procedit de materia non disposita C
[d] arguatur] arguitur TC
[e] electionis] tunc non imponit necessitatem *add. sed del.* O
[f] Si ... electionis *om.* C
[g] tertio] De trinitate *add.* O
[h] ipsum evenire *om.* TC
[i] ipsum *scripsimus*; ipsam O
[j] sua scientia *om.* TC
[k] quod] corpora caelestia *add. sed del.* O

there was a first man and he was the cause of the men coming after him; hence, the same will apply to the generation of motions. Thus, since one can posit a first motion, that motion will be the cause of the motions that come after it, and thus of all the generations here below.

3. Furthermore, it is the case that natural matter has in it all the forms in potency or the causal principles of all the forms. And these are all equally related to an agent, and the agent, namely, a celestial body, acts uniformly in virtue of its illuminating powers; therefore, if it necessarily produces one thing, then this also applies to another thing, and thus to all things. But if we speak of matter without disposition, the argument is the same.

4. If it is said that matter is more disposed to one form than to another, then we can argue about that disposition by asking in what way matter is related to that disposition. And thus in natural things all things are ordained by fate in this way.

5. To the contrary: if rainfalls, droughts, and storms were ordained by fate in this way, then sacrifices and the like that are now done would be done in vain.

6. Furthermore, the first cause is the cause of all the things that are, and this in so far as it is a will; for it acts by the condition of choice and not by the condition of necessity. Therefore, if the first cause acts by the condition of choice alone, then it does not impose necessity on the occurrence of the things.

7. Furthermore, it is easier to destroy than to build; therefore, if the first cause produced all things by its will, then it could equally easily not have made them before it did make them; therefore, there was no necessity in its action.

8. Furthermore, in the third book Augustine says: "All things are subject to the choice of what can be diverted but not changed".[181] And it is said that God's knowledge can in some way be diverted but not changed, and this because, if God knows that something happens, as long as he does not intend it to happen, if it does not happen, then his knowledge is diverted in some way, but not changed. If, however, God intended it to happen and then it did not happen, then his knowledge could be changed.

<Question 27>
The question is whether voluntary things are ordained by fate in such a way that the celestial bodies establish necessity in them.

1. First the arguments that voluntary things receive necessity from the celestial bodies: the soul by being joined to the body is subject to the passions

[181] This statement is not by Augustine himself but seems to be an interpretation of Augustine's discussion about the compatibility between God's foreknowledge of the future and free will in *De lib. arb.* III.2–4, CCSL 29, 276–281.

passiones scilicet et condiciones corporis; ergo corpora caelestia, [a]cum habeant dominium supra complexiones[b] rerum voluntarium, ut videtur, possunt dare necessitatem ipsis.

2. Item, texere et aedificare et omnes actiones voluntariae sunt ipsius coniuncti et non animae per se nec corporis per se. Tunc, cum corpora caelestia habeant dominium supra ipsa coniuncta, potuerunt inducere necessitatem in illis. Et ita videtur quod corpora caelestia inducant necessitatem in rebus voluntariis.

3. Quod autem ipsa intelligentia quae motor est caeli inducat in ipsis necessitatem videtur sic: quae est proportio corporis caeli ad corpus hic inferius, ea erit proportio intelligentiae quae est motrix ipsius corporis caeli ad animam quae est motor corporis hic inferius. Sed talis est proportio corporis caeli ad corpus inferius hic quod caelum inducit sanitatem et aegritudinem in corporibus hic inferius, et ita necessitatem in suis proprietatibus; ergo ea erit proportio intelligentiae ad animam quod inducet necessitatem in suis operationibus et in suis proprietatibus.

4. Item, duplex est intellectus, scilicet agens et possibilis; aut ergo tota anima dicitur intellectus agens aut pars animae aut intelligentia. Sed non tota, tum quia intellectus agens omnia [c]intelligit, intellectus possibilis nihil —tunc, si tota anima esset etc., sequeretur quod anima simul omnia intelligeret et nihil, quod est inconveniens. Tum quia tunc esset unum et idem in ratione dantis sive agentis et patientis simul et semel, quod est inconveniens. Consequentia patet, quia intellectus agens est in ratione dantis et agentis, quia illuminat species receptas in imaginativa, et intellectus possibilis est in ratione recipientis. Recipit enim species illuminatas ab agente. Tum quia, cum[d] semper praesens sit illud quod illuminat, si sit tota anima, tunc semper intelligeret, quod est inconveniens. Et eadem ratione nec erit intellectus agens pars animae, quia, cum illa pars animae semper sit praesens, tunc semper intelligeret.

5. Sed huic rationi respondent quidam sic, quod, per Aristotelem, intellectus possibilis intelligit per species receptas in imaginativa; **[O 107ra]** quia ergo species non sunt semper praesentes imaginativae, ideo nec semper illuminantur ab intellectu agente. Et ita, licet ponamus quod tota anima vel pars animae sit intellectus agens, non tamen sequitur quod semper intelligat.

[a] cum habeant *ex corr.* O
[b] complexiones *scripsimus*; complexiores O
[c] intelligit, intellectus *ex corr.* O
[d] cum *ex corr.* O

and the conditions of the body; therefore, since the celestial bodies have dominion over the dispositions of voluntary things, as seems to be the case, they can establish necessity in them.

2. Furthermore, weaving and building and all voluntary actions belong to the composite and neither to the soul by itself nor to the body by itself. Then, since the celestial bodies have dominion over the composites themselves, they could establish necessity in them. And so it seems that the celestial bodies establish necessity in voluntary things.

3. That the intelligence itself, which is the mover of the heavens, establishes necessity in voluntary things is shown as follows: the relation of the intelligence, which is the mover of the celestial body itself, to the soul, which is the mover of the body here below, is the same as the relation of the celestial body to the body here below; but the relation of the celestial body to the body here below is such that the heavens establish health and illness in the bodies here below, and thus establish necessity in their properties; therefore, the relation of the intelligence to the soul will be such that the intelligence establishes necessity in the actions and the properties of the soul.

4. Furthermore, there are two intellects, namely, the agent and the possible intellect; therefore, an agent intellect is either the whole soul or a part of the soul or an intelligence. But it is not the whole soul for the following reasons: (i) an agent intellect understands all things, and a possible intellect understands nothing—then, if the whole soul were <an agent intellect>, it would follow that the soul would at the same time understand all things and nothing, which is absurd. (ii) Moreover, one and the same thing would at the same time have the role of the giver or agent and of the patient, which is absurd. The consequence is clear because the agent intellect has the role of the giver or agent, since it illuminates the species received in the imaginative power, and the possible intellect has the role of the recipient, for it receives the species illuminated by the agent intellect. (iii) Since the thing that illuminates is always present, if the whole soul were such a thing, then it would always understand, which is absurd. And for the same reason the agent intellect is not a part of the soul, because, since that part of the soul is always present, then the soul would always understand.

5. Some people reply to this argument that, according to Aristotle, the possible intellect understands in virtue of the species received in the imaginative power;[182] therefore, since the species are not always present to the imaginative power, neither are they always illuminated by the agent intellect. And thus, even if we posit that the whole soul or a part of the soul is the agent intellect, it does not follow, however, that it always understands.

[182] Arist. *De anima* III.7, 431b2–5.

6. Contra: si species ipsius albedinis possit deferri ad organum sensus mediante luce, tunc multo fortius ipsa lux. Ergo, a simili, cum ipsa lux quae illuminat species receptas ab imaginativa sit propinquior intellectui possibili quam ipsa species, tunc, cum ipsa species mediante ipsa luce illuminante intelligatur, [est[a]] et multo fortius ipsa lux illuminans. Et ita numquam cessat intelligere, dato etc. Oportet ergo ponere quod illud quod illuminat illas species receptas ab imaginativa sit distans. Oportet igitur ponere quod sit intelligentia, quia non potest esse causa prima, quia, si esset causa prima, tunc intelligere posset sine omni phantastica virtute, quod est contra Aristotelem. Patet igitur[b], quod manifestum est, quod absurdum est dicere causam primam esse illud quod illuminat species etc. Erit tunc necessario intelligentia, et hoc vel movens caelum vel aliqua alia. Sed quod facit hominem intelligere dirigit ipsum in suis operationibus; ergo intelligentia potest inducere necessitatem in actionibus voluntariis istarum rerum.

7. Ad oppositum: nullum ignobilius agit in nobilius, testante Aristotele: "Agens nobilius patiente". Sed virtutes caelestes sunt ignobiliores spiritu humano vel[c] hominis; quare etc.

8. Item, duplex est principium agendi, unum ut natura, aliud ut voluntas. Sed agens per voluntatem nobilius est agente naturaliter. Sed corpora caelestia agunt naturaliter, anima vero agit voluntarie. Et ignobilius non potest agere in nobilius; quare etc.

9. Item, movens aliud immediate est[d] praesens illi; sed intelligentia non est praesens; quare etc.

10. Diceret aliquis quod intelligentia est praesens per virtutem, licet non per suam substantiam.

11. Contra: omnis virtus est a substantia; quare necesse est ponere substantiam esse ibidem semper praesens, quod falsum est.

12. Item, cum intelligentia movens caelum semper uniformiter et continue et uno modo moveat caelum, tunc anima intellectiva, si intelligeret per intelligentiam, uniformiter et continue et uno et eodem modo prorumperetur semper in suis operationibus, quod est inconveniens.

[a] est *delevimus*
[b] igitur *suppl. s.l.* O
[c] vel *scripsimus*; ut O
[d] est *ex corr.* O

6. Against this: if the species of whiteness can be carried to the organ of sense by means of light, then there is a much stronger reason to hold that light itself <is so carried>. Therefore, similarly, since the light itself which illuminates the species received by the imaginative power is closer to the possible intellect than the species itself, then, since the species itself is understood by means of the illuminating light, there is a much stronger reason to hold that the illuminating light itself is understood. And thus the soul never ceases to understand, given that <the whole soul or a part of the soul is the agent intellect>. It is necessary, therefore, to posit that the thing that illuminates the species received by the imaginative power is distant, and so it is necessary to posit that this thing is an intelligence; for it cannot be the first cause, because, if it were the first cause, then the soul could understand without any imaginative power, which is contrary to Aristotle.[183] It is clear, therefore, that it is absurd to say that the first cause is the thing that illuminates the species. Hence that thing will necessarily be an intelligence, and either one that moves the heavens or another. But what makes men understand directs them in their actions; therefore, an intelligence can bring about necessity in their voluntary actions.

7. To the contrary: nothing more ignoble acts on the more noble, according to Aristotle's statement that the agent is nobler than the patient.[184] But the celestial powers are more ignoble than the human spirit, that is, that of men; therefore, <the celestial powers do not act on men>.

8. Furthermore, there are two principles of acting: nature and will. But a voluntary agent is nobler than a natural agent. But celestial bodies act naturally, whereas the soul acts voluntarily. And the more ignoble cannot act on the more noble; therefore, <celestial bodies do not act on men>.

9. Furthermore, the immediate mover of something is present to it; but an intelligence is not present; therefore, <an intelligence is not the immediate mover of men>.

10. One could say that an intelligence is present by its power, although not by its substance.

11. Against this: every power comes from a substance; therefore, it is necessary to posit that a substance is always present there, which is false.

12. Furthermore, since the intelligence that moves the heavens moves them always uniformly, continuously, and in one way, then if the intellectual soul understood in virtue of an intelligence, it would always burst into action uniformly, continuously, and in one and the same way, which is absurd.

[183] Arist. *De anima* I.1, 403a5–10.
[184] Arist. *De anima* III.5, 430a18–19.

13. Solutio. 'Fatum', secundum Augustinum, duobus modis dicitur: uno modo a fando, id est a loquendo, et tunc sumitur locutio pro verbo mentis sive verbo divino, quod omnia disponit, et sic dicuntur omnia fatalia. Alio modo dicitur 'fatum' quidam ordo causalis procedens a stellis et motibus corporum supracaelestium, ut quidam dicunt: aliquis debet esse bonus eo quod natus est in tali constellatione, et aliquis malus eo quod natus est in alia constellatione.

14. Sed videtur Augustinus male[a] dicere quoad primum membrum, cum dicit quod hoc modo omnia sunt necessaria. Unde oportet distinguere sic: eorum quae fiunt a natura vel a voluntate quaedam fiunt cum intentione et quaedam sine. Secundum membrum non intellexit Augustinus, quia huiusmodi proprie sunt casualia. Unde utroque modo non[b] sunt fatalia, sed tantum ea quae fiunt cum intentione. Et Augustinus, nisi hoc modo intelligat, simpliciter falsum dicit. Non enim casualia sunt fatalia ita [c]quod necessario eveniant. Per hoc qualiter intelligit patet.

15. Ad quaestiones utrasque dicendum sic quod eorum quae fiunt [d]quaedam fiunt a voluntate tantum, quaedam et a voluntate et a natura sua, [e]id est a natura cum voluntate. Nec est ponere tertium membrum, scilicet quod quaedam fiant a natura tantum, quia hoc est impossibile. Non enim potest aliquid fieri sine voluntate divina coniuncta. Si primo modo[f], hoc est dupliciter. Aut enim fiunt a voluntate incausata tantum aut a voluntate causata tantum. Similiter quae fiunt a natura et a voluntate aut fiunt a voluntate causata aut a voluntate incausata. Si primo modo fiant, videlicet a voluntate incausata, hoc est dupliciter, quia quaedam sic fiunt a tali voluntate quod ab alia fieri non possunt, ut caelum, stellae, intelligentiae, animae rationales et huiusmodi. Et quaedam econverso[g] fiunt ab ipso ita quod possunt ab alio fieri, ut aer et elementa, quia natura huiusmodi facere potest, ut per transmutationem patet, ut ignis in destruendo aerem et quodlibet elementum in destruendo aliud. Si primo modo, sic sunt necessaria valde et necessario eveniunt, tamen, antequam fecit ea, potuerunt non evenisse. Si secundo modo, sic sunt necessaria, sed non ita necessaria.

[a] male *scripsimus*; melius *ex corr.* O
[b] non *suppl. s.l.* O
[c] quod necessario *inv. sed corr.* O
[d] quaedam fiunt *suppl. mg.* O
[e] id est a *lectio incerta* O
[f] modo] vel secundo *add. sed del.* O
[g] econverso *lectio incerta* O

13. Solution. According to Augustine, 'fate' is said in two ways: (i) from 'to assent', that is from, 'to speak',[185] and then the locution is taken for the mental word or the divine word, which arranges all things, and in this way all things are said to be ordained by fate. (ii) In another way[186] fate is said to be a causal order proceeding from the stars and from the motions of the celestial bodies, as when some people say that one person must be good since he was born in that constellation, and one person bad since he was born in another constellation.

14. But it seems that Augustine speaks wrongly about the first way of speaking (i), when he says that in this way all things are necessary.[187] Accordingly, it is necessary to make a distinction as follows: of the things that happen by nature or by the will (a) some happen with an intention and (b) some without an intention. Augustine, however, did not understand <the issue in terms of> the second case (b), because such things are properly casual. Accordingly, things are not <said to be> ordained by fate in both ways (a) and (b), but only those that happen with an intention. And Augustine says something absolutely false, unless he understands the issue in this way; for casual things are not ordained by fate in such a way that they necessarily happen. With this it is clear in what way he understands the issue.

15. To both questions it must be said that of the things that come to be (i) some come to be by will alone, (ii) some both by will and by their nature, that is, by nature together with will. And a third member is not to be posited, that is, that some things come to be by nature alone, because this is impossible; for something cannot come to be without the conjunction of the divine will. Case (i) is twofold; for these things either (ia) come to be by the uncaused will alone or (ib) by a caused will alone. Likewise, the things that come to be by nature and by will either come to be by a caused will or by the uncaused will. Case (ia) is twofold, because (ia') some things come to be by the uncaused will in such a way that they cannot come to be by another will, for instance, the heavens, the stars, the intelligences, rational souls, and the like. Conversely, (ia'') some things come to be by the uncaused will in such a way that they can come to be by another thing, for instance, the air and the elements, because nature can produce these things, as is clear through their changes, for instance, fire destroys air and every element destroys another element. In case (ia'), things are indeed necessary and necessarily come to be, but before they were produced, they were able not to come to be. In case (ia''), things are necessary but not in the aforesaid way.

[185] See above, q. 25, n. 7.
[186] See above, q. 25, n. 8.
[187] See above, q. 25, nn. 1–4.

16. Si secundo modo fiant, videlicet a voluntate causata, hoc est dupliciter. Aut enim illa voluntas causata eligit quod melius est aut non, sed quandoque bona, quandoque mala. Si primo modo, sic sunt intelligentiae separatae, et tales sunt necessariae et non possunt non evenire. Si secundo modo, sic sunt **[O 107rb]** illa quae eveniunt et fiunt a nobis et a voluntate nostra, et huiusmodi non sunt necessaria, sed sunt contingentia et possibilia[a] non esse.

17. Et intelligendum est quod non est idem esse contingens et casuale, quia casuale est illud quod evenit non intentum, contingens vero est illud quod evenit intentum; haec tunc est differentia. Et sic patet qualiter primum membrum primae distinctionis dividitur, scilicet illa quae fiunt a voluntate tantum.

18. Si secundo modo fiant, scilicet a natura cum voluntate, sive a voluntate causata sive incausata, hoc est dupliciter. Aut enim sunt immediate procedentes cum voluntate tali aut mediate. Si primo modo, tunc impossibile est[b] talia non evenire et sunt necessaria, ut, quando sol est in canicula, raro contingit fieri pluviam, tamen, quando evenit, non potest non evenire, quia Saturnus, qui est stella valde frigida et humida, cum perficit cursum suum in tribus centum annis, si[c] contingat ipsam redire per solem quando est in canicula, tunc propter suam humiditatem et frigiditatem facit pluviam. Et ista pluvia tunc necessario evenit nec potest non evenire quando Saturnus ita redit per ipsam. Si secundo modo, sic non necessario eveniunt, et hoc est propter inoboedientiam et inhabilitatem materiae, verbi gratia herbas crescere. Et huiusmodi sunt mediate procedentes a natura cum voluntate, quia primo oportet ut corpora caelestia faciant calorem et postea humiditatem et multa consimilia quae manifeste patent. Et licet aliquando virtus supracaelestis omnia ista immiserit, tamen ista non eveniunt, et hoc propter inhabilitatem materiae.

19. Iam patet qualiter respondendum est ad quaestiones et de naturalibus et de voluntariis[d] per illam distinctionem.

20. Primae rationi huius ultimae quaestionis dicendum: differunt habilitatio et coactio. Et ratio tunc non plus concludit nisi quod corpora caelestia possunt agere in ista inferiora et habere dominium super ipsa per habilitationem in suis actionibus, non autem per coactionem sive per necessitatem, quia corpus[e] non habet dominium supra animam[f] per coactionem, sed solum per habilitationem, quia aliquis qui de sua natura propria est sanguineus potest esse malus et aliquis qui similiter de sua natura propria est melancholicus

[a] possibilia] possibilis O
[b] est *suppl. s.l.* O
[c] si *ex corr.* O
[d] voluntariis *scripsimus*; violentis O
[e] corpus *scripsimus*; anima O
[f] animam *scripsimus*; corpus O

16. Case (ib), namely that of the things that come to be by a caused will, is twofold; for either that caused will chooses (ib') what is better or (ib'') it does not, but sometimes chooses good things and sometimes bad things. Case (ib') is that of the separated intelligences, and these are necessary and they cannot fail to come about. Case (ib'') is that of the things that come to be and are made by us and by our will, and these things are not necessary, but are contingent and have the possibility of non-being.

17. It must also be understood that to be contingent and to be casual are not the same, because a casual thing is one that happens without being intended, whereas a contingent thing is one that happens as being intended; this then is the difference. It is now clear in what way the first member (i) of the first distinction, namely, the things that come to be by the will alone, is divided.

18. Case (ii), namely, the things that come to be by nature together with will, whether it be a caused will or the uncaused will, is twofold; for either (iia) they proceed immediately together with that will or (iib) mediately. In case (iia) it is impossible for these things not to happen and they are necessary, for instance, when the sun is in the dog-star, it rarely happens that it rains, but when it does happen, it cannot fail to happen, because if Saturn, which is a very cold and humid star, when it completes its orbit in three hundred years, returns through the sun when that is in the dog-star, it produces rain because of its humidity and coldness. And then rain necessarily happens nor can it fail to happen when Saturn returns in this way. In case (iib), things do not necessarily happen, and this because of the disobedience and unfitness of the matter: the growth of grass is a case in point. And this kind of thing proceeds from nature together with will mediately, because it is first necessary that the celestial bodies should produce heat and then humidity and many similar things, which are obvious. And although a celestial power has produced all these conditions, sometimes, however, these things do not happen, and this is due to the unfitness of the matter.

19. It has now become clear how one can use this distinction to reply to the questions about natural things and about voluntary things.

20. To the first argument of the last question[188] it must be said that enablement and compulsion are different. The argument then only proves that celestial bodies can act on things here below and have dominion over them for the enablement of their actions, but not through compulsion or necessity, because the body does not have dominion over the soul through compulsion, but only through enablement. For instance, a man who is sanguine by virtue of his own nature can be bad and a man who is melancholic

[188] See above, n. 1.

potest esse bonus. Hoc autem non est per coactionem, sed per habilitationem et ex bona assuetudine, quia corpus nihil facit per coactionem animae, sed solummodo per eius habilitationem.

21. Secundae rationi dicendum quod corpora caelestia respectu quarundam proprietatum se habent necessitatis condicione et respectu quarundam habilitationis condicione. Respectu enim proprietatum ipsius corporis, cuiusmodi sunt calefacere, frigefacere, habent corpora caelestia necessitatis condicione dominium, respectu autem proprietatum ipsius animae, cuiusmodi sunt intelligere etc., habent dominium solum habilitationis[a] condicione.

22. Tertiae rationi dicendum quod proportio deficit, quia anima est in corpore ut agens[b] sive motor et ut perfectio, intelligentia vero [c]in caelo non nisi ut motor.

23. Vel dicendum quod quoad quid est proportio et quoad quid non. Bene enim verum est quod intelligentia quoad quasdam operationes animae inducit necessitatem et quoad quasdam non. Quo enim ad istas condiciones quae aliquo <modo[d]> sunt ipsius corporis, ut intelligere et movere corpus, eum[e] perficiendo, non inducit necessitatem nec deservit ipsi animae in huiusmodi operationibus. Tamen quoad operationes quae sunt ipsius animae in se, non respiciendo corpus, dum est in corpore, deservit ei. Ponimus enim quod ipsa intelligentia revelet animae in se multa in somniando in nocte quae aliquando enuntiat animali, aliquando non. Et dicimus quod magis in nocte quam in aliquo alio tempore tum quia non est motus ambulantium nec loquentium nec aliorum operum de nocte sicut de die, tum quia ipsa anima tunc intelligit sibi ipsi, tum quia anima non est tunc sollicita circa exteriora sicut de die. Unde ponimus quod, si anima ita in somniando de nocte cogitaret de specie alicuius rei quae fuit in extremis partibus mundi, anima videret illam speciem per muros, per parietes, per ostia et per omnia intermedia, quia perspicuum divisum est per totum mundum. Tamen nos non possumus videre perspicuum quod[f] est in domo, si fuerimus extra domum et domus claudatur, et hoc propter grossitatem sensus. Non sic est a parte animae, quia non habet sensum grossum, sed valde subtilem; et ideo potest videre illam speciem, hoc modo tamen debiliter videt.

[a] habilitationis *scripsimus*; habilitatis O
[b] agens *scripsimus*; actus O
[c] in caelo *suppl. s.l.* O
[d] modo *supplevimus*
[e] eum *ex corr.* O
[f] quod *scripsimus*; quae O

similarly by virtue of his own nature can be good. However, this happens not through compulsion but through enablement and in virtue of good habits, because the body does nothing through the compulsion of the soul, but only through its enablement.

21. To the second argument[189] it must be said that the celestial bodies are related to some properties by the condition of necessity and to others by the condition of enablement; for the celestial bodies have dominion by the condition of necessity in relation to the properties of the body, such as heating, cooling, but have dominion by the condition of enablement alone in relation to the properties of the soul, such as understanding and the like.

22. To the third argument[190] it must be said that the relation fails, because the soul is in the body as agent or mover and as perfection, whereas the intelligence is in the heavens only as mover.

23. Alternatively, it must be said that in one respect there is a relation and in another there is not; for it is indeed true that an intelligence produces necessity in some actions of the soul and in other actions it does not. In what concerns the conditions that in some way belong to the body, such as understanding and moving the body, <thus> giving perfection to it, the intelligence does not produce necessity nor does it serve the soul itself in this kind of action. However, in what concerns the actions that belong to the soul by itself without relation to the body, the intelligence does serve the soul, as long as it is in the body; for we posit that the intelligence reveals many things to the soul alone when it dreams at night, things which the soul sometimes communicates to the animal and sometimes not. And we say that this is more frequent at night than at any other time because at night there is no motion of walking or of talking or of other activities as there is during the day, and because the soul then understands by itself, and also because the soul is not attentive to exterior things then as it is during the day. Accordingly, we posit that if the soul while dreaming at night thought of the species of a thing that was in the most remote parts of the world, the soul would see that species through the outer walls, through the inner walls, through the doors, and through all the media, because what is transparent is diffused throughout the world. However, if we were outside the house and the house was closed, we could not see what is transparent inside the house, and this because of the coarseness of our senses. But this does not happen to the soul, since the soul does not have a coarse sense but a very subtle one, and therefore it can see that species, although in this case rather weakly.

[189] See above, n. 2.
[190] See above, n. 3.

24. Ad argumenta sequentia patebit alias, ubi disputabitur illa quaestio. Non enim posset hic ad argumenta responderi[a] nisi quaestio illa disputaretur. Et locus quaestionis non est disputari hic.

<Quaestio 28>
Quaeritur qualiter Deus praescivit omnia quae eveniunt, an scilicet per conversionem sui supra se, [O 107va] ut intelligeret [[b]sed intelligeret] infinite[c] infinities, sic intelligendo se, aut habeat similitudinem cuiuslibet[d] rei fiendae in mente sua.

1. Et hoc est quaerere an[e] cuilibet rei correspondeat una idea in causa prima vel sit tantum unica idea per quam scit omnia.

2. Et dicunt quidam quod non habet nisi unam ideam, scilicet unum intelligere prout infinities convertit se supra se, intelligendo se.

3. Et dicunt quidam aliter et melius quod cuilibet[f] rei correspondet una idea. Unde melius debemus dicere quod idea respondet <tam[g]> singulari quam universali, quod tamen non dixerunt quidam.

<Quaestio 29>
Quaeritur etiam qualiter prima causa intelligat istas propositiones: 'homo est animal', 'homo est substantia' etc.

1. Et dicendum quod per simplicem ideam hominis scit omnes propositiones ubi aliqua enuntiantur de homine.

2. Et nota quod quaelibet idea est eadem cum causa prima, ideae tamen inter se non sunt eaedem, et hoc est quia in ipso sunt multae rationes causales.

3. Unde sciendum quod in simplicissimo sunt maximae et innumerabiles rationes et respectus, ut patet in puncto et in unitate.

[a] responderi *corr. ex* respondere O
[b] sed intelligeret *delevimus* (intelligeret *ex corr.* O)
[c] infinite *ex corr.* O
[d] cuiuslibet *ex corr.* O
[e] an *ex corr.* O
[f] cuilibet *ex corr.* O
[g] tam *supplevimus cum* TC (*cf. Appendix, Book II, q. 28, n. 3*)

24. The reply to the arguments that follow[191] will be made clear elsewhere, where that question will be discussed,[192] for the reply to these arguments cannot be given here if that question is not discussed, and here is not the place to discuss it.

<Question 28>

The question is in what way God has foreknowledge of all the things that happen, that is, either by his turning inward, so that by understanding himself in this way he would understand infinitely and infinitely many times, or by having in his mind a likeness of each thing to be produced.

1. And this is equivalent to asking whether to each thing there corresponds an idea in the first cause or whether there is only one idea by which the first cause knows all things.

2. Some people say that the first cause has only one idea, namely, one understanding in so far as it turns inward infinitely many times, by understanding itself.

3. Other people have a different and better view, namely that to each thing there corresponds an idea. Accordingly, it is better to say that an idea corresponds both to a singular and to a universal, which, however, some people do not posit.

<Question 29>

Another question is in what way the first cause understands the propositions 'man is an animal', 'man is a substance' <and the like>.

1. It must be said that the first cause knows all the propositions in which something is asserted about man by means of the simple idea of man.

2. And note that every idea is the same as the first cause, but one idea is not the same as another. The reason for this is that in the first cause there are many causal principles.

3. Accordingly, it must be understood that in the simplest thing there are not only the greatest principles and relations, but also countless principles and relations, as is clear in the case of a point and a unit.

[191] See above, nn. 4–6.

[192] As the version of Aspall's Questions on *Physics* II transmitted by MSS T and C indicates (cf. Appendix, Book II, q. 27, n. 27), this is a reference to a discussion of the nature of the agent intellect in a commentary on *De anima* III.5. This reference, however, is not verified by Aspall's Questions on *De anima* preserved in MS O (ed. Čizmić), which ends at *De anima* III.3.

<Quaestio 30>
Quaeritur qualiter Deus scit contingentia, an scilicet contingentia habeant ideam in causa prima.

1. Et videtur quod sic, quia causa prima scit discernere contingens[a] a necessario; hoc autem non posset nisi haberet speciem et similitudinem illorum; quare etc.

2. Contra: si habeat utramque ideam, necessitatis et contingentiae, tunc, cum idea contingentiae sit in ipso, erit necessaria. Sed per necessarium non discernit necessarium a <non[b]> necessario; quare de necessitate oportet ponere quod non habeat ideam.

1*. Quaeritur iuxta hoc de monstris. Et cum monstra fuerint duobus modis, uno modo communiter loquendo de monstris, sicut loquitur Aristoteles sexto decimo *De animalibus*, dicendo quod mulus est animal monstrum, alio modo proprie loquendo de monstris, sicut homo quando habet sex digitos vel arbor quae partim est vitis et partim oliva, de monstris primo modo tria quaeruntur.

<Quaestio 31>
Primum est quare sunt sive dicuntur animalia monstruosa et qualiter.

1. Ad quod dicendum quod huius causa est quia non producitur simile ex simili. Mulus enim neque est similis asino neque equae, quia non est in aliqua specie completa, scilicet neque in specie asini neque in specie equae complete.

2. Dicendum ergo sic quod huiusmodi animal est monstrum quoad productionem similis ex simili, tamen, quantum ad intentionem naturae operantis, quia huiusmodi animal producitur in esse per viam naturae, non est monstruositas. Patet ergo qualiter possumus ponere quod ibi uno modo est monstruositas, alio[c] modo non.

3. Et per istam distinctionem patet solutio ad argumenta quae possunt adduci ad unam partem et ad aliam.

<Quaestio 32>
Secundum est quare huiusmodi animal est sterile.

1. Dicendum quod huius causa est quod, quia natura operatur breviori via qua potest et viis determinatis, ideo ordinavit natura huiusmodi animal produci ex diversis naturis [d]et non sua propria, quia, si sic, tunc individuis

[a] contingens *ex corr.* O
[b] non *supplevimus*
[c] alio *corr. ex* quo O
[d] et ... propria *suppl. s.l.* O

<Question 30>
The question is in what way God knows contingent things, that is, whether there is an idea of contingent things in the first cause.

1. Arguments for the affirmative: the first cause knows how to distinguish the contingent from the necessary; but it could not know this unless it had their species and likeness; therefore, <the idea of contingent is in the first cause>.

2. Against this: if the first cause has both the idea of necessity and that of contingency, then, since it contains the idea of contingency, this idea will be necessary. But it does not distinguish the necessary from the non-necessary through what is necessary; therefore, it is necessary to posit that the first cause does not have an idea <of contingent things>.

1*. Questions about monsters come next. And since monsters are spoken of in two ways –(i) in a broad sense, as Aristotle speaks of them in *On Animals* XVI, saying that the mule is a monstrous animal,[193] and (ii) in the proper sense, as when a man has six fingers or a tree is in part a grape-vine and in part an olive-tree—there are three questions about monsters in the first sense.

<Question 31>
The first question is why animals are or are said to be monstrous and in what way.

1. To this question it must be said that the cause of this is that like is not produced by like; for a mule is similar neither to an ass nor to a mare, because it is not in a complete species, that is, it is neither completely in the species of the ass nor in the species of the mare.

2. Therefore, it must be said that such an animal is a monster in relation to the production of like by like, but in relation to the intention of the nature that acts it is not a monstrosity, since such an animal is brought into being by a natural process. It is then clear how we can posit that in one way there is a monstrosity in this case and in another way there is not.

3. And with this distinction the solution to the arguments that can be adduced in favour of one part and of the other is obvious.

<Question 32>
The second question is why such an animal is sterile.

1. It must be said that the cause of this is that since nature acts by the shortest way that it can and by determined ways, nature therefore ordered that

[193] Arist. *De gen anim.* II.8, 748b16–19.

unius speciei responderet[a] duplex generatio. Quod negat Commentator, <ut[b]> videtur undecimo *Metaphysicae*, ubi quaerit[c] utrum generata per putrefactionem possint generari per propagationem. Et respondet dicens quod non, et hoc ratione praedicta. Dicit enim quod causa quare animalia generata per putrefactionem non possunt generari per propagationem est quia natura operatur breviori via qua potest, et praeter hoc viis ordinatis et determinatis.

<Quaestio 33>

Tertium est de convenientia asini cum equa, per quam convenientiam quae sit inter illa possunt ex se[d] huiusmodi animal producere.

1. Dicendum quod duplex est convenientia, una quae est in genere vel in specie et alia quae est in naturali complexione, verbi gratia ego et omnes homines aequaliter convenimus in genere vel in specie, ego tamen et quidam alius plus convenimus in naturali complexione sicut duo fratres vel aliqui omnino diversi, sicut patet multotiens, et isto modo est in proposito.

2. Unde dicendum quod convenientia inter illa non est convenientia <in[e]> specie vel genere, ita quod per illam convenientiam in genere producant huiusmodi animal, sed est convenientia in naturali complexione propter quam naturaliter producere possunt huiusmodi animal.

1*. De monstro secundo modo, ut quae proprie dicuntur monstra, ut quando producitur aliquod animal quod partim assimilatur homini et partim bovi, sicut asserunt quidam homines se vidisse, tria quaeruntur: primum est utrum possit esse aliquod tale animal; secundum est utrum tale animal sit a natura et exeat in esse per viam naturae; tertium est utrum fuerit unum individuum vel multa et, si unum, in qua specie fuerit.

<Quaestio 34>

1. De primo sic: quod non possit aliquod tale animal esse videtur sic. Dicit Aristoteles libro *Physicorum*[f] sic: “Non autem visum est quod sint vitigenae”, id est aliqua planta quae partim est vitis et partim oliva, sicut posuit Empedocles; sed natura magis sollicita est circa talia animalia quam circa plantas; quare etc.

[a] responderet *scripsimus*; respondet O
[b] ut *supplevimus*
[c] quaerit *ex corr.* O
[d] se *suppl. mg.* O
[e] in *supplevimus*
[f] Physicorum *scripsimus cum* TC; De vegetabilibus O

such an animal should be produced by different natures and not by its proper nature because in that case a double generation would correspond to the individuals of one species. And the Commentator apparently denies this in *Metaphysics* XI, where he asks whether the things generated by putrefaction can be generated by reproduction.[194] He replies by saying that this is not possible, and this for the reason already given; for he says that the cause why animals generated by putrefaction cannot be generated by reproduction is because nature acts by the shortest way that it can and in addition to this <it acts> by ordered and determined ways.

<Question 33>

The third question is about the agreement of an ass with a mare, by which agreement between them they can produce such an animal by themselves.

1. It must be said that the agreement is of two kinds: one of genus and of species and the other of natural constitution; for example, I and all men agree equally in genus and in species, but I and some other man agree more in natural constitution, as do two brothers or even some men who are altogether different, as is clear in many cases, and so it is in the present case.

2. Accordingly, it must be said that the agreement between them is not an agreement of species or of genus, so that they produce an animal of this sort in virtue of that agreement in genus, but it is an agreement in natural constitution in virtue of which they can produce such an animal naturally.

1*. About a monster in the second way, that is, those things that are properly said to be monsters, as when an animal that is in part similar to a man and in part to an ox is produced, as some people claim to have seen, there are three questions: the first is whether there can be an animal of that kind; the second is whether such an animal is by nature and comes into being through the way of nature; the third is whether it is one individual or many, and if one, what species it belongs to.

<Question 34>

1. As to the first question, arguments that there cannot be an animal of that kind: in the *Physics* Aristotle says that nobody has seen anything that is a *vitigena*, that is, a plant that is in part a grape-vine and in part an olive-tree, as Empedocles posited;[195] but nature is more attentive to animals of that sort than it is to plants; therefore, <there cannot be an animal of that kind>.

[194] This question is not found in Averroes' discussion of animals generated by putrefaction in *In Metaph.* XII, t.c. 13, fol. 299rb–vb, and t.c. 18, fols. 303rb–305va.
[195] Arist. *Phys.* II.8, 199b10–13.

2. Item, quod non intenditur a natura non producitur in esse per viam naturae; sed monstrum tale non intenditur a natura; ergo nec producitur in esse per viam naturae. Nec per viam artis; quare nullo modo. Non est ergo ponere quod aliquod huiusmodi sit. **[O 107vb]**

3. Item, ars imitatur naturam in quantum potest; sed ars in suis operationibus non facit aliquod artificiatum cuius una medietas sit domus et alia medietas scamnum[a]; ergo nec natura, ut videtur, faciet aliquod huiusmodi animal.

4. Ad oppositum: nos videmus in plantis quod super eandem radicem fundatur planta cuius una pars est unius speciei, ut in superiori parte, et alia pars alterius speciei, ut in inferiori parte; ergo eadem ratione poterit vegetativa eodem modo operari in productione animalis producendo animal cuius una pars est unius speciei et alia pars alterius speciei.

5. Item, dictum est quod talia videntur.

6. Solutio. Dicendum quod contingit huiusmodi animal evenire.

7. Ad primum in oppositum dicendum quod Aristoteles intelligit sic, quod ex uno semine non fit aliqua planta quae partim est oliva et partim vitis. Si autem per artem inseretur una planta in aliquam radicem plantae alterius speciei, tunc posset talis arbor et talis planta evenire, quod non fit ex semine uno, sed ex semine duplici.

8. Ad alias rationes patebit consequenter.

1*. Sed iuxta hoc quaeritur de insertis, circa quae[b] duo principaliter quaeruntur.

<Quaestio 35>
Primum est utrum ibi sint duae animae vegetativae vel una tantum.

1. Ad quod dicendum quod non sunt ibi duae animae vegetativae nec una anima vegetativa simpliciter.

2. Sed si quaeratur qualiter possunt duae cedere in perfectionem alicuius continuae plantae secundum totum crescentis, dicitur quod, sicut est in divisione anguillae quod, quia anima magis materialiter unitur corpori quam in aliis animalibus, et cum hoc est maior conformitas partium corporis et

[a] scamnum *ex corr.* O
[b] quae *scripsimus*; quas O

2. Furthermore, what is not intended by nature is not brought into being through the way of nature; but such a monster is not intended by nature; therefore, it is not brought into being through the way of nature, nor <is it brought into being> through the way of art; so <it is not brought into being> in any way. It is not to be posited, therefore, that something of that kind exists.

3. Furthermore, art imitates nature in so far as it can; but art in its operations does not produce an artefact one half of which is a house and the other half of which is a bench; therefore it seems that nature does not produce an animal of that kind.

4. To the contrary: in the case of plants we see that a plant, one part of which, namely the superior part, is of one species, and another part of which, namely the inferior part, is of another species, is based on the same root; therefore, by the same argument the vegetative soul can act in the same way in the production of animals, by producing an animal one part of which is of one species and another part of which is of another species.

5. Furthermore, it is said that these things are seen.

6. Solution. It must be said that it does happen that such an animal is produced.

7. To the first argument to the contrary[196] it must be said that Aristotle means that a plant that is in part an olive-tree and in part a grape-vine does not come to be from one seed. But if by art a plant is inserted into the root of a plant of another species, then such a tree and such a plant can come to be, not from one seed, but from a double seed.

8. The reply to the other arguments[197] will be made clear in what follows.[198]

1*. Next are the questions about grafting, and there are two main questions.

<Question 35>
The first is whether there are two vegetative souls there or only one.

1. To this question it must be said that there are neither two vegetative souls there nor one vegetative soul absolutely.

2. But to the question in what way two vegetative souls can submit to the perfection of a continuous plant that grows as a whole, it is said that, just as it happens in the division of an eel, that because the soul is more materially joined to the body than is the case in other animals, and in addition to this there is more conformity and uniformity in the parts of the body, so one thing

[196] See above, n. 1.
[197] See above, nn. 2–3.
[198] See below, q. 37, n. 7.

unigeneitas, ideo per divisionem fit una plures, cum ergo in planta sit materialior alligatio animae et maior partium corporis unigeneitas, tunc, si dividatur unum ab altero, manet vegetativa in utroque et iterum, si inseratur unum alteri, manent diversae rationes animae in utroque, scilicet in stipite et in eo quod insertur[a].

<Quaestio 36>

Secundum est quid[b] illorum convertit alterum in sui naturam.

1. Et quod stipes convertat illud quod insertur in sui naturam videtur, quia radix attrahit nutrimentum a terra et in stipite fit una digestio, et surculus non recipit nisi a stipite et stipes non recipit a surculo; ergo stipes convertit insertum in suam naturam et non econverso.

2. Ad quod dicendum quod surculus convertit stipitem, et hoc quia, cum recipitur nutrimentum a terra in stipite, fit una digestio et postea defertur illud digestum per poros et per distantias et partes connaturales usque ad surculum. Et cum recipitur in surculo, ibi est virtus animae et calor naturalis <et[c]> caelestis, et per istas virtutes in surculo fit ulterior digestio et tolluntur dispositiones quae fuerunt ad formam sive stipitis speciem et convertuntur in speciem propriam, [d]scilicet surculi, et ita totum[e] assimilatur surculo.

<Quaestio 37>

[fg]Dato quod tale animal possit esse, quaeritur secundo utrum sit a natura.

1. Et quod [h]sit a natura videtur, quia generatio est natura [i]eo quod ipsa est via in naturam; sed talia procedunt in esse per generationem; quare etc.

2. Item, forma talis animalis producitur in esse ex massa per principium intra et, si hoc, producitur per naturam.

3. Item, duo sunt principia agendi[j], scilicet ars[k] et natura; sed ars potest facere aliquid cuius una pars sit unius speciei et alia pars alterius speciei[l]; quare etc.

[a] insertur] adhuc quaeritur quid illorum convertit alterum *add. sed del.* O
[b] quid *scripsimus*; quod O
[c] et *supplevimus cum* TC (*cf. Appendix, Book II, q. 36.2, n. 1*)
[d] scilicet surculi *suppl. s.l.* O
[e] totum *scripsimus*; totus O
[f] Dato ... secundo] consequenter quaeritur dato quod tale animal sit T
[g] Dato ... natura] consequenter dato quod talia monstra sint, quaero an sint a natura C
[h] sit a natura] sic C
[i] eo quod ipsa] quia TC
[j] agendi *om.* TC
[k] ars] non *add.* O
[l] speciei *om.* TC

becomes many by division, <similarly> therefore in a plant, since there is a more material bond of the soul and more uniformity of the parts of the body, if one part is divided from the other, a vegetative soul remains in both, and if one part is re-inserted into the other, the different principles of the soul remain in each of them, that is, in the trunk and in what is inserted.

<Question 36>
The second question is which of those two converts the other into its nature.

1. Argument that the trunk converts the inserted thing into its nature: the root attracts nourishment from the earth and digestion occurs in the trunk; now, the twig only receives from the trunk, and the trunk does not receive anything from the twig; therefore, the trunk converts the inserted thing into its nature and not the other way around.

2. To this it must said that the twig converts the trunk <into its own species>, and this is because, when nourishment from the earth is received in the trunk, digestion occurs and afterwards the digested thing is carried through the pores and the openings and the connatural parts up to the twig. And when it is received in the twig, it finds the power of the soul there and both the natural and the celestial heat, and in virtue of these powers another digestion takes place in the twig and the dispositions it had to the form or species of the trunk are removed and they are converted into the species proper to the twig, and thus the whole is assimilated to the twig.

<Question 37>
Granted that an animal of that kind can exist, the second question is whether it exists by nature.

1. Arguments that it exists by nature: generation is a nature in so far as it is a way towards a nature; but such things are brought into being through generation; therefore, <such things exist by nature>.

2. Furthermore, the form of such an animal is brought into being from a mass in virtue of an internal principle, and if so, it is produced by nature.

3. Furthermore, there are two principles of acting, namely, art and nature; but art can produce something one part of which is of one species and another part of another species; therefore, <nature produces such things>.

4. Ad oppositum: quod non intenditur a natura non est a natura; sed tale animal non intenditur a natura, quia natura non intendit ex diversis speciebus aliquod unum producere nisi in certis et determinatis speciebus.

5. Item, secundum Aristotelem in septimo *Metaphysicae*, commune est naturae et arti quod utrumque producat simile ex simili; sed in tali generatione non est sic; [a]quare etc.

6. Item, species impermixtae sunt nisi ubi natura determinate ordinavit; sed in huiusmodi non ordinavit natura; quare etc.

7. Solutio[b]. Dicendum quod duplex est natura, scilicet aut natura universalis ordinans aut natura [c]particularis operans. Si[d] particularis operans, [e]hoc est dupliciter. Aut enim est[f] ordinans secundum naturam et[g] operans secundum naturam aut[h] ordinans contra naturam et[i] operans cum natura sive per naturam. Si autem sumatur natura universalis ordinans [j]et particularis ordinans [k]secundum naturam et[l] operans[m] per naturam[n], [o]sic sunt talia monstra contra naturam, et hoc quia natura nullo modo intendit talia[p]. Si autem sumatur particularis[q] ordinans contra naturam et[r] operans cum natura, sic sunt talia monstra naturalia, [s]et hoc quia[t] contra naturam ordinantur[u], tamen, cum ordinantur, per [v]operationem naturae ad esse producuntur.

[a] quare etc. *om.* T
[b] Solutio *om.* TC
[c] particularis ... talia] particularis ordinans et operans. Natura particularis ordinans et operans aut est ordinans et etiam operans secundum naturam aut est ordinans contra naturam et operans cum natura sive per naturam. Si autem sumatur natura universalis ordinans et particularis ordinans per naturam et similiter operans per naturam, sic sunt huiusmodi C
[d] Si] universalis ordinans et *add. s.l.* O
[e] hoc ... enim] aut T
[f] est] universalis *add. s.l.* O
[g] et] particularis *add. s.l.* O
[h] aut] universalis *add. s.l.* O
[i] et] particularis *add. s.l.* O
[j] et particularis ordinans *scrips. sed del.* O
[k] secundum naturam *om.* TC
[l] et] particularis *add. ex corr.* O
[m] operans] ordinans TC
[n] naturam] et operans *add.* T
[o] sic ... intendit talia] et illud ponit utrumque membrum T
[p] talia] et illud ponit utrumque *add. sed del.* O
[q] particularis] universalis O; *om.* T; natura particularis C
[r] et] particularis *add. s.l.* O
[s] et hoc *om.* TC
[t] quia] licet *add.* TC
[u] ordinantur] ordinentur TC
[v] operationem naturae] naturam C

4. To the contrary: what is not intended by nature is not by nature; but such an animal is not intended by nature, because nature does not intend to produce something that is one thing from different species except in some certain and determinate species.

5. Furthermore, according to Aristotle in *Metaphysics* VII, it is common to nature and art that each of them should produce like from like;[199] but this is not the case in such generation; therefore, <such generation is not by nature>,

6. Furthermore, species are unmixed except where nature has determinately ordered this; but nature did not order this with respect to such things; therefore, <such things do not exist by nature>.

7. Solution. It must be said that nature is of two kinds, either the universal ordering nature or a particular acting nature. A particular acting nature is of two kinds; for either it orders in accordance with nature and acts in accordance with nature or it orders against nature and acts with nature or through nature. If we consider the universal nature as ordering in accordance with nature and a particular nature as ordering and as acting in accordance with nature, then such monsters are against nature, and this is because nature in no way intends such things. But if we consider a particular nature as ordering against nature and as acting with nature, then such monstrosities are natural, and this because, although they are ordered against nature, however, since they are ordered, they are brought into being through the operation of nature.

[199] Arist. *Metaph.* VII.9, 1034a21–25

8. [a]Ad rationes patet, quia procedunt secundum quod huiusmodi monstra sunt intenta a natura et operantur a natura. Et secundum hoc verum concludunt.

<Quaestio 38>
Consequenter quaeritur de tertio, scilicet an sit unum individuum vel duo.

1. Et quod sit unum videtur sic: si[b] duo, aut ergo unius speciei aut plurium. Non plurium specierum, quia non sunt separata ad invicem. Nec unius speciei, quia forma speciei non numeratur nisi in diversis et separatis substantiis.

2. Item, ibi est unum cor; ergo et[c] unum animal.

3. Item, generatum assimilatur generanti; cum ergo[d] generans est unum, [e]et generatum[f].

4. [g]Ad oppositum: ex quo una medietas est hominis et alia bovis, **[O 108ra]** non est maior ratio quare sit sub una specie quam altera[h]. [i]Et constat quod non est sub nulla specie; est ergo sub utraque.

5. Solutio[j]. Dicendum quod est individuum unum, et hoc quia cor eius est unum. Et est in specie aliqua, tamen illa species non est determinata a natura, quia, sicut nec ordinate processit in esse, sic nec [k]a natura ordinatur ei species. Unde, quia mulus [l]est a natura[m] determinatus, ideo ordinatur ei species a natura. [n]Si ergo huiusmodi animal non ordinetur a natura, non debet ei species a natura determinari. Et est in tali animali sensitiva, diversificata tamen per diversificationem partium in suis operationibus.

6. Sed quid est dicendum de his quae habent utrumque sexum? An[o] debeant collocari sub altero sexu, ut sub masculo vel sub femina?

7. Et forte non est reperire aliquod animal quod habeat utrumque sexum essentialiter per praedominium[p], sed necesse est quod unum habeat per praedominium. Et si habeat sexum masculi per praedominium, debet dici masculus, si sexum feminae, femina.

[a] Ad ... concludunt *om.* TC
[b] si] essent *add.* TC
[c] et] est TC
[d] ergo *suppl. s.l.* O
[e] et generatum] animal; ergo et generatum erit unum animal C
[f] generatum] erit unum *add.* T
[g] Ad oppositum] Contra TC
[h] altera] sub alia TC
[i] Et ... utraque] aut ergo est sub utraque aut sub nulla; non sub nulla constat; ergo sub utraque C
[j] Solutio *om.* TC
[k] a ... species] sub aliqua specie ordinata a natura C
[l] est ... species a natura] determinatur a natura ideo a natura ordinatur in specie determinata C
[m] natura] terminatus *add. sed del.* O; ordinatus vel *add.* T
[n] Si ... determinari *om.* C
[o] An *ex corr.* O
[p] praedominium] dominium TC

8. The reply to the arguments[200] is clear because they proceed on the assumption that these monsters are <both> intended by nature and produced by nature. And given that assumption, their conclusion is true.

<Question 38>
The next question is about the third issue, namely, whether it is one individual or two.

1. Arguments that it is one individual: if they are two, then they are either of one species or of many species. They are not of many species because they are not separate one from the other. Nor they are of one species because the form of a species is not numbered except in different and separate substances.

2. Furthermore, there is one heart there; therefore, there is only one animal.

3. Furthermore, the generated is similar to the generator; therefore, since the generator is one, the generated is also one.

4. To the contrary: given that one half is from a man and the other is from an ox, there is no more reason why it is under one species than under the other. And it is apparent that it is not under no species; therefore, it is under both.

5. Solution. It must be said that it is one individual, and this because its heart is one. And it is in some species or other, but that species is not determined by nature, because, just as its coming into being was not ordered, so it is not ordered in a species by nature. Thus, it is because the mule is determined by nature that it is ordered in a species by nature. If, therefore, an animal of this sort is not ordered by nature, it is not placed in a determinate species by nature. And there is a sensitive soul in an animal of this sort, but the soul is diversified through the difference of the parts in their operations.

6. But what should be said about things that have both sexes? Should they be placed under one of the two sexes, that is, under the male or under the female sex?

7. Perhaps an animal cannot be found that has both sexes essentially and predominantly, but it is necessary for it to have predominantly one sex. And if it has predominantly the male sex, it must be said to be a male, if the female sex, a female.

[200] See above, nn. 4–6.

8. Sed quod non sit reperire[a] tale animal probatio. Dicit Aristoteles in decimo *Metaphysicae* quod quaedam sunt passiones materiales, quae non diversificant speciem, sicut mas et femina. Et dicit Commentator quod, si in commixtione seminis dominetur calor, erit masculus, [b]si autem frigus dominatur, generabitur femina. Sed impossibile est utrumque dominari; et ideo impossibile est aliquid utriusque sexus generari.

9. Quod autem sit possibile videtur, quia, si materia fuerit passa caliditate, generatur masculus et, si frigiditate, generatur femina. Sit ergo A terminus ille caliditatis citra[c] quem, si materia fuerit[d], non[e] generatur masculus, et sit B terminus frigiditatis citra[f] quem, si fuerit materia[g], non[h] generatur femina[i]. Constat quod inter A et B sunt multi gradus medii. Si ergo materia fuerit in illis gradibus intermediis, constat quod illud generatum non est[j] sub nullo sexu. Nec erit sub altero determinate, quia est citra[k] utrosque terminos; erit ergo sub utroque sexu. Et ita erit animal vel ad minus contingere potest tale animal [l]quod sit utriusque sexus.

<Quaestio 39>

Quaeritur[m] de assimilatione prolis. Et quia triplex est assimilatio —una scilicet in forma substantiali[n], de qua non est intentio [o]ad praesens, alia in lineationibus corporis et dispositione faciei [p]et huiusmodi membrorum[q], et de tali est intentio, tertia est de accidentibus aliis, ut albedine, scientia et huiusmodi, [r]et de hac adhuc non est intentio—primo ergo quaeritur utrum assimilatio habeat causam naturalem.

1. Et quod non videtur, quia[s] motus est natura eo quod est via in naturam; sed in 'ad aliquid' non est motus, secundum Aristotelem in quinto huius, quia

[a] reperire] invenire TC
[b] si ... femina *om.* T
[c] citra] circa OT
[d] fuerit] passa caliditate *add.* C
[e] non *om.* O
[f] citra] circa OT
[g] materia] passa *add.* C
[h] non *om.* O
[i] femina *suppl. s.l.* O
[j] est] erit TC
[k] citra] circa OT
[l] quod ... sexus] habens utrumque sexum. qualiter hoc sit tu videas C
[m] Quaeritur] consequenter *praem.* TC
[n] substantiali] speciali TC
[o] ad praesens *om.* TC
[p] et huiusmodi membrorum *om.* C
[q] membrorum *om.* T
[r] et ... intentio *om.* T; et de hac nihil ad prasens C
[s] quia] secundum Aristotelem *add.* C

8. Here is the proof that such an animal cannot be found. In *Metaphysics* X, Aristotle says that some passions are material, such as male and female, and these do not diversify the species.[201] And the Commentator says that, if heat dominates in the mixture of the semen, what is generated will be male, and if coldness dominates, it will be female.[202] But it is impossible for both to dominate; therefore, it is impossible for something belonging to both sexes to be generated.

9. It seems, however, that this is possible because if the matter was subject to heat, a male would be generated, and if it was subject to coldness, a female would be generated. Thus, let A be the terminal degree of heat such that if the matter goes beyond it, a male is not generated, and let B be the terminal degree of coldness such that if the matter goes beyond it, a female is not generated. It is clear that between A and B there are many intermediate degrees. If, then, the matter comes under those intermediate degrees, it is clear, on one hand, that that generated thing does not belong to no sex, but, on the other hand, it will not belong to one of them determinately, because the matter comes beyond both terminal degrees; therefore, it will belong to both sexes. And so there is or at least there can be an animal that belongs to both sexes.

<Question 39>

This question is about the resemblance of offspring. And since there are three kinds of resemblance—one in substantial form, which is not that under discussion here; another in features of the body and in the disposition of the face and of such members, which is the subject of the present discussion; the third is in other accidents, as whiteness, knowledge, and the like, and this is not yet under discussion—the first question is whether the resemblance has a natural cause.

1. Arguments for the negative: motion is nature in so far as it is a way towards a nature; but in relatives there is no motion, according to Aristotle in

[201] Arist. *Metaph.* X.9, 1058b21–24.
[202] Aver. *In De gen. anim.* IV, fols. 114vb–115ra.

ibi dicit quod motus est tantum in tribus generibus, [a]et ita nec via naturalis et, si hoc, neque causa naturalis. Sed assimilatio est relatio quaedam; ergo etc.

2. Item, si habet causam, aut[b] ergo materiam aut formam. Non materiam, quia materia non coincidit cum efficiente, et loquor de causa efficiente, et etiam materia est eadem in omnibus, [c]et tamen similitudo non eadem. Non formam, quia, si formam, aut specialem aut individualem. Non specialem, et hoc quia omnia individua illius speciei habent talem formam [d]quae specialis est, [e]et [f]tamen talis assimilatio non est inter omnia ista. Non individualem, quia forma individualis non est nisi in uno tantum, et assimilatio est inter duo vel plura.

3. Solutio[g]. Dicendum quod est loqui de assimilatione duobus modis: aut de[h] ipsa relatione aut de natura quae subest relationi. Si autem primo modo loquamur, sic non est motus in ipsam[i] [j]nisi per accidens [k]et non habet causam nisi per accidens. Per se enim est quod natura producat formam substantialem, sed quod producat alicui simile, hoc accidit. Si autem secundo modo loquamur, habet causam naturalem et [l]viam naturalem in ipsam. Quid autem sit illa natura postea patebit, ubi patebit responsio secundae rationis.

<Quaestio 40>
Quaeritur consequenter quae est causa huius assimilationis.

1. Et dicit Aristoteles quarto decimo *De animalibus* quod, si semen patris fuerit vincens, tunc assimilatur generatum patri et, si semen matris [m]fuerit vincens, tunc assimilatur matri.

2. [n]Et exponunt quidam istam assimilationem in sexu, ita scilicet quod, si semen patris vincat, assimilatur patri in sexu ita quod erit masculus et, si semen matris [o]vincat, assimilatur in sexu matri ita quod erit femina.

[a] et ... etc.] ergo in ad aliquid non est via naturalis; cum ergo assimilatio sit ad aliquid sive relatio quaedam, assimilatio non habebit causam materialem C
[b] aut *ex corr.* O
[c] et ... eadem *om.* C
[d] quae specialis est *om.* T
[e] et ... ista] non tamen sunt similia secundum quod hic intendimus C
[f] tamen talis *om.* T
[g] Solutio *om.* TC
[h] de *ex corr.* O
[i] ipsam] ipsa OC
[j] nisi per accidens *om.* T
[k] et ... accidens *om. hom.* C
[l] viam naturalem] est motus naturalis C
[m] fuerit vincens, tunc *om.* C
[n] Et ... femina *om.* C
[o] vincat ... quod *om.* T

Physics V, since he says there that motion is only in three categories;[203] thus, in relatives there is no natural way, and if so, neither is there a natural cause. But resemblance is a relation; therefore, <resemblance does not have a natural cause>.

2. Furthermore, if resemblance has a cause, this is either matter or form. But it is not matter, because matter does not coincide with the efficient, and I mean the efficient cause; moreover, matter is the same in all things, whereas a likeness is not the same. It is not form, because then it is either a specific form or an individual form. It is not a specific form because all individuals of that species have that specific form, whereas a resemblance of this sort does not hold between all these individuals. It is not an individual form because an individual form exists only in one individual, and resemblance exists between two or more individuals.

3. Solution. It must be said that one can speak of resemblance in two ways: either (i) of the relation itself or (ii) of the nature that is subject to the relation. In case (i), there is no motion in resemblance except accidentally and resemblance does not have a cause except accidentally; for the fact that nature produces a substantial form is per se but the fact that it produces a thing similar to something is accidental. In case (ii), the resemblance has a natural cause and a natural way towards itself. What that nature is will be made clear later, when the reply to the second argument is made clear.[204]

<Question 40>
The next question is what the cause of this kind of resemblance is.

1. In *On Animals* XIV Aristotle says that, if the semen of the father wins, then what is generated is similar to the father, and if the semen of the mother wins, then it is similar to the mother.[205]

2. And some people explain this resemblance as relative to sex, that is, if the semen of the father wins, that the thing generated is similar to the father in sex so that it will be a male, and if the semen of the mother wins, then it is similar to the mother in sex so that it will be a female.

[203] Arist. *Phys.* V.1, 225b5–9; V.2, 225b11–14.
[204] See below, q. 40, n. 13.
[205] Arist. *De gen. anim.* IV.1, 764a9–12, 766b15–18.

3. Sed contra istud est illud quod dicit Aristoteles, quod causa talis sexus vel talis est situs diversus seminis in matrice, [a]quia, si situetur in unam partem, generatur masculus, si in aliam, femina.

4. Et propter hoc quaeritur utrum causa Aristotelis praedicta sit [b]causa sufficiens huius assimilationis praedictae.

5. [c]Et videtur quod non, quia, si causa esset bona, cum semen patris semper vincat, semper assimilaretur generatum patri[d]. Probatio maioris: in omni actione naturali consequente finem intentum, agens habet dominium in patiens; sed secundum Aristotelem, semen patris est agens; quare etc.

6. Item, quod semper assimilatur patri videtur, quia, secundum Aristotelem **[O 108rb]** in *De generatione*, agens assimilat sibi patiens, ita quod in fine convertit patiens in sibi simile sive in suam naturam; sed semen patris est agens in generatione naturali[e]; ergo etc.

7. Item, secundum Aristotelem in septimo [f]*Primae philosophiae*, generans et generatum sunt unum in forma, duo autem in materia.

8. Item, natura semper facit quod melius est[g]; [h]sed melius est assimilari nobiliori quam minus nobili. Sed [i]agens, ut pater, est nobilius quam patiens, ut mater; quare natura magis producit simile patri quam matri.

9. Ad oppositum: [j]quod generatum assimiletur matri videtur sic. Omne receptum est in recipiente per modum recipientis et non[k] per modum recepti. Sed utrumque semen est receptum in matre; ergo[l] per modum recipientis, [m]ut matris; quare etc.

10. Item, illud quod dat totam substantiam et totam naturam ei quod debet produci, ita quod forma inducenda totaliter est in ipso, illud magis assimilat sibi illud quod in esse producitur quam illud quod non. Sed sic est de matre[n], secundum quod dicit Aristoteles in quinto decimo *De animalibus*, ubi dicit quod semen [o]patris est sicut manus figuli in olla; [p]quare etc.

[a] quia ... femina *om.* C
[b] causa ... praedictae] bona C
[c] Et ... non *om.* C
[d] patri] quod falsum est *add.* C
[e] naturali] et semen matris patiens *add.* C
[f] Primae philosophiae *om.* TC
[g] est] semper *add.* OC
[h] sed melius est *suppl. s.l.* O
[i] agens ... matri] pater est agens et ita in ratione nobilioris, mater autem est in ratione patientis et ita in ratione ignobilioris; ergo generatum semper assimilabitur patri C
[j] quod ... sic *om.* C
[k] non *suppl. s.l.* O
[l] ergo] erit in matre *add.* C
[m] ut ... etc.] et ita generatum ex semine utroque magis assimilatur patri quam matri C
[n] matre *lectio incerta* (*ex corr.*) O
[o] patris ... etc.] maris operatur in semine matris sicut manus figuli in olla; quare productum magis debet assimilari matri quam patri C
[p] quare etc. *om.* T

3. But against this is Aristotle's statement that the cause of this or that sex is the different position of the semen in the mother, because if it is situated in one part, a male is generated, and if in another part, a female.[206]

4. And because of this the question arises whether the cause mentioned by Aristotle is the sufficient cause of such a resemblance.

5. Arguments for the negative: if that cause were the right one, since the semen of the father always wins, then what is generated would always be similar to the father. Proof of the major premise: in every natural action that reaches its intended end, the agent dominates the patient; but, according to Aristotle, the semen of the father is the agent;[207] therefore, <the semen of the father dominates>.

6. Furthermore, it seems to be the case that what is generated is always similar to the father because, according to Aristotle in *On Generation*, the agent makes the patient similar to itself, so that in the end it converts the patient into something similar to itself or of the same nature;[208] but the semen of the father is the agent in natural generation; therefore, <what is generated is similar to the father>.

7. Furthermore, according to Aristotle in *Metaphysics* VII, the generator and the generated are one in form but two in matter.[209]

8. Furthermore, nature always produces what is better; but it is better to resemble what is more noble than what is less noble. But the agent, that is, the father, is nobler than the patient, that is, the mother; therefore, nature produces something similar to the father rather than to the mother.

9. To the contrary: it seems to be the case that what is generated resembles the mother: everything received is in the receiver in accordance with the mode of the receiver and not in accordance with the mode of the received. But both semens are received in the mother; therefore, in accordance with the mode of the receiver, that is, that of the mother; therefore, <what is generated resembles the mother>.

10. Furthermore, what gives the entire substance and nature to what it produces, so that the form to be induced is totally in that thing produced, more closely resembles the thing it produces than does the thing which does not <give in that way>. But this applies to the mother, in accordance with Aristotle in *On animals* XV, where he says that the semen of the father is like the hand of a potter in a pot;[210] therefore, <what is generated resembles the mother>.

[206] Arist. *De gen. anim.* IV.1, 763b28–764a23.
[207] Arist. *De gen. anim.* I.21, 730a25–27.
[208] Arist. *De gen. et corr.* I.7, 324a9–14.
[209] Arist. *Metaph.* VII.8, 1033b29–1034a8.
[210] Arist. *De gen. anim.* I.22, 730b9–23.

11. Quidam dicunt quod in semine viri[a] duo sunt, scilicet pars pura et pars impura. Unde, si purum dominetur, tunc assimilatur patri, si autem impurum [b]et illud quod est grossum vincat, non oportet quod assimiletur patri. [c]Et similiter est in semine matris quod, si purum quod est in semine matris dominetur, tunc assimilatur matri, si autem grossum sive impurum, non necesse est quod assimiletur patri vel matri, sed parentibus[d]. Unde [e]dicunt quod auctoritas Aristotelis intelligenda est solum quando purum quod est in semine[f] patris vincit, et similiter est[g] de matre.

12. Mihi videtur quod victoria non est solum causa assimilationis, sed cum hoc exigitur quod sit[h] ibi[i] consimile esse formae substantialis in sua materia. Et hoc patet plane. Videmus enim quod una forma substantialis per diversa esse quae recipit in sua materia est causa diversorum accidentium specie; ergo per oppositum diversae formae substantiales per aliquod consimile esse in sua materia possunt esse causa eiusdem accidentis specie.

13. Et per hoc patet responsio [j]secundo argumento primi problematis, quia forma substantialis filii non unde forma substantialis est, sed[k] unde est sub consimili esse in sua materia, est causa talis assimilationis accidentalis. Et quia pater est generans filium, magis forte potest complexio filii convenire cum patre quam cum alio. Et ideo saepe [l]magis est[m] forma substantialis filii conveniens in esse cum forma substantiali patris quam cum forma alterius, aliquando autem contingit modo contrario.

14. Unde[n] respondendum[o] auctoritati Aristotelis, scilicet quod non solum est causa victoria, sed causa radicalis cum hoc est esse consimile in sua materia.

15. Ad aliud dicendum quod[p] duplex est[q] generatio, scilicet aequivoca, et ibi non[r] generatur simile ex simili, ut patet in putrefactis, et generatio univoca.

[a] viri *om.* O
[b] et ... grossum *om.* TC
[c] Et ... parentibus *om.* TC
[d] parentibus *ex corr.* O
[e] dicunt quod *om.* TC
[f] semine *corr. ex* specie O
[g] est] etiam O; dicunt C
[h] sit] est (*lectio incerta*) O
[i] ibi *suppl. s.l.* O; *om.* C
[j] secundo ... problematis] ad secundam rationem positam ad primum problema C
[k] sed *ex corr.* O
[l] magis est *om.* TC
[m] est *suppl. s.l.* O
[n] Unde] per illud *add.* TC
[o] respondendum] est *add.* TC
[p] quod] cum TC
[q] est] sit TC
[r] non *suppl. s.l.* O

11. Some people say that in the semen of a man there are two things: a pure part and an impure part. Thus, if the pure part dominates, then what is generated resembles the father; if the impure and coarse part wins, it is not necessary for what is generated to resemble the father. And the same applies to the semen of the mother: if the pure part that is in the semen of the mother dominates, then what is generated resembles the mother; if the coarse or impure part dominates, it is not necessary for what is generated to resemble either the father or the mother but <only> its progenitors. Thus, they say that the authoritative text of Aristotle[211] must be understood as applying only when the pure part that is in the father's semen wins, and the same applies to the mother.

12. It seems to me that a victory is not the only cause of resemblance, but together with this it is required that the substantial form should exist in a similar way in its matter. And this is absolutely clear; for we see that one substantial form in virtue of the different beings that it receives in its matter is the cause of accidents which differ in species; therefore, by an opposite argument, different substantial forms in virtue of existing in a similar way in their matter can be the cause of an accident of the same species.

13. On this basis the reply to the second argument of the first problem[212] is clear, because the substantial form of the son is the cause of this resemblance in accidents not in so far as it is substantial but in so far as it exists in a similar way in its matter. And since the father is the generator of the son, it is perhaps the case that the constitution of the son can agree more with the father than with another man. Therefore, it is often the case that the substantial form of the son agrees in its manner of existence more with the father's substantial form than with the form of another man, although sometimes the contrary happens.

14. Accordingly, one must reply to the authoritative text of Aristotle[213] by saying that the cause is not only the victory but together with this, existing in a similar way in its matter is the root cause.

15. To the other argument[214] it must be said that generation is of two kinds: (i) equivocal generation, and in this case like is not generated from like, as is clear in rotten things; and (ii) univocal generation. Univocal generation

[211] See above, n. 1.
[212] See above, q. 39, n. 2.
[213] See above, n. 1.
[214] See above, n. 6.

Sed haec fit dupliciter: aut per propagationem et seminis decisionem aut per mutuam actionem et passionem, ut patet in generatione conversiva. Si autem primo modo sumatur generatio, tunc generans assimilat sibi generatum in forma substantiali constituente speciem et non in omnibus accidentibus. Si secundo modo sumatur generatio, tunc corrumpit contrarium, ut agens primum, et assimilat sibi patiens non solum in forma substantiali, sed etiam in omnibus formis accidentalibus. Et de tali generatione intelliguntur omnes auctoritates[a] Aristotelis [b]in libro *De generatione* et multae alibi dictae.

16. Sed si hoc est verum, scilicet de generatione per contrarietatem, quod [c]generat sibi simile et in forma substantiali et[d] in formis[e] accidentalibus, tunc, cum generans animatum sit agens nobilius quam generans inanimatum et potentior in sua actione, generans animatum non solum generat sibi simile in forma substantiali, sed[f] etiam in omnibus formis accidentalibus.

17. [g]Dicendum quod non sequitur, quia in generatione per contrarietatem ita est quod ibi forma substantialis perfectiva[h] est forma valde materialis et materialiter perficiens. Et propter hoc sunt omnes passiones materiales proximae cum forma substantiali et immediate consequentes formam substantialem. Unde Aristoteles aliquando vocat formas perfectivas elementorum gravitatem et levitatem propter hoc quod sunt proxime[i] consequentes formam substantialem. Sed a parte alia[j] est econverso, quia forma substantialis perfectiva animalis est forma valde[k] remota a materia inter formas materiales. Propter hoc est ita quod passiones materiales non ita immediate se tenent cum suis formis substantialibus. Ex nobilitate ergo suae formae est ita quod passiones materiales non sunt eidem ita immediate consequentes [l]et unitae[m]. Et ita magis debet concludere oppositum quam propositum.

18. Ad argumentum quod est contra Aristotelem, scilicet quod semen viri semper habet dominium, dicendum quod semen patris simul cum semine matris intendit principaliter unum, scilicet producere formam substantialem

[a] auctoritates *ex corr.* O
[b] in ... dictae] et alibi in multis locis C
[c] generat ... et] generans assimilat sibi generatum non solum C
[d] et] sed etiam C
[e] formis *om.* TC
[f] sed *ex corr.* O
[g] Dicendum quod *om.* C
[h] perfectiva] perfectio O *et praem. s.l.* quae est
[i] proxime] sive immediate *add.* C
[j] alia] altera TC
[k] valde] maxime TC
[l] et unitae *om.* C
[m] unitae *ex corr.* O

happens in two ways: either (iia) by reproduction and the separation of the semen or (iib) by mutual action and passion, as is clear in converse generation. In case (iia), the generator makes the generated resemble itself in the substantial form that constitutes the species but not in all the accidents. In case (iib), the generator destroys the contrary, acting as a first agent does, and makes the patient resemble itself not only in substantial form but also in all the accidental forms. And all the authoritative texts of Aristotle in *On Generation*[215] and many found in other works must be understood as relative to case (iib).

16. But if it is true that in generation by contrariety a generator generates something similar to itself both in the substantial form and in the accidental forms, then, since an animate generator is a nobler agent than an inanimate generator and more powerful in its action, an animate generator generates not only something similar to itself in substantial form, but also in all the accidental forms.

17. It must be said that this conclusion does not follow because in generation by contrariety the perfective substantial form is a very material form and gives perfection materially. Because of this all the material passions are close to the substantial form and follow from the substantial form immediately. Accordingly, Aristotle sometimes calls the perfective forms of the elements heaviness and lightness because they follow closely from their substantial form.[216] But in the other case <that is, that of an animate generator> it happens the other way around because among the material forms the perfective substantial form of an animal is a form very remote from the matter. Because of this the material passions are not immediately linked to their substantial forms. Therefore, it follows from the nobility of an animal's form that the material passions are not immediately following from and united with that form. Thus, an opposite conclusion to the one proposed should follow.

18. To the argument that it is against Aristotle to say that the semen of the male always has dominion,[217] it must be replied that the semen of the father together with the semen of the mother has one main aim, that is, to produce a substantial form similar to itself in species, and in this the semen of the male

[215] See above, n. 6.
[216] The reference here could be, for example, to Arist. *Phys.* VIII.4, 255b9–21 and *De caelo* I.3, 269b18–20. Note, however, that in Book I, q. 39.2, n. 4 a similar claim is attributed to Averroes.
[217] See above, n. 5.

sibi[a] similem in specie, et in hoc dominatur semen viri. Alia autem sunt[b] cointenta, ut accidentia. Unde, si loquamur de accidentibus absolute, adhuc est semen patris vincens. Si autem sumatur accidens tale vel tale, non semper est semen patris vincens, sed aliquando semen matris.

<Quaestio 41>
Iuxta hoc est quaerendum utrum aliqua pars quantitativa [O 108va] concepti sit a patre.

1. Et videtur quod sic, quia, si non, cum, secundum Aristotelem, pater dat formam, tunc forma substantialis transferret se a subiecto in subiectum, ut a patre in filium, nulla parte materiali simul transita cum ipsa[c].

2. [d]Et etiam tunc non esset aliquo modo cognatus patri.

3. Et etiam nec haberet animam a patre, quia anima non transfert[e] se a corpore in corpus.

4. Item, nec potest dici quod totum secundum speciem sit a patre et secundum materiam a matre. Cum caro secundum materiam continue fluat et refluat[f], tunc in parvo tempore totum quod esset receptum a matre esset consumatum, et ita non esset [g]cognatus matri.

5. Mihi videtur quod [h]in utraque decisione, scilicet et patris et matris, descinditur aliquid de puro et aliquid de impuro. Maior tamen pars est impura, secundum quod dicit Aristoteles quod maior pars spermatis est sordes, qua non[i] indigetur.

6. Pono etiam quod illud purum quod est [j]a patre est multo purius quam illud purum quod est a matre, et illud impurum [k]quod est a matre est impurius quam illud impurum quod est a patre.

7. Pono etiam quod [l]duo pura cedunt in radicale principium generati et cedunt simul loco et quod duae partes impurae cedunt similiter simul loco et fiunt pars materialis generati. Unde illa pars materialis cedit simul loco cum parte secundum speciem generata ex puriori parte utriusque seminis, quatenus

[a] sibi *ex corr.* O; *om.* TC
[b] sunt *suppl. s.l.* O
[c] ipsa] quod falsum est *add.* C
[d] Et ... patri] Et praeter hoc si hoc tunc non esset simul de cognatione patris C
[e] transfert *iter. sed corr.* O; *corr. ex* transfertur T
[f] refluat] et *add.* TC
[g] cognatus matri] proles de cognatione sive parentela matris C
[h] in ... matris] tam a patre quam a matre C
[i] non *iter.* O
[j] a ... quod est *om. hom.* C
[k] quod ... matre est] similiter et tam illud C
[l] duo ... et] illud quod est purum utriusque cedit in radicale principium generati ita quod illa duo pura C

dominates, whereas other things, namely the accidents, are aimed at concomitantly. Thus, if we talk about accidents absolutely, the semen of the father still wins. If, however, we take this or that accident, the father's semen does not always win but sometimes the mother's semen does.

<Question 41>
The next question is whether a quantitative part of the embryo comes from the father.

1. Arguments for the affirmative: if that is not the case, then, since according to Aristotle the father gives the form,[218] the substantial form would pass from subject to subject, that is, from the father to the son, but no material part would pass along with it.

2. Also, the embryo would not be related by birth to the father in some way.

3. Again, the embryo would not receive a soul from the father because a soul does not pass from body to body.

4. Furthermore, it cannot be maintained that the whole with respect to the species comes from the father and with respect to the matter from the mother. Since flesh in accordance with its matter continuously comes and goes, then in a short time the whole that is received from the mother would be consumed, and so the embryo would not be related by birth to the mother.

5. It seems to me that in both separations, that is, that of the father and that of the mother, both something pure and something impure is separated. The major part, however, is impure, according to Aristotle's statement that the major part of sperm consists of impurities, which are not needed.[219]

6. I also posit that the pure part that comes from the father is much purer than the pure part that comes from the mother, and that the impure part that comes from the mother is more impure than the impure part that comes from the father.

7. Further, I posit that two pure things yield to a root principle of the generated thing and come to be in the same place, and that likewise two impure parts come to be in the same place and become a material part of what is generated. Accordingly, that material part comes to be in the same place as the part generated in accordance with the species from the purer part of each of the two semens, so that the action of the active and passive qualities

[218] Arist. *De gen. anim.* I.20, 729a8–11.
[219] Arist. *De gen. anim.* II.4, 739a7–8.

possit [a]fieri actio qualitatum activarum et passivarum, ita ut[b] non fiat consumptio humoris radicalis [c]sive radicalis principii, sed humoris fluentis a parte principii materialis se tenentis, quia, si fiat eius deperditio, possibilis est restauratio. Et est exemplum de lampade et lichinio, quia, si fiat consumptio lichinii, non potest fieri restauratio, si autem [d]olei, bene potest.

8. Sed si quaeratur quid[e] illorum dat animam, mihi videtur quod pater. Et hoc quia illud quod est de patre in generato est nobilius, et quod dat nobilius, dat et animam.

9. Tamen audivi dicere[f] de quadam muliere quae numquam habuit rem cum viro quod in matrice sua generabatur massa quaedam carnea et post eiectionem illius[g] [h]a matrice movebatur per aliquod tempus [i]et, si pungeretur acu, moveret se.

<Quaestio 42>
Hic potest quaeri utrum unus homo per plures accessus possit super unum generatum.

1. Et videtur quod sic, quia, si agens per unam actionem potest super effectum multiplicatum, tunc eadem ratione pluribus actionibus[j] poterit super unum. Sed Aristoteles[k], tertio decimo *De animalibus*, [l]dicit quod homo unico accessu potest super [m]plures[n] filios; quare etc.

2. Item, sol est adeo potens in suis generationibus sicut animal in suis; sed unico accessu, ut unica revolutione, non potest sol super suum generatum; ergo [o]similiter erit de animali quod unico accessu non poterit super suum generatum. [p]Oportet ergo quod hoc fuerit per plures accessus.

[a] fieri ... passivarum] sustentare actionem et passionem qualitatum contrariarum C
[b] ut] quod TC
[c] sive ... quia] quod est principium radicale immo humoris materialis cuius C
[d] olei, bene potest] fiat consumptio humoris olei existentis in lichinio potest fieri restauratio C
[e] quid *scripsimus*; quod OTC
[f] dicere *om.* TC
[g] illius] eius TC
[h] a matrice *om.* TC
[i] et ... se *om.* TC
[j] actionibus] potest super *add. sed del.* O
[k] Aristoteles] secundum Aristotelem TC
[l] dicit quod *om.* TC
[m] plures filios] generationem plurium filiorum C
[n] plures *om. sed suppl. s.l.* duos O
[o] similiter ... quod] animal C
[p] Oportet ... accessus *om.* TC

can take place, in such a way that there is no consumption of the basic fluid or the basic principle, but only of the fluid that belongs to the material principle, because, if that is lost, then its replacement is possible. An example is that of the torch and the wick because if the wick is consumed, it cannot be replaced, but if the oil is consumed, it can easily be replaced.

8. But if it is asked which of the two gives the soul, it seems to me that this is the father. The reason for this is that what comes from the father in the generated thing is nobler, and what gives the nobler thing also gives the soul.

9. I heard, however, of a woman who never had a relation with a man but a fleshy mass was generated in her womb, and after its ejection from the womb it kept moving for some time and if it were pierced with a needle, it would move itself.

<Question 42>
Here one can raise the question whether one man can generate one thing through many sexual acts.

1. Arguments for the affirmative: if an agent can produce multiple effects through one action, then for the same reason it will be able to produce one effect with many actions. But in *On Animals* XIII, Aristotle says that a man can produce many sons with one sexual act;[220] therefore, <he can produce one son with many sexual acts>.

2. Furthermore, the sun is as powerful in its generations as is an animal in its generations; but the sun cannot produce the thing generated with only one act, that is, only one revolution; therefore, likewise an animal will not be able to produce the thing generated by it with only one sexual act. It is necessary, therefore, that this should be done through many sexual acts.

[220] Arist. *De gen. anim.* I.20, 728b36–729a5.

3. [a]Ad oppositum: unico effectui numero respondet unum agens numero, et ita unico filio numero respondet unum semen numero; [b]quare etc.

4. Item, matrix[c], secundum Aristotelem, non solum attrahit semen, sed et [d]semen superfluum vel non aptum eicit. Si ergo primum semen sit sufficiens, tunc eicit alterum, si insufficiens, tunc, antequam aliud semen receperit, ipsum eicit. [e]Quocumque ergo modo dicatur, semper sequitur quod non faciet eundem generatum.

<Quaestio 43>
Quaeritur consequenter utrum duo patres possunt super generationem unius filii.

1. Et videtur quod sic: sicut se habet filius ad patrem, sic se habet econverso. Sed plures possunt[f] esse filii eiusdem patris; ergo plures poterunt[g] esse patres eiusdem filii.

2. Item, duo semina possunt sic se habere quod unum potest esse insufficiens et alterum consimilis complexionis potest supervenire insufficientiam prioris[h] complens; potest ergo illud superveniens ad generationem conferre; quare etc.

3. Item, in actione naturali per contrarietatem ita est quod duo agentia possunt immittere speciem in aliquo uno et producere effectum unum tantum, ut patet si ignis immittat speciem in aere et aqua similiter. Ergo similiter erit a parte generantis per propagationem quod duo poterunt super unum generatum.

4. Item, in artificialibus ita est quod duo artifices possunt super unum artificiatum; ergo similiter in actione naturali erit quod duo patres possunt generare unum filium[i], [j]cum ars imitatur naturam etc.[k]

5. [l]Ad oppositum: unico effectui numero respondet unicum agens numero; [m]quare etc.

[a] Ad oppositum] Contra TC
[b] quare etc. *om.* C
[c] matrix *scripsimus*; vel *sed corr. s.l. in* vulva vel matrix O; mater TC
[d] semen ... eicit] eicit superfluum C
[e] Quocumque ... generatum *om.* TC
[f] possunt] poterunt TC
[g] poterunt *ex corr.* O
[h] prioris] eius TC
[i] filium] generatum sive duo patres super unum filium C
[j] cum ... etc. *om.* TC
[k] etc. *scripsimus*; et circa *scrips. et* item *add. sed del.* O
[l] Ad oppositum] Contra TC
[m] quare etc. *om.* C

3. To the contrary: to numerically one effect there corresponds numerically one agent; hence, to numerically one son there corresponds numerically one semen; therefore, <a man cannot produce one son with many sexual acts>.

4. Furthermore, according to Aristotle, the mother does not only attract the semen, but she also ejects the superfluous or non-fitting semen.[221] Therefore, if the first semen is sufficient, then she ejects the other; if it is insufficient, then she rejects it before receiving another semen. Whichever is the case, then, it always follows that same generated thing is not produced <by the two sexual acts>.

<Question 43>
The next question is whether two fathers can generate one son.

1. Arguments for the affirmative: as the son is related to the father so also the other way around. But there can be many sons of the same father; therefore, there can also be many fathers of the same son.

2. Furthermore, two semens can be such that one can be insufficient and the other of a similar constitution can supervene on the first and complete the insufficiency of the first; therefore, the supervening semen can contribute to the generation <of one thing>; therefore, <two fathers can generate one son>.

3. Furthermore, in natural action by contrariety two agents can introduce a species into something that is one and can produce just one effect, as is clear when fire introduces its species into the air and water does the same. Therefore, in the case of a generator through reproduction the case will be similar so that two generators can generate one thing.

4. Furthermore, in artificial things it is the case that two artisans can produce one artefact; therefore, similarly in natural action it will be the case that two fathers can generate one son, since art imitates nature <in so far as it can>.

5. To the contrary: to numerically one effect there corresponds numerically one agent; therefore, <two fathers cannot generate one son>.

[221] Arist. *De gen. anim.* II.4, 739a36-b19

6. Item, unico actui numero respondet una potentia numero; sed in seminibus diversis sunt duae potentiae numero; ergo non erunt in unum actum numero.

7. Quidam dicunt quod homo per plures accessus potest super [a]unum filium, et hoc quia Aristoteles dicit quod, cum gallus multotiens accedit, generantur in uno ovo duo vitella et ex illo ovo generatur animal cum uno capite et una cauda et cum quattuor alis.

8. [b]Qualiter hoc sit quaere.

[a] unum filium] numerum animalium (numerum *ex corr.*) O; *inv.* T

[b] Qualiter ... quaere] sed quid dicendum sit de illis duabus (*sic*) seminibus tu videas C

6. Furthermore, to numerically one act there corresponds numerically one potency; but in different semens there are numerically two potencies; therefore, they will not be directed to numerically one act.

7. Some people say that a man can produce one son by many sexual acts, and this because Aristotle says that, when a cock performs many sexual acts, two yolks are generated in one egg and an animal with one head, one tail, and four wings is generated from that egg.[222]

8. Ask in what way this happens.

[222] Arist. *De gen. anim.* IV.4, 770a5–22.

LIBER III

1*. [a]*Quoniam autem natura quidem est principium motus* [b]*et status et mutationis*

2*. [c]Quaeritur[d] de motu.

<Quaestio 1>
Et primo quaeritur utrum motus sit.

1. Et videtur primo quod non: signemus in spatio primum punctum et medium et finem, et vocetur primum punctum[e] A, medium B, finis C[f]. Quando mobile est in B non est in aliqua parte motus, quia non sub praeterita, cum ipsa non sit, nec sub futura, cum ipsa [g]adhuc non generetur. Et ita [h]signetur spatium alibi et eodem modo arguatur. Erit ergo mobile in omni parte spatii, et tamen sub nulla parte motus, quod non contingeret si motus esset; quare etc.

2. Item, partes motus[i] [j]solum imaginantur sub priori et posteriori; ergo non habet partes nisi sub priori et posteriori, et ita nihil motus[k] praesens est; sed quod prius est non est [l]nec quod posterius est; ergo motus non est.

3. Item, si motus esset, esset in mobili; [m]sed non est in mobili, ut videtur, quia mobile est in omni instanti temporis, quia, secundum Aristotelem, instans respondet mobili, et motus est in tempore et non in instanti; quare etc.

4. Item, secundum Aristotelem, motus est entis in potentia, et etiam motus est in potentia; sed quod est in potentia non est; quare etc.

5. Dicitur sic quod quaedam sunt quorum esse est permanens et partes sunt praesentialiter et quorum substantia facta est, alia sunt quorum esse est in fieri et successione. Ad esse autem rerum primo modo dictarum exigitur permanentia partium et quod partes sint in aliquo praesenti in actu; ad esse autem rerum secundo modo dictarum[n] non exigitur nisi successio partium ita

[a] Quoniam … mutationis *om.* T
[b] et ... mutationis] etc. C
[c] Quaeritur ... sit] In hoc libro tertio principaliter determinatur de motu et de his quae exiguntur ad motum, dans primo intentionem respectu motus. Sit ergo prima quaestio an sit motus C
[d] Quaeritur] Super tertium Physicorum *praem.* T
[e] punctum *om.* TC
[f] C] Arguitur (*lectio incerta*) sic *add.* C
[g] adhuc non] nondum adhuc O
[h] signetur ... arguatur] arguitur ubicumque signetur punctus in spatio C
[i] motus *scripsimus*; temporis OTC
[j] solum imaginantur] non possunt imaginari nisi C
[k] motus *scripsimus*; temporis OC; ipsis (*lectio incerta*) T
[l] nec ... sed non est (n. 3) *om. hom.* O
[m] sed ... mobili *om. hom. sed suppl. mg.* (*post* videtur) sed quod non sit in mobili videtur T
[n] dictarum *om.* O; *suppl. mg.* T

BOOK III

1*. *Since nature is a principle of motion and rest and change <and it is the subject of our inquiry, we must therefore see that we understand what motion is>.*[1]

2*. Questions about motion.

<Question 1>
The first question is whether motion exists.

1. First the arguments for the negative: let us mark off a first point, a middle point, and an end point in a space, and let us call the first point A, the middle point B, and the end point C. When a mobile thing is in B it is not in a part of motion because it is not in the past part, since the past part does not exist, nor is it in the future part, since the future part has not yet been generated. And if another region of space is marked off in the same way, the same argument holds. Therefore, the mobile thing will be in all parts of space, and yet in no part of motion, which would not happen if motion existed; hence, <motion does not exist>.

2. Furthermore, the parts of motion are only imagined as being prior and posterior; therefore, motion only has parts that are taken as prior and posterior, and thus nothing belonging to a motion is present; but what is prior does not exist nor does what is posterior; therefore, motion does not exist.

3. Furthermore, if motion existed, it would exist in a mobile thing; but it seems that motion does not exist in a mobile thing, because a mobile thing exists in every instant of time, since, according to Aristotle, an instant corresponds to a mobile thing,[2] and motion is in time and not in an instant; hence, <motion does not exist in a mobile thing>.

4. Furthermore, according to Aristotle, motion belongs to a being in potency,[3] and motion itself is in potency; but what is in potency does not exist; hence, <motion does not exist>.

5. It is said that (i) some things are such that their being is permanent, their parts exist in the present, and their substance is fully realized, whereas (ii) other things are such that their being is in becoming and in succession. The existence of things said to be of kind (i) requires the permanence of their parts and the existence of their parts in something present in act; but the existence of things said to be of kind (ii) only requires the succession of their

[1] Arist. *Phys.* III.1, 200b12–13.
[2] Arist. *Phys.* IV.11, 219b22–23.
[3] Arist. *Phys.* III.1, 201a10–11.

quod una pars sit praeterita et corrupta, alia autem pars futura et generanda. [a]Et motus est aliquid tale in fieri.

6. Contra: si motus est huiusmodi cuius pars aliqua praeterita est et alia futura et nihil[b] in actu, tunc sic: cum nihil sit praeteritum nisi quod aliquando fuit praesens[c] in actu, si aliqua pars motus est praeterita, illa pars motus aliquando fuit praesens in actu; sed hoc non contingit quia numquam fuit aliquod tempus praesens in actu in quo posset motus fieri, sed solum instans, [d]in quo non est motus.

7. Item, si [e]motus est tantum [f]in successione et in fieri, cum cuiuslibet fieri sit motus, tunc motus erit motus.

8. Item, si motus est, aliquando incepit esse; sed hoc non contingit, quia quod incepit esse transmutatur a non esse in esse, et quod exit[g] a non esse in esse aut exit motu aut mutatione; si motus exivit in esse per motum, erit eadem quaestio de motu illo; si mutatione, illa mutatio exivit in esse. Aut ergo motu aut mutatione, et sic in infinitum. Aut standum est in primo, et dicendum quod non exivit in esse, et ita non est. Sic enim arguit Aristoteles in principio octavi.

9. Item, si motus non sit [h]nisi in successione et in[i] fieri propter hoc quod in motu est accessus ad terminum et semper acceptio partis post partem, tunc cum accessui ad terminum respondet recessus a termino et acceptioni[j] partis post partem respondeat corruptio partium praecedentium, qua ergo ratione est successivum et in fieri [k]est et corruptivum et recessivum et[l] non in fieri.

10. Solutio[m]. Dicendum quod motus est, sed distingue[n] esse sicut prius, scilicet quod quaedam sunt[o] ut substantia facta et quaedam ut in fieri et successione.

11. [p]Argumentis in contrarium dicendum.

[a] Et ... tale] motus autem est huiusmodi quia eius esse est C
[b] nihil] nulla pars C
[c] praesens] prius OC (*ante* fuit C); *corr. mg. ex* penes T
[d] in quo] sed in instanti C
[e] motus est tantum] motus tantum esset *scrips. sed* tantum *del.* T
[f] in … quod (n. 8) *om. sed* incepit esse sed quod *suppl. mg*. T
[g] exit] in esse *add. sed del.* O
[h] nisi in] sine O
[i] in *scrips. sed del.* O
[j] acceptioni] receptioni O; *corr. s.l. ex* recessio T
[k] est et *suppl. s.l.* O; et TC
[l] et *suppl. s.l.* O
[m] Solutio *om.* TC
[n] distingue] distinguo TC
[o] sunt] sive habent esse *add.* C
[p] Argumentis ... dicendum *om.* C

parts, so that one part is past and corrupted whereas another part is future and to be generated. And motion is something in becoming of that sort.

6. Against this: if motion is such that one of its parts is past and another is future and nothing of it is in act, then, since nothing is past unless it was at some time present in act, if a part of motion is past, that part of motion was at some time present in act; but this does not happen because there was never a time present in act in which motion could occur, but only an instant, in which there is no motion.

7. Furthermore, if motion is only in succession and in becoming, since of any becoming there is a motion, then there will be a motion of a motion.

8. Furthermore, if motion exists, it began to exist at some time; but this is not possible, because what begins to exist passes from non-being into being, and what comes from non-being into being does that either by a motion or by a sudden change. If motion came into being by a motion, the same question will arise about that motion. If by a sudden change, that change also came into being; therefore<it did so> either by a motion or by a sudden change, and so on to infinity. Alternatively, one must stop at the first step and say that motion did not come into being and thus it does not exist. Indeed, Aristotle argues in this way at the beginning of *Physics* VIII.[4]

9. Furthermore, if it is because of the fact that in motion there is an accession to a terminus and one part is always taken after another that motion exists only in succession and in becoming, then since to an accession to a terminus there corresponds a recession from a terminus and to the taking of one part there corresponds the corruption of the preceding parts, therefore, for the same reason by which motion is successive and in becoming it is also corruptive and recessive and not in becoming.

10. Solution. It must be said that motion exists, but one must make the same distinction as before[5] and say that some things exist as fully realized substances and others exist in becoming and in succession.

11. Reply to the arguments to the contrary:

[4] Arist. *Phys*. VIII.1, 251a17–20.
[5] See above, n. 5.

12. [a]Et primo argumento sic: sicut in tempore est ita[b] quod non est aliqua pars sui praesens sed instans, sic est in motu quod non est aliqua pars motus praesens sed tantum mutatio quae respondet instanti. Unde cum dicit 'illud quod praeteritum est aliquando fuit praesens', hoc est verum in illis quae se habent ut substantia facta, cuiusmodi sunt permanentia.

13. Sed qualiter potest esse quod motus sit, et tamen non potest aliqua eius pars esse, dicendum sicut dicit Aristoteles de tempore quia, secundum ipsum, instans fluens facit tempus et non est tempus directa praedicatione, est tamen tota substantia temporis quia, secundum ipsum, instans replicatum per diversa facit tempus. Unde licet tempus secundum partes suas non fuerit praesens, est tamen praesens in sua substantia ut in instanti replicante se. Unde licet temporis continuitas non fuerit praesens sive ipsum tempus continuum, tamen[c] fluxus eius continuus super idem esse instantis ut praesens, quod substantialiter defert totum tempus et est tota substantia temporis, est[d] praesens. Unde tota continuitas temporis est a continuitate illius fluxus continui super idem indivisibile ut instans[e], quod non est nihil temporis sed tota substantia eius. Unde isto modo ponitur de motu, scilicet quod fluxus motus transit per mutationes praesentes. Licet ergo continuitas motus non sit praesens, continuus tamen transitus eius est super aliquid, ut super mutationem, quae praesens est, et non est nihil ipsius motus, sed forte tota substantia ipsius motus.

14. [f]Argumento secundo dicendum[g], ut quidam dicunt, quod motus est ipsa successio [h]et forma acquirenda est illud quod successive inducitur. Unde tempus mensurat successionem ipsam [i]et ita est eius quod est in fieri, ut formae, per se[j].

15. Vel sic: motus est in fieri, sed illud fieri non est per se fieri sui sed alterius, ut formae[k] inducendae. Et quia per se est fieri in formam, per accidens est fieri in fieri, et ita motus per se est in terminum, per accidens autem est motus in motum.

[a] Et ... sic] Ad primum in contrarium quod C
[b] ita *scrips. sed del.* O
[c] tamen *ex corr.* O
[d] est *suppl. mg.* O; *om.* TC
[e] instans] instantis OT
[f] Argumento ... quod] Ad secundum argumentum dico quod, secundum quosdam C
[g] dicendum] quod *add. sed del.* O
[h] et ... est] in formam acquirendam et O (in *corr. ex* et)
[i] et ... se *om.* C
[j] se] et hoc sequitur *add.* OT *sed del.* O
[k] formae] acquirendae vel *add.* C

12. To the first argument[6] it must be said that just as it happens in time that no part of time but only an instant is present, so it happens in motion that no part of motion but only the sudden change which corresponds to an instant, is present. Thus, when it is said 'that which is past was at some time present', this is true of the things that have being as a fully realized substance, and permanent things are of this kind.

13. To the question of how it can be that motion exists, even though no part of it can exist, it must be replied as Aristotle does about time, because, according to him, the flowing instant produces time; and yet it is not time by direct predication. It is, however, the whole substance of time because, according to Aristotle, the instant by being replicated through its different existences produces time.[7] Thus, although time is not present in its parts, it is however present in its substance, that is, in the instant that replicates itself. Thus, although the continuity of time or continuous time itself is not present, its continuous flowing over the same existence of the instant as present, which substantially carries time as a whole and is the whole substance of time, is present. Thus, the whole continuity of time derives from the continuity of that continuous flowing over the same indivisible, namely the instant, which is not nothing in relation to time but is its whole substance. Accordingly, the same kind of claim can be made about motion, namely, that the flowing of motion passes through the present sudden changes. Therefore, although the continuity of motion is not present, its continuous passage, however, occurs over something that is present, that is, the sudden change, and that is not nothing in relation to motion itself but is perhaps its whole substance.

14. To the second argument[8] it must be said, as some people say, that motion is that very succession and the form to be acquired is that which is successively induced. Thus, time measures that very succession, and so it belongs to the thing which is per se in becoming, namely, to the form.

15. Another reply is that motion exists in becoming, but that that becoming is not per se the becoming of motion itself but of something else, that is, of the form to be induced. And since the becoming is per se directed towards the form and accidentally towards the becoming, thus motion is per se directed towards its terminus and only accidentally towards a motion.

[6] See above, n. 6.
[7] Arist. *Phys*. IV.11, 219b9–33.
[8] See above, n. 7.

16. Ad tertium dicendum quod motus exivit in esse. Et **[O 109ra]** cum quaeritur: aut motu aut mutatione, dico quod mutatione. Et cum quaeritur[a] de illa [b]utrum alia mutatione exivit in esse, dicendum quod ipsa[c] se ipsa exivit in esse, et hoc quia motus sive mutatio sunt viae in formam, et viae non est via. Unde illa prima mutatio non exivit in esse per aliquid quod sit via eius sed per aliquid ut efficiens eius[d] tantum. Via enim[e] est in formam[f] per se, via non est viae nisi per accidens.

17. Ad ultimum dicendum quod, licet motus sit a termino in terminum, tamen per se est in terminum in quem, quia per se intenditur terminus in quem. Non enim fit recessus a termino a quo nisi[g] per se intendatur terminus in quem, et ita per accidens est a termino a quo, quia sic considerando motum, facit distare quod est. Unde quia comparando ad terminum in quem est[h] successivum et acceptivum et productio [i]eius quod non est, et illud est quod est intentum[j] essentialiter in motu, ideo ab illo fit tota denominatio. Et praeter hoc successivum commune recessivo et accessivo, nec opponitur successivum recessivo, sed accessivum opponitur recessivo.

1*. Consequenter quaeritur de hoc quod dicit in littera, scilicet[k] quod est loqui de infinito propter continuitatem in motu.

<Quaestio 2>
Et primo de causa[l].

1. Et videtur quod nulla sit sua causa quia quod repugnat infinito non est causa eiusdem; sed continuitas repugnat infinito; quare etc. Probatio minoris: Aristoteles improbat infinitum esse, et hoc per naturam corporis unde continuum est.

[a] quaeritur] quaerit TC
[b] utrum ... esse] mutatione C
[c] ipsa *om.* TC
[d] eius *om.* TC
[e] enim *corr. ex* non O; non C
[f] formam] forma O
[g] nisi] quia *add. s.l.* O
[h] est *ex corr.* O
[i] eius ... denominatio *om.* C
[j] intentum] in motu O
[k] scilicet *om.* TC
[l] causa *corr. ex* consequentia O; consequentia TC

16. To the third argument[9] it must be said that motion came into being. And when it is asked whether <this was> by a motion or by a sudden change, I reply that it was by a sudden change. And when it is asked about that sudden change whether it came into being by another sudden change, it must be said that that sudden change came into being by itself, and this happens because motion and sudden change are ways towards a form, and there is no way of a way. Thus, that first sudden change did not come into being through something which was a way towards it, but through something which was only its efficient cause; for a way is per se directed towards a form, and there is no way of a way except accidentally.

17. To the last argument[10] it must be said that, although motion is from a terminus towards a terminus, it is, however, per se towards the terminus towards which, because the terminus towards which is aimed at per se. The recession from the terminus from which only occurs if the terminus towards which is aimed at per se, and thus motion is accidentally from the terminus from which, given that motion considered in this way removes what exists. Thus, because motion related to the terminus towards which is successive and receiving and is the production of something that does not exist, and this is what is essentially aimed at in motion, the whole denomination derives from the terminus towards which. Moreover, being successive is common to being recessive and to being accessive, and being successive is not opposite to being recessive, but being accessive is opposite to being recessive.

1*. The next questions are about what Aristotle says in the text, namely that one needs to speak about the infinite because of the continuity in motion.[11]

<Question 2>
And first about the cause.

1. Arguments that the cause Aristotle postulates is nothing: what is inconsistent with the infinite is not its cause; but continuity is inconsistent with the infinite; hence, <continuity is not the cause of the infinite>. Proof of the minor premise: Aristotle disproves the existence of the infinite on the basis of the nature of body in so far as it is continuous.[12]

[9] See above, n. 8.
[10] See above, n. 9.
[11] Arist. *Phys.* III.1, 200b16–20.
[12] Arist. *Phys.* III.5, 204b1–7.

2. Item, omne continuum est totum, et totum unde totum repugnat infinito; quare etc.

3. Solutio[a]. Dicendum quod bonam causam dat secundum hoc quod intendit de infinito, quia est acceptio infiniti dupliciter: aut actu aut potentia. Si potentia, hoc potest esse vel in appositione[b] numeri vel in[c] divisione continui. Sed hoc est dupliciter: vel continui permanentis vel continui successivi. [d]Unde Aristoteles: quia motus est continuus, respondet ei infinitum in potentia secundum divisionem, et hoc semper successive accipiendo partem post partem. Unde infinitas in actu repugnat continuitati.

<Quaestio 3>

Consequenter [e]quaeritur utrum omnis motus sit continuus.

1. Et videtur quod non, quia, si sic, tunc, cum[f] omne continuum sit quantum, omnis [g]motus esset quantus, et quantus[h] non est nisi[i] ad quantitatem; ergo omnis motus est in quantitate, et ita generatio et alteratio non sunt motus.

2. Item, si sit quantus, et mathematici est considerare de quantitate et non naturalis, [j]quare etc.

3. Item, si[k] continuitas sit in motu, aut ergo a continuitate mobilis aut a continuitate spatii. Non mobilis, quia tunc quantitas motus esset a quantitate mobilis, et ita maior est quantitas motus magni lapidis quam parvi, si simul inciperent et simul terminarentur. Nec a continuitate spatii, cum multi sunt motus qui non [l]a spatio continuantur[m] nec in spatio sunt, quia[n] sola loci mutatio fit in spatio.

4. Item, potest probari quod motus est indivisibilis per argumenta habita.

5. Solutio[o]. Dicendum quod motus est continuus. Unde [p]dico quod generalis causa continuitatis motus est continuitas in gradibus inter terminum

[a] Solutio *om.* TC
[b] appositione] potentia O
[c] in *om.* O
[d] Unde Aristoteles] vult ergo Aristoteles quod C
[e] quaeritur utrum] de hoc quod vult quod C
[f] cum *om.* O; *suppl. s.l.* T
[g] motus esset *inv. sed corr.* O
[h] quantus] quantum TC
[i] nisi] respectu (*lectio incerta*) *add. s.l.* O
[j] quare etc.] ergo nec de motu C
[k] si *suppl. s.l.* O; quantitas vel *add.* C
[l] a ... nec *om.* TC
[m] continuantur *scripsimus*; continentur O
[n] quia *ex corr.* O
[o] Solutio *om.* TC
[p] dico quod *om.* TC

2. Furthermore, every continuum is a whole, and the whole in so far as it is a whole is inconsistent with the infinite; hence, <the continuum is inconsistent with the infinite>.

3. Solution. It must be said that Aristotle assigns a good cause, given his understanding of the infinite; for the infinite can be taken either (i) in act or (ii) in potency. Case (ii) is twofold: either (iia) by the addition of number or (iib) by the division of a continuum. Case (iib) is twofold: either (iib') by the division of a permanent continuum or (iib") by that of a successive continuum. Thus, according to Aristotle, it is because motion is continuous that the infinite in potency by division corresponds to it, that is, if one always takes one part after another successively.[13] Thus, infinity in act is inconsistent with continuity.

<Question 3>
The next question is whether every motion is continuous.

1. Arguments for the negative: if this is the case, then since every continuum is a quantity, every motion would be a quantity; and only that motion which is directed towards a quantity is a quantity; therefore, every motion is in the category of quantity, and thus generation and alteration are not motions.

2. Furthermore, if motion is a quantity, and the discussion of quantity pertains to the mathematician and not to the natural philosopher, then <the discussion of motion does not pertain to the natural philosopher>.

3. Furthermore, if continuity exists in motion, it derives either (i) from the continuity of a mobile thing or (ii) from the continuity of space. Not (i), because then the quantity of motion would derive from the quantity of the mobile thing, and thus the quantity of the motion of a big stone would be greater than that of the motion of a small stone, even if these two motions began and ended simultaneously. Not (ii) either, because there are many motions that are not continuous in virtue of space nor are in space, since it is only locomotion that occurs in space.

4. Furthermore, by the arguments already given[14] it can be proved that motion is indivisible.

5. Solution. It must be said that motion is continuous. Thus, I say that the general cause of the continuity of motion is the continuity of stages between the terminus from which and the terminus towards which. Accordingly,

[13] Arist. *Phys*. III.6, 206a25–b3.
[14] See above, nn. 1–3.

a quo et terminum in quem. Unde licet in loci mutatione sit motus super spatium et continuitas motus a continuitate spatii, non tamen ascribitur continuitas motus spatio absolute, sed in comparatione ad gradus inter terminum a quo et terminum in quem. Si autem debeat fieri motus ab albedine in nigredinem ita quod debeat induci nigredo, continuitas graduum[a] inter terminos est causa continuitatis totius motus; nec est maior quantitas motus sive maius corpus alteretur sive minus, dummodo simul incipiant et terminentur simul. [b]Possem enim dicere: in loci mutatione quantum est tempus tantus est motus, et similiter quantum est spatium tantus est motus, et hoc quia quantum est tempus et duratio motus, quia tempus mensurat[c] eius durationem[d], spatium autem mensurat substantiam motus, et ideo quantum est spatium etc. Si autem in eodem tempore pertranseatur maius spatium et minus, maior est continuitas in motu transeunte maius spatium quam minus.

<Quaestio 4>

Consequenter [e]quaeritur in quo genere sit motus sive in quo praedicamento.

1. Et dicit Aristoteles in hoc tertio quoniam in quattuor.

2. Contra: in quinto huius dicit quod [f]non est nisi tantum in tribus[g], scilicet qualitate, quantitate et 'ubi'.

3. [h]Item, Aristoteles in *Praedicamentis*: "Motus est quantus per accidens quia tempus est quantum per se".

4. Item, Aristoteles in quinto[i] *Metaphysicae*: "Tempus est quantum per accidens[j], motus autem per se[k]".

5. Item, Commentator dicit super quintum huius quod Aristoteles in *Praedicamentis* dicit quod motus est in praedicamento passionis.

6. Item, motus est idem cum actione et passione; ergo est in genere cum utroque.

7. Item, auctor *Sex principiorum*: "Motus est quale[l]".

[a] graduum *scripsimus*; gradus OTC
[b] Possem enim dicere *om.* C
[c] mensurat] motum et *add.* C
[d] durationem *ex corr.* O
[e] quaeritur ... quattuor (n.1)] quia Aristoteles ponit hic motum esse in quattuor generibus et si hoc quattuor esse species motus quaeritur primo in quo genere sive in quo praedicamento sit; secundo utrum species motus <quas> enumerat Aristoteles sufficiant. De primo Aristoteles dicit quod motus est in quattuor generibus C
[f] non est nisi] est TC
[g] tribus] generibus *add.* TC
[h] Item ... se *om.* T
[i] quinto *ex corr.* O
[j] accidens *corr. ex* se O; se T
[k] se *corr. s.l. ex* accidens O; accidens T
[l] quale] aequale O

although in the case of locomotion motion takes place in space and the continuity of motion derives from the continuity of space, this continuity, however, is ascribed to space not absolutely but in relation to the stages between the terminus from which and the terminus towards which. But if there must be a motion from whiteness towards blackness in such a way that blackness must be induced, the continuity of the stages between the termini is the cause of the continuity of the whole motion. Also, the quantity of a motion is not greater when a greater or a smaller body is altered, as long as their motions begin and end simultaneously; for I can say that in the case of locomotion the amount of motion is the same as the amount of time, and similarly the amount of that motion is the same as the amount of space. And the reason for this is that the duration of a motion is the same as the amount of time, since time measures the duration of motion, whereas space measures the substance of motion, and therefore <the amount of motion is the same as> the amount of space. If, however, a greater and a smaller space are traversed in the same time, then the continuity of the motion traversing the greater space is greater than that of the motion traversing the smaller space.

<Question 4>
The next question is in what genus or in what category motion is.

1. And in *Physics* III Aristotle says that motion is in four genera.[15]

2. Against this: in *Physics* V Aristotle says that motion is only in three genera, that is, in quality, in quantity, and in 'where'.[16]

3. Furthermore, in the *Categories* <he says>: "Motion is a quantity accidentally, because time is a quantity per se".[17]

4. Furthermore, in *Metaphysics* V <he says>: "Time is a quantity accidentally, whereas motion is a quantity per se".[18]

5. Furthermore, in his comment on *Physics* V the Commentator says that in the *Categories* Aristotle says that motion is in the category of passion.[19]

6. Furthermore, motion is the same as action and passion; therefore, it is in the same genus with each of them.

7. Furthermore, the author of the book of *Six principles* <says>: "Motion is a quality".[20]

[15] Arist. *Phys.* III.1, 200b32–34.
[16] Arist. *Phys.* V.1, 225b5–9.
[17] Arist. *Cat.* 6, 5a37–b4.
[18] Arist. *Metaph.* V.13, 1020a29–32.
[19] Aver. *In Phys.* V, t.c. 9, fol. 214vb; Arist. *Cat.* 9, 11b1–8.
[20] Anon. *Sex princ.* 2, AL I.6–7, 39.18.

8. Item, in fine *Praedicamentorum*: "Sex sunt species motus", et sequitur ad omnes et ita est in omnibus.

9. Item, Aristoteles in principio octavi: "Movens et motum[a] sunt relativa", et ita in genere relationis.

10. Solutio[b]. Nos possumus loqui communiter de motu extendendo motum ad omnem transmutationem et mutationem, et hoc prout est via in aliquid[c] **[O 109rb]** per se vel per accidens. Et sic est[d] sequens omnia praedicamenta, per se tamen est in substantiam, qualitatem, quantitatem, et 'ubi'. Sic ergo potest motus poni in omni praedicamento. Si autem proprie loquamur de motu et loquamur de generatione, tunc potest ipsa generatio nominare mutationem ultimam et non transmutationem[e], sic non est motus. [f]Si autem non nominet solum mutationem, sed cum hoc omnes dispositiones praecedentes, sic est motus in substantiam. Possumus etiam considerare motum aut ut est via in rem praedicamenti aut ut est species in genere. [g]Si ut est via in rem praedicamenti, tunc[h] quia[i] est via in[j] 'ubi[k]' ibi ponitur. Si ut est species in genere, tunc potest comparari ad mobile cuius est denominativum et qualificativum[l], et sic est in praedicamento qualitatis. Si autem comparetur ad distantiam a qua recipit quantitatem, sic est in praedicamento quantitatis, et isto modo ponitur in tribus praedicamentis. Et sic[m] patet qualiter habet poni in tribus et qualiter in quattuor.

11. Ad hoc quod Aristoteles in *Praedicamentis* dicit quod motus est quantus per accidens quia tempus per se, dicendum quod logicus sumit quantitatem ut mensura est, et tempus est mensura motus. Propter hoc dicit logicus quod tempus est quantum per se, motus autem[n] per accidens. [o]Considerando autem continuitatem et quantitatem in se et non [p]in ratione mensurae, sic dicit Aristoteles [q]in quinto *Metaphysicae* quod motus est quantus per se et tempus per accidens.

[a] motum] mobile TC
[b] Solutio *om.* TC
[c] aliquid] aliud OT
[d] est] accidens et sic est *add.* C
[e] transmutationem] et *add.* TC
[f] Si ... praecedentes] vel potest nominare ipsam mutationem cum omnibus dispositionibus et transmutationibus praecedentibus et C
[g] Si ... genere *om. hom.* T
[h] tunc *lectio incerta* (*ex corr.*) O
[i] quia *om.* O
[j] in *om.* O
[k] ubi *ex corr.* O
[l] qualificativum *ex corr.* O
[m] sic] habet *add. sed del.* O
[n] autem] aut O
[o] Considerando autem] quia autem metaphysicus considerat C
[p] in ratione] quantitatem O; ratione T
[q] in quinto Metaphysicae *om.* C

8. Furthermore, at the end of the *Categories* <Aristotle says>: "There are six species of motion",[21] and motion follows from all categories and so it is in all of them.

9. Furthermore, at the beginning of *Physics* VIII Aristotle <says>: "The mover and the moved are relative",[22] and thus motion is in the genus of relation.

10. Solution. (i) We can speak of motion in a broad sense by extending motion to every transmutation and sudden change, and this in so far as motion is a way towards something either per se or accidentally. In this way motion follows from all the categories, but it exists per se in the categories of substance, of quality, of quantity, and of 'where'. Therefore, in accordance with this way, motion can be placed in every category. (ii) But if we speak of motion in the proper sense, and we take the case of generation, then generation can designate the last sudden change and not the transmutation, and in this way it is not a motion. However, if generation designates not only the sudden change but along with this all the preceding dispositions as well, in this way it is a motion towards a substance. We can also consider motion either (a) as a way towards a thing belonging to a category or (b) as a species of a genus. In case (a), then if a motion is a way towards a 'where', it is placed in that category. In case (b), then motion can be taken in relation (b') to the mobile thing, which it denominates and qualifies, and in this way motion is in the category of quality. (b") If motion is taken in relation to the distance from which it receives its quantity, then it is in the category of quantity. In accordance with this way, motion is placed in three categories. It is then clear in what way motion is to be placed in three categories and in what way in four.

11. In reply to Aristotle's statement in the *Categories* that motion is a quantity accidentally because time is a quantity per se,[23] it must be said that the logician considers quantity in so far as it is a measure, and time is the measure of motion. And this is why the logician says that time is a quantity per se, whereas motion is a quantity accidentally. But if we consider continuity and quantity in themselves and not as types of measure, it is in this way that Aristotle says in *Metaphysics* V that motion is a quantity per se and time is a quantity accidentally.[24]

[21] Arist. *Cat.* 14, 15a13–14.
[22] Arist. *Phys.* VIII.1, 251b5–9.
[23] See above, n. 3.
[24] See above, n. 4.

12. Aliter exponit[a] Commentator dicens quod motus potest considerari ut est via in formam, et sic est in praedicamento passionis. Et dicit quod iste est modus eius famosus. Vel potest considerari motus secundum quod non differt[b] a forma in quam vadit nisi secundum magis[c] et minus. Et sic sicut sunt quattuor formae per se in quibus est motus, et[d] sic habet poni in quattuor generibus.

<Quaestio 5>

Propter hoc quod Commentator ponit hic quod motus sit ipsa forma inducenda non differens [e]ab ipsa nisi secundum magis[f] et minus, quaeritur utrum motus sit illa forma.

1. Et videtur quod non, quia si sic, cum tota forma sit prius[g] in potentia[h] et tota secundum gradus completionis et incompletionis vadit ad actum, tunc totus motus prius erit in actu et totus postea magis in actu, quod est contra rationem successivi, quia successivum semper acquirit partem post partem, et non est totum[i] prius in actu minus et postea totum magis in actu.

2. Item, si ita esset ut dicit Commentator, tunc cum ex illo incompleto fiat forma in actu et non per illud incompletum, ita erit in motu quod ex illo motu fiet forma et non per illum motum, quod[j] scimus esse falsum, quia per motum acquiritur forma.

3. Item, sicut se habet forma in actu ad suum movere, sic forma in potentia ad suum moveri; sed forma in actu differt a suo movere ita quod sunt in diversis praedicamentis; ergo [k]ita erit de forma in potentia ad suum moveri quod non erunt in eodem praedicamento.

4. Contra: quod est terminus et finis alicuius[l] ponit rem in specie, ut patet descendendo a subalterno in subalternum; sed forma est finis et terminus motus; quare etc.

[a] exponit] respondet TC
[b] differt *corr. ex* dicitur OT
[c] magis] maius TC
[d] et *om.* TC
[e] ab ipsa *om.* OT
[f] magis] maius TC
[g] prius] potentia *add. sed del.* O
[h] potentia *ex corr.* O
[i] totum *om.* O
[j] quod *ex corr.* O
[k] ita ... moveri] et forma in potentia differt a suo movere ita C
[l] alicuius *lectio incerta* (*ex corr.*) O

12. The Commentator gives another explanation saying that motion can be considered either (i) as a way towards a form, and in this way motion is in the category of passion. And he says that this is the well-known way of considering motion. Or (ii) motion can be considered in so far as it only differs from the form towards which it is directed in accordance with more and less. And in this way there are four forms per se in which motion exists, and in this way motion is to be placed in four genera.[25]

<Question 5>
Given that the Commentator posits here that a motion is the form itself to be induced, only differing from that form by more and less,[26] the question arises whether a motion is that form.

1. Arguments for the negative: if it is, then since a whole form is first in potency and subsequently moves towards act in accordance with degrees of completeness and incompleteness, a whole motion will first be in act and subsequently will be even more in act, which is contrary to the nature of a successive thing, because a successive thing always acquires one part after another, and is not a whole which is first less in act and then more in act.

2. Furthermore, if it were as the Commentator says, then since a form becomes in act by starting from an incomplete thing and not through that incomplete thing, the same will apply to motion, so that a form will come to be by starting from a motion and not through that motion, which we know to be false, since a form is acquired through a motion.

3. Furthermore, in the way that a form in act is related to its causing motion, so a form in potency is related to its being moved; but a form in act differs from its causing motion in such a way that they are in different categories; therefore, the same will apply to a form in potency and its being moved, namely that they will not be in the same category.

4. Against this: whatever is the terminus and end of a thing places that thing in its species, as is clear when one descends from one subaltern genus to another; but a form is the end and the terminus of a motion; hence, <it places a motion in its species>.

[25] Aver. *In Phys.* III, t.c. 4, fol. 87ra–b.
[26] See above, q. 4, n. 12.

5. Item, essentiale est motui ut compleatur et perficiatur ad terminum in quem; sed quod est terminus essentialis et completivus[a] ponit rem in specie; quare etc.

6. Solutio[b]. Dicendum, ut mihi videtur, quod Commentator aut aliter intellexit quam expressit aut falsum intellexit.

7. Unde [c]primae rationi [d]pro ipso dicendum quod est duplex finis, scilicet essentialis et accidentalis. Et finis essentialis ponit rem in specie, finis autem accidentalis non. Unde finis essentialis motus est mutatio, finis autem accidentalis est ipsa forma. Et est huius simile illud quod dicit Aristoteles quod puncti et lineae sunt termini materiae, et ipsa forma est terminus materiae, et constat quod non univoce, quia puncti et lineae et[e] superficies sunt termini accidentales, et ideo non ponunt materiam in specie, forma autem est terminus eius essentialis, et ideo ponit ipsam in specie[f].

8. Ad secundum dicendum quod aliquid esse essentiale alteri contingit dupliciter. Vel ita essentiale ut non derelinquens[g] ipsum, et sic est forma essentialis motui, et tale essentiale non ponit rem in specie, sicut dicimus quod per se passio est essentiale subiecto suo, et tamen non ponit ipsum in specie. Vel ita essentiale quod sit ingrediens definitionem ipsius et pars suae quidditatis et essentiae, et isto modo ipsum[h] ponit in specie, [i]primo modo non, nec sic est essentiale.

9. Prius positum est quod, si motus comparetur ad mobile ut ipsum qualificans, sic est ut species in genere qualitatis. Arguo tunc sic. Transmutetur[j] Socrates ab albedine in nigredinem. Ille motus habet duplicem comparationem, scilicet ad mobile et ad terminum motus, et adeo essentialiter comparatur ad terminum motus sicut ad mobile. Si ergo per respectum quem habet motus ad mobile [k]possit poni species in praedicamento, tunc per respectum quem habet ad terminum poterit poni aliqua species in genere.

[a] completivus *lectio incerta* OC *sed corr. in* completio rei (*lectio incerta*) O
[b] Solutio *om.* TC
[c] primae ... ipso *om.* C
[d] pro ipso *om.* T
[e] et *om.* OC
[f] specie] Iam patet ad primum argumentum. Patet etiam qualiter salvari potest dictum Commentatoris *add.* C
[g] derelinquens] derelinquat (*ex corr.*) O
[h] ipsum *suppl. mg.* O; *om.* T
[i] primo ... essentiale *om.* C
[j] Transmutetur] transmutatur OC
[k] possit ... praedicamento] ponatur in praedicamento ut species in genere C

5. Furthermore, it is essential to a motion that it should be completed and perfected in the terminus towards which; but whatever is the essential terminus and the one that gives completion places that thing in its species; hence, <the terminus towards which places a motion in its species>.

6. Solution. In my view, it must be said that the Commentator either understood this matter differently from the way he presented it or he understood it falsely.

7. Thus, to the first argument in favour of the Commentator[27] it must be said that there are two kinds of end: (i) an essential end and (ii) an accidental end. And an essential end places a thing in its species, whereas an accidental end does not. Thus, the essential end of a motion is a sudden change, whereas its accidental end is the form itself. And similar to this distinction is Aristotle's statement that points and lines are the termini of matter,[28] and form itself is the terminus of matter, but clearly these things are not univocally termini, because points, lines, and surfaces are accidental termini, and therefore they do not place matter in its species, whereas a form is matter's essential terminus, and therefore it places matter in its species.

8. To the second argument[29] it must be said that something can be essential to another thing in two ways. Either (i) it is essential in such a way that it does not leave that thing; and in this way a form is essential to motion; and what is essential in this way does not place a thing in its species, as when we say that a per se property is essential to its subject, yet it does not place it in its species. Or (ii) something is essential to another thing in such a way that it is included in the definition of that thing and is part of its quiddity and essence; and what is essential in this way places that thing in its species, but what is essential in the first way does not, nor is it essential in this second way.

9. It has earlier been posited that, if motion is taken in relation to a mobile thing as something that qualifies the mobile thing, it is a species in the genus of quality.[30] I then argue as follows: suppose that Socrates is changed from whiteness to blackness. That motion is taken in relation to two things: to the mobile thing and to the terminus of the motion; and it is taken in relation to the terminus of the motion as essentially as it is taken in relation to the mobile thing. Therefore, if a <type of> motion can be placed in a category as a species in virtue of its relation to a mobile thing, then it can also be placed in a category as a species in virtue of its relation to a terminus.

[27] See above, n. 4.
[28] Arist. *De gen. et corr.* I.5, 320b16–17.
[29] See above, n. 5.
[30] See above, q. 4, n. 10.

10. Ad illud dicendum quod illud quod ponitur in praedicamento aut [a]ponitur secundum quod habet rationem substantiae aut secundum quod habet rationem accidentis. Si ergo motus debeat poni in praedicamento, hoc erit secundum quod habet rationem accidentis; sed in comparatione ad subiectum[b] habet rationem accidentis, et ita in comparatione ad illud[c] ponitur in genere; in comparatione autem[d] ad terminum in quem non capit rationem accidentis **[O 109va]** sed tantum rationem viae; ergo[e] in comparatione ad ipsum non ponitur in praedicamento.

<Quaestio 6 >

Quaeritur consequenter [f]utrum sufficiant species enumeratae ab Aristotele.

1. Et videtur quod non, quia rarefactio et condensatio motus sunt, et tamen non sunt sub speciebus motus praeenumeratis, quia non sub generatione, [g]quia idem manens sub specie sua potest rarefieri et condensari. Nec est[h] augmentatio quia augmentatio non fit proprie nisi in animatis, et etiam augmentatio fit aliquo adveniente, condensatio fit nullo [i]recedente, et rarefactio nullo adveniente. Nec alteratio, quia alteratio est respectu qualitatum diversarum, [j]haec autem est in maiorem locum tantum et non ratione qualitatum aliquarum. Nec est motus localis, quia, si sic, aut est rectus aut circularis. [k]Non rectus, quia non est tantum [l]a medio vel ad medium, sed[m] in omnem partem[n] per lineam rectam. Non circularis, cum talis motus sit caelo proprius et supracaelestibus, [o]et etiam est in omnem partem secundum lineam rectam a centro corporis rarefaciendi[p] ad extimum eiusdem.

2. Item, motus nutritionis est motus et non est sub aliqua specie numerata. Non generatio nec augmentatio, quia nutritio dividitur per oppositum contra ipsa, cum sit tertia operatio vegetativae divisa contra illas operationes. Nec alteratio nec loci mutatio, cum sit tantum circa partes materiae.

[a] ponitur ... accidentis] in ratione substantiae aut accidentis C
[b] subiectum *corr. ex* substantiam O
[c] illud] illum O
[d] autem *suppl. s.l.* O; *om.* T
[e] ergo] ideo TC
[f] utrum ... Aristotele] de secundo, scilicet an species motus enumeratae ab Aristotele sufficiant C
[g] quia idem] eadem *et* enim forma *add. s.l.* O
[h] est] argumentatio *add. sed del.* O
[i] recedente ... adveniente] adveniente vel recedente et rarefactio similiter C
[j] haec ... tantum] rarefactio autem et condensatio fiunt in maiorem locum vel minorem C
[k] Non ... tantum] rectus est C
[l] a ... talis] ad T
[m] sed] rarefactio est C
[n] partem] indifferenter *add.* C
[o] et ... eiusdem *om.* C
[p] rarefaciendi *ex corr.* O

10. To this objection[31] it must be said that something is placed in a category either because it has the *ratio* of a substance or because it has the *ratio* of an accident. Therefore, if a motion must be placed in a category, this will be because it has the *ratio* of an accident; but a motion has the *ratio* of an accident when it is taken in relation to its subject, and thus a motion is placed in a genus in so far as it is taken in relation to its subject; in so far as a motion is taken in relation to a terminus towards which, however, a motion does not receive the *ratio* of an accident, but only the *ratio* of a way; therefore, a motion is not placed in a category in so far as it is taken in relation to a terminus.

<Question 6>
The next question is whether the species listed by Aristotle are sufficient.[32]

1. Arguments for the negative: rarefaction and condensation are motions, but do not fall under any of the species of motion listed above, because they do not fall under generation, given that one and the same thing persisting in its species can be both rarefied and condensed. Nor are they an increase, because increase properly occurs only in animated things; moreover, increase occurs through the acquisition of something, but condensation occurs without the loss of anything and rarefaction without the acquisition of anything. Nor are they an alteration, because alteration occurs with respect to different qualities, but rarefaction is only directed towards a greater place and does not occur because of any qualities. Nor are they a locomotion, because if so, they are either (i) a rectilinear locomotion or (ii) a circular one. Not (i), because rarefaction takes place not only from the middle or towards the middle but towards all parts in a straight line. Not (ii), since circular motion is proper to the heavens and to the celestial bodies, and because rarefaction occurs in all the parts in a straight line from the centre of the body to be rarefied to its boundary.

2. Furthermore, the motion of nutrition is a motion but does not fall under any of the species listed. Nutrition is not a generation or an increase, since nutrition is classified as opposite to generation and increase, since it is classified as the third operation of the vegetative soul in opposition to those two operations. Nor is nutrition an alteration or a locomotion, since it only occurs in the parts of matter.

[31] See above, n. 9.
[32] Arist. *Phys*. III.1, 200b32–34.

3. Contra[a], sicut[b] augmentatio ordinatur ad quantitatem rei alitae conservandam sic[c] nutritio ordinatur ad conservandum[d] substantiam rei alitae. Sicut ergo augmentatio ponitur in genere quantitatis[e] sive in specie motus, sic nutritio ponetur in praedicamento substantiae sive in aliqua specie motus.

4. Ad primum dicitur quod rarefactio et condensatio non merentur nomen motus, et hoc quia omnis motus proprie dictus est in aliquam formam sive nominat viam in formam; sed rarefactio et condensatio sunt tantum circa partes [f]materiae; [g]quare etc.

5. Et consimiliter dicunt de nutritione quod illud proprie nutritur quod continue deperditur, et haec est caro secundum materiam.

6. Contra, secundum Aristotelem, illud quod augetur manet, et illud quod nutritur similiter; sed solum caro secundum speciem manet; ergo solum caro secundum speciem est illud quod augeri debet, et [h]ita non tantum caro secundum materiam.

7. Alia est instantia Aristotelis libro *De anima*. Dicit enim[i] quod est motus ad animam et est motus ab anima. Motus ad animam ut[j] motus specierum sensibilium[k], ab anima sicut in somniando simulacra[l] movent animam. Sed motus ad animam sub nulla specie motus continetur, quia omnis motus est in rem alicuius praedicamenti, sed motus ad animam [m]non est in rem praedicamenti [n]quia in anima non recipitur res praedicamenti sed intentio sola.

8. Item, motus specierum extra sua obiecta est motus, secundum quod dicit Aristoteles in *De sensu et sensato* quod quaedam sensibilia movent suum sensum[o] subito, sicut obiectum visus, et quaedam successive, sicut sensibilia omnium aliorum[p] sensuum, quia omnia movent in tempore, ut patet de sono, quoniam prius est in una parte aeris quam in alia[q].

[a] Contra] item (*ex corr.*) O
[b] sicut *om.* O
[c] sic] sed O
[d] conservandum *corr. ex* considerandum O
[e] quantitatis *ex corr.* O
[f] materiae; quare *ex corr.* O
[g] quare etc. *om.* C
[h] ita ... materiam] nutriri C
[i] enim *om.* TC
[j] ut] est TC
[k] sensibilium *ex corr.* O; motus *add.* C
[l] simulacra *ex corr.* O
[m] non est *suppl. s.l.* O
[n] quia ... intentio] immo intentione C
[o] sensum *scripsimus*; obiectum OC; obiecta T
[p] aliorum] aliarum O
[q] alia] responde *add.* C

3. Against this: just as increase is ordered to the conservation of the quantity of a nourished thing so nutrition is ordered to the conservation of the substance of a nourished thing. Therefore, just as increase is placed in the genus of quantity or in a species of motion, so nutrition will be placed in the category of substance or in some species of motion.

4. To the first argument[33] it is said that rarefaction and condensation do not deserve the name 'motion', and the reason for this is that every motion in the proper sense <of the term> is towards a form or designates the way towards a form; but rarefaction and condensation only occur in the parts of matter; hence, <they do not deserve the name 'motion'>.

5. And these people give a similar reply in the case of nutrition,[34] saying that the thing that is nourished in the proper sense is that which is continuously corrupted, and this thing is flesh with respect to its matter.

6. Against this: according to Aristotle, a thing that is increased persists, and similarly a thing that is nourished persists;[35] but only flesh in accordance with its species persists; therefore, only flesh in accordance with its species is that which should be increased, and thus <it is> not merely flesh in accordance with its matter.

7. Another counterexample is that of Aristotle in *On the Soul*; for he says that there is a motion towards the soul and a motion from the soul. For example, a motion towards the soul is that of the sensible species, and a motion from the soul is that produced by the images that move the soul in dreams.[36] Motion towards the soul, however, is not contained under any species of motion, because every motion is towards a thing belonging to a category, whereas a motion towards the soul is not towards a thing belonging to a category, because what is received in the soul is only an intention and not a thing belonging to a category.

8. Furthermore, the motion of species outside their objects is a motion, according to Aristotle's statement in *Sense and sensibilia* that some sensible things move the corresponding sense instantaneously, such as the object of sight, and some successively, such as the sensibles of all the other senses, because they all move in time, as is clear in the case of sound, given that a sound is in one part of air before it is in another.[37]

[33] See above, n. 1.

[34] See above, n. 2.

[35] Arist. *De gen. et corr.* I.5, 321a22–25.

[36] Arist. *De anima* I.4, 408b15–18.

[37] Arist. *De sensu* 6, 446a20–447a11.

1*. Consequenter quaeritur qualiter species motus inter se differant. Et [a]patet super primum *De anima*.

<Quaestio 7>
Quaeritur consequenter utrum motus possit intelligi.

1. Et videtur quod non, quia dicit Aristoteles in *Ethicis* quod operationes et motus sunt singularium et non universalium, et ita motus non est universale.

2. Item, motus est hic et nunc, et universale non; quare motus non est universale. Et quod non est universale non intelligitur, [b]quia singulare sentitur et non aliud, quia[c] secundum[d] Aristotelem, singulare dum sentitur et universale dum intelligitur. Et etiam intellectus agens facit universalitatem in rebus.

3. Item, Aristoteles: "Singulare imaginatur et non intelligitur[e]".

4. Item, intellectus possibilis intelligit per speciem similem rei extra; sed impossibile est quod motus habeat speciem sibi similem in anima. Et hoc quia species in anima est indivisibilis, quia est in substantia indivisibili; sed motus est divisibilis et cum successione; quare etc.

5. Item, species obiecti extra multiplicat se subito ab illo obiecto usque ad animam et sine omni successione. Quod ergo non est in instanti sed in tempore et successione solum non potest speciem in instanti facere sive subito usque ad animam; quare etc.

6. Item, nihil multiplicat speciem ad intellectum nisi illud quod praesens est et in actu, sicut color corruptus non facit speciem [f]nec color generandus[g] facit speciem, sed color solum in actu praesens. Cum igitur nulla pars motus sit praesens, sed una pars est[h] praeterita et corrupta et alia pars generanda et nondum generata [i]ita quod nulla pars praesens est, tunc motus non poterit facere speciem; quare etc.

7. Item, multiplicatio speciei ad animam fit in instanti; sed instanti in tempore respondet mutatio in motu. Si ergo aliquid faciat speciem usque ad

[a] patet ... anima] de hoc dictum est in primo De anima et ideo non hic C
[b] quia ... aliud] sed sentitur T; *om.* C
[c] quia] et O
[d] secundum] singulare *add. sed del.* O
[e] intelligitur] quare etc. *add.* C
[f] nec ... speciem *om. hom.* T
[g] generandus *corr. ex* generatus O; generatus C
[h] est *om.* TC
[i] ita ... tunc *om.* C

1*. The next question is in what way the species of motion differ one from the other. And this is made clear in the comment on *On the Soul* I.[38]

<Question 7>
The next question is whether motion can be understood.

1. Arguments for the negative: in the *Ethics* Aristotle says that operations and motions belong to singulars and not to universals,[39] and thus motion is not a universal.

2. Furthermore, motion is here and now, and a universal is not; hence, motion is not a universal. And what is not a universal is not understood, because a singular is sensed and not anything else, because, according to Aristotle, something is singular as long as it is sensed and universal as long as it is understood.[40] Moreover, the agent intellect is what produces universality in things.

3. Furthermore, Aristotle <says>: "The singular is imagined and is not understood".[41]

4. Furthermore, the possible intellect understands in virtue of the species which is similar to the thing outside; but it is impossible for motion to have a species similar to itself in the soul. The reason for this is that a species in the soul is indivisible, because it is in an indivisible substance; but motion is divisible and has succession; hence, <motion does not have a species similar to itself in the soul>.

5. Furthermore, the species of an external object multiplies itself suddenly from that object all the way to the soul and without any succession. Therefore, what does not exist in an instant but only in time and in succession cannot produce its species at an instant or suddenly all the way to the soul; hence, <motion does not produce its species in the soul>.

6. Furthermore, nothing multiplies its species all the way to the intellect except what is present and in act, just as a colour already corrupted does not produce its species nor does a colour to be generated, but only a colour that is present in act. Therefore, since no part of motion is present, for one part of it is past and corrupted and the other part is to be generated and has not yet been generated, so that no part of motion is present, it follows that motion cannot produce its species; hence, <motion cannot be understood>.

7. Furthermore, the multiplication of a species towards the soul occurs at an instant; but to an instant in time there corresponds a sudden change in motion. Therefore, if something <belonging to motion> produces its species

[38] Cf. Aspall, *Qq. in De anima* I, ed. Čizmić, 80–82.
[39] Arist. *Eth. Nic.* II.7, 1107a31.
[40] Arist. *Phys.* I.5, 189a5–7.
[41] Arist. *An. post.* I.18, 81b6–7

animam, hoc erit ipsa mutatio; sed in minimo motu sunt infinitae mutationes. Si ergo minimus motus debeat intelligi, intelligetur per species infinitas; sed hoc est impossibile; quare etc.

8. Ad oppositum: quod sensu comprehenditur ab intellectu intelligitur, [a]et hoc vel in se vel in suo universali; sed motus apprehenditur[b] sensu, cum sit sensibile commune[c]; quare etc.

9. Item, intelligentia intelligit motum, et hoc quia causat motum; quare etc.

10. Dicendum quod motus potest intelligi et etiam quod habet universale.

11. Unde primae rationi dicendum quod haec est falsa 'singulare non intelligitur', quia per se intelligibile est singulare, per accidens autem universale, quia propter defectum intellectus et etiam[d] propter eius impotentiam est quod nos non intelligimus nisi per universale. Summe enim intelligibile est summe singulare, sicut causa prima.

12. Item, quaelibet intel **[O 109vb]** ligentia cognoscit aliam intelligentiam, et hoc quoad eius omnes condiciones, et ita cognoscit singulare.

13. Item, causa prima cognoscit singulare [e]secundum quod singulare. Haec ergo propositio 'solum universale intelligitur' debet condicionari[f] a parte intellectus ita quod intelligatur de intellectu qui prius est in potentia et postea in actu, et hoc propter eius impotentiam. Eius autem impotentia causatur propter hoc quod unitur virtutibus materialibus, et ita non intelligit nisi per receptionem specierum[g] ab extra, et tunc per decursum rationis decurrit ad universale [h]et abstrahit universale. Unde cum dicit quod motus non est universale, falsum dicit, quia haec est vera 'iste motus est motus et ille motus est motus', et ita praedicatur motus de motibus particularibus sicut commune sive universale de suis particularibus.

14. Ad illud quod dicit quod operationes sunt circa singularia, hoc est verum per se. Sunt tamen circa universalia per accidens.

[a] et ... universali *om.* C
[b] apprehenditur *ex corr.* O
[c] commune *om.* O
[d] etiam *om.* TC
[e] secundum quod singulare *om. hom.* T
[f] condicionari *ex corr.* O
[g] specierum *ex corr.* O
[h] et abstrahit universale *om. hom.* T

all the way to the soul, this will be a sudden change; but in a minimal motion there are infinitely many sudden changes. Therefore, if a minimal motion is to be understood, it will be understood in virtue of infinitely many species; but this is impossible; hence, <motion cannot be understood>.

8. To the contrary: what is grasped by the senses is understood by the intellect, either in itself or in its universal; but motion is grasped by the senses, since it is a common sensible; hence, <motion is understood by the intellect>.

9. Furthermore, an intelligence understands motion, and the reason for this is that an intelligence causes motion; hence, <motion can be understood>.

10. It must be said that motion can be understood and also that it has a universal.

11. Thus, to the first argument[42] it must be said that the proposition 'the singular is not understood' is false because the singular is intelligible per se whereas the universal is intelligible accidentally, given that it is because of the defect of the intellect and its weakness that we only understand in virtue of the universal; for what is maximally intelligible is maximally singular, namely the first cause.

12. Furthermore, any intelligence knows another intelligence, and does this with respect to all its conditions, and thus it knows a singular.

13. Furthermore, the first cause knows a singular in so far as it is singular. Therefore, the proposition 'only the universal is understood' must be qualified on the side of the intellect so that it is understood as relating to the intellect that is first in potency and then in act, and this because of its weakness. But the weakness of the intellect is caused by the fact that the intellect is joined to material powers, and thus the only way the intellect understands is through the reception of species from outside, and then through a rational process by which the intellect reaches the universal and abstracts it. Therefore, when he says that motion is not a universal,[43] he says something false because the proposition 'this motion is motion and that motion is motion' is true, and thus motion is predicated of particular motions in the way that something common or universal is predicated of its particulars.

14. To the claim that operations are about singulars,[44] the reply is that this claim is true if it is understood per se; but operations are about universals accidentally.

[42] See above, nn. 2–3.
[43] See above, n. 2.
[44] See above, n. 1.

15. Ad aliud, cum dicit quod illud quod intelligitur per speciem sibi similem intelligitur, dicitur sic, sicut dicit Aristoteles in capitulo de tempore ubi docet comprehendere tempus, quod percipimus speciem mobilis et spatii, et cum hoc percipimus mobile prius esse in una parte spatii et postea in alia parte, et ita prius et posterius et etiam continuationem prioris et posterioris, et ita demum per decursum rationis percipimus tempus. Unde tempus et motus non percipiuntur speciebus propriis, sicut[a] permanentia, sed speciebus alienis, sicut specie mobilis et spatii.

16. Aliter dicitur sic quod motus percipitur per fluxum mutationis fluentis sicut tempus per fluxum instantis fluentis.

17. Per hoc patet aliis.

18. [bc]Ad idem in *De memoria et reminiscentia*[d] in illa quaestione 'Utrum [e]tempus sit sensibile'.

1*. [f]Quaeritur consequenter de definitione motus.

<Quaestio 8>
Et primo de prima parte, cum dicit "Motus est actus etc.".

1. Et videtur quod non [g]sit actus, quia omnis actus aut est actus primus aut secundus. Actus primus est forma sine actuali operatione, actus secundus est forma cum actuali operatione[h], et ita omnis[i] actus est forma; sed motus non est forma; quare etc.

2. Item, omnis actus terminat potentiam; sed solum forma terminat potentiam; ergo solum forma est actus; sed motus non est forma; ergo [j]nec actus.

3. Item, si motus est actus, et actus et potentia non diversificant essentiam, cum ergo mobile sit illud quod est in potentia, tunc motus erit forma substantialis mobilis et non diversum in essentia a mobili, quod est in potentia; sed hoc est falsum; ergo primum.

[a] sicut *scrips. sed del. et add.* ut O
[b] Ad ... sensibile *om.* C
[c] Ad idem *om.* T
[d] reminiscentia *ex corr.* O
[e] tempus sit sensibile] sensit tempus T
[f] Quaeritur ... etc.] Divisum autem unumquodque genus etc. Definit motum intendens istam definitionem: motus est actus entis in potentia secundum quod huiusmodi. Quaeratur ergo primo an motus sit actus C
[g] sit actus *om.* C
[h] operatione] consideratione O; vel consignatione *add.* T
[i] omnis *om.* O
[j] nec actus] etc. C

15. To the other argument, when he says that what is understood is understood through a species similar to it,[45] it is replied as Aristotle does in the chapter on time where he teaches how to grasp time, that is, that we perceive the species of a mobile thing and of space, and together with this we perceive that the mobile thing is first in one part of space and later in another part, and thus we perceive before and after and also the continuity of before and after, and thus we eventually perceive time through a rational process.[46] Accordingly, time and motion are not perceived by means of their own species, in the way that permanent things are perceived, but by means of extraneous species, namely the species of the mobile thing and of space.

16. Another reply is that motion is perceived through the flowing of a flowing sudden change in the way that time is perceived through the flowing of a flowing instant.

17. On this basis the reply to the other arguments[47] is clear.

18. On the same topic see the question 'Whether time is sensible' in *On Memory*.[48]

1*. Questions about the definition of motion.

<Question 8>
And the first question is about the first part of the definition, when Aristotle says "Motion is an act".[49]

1. Arguments that motion is not an act: every act is either a first act or a second act. A first act is a form without an actual operation, and a second act is a form with an actual operation, and thus every act is a form; but motion is not a form; hence, <motion is not an act>.

2. Furthermore, every act terminates a potency; but only a form terminates a potency; therefore, only a form is an act; but motion is not a form; therefore, motion is not an act either.

3. Furthermore, if motion is an act, and act and potency do not diversify an essence, then, since a mobile thing is that which is in potency, motion will be the substantial form of the mobile thing and not different in its essence from the mobile thing which is in potency; but this is false; therefore, the first premise <is also false>.

[45] See above, n. 4.
[46] Arist. *Phys*. IV.11, 219a22–b1.
[47] See above, nn. 5–7.
[48] A question with this title is found in the anonymous collection of questions on *De memoria et reminiscentia* preserved in MS N (at fol. 193rb). On Aspall's authorship of this work, see Introduction, p. XII.
[49] Arist. *Phys*. III.1, 201a10–11.

4. Dicendum quod actus rei est dupliciter. Aut perfectio rei ponens rem in specie; et talis actus est[a] forma, et sic non est motus actus. Alio modo est actus rei, hoc est rei operatio. Sed [b]hoc contingit dupliciter: vel ita quod complete sit operatio rei, et ita est motus actio; vel incomplete, et ita est motus passio, quia sic est in mobili.

5. Vel sic: mobile est in potentia duplici quia est in potentia ad formam et est in potentia ad[c] viam in[d] formam. Si loquamur de potentia primo modo dicta[e], sic est forma terminus potentiae, et talis[f] actus non est motus. Si loquamur de potentia secundo modo, sic non est forma terminus potentiae sed via in formam, et haec via est actus, [g]et ita est motus actus.

<Quaestio 9>

1. Quaeritur consequenter [h]super hoc quod dicit quod est actus entis in potentia secundum quod huiusmodi[i], quia, si descriptio convertitur cum descripto, tunc omnis[j] quod est in potentia secundum quod [k]est in potentia est motus. Consequens falsum; ergo antecedens.

2. Item, nullum[l] oppositum dicitur de suo opposito cum reduplicatione sui; [mn]sed in hac descriptione dicitur oppositum de suo opposito cum reduplicatione sui[o], ut cum dicitur 'motus est actus etc.'.

3. Item, si motus est entis in potentia, tunc eius quod est maxime in potentia maxime erit motus, et tale est materia prima in sua substantia; sed consequens falsum; ergo antecedens.

4. Solutio[p]. Dicendum quod omne illud quod movetur est in potentia duplici: vel in potentia respectu operationis aut respectu formae. Unde motus est [q]entis in potentia sive imperfecti non simpliciter sed respectu formae.

[a] est *suppl. s.l.* O
[b] hoc contingit dupliciter] talis actus C
[c] ad] in (*ex corr.*) O; *corr. mg. ex* in T
[d] in] ad O
[e] dicta *ex corr.* O; dictum T
[f] talis] forma sive talis *add.* C
[g] et ... actus] quae dicitur motus C
[h] super ... actus] an sit C
[i] huiusmodi] Et videtur quod non *add.* C
[j] omnis *scripsimus*; omne OTC
[k] est in potentia] huiusmodi C
[l] nullum *ex corr.* O
[m] sed ... etc.] quare etc. C
[n] sed ... sui *om. hom.* T
[o] sui *suppl. s.l.* O
[p] Solutio *om.* TC
[q] entis ... est *om. hom.* TC

4. It must be said that a thing has two kinds of act: either (i) a perfection placing the thing in its species; and an act of this kind is a form, and motion is not an act of this kind. (ii) Another kind of act had by a thing is the operation of the thing. Case (ii) is twofold: either (iia) the act is the operation of the thing in a complete way; and in this way motion is an action; or (iib) the act is the operation of the thing in an incomplete way; and in this way motion is a passion, because in this way motion is in the mobile thing.

5. Another reply is that a mobile thing has two kinds of potency because (i) it is in potency to a form and (ii) it is in potency to the way towards a form. If we speak of a potency of kind (i), then the terminus of this potency is a form, and the corresponding act is not a motion. If we speak of a potency of kind (ii), then the terminus of this potency is not a form but the way towards a form, and this way is an act, and motion is an act of this kind.

<Question 9>

1. The next question is about Aristotle's statement that motion is the act of what is in potency in so far as it is such,[50] because if a description is convertible with what is described, then motion belongs to everything which is in potency in so far as it is in potency. The consequent is false; therefore, the antecedent also is false.

2. Furthermore, no opposite is said of its opposite when that is qualified by a reduplicative phrase; but in this description an opposite is said of its opposite when it is qualified by a reduplicative phrase, that is, when it is said 'motion is the act <of what is in potency in so far as it is in potency>'.

3. Furthermore, if motion belongs to a being in potency, then motion will maximally belong to the thing that is maximally in potency, and such a thing is prime matter in its substance; but the consequent is false; therefore, the antecedent also is false.

4. Solution. It must be said that everything that is moved has two kinds of potency: (i) potency with respect to an operation, and (ii) potency with respect to a form. Accordingly, motion belongs to a being that is in potency

[50] Arist. *Phys*. III.1, 201a10–11.

Unde motus est actus etc., hoc est[a], est via media per quam mobile acquirit formam, et sic sumitur[b] 'actus' pro operatione sive via ad formam.

5. [c]Et [d]per iam dicta patet[e] rationibus.

<Quaestio10>

[f]Quaeritur de hoc quod dicit quod motus est actus mobilis[g] utrum sit ita actus mobilis ut mobile sit subiectum motus, et hoc est quaerere[h] in quo est motus ut in subiecto.

1. Et videtur quod sit in movente, quia dupliciter comparatur motus, scilicet ad moventem et ad mobile; sed in comparatione ad moventem est actio et in comparatione ad mobile dicitur passio. Sicut igitur passio est patientis, et hoc ut subiecti, sic erit actio agentis, et hoc ut subiecti.

2. Item, motus et quies sunt contraria; sed quies est in quiescente ut in subiecto; [i]ergo motus erit in movente [j]ut in subiecto.

3. Item, idem significatur per concretum et per abstractum, sed diversimode, quia illud idem quod significatur per abstractum ut absolutum significatur [k]in concreto ut inhaerens, et hoc subiecto proprio; [l]sed sic se habent 'movens' et 'motus' quod 'motus' significat in abstractione et 'movens' in concretione, et hoc ut inhaerens subiecto proprio; sed significat motum ut[m] qualitatem et moventem ut subiectum eiusdem. [n]Erit ergo in movente.

4. Item, actio et passio sunt rationes contrariae ipsius motus; sed huiusmodi contrarietas non simul et semel [o]inest eidem; sed passio est in mobili sive[p] patiente **[O 110ra]** ut in subiecto; ergo actio [q]erit in agente sive motus in movente.

[a] est *suppl. s.l.* O
[b] sumitur] sumit (*lectio incerta*) O
[c] Et ... rationibus] Iam patet ad argumenta C
[d] per iam dicta] ita T
[e] patet] responsio *add. s.l.* O
[f] Quaeritur … quod] Consequenter quia C
[g] mobilis] quaero *add.* C
[h] quaerere *om.* O
[i] ergo ... subiecto *om. hom. sed suppl. mg.* T
[j] ut in subiecto *om.* C
[k] in concreto] per concretum TC
[l] sed ...quod 'motus'] sed movens et motum se habent sicut concretum et abstractum. Motus enim C
[m] ut *ex corr.* O
[n] Erit ... movente] quare etc. C
[o] inest eidem] est in eodem subiecto C
[p] sive] in *add.* TC
[q] erit ... motus] sive motus erit in agente sive TC

or imperfect not absolutely but with respect to a form. Thus, motion is the act <of what is in potency in so far as it is such>, that is, motion is the middle way through which a mobile thing acquires a form, and thus 'act' is taken for the operation or way towards a form.

5. On the basis of what has been said the reply to the arguments[51] is clear.

<Question 10>
Given Aristotle's statement that motion is the act of a mobile thing,[52] the next question is whether motion is the act of a mobile thing in such a way that the mobile thing is the subject of motion, that is, the question is what motion is in as its subject.

1. Arguments that motion is in the mover: motion can be taken in relation to two things, that is, to the mover and to the mobile thing; but in relation to the mover motion is an action and in relation to the mobile thing it is said to be a passion. Therefore, just as a passion belongs to a patient as to its subject, so an action will belong to an agent as its subject.

2. Furthermore, motion and rest are contraries; but rest is in a thing at rest as its subject; therefore, motion will be in a mover as its subject.

3. Furthermore, the same thing is signified by a concrete term and the corresponding abstract term, but in different ways, because the same thing which is signified by the abstract term as absolute is signified by the concrete term as inhering, and as <inhering> in its proper subject; but the terms 'mover' and 'motion' are so related that 'motion' signifies in abstraction and 'mover' in concretion, and <signifies motion> as inhering in its proper subject; so 'mover' signifies motion as a quality and the mover as the subject of that quality. Therefore, motion will be in the mover.

4. Furthermore, action and passion are contrary aspects of motion itself; but such contrariety does not simultaneously inhere in the same thing; but passion is in the mobile thing or patient as its subject; therefore, action will be in the agent or motion in the mover.

[51] See above, nn. 1–3.
[52] Arist. *Phys*. III.1, 201a27–29.

5. Item, si [a]de non movente[b] fiat movens, tunc movens acquirit aliquam novam dispositionem; sed illa dispositio non acquiritur nisi [c]mediante transmutatione et motu[d]; ergo ante primum motum fuit alius motus; sed hoc non contingeret nisi motus esset in movente; quare[e]. Sic enim arguit Aristoteles in principio octavi *Physicorum*[f].

6. Item, scientia est fluxus medius inter sciens et scitum, sicut motus inter movens et mobile; sed scientia est in sciente ut in subiecto; quare motus erit in movente ut in subiecto.

7. Ad oppositum: si motus esset[g] in movente ut in subiecto, tunc omne movens moveretur; sed instantia est in motore primo. Sic arguit Aristoteles in secundo[h] huius.

8. Item, motus est [i]positio mobilis in speciem novam; sed non ponitur aliquid in specie nova nisi ipsi fiat acquisitio aliqua. Cum ergo solum mobili fiat acquisitio, solum erit[j] in mobili ut in subiecto.

9. Dicendum quod motus est in mobili ut in subiecto et est[k] in movente ut in efficiente.

10. Unde dicendum primae rationi quod actio est agentis et patientis, [l]sed diversimode, sive motus est moventis et mobilis, sed diversimode, quia est moventis ut efficientis et mobilis ut subiecti. Et sic fallit[m] prima proportio. Unde quia tam actione quam passione fit acquisitio tota mobili sive patienti, [n]ideo ponitur utrumque in mobili [o]sive patiente ut in subiecto.

11. Ad secundum dicendum quod, cum dicit 'motus et quies sunt contraria', non valet[p] nisi [q]ubi imponat contrarietatem in aliis extremis [r]sic: movens et quiescens sunt contraria; sed haec non sunt contraria —movens et quiescens—, quia contraria nata sunt fieri circa idem nisi determinetur ad

[a] de non *inv.* OT
[b] movente] movens *add. sed del.* O
[c] mediante ... motu] per motum et transmutationem C
[d] motu *corr. ex* moto O; moto T
[e] quare] etc. *add.* TC
[f] Physicorum *om.* TC
[g] esset *corr. ex* est O; est TC
[h] secundo] primo (?) C; *recte* tertio
[i] positio mobilis] ponens mobile C
[j] erit] quaeretur O
[k] est *om.* TC
[l] sed diversimode *om.* C
[m] fallit *ex corr.* O; *om. sed* tenet *suppl. s.l.* T; fit C
[n] ideo ponitur utrumque] item tam actio quam passio ponitur C
[o] sive patiente *om.* C
[p] valet] procedit argumentum C
[q] ubi imponat *corr. in* ponat O; imponat T; ponat C
[r] sic ... contraria] sicut in movente et quiescente C

5. Furthermore, if something passes from being a non-mover to being a mover, then the mover acquires a new disposition; but this disposition is only acquired through a transmutation and a motion; therefore, before the first motion there was another motion; but this conclusion would not follow unless motion were in the mover; hence, <motion is in the mover>. Aristotle indeed argues in this way at the beginning of *Physics* VIII.[53]

6. Furthermore, knowledge is a flowing intermediate between the knower and the known, just as motion is a flowing intermediate between the mover and the mobile thing; but knowledge is in the knower as its subject; therefore, motion will be in the mover as its subject.

7. To the contrary: if motion were in the mover as its subject, then every mover would be moved; but the first mover is a counterexample. Aristotle argues in this way in *Physics* II.[54]

8. Furthermore, motion is the placing of a mobile thing in a new species; but a thing is not placed in a new species unless it acquires something. Therefore, since only a mobile thing acquires something, motion will only be in the mobile thing as its subject.

9. It must be said that motion is in the mobile thing as its subject and in the mover as its efficient cause.

10. Thus, to the first argument[55] it must be said that action belongs to both agent and patient, but in different ways, or motion belongs to both mover and mobile thing, but in different ways, because it belongs to the mover as its efficient cause and to the mobile thing as its subject. And thus the first analogy fails. Accordingly, since it is both by action and by passion that the mobile thing or patient acquires something fully, it follows that both action and passion are placed in the mobile thing or patient as their subject.

11. To the second argument[56] it must be said that when he says 'motion and rest are contraries' the inference is not valid unless it establishes contrariety in the other extremes, for instance, that the mover and the thing at rest are contrary. But these <extremes>—the mover and the thing at rest—are not contrary, because contraries are apt by nature to exist in the same thing, unless that thing is determined to one of them; but there is a mover with no

[53] Arist. *Phys*. VIII.1, 251a23–28.
[54] The correct reference is Arist. *Phys*. III.3, 202a28–30.
[55] See above, n. 1.
[56] See above, n. 2.

alterum; sed[a] aliquod est movens in quo non est aliqua aptitudo respectu quietis; quare non sunt contraria. Sed illa sunt contraria—mobile et quiescens—, quia apta nata sunt esse circa idem. Arguatur ergo in his, et bene sequetur: quies est in quiescente; ergo motus est in mobili[b].

12. Ad tertium dicendum quod concretum significat accidens sub inclinatione ad aliud quod denominat; sed denominare aliud est multipliciter, scilicet vel ut subiectum, ut corpus sanum, vel ut efficiens, ut medicina sana, vel ut signum, ut urina sana, [c]vel ut conservativum, ut dieta sana. Unde motus denominat movens ut efficiens tantum.

13. Ad quartum dicendum quod contraria secundum rem[d] et rationem[e] non sunt simul, sed quae contrariantur per diversos respectus[f] tantum bene possunt[g], sicut idem motus secundum rem est actio et passio, diversificatus tamen per diversos respectus ad agens et ad patiens, et tamen ponitur totum in patiente ut in[h] subiecto.

14. Ad quintam[i] dicendum quod novam dispositionem acquiri alicui contingit duobus modis, scilicet vel per se vel per accidens. Unde per hoc quod acquiritur aliquid per se mobili acquiritur aliquid per accidens moventi ut aliqua nova relatio. Et potest nova relatio acquiri in aliquo, nulla transmutatione facta in eodem[j].

15. Ad ultimum dicendum quod semper fluxus medius inter duo ponitur in eo ut in subiecto cui fit acquisitio. Unde, quia per scientiam fit acquisitio scienti ponitur scientia in sciente, et ita, quia per motum fit acquisitio mobili et non moventi, ponitur motus in mobili et non in movente[k].

<Quaestio 11>
Quaeritur consequenter super hoc quod ipse dicit, scilicet[l] quod actio et passio sunt motus unus.

1. Et videtur quod non: movens et [m]motum sive mobile non sunt idem, ut patet. Primum enim movens et primum mobile non sunt idem; ergo absolute

[a] sed] scimus quod *add.* C
[b] mobili] et illud est verum *add.* C
[c] vel ... sana *om.* C
[d] rem *corr. ex* rationem O
[e] rationem *ex corr.* O
[f] respectus] sive per diversas rationes *add.* C
[g] possunt] esse simul *add.* C
[h] in *om.* O
[i] quintam] quintum TC
[j] eodem] sicut tu sis (stis *ms.*) albus et Socrates iam fiat albus, Socrates fit similis tibi nulla facta transmutatione in te *add.* C
[k] movente] ut in subiecto *add.* TC
[l] scilicet *om.* TC
[m] motum sive *om.* C

aptitude with respect to rest; hence, the mover and the thing at rest are not contrary. But these <other extremes>—the mobile thing and the thing at rest—are contrary, because they are apt by nature to exist in the same thing. Therefore, let us formulate the argument in the case of the mobile thing and the thing at rest, and then this <conclusion> will follow correctly: rest is in the thing at rest; therefore, motion is in the mobile thing.

12. To the third argument[57] it must be said that a concrete term signifies an accident together with its inclination towards another thing which it denominates; but to denominate another thing happens in many ways: either as a subject, for instance, the body is healthy, or as an efficient cause, for instance, the medicine is healthy, or as a sign, for instance, the urine is healthy, or as something preservative, for instance, the diet is healthy. Thus, motion denominates the mover only as its efficient cause.

13. To the fourth argument[58] it must be said that contraries in reality and in accordance with reason do not exist simultaneously, but contraries in accordance with different relations can indeed exist simultaneously. For instance, the same motion in reality is action and passion, although it is diversified by its different relations to the agent and to the patient, but it is placed as a whole in the patient as its subject.

14. To the fifth argument[59] it must be said a thing can acquire a new disposition in two ways: either per se or accidentally. Thus, in virtue of the fact that the mobile thing per se acquires something, the mover also acquires something accidentally, that is, a new relation. And a new relation can be acquired by something without any change in it.

15. To the last argument[60] it must be said that the intermediate flowing between two things is always placed in the thing that acquires something as its subject. Thus, because a knower acquires something in virtue of knowledge, knowledge is placed in the knower, and because the mobile thing but not the mover acquires something in virtue of motion, motion is placed in the mobile thing and not in the mover.

<Question 11>
The next question is about Aristotle's statement that action and passion are one motion.[61]

1. Arguments for the negative: the mover and the moved or mobile thing are not the same, as is clear; for the first mover and the first mobile thing are

[57] See above, n. 3.
[58] See above, n. 4.
[59] See above, n. 5.
[60] See above, n. 6.
[61] Arist. *Phys*. III.3, 202a13–21.

loquendo movens et mobile non sunt idem. Sed actio est in comparatione ad [a]moventem sive agentem, et passio in comparatione ad mobile; ergo[b] unum et idem non erit actio et passio, et ita nec motus unus.

2. Item, actio et passio sunt contraria, et contraria non sunt simul in eodem; ergo nec in motu, et ita [c]non sunt idem motus.

3. Item, secundum Aristotelem, actio [d]debetur formae, pati vero est[e] alterius potentiae ut materiae[f]; sed materia et forma sunt diversa principia secundum rem et sunt naturae diversae, ut patet in secundo[g] huius, et essentiae diversae, secundum quod dicit Aristoteles quod essentia materiae est alia ab essentia cuiuslibet praedicabilis. Et si hoc, eadem diversitas respondet actioni et passioni, scilicet[h] quod sunt naturae diversae et essentiae diversae.

4. Item, per exemplum Aristotelis patet illud, quia in [i]hoc tertio comparat actionem et passionem ad ire et redire; sed ire et redire non sunt simul neque unus motus; ergo similiter erit de actione et passione.

5. Ad oppositum, Aristoteles arguit sic: si actio est in agente et actio est motus, ergo motus est in agente. Supponit ergo Aristoteles quod actio in agente est motus. Et constat quod <passio[j]> in patiente est motus; ergo secundum Aristotelem utrumque est motus.

6. Item, Aristoteles: "Idem est actus motivi et mobilis"; et actus motivi est actio et actus mobilis est passio; quare idem est motus actionis et passionis.

7. Item, Aristoteles: "Cui autem haec insunt et motus inest".

8. Solutio[k]. **[O 110rb]** Dicendum quod actio uno modo est[l] motus comparatus ad agentem, alio modo est excertio ipsius agentis super patiens vel extensio virtutis agentis super patiens, quod idem est. Primo modo sunt actio et passio idem secundum rem et differunt tantum[m] per rationes superadditas, ut patet. Si autem sumatur secundo modo, tunc aut est

[a] moventem sive *om.* C
[b] ergo *ex corr.* O
[c] non ... motus] nec motus unus C
[d] debetur formae] formae est T
[e] est *om.* TC
[f] materiae *corr. ex* materia O
[g] secundo *corr. s.l. in* primo T
[h] scilicet *ex corr.* O
[i] hoc tertio] secundo huius O
[j] passio *supplevimus*
[k] Solutio *om.* TC
[l] est *suppl. s.l.* O
[m] tantum] tamen TC

not the same; therefore, absolutely speaking the mover and the mobile thing are not the same. But something is an action in relation to a mover or agent, and something is a passion in relation to a mobile thing; therefore, an action and a passion will not be one and the same thing, and hence will not be one motion.

2. Furthermore, action and passion are contraries, and contraries do not exist simultaneously in the same thing; therefore, neither do they exist simultaneously in a motion, and thus they are not the same motion.

3. Furthermore, according to Aristotle, action is due to a form, whereas being subject to passion belongs to another potency, that is, to matter;[62] but matter and form are different principles in reality and different natures, as is clear from *Physics* II,[63] and different essences, according to Aristotle's statement that the essence of matter is different from the essence of any predicable thing.[64] If so, the same diversity holds between action and passion, that is, they are different natures and different essences.

4. Furthermore, the example of Aristotle makes this clear because in *Physics* III he compares action and passion to going and coming back;[65] but going and coming back do not exist simultaneously and are not one motion; therefore, the same will apply to action and passion.

5. To the contrary: Aristotle argues that, if an action is in an agent and the action is a motion, it follows that a motion is in the agent.[66] Therefore, Aristotle assumes that an action in an agent is a motion. And it is manifest that a passion in a patient is a motion; therefore, according to Aristotle, each of them is a motion.

6. Furthermore, Aristotle <says>: "The act of a mover and of the mobile thing is the same",[67] and the act of the mover is an action and the act of the mobile thing is a passion; therefore, the motion of an action and of a passion is the same.

7. Furthermore, Aristotle <says>: "In the thing in which an action and a passion inhere motion also inheres".[68]

8. Solution. It must be said that an action is (i) in one way a motion taken in relation to an agent, and (ii) in another way the operation of the agent itself on a patient or, which is the same, the extension of the power of the agent as applied to the patient. In case (i) action and passion are the same in reality and differ only by additional *rationes*, as is clear. In case (ii), then an action is

[62] Arist. *De gen. et corr.* II.9, 335b29–31.
[63] Arist. *Phys.* II.1, 193a28–31.
[64] Arist. *Metaph.* VII.3, 1029a20–23.
[65] Arist. *Phys.* III.3, 202a19–20, b13–14.
[66] Arist. *Phys.* III.3, 202a25–29.
[67] Arist. *Phys.* III.3, 202a14–15.
[68] Arist. *Phys.* III.3, 202b20–21.

voluntaria conversio [a]quatenus moveat, sicut se habet causa prima ad movendum supercaelestia et inferiora et sicut anima ad movendum corpus et sicut intelligentia ad movendum caelum; vel virtutis agentis naturalis influxio, ut est in agentibus naturalibus. Agens enim naturale influit virtutem in suum passivum ad hoc quod moveat ipsum. Et secundum istum modum sic divisum non est actio et passio idem motus, sed actio est virtus sive potentia agentis comparata ad mobile per se vel per accidens. Et dico 'per se vel per accidens' quia[b] virtus sive potentia naturaliter agentis per se comparatur ad suum passivum sive ad aliud, voluntaria vero conversio quatenus moveat per accidens comparatur ad aliud, et hoc quia creatura refertur ad creatorem per se, creator autem ad creaturam per accidens.

9. Per hoc patet responsio ad primam rationem, quia, si sumatur actio pro exercitione agentis super patiens, sic non est idem cum passione. Si autem sumatur actio pro motu comparato ad agentem, sic est idem cum passione. Sumendo actionem primo modo sequitur sua conclusio.

10. Ad secundum dicendum quod, si sumatur actio pro motu comparato etc.[c], sic non sunt contraria [d]nisi secundum contrarias comparationes et rationes diversas, et sic contraria possunt inesse eidem.

11. Ad tertium dicendum sicut ad primum.

12. Ad quartum dicendum quod Aristoteles non vult quod actio et passio sint[e] diversa sicut ire et redire, sed intendit quod sicut eadem est via, et comparationes diversae, sic est in actione et passione quod idem est motus utriusque, recipiens tamen comparationes diversas, quia in comparatione ad agens dicitur actio et in comparatione ad patiens dicitur passio.

<Quaestio 12>

[f]Quaeritur consequenter utrum actio et passio diversificent praedicamenta.

1. Et videtur quod non, quia dicit Boethius super principium *Praedicamentorum* quod iste liber est de decem[g] vocibus[h] decem prima[i] genera rerum significantibus. Secundum[j] ipsum, ergo ad diversitatem praedicamentorum exigitur diversitas realis sive rerum. Cum ergo eadem sit

[a] quatenus moveat] alicuius super aliud quatenus moveat ipsum C
[b] quia *lectio incerta* (*ex corr.*) O
[c] etc.] ad agens C
[d] nisi ... possunt *om.* C
[e] sint] sunt TC
[f] Quaeritur ... diversificent] Consequenter gratia istorum potest quaeri qualiter actio et passio sunt diversa C
[g] decem *ex corr.* O
[h] vocibus *lectio incerta* O
[i] prima *om.* TC
[j] Secundum *ex corr.* O

either (iia) a voluntary conversion of an agent in so far as the agent moves, as in the case of the first cause in relation to its moving the celestial bodies and the things here below, and of a soul in relation to its moving a body, and of an intelligence in relation to its moving the heavens; or (iib) the flowing of the power of a natural agent, as in the case of natural agents; for a natural agent pours its power into the corresponding passive thing in order to move it. In cases (iia) and (iib) action and passion are not the same motion, but the action is the power or potency of an agent in relation to a mobile thing, either per se or accidentally. And I say 'either per se or accidentally' because the power or potency of something acting naturally is per se related to the corresponding passive thing, that is, to something distinct from itself, whereas the voluntary conversion of an agent in so far as the agent moves is accidentally related to a thing distinct from the agent; and the reason for this is that a creature is per se related to the creator, but the creator is accidentally related to a creature.

9. With this the reply to the first argument[69] is clear because, if action is taken for the exercise of an agent on a patient, then it is not the same as passion. But if action is taken for a motion related to an agent, then it is the same as passion. His conclusion follows if we take action in the first way.

10. To the second argument[70] it must be said that, if action is taken as a motion related <to an agent>, then action and passion are not contraries except by contrary relationships and different *rationes*, and things which are contraries in this way can inhere in the same thing.

11. To the third argument[71] it must be replied just as it was replied to the first one.

12. To the fourth argument[72] it must be said that Aristotle does not maintain that action and passion differ as going and coming back, but he means that just as the way is the same, but the relationships are different, so it is in the case of an action and a passion that the same motion belongs to both, but it is taken in different relationships, because in relation to an agent motion is said to be an action and in relation to a patient it is said to be a passion.

<Question 12>

The next question is whether action and passion produce different categories.

1. Arguments for the negative: in his commentary on the beginning of the *Categories*, Boethius says that this treatise is about the ten spoken words that signify the first ten genera of things.[73] Therefore, according to Boethius, a real

[69] See above, n. 1.
[70] See above, n. 2.
[71] See above, n. 3.
[72] See above, n. 4.
[73] Boethius, *In Cat.* I, PL 64, 160.

res actionis et passionis, quia idem motus, tunc actio et passio non sunt[a] diversa praedicamenta.

2. Item, Aristoteles in libro *Posteriorum* dicit quod abnegare rem unius praedicamenti de re alterius facit propositionem veram et immediatam. Si ergo actio et passio sunt praedicamenta diversa, tunc erit haec vera et immediata 'actio non est passio'; [b]sed haec non est vera, quia res unius non potest removeri[c] a re alterius, cum utriusque sit res eadem, cum sint motus unus; nec sufficit inter ipsa diversitas secundum rationem, quia tunc esset haec vera 'homo non est albus', et hoc propter diversam rationem subiecti et praedicati, quod accidens est.

3. Item, nulla duo praedicamenta diversa praedicantur in abstractione de eodem numero; sed actio et passio in abstractione praedicantur de motu; quare etc.

4. Item, si actio et passio essent genera diversa, tunc differrent per essentias, et [d]quae differunt per essentias differunt in naturis, et quae sic differunt sunt res diversae; [e]sed actio et passio non sunt res diversae, cum utrumque sit motus; quare non erunt genera diversa.

5. Item, actio est correlatum ad agentem et passio ad patientem, et ita sunt in genere[f] relationis.

6. Quidam dicunt quod actio et passio non sunt praedicamenta diversa, sed agere et pati in concretione sunt diversa praedicamenta.

7. Contra: Aristoteles in principio *Praedicamentorum* definit univoca [g]quia univocatio maxime convenit[h] naturae generis, et definit aequivoca ut excludat ea a natura generis, et denominativa ut similiter excludat; sed concretum significat ut denominativum; concretum ergo removetur a natura generis.

8. Item, Aristoteles dicit quod qualitas et quantitas sunt genera, et non quantum et quale.

9. Item, concretum non ponitur in genere uno.

10. Solutio[i]. Dicendum quod, cum dicitur 'praedicamenta significant res diversas', hic sumitur 'res' [j]extendendo ad rem veram et ad rerum circumstantias, sicut dicimus quod tempus secundum quod res est [k]quantum

[a] sunt] erunt TC
[b] sed ... vera] quod tamen falsum est C
[c] removeri] moveri *et add. s.l.* vel negari O
[d] quae ... essentias *om.* C
[e] sed ... diversae *om. hom. sed suppl. s.l.* sed non sic actio et passio O
[f] genere] eodem genere sicut in genere C
[g] quia univocatio *om. sed* ut *suppl. s.l.* O
[h] convenit] convenientis *ex corr.* O; convenientia (*lectio incerta*) C
[i] Solutio *om.* TC
[j] extendendo …veram] communiter extendendo nomen rei ad rem C
[k] quantum est] est in praedicamento quantitatis C

difference or a difference in things is required for a difference in category. Therefore, since the thing of an action and of a passion is the same because it is the same motion, then action and passion are not different categories.

2. Furthermore, in the *Posterior Analytics* Aristotle says that to negate a thing of one category of a thing of another category produces a true and immediate proposition.[74] Therefore, if action and passion are different categories, then the proposition 'an action is not a passion' will be true and immediate; but this proposition is not true because the thing of a passion cannot be removed from the thing of an action, since each of them belong to the same thing, given that they are one motion. Nor is the difference in *ratio* between action and passion sufficient because if it were, then the proposition 'a man is not white' would be true, and this because of the different characteristics of the subject and of the predicate, which is an accident.

3. Furthermore, no two predicates belonging to different categories are predicated in abstraction of numerically the same thing; but action and passion are predicated in abstraction of motion; hence, <action and passion do not belong to different categories>.

4. Furthermore, if action and passion were different genera, then they would differ in their essences, and things that differ in their essences differ in their natures, and things that differ in their natures are different things; but action and passion are not different things, since each of them is a motion; hence, they are not different genera.

5. Furthermore, an action is correlative with an agent and a passion with a patient, and thus they both are in the genus of relation.

6. Some people say that action and passion are not different categories, but to act and to be subject to passion in concretion are different categories.

7. Against this: at the beginning of the *Categories*, Aristotle defines univocal things because univocation is maximally in agreement with the nature of a genus, and he defines equivocal things in order to exclude them from forming a genus, and similarly for denominative things;[75] but a concrete term signifies denominatively; therefore, concrete things do not form a genus.

8. Furthermore, Aristotle says that quality and quantity are genera, and not a quantum and a quale.[76]

9. Furthermore, a concrete thing is not placed in one genus.

10. Solution. It must be replied that in the proposition 'the categories signify different things'[77] the term 'thing' is taken as including the genuine thing and the circumstances of things, as when we say that time in so far as it

[74] Arist. *An. post.* I.15, 79b6–22.
[75] Arist. *Cat.* 1, 1a1–15.
[76] Arist. *Metaph.* V.7, 1017a23–28.
[77] See above, n. 1.

est, secundum tamen quod adicitur circumstantia aliqua, ut respectus ad rem temporalem, sic facit 'quando'. Similiter, locus quantus est sive quantitas; [a]si tamen sumamus circumstantiam[b] inter locum et locatum, sic facit[c] 'ubi'. Unde actio et passio sunt idem motus secundum rem, tamen rationes formales adiectae super ipsum motum diversificant sive faciunt diversa praedicamenta. Et secundum quod rationes diversae adiciuntur et diversificantur praedicamenta, sic non praedicatur unum de altero, licet secundum quod unum sunt, ut secundum rem et naturam[d], possit unum de altero[e] praedicari.

11. Contra: si haec ratio diversa adiecta faceret diversitatem praedicamentalem, tunc ire et redire sive accessus et recessus, cum super accessum et recessum diversae adiciantur[f] rationes, erit diversitas praedicamentalis.

12. [g]Quidam dicunt sic ad istud, scilicet quod quaedam habent esse absolutum [h]et quaedam esse comparatum. In his autem quae habent esse absolutum exigitur diversitas realis ad hoc quod sint praedicamenta diversa. Ea autem quae habent **[O 110va]** esse comparatum[i] aut praeter hoc quod comparantur non sunt aliquae naturae in se, sicut paternitas et filiatio, et in talibus non potest respectus diversificare praedicamenta. Alia autem sunt quae praeter hoc quod comparantur sunt res et naturae in se, et huiusmodi per diversum respectum diversificant[j] praedicamenta, et ita est de actione et passione.

13. Contra: accessui et recessui praeter hoc quod sunt sub diversis rationibus subest aliqua res, sicut motus aliquis.

14. Mihi videtur quod rationes formales adiectae super motum sufficiunt ad diversificandum praedicamenta. Unde ad illud quod dicitur de accessu et recessu dicendum quod accessus et recessus comparantur ad terminos diversos, et actio et passio comparantur ad terminos [k]diversos; sed hoc est aequivoce, quia actio et passio sunt inter terminos ut inter agens et patiens, et hoc ut[l] inter causam et suum subiectum. Et istae sunt diversae rationes formales quae sufficiunt ad diversificandum praedicamenta. Accessus autem et recessus sunt inter terminos [m]ut tantum inter terminos motus quae sunt materiales condiciones adiectae, et haec non diversificant praedicamenta.

[a] si tamen sumamus] respiciendo tamen C
[b] circumstantiam] causatam *add.* TC
[c] facit *ex corr.* O
[d] naturam] materiam O
[e] altero] alio O
[f] adiciantur] condiciones sive *add.* C
[g] Quidam ... scilicet] Ad illud quod dicitur C
[h] et ... absolutum *om. hom. sed* ad quorum *suppl. s.l. et* esse *add.* T
[i] comparatum] corparatum O
[j] diversificant] diversificat O
[k] diversos ... terminos *om. hom.* T
[l] ut *suppl. s.l.* O
[m] ut ... terminos *om. hom.* O

is a thing is a quantity, but in so far as a circumstance is added to it, such as the relation to a temporal thing, it produces the category 'when'. Similarly, place is a quantum or a quantity; but if we consider the circumstantial relationship between place and a located body, it produces the category 'where'. Accordingly, an action and a passion are the same motion in reality, but the formal *rationes* added to the motion diversify or produce different categories. And in so far as different *rationes* are added to a motion and the categories are diversified, one is not predicated of the other, although in so far as they are one, that is, in reality and in nature, one can be predicated of the other.

11. Against this: if this different added *ratio* produced a categorical difference, then there would be a categorical difference between going and coming back or accession and recession, since different *rationes* are added to accession and recession.

12. Some people reply to this by distinguishing between (i) things that have an absolute being and (ii) things that have a relative being. In case (i) a real difference is required in order to have different categories. In case (ii), either (iia) these things, apart from having a relative being, are not natures in themselves, as in the case of paternity and filiation, and for such things the relation cannot diversify the categories. (iib) There are other things, however, that apart from having a relative being, are also things and natures in themselves, and such things can diversify the categories through a different relation, and this is the case of action and passion.

13. Against this: apart from being under different *rationes*, accession and recession have a thing underlying them, that is, a motion.

14. It seems to me that the formal *rationes* added to a motion are sufficient to diversify the categories. Accordingly, to the counterexample of accession and recession[78] it must be said that accession and recession are related to different termini, and so too are action and passion related to different termini; but this happens equivocally because action and passion are between termini, namely between agent and patient, that is, between a cause and what it acts on. And these different formal *rationes* are such that they are sufficient to produce different categories. Accession and recession, however, are between termini only as between the termini of motion, which are material conditions added to motion, and these do not diversify the categories.

[78] See above, n. 11.

15. Sed tunc quaeritur[a] quare istae rationes ‘ab hoc’, ‘in hoc’ ut ‘ab agente’, ‘in patiens’, diversificant praedicamenta, et ‘ab hoc’, ‘in hoc’ ut ‘a termino’, ‘in terminum’, non diversificant[b].

1*. [c]Quia[d] finitum est dupliciter, scilicet aut extensione, ut[e] in continuo, aut numeratione, ut in discretis, ideo potest infinitum dupliciter accipi, scilicet aut numeratione, ut in discretis, aut extensione[f], ut in continuo.

<Quaestio 13>
Propter hoc quaeritur primo an sit infinitum in discretis[g].

<Quaestio 13.1>

1. [h]Et videtur primo quod sic quia, secundum Aristotelem, mundus non incepit neque incepit generatio, et ita non fuit primus homo, et si hoc, infiniti homines praecesserunt. Et secundum Aristotelem, anima separatur a corpore sicut perpetuum a corruptibili; ergo manent infinitae animae; aliquid ergo est infinitum numeratione.

2. Item, secundum Aristotelem, demonstrationes sunt infinitae et scientiae sunt infinitae.

3. Item, demonstrata linea quacumque, in ipsa sunt infiniti puncti in potentia, et si hoc, cum de quolibet puncto in potentia possum dicere quod est punctus in potentia, et oratio actualiter vera, tunc sunt infinitae veritates actuales[i].

4. Item, causa prima, secundum Augustinum, quia bonus, ideo[j] communicat suam bonitatem cum his inferioribus. Cum ergo sit bonitatis infinitae, communicat cum creaturis infinitis in numero[k].

5. Contra, Aristoteles probat in hoc tertio[l] quod non est infinitum nisi in potentia tantum[m].

[a] quaeritur] adhuc potest quaeri C
[b] diversificant] stude responde *add.* C
[c] Quia ... continuo] Quoniam autem de natura etc. Capitulum de infinito, utrum primo quaerendum est an sit infinitum. Et quia infinitum dupliciter potest accipi, scilicet aut secundum numerationem ut in discretis aut secundum extensionem ut in continuis C
[d] Quia] De infinito *praem.* T
[e] ut *om.* O
[f] extensione *corr. ex* intentione O
[g] discretis] secundo an in continuis *add.* C
[h] Et videtur primo] De primo videtur C
[i] actuales] ubituales O
[j] ideo *om.* TC
[k] numero] quare etc. *add.* C
[l] tertio *corr. mg. ex* secundo O; secundo C
[m] tantum] Item, angulus contingentiae est indivisibilis; sed non propter solum contactum qui fit (fit *om.* C) in puncto, quia sic esset angulus rectus indivisibilis; ergo propter spatium interceptum

15. But then the question arises why it is the case that these *rationes* 'from this', 'to this' as in 'from the agent', 'to the patient', diversify the categories, whereas these *rationes* 'from this', 'to this' as in 'from the terminus', 'to the terminus' do not diversify them.

1*. Since the finite is of two kinds, that is, either by extension, as in a continuum, or by numeration, as in discrete things, it follows that an infinite can be taken in two ways, that is, either by numeration, as in discrete things, or by extension, as in a continuum.

<Question 13>
Given this, the first question is whether an infinite exists in discrete things.

<Question 13.1>

1. First the arguments for the affirmative: according to Aristotle, the world did not begin nor did generation, and so there was no first man, and if so, infinitely many men preceded.[79] And according to Aristotle, the soul is separated from the body in the way that a perpetual thing <is separated> from a corruptible thing;[80] therefore, infinitely many souls persist; therefore, something is infinite by numeration.

2. Furthermore, according to Aristotle, demonstrations and sciences are infinitely many.[81]

3. Furthermore, given any line, there are infinitely many potential points in it, and if so, since of any point in potency I can say that it is a point in potency, and that utterance is actually true, there are infinitely many actual truths.

4. Furthermore, according to Augustine, the first cause, because it is good, communicates its goodness to the things here below.[82] Therefore, since the first cause has infinite goodness, it communicates it to an infinite number of creatures.

5. Against this: in *Physics* III Aristotle proves that the infinite exists only in potency.[83]

[79] Arist. *De caelo* I.10-II.1; *Phys*. VIII.1.
[80] Arist. *De anima* II.2, 413b26–27.
[81] Arist. *An. post*. I.32, 88b7.
[82] Aug. *De doctr. christ.* I.32, CCSL 32, 26.3–4.
[83] Arist. *Phys*. III.6, 206a9–18.

<Quaestio 14>

[a]Quaeritur consequenter an sit infinitum in quantitate continua.

<Quaestio 14.1>

1. Et videtur quod sic quia [b]linea est infinita; quare etc. Probatio praemissae. Per[c] Aristotelem, sic se habet punctus ad lineam sicut instans ad tempus; sed secundum Aristotelem, omne instans est medium duorum temporum; ergo omnis punctus erit medium duarum linearum. Si ergo ponamus punctum in[d] extremo lineae, adhuc erit medium, [e]et ita in infinitum, et ita linea erit infinita[f].

2. Item, est intelligere unumquodque praeter illud quod se habet per privationem[g] in eodem; sed punctus se habet per privationem in linea; est ergo intelligere lineam sine punctis; sed extrema lineae sunt duo puncta; erit ergo intelligere lineam sine extremis, [h]et ita poterit esse linea sine extremis.

3. Item, secundum Aristotelem, potentia infinita non est nisi corporis infiniti; sed est potentia corporis infinita, ut[i] solis, quia, si extenderetur mundus in infinitum, sol multiplicaret suam lucem in infinitum, sicut patet per Augustinum dicentem quod lux statim cum generatur gignit, et ita, secundum ipsum, in eodem instanti fiunt infinitae[j] gignitiones; sed si non posset se multiplicare in infinitum, alicubi staret, [k]et tunc non posset gignere se, quod est contra Augustinum.

sub cono; sed angulus rectus potest dividi (addi *pro* dividi T) in angulos contingentiae; quare aliquod continuum dividitur in indivisibilia et ita erit aliquod actu infinitum in numero. Item, plus excedit substantia incorporea substantiam corpoream (quam substantia corporea *pro* substantiam corpoream T) quam instans tempus; et patet argumentum superius. Item, cuilibet potentiae passivae respondet potentia activa; sed potentiae passivae sunt infinitae; quare etc. (et patet argumentum superius *pro* sed ... etc. C). Ad hoc dicitur quod maior est falsa quia creatura de se potest in infinitum in deficiendo, non autem in perficiendo (Ad...perficiendo *om.* C) *add.* TC

[a] Quaeritur ... sic (n.1)] De secundo. Et videtur quod sit ponere infinitum in continuis C

[b] linea ... praemissae *om.* C

[c] Per] secundum TC

[d] in *suppl. s.l.* O

[e] et ... infinitum] inter etc. T

[f] infinita] Ad hoc dicitur quod inter puncta diversorum continuorum non (non *om.* C) cadit linea media. Contra: tangant (tangunt *pro* tangant C) se extremitates duarum linearum et superponatur corpus (et punctus *add.* T) et protrahantur a punctis tangentibus (tangibilibus *pro* tangentibus T) duae lineae per medium corporis; lineae sunt diversae et (et *om.* C) inter illas cadit (cadat *pro* cadit T) medium, cum sint eiusdem continui; ergo inter illa puncta quae sunt termini linearum debet (oportet *pro* debet C) cadere medium, quia sicut se habet linea ad lineam sic terminus ad terminum *add.* TC

[g] privationem] modum privationis TC

[h] et ... extremis] quare etc. C

[i] ut] potentia corporis *add.* TC

[j] infinitae] multae O

[k] et ... Augustinum] illa gignitio, sed hoc est inconveniens Augustino C

<Question 14>
The next question is whether an infinite exists in continuous quantity.

<Question 14.1>

1. Arguments for the affirmative: a line is infinite; hence, <an infinite exists in continuous quantity>. Proof of the premise: according to Aristotle, a point is to a line as an instant is to time;[84] but according to Aristotle, every instant is a mid-point between two times;[85] therefore, every point will be a mid-point between two lines. Therefore, if we posit a point at the extremity of a line, it will still be a mid-point between two lines, and so on to infinity, and thus the line will be infinite.

2. Furthermore, it is possible to understand any thing apart from what is a privation in that thing; but a point is a privation in a line; therefore, it is possible to understand a line without points; but the extremities of a line are two points; therefore, it will be possible to understand a line without its extremities, and thus it will be possible that there exists a line without extremities.

3. Furthermore, according to Aristotle, an infinite power only belongs to an infinite body;[86] but there is a body with an infinite power, namely the sun, because if the world were extended to infinity, the sun would multiply its light to infinity, as is clear from Augustine who says that light generates as soon as it is generated,[87] and thus, according to him, infinitely many generations occur in the same instant; but if light could not multiply itself to infinity, light would stop somewhere, and then it could not generate itself, which is against Augustine.

[84] Arist. *Phys*. IV.11, 220a9–20.
[85] Arist. *Phys*. VIII.1, 251b20–23.
[86] Arist. *Phys*. VIII.10, 266a10–b6.
[87] The text of Augustine alluded to here could be *De genesi ad litt.* I.10, CSEL 28.1, 14.12–20.

4. Item, [a]si sit finitae multiplicationis, imaginemur duos soles, unum in meridie, alium in septentrione, et [b]imaginemur quod incipiant distare in infinitum. Tunc, si sint finitae multiplicationis, aliquando in tantum distabunt quod lumina illorum duorum corporum coniungunt se secundum superficiem; hoc autem negat Aristoteles in principio *De sensu et sensato*.

5. Item, quantitati potentiae agentis respondet quantitas in suo effectu, si non fuerit aliquod impedimentum. Cum ergo causa prima producat mundum in esse, et ipsa est potentiae infinitae[c], tunc mundus erit infinitus.

6. <[d]Item, per Aristotelem, motus caeli est infinitus; quare spatium circa quod fit ille motus est infinitus; ergo etc.>

<Quaestio 13.2>

1. [e]Ad primum problema dicendum quod in discretis non est infinitum nisi in potentia tantum.

2. [f]Unde ad primam rationem dicitur sic. Si velimus ponere, secundum Aristotelem, quod mundus non[g] incepit, dicendum sic, secundum quosdam, quod Aristoteles non[h] improbat numerum esse infinitum nisi de illo numero qui resultat ex divisione continui, [i]quia, secundum ipsum, solum per divisionem continui crescit appositio numeri. Quod autem animae sint infinitae non haberet Aristoteles pro inconvenienti; stant ergo simul quod sint

[a] si ... multiplicationis *om.* C
[b] imaginemur quod *om.* C
[c] infinitae] nec patitur aliquod impedimentum in sua actione *add.* C
[d] Item ... etc. *supplevimus cum* C; *om.* OT *sed* Item, potentia infinita non est nisi corporis infiniti; sed in caelo est potentia infinita movendi; ergo etc. Diceres quod potentia movendi caelum non est corporis caeli sed intelligentiae moventis caelum. Contra cum caelum (Contra…caelum *suppl. mg.*) posset de sui natura habere esse sine intelligentia illa, cum non sit aliquid sui, sed non posset esse caelum nisi motus ei insit (caelum ... insit *suppl. mg.*), cum motus insit caelo sicut vita animali secundum Commentatorem, potentia ergo movendi caelum est corporis caeli et non intelligentiae (potentia…intelligentiae *suppl. mg.*). Item, supra quod potest virtus inferior supra idem potest virtus superior; sed virtus imaginativa quae est inferior potest supra infinitum; quare virtus intellectiva quae est superior etc.; sed nihil intelligitur nisi quod vere est; quare etc. Dicendum quod super quod potest virtus inferior ratione nobilitatis et perfectionis potest superior etc., non tamen supra quod et cum ratione imperfectionis quia sensitiva comprehendit sensibile secundum quod est hic et nunc, quod tamen non potest intellectiva. Item, si caelum sit finitum, ergo est unum; sed non est unum; quare est infinitum. Minor patet quia si est unum est in se indivisum et ab aliis divisum (et ab aliis divisum *suppl. mg.*) et hoc ab aliis eiusdem generis; sed caelum non est sic divisum quia sic essent plures (quia sic essent plures *suppl. s.l.*); quare etc. *add.* T
[e] Ad ... est] Ad ista dico quod nec in discretis nec in continuis est ponere C
[f] Unde … nisi] ad primam rationem positam ad primum problema dicunt quidam ponentes secundum Aristotelem mundum non incepisse quia Aristoteles non intendit numerum procedere in infinitum nisi loquendo C
[g] non *suppl. s.l.* O
[h] non *ex corr.* O
[i] quia ... continui *om. hom. sed* quia per divisionem continui *suppl. mg.* T

4. Furthermore, if the multiplication of light is finite, let us imagine two suns, one in the south and the other in the north, and let us imagine that they begin to move apart to infinity. Then, if the multiplication of their lights is finite, at some point they will be so distant that the lights of those two bodies will join together in a surface; but Aristotle denies this at the beginning of *Sense and sensibilia*.[88]

5. Furthermore, to the quantity of an agent's power there corresponds the quantity of its effect, provided that there is no hindrance. Therefore, since the first cause brings the world into being, and it has an infinite power, then the world will be infinite.

6. Furthermore, according to Aristotle, the motion of the heavens is infinite;[89] hence, the space around which that motion occurs is infinite; therefore, <an infinite exists in continuous quantity>.

<Question 13.2>

1. To the first problem[90] it must be said that in discrete things an infinite exists only in potency.

2. Thus, to the first argument[91] it is replied as follows. If, following Aristotle, we are willing to posit that the world did not begin, then according to some people, it must be said that Aristotle does not disprove the existence of an infinite number with the exception of the number produced from the division of a continuum, because, according to him, an increase of number only occurs through the division of a continuum.[92] That there are infinitely many souls, however, is not regarded as absurd by Aristotle. Therefore, the claim that

[88] Arist. *De sensu* 2, 438a29–b1.
[89] Arist. *Phys*. VIII.8, 261b27–29.
[90] That is, q. 13.
[91] See above, q. 13.1, n. 1.
[92] Arist. *Phys*. III.7, 207b7–15.

infinitae animae, et tamen[a], secundum intentionem Aristotelis, non sit numerus infinitus.

3. [b]Vel [c]sic. Secundum Commentatorem, intellectus est unus in omnibus. Unde secundum diversas intentiones imaginatas diversis copulatur.

4. [d]Si autem velimus dicere quod mundus incepit, tunc dicendum est, secundum Aristotelem, quod non improbat mundum incepisse, improbat tamen ipsum incepisse per generationem.

5. Ad secundum dicendum sic vel quod scientiae[e] non sunt infinitae simpliciter, **[O 110vb]** sed quoad nos tantum.

6. Vel sic. Sicut dicit Aristoteles libro *Elenchorum*, res sunt infinitae, nomina finita. Intelligit[f] sic quod res sunt infinitae, hoc est multo plures quam nomina[g], et ita sunt scientiae infinitae, hoc est multo plures quam sciuntur.

7. [h]Vel sic. Sunt infinitae potentia solum, et similiter demonstrationes [i]sunt infinitae.

8. Ad tertium dicendum quod non sunt ibi infinitae veritates in actu, et hoc quia non est[j] veritas aliqua in actu nisi ubi est aliquid in actu conceptum; unde quia non est actualiter concipere infinita puncta in linea, ideo non sunt ibi actuales veritates infinitae. Sed tot sunt ibi veritates in actu quot contingit accipere et concipere puncta in actu, et hoc per signationem, et [k]sic non est infinita puncta concipere in aliquo.

9. Vel sic. Non est inconveniens infinitas esse veritates enuntiantes[l] potentias infinitas de infinitis in potentia, et hoc contingit in proposito; sed enuntiare[m] infinitas actualitates[n] de infinitis in actu est impossibile.

10. Ad ultimum dicendum quod aliter est de primo agente et de inferioribus agentibus, quia primum agens, cum sit voluntarie agens, licet[o] sit infinitae potentiae [p]et agat secundum suam totam potentiam, potest tamen mitigare in effectu.

[a] tamen *ex corr.* O

[b] Vel ... Commentatorem] Aliter dicunt quidam ponentes secundum opinionem Commentatoris quod C

[c] sic. Secundum Commentatorem] sicut dicit Commentator quod T

[d] Si ... generationem] qui autem ponunt mundum incepisse dicunt quod Aristoteles non improbat mundum non incepisse, immo probat ipsum non incepisse per generationem C

[e] scientiae] vel demonstrationes *add.* C

[f] Intelligit] intendit TC

[g] nomina] solum *add. sed del.* O

[h] Vel ... demonstrationes sunt infinitae] Vel aliter sic: scientiae sunt infinitae potentia, non actu vel natura quia non est actualiter concipere infinita puncta in linea et similiter demonstrationes C

[i] sunt infinitae *om.* T

[j] est] actualiter *add. sed del.* O

[k] sic non est] hoc sic non est *scrips. sed* sic *del. et* scilicet *add. s.l.* O; sic non sunt TC

[l] enuntiantes] actualitates de infinitis in actu *add. sed del.* O

[m] enuntiare] enuntiandae O

[n] actualitates] enuntiationes actuales O

[o] licet] cum *add. s.l.* O

[p] et agat] si agat (*ex corr.*) O

there are infinitely many souls and the claim that, in accordance with Aristotle's intention, there is no infinite number, are compatible.

3. Another reply is that, according to the Commentator, the intellect is one in all men.[93] Thus, the intellect is joined to different men in accordance with different imagined intentions.

4. If, however, we want to maintain that the world began, then it must be said, following Aristotle, that he does not disprove that the world began, but he does disprove that it began through generation.

5. To the second argument[94] one reply is that sciences are not infinitely many absolutely, but only with respect to us.

6. Another reply is the following. As Aristotle says in the *Sophistical Refutations*, things are infinite in number, but names are finite in number.[95] What he means is that things are infinite in number in the sense that there are many more of them than of names, and in this way sciences are infinitely many, that is, there are many more of them than of those that are known.

7. Another reply is that sciences are infinitely many only in potency, and the same applies to demonstrations.

8. To the third argument[96] it must be said that in that case there are not infinitely many truths in act, and this because a truth is in act only when something is conceived in act; thus, since it is not possible to actually conceive infinitely many points in a line, therefore, in that case there are not infinitely many actual truths. But there are as many actual truths as there are actual points that it is possible to take and conceive by designating them; and in this way it is not possible to conceive infinitely many points in something.

9. Another reply is that it is not absurd for there to exist infinitely many truths that enunciate infinitely many potencies of infinitely many things in potency —and this happens in the present case— but it is impossible to enunciate infinitely many actualities of infinitely many things in act.

10. To the last argument[97] it must be said that the case of the first agent is different from that of the agents here below, because, since the first agent is a voluntary agent, although it has infinite power and acts in accordance with the whole of its power, it can nevertheless mitigate its effect.

[93] Aver. *In De anima* III, t.c. 5, ed. Crawford, 392.158–413.758.
[94] See above, q. 13.1, n. 2.
[95] Arist. *Soph. el.* I, 165a10–13.
[96] See above, q. 13.1, n. 3.
[97] See above, q. 13.1, n. 4.

<Quaestio 14.2>

1. [a]Ad secundum problema [b]dicendum quod non est infinitum in quantitate continua nisi in potentia tantum.

2. Unde ad primum argumentum, si ponatur omne instans esse medium duorum temporum, [c]dicendum tunc quod, sicut se habet instans ad tempus, sic se habet punctus [d]proportionalis instanti ad lineam; et punctus proportionalis est solum in linea circulari. Si autem non ponatur omne instans esse[e] medium [f]duorum temporum, tunc [g]sic: omne instans infra continuitatem temporis est medium etc., et hoc verum est, et ita omnis punctus infra continuitatem lineae.

3. [h]Ad secundum, dicendum quod punctus non dicit solam privationem sed aliquid per positionem.

4. Ad tertium [i]dicendum, secundum quosdam, quod potentia lucis non est infinita, quia quod potest debilitari potest corrumpi; sed lux [j]potest debilitari, ut patet de luce candelae. Et respondent Augustino dicentes quod Augustinus non dicit propter hoc quod multiplicat se in infinitum, sed propter sui subitam[k] immutationem. Unde cum dicit quod quilibet punctus gignit alium punctum subito, [l]hoc est intelligendum dum durat potentia gignendi.

5. [m]Aliter dicitur sic quod lux solis multiplicaret se in infinitum si haberet in quod. Et respondent ad illud quod potentia infinita non est nisi corporis infiniti dicentes quod non est potentia una qua fit ista multiplicatio, quia

[a] Ad secundum *lectio incerta* O
[b] dicendum … argumentum (n.2)] dico ad primam rationem quod C
[c] dicendum tunc quod] tunc C
[d] proportionalis … solum] ad lineam, talis autem punctus non reperitur nisi C
[e] esse *om.* O; *suppl. s.l.* T
[f] duorum ... instans *om. sed* etc. sed omne instans *suppl. mg.* T
[g] sic ... lineae] dicendum quod Aristoteles intendit de instanti quod est infra continuitatem temporis et eodem arguitur de puncto infra continuitatem lineae et sic non sequitur conclusio C
[h] Ad secundum] Item O
[i] dicendum, secundum quosdam] respondent quidam dicentes C
[j] potest ... immutationem] in sua multiplicatione debilitatur sicut patet in luce. Et ad auctoritatem Augustini dicendum nec intelligit quod lux se multiplicet in infinitum immo quod subito immutet C
[k] subitam *om.* O
[l] hoc est intelligendum] et hoc *scrips. et post* hoc *suppl. s.l.* est verum O; hoc est T
[m] Aliter ... sunt in illo (n.7)] Alii dicunt quod lux solis multiplicat se in infinitum si esset diaphanum in infinitum quod reciperet istam multiplicationem. Dicunt ergo quod illa potentia multiplicandi non est una potentia alicuius unius corporis, immo sunt infinitae potentiae infinitorum punctorum lucis successive se gignentium in infinitum sicut patet et ita non est contra quod dicit Aristoteles quod nullius corporis finiti est potentia infinita. Si sic arguitur: omnes potentiae illae processerunt a corpore finito; ergo prius fuerunt in illo et ita idem quod pridem, dicendum quod esse in corpore potest esse dupliciter: vel ut congregatae in illo et sic non,vel ut in sua causa efficiente et sic fuerunt in illo C

<Question 14.2>

1. To the second problem[98] it must be said that in a continuous quantity an infinite exists only in potency.

2. Thus, on the assumption that every instant is a mid-point between two times, to the first argument[99] it must be said that an instant is to time as a point analogous to the instant is to the line; and a point analogous to the instant only exists in a circular line. If we do not make the assumption that every instant is a mid-point between two times, it must then be said that every instant within the continuity of time is a mid-point <between two times>, and this claim is true, and hence every point within the continuity of the line <is a mid-point between two lines>.

3. To the second argument[100] it must be said that a point expresses not only a privation but also something positive.

4. To the third argument[101] some people reply that the power of light is not infinite because what can be weakened can be corrupted; but light can be weakened, as is clear <when we consider> the light of a candle. And they reply to Augustine by saying that Augustine does not speak as he did because light multiplies itself to infinity, but because of its sudden change. Thus, when he says that any point of light immediately generates another point, this must be understood as long as the generating power lasts.

5. Another reply is that the light of the sun would multiply itself to infinity, if there were something in which <this multiplication could occur>. And these people reply to the claim that an infinite power only belongs to an infinite body by saying that there is not only one power by which the

[98] That is, q. 14.
[99] See above, q. 14.1, n. 1.
[100] See above, q. 14.1, n. 2.
[101] See above, q. 14.1, n. 3.

potentia corporis luminosi est gignere iuxta se lucem, et illa lux ibi genita gignit lucem in aere sibi proximo, et ita sunt ibi potentiae infinitae.

6. Contra: omnes illae potentiae processerunt a corpore; ergo fuerunt prius in illo.

7. Dicendum quod non valet, quia esse in illo corpore contingit dupliciter: vel ut aggregatae in illo, et sic non sunt in illo, vel ut in sua causa efficiente, et sic sunt in illo.

8. Ad argumentum de motu[a] dicendum quod, si[b] [c]motus caeli esset[d] motus unus [e]et non iteratus, et cum hoc infinitus, [f]tunc necessario fieret super spatium infinitum; [g]sed[h] motus qui est infinitus per iterationem potest esse super spatium finitum.

<Quaestio 15>

[i]Quaeritur consequenter utrum infinitum in actu repugnet magnitudini unde magnitudo sive magnitudini unde naturale est.

1. Quod non magnitudini simpliciter probatio, quia infinitum quantitati congruit, et ita non repugnat quanto; quare etc.

2. Item, magnitudo simpliciter potentia permittit quantamcumque divisionem in infinitum; ergo eadem ratione et appositionem. Et si hoc, nullo modo repugnat[j] infinitati.

3. Item, dicit Aristoteles in secundo *De anima* quod, si calor naturalis esset solum agens in augmentatione naturali, augmentatum cresceret[k] in infinitum. Et propter hoc dicit quod oportet ponere animam quae sit terminans et regulans actionem caloris; sed primum non contingeret si infinitum repugnaret magnitudini naturali; quare etc.

4. Ad oppositum: Aristoteles in capitulo de infinito improbat infinitum esse, et hoc per naturam magnitudinis absolute unde magnitudo est; quare etc.

5. Item, ponere partem et non ponere totum ei respondens est ponere impossibile; sed ponere infinitum in magnitudine est sic ponere; quare etc.

[a] motu] caeli *add.* C
[b] si *corr. s.l. in* quia O
[c] motus caeli esset] esset motus C
[d] esset *corr. s.l. in* est O
[e] et non *scrips. sed del.* O
[f] tunc … infinitus *om. hom.* O
[g] sed … esse] motus autem iteratus infinities potest fieri C
[h] sed *scripsimus*; quia T
[i] Qq. 15–17 *om.* C
[j] repugnat *ex corr.* O
[k] cresceret *ex corr.* O

multiplication of light is produced, because the power a luminous body has is that of generating light next to itself, and the light generated there generates <another> light in the air proximate to itself, and thus there are there infinitely many powers.

6. Against this: all those powers proceeded from a body; therefore, they existed in that body beforehand.

7. It must be said that this objection is not valid because there are two ways in which those powers can exist in that body: either (i) as aggregated in it, and in this way the powers are not in that body; or (ii) as in their efficient cause, and in this way the powers are in that body.

8. To the argument about motion[102] it must be said that, if the motion of the heavens were a motion that is one, non-iterated, and also infinite, then it would necessarily occur over an infinite space; but a motion that is infinite by iteration can occur over a finite space.

<Question 15>
The next question is whether the infinite in act is inconsistent with magnitude in so far as it is a magnitude or in so far as it is natural.

1. Arguments that an infinite is not inconsistent with magnitude taken absolutely: an infinite is consistent with quantity, and thus it is not inconsistent with quantity; hence, <it is not inconsistent with magnitude taken absolutely>.

2. Furthermore, magnitude taken absolutely potentially allows a division of any size to infinity; therefore, for the same reason it also allows an addition <of any size> to infinity. And if so, then magnitude is in no way inconsistent with infinity.

3. Furthermore, in *On the Soul* II, Aristotle says that, if natural heat were the only agent in a natural increase, then the thing increased would increase to infinity.[103] And he says that because of this it is necessary to posit a soul which limits and regulates the action of heat; but the initial situation would not be possible if the infinite were inconsistent with a natural magnitude; hence, <the infinite is not inconsistent with a natural magnitude>.

4. To the contrary: in the chapter on the infinite Aristotle disproves the existence of the infinite on the basis of the nature of magnitude taken absolutely in so far as it is a magnitude;[104] hence, <the infinite is inconsistent with magnitude taken absolutely>.

5. Furthermore, to posit a part without positing the whole corresponding to it is to posit something impossible; but to posit an infinite in magnitude is

[102] See above, q. 14.1, n. 6.
[103] Arist. *De anima* II.4, 416a9–18.
[104] Arist. *Phys*. III.5, 204b1–7.

Minor patet sic. Si linea est infinita, ergo non contingit sumere ibi aliquod totum, quia totum est ubi nihil derelinquitur extra; contingit tamen sumere partem ipsius lineae. Erit ergo pars sine toto, quod est impossibile.

6. Item, magnitudo unde[a] talis habet istas differentias 'sursum', 'deorsum', 'ante' et 'retro' etc., secundum Aristotelem in libro *Caeli et mundi*; sed habens has differentias habet finitatem; quare etc.

7. Item, forma cuiuslibet corporis est cum tota materia eiusdem; sed si ponatur infinitum, [b]forma non est cum tota materia; quare etc. Probatio minoris: quod solum est in profundo alicuius non est cum tota materia eius; sed si ponatur infinitum, forma erit solum in profundo, quia, si esset in exteriori parte, terminaret et finitaret, et ita esset finitum, quod est contra hypothesim. Si ergo ponatur infinitum, non erit forma cum tota materia; sed hoc est inconveniens in re naturali; quare etc.

8. Solutio. Dicendum quod infinitas in actu repugnat magnitudini **[O 111ra]** et unde magnitudo et unde naturalis est.

9. Ad primam rationem in oppositum, dicendum quod propositio est intelligenda eo modo quo infinitum habet esse, et hoc est solum potentia.

10. Ad secundum dicendum quod magnitudo habet quod possit dividi in infinitum gratia alicuius quod est intra magnitudinem, ut gratia partium infinitarum materiae, et ita in magnitudine est causa essentialis suae divisionis in infinitum. Si autem imaginemur aliquam magnitudinem crescere in infinitum per additionem, hoc non erit per aliquam intraneam naturam ipsius magnitudinis sive corporis, sed per naturam additamenti, et ita per naturam extrinsecam. Unde non potest esse causa aliquo modo essentialis in appositione magnitudinis in infinitum, sed solum accidentalis, et ita non dicimus quod magnitudo possit crescere in infinitum [c]per additionem quia hoc non est ex natura sua intranea.

11. Contra, si hoc est verum, videtur quod numerus non possit crescere in infinitum.

[a] unde *lectio incerta* O
[b] forma ... hypothesim] non erit forma in profundo *et* solum *add. sed del.* T
[c] per additionem *om.* T

to posit something of this kind <that is, a part without the whole>; hence, <it is impossible to posit the infinite in magnitude>. The minor premise becomes clear as follows: if a line is infinite, it is not possible to take any whole, because a whole is found when nothing is left outside it; but it is possible to take a part of the line itself; therefore, there will be a part without the whole, which is impossible.

6. Furthermore, a magnitude in so far as it is a magnitude has the differences 'up', 'down', 'front', 'back', <'left', 'right'>, according to Aristotle in *On the Heavens*;[105] but what has these differences is finite; hence, <magnitude is finite>.

7. Furthermore, the form of any body exists together with the whole matter of that body; but if an infinite body is posited, its form does not exist together with its whole matter; hence, <an infinite body cannot be posited>. Proof of the minor premise: what exists only in the depths of something does not exist together with its whole matter; but if an infinite body is posited, its form will exist only in the depths, because if it also existed in the exterior part, it would limit the body and make it finite, and then the body would be finite, which is contrary to the hypothesis. Therefore, if an infinite body is posited, its form will not exist together with its whole matter; but this is absurd in a natural thing; hence, <an infinite body cannot be posited>.

8. Solution. It must be said that infinity in act is inconsistent with a magnitude both in so far as it is a magnitude and in so far as it is natural.

9. To the first argument to the contrary[106] it must be said that the proposition must be understood in accordance with the way in which an infinite exists, and this is only in potency.

10. To the second argument[107] it must be said that a magnitude has the property of being able to be divided to infinity in virtue of something intrinsic to it, that is, in virtue of the infinitely many parts of matter, and thus the essential cause of its division to infinity exists in a magnitude. If we imagine, however, that a magnitude increases to infinity by addition, this will not be in virtue of some intrinsic nature of the magnitude itself or of body, but in virtue of the nature of what is added, and so in virtue of an extrinsic nature. Thus, there cannot be a cause of a magnitude's increase to infinity that is in some way essential, but only an accidental cause, and thus we do not say that a magnitude can increase to infinity by addition because this does not happen in virtue of its intrinsic nature.

11. Against this: if this is true, it seems that number cannot increase to infinity.

[105] Arist. *De caelo* II.2, 284b20–25.
[106] See above, n. 1.
[107] See above, n. 2.

12. Dicendum quod quilibet numerus est in potentia de se respectu additionis super ipsum, et quaelibet pars numeri ordinatur ad aliam in infinitum, et ideo dicimus quod numerus per additionem crescit in infinitum, scilicet quia quilibet numerus de se est in potentia ad alium numerum. Et praeter hoc, cuilibet parti numeri debetur forma propria, et ita numerus habet diversas potentias quarum quaelibet ordinatur ad aliam, et ideo crescit sive est in potentia ut crescat etc. Sed in magnitudine non est ita, quia una magnitudo non ordinatur ad aliam, et etiam quando una magnitudo additur alteri magnitudini, totum fit sub una forma sive sub uno actu existendi. Unde totum est terminatum sicut fuit prius, et ideo non crescit in infinitum per additionem.

13. Ad tertium dicendum quod Aristoteles non intendit quod cresceret in infinitum simpliciter, sed ultra magnitudinem determinatam tali speciei.

<Quaestio 16>

[a]Quaeritur consequenter super hoc quod dicit in littera quod, si ponatur unum elementum esse infinitum, illud elementum corrumperet omnia alia.

1. Contra: licet elementa[b] secundum partes suas [c]sint corruptibilia, secundum totum tamen [d]sunt incorruptibilia, et ita qualitercumque ponatur unum elementum, alia secundum se tota manent incorruptibilia.

2. Item, si ponamus unum elementum esse infinitum, illud continebit omnia alia, et erit locus eorum; sed locus est salvans et non corrumpens; quare etc.

3. Dicendum quod Aristoteles verum dicit, quia[e] nullum[f] continens inferius est salvans et continens nisi per virtutes superiores. Unde non solum salvant et continent per naturam propriam particularem[g], sed per virtutem caelestem. Unde si ponamus unum infinitum, nos non ponemus aliquod corpus caeleste, et ita nec salvans. Et ita patet secundo argumento.

4. Ad primum patet per hoc quod elementa non sunt incorruptibilia in[h] naturis propriis, sed per naturam superiorum.

5. Sed videtur quod dicat falsum in responsione primae rationis[i], cum dicit quod supracaelestia influunt suas virtutes in haec inferiora ad hoc quod

[a] Qq. 16–18 *post* inconveniens (Q. 20.1, n. 11) *transp.* T
[b] elementa] elementum T
[c] sint corruptibilia] fuerit corruptibile T
[d] sunt incorruptibilia] est incorruptibile T
[e] quia] et sciendum quod T
[f] nullum *ex corr.* O
[g] particularem *ex corr.* O
[h] in *suppl. s.l.* O; *om.* T
[i] rationis] quaestionis O

12. It must be said that any number is in itself in potency with respect to an addition to it, and any part of a number is ordered to another part to infinity, and so we say that a number increases to infinity by addition because any number is in itself in potency to another number. Moreover, any part of a number has its own form, and thus a number has different potencies each of which is ordered to another, and this is why number increases or has the potency to increase <to infinity>. But it is not like that in a magnitude because one magnitude is not ordered to another magnitude, and when one magnitude is added to another magnitude, the whole comes to be under one form or under one act of existing. Thus, the whole is limited as it was beforehand, and therefore it does not increase to infinity by addition.

13. To the third argument[108] it must be said that Aristotle does not mean that the thing subject to increase would increase to infinity in an absolute sense, but that it would increase beyond the magnitude fixed for that species.

<Question 16>
The next question is about Aristotle's statement in the text that, if one element is posited to be infinite, that element would destroy all the others.[109]

1. Against this: although the elements are corruptible in their parts, they are, however, incorruptible as a whole, and thus in whatever way one element is posited <to exist>, the others remain incorruptible as wholes.

2. Furthermore, if we posit that one element is infinite, that element will contain all the others and it will be their place; but place is preserving and not corruptive; hence, <an infinite element would not destroy the others>.

3. It must be said that Aristotle's statement is true because no container here below is preserving and containing except in virtue of the superior powers. Thus, the elements do not preserve and contain only in virtue of their particular nature, but in virtue of celestial power. Thus, if we posit that one element is infinite, we do not posit any celestial body, and hence no preserving body. And with this the reply to the second argument[110] is clear.

4. The reply to the first argument[111] is clear too, given that the elements are not incorruptible in their own natures, but in virtue of the nature of the superior bodies.

5. But it seems that he says something false in the reply to the first argument, when he says that the celestial bodies pour their powers into the

[108] See above, n. 3.
[109] Arist. *Phys.* III.5, 204b28–29.
[110] See above, n. 2.
[111] See above, n. 1.

salvent[a] et contineant, quia secundum eandem viam est influxio virtutis ab aliquo secundum quam inclinatur ad motum, ut patet. Grave solum influit suam virtutem deorsum quia solum movetur in illam partem, et leve sursum. Cum ergo motus caeli sit circularis tantum, tunc erit influxio suae virtutis secundum circulum tantum, et non secundum rectum, et ita non in haec inferiora.

6. Item, si corpora caelestia influant suas virtutes hic inferius secundum lineam rectam, et haec inferiora influunt suas virtutes ut multiplicant suas species sursum secundum lineas rectas, tunc cum eadem sit proportio elementorum ad caelum et caeli ad elementa, tunc si caelum influit secundum lineam rectam, et haec inferiora[b] similiter in illa influent et ita, si homo esset ultra solem, posset videre haec inferiora nisi esset defectus ex parte sua. Hoc tamen habet quaestionem.

<Quaestio 17>

Consequenter quaeritur super hoc quod consequenter dicit quod, si unum excedit aliud et illud quod exceditur crescat in infinitum, excedet illud quod prius excessit ipsum.

1. Contra: anima et corpus proportionantur invicem ita quod anima excedit corpus in suis operationibus; sed si corpus crescat in infinitum, [c]adhuc excedet, ut videtur, quia prius fuit anima tota in qualibet parte corporis. Ergo, si crescat in infinitum, cum sit corpus animatum, adhuc erit[d] ubique, aut plures erunt animae in illo corpore, quod non contingit; et ita idem erit excessus animae si corpus sit infinitum et si [e]corpus sit finitum.

2. Ad hoc dicitur sic sicut dicit Aristoteles in secundo *De anima* quod omnium constantium natura determinata est ratio magnitudinis et augmenti, et hoc natura animae. Unde determinatur in natura animae quantum potest esse corpus cuius potest esse perfectio. Unde si corpus excedat, non poterit anima [f]in perfectionem totius, et ita si crescat in infinitum, non poterit in[g] perfectionem eius.

3. Vel sic: consequentia Aristotelis tenet in his quae proportionantur in actione et passione naturali, et non alibi[h].

[a] salvent *corr. ex* solvent O
[b] inferiora] et ita *add. sed del.* O; et *add.* T
[c] adhuc ... infinitum *om. hom.* T
[d] erit] anima *add.* T
[e] corpus sit *om.* T
[f] in perfectionem] esse perfectio T
[g] in] esse T
[h] alibi *ex corr.* O; Contra maioris potentiae est anima quam forma ignis, et si apponantur ligna in infinitum, comburet ignis in infinitum; ergo etc. Dicitur quod si ligna apponantur etc., et hoc verum est, sed ista infinitas non est ab una potentia sive a potentia unius corporis et ideo non

things here below in order to preserve and contain them, because the direction in which a thing pours its power is that in which that thing has an inclination towards motion, as is clear: a heavy thing only pours its power downwards because it only moves in that direction, and a light thing <does so> upwards. Therefore, since the motion of the heavens is only circular, the pouring of their power will only happen in a circle, and not in a straight line, and thus <will not be directed> towards the things here below.

6. Furthermore, if the celestial bodies pour their powers here below in a straight line, and the things here below pour their powers in order to multiply their species in straight lines, then since the relation of the elements to the heavens is the same as that of the heavens to the elements, if the heavens pour their power in a straight line, the things here below will likewise pour their power into the celestial bodies in a straight line, and thus if a man were beyond the sun, he could see the things here below, unless there were a defect on his part. But this is questionable.

<Question 17>
The next question is about Aristotle's next statement that, if one thing exceeds another and the thing exceeded increases to infinity, it will exceed the thing that exceeded it beforehand.[112]

1. Against this: the soul and the body are proportionate one to the other in such a way that the soul exceeds the body in its operations; but if the body increases to infinity, the soul would still exceed it, as it seems, because beforehand the soul existed as a whole in every part of the body, therefore, if the body increases to infinity, since it is an animate body, the soul will still be everywhere, or there will be many souls in that body, which does not happen; and thus the excess of the soul over the body will be the same both if the body is infinite and if it is finite.

2. The reply to this argument is in accordance with what Aristotle says in *On the Soul* II, namely that for all the things existing by nature there is a definite ratio of magnitude and of increase,[113] and this is in virtue of the nature of the soul. Thus, the nature of the soul determines how large the body that the soul may perfect can be. Accordingly, if the body exceeds these bounds, the soul cannot confer perfection on the whole of it, and thus if the body increases to infinity, the soul cannot confer perfection on it.

3. Another reply is that Aristotle's inference is valid for the things that are proportional in accordance with natural action and passion, and not in other cases.

[112] Arist. *Phys*. III.5, 204b14–19.
[113] Arist. *De anima* II.4, 416a16–17.

<Quaestio 18>
Consequenter quaeritur utrum infinitum possit esse acceptum.

1. Et videtur quod non, quia semper est in fieri et accipiendum sicut dies et agon, sicut [a]dicit Aristoteles in libro hoc tertio[b].

2. Item, omne acceptum est[c] totum, et omne totum est finitum; **[O 111rb]** [d]quare etc.

3. Item, omne acceptum habet ultimum; aut ergo habet[e] primum aut non. [f]Si non habet primum, non fuit acceptum. Habet ergo primum, et ita est finitum.

4. [g]Ad oppositum: Aristoteles vult quod tempus pertransitum est infinitum, et tamen totum acceptum est.

5. Item, totum esse Dei est acceptum et est infinitum.

6. Item, in aliquo plano sunt infinitae figurae[h].

7. Solutio[i]. Dicendum quod infinitum in actu non potest esse acceptum, infinita tamen in aliquo finito possunt accipi et pertransiri[j], sicut infinita puncta in uno continuo finito pertransiri possunt, in se autem nec possunt pertransiri nec accipi.

8. Ad primum patebit super octavum[k].

9. Ad secundum dicendum quod esse Dei est infinitum non extensione, sed simplicitate et indivisibilitate; tempus autem est infinitum in extensione.

<Quaestio 19>
Consequenter quaeritur an sit infinitum in actu.

<Quaestio 19.1>

1. Et videtur quod sic: suppono istam propositionem, scilicet quod omnis numerus habet esse in actu et in factum esse. Tunc sic: aliquis numerus est infinitus, et omnis numerus habet esse in factum esse; quare aliquid in factum esse est infinitum, et ita aliquid in actu. Probatio maioris. In continuo sunt

---

sequitur. Sed si tantum potentia una corporalis posset in infinitam actionem, bene sequeretur quod anima similiter posset *add.* T *et* Q. 17a (*Cf. Appendix*)

[a] dicit ... tertio] vult hic Aristoteles C

[b] tertio] non potest esse acceptum; quare etc. *add.* T

[c] est *corr. ex* et O

[d] quare etc. *om.* C

[e] habet *om.* O

[f] Si ... finitum] Si sic, ergo est finitum. Si non, ergo non fuit acceptum; sed infinitum non habet primum et ultimum; ergo etc. C

[g] Ad oppositum] Contra TC

[h] figurae] lineae vel figurae T; lineae C

[i] Solutio *om.* TC

[j] pertransiri *ex corr.* O

[k] octavum] conclusionem (*lectio incerta*) O

<Question 18>
The next question is whether an infinite can be taken as completed.

1. Arguments for the negative: an infinite is always in becoming and in going to be taken, just like a day and a contest, as Aristotle says in *Physics* III.[114]

2. Furthermore, every thing that is taken as completed is a whole, and every whole is finite; hence, <the infinite cannot be taken as completed>.

3. Furthermore, every thing that is taken as completed has a final element; therefore, it either has a first element or it does not. If it does not have a first element, then it was not taken as completed. Therefore, it has a first element, and thus it is finite.

4. To the contrary: Aristotle wants to say that past time is infinite,[115] yet the whole of it is taken as completed.

5. Furthermore, the whole being of God is taken as completed and is infinite.

6. Furthermore, there are infinitely many figures in a plane.

7. Solution. It must be said that an infinite in act cannot be taken as completed but infinites existing in something finite can be taken as completed and traversed, in the way that the infinitely many points in a finite continuum can be traversed, but in themselves they cannot be traversed or taken as completed.

8. The reply to the first argument[116] will be clear in the comment on *Physics* VIII.[117]

9. To the second argument[118] it must be said that the being of God is infinite not in extension but in simplicity and indivisibility, whereas time is infinite in extension.

<Question 19>
The next question is whether an infinite exists in act.

<Question 19.1>

1. Arguments for the affirmative: I make the assumption that every number has an actual and fully realized being. Then I argue as follows: a number is infinite, and every number has a fully realized being, hence something with a fully realized being is infinite, and thus something <infinite> is in act. Proof of the major premise: there are infinitely many halves in potency in a

[114] Arist. *Phys*. III.6, 206a18–23.
[115] Arist. *Phys*. VIII.1, 251b19–26.
[116] See above, n. 4.
[117] See below, Book VIII, q. 5, n. 18.
[118] See above, n. 5.

medietates infinitae in potentia, et quot sunt medietates in potentia sive[a] partes in potentia tot erunt in actu, si dividatur. Cum ergo possibile sit dividi, possibile est aliquid esse actu infinitum.

2. Item[b], infinitae sunt medietates in potentia in continuo, et quaelibet medietas est maior medietate sua; infinities ergo potest enuntiari medietatem esse maiorem sua medietate, et ita sunt infinitae veritates, et etiam numerus medietatum est infinitus, et ita aliquid est actu infinitum.

3. Diceres quod non est huiusmodi veritas veritas actualis, quia veritas causatur a re, quia sicut se habet res etc. Si ergo sumptio medietatum sit solum in potentia et excessus similiter, tunc respondebit ei veritas potentialis solum et non actualis, et ita sequitur infinitas esse veritates in potentia et non in actu.

4. Contra, veritas est adaequatio[c] rerum [d]et intellectuum, et veritas actualis est actualis adaequatio[e]; sed tanta[f] est adaequatio[g] rei ad intellectum si enuntietur res in potentia de re existente in potentia et adeo actualis adaequatio sicut si enuntietur res existens in actu de re existente in actu. Sicut ergo in una est actualis veritas sic et in altera.

5. Item, ad principale, aeternitas est nobilis vitae possessio tota simul, secundum Boethium, et loquitur de aeternitate creata; sed haec aeternitas est infinita duratione et actualiter existens; quare aliquid actu est infinitum.

6. Probatio propositionis suppositae in primo argumento, scilicet quod omnis numerus habet esse in factum esse et in actu, quia quod[h] indifferenter est rerum existentium [i]in actu et rerum existentium in potentia a neutro trahit suam actualitatem; sed indifferenter contingit numerare res existentes in potentia et res existentes in actu; quare etc.

7. Item, secundum Aristotelem in quarto huius, idem est numerus numero[j] decem hominum et decem equorum; ergo pari ratione idem erit numerus numero decem actuum et decem potentiarum respectu illorum actuum; quare etc.

8. Ad oppositum, omne existens in actu sub aliquo genere potest comparari ad res eiusdem generis; ergo, si esset numerus infinitus vel aliquid infinitum in actu, posset comparari; sed hoc non contingit quia, secundum Aristotelem[k], infinitum nulli proportionatur.

[a] sive] aliquid esse actu infinitum. Item infinitae *add. sed del.* O; quot *add.* T; suae C
[b] Item *om.* O
[c] adaequatio] coaequatio O
[d] et intellectuum] etc. T
[e] adaequatio] coaequatio O
[f] tanta] tam O
[g] adaequatio] rerum *add. sed. del.* O
[h] quod *om.* O
[i] in ... existentium *om. hom.* T
[j] numero *om.* O; *ante* numerus *transp.* T
[k] Aristotelem] dicentem quod *add.* T

continuum, and as many halves or parts that exist in potency will also exist in act, if the continuum is divided. Therefore, since it is possible for a continuum to be divided, it is possible that something is infinite in act.

2. Furthermore, there are infinitely many halves in potency in a continuum, and every half is greater than its own half; therefore, it is possible to assert infinitely many times that a half is greater than its own half, and thus there are infinitely many truths; moreover, the number of halves is infinite, and thus there is something infinite in act.

3. You might say that such a truth is not an actual truth, because a truth is caused by a thing, because just as a thing is related <to being so it is related to truth>. Therefore, if the taking of the halves is only potential and similarly for the excess, then only a potential truth and not an actual truth will correspond to this process, and thus it follows that there are infinitely many truths in potency and not in act.

4. Against this: truth is the adequation of things and intellects, and an actual truth is an actual adequation; but the adequation of a thing to an intellect when a thing in potency is asserted of a thing in potency is as much and as actual as when a thing in act is asserted of a thing in act. Therefore, just as there is an actual truth in one case, so also in the other.

5. Furthermore, about the principal claim, eternity is the whole simultaneous possession of a noble life, according to Boethius,[119] and he speaks of a created eternity; but this kind of eternity is infinite in duration and is actually existent; therefore, something is infinite in act.

6. Proof of the assumption made in the first argument,[120] namely, that every number has a fully realized and actual being: what belongs indifferently to things existing in act and to things existing in potency receives its actuality from neither of them; but things existing in potency and things existing in act are numbered indifferently; hence, <a number does not receive its actuality from the things numbered>.

7. Furthermore, according to Aristotle in *Physics* IV, the number of ten men and of ten horses is the same number;[121] therefore, by the same reasoning the number of ten acts and that of the ten potencies relative to those acts will be the same number; hence, <a number does not receive its actuality from the things numbered>.

8. To the contrary: everything that exists in act in a genus can be compared to the <other> things in that same genus; therefore, if there were an infinite number or something infinite in act, it could be compared <to the other things>; but this does not happen because, according to Aristotle, an infinite is not proportionate to anything.[122]

[119] Boethius, *De cons.* V.6.4, CCSL 94, 102.8–9.
[120] See above, n. 1.
[121] Arist. *Phys.* IV.14, 223b4–6, 224a2–3.
[122] Arist. *Phys.* VIII.1, 252a13.

<Quaestio 20>
Et gratia huius quaeratur an infinitum comparetur sive habeat comparationem alicui.

<Quaestio 20.1>

1. Et videtur quod sic, quia incipiat numerus a denario in infinitum et alius a millenario in infinitum. Isti duo numeri sunt comparabiles secundum maioritatem et minoritatem; quare etc.

2. [a]Item, imaginemur[b] numerum infinitum. In illo numero sunt infiniti millenarii et infinitae unitates, et tamen multo plures sunt ibi unitates quam millenarii, et ita est ibi comparatio secundum maioritatem[c] et minoritatem[d].

3. [e]Item, accipiamus duas lineas aequales in omni parte. In utraque[f] sunt puncta infinita, et sicut lineae proportionantur[g], sic et puncta in ipsis lineis proportionari debent[h], et ita infinitum proportionatur infinito.

4. Item, plura sunt puncta in tota linea quam in sua medietate, et tamen in utraque[i] sunt puncta infinita.

5. [j]Item, nisi infiniti ad infinitum esset proportio, non esset una creatura nobilior alia[k], quia inter creatorem et creaturam est infinita distantia, et tamen quaedam sunt meliora quia sunt immediatiora, et ita infinitum est propinquius primo quam aliud[l] infinitum.

6. Contra, imaginemur[m] triangulum et inter basim et conum imaginemur lineam transversalem, et a cono trianguli protrahantur lineae quot possunt usque ad basim. Istae lineae solum tangent[n] se in cono et proportionaliter distabunt per approximationem ad basim; et non est sumere punctum in basi ad quem non possit protrahi linea nec concurrent lineae nisi in cono; et ita quot possunt protrahi ad basim **[O 111va]** sine concursu et per transversalem

[a] Item ... minoritatem *om.* T
[b] imaginemur *corr. ex* imaginemus O
[c] maioritatem *corr. ex* maiorem O
[d] minoritatem *corr. ex* minorem O
[e] Item ... infinita (n. 4)] *post* aliud infinitum (n. 5) *transp. et* Item, plures sunt binarii in numero infinito quam denarii, tamen cum sint ibi infiniti denarii sicut et (sicut et *corr. mg. ex* quam) binarii et ita est comparare infinitum infinito secundum maioritatem *add.* T
[f] utraque] utroque O; utriusque T
[g] proportionantur] infinito *add. sed del.* O
[h] debent] deberent O
[i] utraque *scripsimus*; utroque OTC
[j] Item ... aliud infinitum *post* etc. (n. 1) *transp.* T
[k] alia] altera O
[l] aliud *ex corr.* O
[m] imaginemur *corr. ex* imaginamur O
[n] tangent] tangunt O

<Question 20>
Because of this, it is asked whether an infinite is compared to something or has a comparison to something.

<Question 20.1>

1. Arguments for the affirmative: suppose that one number begins at ten and goes to infinity and another number begins at one thousand and goes to infinity. These two numbers are comparable in accordance with greater and smaller; hence, <an infinite is compared to something>.

2. Furthermore, let us imagine an infinite number. In that number there are infinitely many thousands and infinitely many units; however, there are many more units than thousands in it, and thus there is a comparison there in accordance with greater and smaller.

3. Furthermore, let us take two lines equal in every part. In both lines there are infinitely many points; and as the lines are proportionate, so the points in those lines also have to be proportionate, and thus an infinite is proportionate to an infinite.

4. Furthermore, there are more points in the whole line than in its half; however, there are infinitely many points in both.

5. Furthermore, if there were no relation of an infinite to an infinite, one creature would not be nobler than another because there is an infinite distance between the creator and a creature; some creatures, however, are better because they are more immediately related <to the creator>, and thus one infinite is closer to the first cause than is another infinite.

6. Against this: let us imagine a triangle and a transversal line between the basis and the vertex, and let us draw as many lines as possible from the vertex of the triangle to the basis. These lines will meet only in the vertex and their distance will be proportionate to their closeness to the basis; also, it is not possible to take a point in the basis to which a line cannot be drawn, and these lines intersect only in the vertex. Accordingly, the lines that can be drawn to the basis without intersection are as many as the lines that can be drawn

sine concursu, quod[a] non contingeret si plura essent puncta in maiori linea[b] quam in minori; sed infinita sunt in[c] utraque; non ergo comparantur infinita secundum maius aut minus.

7. Idem argumentum est si describatur magnus circulus [d]et infra iuxta centrum parvus circulus, et protrahatur linea a centro ad maiorem circulum[e].

8. Item, Aristoteles in primo *Caeli et mundi* probat quod infinitum non comparatur infinito.

9. Istis rationibus respondetur sic, scilicet quod infinitum comparatur infinito non a parte qua procedit in infinitum, sed a parte illa qua incipit.

10. Contra, imaginemur[f] duas lineas finitas a parte una et infinitas a parte altera et resecentur ad aequalitatem a parte illa [[g]a parte] qua finitae sunt, et vocentur termini resectionis A, B. Et protrahatur linea recta ab A ad B ita quod in utraque parte faciat angulum rectum. Deinde protrahatur linea[h] ab A in infinitum secans lineam B, et vocetur illa linea C. Si ergo dicas quod linea A et linea B proportionentur secundum aequalitatem, probatio quod triangulus constat ex pluribus quam duobus rectis, quia linea A et linea C[i] sunt aequales, quia incipiunt ab eodem termino [j]et procedunt in infinitum, et linea A et linea B sunt aequales, [k]ut patet per resectionem; ergo a primo: linea B et linea C sunt aequales. Et si ab aequalibus aequalia demas, quae relinquuntur erunt aequalia; et ita cum in C[l] puncto resecetur utraque linea A[m], C, usque in infinitum sunt aequales, quia incipiunt in eodem puncto et procedunt in infinitum, et ita quae remanent sunt aequalia, et sic probatur triangulum constare ex pluribus rectis quam duobus.

11. Iterum, si erigatur linea ab A ad B in infinitum, probatur quod pars aequalis toti, [n]et hoc est aliud inconveniens[o].

[a] quod *ex corr.* O
[b] linea *suppl. s.l.* O; *om.* TC
[c] in *om.* O
[d] et ... circulus *om. hom. sed* parvus et *post* describatur *suppl. mg.* T
[e] circulum] et est idem argumentum *add.* OT
[f] imaginemur *corr. ex* imaginamur O
[g] a parte *delevimus*
[h] linea] recta ab A ad B ita quod in utraque parte faciat angulum rectum. Deinde protrahatur linea *add.* C
[i] C *ex corr.* O
[j] et ... infinitum *suppl. s.l.* T
[k] ut ... aequales *om. hom. sed* ut patet *suppl. s.l.* T
[l] C] F (*lectio incerta*) O
[m] A] *fortasse scribendum* B
[n] et ... inconveniens *om.* TC
[o] Qq. 16–18 *hic transp.* T

through the transversal line without intersection, which would not happen if there were more points in a greater line than in a smaller line; but there are infinitely many points in both lines; therefore, the infinites are not compared in accordance with more and less.

7. The same argument applies if one sketches a big circle with a small circle inside it close to the centre, and <then> draws a line from the centre to the bigger circle.

8. Furthermore, in *On the Heavens* I Aristotle proves that an infinite is not compared to an infinite.[123]

9. A reply to these arguments[124] is that an infinite is compared to an infinite not in accordance with the direction in which it proceeds to infinity, but in accordance with the direction from which it begins.

10. Against this: let us imagine two lines that are finite in one direction and infinite in the other direction, and let us cut equal sections in accordance with the direction in which they are finite, and let the termini of the sections be A and B. Draw a straight line from A to B in such a way that it produces a right angle with each line. Then draw a line from A to infinity intersecting line B, and let that line be C. Therefore, if you say that line A and line B are proportionate in accordance with equality, it can be proved that a triangle has more than two right angles as follows. Line A and line C are equal, since they begin from the same terminus and proceed to infinity, and line A and line B are equal, as is clear from the way they were cut; therefore, by transitivity, line B and line C are equal. And if you subtract equals from equals, the things that are left are equal; thus if we cut off the parts of each line A and C that begin from point C and proceed to infinity, these two parts are equal, because they begin from the same point and proceed to infinity, and thus the lines that remain <when these two parts are subtracted> are equal, and in this way it is proved that a triangle has more than two right angles.

11. Again, if a line is raised from A to B and prolonged to infinity, it is proved that a part is equal to its whole, and this is another absurdity.

[123] Arist. *De caelo* I.6, 274a7–8.
[124] See above, nn. 6–8.

<Quaestio 19.2>

1. Solutio[a]. Dicendum ad primum problema quod infinitum sub actuali existentia non potest esse.

2. Unde ad suppositum in primo argumento dicendum quod numerum esse in actu contingit dupliciter. Vel ita quod sit actualiter in rebus signatis et distinctis invicem; et sic vult Aristoteles quod numerus sit in actu, quia dicit numerum crescere secundum actualem divisionem continui; et illo modo non est numerus infinitus. Alio modo est numerus in actu quia est in actu nostrae signationis sive numerationis, et hoc contingit dupliciter. Vel sub una summa numerando, et sic potest esse infinitum. Bene enim dicitur 'infinita puncta sunt in hoc continuo'. Vel sub singulis discretionibus numerando, ita quod quaelibet sub proprio numerabili signetur[b] ab actu nostro, et sic est impossibile aliquid esse infinitum. Per hoc patet argumentis.

<Quaestio 20.2>

1. Solutio[c]. Ad secundum problema dicendum [d]sicut prius, scilicet quod in discretis comparantur infinita[e] a parte illa [[f]a parte] qua finita sunt; in continuis nec proportionantur sive comparantur a parte illa [[g]a parte] qua finita sunt nec a parte illa[h] qua infinita sunt. Et patebit ratio postea.

2. Unde est considerare numerum infinitum dupliciter: aut solum secundum quod [i]mensura est aut secundum quantitatem et congregationem unitatum. Primo modo non est numerus unus infinitus maior alio; secundo autem modo est unum infinitum maius alio, quia in infinito maior est congregatio unitatum quam millenariorum.

3. Et quod contingat[j] ita diversimode considerare numerum patet, quia Aristoteles in quinto[k] *Metaphysicae* dicit quod unum opponitur privative multo, et hoc secundum suam substantiam, et etiam opponitur ei relative, et hoc secundum quod est mensura.

[a] Solutio *om.* TC
[b] signetur] signatione O
[c] Solutio *om.* TC
[d] sicut prius *om.* C
[e] infinita *om.* O
[f] a parte *delevimus*
[g] a parte *delevimus*
[h] illa *om.* TC
[i] mensura est *corr. ex* mensurato O
[j] contingat] contingit TC
[k] quinto] *recte* decimo

<Question 19.2>

1. Solution. To the first problem[125] it must be said that an infinite cannot have actual existence.

2. Accordingly, to the assumption in the first argument[126] it must be said that a number can be in act in two ways: either (i) in such a way that it actually exists in the things that are marked off and distinct one from another; and Aristotle wants to say that a number exists in act in this way, because he says that a number increases in accordance with the actual division of a continuum;[127] but a number is not infinite in this way. Or (ii) a number is in act because it exists through our act of marking off or numeration. And this happens in two ways: either (iia) by numbering under one total, and in this way a number can be infinite; for it is rightly said that there are infinitely many points in this continuum. Or (iib) by numbering things as distinct singulars, in such a way that each thing is marked off by its own number through our act; and in this way it is impossible for anything to be infinite. By virtue of these remarks the reply to the arguments is clear.

<Question 20.2>

1. Solution. To the second problem[128] it must be replied as before,[129] namely, that in discrete things the infinites are compared in accordance with the direction in which these things are finite, whereas in continua the infinites are proportionate or compared neither in accordance with the direction in which they are finite nor in accordance with that in which they are infinite. And the reason for this will be made clear below.[130]

2. Thus, an infinite number can be considered in two ways: either (i) only in so far as it is a measure or (ii) in accordance with the quantity and aggregation of the units. In case (i) one infinite number is not greater than another; in case (ii) one infinite number is greater than another because in an infinite number there is a greater aggregation of units than of thousands.

3. And that it is legitimate to consider a number in these different ways is clear because in *Metaphysics* V Aristotle says that the one is privatively opposite to the many, when taken in accordance with its substance, and it is also relatively opposite to the many, in so far as it is a measure.[131]

[125] That is, q. 19.
[126] See above, q. 19.1, n. 1.
[127] Arist. *Phys.* III.7, 207b10–14.
[128] That is, q. 20.
[129] See above, q. 20.1, n. 9.
[130] See below, n. 5.
[131] The correct reference is Arist. *Metaph.* X.6, 1056b17–35.

4. Item, aliud exemplum, in numero ita est quod omnes numeri sunt eiusdem speciei considerando ad solam aggregationem unitatum, quia unitas unitati est eadem specie. Si autem consideremus rationem discretionis et certitudinem mensurae, sic, quia quaelibet unitas adiecta numero novam facit discretionem secundum speciem, sic differt quilibet numerus ab alio numero.

5. Sed quare est ita quod in continuis non comparantur sive non proportionantur duo continua infinita a parte una, licet sint finita a parte altera, et in discretis comparantur a parte illa [[a]a parte] qua terminati et finiti sunt numeri? Ratio est quia continuum unde continuum incipit indifferenter a parte una et a parte altera, et ideo non proportionantur duo continua aliquo modo nisi sint in omni parte terminata; sed in discretis est inceptio a parte determinata, ut ab unitate vel binario, et ideo licet procedant[b] in infinitum a parte finis, a parte tamen principii possunt comparari secundum aequalitatem vel inaequalitatem.

6. Sed nonne finis et principium [c]sunt correlata? Si ergo in numeris est necessario principium, erit et [d]in numero necessario finis.

7. Solutio[e]. Dicendum quod duplex est principium: vel principium existentiae durationis vel principium dimensionis[f]. Primo modo non necesse est, si habeat principium, quod habeat finem. Si loquamur de principio dimensionis[g], tunc possumus loqui de dimensione[h] vera et propria, et talis est in continuis solum, et in tali est necessarium quod, si habeat principium, quod habeat finem[i]. Et huius[j] ratio est quia non est maior ratio quare incipit plus a parte una quam a parte altera. Si loquamur de dimensione proportionali, cuiusmodi est in numeris, non est necesse **[O 111vb]** quod, si habeat principium, quod habeat finem, et hoc quia incipit a parte determinata.

8. Generalis ratio quare non est necesse quod, si aliquid habeat principium, quod habeat finem est quia posterius dependet a priore et non econverso, et ideo[k] ubi est finis erit principium necessario, et ubi est principium non necessario finis.

9. Sed quare est ita in numeris quod apposita unitate alicui numero necessario crescit nova species numeri, et in continuo non resultat nova species propter aliquam additionem?

[a] a parte *delevimus*
[b] procedant *scripsimus*; procedat OTC
[c] sunt correlata] dicuntur correlative T
[d] in numero necessario *om.* T
[e] Solutio *om.* TC
[f] dimensionis] divisionis O
[g] dimensionis] divisionis O
[h] dimensione] divisione O
[i] finem] et ratio est *add. sed del.* O
[j] huius *om.* TC
[k] ideo] non est finis *add. sed del.* O

4. Furthermore, another example: in the case of numbers, if we consider the aggregation of units alone, all numbers are of the same species because one unit is of the same species as another unit. But if we consider the *ratio* of discreteness and the determination of measure, then since any unit added to a number produces a new discreteness different in species, in this way any number differs from any other number.

5. But why is it the case that in continua two continua that are infinite in accordance with one direction are not compared or not proportionate, although they are finite in accordance with the other direction, whereas in discrete things numbers are compared in accordance with the direction in which they are limited and finite? The reason for this is that a continuum in so far as it is a continuum begins from one direction or the other indifferently, and therefore two continua are in no way proportionate unless they are limited in every direction; but in discrete things the beginning is from a determinate part, such as the number one or the number two, and therefore although they proceed to infinity when directed towards their end, they can, however, be compared in accordance with equality and inequality when taken in relation to their beginning.

6. But isn't it the case that the end and the beginning are correlative? Therefore, if in numbers there is necessarily a beginning, there will also necessarily be an end.

7. Solution. It must be said that a beginning is of two kinds: either (i) the beginning of the duration of existence or (ii) the beginning of the extent. In case (i) it is not necessary that if something has a beginning, it also has an end. In case (ii), we can speak of (iia) a true and proper extent, and this extent only exists in continua, and with respect to such an extent it is necessary that if something has a beginning, it also has an end. And the reason for this is that there is no stronger reason why something begins from one part rather than from another. (iib) If we speak of proportional extent, such as that which exists in numbers, it is not necessary that if something has a beginning, it also has an end, and this is because it begins from a determinate part.

8. The general reason why it is not necessary that if something has a beginning, it also has an end is that the posterior depends on the prior but not the other way around; therefore, where there is an end there is necessarily a beginning, and where there is a beginning there is not necessarily an end.

9. But why is it the case that in numbers a new species of number arises when a unit is added to a number, whereas in a continuum a new species does not come to be by addition?

10. Dicitur quod, quia in numero non est diminutum cum non sit ibi motus, ideo quocumque apposito vel diminuto statim cedit in speciem completam. In continuo autem non est ita.

11. Contra, in linea mathematica non est motus, et ita non est ibi incompletum, et tamen per additionem alicuius partis lineae non resultat nova species.

12. Mihi videtur quod ratio est quia numerus de se maxime sonat in discretionem et rationem mensurae, secundum quod dicit Commentator quod tempus secundum quod est mensura est numerus. Propter hoc diversitas discretionum et mensurarum facit diversitatem specialem, et ita ratio discretionis et mensurae est formale[a] in ipso. Cum ergo per ablationem unitatis sive per appositionem unitatis causetur nova discretio et nova ratio mensurae, et hoc in se, ideo appositio vel ablatio diversificat speciem. Sed in linea est ita quod solum mensurat[b] ex positione nostra, et ita accidit ipsi lineae quod mensuret, et ideo penes diversam mensurationem non diversificatur in ratione lineae, et ita non causat diversitatem in linea, sed solum in ratione mensurae.

1*. In hoc capitulo consequenter dicit quod omne corpus sensibile est in loco per se. Et exponit Commentator 'sensibile', hoc est tangibile.

<Quaestio 21>
Propter hoc quaeritur an sit caelum tangibile.

1. Et videtur quod sic, quia, secundum Aristotelem in secundo *De anima*[c], quae est proportio[d] virtutum, eadem est proportio obiectorum; quae ergo est proportio visus ad tactum, eadem est proportio visibilis ad tangibile. Sed omne videns est tangens; ergo omne visibile est tangibile; caelum est visibile; ergo et tangibile.

2. Item, si manus apponeretur caelo, non cederet neque finderet; resisteret ergo, et ita tangibile.

3. Item, maioris soliditatis est illud[e] quod non patitur fissionem aliquam quam[f] illud quod patitur; sed omne elementum patitur fissionem; sed corpus caeli non, quia tunc esset corruptibile, secundum Aristotelem in libro[g] *De caelo et mundo*.

[a] formale *ex corr.* O
[b] mensurat *ex corr.* O
[c] anima] quod *add.* OTC *sed del.* O
[d] proportio] visus ad *add. sed del.* O
[e] illud *om.* O
[f] quam *suppl. s.l.* O
[g] libro *om.* TC

10. It is replied that in numbers there is nothing that is diminished, since there is no motion there, and because of this any addition or subtraction to a number produces a complete species. In a continuum, however, it is not like this.

11. Against this: in a mathematical line there is no motion, and thus there is nothing incomplete there; however, a new species does not come to be by the addition of a part of a line.

12. It seems to me that the reason for this is that a number in itself is especially thought of as a discreteness and a *ratio* of measure, in accordance with the Commentator's statement that time is a number in so far as it is a measure.[132] Because of this the difference in discreteness and measure produces a specific difference, and thus the *ratio* of discreteness and measure is something formal in a number. Therefore, since by the subtraction or addition of a unit a new discreteness and a new *ratio* of measure is produced, and this in itself, it follows that addition or subtraction produces a difference of species. But a line is such that it only measures by our convention, and thus it is accidental to a line that it measures; therefore, a line is not diversified in the *ratio* of line by a different measure, and thus a different measure does not cause a difference in the line, but only in the *ratio* of measure.

1*. In this chapter Aristotle next says that every sensible body is per se in a place.[133] And the Commentator interprets 'sensible' as 'tangible'.[134]

<Question 21>

Because of this the question arises whether the heavens are tangible.

1. Arguments for the affirmative: according to Aristotle in *On the Soul* II, the relation between objects is the same as the relation between powers;[135] therefore, the relation of the visible to the tangible is the same as the relation of sight to touch. But every thing that has sight also has touch; therefore, every visible is tangible; the heavens are visible; therefore, they also are tangible.

2. Furthermore, if a hand were placed on the heavens, the heavens would not recede or split; therefore, they would resist, and so they are tangible.

3. Furthermore, a thing that is not subject to splitting has more solidity than a thing subject to it; but every element is subject to splitting, whereas the body of the heavens is not, because if it were, then it would be corruptible, according to Aristotle in *On the Heavens*.[136]

[132] Aver. *In Phys.* IV, t.c. 16, fol. 182ra.
[133] Arist. *Phys*. III.5, 205b31.
[134] Aver. *In Phys.* III, t.c. 54, fol. 110ra.
[135] Arist. *De anima* II.4, 415a14–22.
[136] Arist. *De caelo* II.8, 289b31–33 (cf. Aver. *In De caelo* II, t.c. 46, ed. Carmody, II, 363.52–54; t.c. 47, 364.5–7).

4. Solutio[a]. Dicitur quod in elementis et elementatis est ita quod, quanto[b] partes magis aggregantur et uniuntur, tanto est ibi maior soliditas; sed in caelo non est ita, quia in caelo est consolidatio[c] [d]partium sive soliditas; nec tamen resisteret sua soliditas tactui, et hoc quia soliditas eius est cum maxima partium extensione propter nobilitatem formae perficientis ipsam. Unde propter naturalem unionem partium et propter partium[e] unigeneitatem et etiam propter formae incorruptibilitatem causatur ibi tanta soliditas quod non potest ibi esse aliqua partibilitas.

<Quaestio 22>
Secunda suppositio Aristotelis est quod idem est locus totius et partis.

1. Contra, si hoc est verum, tunc, si scintilla ignis esset in medio mundi, moveretur[f] aut ad omnem partem regionis ignis aut ad nullam; sed constat quod non ad omnem partem; ergo ad nullam.

2. Item, si idem est locus etc., tunc posita parte terrae in quacumque parte[g] loci totius terrae quiesceret naturaliter; sed hoc videtur falsum, quia amota inferiore sustinente[h] descendit naturaliter; quare[i] etc.

3. Solutio[j]. Dicitur quod duplex est locus: aut in quo[k] continetur[l] aut ad quem[m] movetur. Unde locus ad quem[n] movetur terra est centrum terrae sive medium mundi[o], et tamen illud non est eius continens; locus autem ad quem movetur ignis est intima superficies concavi orbis lunae, et tamen non[p] est eius continens. Unde idem est locus totius[q] et partis, [r]verum est de loco ad quem moventur, de loco autem continente falsum est.

4. Vel sic: idem est locus quo ad salutem, non idem quoad quantitatem.

[a] Solutio *om.* TC
[b] quanto *ex corr.* O
[c] consolidatio] consideratio *scrips. sed corr. s.l. in* condensatio O
[d] partium sive soliditas *om. sed suppl. mg.* et si esset ibi soliditas partium T
[e] partium *om.* TC
[f] moveretur] incederetur (*lectio incerta*) *ex corr.* O
[g] parte] terrae *add. sed del.* O
[h] sustinente] sufficiente *scrips. sed corr. s.l. in* superior O
[i] quare *om. sed suppl. mg.* ergo O
[j] Solutio *om.* TC
[k] quo] quem O; *corr. ex* quod T; quod C
[l] continetur *scripsimus*; quiescit OTC; aliud *add.* TC
[m] quem] quod TC
[n] quem] quod TC
[o] mundi om. O
[p] non *fortasse delendum*
[q] totius *corr. s.l. ex* continuus O
[r] verum ... locus (n. 4) *om.* T

4. Solution. It is said that in the elements and in things derived from the elements the more the parts are aggregated and united the greater is their solidity; but in the heavens it is not like that. In the heavens there is consolidation or solidity of parts, but the solidity of the heavens does not offer resistance to touch, and the reason for this is that their solidity is accompanied by the maximal extension of their parts owing to the nobility of the form that gives perfection to their solidity. Accordingly, because of the natural union of the parts, because of the homogeneity of the parts, and also because of the incorruptibility of the form, such a great solidity is caused in the heavens that there cannot be any divisibility into parts there.

<Question 22>
Aristotle's second assumption is that the place of a whole and of a part is the same.[137]

1. Arguments for the negative: if this is true, then if a spark were in the middle of the world, it would either move towards every part of the region of fire or towards none of them; but it is apparent that it does not move towards every part; therefore, it moves towards no part.

2. Furthermore, if the place <of a whole and of a part> is the same, then a part of earth placed in any part of the place of earth would be naturally at rest there; but this seems to be false because when the part below it and supporting it is removed, it moves naturally downwards; hence, <it is not naturally at rest there>.

3. Solution. It must be said that there are two kinds of place: either (i) the place in which something is contained or (ii) that towards which something moves. Thus, the place towards which the earth moves is the centre of the earth or the middle of the world, but that thing is not its container. The place towards which fire moves is the internal surface of the concavity of the lunar orb, but this is not its container. Thus, the claim that the place of a whole and of a part is the same is true of the place towards which something moves, but it is false of the containing place.

4. Another solution is that the place is the same in relation to safety but not in relation to quantity.

[137] Arist. *Phys.* III.5, 205b20–22.

5. Argumento dicendum quod, si imaginetur scintilla in medio mundi, undique distendet se et movebitur undique per lineam brevissimam quousque corrumpatur.

6. Contra[a]: nonne videmus de flamma quod in ascensu suo in conum tendit, et ita non ascendit per lineam brevissimam?

7. [b]Dicendum quod ignis, si[c] non impediatur, semper ascendit per lineam brevissimam. Unde aliquid tendere ad conum potest esse duplici de causa: vel quia partes exteriores tendant in conum de se, et hoc non contingit in flamma, vel quia citius corrumpuntur partes exteriores et diutius durant partes in medio, et propter istam causam ascendit flamma sub figura conoidali. Partes enim exteriores citius corrumpuntur propter frigiditatem aeris circumstantis.

8. Ad secundum dicendum quod simul stant quod quiescat naturaliter et tamen, amota parte inferiore, descendat naturaliter, et hoc[d] quia descendit et movetur in suo loco. Si autem esset extra suum locum et moveretur naturaliter ad suum locum ita quod moveretur de loco ad locum, tunc quievit ibi violenter.

<Quaestio 23>

Tertia suppositio est quod omne finitum[e] consumitur per abstractionem finiti.

1. Contra: corpus solis est finitum, non tamen consumitur per subtractionem lucis.

2. Item, ad idem habetur argumentum de angulo contingentiae.

3. Dicendum quod Aristoteles intendit de subtractione quanti, et lux non est quantum; quare non est instantia. Item, Aristoteles loquitur de subtractione, et lux non subtrahitur sed multiplicatur. **[O 112ra]**

[a] Contra *suppl. s.l.* O; *om.* C
[b] Dicendum ... brevissimam *om. hom.* C
[c] si *suppl. s.l.* O
[d] hoc *suppl. s.l.* O; *om.* TC
[e] finitum] infinitum O

5. To the argument[138] it must be said that, if a spark is imagined in the middle of the world, it will expand itself in all directions and will move in all directions through the shortest line until it is corrupted.

6. Against this: don't we see that a flame in its motion upwards tends towards the conical apex, and so it does not move upwards through the shortest line?

7. It must be said that in the absence of obstacles fire always moves upwards through the shortest line. Thus, there can be two reasons why something tends towards the conical apex: either (i) because the exterior parts by themselves tend towards the conical apex, and this is not what happens in a flame; or (ii) because the exterior parts are corrupted more quickly whereas the parts in the middle last longer, and this is the cause of the fact that a flame moves upwards in a conical shape. The exterior parts of a flame are indeed corrupted more quickly because of the coldness of the surrounding air.

8. To the second argument[139] it must be said that the fact that a part of earth is at rest naturally and the fact that when the part below it is removed it moves downwards naturally are compatible. The reason for this is that that part of earth moves downwards within its place. But if a part of earth were outside its place and naturally moved towards its place so that it moved from a place to a place, then it would be at rest there by violence.

<Question 23>
The third assumption is that every finite thing is gradually destroyed by the subtraction of something finite.[140]

1. Against this: the body of the sun is finite, but it is not gradually destroyed by the subtraction of light.

2. Furthermore, the argument about the angle of contact proves the same point.

3. It must be said that Aristotle refers to the subtraction of a quantity, and light is not a quantity; therefore, this is not a counterexample. Also, Aristotle refers to subtraction, and light is not subtracted but multiplied.

[138] See above, n. 1.
[139] See above, n. 2.
[140] Arist. *Phys.* III.6, 206b9–12.

LIBER IV

1*. [a]*Similiter autem necesse est physicum et de loco sicut de infinito considerare et cognoscere si est aut non est* etc.

2*. Quaeritur[b] de loco.

<Quaestio 1>

<Quaestio 1.1>

1. Et quia Aristoteles probat locum esse, sufficiant rationes eius ad istam partem. Prima est: omne ens [c]videtur esse in loco; quare etc. Secunda: nos videmus corpus transmutari de loco ad locum.

2. Ad oppositum: si locus sit, aut est substantia aut accidens; sed non est substantia, quia, si sic, tunc transmutatio secundum locum esset transmutatio secundum substantiam, et talis est generatio et corruptio, et ita generatio [d]et corruptio essent loci mutatio.

3. Item, si sit substantia, aut ergo est corporea aut incorporea. Non corporea, quia, cum locus et locatum sunt simul, tunc duo corpora essent simul. Si sit incorporea, tunc aut est intelligentia aut anima rationalis aut causa prima, quorum quodlibet est impossibile.

4. Quod non sit accidens videtur, quia nullum accidens unum et idem numero potest esse in diversis subiectis numero; sed recedente uno corpore et adveniente altero, est unus et idem locus; quare non est accidens.

5. Item, nullum accidens est prius suo subiecto; sed locus est prius locato, cum sit receptivus eiusdem; quare etc.

6. Item, nullum accidens est unigeneum cum[e] substantia, sed est alterius naturae et specialis et generalis; sed locus est unigeneus locato, secundum quod dicit Aristoteles quod res movetur ad locum ut ad suam formam[f].

[a] Similiter ... etc. *om.* C
[b] Quaeritur] In principio quarti huius *praem.* C
[c] videtur esse *om. sed* est *suppl. s.l.* O
[d] et corruptio essent] esset C
[e] cum *suppl. mg.* O
[f] formam] quare etc. *add.* C

BOOK IV

1*. *The physicist must investigate place, too, as well as the infinite, and must know whether it exists or not <and the manner of its existence and what it is>.*[1]

2*. Questions about place.

<Question 1>

<Question 1.1>

1. Given that Aristotle proves that place exists, let his arguments be sufficient for this conclusion. The first argument is: every being seems to be in a place; hence, <place exists>.[2] The second: we see that a body moves from place to place.[3]

2. To the contrary: if place exists, it is either a substance or an accident; but place is not a substance, because, if it were, then a change with respect to place would be a change with respect to substance, and the latter kind of change is generation and corruption, and thus generation and corruption would be a change of place.

3. Furthermore, if place is a substance, then it is either a corporeal substance or an incorporeal substance. It is not a corporeal substance because, since place and the located body are in the same place, then two bodies would be in the same place. If it is an incorporeal substance, then it is either an intelligence or a rational soul or the first cause, each of which cases is impossible.

4. Arguments that place is not an accident: no accident that is one and the same in number can be in numerically distinct subjects; but when one body moves out of a place and another body comes in, place remains one and the same; therefore, place is not an accident.

5. Furthermore, no accident is prior to its subject; but place is prior to a located body, since it receives the located body; therefore, <place is not an accident>.

6. Furthermore, no accident is of the same kind as a substance, but is of another nature, both specific and general; but place is of the same kind as the located body, according to Aristotle's statement that a thing moves towards its place as if towards its form.[4]

[1] Arist. *Phys*. IV.1, 208a27–28.
[2] Arist. *Phys*. IV.1, 208a29–31.
[3] Arist. *Phys*. IV.1, 208a31–32.
[4] Arist. *De caelo* IV.3, 310a33–b1.

7. Et praeter hoc, secundum Aristotelem, quod est principium motus localis est forma substantialis; [a]sed forma substantialis non movet per se nisi ad formam substantialem, et movet ad locum; quare etc.

8. Item, Commentator dicit quod corpora caelestia sunt causae suorum locorum, et ista inferiora causantur a suis locis; sed [b]accidens non est causa substantiae; quare etc.[c]

9. Item, si est accidens, aut[d] corporis locantis aut locati.

<Quaestio 2>
Et sit hoc pro secundo problemate.

1. Quod non [e]sit locantis[f] accidens videtur, quia omne accidens quantum distenditur a distensione proprii subiecti in quo est; sed locus non recipit distensionem sive impressionem a locante, sed a locato; quare etc.

2. Item, si sit in locante, [g]tunc amoto locante, amovetur et locus; sed hoc est falsum, quia contingit aliquid manens in eodem loco prius locari in aere, [h]et illo recedente, locari in aqua, et tamen[i] manet idem locus, quia, si non, tunc transmutat locum, et quiescit secundum totum et secundum partes; ergo aliquod quiescens secundum locum transmutaret locum, [j]quod est inconveniens.

3. Quod non sit accidens locati videtur, quia nihil quod est superficies unius corporis est accidens alterius; sed, secundum Aristotelem, locus est ultimum corporis continentis; quare non est accidens locati.

[a] sed forma substantialis] et C
[b] accidens non est] est accidens non est *scrips. sed corr. in* si esset accidens, non esset O
[c] etc.] locus non est accidens C
[d] aut *suppl. mg.* O; est accidens *add.* C
[e] sit locantis accidens] locantis C
[f] locantis *corr. ex* localis O
[g] tunc amoto locante *om. hom. sed* amoto locante *suppl. s.l.* O
[h] et illo] in illo *scrips. et* aere *add. s.l.* O
[i] tamen *ex corr.* O
[j] quod est inconveniens *om.* C

7. Also, according to Aristotle, the principle of locomotion is a substantial form;[5] but a substantial form per se causes motion only towards a substantial form; and it causes a motion towards a place; therefore, <place is a substantial form>.

8. Furthermore, the Commentator says that the celestial bodies are the causes of their places, whereas things here below are caused by their places;[6] but an accident is not the cause of a substance; therefore, <place is not an accident>.

9. Furthermore, if place is an accident, it is an accident either of a locating body or of a located body.

<Question 2>
Let us raise this as a second problem.

1. Arguments that place is not an accident of the locating body: every quantified accident is extended in virtue of the extension of its own subject, that in which it exists; but place does not receive its extension or impression from the locating body, but from the located body; therefore, <place is not an accident of the locating body>.

2. Furthermore, if place is in the locating body, then when the locating body is removed, place is also removed; but this is false because it can happen that something remaining in the same place is first located in air, and when air recedes, it is located in water, but its place remains the same; for, if its place is not the same, it would change place, yet it is at rest as a whole and in its parts; therefore, something at rest with respect to place would change place, which is absurd.

3. Arguments that place is not an accident of the located body: nothing that is the surface of one body is an accident of another body; but according to Aristotle, place is the boundary of the containing body;[7] therefore, it is not an accident of the located body.

[5] Arist. *De caelo* I.8, 276b23–26 (cf. Aver. *In De caelo* I, t.c. 81, ed. Carmody, I, 151–152).
[6] The contrast between the place of the celestial bodies and the place of the sublunar bodies that Aspall ascribes to Averroes does not appear in Averroes' specific discussion of the place of the heavens in *In Phys.* IV, t.c. 43–46. In a passage of his commentary on *De caelo*, however, Averroes proposes a similar contrast between the place of the elements and the place of the composite bodies. Cf. Aver. *In De caelo* IV, t.c. 40, ed. Carmody, II, 747.89–96: "Et iam apparet hic quod non est de locis istorum sicut de locis aliorum corporum naturalium: loca enim in istis elementis sunt cause locatorum, in aliis autem corporibus est econverso, scilicet quod locata sunt cause locorum, licet hoc inveniatur in eis secundum prius et posterius et non equaliter, scilicet quod compositorum illud quod est vicinius elementis, locus est ei magis necessarius et indiget plus loco secundum illud esse, et quanto remotius tanto minus... ".
[7] Arist. *Phys.* IV.4, 212a20–21.

4. Item, locus est prius locato, et accidens est posterius suo subiecto; quare etc.

5. Ad idem est argumentum quod prius, scilicet quod recedente uno corpore etc.

6. Propter hoc ultimum argumentum [a]commune dicitur diversimode. Et dicitur sic quod locus est accidens, non tamen est accidens absolutum alterius istorum, sed est accidens unius[b] per respectum ad alterum, ut[c] per respectum locati ad locans proximum, et cum hoc ad centrum et circumferentiam mundi, et[d] est haec ultima condicio completio loci. Et per istum respectum est omne continens continens et omnis locus locus, et ab identitate huius respectus sive[e] huius distantiae est [f]identitas loci [g]et a diversitate diversitas loci. Unde non est accidens huius corporis vel illius, sed est accidens corporis absolute secundum quod est sub respectu tali.

7. Contra: iste respectus aut est substantia aut accidens. Non substantia, quia est relatio quaedam quae est de genere accidentium. Si accidens, tunc cum nullum accidens maneat idem numero in diversis subiectis, si removeatur unum corpus et succedat ei aliud, non maneret idem locus. Aut ergo dabis aliam causam identitatis aut procedetur in infinitum.

8. Item, cum caelum sit in continuo motu et ita partes eius transmutant situm, tunc est impossibile quod eadem linea semper habeat respectum eundem[h] ad eandem partem caeli sed ad partes infinitas, et ita non potest manere idem respectus numero.

<Quaestio 1.2>

1. Dicendum quod locus est. Unde quia propositio Aristotelis per quam probat locum esse sapit falsitatem[i] nisi glossetur, ideo dico[j] quod sic intelligenda est: omne ens videtur esse in loco, hoc est omne quod est quantum de se est in loco. Et hoc dico propter causam primam et intelligentiam et animam, quae non sunt in loco proprie, tamen non est negandum ipsa esse [k]in loco omni modo, quia esse in loco contingit dupliciter. Vel ita quod circumscribatur loco, hoc est ita quod partes loci commetiantur se partibus

[a] commune ... Et *scrips. sed del.* O (et cum *post* commune *add.*)
[b] unius *ex corr.* O
[c] ut *corr. ex* unde O; unde C
[d] et *ex corr.* O
[e] sive *om.* O
[f] identitas ... diversitate *suppl. mg.* O
[g] et ... loci *om. hom.* C
[h] eundem *corr. s.l. ex* semper O; *om.* C
[i] falsitatem *ex corr.* O
[j] dico *suppl. s.l.* O; dicendum C
[k] in loco *suppl. s.l.* O

4. Furthermore, place is prior to the located body, and an accident is posterior to its subject; therefore, <place is not an accident of the located body>.

5. The same point is shown by a preceding argument,[8] that is, when one body moves out <of a place and another body comes in, place remains one and the same>.

6. This last standard argument gives rise to various views. One view is that place is an accident, but it is not an absolute accident of either the located body or the locating body, but it is an accident of one of them in virtue of the relation this one has to the other, that is, in virtue of the relation of the located body to its proximate locating body, and in addition to this <in virtue of the relation of the locating body> to the centre and to the circumference of the world. This last condition completes the notion of place. It is in virtue of this relation that every container is a container and that every place is a place, and the identity and diversity of place derive from the identity and diversity of this relation or distance. Accordingly, place is not an accident of this or that body, but is an accident of a body absolutely in so far as it has such a relation.

7. Against this: this relation is either a substance or an accident. It is not a substance because it is a relation belonging to the genus of accidents. If it is an accident, then since no accident remains numerically the same in different subjects, if one body is removed and another one succeeds it, place would not remain the same. Therefore, either you will assign another cause of the identity of place or there will be an infinite regress.

8. Furthermore, since the heavens are in perpetual motion and so their parts change their position, it is impossible for the same line always to have the same relation to the same part of the heavens, but <it has relations> to infinitely many parts, and thus that relation cannot remain numerically the same.

<Question 1.2>

1. It must be said that place exists. Thus, since the proposition that Aristotle uses to prove that place exists[9] sounds false, unless it is glossed, therefore, I say that it must be understood as follows: 'every being seems to be in a place', that is, every thing that is a quantity by itself is in a place. I say this because of the first cause, intelligences, and souls, which are not properly in a place, but it is not to be absolutely denied that they are in a place, because something can be in a place in two ways. Either (i) in such a way as to be

[8] See above, q. 1.1, n. 4.
[9] Arist. *Phys*. IV.1, 208a29.

locati; et sic est corpus in loco, et tale est proprie in loco. Vel ita quod determinetur loco, et sic [a]sunt [b]formae illae in loco, scilicet[c] anima et intelligentia. Anima enim mea determinatur ita corpori meo quod non alteri corpori. Sed utroque modo negatur locus a causa prima. Tertio modo est aliquid in loco quia praesens est loco; et ita est causa prima in loco, quia est ubique praesentialiter et essentialiter et potentialiter, secundum Augustinum.

2. Argumentis in oppositum dicendum quod est loqui de loco dupliciter, scilicet aut de loco mathematice dicto, et hoc secundum quod respicit solam continentiam; et isto modo est locus accidens. Alio modo est considerare locum secundum quod naturalis est, et ista consideratio addit priori considerationi influentiam virtutis et salutis. Unde locus isto modo est ultima superficies corporis continentis cum influentia virtutis corporis locantis proximi[d] et etiam omnium corporum superiorum; et sic est locus aliquo modo substantia, ut scilicet[e] consideratur in ratione influxionis virtutis et salutis.

3. Adhuc quod locus non sit accidens videtur, quia omne accidens dependet ab aliquo subiecto; **[O 112rb]** locus non dependet ab aliquo subiecto; quare etc. Probatio minoris. Si continens et contentum utrumque amoveatur [et[f]] advenientibus aliis duobus, ut continente et contento, aequalibus, manet locus idem numero, quod non contingeret si ab aliquo illorum dependeret.

4. Item, quia lux sive radius lucis manet idem numero, uno aere recedente et alio succedente, ideo dicimus quod lux non est in aere ut in subiecto, cum a neutro dependeat. Cum ergo locus similiter se habeat respectu locantis et locati, tunc in neutro erit ut in subiecto, nec ab altero istorum poterit dependere.

5. Item, ita est quod, quia materia manet eadem sub formis diversis, dicimus[g] quod materia est substantia et quod non dependet ab istis formis. Cum ergo locus maneat idem numero diversificatis locantibus et locatis[h], videtur quod locus sit substantia, et ita a nullo illorum dependet, nec est accidens alicuius illorum.

[a] sunt ... anima] est anima in loco C
[b] formae illae *ex corr.* O
[c] scilicet *suppl. s.l.* O
[d] proximi *lectio incerta* O
[e] scilicet *suppl. mg.* O; *om.* C
[f] et *delevimus*
[g] dicimus *illegibile scrips.* O
[h] locatis *ex corr.* O

circumscribed by place, that is, in such a way that the parts of place are commensurate to the parts of the located thing; and such a thing is properly in a place. Or (ii) in such a way as to be determined by place; and in this way those forms, namely souls and intelligences, are in a place; for my soul is so determined to my body that it is not determined to another body. But the first cause is not in a place in either of these ways. (iii) In a third way something is in a place because it is present to a place; and in this way the first cause is in a place because it is everywhere by its presence, its essence, and its power, according to Augustine.[10]

2. To the arguments to the contrary[11] it must be said that we speak of place in two ways: either (i) of place in a mathematical sense, and this in so far as place concerns containment alone; and in this way place is an accident. Or (ii) of place in so far as it is natural. The second consideration (ii) of place adds to the first consideration (i) the influence of power and of safety. Thus, place as in (ii) is the extreme surface of the containing body together with the influence of the power of the proximate locating body and of all of the superior bodies as well. And thus place is in some way a substance, that is, when it is considered in relation to the pouring of its power and to safety.

3. Further arguments that place is not an accident: every accident depends on a subject; but place does not depend on a subject; therefore, <place is not an accident>. Proof of the minor premise. If the containing and the contained bodies are both removed and are replaced by two other bodies—one containing and the other contained—equal to them, place remains numerically the same, which would not happen if place depended on one of them.

4. Furthermore, it is because light or a ray of light remains numerically the same, when one part of air recedes and another one succeeds, that we say that light is not in the air as its subject, since it depends on neither of the two parts of air. Therefore, since the same is true of place in relation to the locating and the located body, then place will be in neither of them as its subject, and it will depend on neither of them.

5. Furthermore, it is because matter remains the same under different forms that we say that matter is a substance and that it does not depend on these forms. Therefore, since place remains the same in number when the locating and the located bodies vary, it seems that place is a substance, and thus it does not depend on any of them nor is it an accident of one of them.

[10] The correct reference is Peter Lombard, *Sententiae* I, dist. 37 c.1 § 5 (ed. Coll. S. Bonav. IV, 265.2–4); based on *Glossa ord. in Cant*. 5.17 (ibid., § 3, 264.6–8 and critical apparatus).
[11] See above, q. 1.1, nn. 2–8.

6. Dicendum, sicut prius, quod considerando influxionem virtutis, sic aliquo modo est substantia. Considerando mathematice sive solam distantiam, sic est accidens. Sed intellige quod non est accidens absolute unius illorum, sed unius per comparationem ad alterum. Dicendum ergo huic ultimo quod est in aliquo ut in subiecto.

7. Unde primae rationi dicendum quod, licct locus non dependeat ab hoc corpore vel ab illo, dependet tamen a corpore simpliciter.

8. Ad secundum dicendum quod lux non manet eadem numero, recedente uno aere et succedente alio, et adhuc etiam, si maneret aer, non maneret eadem lux numero per aliquod tempus, quia lux semper est in continua gignitione, nec etiam potest lux aliquo modo manere[a]. Esto tamen[b] quod maneat eadem lux. Dicendum argumento quod alio modo est lux in aere quam locus in suo susceptibili, quia lux est in aere per diffusionem et expansionem, et est aliquid in se praeter hoc quod est illuminans medium, quod patet, quia aliter non numerarentur diversae luces in eadem parte medii, quod non contingit de aliquo accidente sive aliquo quod sit forma tantum, et ita non dependet lux etc. Sed locus praeter hoc quod est accidens non est aliquid in se nec est simile etc.

9. Ad tertium respondet Aristoteles dicens quod materia est ita sub formis diversis quod est pars rei; sed locus non est pars rei; quare etc.

<Quaestio 3>
Sed quaeritur in quo est locus ut in subiecto.

1. Et dicit auctor *Sex principiorum* quod 'ubi' est in circumscripto, locus autem in circumscribente.

2. Alii dicunt quod locus nec est absolute in locante neque in locato absolute, sed est in utroque quia causatur ab utroque. Et inducunt simile ad hoc de sono quod nec est in percusso nec in percutiente sed est in utroque, et hoc quia utrumque percussum est.

[a] manere *scripsimus*; moveri OC
[b] tamen *suppl. s.l.* O

6. It must be said as before[12] that if we consider the pouring of power, then place is in some way a substance. But if we consider place mathematically or distance alone, then it is an accident. You have to understand, however, that place is not absolutely an accident of either the locating or of the located body, but of one of the two in relation to the other. Therefore, to the last argument it must be said that place is in something as its subject.

7. Accordingly, to the first argument[13] it must be said that, although place does not depend on this or that body, however it depends absolutely on body.

8. To the second argument[14] it must be said that light does not remain numerically the same when one part of air recedes and another succeeds, and even if air remained, light would not remain numerically the same for some period of time because light is generated continuously, nor can light remain in some way or other. Suppose, however, that light remains the same. To the argument it must then be said that light is in the air in a different way from that in which place is in its recipient, because light is in the air by diffusion and expansion, and it is something in itself in addition to being what illuminates the medium. This is obvious, because otherwise different lights would not be numbered in the same part of the medium, which does not happen to an accident or something that is only a form, and thus light does not depend <on the medium>. But place is not something in itself in addition to being an accident nor is the case of place similar <to that of light>.

9. To the third argument[15] Aristotle replies that matter is under different forms in such a way that it is a part of the thing; but place is not a part of the thing;[16] hence, <the case of place is not similar to that of matter>.

<Question 3>
The question is what place is in as its subject.

1. The author of the *Six principles* says that the 'where' is in the circumscribed thing, whereas place is in the circumscribing thing.[17]

2. Other people say that place is neither in the locating body nor in the located body absolutely, but it is in both because it is caused by both. And they put forward a similarity with sound, which is neither in the thing struck nor in the thing striking, but it is in both of them, and this because both of them are struck.

[12] See above, n. 2.
[13] See above, n. 3.
[14] See above, n. 4.
[15] See above, n. 5.
[16] Arist. *Phys*. IV.2, 209b22–24; IV.4, 211b31–212a2.
[17] Anon. *Sex princ*. 5, AL I.6–7, 45.8–9.

3. Sed nonne sonus accidens est? Qualiter potest esse in pluribus ut in subiecto?

4. Et respondent[a] sic: sicut est de numero quaternario quod habet omnia numerata non ut plura subiecta sed ut unum subiectum, sic est de sono quod habet utrumque ut unum subiectum[b].

5. [c]Sed contra videtur quod simile non sufficiat, quia quattuor[d] non sunt unum subiectum secundum[e] quod unum est numerans et aliud est numeratum, sed secundum quod omnia sunt numerata. Cum ergo locans[f] et locatum solum se habeant ita quod unum est locans et aliud locatum, et non secundum quod utrumque est locatum, cum ergo non conveniant in aliqua eadem ratione, non possunt esse unum subiectum.

6. Similiter est de alio exemplo, quia in generatione soni est utrumque percussum, et ideo possunt esse ut unum subiectum soni; sed si unum esset tantum in ratione percutientis et aliud tantum in ratione percussi, non possent esse unum subiectum respectu alicuius accidentis; sed sic est in locante et locato.

7. Quidam autem dicunt quod locus est in locante ut in subiecto, et hoc quia accidens solum est in eo ut in subiecto ex cuius principiis egreditur. Cum ergo egrediatur ex principiis locantis, cum sit eius ultimum, erit locus in locante ut in subiecto; sed[g] non absolute sed per comparationem ad locatum.

8. Et dicunt argumento de distensione loci: sicut est si fuerit aliquid congregatum et densum, calor agens in ipsum distendit et rarefacit illud congregatum[h]. Unde illa distensio est effective in calore, subiective tamen est in illo rarefacto. Unde ita est quod effective est distensio in locato, subiective est in locante.

9. Aliter dicitur sic quod locus proprie non distenditur distensione locati. Et dicunt quod nihil[i] est in loco proprie nisi ignis, et hoc est quia suum continens est terminatum et distensum dimensionibus propriis et non dimensionibus alienis, sicut concavum orbis lunae non distenditur distensione ignis.

10. Sed quod non sit in locante videtur, quia, quando aliquis habitus est causatus inter duo, est in eo ut in subiecto cui acquiritur ille habitus, ut patet

[a] respondent *ex corr.* O
[b] subiectum *ex corr.* O
[c] Sed contra *om.* C
[d] quattuor *scripsimus*; decem O; quinque C
[e] secundum *ex corr.* O
[f] locans *scripsimus*; locus OC
[g] sed] nisi O
[h] congregatum] aggregatum O
[i] nihil *ex corr.* O

3. But isn't it the case that sound is an accident? How then can sound be in many things as one subject?

4. They reply as follows: just as the number four has all the things numbered by it as its one subject and not as many subjects, so sound has both things as one subject.

5. Against this you can argue that this similarity is not sufficient, because four things are not the one subject of number four in so far as one of them is numbering and another is numbered, but in so far as all of them are numbered. But since the locating and the located bodies are such that one of them is locating and the other located, and they are not both located, therefore they cannot be one subject, since there is not one and the same *ratio* in which they agree.

6. The same applies to the other example, because in the generation of sound both things are struck and, therefore, they can be the one subject of sound; but if one of the two things only had the *ratio* of what strikes and the other only had the *ratio* of what is struck, they could not be one subject with respect to an accident; but this is what happens in the case of the locating and the located bodies.

7. Some people say that place is in the locating body as its subject, and this because an accident is only in a thing as its subject when it emerges from the principles of that thing. Therefore, since place emerges from the principles of the locating body, since it is the boundary of this body, place will be in the locating body as its subject; not absolutely, however, but in comparison to the located body.

8. And to the argument about the extension of place[18] they reply that the case of place is like that of a thing that is an aggregate and dense when heat acting on it makes that aggregate extended and rarefied. Hence, that extension is effectively in heat, but in the rarefied thing as its subject. Accordingly, the extension of place is effectively in the located body and in the locating body as its subject.

9. Another reply is that place is not properly extended by the extension of the located body. And these people say that nothing is properly in a place except fire, and this because the container of fire is limited and extended in virtue of its own dimensions and not in virtue of extraneous dimensions, that is, the concavity of the lunar orb is not extended by the extension of fire.

10. Arguments that place is not in the locating body: when a habit is caused between two things, it is in the one of the two things that acquires that habit as its subject, as is clear in the case of knowledge, which is in the knower.

[18] See above, q. 2, n. 1.

de scientia, [a]quae est in sciente. Cum ergo locus sic causetur et sit[b] habitus acquisitus locato, erit in locato ut in subiecto.

11. Item, est movens et motum[c] et est causatum medium ex istis, ut motus, et est in moto ut in subiecto; ergo locus erit in locato ut in subiecto.

12. Item, actio et passio, et medium inter ista est in recipiente. Tunc, ut videtur, erit locus in locato, cum sit recipiens.

13. Item, caelum habet locum, nec tamen habet locans extra; ergo locus eius non est in locante ut in subiecto[d] simpliciter loquendo.

14. Solutio[e]. Dicendum quod est considerare locum dupliciter. Et loquamur de loco inferiorum. Aut secundum quod est quiddam extra continens et ambiens ipsum extra, ut secundum quod est terminus corporis continentis, egrediens ex principiis eius; et sic est in locante ut in subiecto, et sic vult auctor *Sex principiorum* quod sit in continente. Aut non solum ut **[O 112va]** est extra continens, sed ut est habitus quidam acquisitus locato, profundans se usque in centrum corporis ut quodammodo mensura intra, quia partes loci correspondent partibus corporis, non sub dimensione propria sed sub dimensione corporis locati, [f]et sic est in locato ut in subiecto. Et quod sit possibilis ista consideratio videtur. Scimus quod una relatio per diversos respectus diversa sortitur nomina et subiecta, ut paternitas et filiatio una relatio est secundum rem.

<Quaestio 4>

Sed ex quo ita est quod est in utroque ut in subiecto, et hoc per diversam considerationem, qualiter est quod, utroque recedente et duobus aliis aequalibus succedentibus, adhuc manet idem locus numero? [g]Et hoc est quaerere de immobilitate[h] loci.

1. Quidam dicunt quod, cum dicimus quod locus est immobilis, hoc est verum de distantia mathematica. Locus enim nominat distantiam, et hoc potest esse vel mathematicam vel naturalem. De distantia autem unde naturalis est non est verum.

[a] quae est *suppl. s.l.* O; *om.* C
[b] sit *suppl. s.l.* O; *om.* C
[c] motum *ex corr.* O
[d] subiecto] et ita non est locante *add.* C
[e] Solutio *om.* C
[f] et ... locato *om.* C
[g] Et ... loci *om.* C
[h] immobilitate *corr. s.l. ex* identitate O

Therefore, since place is caused in this way and is a habit acquired by the located body, it will be in the located body as its subject.

11. Furthermore, consider the mover and the moved: what results from them, that is, motion, is intermediate between them; and motion is in the moved as its subject; therefore, place will be in the located body as its subject.

12. Furthermore, consider action and passion: the intermediate between them is in the recipient. Then, it seems, place will be in the located body, since the located body is the recipient.

13. Furthermore, the heavens have a place, but they do not have an external locating body; therefore, their place is not in a locating body as its subject absolutely speaking.

14. Solution. It must be said that place can be considered in two ways. And we are talking about the place of the things here below. Either (i) in so far as it is something that externally contains and surrounds the located body, that is, in so far as it is the terminus of the containing body and emerges from its principles; and in this way place is in the locating body as its subject, and it is in this way that the author of the *Six principles* posits that place is in the containing body.[19] Or (ii) place can be considered not only as an external container but also as a habit acquired by the located body which reaches down into the centre of the body as a kind of internal measure, because the parts of place correspond to the parts of the body, not in accordance with their own dimension but in accordance with the dimension of the located body; and in this way place is in the located body as its subject. And that it is possible to consider place as in (ii) is shown by the case of relation, because we know that one relation in virtue of different aspects acquires different names and different subjects, in the way that paternity and filiation are one relation in reality.

<Question 4>

Since place is in both the locating and the located body as its subject, and this in virtue of different considerations, the question arises of how it is possible that place still remains numerically the same when both bodies recede and two other equal bodies succeed. And this question is about the immobility of place.

1. Some people say that the claim that place is immobile is true of mathematical distance; for place designates a distance, and this can be either mathematical or natural. But the claim is not true of distance in so far as it is natural.

[19] See above, n. 1.

2. Contra: mathematica secundum actum existendi sunt in naturalibus, et tantum[a] sunt abstracta secundum considerationem, [b]et intelligentia et anima rationalis sunt abstracta et secundum esse et secundum[c] considerationem; sed anima rationalis, quae ultimo modo abstrahitur, movetur per accidens, secundum Aristotelem; quare multo fortius illa distantia mathematica aliquo modo erit mobilis secundum accidens, et ita locus, quod est contra Aristotelem.

3. Aliter dicitur sic quod extra orbem stellatum est orbis non motus, diffundens suas virtutes luminares per totum universum, ordinans universa in naturis propriis, et est sub minus nobili esse in terra quam in aere, et sic deinceps. Et quia ista natura lucis caeli est immobilis, ab identitate huius attribuimus loco identitatem et immobilitatem.

4. Contra: illa lux aut est semper in fieri aut in factum esse. Si in fieri, tunc non est causa identitatis. Si in factum esse, contra: lux ista aut est corpus aut virtus defluxa a corpore. Non corpus, quia sic essent duo corpora in eodem loco. Si non sit corpus, ut virtus aliqua, sed virtus aliqua non dat realem identitatem ipsi rei; lux ergo non erit causa identitatis loci.

5. Aliter dicitur sic quod, licet caelum moveatur secundum partes, non tamen movetur secundum totum. Unde identitas loci non est ab identitate respectus ad aliquam partem caeli, sed ad totum caelum unde totum est.

6. Contra: imaginemur quod aliquid quiescat[d] in aere et quod caelum aliquantulum moveatur a loco suo ita quod transmutet centrum. Illud quiescens manet in eodem loco, cum non moveatur de loco ad locum, et tamen non manet idem respectus ad totum caelum.

7. Item, aliquid potest mutari continue[e] localiter, et tamen habere semper eundem respectum ad circumferentiam totius caeli, ut si aliquid moveatur continue et aliquid[f] moveatur super illud motu regressionis ita quod non sit maior velocitas unius quam alterius, illud motum supra movens non transmutat respectum ad circumferentiam [g]et tamen etc.

8. Ad primum dicendum quod imaginatio ponit incompossibilia[h], scilicet quod caelum locum transmutet secundum totum et [i]tamen quod aliquid quiescat in loco.

[a] tantum *scripsimus*; tamen O; quando C
[b] et ... considerationem *om. hom.* C
[c] secundum *ex corr.* O
[d] quiescat] crescat O
[e] continue] in continue *scrips. sed* continue *del. et corr. s.l. in* aere alio O
[f] aliquid] ut musca *add. s.l.* O
[g] et tamen *ex corr.* O; ergo per oppositum C
[h] incompossibilia] impossibilia O
[i] tamen quod *inv. sed corr.* O

2. Against this: mathematical things are in natural things through their act of existing, and they are only abstract through our consideration <of them>, whereas an intelligence and a rational soul are abstract both in being and in consideration; but a rational soul, which is abstract in the latter way, moves accidentally, according to Aristotle;[20] hence, there is a much stronger reason to say that mathematical distance is in some way accidentally mobile, and so too is place, which is contrary to Aristotle.[21]

3. Another opinion is that outside the orb of the stars there is an immobile orb, which spreads its illuminating powers throughout the whole universe, and orders all things in their proper natures, and <this nature of the celestial light> has a less noble existence in earth than in air, and so on. And since this nature of the celestial light is immobile, it is from the identity of this nature that we ascribe identity and immobility to place.

4. Against this: that light is either always in becoming or in fully realized being. If it is in becoming, then it is not the cause of the identity of place. If it is in fully realized being, one can argue against this: this light is either a body or a power pouring out of a body. It is not a body, because, if so, two bodies would be in the same place. If it is not a body but a power, then the problem is that a power does not give real identity to a thing; therefore, light will not be the cause of the identity of place.

5. Another opinion is that, although the heavens move as to their parts, they do not, however, move as a whole. Thus, the identity of place does not derive from the identity of its relation to a part of the heavens, but to the whole heavens in so far as they are a whole.

6. Against this: let us imagine that something is at rest in air and that the heavens are slightly displaced so that the centre changes. The thing at rest remains in the same place, since it does not move from a place to a place, but its relation to the whole heavens does not remain the same.

7. Furthermore, something can change place continuously, yet it can always have the same relation to the circumference of the whole heavens, for instance, if a thing moves continuously and another thing moves over it in the opposite direction in such a way that the speed of one motion is not greater than the speed of the other motion, then the thing that moves over the thing carrying it does not change its relation to the circumference, yet <it does change its place>.

8. To the first argument[22] it must be said that the imagination posits things that are not simultaneously possible, that is, that the heavens change their place as a whole, and yet that something is at rest in a place.

[20] Arist. *De anima* I.4, 408a30–33.
[21] Arist. *Phys*. IV.4, 212a18–20.
[22] See above, n. 6.

9. Ad secundum dicitur quod illud motum movebitur [a]in loco, [b]non de loco ad locum.

10. Contra: nihil movetur in loco nisi moveatur circulariter.

11. Dicendum quod nihil movetur in loco ita quod super spatium immobile nisi circulariter, aliquid tamen potest moveri in loco super spatium mobile, licet non moveatur circulariter.

<Quaestio 5>

1. Ad argumentum de caelo quod est in loco, versus finem huius capituli quaerit Aristoteles qualiter caelum est in loco. Et innuit quod caelum secundum se totum non est in loco, sed solum partes eius sunt in loco.

2. Et obicit Commentator contra illud sic: si partes caeli sunt in loco, aut ergo sectores circuli aut partes quarum una continet aliam; non sectores, quia in illis una est iuxta aliam et non ut continens aliam; non partes quarum una est sub alia, quia sic non esset ultima pars, ut quae continet omnes, in loco.

3. Item, totum caelum est primum mobile et non pars eius; ergo toto caelo debetur locus prius quam parti eius.

4. Commentator respondet sic. Totum non est in loco nec per partes sectores nec per alias partes, sed propter partem eius quae centrum est, quae est in loco per se.

5. Contra: Commentator dicit quod caelum est primum corporum, et prius per naturam quam centrum; ergo prius debetur ei locus quam centro.

6. Item, quies est privatio motus, et ita motus est prior quam quies; et caelum prius est terra; ergo prius per naturam est corpus caeli motum quam terra quiescat, et ita prius debetur ei locus quam terrae, et ita quam centro.

[a] in loco, non *om.* C
[b] non de *inv. sed corr.* O

9. To the second argument[23] it is said that the moved thing will move in a place, but not from a place to a place.

10. Against this: nothing moves in a place unless it moves in a circle.

11. It must be said that nothing moves in a place in such a way that it moves over an immobile space unless it moves in a circle, but something can move in a place over a mobile space without moving in a circle.

<Question 5>

1. As to the argument about the heavens which shows that they are in a place,[24] towards the end of the chapter Aristotle asks in what way the heavens are in a place. And he suggests that the heavens are not in a place as a whole, but that only their parts are in a place.[25]

2. And the Commentator raises the following objection to Aristotle's claim: if the parts of the heavens are in a place, then either (i) these are the sectors of a circle or (ii) the parts that are such that one of them contains another. Not (i), because sectors are such that one is next to the other instead of containing the other. Not (ii), because in this way the last part, that is, the one that contains all the others, would not be in a place.[26]

3. Furthermore, the whole heavens and not one of their parts constitute the first mobile thing; therefore, the whole heavens must have a place before one of their parts has a place.

4. The Commentator replies as follows. The whole heavens are in a place neither in virtue of the sector-parts nor in virtue of the other kind of parts, but in virtue of that part of them which is the centre, which is in a place per se.[27]

5. Against this: the Commentator says that the heavens are the first among bodies,[28] and are prior by nature to the centre; therefore, the heavens must have a place before the centre does.

6. Furthermore, rest is a privation of motion, and thus motion is prior to rest; and the heavens are prior to the earth; therefore, the being in motion of the body of the heavens is prior by nature to the being at rest of the earth, and the heavens must have a place before the earth does, and hence before the centre does.

[23] See above, n. 7.
[24] See above, q. 3, n. 13.
[25] Arist. *Phys*. IV.5, 212a31–b11.
[26] This objection is not found in Averroes' discussion of the place of the heavens in *In Phys*. IV, t.c. 43–46, but is popular among contemporary English commentators and some of them ascribe it to Averroes. See Trifogli, 2007, 152.
[27] Aver. *In Phys*. IV, t. c. 43, fol. 142va.
[28] Aver. *In De caelo* I, t.c. 16, ed. Carmody, I, 29.

7. Mihi videtur quod caelum non est in loco per accidens quia centrum per se, sed dicitur caelum esse in loco per accidens quia est in loco cum non indigeat loco. Et propter hoc dicit Commentator quod corpora caelestia causant sibi loca. Et ita propter perfectionem eius et nobilitatem est quod loco non indiget. Et omne quod proprie est in loco indiget loco. Alio modo est aliquid in loco per accidens quia exigit necessario intra se aliquid fixum circa quod moveatur, quod est in loco per se, dummodo non sit aliquid extra[a] illud, et sic est caelum [b]in loco per accidens. Debemus enim imaginari quod extima superficies caeli non unde tantum superficies est, sed cum hoc unde profundans[c] se in profundum totius universi, ut ad centrum universi, est locus caeli. Et deficit locus caeli a potentia loci[d], et hoc quia in caelo **[O 112vb]** non est aliqua indigentia loci[e].

<Quaestio 6>
Quaeritur consequenter utrum debeatur loco alia distensio a distensione locati.

1. Et videtur quod sic: omne quod recipit aliud sub aliqua dispositione sine qua impossibile est ipsum recipere prius fuit sub illa dispositione; sed locus recipit locatum sub distensione et dimensione sine qua ipsum non potest recipere; quare etc.

2. Item, nihil capit aliquid ab aliquo si insit illi, ipso non existente; sed distensio inest loco remoto uno et alio succedente locato; quare etc.

3. Ad oppositum: omne adveniens alii, causans aliquam dispositionem in illo sine cuius adventu non causatur, est causa dispositionis eius cui advenit; sed locatum sic se habet respectu loci; quare etc.

4. Dicendum quod est loqui de loco dupliciter: aut de loco partium essentialium mundi aut partium accidentalium. Si primo modo, sic habent distensionem a se ipsis, ut si imaginetur quod totus ignis amoveatur a loco suo <et[f]> succedat ei aliud corpus, adhuc retinet locus eius suam distensionem, ut concavum orbis[g] lunae. Si secundo modo loquamur, scilicet de partibus accidentalibus sive de corporibus pulsis[h], ut aere et aqua et huiusmodi, sic est

[a] extra] *illegibile add.* O
[b] in loco *suppl. s.l.* O
[c] profundans *ex corr.* O
[d] loci] caeli O
[e] loci] *hic explicit* C
[f] et *supplevimus*
[g] orbis *lectio incerta* O
[h] pulsis *lectio incerta* O

7. It seems to me that the heavens are not in a place accidentally because the centre is in a place per se, but the heavens are said to be in a place accidentally because they are in a place but they do not require a place. And this is why the Commentator says that the celestial bodies cause their own places.[29] Thus, it is because of their perfection and their nobility that the heavens do not require a place. And everything that is properly in a place does require a place. In another way something is in a place accidentally because it necessarily requires something fixed inside itself around which it moves and which is in a place per se, provided that there is nothing outside it, and in this way the heavens are in a place accidentally. For we have to imagine that the extreme surface of the heavens is the place of the heavens not in so far as it is only a surface, but in so far as, in addition to this, it reaches down into the depths of the whole universe, that is, down to the centre of the universe. The place of the heavens lacks the power of place, and this is because the heavens have no need of a place.

<Question 6>
The next question is whether an extension different from the extension of the located body should be ascribed to place.

1. Arguments for the affirmative: everything that receives another thing in accordance with a disposition without which it is impossible for it to receive that thing was under that disposition before <receiving that thing>; but place receives a located body in accordance with extension and dimension, without which place cannot receive the located body; hence, <place has extension before receiving the located body>.

2. Furthermore, a thing <A> does not receive a thing <B> from another thing <C> if B inheres in A when C does not exist; but extension inheres in place when one located body recedes and another succeeds; hence, <place does not receive its extension from the located body>.

3. To the contrary: a thing <A> that comes to be in another thing <B> and causes a disposition in B that cannot be caused without A's coming to be in B is the cause of that disposition in B; but the located body is related in this way to place; hence, <the located body causes the extension of place>.

4. It must be said that we can speak of place in two ways: either (i) of the place of the essential parts of the world or (ii) of the place of the accidental parts. In case (i), these places have extension by themselves, for instance, if we imagine that the whole fire is removed from its place, and that another body replaces it, its place, that is, the concavity of the lunar orb, still retains its extension. In case (ii), that is, in the case of the accidental parts or of bodies

[29] Cf. above, note 6.

distensio sive dimensio loci a distensione locati. Sed potest esse locati dupliciter: vel ita quod sit ab eo ut accidens a subiecto; sic non est distensio sua a locato, quia est ultima superficies corporis continentis, et est in locante ut in subiecto; aut ita quod sit ab eo ut a causante, et sic est dimensio a locato[a].

5. Primae rationi dicendum quod simul tempore recipit locus distensionem suam et locatum, et a locato habet suam distensionem ut a causante, tamen prius per naturam est aliqua distensio in loco quam recipiatur locatum, sicut dicimus quod, ad[b] hoc quod[c] aliquid moveatur in aere, prius pellit partes aeris per naturam et facit ipsas distare quam moveatur inter ipsas partes, et tamen simul tempore fit totum et est totum a movente ut ab efficiente sive causante. Stant ergo simul quod prius per naturam est distensio in loco quam recipiat locatum, et tamen tota sua distensio est a locato ut a causante.

6. Ad secundum respondetur per causam identitatis loci.

7. Vel sic, secundum quosdam: corrupto agente sive causante, adhuc manet accidens sive manere potest.

<Quaestio 7>
Quaeritur consequenter utrum locus agat in locatum.

1. Et videtur quod non, quia receptio est condicio passionis, et locus recipit locatum; quare respectu eius habet condicionem passionis.

2. Item, anima sensitiva est aliquid compositivum[d], et tamen est de genere passivorum, secundum Aristotelem; quare multo fortius est et locus.

3. Contra: locus continet locatum et ipsum salvat; quare agit in ipsum.

4. Item, Commentator dicit quod in his inferioribus locus est causa locati et in superioribus econverso; sed causa agit in causatum secundum quod tale; quare etc.

[a] locato *ex corr.* O
[b] ad] aliquid *add. sed del.* O
[c] quod] ad *add. sed del.* O
[d] compositivum *scripsimus*; compositum O

that are pushed, such as air, water, and the like, the extension or dimension of place derives from the extension of the located body. The extension of place, however, can derive from that of the located body in two ways: either (iia) as an accident derives from its subject: the extension of place does not derive from the located body in this way, because place is the extreme surface of the containing body, and so it is in the locating body as its subject. Or (iib) as something derives from what causes it: the dimension of place derives from the located body in this way.

5. To the first argument[30] it must be said that place receives both its extension and the located body at the same time, and it receives its extension from the located body, that is, from its cause. However, an extension's existence in place is prior by nature to place's reception of the located body, as when we say that, if something moves through air, its pushing the parts of air and separating them is prior by nature to its moving through them, yet this whole event occurs at the same time and the whole event derives from the mover, namely from what produces or causes it. Therefore, <the claims> that extension's existence in place is prior by nature to the place's reception of the located body and that the whole extension of place derives from the located thing, namely from its cause, are compatible.

6. To the second argument[31] it is replied by appealing to the cause of the identity of place.

7. Some people give another reply: once the agent or cause is destroyed, the accident still remains or can remain.

<Question 7>
The next question is whether place acts on the located body.

1. Arguments for the negative: reception is the state of being passive, and place receives the located body; hence, place has the state of being passive relative to the located body.

2. Furthermore, the sensitive soul is something capable of making a composition, yet it belongs to the genus of passive things, according to Aristotle;[32] hence, there is a much stronger reason to say that place is passive.

3. Against this: place contains the located body and preserves it; hence, place acts on it.

4. Furthermore, the Commentator says that in the things here below place is the cause of the located body, whereas in the superior things it is the other way around;[33] but a cause acts on the caused thing in so far as it is caused; hence, <place acts on the located body>.

[30] See above, n. 1.
[31] See above, n. 2.
[32] Arist. *De anima* II.5, 416b33–34.
[33] Cf. above, note 6.

5. Item, forma est agens, et locus est forma locati, secundum Aristotelem in pluribus locis.

6. Dicendum quod locus consideratur dupliciter: aut secundum quod naturalis est, et sic agit; aut secundum quod mathematicus est, et sic non.

7. Unde dicendum argumento quod prima condicio loci est receptio; secunda autem virtutis influxio, et quoad hanc condicionem agit.

8. Ad secundum dicendum quod sensus considerando[a] receptionem suam sic vult Aristoteles quod sit de genere passivorum. Considerando tamen condiciones adiectas, quae sunt discretionem[b] et iudicium, est de genere activorum. Et sic est de loco.

<Quaestio 8>
Quaeritur consequenter utrum 'ubi' sit praedicamentum distinctum.

1. Et videtur quod non, quia decem praedicamenta sunt decem voces decem prima genera rerum significantes. Significant ergo praedicamenta res primas. Sed locus est prior 'ubi', et locus non est praedicamentum; ergo nec 'ubi'.

2. Item, si propter respectum loci ad locatum causatur praedicamentum, tunc quicumque alius respectus diversificaret praedicamenta, et ita duplum et dimidium diversificarent, et etiam tunc accessus et recessus diversificarent praedicamenta.

3. Item, mobile habet esse in tempore et in motu et in loco; sed mobile esse in tempore facit 'quando' et mobile esse in loco facit 'ubi'; ergo mobile esse in motu faciet aliud praedicamentum diversum, quod est falsum; ergo primum.

4. Item, si habitudo loci ad locatum facit praedicamentum, sicut[c] ergo est in motu, qui[d] est habitudo inter agens et patiens, sic erit ibi; sed in motu ita[e] est quod comparatus ad unum extremum[f] facit unum praedicamentum et ad aliud aliud; ergo ita erit in habitudine loci ad locatum quod per comparationem ad locum erit unum praedicamentum et per comparationem ad locatum erit aliud[g] <...>

[a] considerando] tamen condiciones *add. sed del.* O
[b] discretionem *scripsimus*; discretam O
[c] sicut *scripsimus*; sic O
[d] qui *lectio incerta* O
[e] ita *suppl. s.l.* O
[f] extremum] loci ad locatum *add. sed del.* O
[g] aliud] *Hic explicit ex abrupto* O

5. Furthermore, a form is an agent, and place is the form of the located body, according to Aristotle in many places.[34]

6. It must be said that place is considered in two ways: either (i) in so far as it is natural, and in this way place acts; or (ii) in so far as it is mathematical, and in this way place does not act.

7. Accordingly, to the argument[35] it must be said that the first state of place is reception, but the second state is a pouring of power, and place acts in accordance with this second state.

8. To the second argument[36] it must be said that Aristotle wants the senses to belong to the genus of passive things when one considers reception. However, if one considers the additional states of discernment and judgment, the senses belong to the genus of active things. And so it is in the case of place.

<Question 8>
The question is whether 'where' is a distinct category.

1. Arguments for the negative: the ten categories are the ten spoken words signifying the first ten genera of things. Therefore, the categories signify the first things. But place is prior to 'where', and place is not a category; therefore, neither is 'where'.

2. Furthermore, if the relation of place to a located body produces a category, then any other relation would diversify the categories, and so double and half, and also accession and recession, would diversify the categories.

3. Furthermore, a mobile thing has being in time, in motion, and in place; but a mobile thing's being in time produces 'when', and the mobile thing's being in place produces 'where'; therefore, the mobile thing's being in motion will produce another distinct category, which is false; therefore, the first premise <is also false >.

4. Furthermore, if the relationship of place to a located body produces a category, then the case of place will be like the case of motion, which is a relation between agent and patient; but motion is such that when related to one extreme of the relation it produces one category and when related to the other extreme it produces another category; therefore, the same will apply to the case of the relationship of place to a located body, that is, when related to place it will be in one category and when related to the located body it will be in another category <...>

[34] Arist. *Phys*. IV.5, 213a1–10. Cf. also Arist. *De caelo* IV.3, 310a33–b1.
[35] See above, n. 1.
[36] See above, n. 2.

LIBER VIII

[O 113ra] 1*. [a]*Utrum autem factus sit aliquando motus, cum non esset prius, et corrumpitur iterum sicque moveri non sit aut neque* etc.

<Quaestio 1>
Quaeritur[b] primo an sit motus primus.

1. Et videtur quod non, quia Aristoteles in sexto huius probat pluribus demonstrationibus quod non est motus primus.

2. Item, non est mobile primum; ergo nec motus primus. Probatio praemissae, scilicet quod caelum non sit mobile primum, quia caelum est ab alio contentum, quia quod per se est in loco per se continetur a locante[c]. Et quod caelum[d] sit per se in loco videtur, quia quod[e] per se movetur per se est in loco; [f]caelum per se movetur; ergo per se est in loco. Et omne tale habet corpus continens extra, et sit eadem quaestio de illo continente, et sic in infinitum.

3. Item, si corpus caeli sit terminatum, aut ergo est terminatum ad aliud extra aut sua quantitas propria est infinita; sed hoc ultimum non contingit; est ergo ad aliud terminatum, et ita ut prius.

4. Ad oppositum: in septimo probat Aristoteles quod est motus primus.

5. Solutio[g]. Motus primus [h]dicitur dupliciter[i]: causalitate et[j] divisione. Primo modo est motus primus, ut motus caeli, qui est causa omnium motuum inferiorum. Secundo modo non est sumere primum, et hoc probat Aristoteles in sexto.

6. Et sic patet ad primum argumentum.

7. Ad secundum dicendum quod aliquid est per se in loco dupliciter: vel ita quod sit causa sui loci, et sic sunt corpora caelestia, ut caelum, in loco. Et quod sic est in loco non indiget continente extra, quia non exigitur aliquid extra continens nisi propter imperfectionem locati. Et non valet: ‘per se

[a] Utrum ... etc. *om.* LT
[b] Quaeritur] <S>uper octavum Physicorum *praem.* T
[c] locante] etc. *add. mg.* O
[d] caelum *suppl. s.l.* OT; *om.* L
[e] quod *suppl. s.l.* O
[f] caelum ... loco *om. hom.* L
[g] Solutio] dicendum quod T
[h] dicitur dupliciter: causalitate] est dupliciter causa *ex corr.* O
[i] dupliciter] aut *add.* LT
[j] et] aut primus motus LT

BOOK VIII

1*. *Was there ever a becoming of motion before which it had no being, and is it perishing again so as to leave nothing in motion? Or <are we to say that it> never <had any becoming and is not perishing, but always was and always will be?>*[1]

<Question 1>
The first question is whether there is a first motion.

1. Arguments for the negative: in *Physics* VI Aristotle proves with many demonstrations that there is no first motion.[2]

2. Furthermore, there is no first mobile thing; therefore, neither is there a first motion. The proof of the premise—that is, that the heavens are not the first mobile thing—is that the heavens are contained by something else, because what is per se in a place is per se contained by a locating body. And it seems that the heavens are per se in a place because what moves per se is per se in a place; but the heavens move per se; therefore, they are per se in a place. And everything of this kind has an external containing body. Let us then ask the same question of that containing body, and so on to infinity.

3. Furthermore, if the body of the heavens is limited, therefore it is either (i) limited by something outside it or (ii) its proper quantity is infinite; but case (ii) does not happen; therefore, the heavens are limited by something else, and thus <we argue> as before.

4. To the contrary: in *Physics* VII Aristotle proves that there is a first motion.[3]

5. Solution. A motion is said to be first in two ways: (i) by causality and (ii) by division. There is a first motion in way (i), namely the motion of the heavens, which is the cause of all motions here below. It is not possible for there to be a first motion in way (ii), and Aristotle proves this in *Physics* VI.

6. And with this the reply to the first argument[4] is clear.

7. To the second argument[5] it must be said that something can be per se in a place in two ways: either (i) in such a way that it is the cause of its place; and in this way the celestial bodies, namely the heavens, are in a place. And what is in a place in this way does not require an external container, given that an external container is only required because of the imperfection of the located body. And the inference 'it moves per se; therefore, it is per se in a place' is not

[1] Arist. *Phys*. VIII.1, 250b11–12.
[2] Arist. *Phys*. VI.6, 236b32–237b22.
[3] Arist. *Phys*. VII.1, 243a27–28.
[4] See above, n. 1.
[5] See above, n. 2.

movetur; ergo per se est in loco', sumendo proprie esse per se in loco, nisi in illis quae moventur de loco ad locum; in his autem quae moventur in spatio sive in suo[a] loco non valet. Alio modo est esse per se in suo loco ita quod causetur a suo loco, ut sunt elementa et elementata, et huiusmodi propter suam imperfectionem indigent continente extra.

8. Ad ultimum dicendum quod caelum est terminatum, non tamen ad aliud extra nec exigitur hoc nisi propter imperfectionem terminati.

9. Quid autem sit terminans alias patebit.

<Quaestio 2>
Quia supponitur quod caelum est mobile primum, propter hoc quaeritur utrum caelum moveatur.

1. Et videtur quod non, quia mobile non movetur nisi propter resistentiam mobilis ad motorem, sicut patet in suo simili, quia illud ex quo debet generari aliud indiget forma contraria ad hoc ut resistat actioni, ita quod fiat generatio illa in tempore; sed in caelo non est resistentia, quia, ubi est resistentia, ibi est violentia; sed in caelo non est violentia; quare etc.

2. Item, si caelum moveatur, aut ergo naturaliter aut violenter. Non violenter—constat; nec naturaliter, quia in omni motu naturali fit aliqua nova dispositio acquisita per motum; sed caelum in una revolutione non acquirit quod non habuit in alia; quare etc.

3. Solutio. Dicendum quod caelum movetur, et hoc motu diurno, [b]de quo quaeritur[c] aeternitas.

4. Unde ad primum dicendum quod maior falsa est. Non enim est resistentia causa motus per se, sed pluralitas graduum medii qui sunt inter terminos, et hoc, quia impossibile est in eodem instanti aliquid esse in diversis locis, necesse est pertransitum cuiuslibet spatii fieri in tempore. Et ita, si imaginetur quod medium non resistat aliquo modo, adhuc non minus erit motus in tempore. Et est simile in suo exemplo quod, ad hoc quod fiat actio naturalis, exigitur quod in patiente sit forma contraria quatenus fiat

[a] suo] illo O
[b] de ... aeternitas *om.* L; *scrips. sed del.* T
[c] quaeritur *lectio incerta* O

valid when one takes being per se in a place in the proper sense, except where those things that move from a place to a place are concerned. Where the things that move in a space or in their place are concerned, the inference is not valid. Or (ii) something is per se in a place in such a way that it is caused by its place, as are the elements and the things derived from the elements, and these things because of their imperfection require an external container.

8. To the last argument[6] it must be said that the heavens are limited, but not by something outside them nor is this required except because of the imperfection of the thing limited.

9. By what kind of thing the heavens are limited will be made clear elsewhere.[7]

<Question 2>
Since it is assumed that the heavens are the first mobile thing, the question is whether the heavens move.

1. Arguments for the negative: a mobile thing does not move except in virtue of the resistance of the mobile thing to the mover, as is clear in a similar case, because the thing from which another thing is to be generated must have a contrary form so that it offers resistance to action in such a way that the generation occurs in time; but there is no resistance in the heavens, because, where there is resistance, there is violence; but there is no violence in the heavens; hence, <the heavens offer no resistance and so do not move>.

2. Furthermore, if the heavens move, they move either by nature or by violence. Not by violence, as is manifest; nor by nature, because in every natural motion there is a new disposition acquired through that motion; but the heavens do not acquire something in one of their revolutions that they did not have in another revolution; hence, <the heavens do not move by nature>.

3. Solution. It must be said that the heavens move, and that they move with diurnal motion, whose eternity is in question.

4. Thus, to the first argument[8] it must be said that the major premise is false; for the per se cause of a motion is not resistance, but the plurality of the intermediate stages between the termini of the motion, because it is impossible for something to be in different places in the same instant, and therefore it is necessary that the traversal of any space should take time. Thus, if we imagine that the medium does not offer resistance in any way, motion will nonetheless still be in time. Similarly, in Aristotle's example we find that in order for a natural action to occur a contrary form is required in the patient,

[6] See above, n. 3.
[7] This could be a reference to Book VI, q. 11, n. 20.
[8] See above, n. 1.

resistentia, ita quod fiat motus, et quatenus habeat condiciones quae exiguntur ad passionem naturalem. Sed non exigitur forma contraria ut per se causa, sed ut occasio necessaria, et hoc quia agens intendit per se excitare potentiam diminutam repertam in patiente et ipsam ad actum promovere; tunc, propter hoc quod reperit ipsam sub forma contraria, per accidens corrumpit eam in per se excitando et promovendo potentiam illam ad actum. Si autem imaginetur sic per impossibile quod sit materia per se existens cum sua potentia diminuta ad ignem, ita quod non sit actualiter sub forma contraria, et cum hoc imaginatur quod ignis immittat speciem in illud diminutum, adhuc videtur mihi quod generatio illa fieret in tempore, et [a]hoc propter gradus in illo diminuto[b] antequam fiat sub ultima actualitate. Nec potest aliqua virtus materialis transire per diversos gradus nisi in tempore.

5. Et sic patet quod, etsi[c] non fuerit resistentia, dummodo fuerit multitudo mediorum, necesse est motum fieri. Nec in caelo oportet [d]totum moveri in tempore propter se totum, sed ut partes caeli possint se successive cum omnibus partibus spatii commetiri[e].

6. Ad secundum dicendum quod non oportet in omni motu naturali acquirere dispositionem[f] novam, sed tantum in illo motu qui fit de loco ad locum.

7. Vel sic: sicut movetur secundum partes tantum, [g]sic secundum partes acquirit novas dispositiones, ut novos locos, et completa una revolutione remanet idem appetitus qui prius.

8. Vel sic: est quidam motus qui est propter aliud et non[h] propter se, et sic movetur caelum, et in talibus non acquiritur nova dispositio in mobili[i]. Alius est motus qui fit propter se et non propter **[O 113rb]** aliud, et in tali motu, quia mobile movetur[j] propter se, acquirit in se aliquam novam dispositionem, sicut patet in istis inferioribus.

[a] hoc propter *inv.* OL
[b] diminuto *om.* O
[c] etsi *ex corr.* O
[d] totum moveri *corr. s.l. in* motum esse O
[e] commetiri *corr. s.l. ex* contineri O; commensurari L
[f] dispositionem *ex corr.* O
[g] sic secundum partes *om. sed* sic *suppl. s.l.* T
[h] non] quam *scrips. sed corr. s.l. in* numquam O
[i] mobili *scripsimus*; movente OLT
[j] movetur] movet OL

both so that there is resistance in such a way that motion occurs, and so that the patient has the conditions required for a natural passion. However, this contrary form is not required as a per se cause, but as a necessary occasion. The reason for this is that the agent per se intends to stimulate the diminished potency found in the patient and to bring it into act; then, because the agent finds that potency characterized by a contrary form, it destroys that form accidentally by per se stimulating and bringing that potency into act. If, however, we imagine the impossible situation in which matter exists by itself together with a diminished potency to become fire in such a way that matter is not actually characterized by a contrary form, and together with this we imagine that fire introduces its species into that diminished thing, it still seems to me that that the generation <of fire> would occur in time, and this because of the stages that exist in that diminished thing before it acquires its ultimate actuality. Nor can a material power traverse different stages except in time.

5. Accordingly, it is clear that, even if there is no resistance, as long as there is a multitude of media, it is necessary for motion to occur. Nor is it necessary in the heavens for the whole to move in time for the sake of the heavens as a whole, but in order for the parts of the heavens to commensurate themselves successively to all the parts of space.

6. To the second argument[9] it must be said that it is not necessary for there to be the acquisition of a new disposition in every natural motion, but only in the motion that occurs from a place to a place.

7. Another reply is that, just as the heavens move only in their parts, so they acquire new dispositions, that is, new places, only in their parts, and when one revolution is completed the same appetite as before remains.

8. Another reply is that (i) one kind of motion occurs for the sake of something else and not for the sake of itself; and the heavens move in this way, and in motions of this kind a mobile thing does not acquire a new disposition. (ii) Another kind of motion occurs for the sake of itself and not for the sake of something else, and in a motion of this kind a mobile thing acquires a new disposition because the mobile thing moves for the sake of itself, as is clear where the things here below are concerned.

[9] See above, n. 2.

<Quaestio 3>
Consequenter quaeritur super hoc quod dicit quod inest rebus naturalibus motus sicut vita in animalibus.

1. Contra: vita inest animalibus ita quod non quandoque et quandoque non, sed semper et continue; ergo ita erit motus in naturalibus quod semper et continue, et ita non inerit quies.

2. Item, videtur quod deberet dicere de quiete quod dicit de motu, et hoc quia in inferioribus non fit motus nisi propter quietem; maior ergo est continuitas in quiete quam in motu; quare etc.

3. Dicendum quod non est omnimoda similitudo, sed quia sic est motus in naturalibus ab intra sicut in animalibus est vita ab intra.

1*. Antequam quaeratur de principali conclusione Aristotelis, quaeratur primo de opinionibus. Et fuit una quod mundus infinities generatur et infinities corrumpitur.

<Quaestio 4>
Propter hoc quaeratur an mundus sit generabilis et corruptibilis.

1. Et videtur quod sic, quia omne habens qualitates[a] elementares est corruptibile vel secundum totum vel secundum partem; sed caelum est huiusmodi, quia quaedam partes eius sunt calidae, ut stellae quaedam[b], et quaedam sunt[c] frigidae; quare etc.

2. Si dicatur sic quod, ubi est calidum vel[d] frigidum ut passio naturalis vel passibilis qualitas, illud est corruptibile; sed in stellis non est calidum vel frigidum nisi effective tantum, et ita non sequitur, contra: si effectus univocus, et causa univoca. Sed calor in sole et calor in igne univocum habent effectum, quia utrumque constringit lutum et dissolvit glaciem et huiusmodi; quare calor est univoce in igne et in sole. Sed in igne est passio vel passibilis qualitas; ergo et in sole; quare etc.

3. Item, omne quod est grave vel leve est corruptibile; sed caelum est huiusmodi; quare etc. Probatio minoris: ita est in elementis quod, si pars sit distans a suo toto, necessario appetit esse cum suo toto, [e]et hoc propter conformitatem quam habet[f] cum toto. Sed maior est conformitas partium cum toto in caelo quam in elementis; ergo si pars caeli, ut stella aliqua, fuerit

[a] qualitates] quantitates O
[b] quaedam *om.* LT
[c] sunt *om.* LT
[d] vel *om. sed* et *suppl. s.l.* O; *om.* L
[e] et ... toto *om. hom.* T
[f] habet *suppl. s.l.* O

<Question 3>
The next question is about Aristotle's statement that motion exists in natural things in the way that life exists in animals.[10]

1. Against this: life exists in animals not in such a way that it sometimes exists and sometimes does not, but always and continuously; therefore, motion will exist in natural things always and continuously, and thus rest will not exist in them.

2. Furthermore, it seems that Aristotle should say about rest what he says about motion, because in the things here below motion occurs only for the sake of rest; therefore, there is a greater continuity in rest than in motion; hence, <Aristotle's claim is not correct>.

3. It must be said that the similarity is not complete, but it holds in so far as motion exists intrinsically in natural things in the way that life exists intrinsically in animals.

1*. Before raising questions about Aristotle's main conclusion, let us first raise questions about other views. And one such view was that the world is generated and corrupted infinitely many times.[11]

<Question 4>
The question then is whether the world is generable and corruptible.

1. Arguments for the affirmative: everything that has elementary qualities is corruptible either as a whole or in part; but the heavens are of this kind, because some of their parts are hot, such as some of the stars, and others are cold; hence, <the heavens are corruptible>.

2. One might say that only the thing in which hot and cold exist as natural passions or passive qualities is corruptible; but in the stars hot and cold exist only as in their efficient cause, and thus the conclusion does not follow. Against this one may argue: if the effect is univocal, the cause also is univocal. But the heat of the sun and the heat of fire have a univocal effect, because they both consolidate clay and melt ice and the like; therefore, heat exists univocally in fire and in the sun. But in fire heat is a passion or a passive quality; therefore, <it is a passion> in the sun as well; hence, <the reply is not valid>.

3. Furthermore, everything that is heavy or light is corruptible; but the heavens are of this kind; hence, <the heavens are corruptible>. Proof of the minor premise: in the elements it is the case that if a part is distant from its whole, it necessarily desires to be with its whole, and this because of its

[10] Arist. *Phys*. VIII.1, 250b14–15.
[11] This is a reference to the opinion of Empedocles reported in Arist. *Phys*. VIII.1, 250b27–251a5.

hic inferius, necessario ascendet et movebitur ad suum totum. Et nihil ascendit nisi leve; quare etc.

4. Item, omne quod est grave vel leve est corruptibile; sed caelum est huiusmodi; quare etc. Probatio minoris: omne quod resistit suo motori aut resistit quia grave aut quia leve; sed caelum resistit, quia Aristoteles dicit quod, si apponeretur una stella, [a]magis tunc moveret cum labore et poena, et hoc non contingeret nisi propter resistentiam[b]; quare etc.

5. Item, in quocumque est passio vel passibilis qualitas, illud est corruptibile; sed caelum est huiusmodi; quare etc. Minor[c] patet, quia omnia sensibilia eiusdem sensus sunt eiusdem contrarietatis; sed caelum est visibile et haec inferiora sunt visibilia; quare sunt eiusdem contrarietatis; sed illud per quod inferiora sunt visibilia est passio vel passibilis qualitas; quare etc.

6. Item, omne tangibile habet qualitatem tangibilem, et tale est corruptibile; sed caelum est tangibile, quia, si manus apponeretur caelo, resisteret; quare etc. Quod caelum[d] resistat videtur, quia dicit in quarto[e] *Caeli et mundi* quod, si stella moveatur in suo orbe findendo ipsum[f], tunc est ipsum corruptibile, si volvendo, tunc non appareret semper macula lunae et, si hoc, caelum non cederet manui; quare etc.

7. Item, secundum Aristotelem, necesse est imaginari[g] materiam in re mota; ergo in caelo; sed secundum ipsum, materia est qua res potest esse et non esse; [h]quare etc.

8. Diceres quod materia non in sua substantia est qua res potest esse et non esse, sed secundum quod est sub forma transmutabili.

[a] magis tunc *om.* T
[b] resistentiam] unde resisteret *add. sed del.* O; ut resisteret *add.* L; vel nisi resisteret *add.* T
[c] Minor] probatio minoris L
[d] caelum *om.* LT
[e] quarto] *recte* secundo
[f] ipsum] corruptibile *add. sed del.* O
[g] imaginari *corr. ex* imaginare O; imaginare LT
[h] quare ... non esse (n. 8) *om. hom.* L

homogeneity with the whole. But there is a greater homogeneity of parts with their whole in the heavens than in the elements; therefore, if a part of the heavens, such as a star, were here below, it would necessarily move up towards its whole. But nothing moves upwards unless it is light; hence, <the heavens are light>.

4. Furthermore, everything that is heavy or light is corruptible; but the heavens are of this kind; hence, <the heavens are corruptible>. Proof of the minor premise: everything that offers resistance to its mover does so either because it is heavy or because it is light; but the heavens offer resistance, since Aristotle says that, if even one star were added to the heavens, the heavens would then move with more effort and pain,[12] and this could only happen because of resistance; hence, <the heavens are heavy or light>.

5. Furthermore, anything in which there is a passion or a passive quality is corruptible; but the heavens are of this kind; hence, <the heavens are corruptible>. The minor premise is clear because all the things that can be sensed by one sense are of the same contrariety; but the heavens are visible and the things here below are also visible; hence, the heavens and the things here below are of the same contrariety; but that in virtue of which the things here below are visible is a passion or a passive quality; hence, <the heavens also are visible in virtue of a passion or passive quality>.

6. Furthermore, every tangible thing has a tangible quality, and something of this kind is corruptible; but the heavens are tangible, because if a hand were placed on the heavens, the heavens would resist; hence, <the heavens are corruptible>. That the heavens resist is shown by Aristotle's statement in *On the Heavens* IV that, (i) if a star moves in its orb by splitting it, then the heavens are corruptible,[13] (ii) if <a star moves> by revolving, then the dark spot of the moon would not always appear[14] and, if so, the heavens would not yield to the hand; hence, < the heavens do resist>.

7. Furthermore, according to Aristotle, it is necessary to imagine matter in a thing subject to motion;[15] therefore, <it is necessary> in the heavens as well; but according to him, matter is that by which a thing can be and not be;[16] hence, <the heavens can be and can not be>.

8. You might say that matter in its substance is not that by which a thing can be and not be, but matter in so far it is under a transmutable form.

[12] Arist. *De caelo* II.12, 293a10–11. Cf. the Latin-Arabic translation of Aristotle's *De caelo* in Aver. *In De caelo* II, t.c. 71, ed. Carmody, II, 407.1–3.

[13] The correct reference is Arist. *De caelo* II.8, 289b31–33 (cf. Aver. *In De caelo* II, t.c. 46, ed. Carmody, II, 363.52–54; t.c. 47, 364.5–7).

[14] The correct reference is Arist. *De caelo* II.8, 290a25–29.

[15] Arist. *Metaph.* II.2, 994b25–26.

[16] Arist. *Metaph.* VII.7, 1032a20–22.

9. Contra: si sub maiori perfectione est causa corruptionis, ergo sub minori erit; sed quando est sub forma transmutabili est magis perfecta quam quando est considerata in sua essentia; ergo, si sub forma transmutabili est causa corruptionis, ergo multo fortius in sua essentia erit causa corruptionis.

10. Item, omne divisibile est corruptibile; sed caelum est divisibile, quia dividitur per gradus et minuta[a] et huiusmodi et per sectores; quare etc.

11. Dicendum quod mundus non generatur, habuit tamen initium nec forte habebit finem, sed in fine saeculi, sicut dicimus, habebit finem, scilicet quoad transmutationem et actionem elementorum sublimabitur natura universi in tantum quod neque fiet actio neque passio neque generatio neque corruptio. Et forte tunc erit natura elementorum nobilior quam modo est natura corporis caelestis.

12. Ad primum argumentum dicatur sicut prius dicebatur. Et cum arguit in contrarium sic: 'si effectus univocus, et causa', dicendum quod, si sumatur causa pro subiecto proximo, sic est propositio vera, secundum quod dicit Aristoteles in secundo *Posteriorum* quod, si passio univoca, et subiectum. Si autem sumatur causa pro efficiente, sic est propositio falsa, quia prima causa creat animam et hic inferius generatur substantia aliqua, et ponitur communiter quod illud hic generatum et anima a Deo creata[b] sunt univoce substantia, et tamen agentia sive efficientia eorum non sunt efficientia univoce.

13. Ad secundum dicendum, secundum quosdam, quod, si stella esset hic in medio, moveretur sursum, nec tamen est leve, quia moveri sursum **[O 113va]** est dupliciter: vel ita quod solum continuetur suo toti, et tale non movetur propter levitatem, sed propter conformitatem quam habet cum suo toto, et sic movens sursum non est leve; alio modo movetur aliquid sursum per se, ut ad suum locum, et tale est leve. Unde si dicam: pars ignis movetur sursum et stella movetur sursum, hoc est aequivoce, quia ignis movetur ad sursum elementare, stella ad supra sursum.

[a] minuta *ex corr.* O
[b] creata] creatum OL

9. Against this: if matter is the cause of corruption when characterized by a greater perfection, then it will also be such a cause when characterized by a minor perfection; but matter characterized by a transmutable form is more perfect than when it is considered in its essence; therefore, if matter is the cause of corruption when characterized by a substantial form, there is a much stronger reason to say that matter is the cause of corruption in its essence.

10. Furthermore, every divisible thing is corruptible; but the heavens are divisible, because they are divided into degrees, minutes and the like, and into sectors; hence, <the heavens are corruptible>.

11. It must be said that the world is not generated, but it had a beginning, and perhaps it will not have an end, but at the end of times, as we say, it will have an end, that is, where the transmutation and the action of the elements are concerned, the nature of the universe will be elevated to the extent that neither action nor passion nor generation nor corruption will occur. And then perhaps the nature of the elements will be nobler than the nature of a celestial body is now.

12. To the first argument[17] let it be replied as before.[18] And to the argument to the contrary 'if the effect is univocal, the cause is as well',[19] it must be said that, if cause is taken as the proximate subject, then the proposition is true, according to Aristotle's statement in the *Posterior Analytics* II that, if a passion is univocal, the subject also is.[20] But if cause is taken as the efficient cause, then the proposition is false, because the first cause creates the soul and a substance is generated here below, and it is commonly posited that the thing generated here below and the soul created by God are univocally substances, but their agents or efficient causes are not univocally efficient causes.

13. To the second argument[21] it must be said, according to some people, that if a star were here in the middle <of the world>, it would move upwards, but a star is not light, because something can move upwards in two ways: either (i) only in order to become continuous with its whole; and a thing of this kind does not move because of its lightness, but because of the conformity which it has with its whole, and what moves upwards in this way is not light. Or (ii) something moves upwards per se, that is, towards its place; and a thing of this kind is light. Thus, the claims 'a part of fire moves upwards' and 'a star moves upwards' are equivocal, because fire moves towards the heights of the elementary region, whereas a star moves towards the heights of the superior region.

[17] See above, n. 1.
[18] See above, n. 2.
[19] See above, n. 2.
[20] Arist. *An. post.* II.17, 99a1–17.
[21] See above, n. 3.

14. Ad aliud dicendum quod resistentia non est causa quare movetur cum labore et poena, si apponeretur una stella, sed improportio[a] motoris ad mobile est causa, quia, si apponeretur una stella, esset maior inclinatio ad motum quam nunc est, et ita non proportionaretur suo motori.

15. Ad aliud dicendum[b] quod sensibilia eiusdem sensus sunt eiusdem contrarietatis, hoc verum est si sentiantur univoce; sed lux et color non sentiuntur univoce, quia lux sentitur de se et sine medio, color sentitur per aliud et per medium. Unde, cum dicit: 'sensibilia visus sunt in tertia specie qualitatis', hoc verum est de illis quae sentiuntur ut color.

16. Vel sic: Aristoteles non intendit de contrarietate quae est passio vel passibilis qualitas, sed in visu de hac contrarietate 'visibile-non visibile', in auditu de hac contrarietate[c] 'audibile-non audibile', et ita de aliis.

17. Vel sic: Aristoteles, cum dicit[d] quod sensibilia unius sensus sunt[e] unius contrarietatis, hoc est intelligendum tantum de sensibilibus tactus. Est tamen splendor lucis aliquo modo passio vel passibilis qualitas.

18. Ad aliud dicitur sic quod caelum non resisteret, nec tamen posset manus aliqua intrare. Et huius[f] est aliqua causa per privationem et non per positionem. Quia enim in caelo non est locus proportionalis manui et etiam extra caelum non est locus aliquis, ideo non intraret manus, si caelo apponeretur.

19. Aliter dicitur distinguendo tangibile, quia est tangibile tactu perceptibile et tactum immutans, vel tangere tactu percipere vel tactum immutare etc.

20. Ad aliud dicendum quod materia in sua substantia non est qua res potest esse etc., nec unde sub forma transmutabili, sed unde privata est.

[a] improportio] proportio OL; *corr. s.l. ex* proportio T
[b] dicendum] dicitur LT
[c] contrarietate] accidentale *add. sed del.* O; *om.* T
[d] dicit *corr. ex* dicat O; dicat L
[e] sunt] modo passio *add. sed del.* O
[f] huius *ex corr.* O

14. To the other argument[22] it must be said that resistance is not the cause of the fact that the heavens would move with effort and pain if a star were added to them, but the cause of this is the disproportion of the mover to the mobile thing, because, if a star were added, there would be a greater inclination towards motion than there is now, so that the heavens would not be proportionate to their mover.

15. To the other argument[23] it must be said that the claim that all the things that can be sensed by the same sense are of the same contrariety is true of the sensibles that are sensed univocally; but light and colour are not sensed univocally, because light is sensed by itself and without a medium, whereas a colour is sensed in virtue of something else and of the medium. Thus, when he says that the sensibles of sight are in the third species of quality, this is true of the things that are sensed in the way a colour is.

16. Another reply is that Aristotle does not refer to the contrariety that is a passion or a passive quality, but to the contrariety 'visible-non visible' when speaking of sight, to the contrariety 'audible-non audible' when speaking of hearing, and similarly for the other senses.[24]

17. Another reply is that, when Aristotle says that the things that can be sensed by one sense are of one contrariety,[25] this is to be understood only of the sensibles of touch. The brightness of light, however, is in some way a passion or a passive quality.

18. To the other argument[26] it is said that the heavens would not resist, nor could a hand get inside them. And the cause of this is a privation and not something positive; for, since inside the heavens there is no place proportionate to a hand and outside the heavens there is no place at all, this is why a hand would not get inside, if it were placed on the heavens.

19. Another reply makes a distinction concerning the tangible, namely that the tangible is either what can be perceived by touch and what can modify touch, or that to touch is either to perceive by touch or to modify touch.

20. To the other argument[27] it must be said that matter in its substance is not that by which a thing can be <and not be> nor is matter such a thing in so far as it is under a transmutable form, but in so far as it is deprived.

[22] See above, n. 4.
[23] See above, n. 5.
[24] Arist. *De anima* II.11, 422b23–24.
[25] See above, n. 16.
[26] See above, n. 6.
[27] See above, n. 7.

21. Ad ultimum dicendum quod est divisio mathematica, et sic dividitur caelum per gradus et per minuta et per sectores, et non omne sic divisibile est corruptibile. [a]Alia est divisio quae est actualis separatio partis a parte, et omne sic divisibile est[b] corruptibile, secundum quod dicit Aristoteles quod, si stella findat orbem, est corruptibile.

<Quaestio 5>
Quaeritur consequenter utrum tempus sit aeternum.

1. Et videtur quod sic, quia, si incepit, tunc suum esse et suum non esse aut sunt simul in eodem instanti aut non. Si sic, hoc est impossibile, quia inter quaelibet indivisibilia cadit aliquod medium. Si non, adhuc cadit tempus medium.

2. Dicitur sic, quod inter quaelibet duo indivisibilia eiusdem continui cadit aliquid medium, sed inter duo indivisibilia[c] diversorum continuorum non necesse est aliquid cadere medium. Unde instans sui non esse est instans aeternitatis et instans sui esse est instans temporis, et ita sunt diversorum continuorum, et ita inter ista non cadit medium.

3. Contra: si hoc est verum, scilicet quod ordinantur indivisibilia simul sine medio, tunc simul sunt illa indivisibilia et, si hoc, simul est sub non esse aeternitatis et sub esse temporis; sed hoc est falsum; ergo primum.

4. Item, si mundus creetur et motus et tempus similiter, tunc prius fuit mobile quam moveatur, sicut dicit Aristoteles quod prius est combustibile quam comburatur. Tunc in illo priori mobile habuit esse; sed non in instanti temporis; est ergo in instanti aeternitatis, et ita primum instans sui esse et ultimum sui non esse, utrumque erit in instanti aeternitatis. Et tunc arguatur ut in primo argumento.

5. Item, causa prima scivit mundum esse quando fuit et scivit mundum non[d] esse quando non fuit; et in scientia eius non cadit prius et posterius; ergo

[a] Alia ... corruptibile *om. hom.* L
[b] est *lectio incerta* (*ex corr.*) O
[c] indivisibilia *corr. s.l. ex* divisibilia O
[d] non *suppl. s.l.* O

21. To the last argument[28] it must be said that (i) one kind of division is mathematical, and it is by a division of this kind that the heavens are divided into degrees, minutes, and sectors; and not everything that is divisible in this way is corruptible. (ii) Another kind of division is the actual separation of a part from a part; and everything that is divisible in this way is corruptible, according to Aristotle's statement that if a star splits an orb, then the heavens are corruptible.[29]

<Question 5>
The next question is whether time is eternal.

1. Arguments for the affirmative: if time began, then its being and its non-being either occur together in the same instant or they do not. It is impossible to adopt the first alternative because between any two indivisibles there is something intermediate. If they are not simultaneously together, there is still an intermediate time between them.

2. It is replied to the first alternative by saying that there is something intermediate between any two indivisibles of the same continuum, but there is not necessarily anything intermediate between two indivisibles of different continua. Thus, an instant of the non-being of time is an instant of eternity and an instant of its being is an instant of time, and thus these instants belong to different continua, so that there is nothing intermediate between them.

3. Against this: if this is true, that is, that some indivisibles are ordered together without an intermediate, then those indivisibles are simultaneous, and if so, something comes simultaneously under the non-being of eternity and the being of time; but this is false; therefore, the first premise <is false>.

4. Furthermore, if the world is created and motion and time are likewise created, then a mobile thing existed before it was in motion, just as Aristotle says that a combustible thing exists before it is burnt.[30] Then in that prior state the mobile thing had being; but not in an instant of time; therefore, in an instant of eternity, and so the first instant of its being and the last instant of its non-being will both occur at an instant of eternity. And let us then argue as we did in the first argument.[31]

5. Furthermore, the first cause knew that the world existed when it existed and knew that the world did not exist when it did not exist; and in the knowledge of the first cause there is no before and after; therefore, the first cause knew that the world simultaneously exists and does not exist. But this

[28] See above, n. 10.
[29] Arist. *De caelo* II.8, 289b31–33 (cf. Aver. *In De caelo* II, t.c. 46, ed. Carmody, II, 363.52–54; t.c. 47, 364.5–7).
[30] Arist. *Phys.* VIII.1, 251a15–16.
[31] See above, n. 1.

scivit mundum simul esse et non esse. [a]Et hoc falsum est; mundus ergo non habuit prius non esse quam esse, et ita ab aeterno fuit.

6. Item, [b]in omni potentia ordinata[c] ad actum actus est melior quam potentia, ut dicit Aristoteles in nono; [d]cum ergo causa prima sit semper in dispositione optima, tunc non fuit factus de potentia agente actu agens, et ita semper fuit actu agens, et ita semper sive coaeternum sibi fuit suum actum.

7. Et ita sequitur imaginatio antiquorum, scilicet quod mundus est coaeternus Deo, et tamen est efficiens mundum, sicut imaginatur de vestigio pedis posito in pulvere: si ab infinito extitisset, semper esset pes causa efficiens vestigii.

8. Item, per Aristotelem in perpetuis non differt esse a posse, et ita nec actus et potentia; sed ab infinito potuit[e] prima causa fecisse mundum; ergo ab aeterno fecit mundum.

9. Item, si produxit mundum in esse, tunc factus est de non efficiente efficiens et, si hoc, aliqua transmutatio facta est in ipso efficiente. Consequentia patet in principio octavi. Arguit enim Aristoteles sic: si mobile incepit moveri, facta fuit transmutatio aliqua in mobili.

10. Item, si causa prima fecit mundum, tunc refertur ad illum effectum, et ita aliqua est in ipso innovatio.

11. Item, omnis rationabiliter agens exiens de potentia ad actum in agendo habet rationem quare tunc facit ita quod non in alio tempore; sed a parte eius ex quo non est aliqua ratio; ergo a parte agentis, et ita a parte primi; sed non videtur ratio.

12. Item, Aristoteles in octavo dicit quod omne instans est medium duorum temporum et, si hoc, non est sumere primum instans.

13. Ad oppositum: omne acceptum est finitum; totum tempus praeteritum[f] est acceptum; quare etc.

14. Item, quantumcumque aliquid fuit, finitum fuit; sed totum tempus praeteritum aliquando fuit praesens; quare etc.

[a] Et ... est *om.* L
[b] in omni] omnis OLT, *sed corr.* T
[c] ordinata *corr. s.l. ex* ordina O
[d] cum ... prima] et cum causa prior L
[e] potuit *ex corr.* O
[f] praeteritum *corr. s.l. ex* praedictum O

is false; therefore, the world did not have non-being before being, and so it existed from eternity.

6. Furthermore, in every potency ordered to an act, the act is better than the potency, as Aristotle says in *Metaphysics* IX;[32] therefore, since the first cause is always in the best disposition, it did not pass from being an agent in potency to being an agent in act, and so it always was an agent in act, and its act always was, that is, was coeternal with it.

7. And this leads to the view imagined by the ancient philosophers that the world is coeternal with God, yet God is the efficient cause of the world. Thus one can imagine of a footprint in the dust that even if the footprint existed from eternity, the foot would still be the efficient cause of the footprint.[33]

8. Furthermore, according to Aristotle, in eternal things there is no difference between being and being possible,[34] and hence neither is there a difference between act and potency; but the first cause could have made the world from eternity; therefore, it did make the world from eternity.

9. Furthermore, if the first cause brought the world into being, then the first cause passed from being a non-efficient cause to being an efficient cause, and if so, some transmutation occurred in the efficient cause itself. This consequence is made clear at the beginning of *Physics* VIII, where Aristotle argues as follows: if a mobile thing began to be moved, a transmutation occurred in the mobile thing.[35]

10. Furthermore, if the first cause made the world, then it has a relation to that effect, and thus something new comes to be in the first cause.

11. Furthermore, every rational agent that passes from potency to act in acting has a reason why it does this then and not at another time; but where what it comes out of is concerned there is no such reason; therefore, the reason concerns the agent, and thus it concerns the first cause; but there does not seem to be such a reason.

12. Furthermore, in *Physics* VIII Aristotle says that every instant is intermediate between two times,[36] and if so, a first instant cannot be assumed.

13. To the contrary: everything that is taken as completed is finite; but the whole of past time is taken as completed; hence, <it is finite>.

14. Furthermore, whatever quantity something had, it was finite; but the whole of past time was present at some point; hence, <the whole of past time is finite>.

[32] Arist. *Metaph.* IX.9, 1051a4–5.
[33] Aug. *De civ. Dei* X.31, CCSL 47, 308–309. 4–18.
[34] Arist. *Phys.* III.4, 203b30.
[35] Arist. *Phys.* VIII.1, 251a23–28.
[36] Arist. *Phys.* VIII.1, 251b20–22.

15. Item, si tempus praeteritum esset infinitum [a]a parte ante, **[O 113vb]** non pervenissemus ad hodiernum diem, cum non contingat transire infinita.

16. Item, si hoc esset, tunc generatio esset infinita, et tunc animae essent infinitae in actu.

17. Item, ubi non est primum, ibi non est ultimum; si ergo in esse temporis non fuit primum, nec est ultimum, quod falsum est.

18. Solutio. Dicendum quod tempus non est[b] aeternum, sed incepit tempus et motus et fuit primus homo. Nec vult Aristoteles oppositum, si recte exponatur.

19. Unde primae rationi dicendum quod ultimum sui non esse et primum sui esse non sunt in eodem instanti, quia ultimum sui non esse est in instanti aeternitatis et primum instans sui esse est in primo instanti temporis. Nec est illud instans aliquid infra tempus, sed est principium indivisibile a quo fluit[c] totum tempus, sicut est in motu quod totus motus fluit a mutatione prima et tota linea a puncto indivisibili primo. Unde, cum dicit quod inter quaelibet duo indivisibilia cadit medium, dicendum quod hoc verum est de indivisibilibus eiusdem continui. Et huius ratio est: necessario, quando duo indivisibilia sunt simul, unum sunt, quia eis superadditur continuitas, quae facit unitatem. Super indivisibilia autem diversorum continuorum superadditur discontinuitas, quae facit diversitatem.

20. Sed adhuc stat argumentum probans quod contradictoria sunt simul.

21. Propter hoc mihi videtur quod 'simul esse' est aequivocum, quia uno modo privat medium faciens extrema distare ad invicem, ponens tamen immediatam existentiam [d]unius post aliam, ut si duae lineae applicentur simul in extremis, illae duae lineae sunt simul quia nihil est medium inter earum extrema. Aut sunt aliqua simul quia simul existentiam habent et entitatem. Primo autem modo simultatis sunt contradictoria simul, quia ordinantur invicem sine medio, et ita instans aeternitatis quod mensurabat ultimum sui non esse et instans temporis mensurans primum sui esse sic sunt[e] simul[f].

22. Contra: nonne verum est dicere in primo instanti temporis quoniam aeternitas creata[g] est, dum hoc instans temporis est? Et aeternitas creata sive aevum est totum simul; ergo tota aeternitas creata est simul cum primo nunc temporis, et hoc tam primo modo quam secundo.

[a] a parte ante *om.* L
[b] est] esse *add. sed del.* O
[c] fluit *corr. ex* fuit O
[d] unius post aliam *om.* T
[e] sunt] secundo modo non *add.* L
[f] simul] et non alio modo *add. s.l.* O; secundo modo non sunt simul *add.* T
[g] creata] creatura OL

15. Furthermore, if past time were infinite into the past, we would not have arrived at the present day, since it is not possible to traverse the infinite.

16. Furthermore, if this were the case, then generation would be infinite, and then souls would be infinitely many in act.

17. Furthermore, where there is no first there is no last; therefore, if there was no first instant of time, neither is there a last, which is false.

18. Solution. It must be said that time is not eternal, but it began, as did motion, and there was a first man. And Aristotle does not want to claim the opposite, if his view is explained correctly.

19. Thus, to the first argument[37] it must be said that the last instant of <time's> non-being and the first instant of its being are not the same instant, because the last instant of its non-being is an instant of eternity and the first instant of its being is the first instant of time. Also, that instant is not something within time, but is an indivisible principle from which the whole of time flows, just as in the case of motion the whole motion flows from a first sudden change and the whole line from a first indivisible point. Accordingly, when he says that between any two indivisibles there is something intermediate, it must be said that this is true about indivisibles of the same continuum. The reason for this is that it is necessarily the case that when two indivisibles are together they are one, because continuity, which produces unity, is added to them. But discontinuity, which produces diversity, is added to indivisibles of different continua.

20. However, the argument showing that contradictories are simultaneous still applies.[38]

21. Because of this it seems to me that 'to be together' is equivocal: (i) in one sense <being together> takes away the intermediate that makes extremes distant one from the other, while positing the immediate existence of one extreme after that of the other. For instance, if two lines are together joined at their extremes, those two lines are together because there is no intermediate between their extremes. (ii) Some things are together because they have their existence and their entity simultaneously. Contradictories are together in sense (i) because they are ordered one to the other without an intermediate, and in this way the instant of eternity that measured the last non-being of time, and the instant of time that measures the first being of time are together.

22. Against this: isn't it true to say at the first instant of time that created eternity exists, when this instant of time exists? And created eternity or the *aevum* is together as a whole; therefore, created eternity as a whole is together with the first 'now' of time, and this both in way (i) and in way (ii).

[37] See above, n. 1.
[38] See above, n. 3.

23. Dicendum quod tota aeternitas creata fuit ante primum instans temporis et tota fuit cum primo instanti temporis, sicut est dicere de anima, quoniam tota est in qualibet parte corporis. Verumtamen, licet totum sit in capite et totum in pede, in quantum tamen totum in capite non est totum in pede. Eodem modo est de aeternitate quod tota est cum priore et tota cum posteriore, in quantum tamen est cum priore, non est cum posteriore, et ita ratione diversorum diversatur.

24. Ad secundum argumentum ad principale dicendum quod prius per naturam non ponit prioritatem actualis existentiae, et ita in illo priori per naturam non respondet ei aliquod[a] esse instantis[b].

25. Quidam tamen dicunt quod in instanti sunt plures gradus. Et ratio eorum est quia omne quod est[c] ex nihilo potest cedere in nihil; unde intelligentia, cum[d] sit ex nihilo, potest cedere in nihil. [e]Sed per motum non cedit in nihil, cum impartibile non moveatur; ergo cessioni[f] suae in nihil non respondet[g] tempus, cum motus respondeat tempori; si ergo debeat cedere in nihil, cedet in instanti. Sed respectu eiusdem indivisibilis non potest esse et non esse; necesse est ergo ponere in instanti gradus, quorum unus respondeat suo esse et alius suo non esse.

26. Contra eos arguitur sic: nonne intelligentia habet esse in toto instanti? Ergo non est in parte corruptibilis et in parte non corruptibilis.

27. Item, nihil est velocioris pertransitus quam sit instans temporis; sed impossibile est instans esse et non esse in eodem instanti; aliud[h] ergo quam instans non poterit esse et non esse in eodem instanti.

28. Ad aliud de scientia Dei dicendum [i]quod Deus scivit simul ipsum esse quando fuit et ipsum non esse quando non fuit. Et bene sequitur quod scivit simul mundum esse et non esse, ita quod li[j] 'simul' determinet li 'scire'. Si autem determinet li 'esse', sic non sequitur et est propositio falsa.

[a] aliquod] aliquid OT
[b] instantis] instans O
[c] est] quia est *add.* LT
[d] cum *suppl. s.l.* O
[e] Sed ... nihil *om. hom.* T
[f] cessioni *corr. ex* cessione O
[g] respondet] nihil *add. sed del.* O
[h] aliud] aut LT
[i] quod Deus *om.* LT
[j] li] si O

23. It must be said that the whole of created eternity existed before the first instant of time and the whole of it existed together with the first instant of time, just as one can say of the soul that the whole soul is in every part of the body. Nonetheless, although the whole soul is in the head and the whole soul is in the foot, the whole soul in so far as it is in the head is not the whole soul in the foot. Similarly in the case of eternity, the whole exists with the prior state and the whole exists with the posterior state, but the whole in so far as it exists with the prior state does not exist with the posterior state, and thus the whole is diversified in virtue of different things.

24. To the second argument for the main conclusion[39] it must be said that the state prior by nature does not involve the priority of actual existence, and thus no actual instant of time corresponds to the mobile thing in that state <which is> prior by nature.

25. Some people, however, say that there are many stages in an instant. Their reason for this is that every thing that comes from nothing can pass into nothing; hence, since an intelligence comes from nothing, it can pass into nothing. But an intelligence does not pass into nothing through a motion, since an indivisible thing is not subject to motion; therefore no time corresponds to its passing into nothing, since motion corresponds to time; therefore, if an intelligence is to pass into nothing, it will do this in an instant. But <an intelligence> cannot be and not-be with respect to the same indivisible; therefore, it is necessary to posit stages in an instant, so that one stage corresponds to the being of the intelligence and another stage to its non-being.

26. One can argue against these people like this: isn't it the case that an intelligence has being in the whole of an instant? Therefore, it is not the case that an intelligence is corruptible in one part of an instant and is incorruptible in another part.

27. Furthermore, nothing is faster in its passage than is an instant of time; but it is impossible for an instant to exist and not to exist in the same instant; therefore, something other than the instant cannot exist and not exist in the same instant.

28. To the other argument about God's knowledge[40] it must be said that God knew simultaneously that the world existed when it existed and that the world did not exist when it did not exist. And it does follow that he simultaneously knew that the world existed and did not exist, if 'simultaneously' qualifies 'to know'. But if 'simultaneously' qualifies 'to exist', then the conclusion does not follow and the proposition is false.

[39] See above, n. 4.
[40] See above, n. 5.

29. Ad [a]illud quod actus melior potentia dicendum quod, simpliciter loquendo, actus est melior potentia; respectu tamen condicionum loci, temporis et aliorum, potest potentia esse melior actu.

30. Ad illud quod in perpetuis non differt etc., hoc est verum [b]quantum ad esse, quia non differt esse et posse esse in ipsis, tamen differt agere et posse agere. [c]Et etiam in primo differt agere in extrinsecus, ut creare, dominari et huiusmodi, et posse agere.

31. Ad illud quod dicit 'factus est de non efficiente efficiens', dicendum quod non sequitur, licet de non agente fiat agens, quod in agente fiat transmutatio aliqua, nisi ipsum agens [d]fuerit agens naturale. In voluntarie agentibus etiam[e] inferioribus non valet, quia eadem voluntate qua volo aliquid fieri cras volo illud idem non fieri hodie et, eadem voluntate non mutata, fit res cras. Et ita fit de non volente volens nulla mutatione facta in volente.

32. Sed nonne sic arguit Aristoteles in principio octavi: si de non movente fiat movens, tunc mobile refertur ad movens[f] cum prius ad ipsum non fuit relatum, et si hoc, in mobili fit aliqua transmutatio?

33. Item, arguit sic absolute: si de non movente fiat movens, accidit transmutatio in mobili.

34. Solutio. Ad primum dicendum quod Aristoteles bene arguit nec posset aliter[g] arguere, quia supponit[h] quod mobile fuit prius in actu existendi et, quando duo sunt in actu et[i] inter ipsa causetur relatio, necesse est quod fiat transmutatio in uno illorum vel in utroque. Unde sic est in mobili: cum sit prius in actu, necesse est acquirere novam dispositionem. Si autem unum extremum subito producatur in esse, non est necesse.

[a] illud ... melior] aliud quod dicit quod actus melior est quam T
[b] quantum ad esse *om.* T
[c] Et ... agere *om. hom.* L
[d] fuerit agens *om. sed suppl. mg.* sit agens T
[e] etiam *suppl. s.l.* O; *om.* LT
[f] movens] motum LT
[g] aliter *ex corr.* O
[h] supponit] si ponit O
[i] et *suppl. s.l.* O

29. To the argument that act is better than potency[41] it must be said that act is better than potency absolutely speaking; but in relation to the conditions of place, time and other things, potency can be better than act.

30. In reply to the argument that in perpetual things <there is no difference between being and being possible>[42]: this is true with respect to their being, since being and being possible do not differ in them, but acting and being able to act are different in them. Where the first being is concerned, to act on something extrinsic by creating, dominating, and the like, and to be able to act <in those ways> are also different.

31. To the argument saying 'it has passed from being a non-efficient cause to being an efficient cause',[43] it must be replied that it does not follow that there is a change in the agent, even if the agent passes from being a non-agent to being an agent, unless the agent is a natural agent. The argument is not valid for voluntary agents, even for those here below, because by the same will by which I will something to happen tomorrow I will the same thing not to happen today, and the thing does happen tomorrow without a change in my will. And in this way something passes from being non-willing to being willing without any change in the willing agent.

32. But isn't it the case that Aristotle argues as follows at the beginning of *Physics* VIII: if something passes from being a non-mover to being a mover, then the mobile thing comes to be related to the mover, and it was not related to it before, and if so, there is some change in the mobile thing?[44]

33. Furthermore, Aristotle formulates his argument in absolute terms: if something passes from being a non-mover to being a mover, a change occurs in the mobile thing.[45]

34. Solution. To the first argument[46] it must be said that Aristotle argues correctly and he cannot argue in another way, because he assumes that the mobile thing actually existed beforehand, and when two things are in act and a relation between them is brought about, it is necessary for a change to occur in one of them or in both. Thus, in the case of the mobile thing, since it is in act beforehand, it is necessary for it to acquire a new disposition. But if one of the two extremes is suddenly brought into being, it is not necessary <for a change to occur>.

[41] See above, n. 6.
[42] See above, n. 8.
[43] See above, n. 9.
[44] Arist. *Phys*. VIII.1, 251b5–7.
[45] Arist. *Phys*. VIII.1, 251a23–28.
[46] See above, n. 32.

35. Ad secundum similiter patet quod bene arguit, quia supponit quod mobile prius sit secundum actum existendi ante motum, et ita supponit quod sit quiescens. Et si de quiescente **[O 114ra]** fiat aliquid motum, necessario transmutatur[a]. Et ita non est a parte ista.

36. Sed caditne causa prima sub relatione?

37. Mihi videtur quod non, quia alia secundum totam suam substantiam ad ipsum referuntur.

38. Item, si arguatur sic, secundum Augustinum: quia Deus bonus, ideo sumus, cum ergo ab aeterno fuit aeque bonus, aequaliter ab aeterno fuimus, dicendum quod maior habet duos sensus. Unus est: quia Deus voluit, cui coniungitur sua bonitas, ideo sumus, et est verus sensus. Alius est sensus, quandocumque ipse est, quod nos sumus, et iste sensus est falsus.

39. Contra imaginationem antiquorum qua[b] imaginantur vestigium pedis in pulvere etc.: causa prima[c] cadit in triplici genere causae, scilicet formalis, finalis et efficientis; [d]causa prima ergo est causa efficiens mundum. Sed duae sunt condiciones efficientis sufficientis, scilicet[e] rem in esse producere et [f]productum in esse conservare. Sed prima condicio est generalis cuilibet efficienti; si ergo causa prima est efficiens mundum, mundum produxit in esse, et ita non est imaginari[g] quod mundus[h] sit coaeternus[i] sibi, et tamen quod sit efficiens. Eodem modo arguitur si pes sit causa efficiens vestigii.

40. Ad illud quod [j]dicit Aristoteles, quod omne[k] nunc est medium duorum temporum, dicendum quod bene stant simul quod[l] tempus incepit et tamen non fuit primum praesens temporis post exitum [m]temporis in esse. Unde[n] fuit primum instans a quo fluit totum tempus sequens. Et quando aliquid fluit ab aliquo indivisibili, post illud indivisibile non est sumere primum, quia, si sumatur, cum sit continuum, est divisibile in infinitum, et ita fuit aliquid prius. Est ergo considerare nunc dupliciter: vel secundum quod nunc est principium temporis et non infra continuitatem temporis, et sic non intendit Aristoteles quod sit medium etc.; aut ut est infra continuitatem temporis, et sic intendit Aristoteles quod est medium duorum temporum.

[a] transmutatur] transitur O
[b] qua *ex corr.* O
[c] prima] est vel *add.* T
[d] causa ... sufficientis *om.* T
[e] scilicet] efficientis est (est *s.l.*) T
[f] productum in esse] rem in esse productam T
[g] imaginari *corr. ex* imaginare O; imaginare LT
[h] mundus *corr. ex* mundum O
[i] coaeternus *corr. ex* coaeternum O
[j] dicit Aristoteles, quod *om. hom.* L
[k] omne *corr. ex* esse O; esse LT
[l] quod *corr. ex* quia O
[m] temporis in esse *corr. ex* in esse temporis O
[n] Unde] enim *add. s.l.* O

35. Similarly, it is clear that the second argument[47] is correct, because it assumes that the mobile thing actually exists before its motion begins, and thus it assumes that the mobile thing is at rest. And if something passes from being at rest to being in motion it necessarily changes. But that does not happen in the present case.

36. But isn't it the case that the first cause falls under a relation?

37. It seems to me that this is not the case, because other things are related to it in accordance with their whole substance.

38. Furthermore, if, following Augustine,[48] someone argues as follows: since God is good, we exist; therefore, since God was equally good from eternity, we equally existed from eternity, it must be replied that the major premise has two senses. (i) One is: because God, whose goodness is united with him, willed, we exist; and this sense is true. (ii) The other sense is that we exist whenever God exists; and this sense is false.

39. Against the view imagined by the ancients about the footprint in the dust <and its implications>:[49] the first cause falls under three kinds of cause, that is, formal, final, and efficient; therefore, the first cause is the efficient cause of the world. But there are two conditions sufficient for an efficient cause: (i) to bring a thing into being and (ii) to preserve the being of the thing produced. Condition (i) is common to every efficient cause; therefore, if the first cause is the efficient cause of the world, it brought the world into being, and thus we should not imagine that the world is coeternal with the first cause and yet that the first cause is its efficient cause. The same argument holds when a foot is the efficient cause of a footprint.

40. To Aristotle's statement that every 'now' is intermediate between two times,[50] it must be said that the claim that time began and the claim that there was no first present time after time came into being are indeed compatible. Thus, there was a first instant from which the whole of subsequent time flows. And when something flows from an indivisible, one cannot take something after that indivisible to be first, because if something is taken to be first, given that it is a continuum, it is divisible to infinity, and thus there was something before <the first>. Therefore, the instant can be considered in two ways: either (i) in so far as the 'now' is the beginning of time and not within the continuity of time; and it is not in this way that Aristotle intends the instant to be intermediate <between two times>. Or (ii) in so far as the instant is within the continuity of time; and it is in this way that Aristotle intends the instant to be intermediate between two times.

[47] See above, n. 33.

[48] Aug. *De doctr. christ.* I.32, CCSL 32, 26. 3–4.

[49] See above, n. 7.

[50] See above, n. 12.

<Quaestio 6>
Consequenter [a]quaeritur de hoc quod dicit quod nihil est inordinatum a natura.

1. Contra: aliquando producuntur sex digiti in una manu, quod est inordinatum. Quod autem producatur per naturam videtur, quia propria forma perficit propriam materiam per naturam; si ergo ibi sit materia ordinata ad sextum digitum, hoc erit a natura quod producatur.

2. Item, determinatum est in natura quanto de materia quantum respondeat de forma; cum ergo plus sit de materia ubi est sextus digitus, necesse est quod respondeat ei plus[b] de forma.

3. Item, a natura est quod [c]per viam naturae sive per naturam in esse producitur; sed talis est sextus digitus.

4. Item, quod est a principio intra per se et non per accidens est a natura et, praeter hoc, principium producens sextum digitum in esse est natura.

5. Contra: quod est praeter naturae intentionem non est a natura; sed sextus digitus est praeter intentionem naturae; quare etc.

6. Dicitur quod sextus digitus, id quod est, a natura est, sextus tamen in ratione sexti non est a natura.

7. Aliter [d]dicitur sic quod est loqui de natura dupliciter: aut de natura universali intendente, et sic non est a natura; aut de natura operante. Et hoc dupliciter, quia aut ut operetur ad productionem superflui in materia, et sic non est a natura, [e]aut ut operetur post productionem superflui in esse, et sic est a natura.

8. Aliter dicitur quod monstruositas habet duplex significatum, scilicet unum a quo imponitur, et est ipsa deformitas, et isto modo non est monstrum a natura; aliud est significatum, ut cui imponitur, et est illud quod sustinet ipsam deformitatem, et illud est a natura.

[a] quaeritur ... quod *om.* T
[b] plus *suppl. s.l.* O; *om.* LT
[c] per ... sive *om.* T
[d] dicitur sic] dicendum scilicet O
[e] aut ... natura *om. hom.* T

<Question 6>
The next question is about Aristotle's statement that nothing disordered comes from nature.[51]

1. Against this: sometimes six fingers are produced on one hand, and this is something disordered. That it is produced through nature is shown because a proper form gives perfection to its proper matter through nature; therefore, if the matter there is ordered to a sixth finger, the production of a sixth finger will come from nature.

2. Furthermore, nature determines what quantity of form corresponds to a quantity of matter; therefore, since there is more matter where there is a sixth finger, it is necessary for more form to correspond to it.

3. Furthermore, what is brought into being through the way of nature or through nature comes from nature; but a sixth finger is of this kind.

4. Furthermore, what derives from an internal principle per se and not accidentally comes from nature, and in addition to this, the principle that brings a sixth finger into being is nature.

5. Against this: what is beyond the intention of nature does not come from nature; but a sixth finger is beyond the intention of nature; hence, <it does not come from nature >.

6. It is said that a sixth finger as such comes from nature, but in so far as it is the sixth it does not come from nature.

7. Another reply is that we speak of nature in two ways: either (i) of a universal intending nature; and in this way a sixth finger does not come from nature; or (ii) of an operative nature. And case (ii) is twofold: either we speak (iia) of nature as working towards the production of a superfluity of matter; and in this way a sixth finger does not come from nature; or (iib) of nature as operating after the production of something actually superfluous; and in this way a sixth finger does come from nature.

8. Another reply is that the term 'monstrosity' has two significates: (i) one from which the name is imposed, and this is the deformity itself, and in this sense a monster does not come from nature; (ii) the other significate is that on which the name is imposed, and this is the thing that has the deformity, and this thing does come from nature.

[51] Arist. *Phys*. VIII.1, 252a11–12.

<Quaestio 7>
Consequenter [a]quaeritur de hoc quod dicit quod natura est causa omnis ordinationis.

1. Contra: natura aut est materia aut forma; sed nec materia nec forma est causa omnis ordinationis[b], et hoc quia[c] dicit Aristoteles in undecimo quod duplex est ordo rerum, unus scilicet inter se et alius ad causam primam; et non est primus ordo nisi propter secundum; quare causa prima erit causa [d]omnis et totius ordinationis.

2. Item[e], intelligentiae motrices orbium praecedunt omnem naturam et maximum habent ordinem inter se, secundum quod dicit Aristoteles quod corpora caelestia ordinantur inter se, ita quod quaedam sunt priora et quaedam posteriora, et ita idem est ordo moventium et mobilium, et ita, cum praecedant naturam, natura non est principium ordinationis in ipsis.

3. Item, est ordinatio quaedam causalis, quae est rationum causalium in mente divina et est incausata, et ita natura non est principium ordinationis ibi.

4. Alia est instantia quia ars ordinat artificialia, et ita natura non est causa omnis ordinationis.

5. Item, intellectus ordinat praemissam ad conclusionem, et intellectus non est natura; quare etc.

6. Dicendum quod omnis ordinatio aut est a[f] natura aut ab aliquo quod ordinat sicut natura ordinaret si in illud ageret, [g]secundum quod dicit Aristoteles [h]quod ars facit domum eodem modo sicut natura faceret.

7. Dicendum igitur[i] quod est ordinatio causalis, et est in causa prima, et haec ordinatio est causa totius ordinationis[j] inferioris, et haec est incausata, et in hac non est natura principium ordinationis. Alia est ordinatio causata. Et tunc distinguo naturam quod est natura communiter dicta pro natura superiorum et inferiorum causatorum, et sic est natura talis causa ordinationis omnium causatorum. Aut sumatur natura proprie dicta, ut pro illa quae est in generabilibus et corruptibilibus, et sic est natura causa omnium ordinationum inferiorum. Si autem loquamur de natura naturante et non naturata, ista est causa universalis omnium[k].

[a] quaeritur ... quod *om.* T
[b] ordinationis] quare etc. *add.* T
[c] quia] quod LT
[d] omnis et *om.* T
[e] Item *om.* O
[f] a *suppl. s.l.* O; *om.* L
[g] secundum quod] sicut T
[h] quod ars *om. sed suppl. s.l.* ars O
[i] igitur *suppl. s.l.* O; *om.* LT
[j] ordinationis *suppl. mg.* O
[k] omnium] *Hic quaestionem de comparatione finiti ad infinitum add. T. Cf. Appendix, Liber VIII, q. 7a.*

<Question 7>
The next question is about Aristotle's statement that nature is the cause of every ordering.[52]

1. Against this: nature is either matter or form; but neither matter nor form is the cause of every ordering, and this because in *Metaphysics* XI Aristotle says that there is a twofold order of things: one among themselves and the other to the first cause; and the first order is only for the sake of the second;[53] therefore, the first cause will be the cause of the whole of every ordering.

2. Furthermore, the intelligences moving the orbs precede every nature and have maximal order among themselves, according to Aristotle's statement that the celestial bodies are ordered among themselves so that some are prior and others are posterior;[54] thus, the order of movers and mobile things is the same. And so, since the celestial movers precede nature, nature is not a principle of their ordering.

3. Furthermore, there is a causal ordering, which is that of the causal principles in the divine mind, and this ordering is uncaused, and so nature is not a principle of the ordering in this case.

4. Another counterexample is that art gives order to artefacts, and thus nature is not the cause of every ordering.

5. Furthermore, intellect orders a premise to its conclusion, and an intellect is not a nature; hence, <nature is not the cause of every ordering>.

6. It must be said that every ordering either comes from nature or from something that orders a thing in the way that nature would order it if nature were to act on it, in accordance with Aristotle's statement that art produces a house in the same way that nature would produce it.[55]

7. Therefore, it must be said that (i) there is a causal ordering, and this is in the first cause, and this ordering is the cause of the whole ordering here below, and this ordering is uncaused; and nature is not a principle of this ordering. (ii) Another ordering is caused. And then I distinguish between: (iia) nature said in a broad sense of the nature of superior things and of things here below, and this nature is the cause of the ordering of all the things that are caused. And (iib) nature properly speaking, as that which exists in generable and corruptible things; and in this way nature is the cause of all the orderings of things here below. But if we speak of creative nature and not of created nature, this kind of nature is the universal cause of all things.

[52] Arist. *Phys.* VIII.1, 252a12.
[53] Arist. *Metaph.* XII.10, 1075a11–15.
[54] Arist. *Metaph.* XII.8, 1073a14–b3.
[55] Arist. *Phys.* II.8, 199a12–13.

<Quaestio 8>

[a]Quaeritur consequenter de hoc quod dicit Anaxagoras, qui posuit quod elementa quieverunt per infinitum tempus in quodam chaos et tunc venit intellectus et separavit ea.

1. Et arguit contra ipsum sic: si ita quieverunt, tunc **[O 114rb]** non est maior ratio quare veniat intellectus separans in uno tempore [b]quam in alio; aut ergo in omni tempore separat aut in nullo. Si Aristoteles bene arguit, arguo tunc eodem modo: causa prima fuit ab aeterno; si ergo debuisset producere aliquid, non est maior ratio quare in uno tempore quam in alio; aut ergo in nullo aut in omni.

2. Item, dicit Aristoteles[c] quod, si quiescat illud chaos, non est maior ratio quare magis in uno quam in alio tempore terminetur. Arguo tunc de motu, quia, si aliquid movetur, non est maior ratio quare moveatur magis in uno tempore quam in alio, et ita non est ratio quare terminetur in una hora magis quam in alia.

3. Solutio. Dicendum quod Aristoteles bene arguit, quia quies est privatio motus et prius per naturam est motus quam quies; qui ergo ponit prius quietem quam motum ponit ordinis perversionem et, cum perversionis non sit aliqua causa, bene arguit Aristoteles. In his autem quae creantur[d] non esse simpliciter praecedit esse, et ita est ibi rectitudo ordinis[e]. Et ista est causa quare tunc et[f] non alias. Sed haec causa non est nisi voluntas Dei, secundum[g] quod dicit Augustinus, qui[h] quaerit tres quaestiones. Prima est quare fecit mundum tam parvum. Secunda est quare fecit mundum plus hic quam alibi. Tertia est quare fecit mundum magis tunc quam prius [i]vel post. Et dicit quod duae primae quaestiones sunt ridiculosae, tertia vero est difficilis, et non est causa nisi voluntas Dei.

4. Ad secundum dicendum quod Aristoteles bene dicit de quiete, de motu non valet. Et ratio est quia motus determinatur tribus modis, ut per potentiam motoris et per terminum a quo et per terminum in quem, et ita motus determinatur per terminum. Et ista est ratio quare terminetur. Sed ex parte quietis non est ita, quia quies solum determinat sibi aliquid [j]ut subiectum et non determinat sibi aliquid ut terminum, et ideo non est maior ratio quare

[a] Quaeritur ... qui] Consequenter dicit quod Anaxagoras T
[b] quam in alio *suppl. s.l.* O; *om.* L
[c] Aristoteles *scripsimus*; Anaxagoras OLT
[d] creantur *ex corr.* O
[e] ordinis] corporis O
[f] et *om.* O
[g] secundum *ex corr.* O
[h] qui *corr. ex* quia O; quod L; quia T
[i] vel post *om.* T
[j] ut ... aliquid *om. hom. sed suppl. mg.* O

<Question 8>
The next question is about Anaxagoras' assumption that all the elements were at rest for an infinite time in some chaos and then intellect arrived and separated them.[56]

1. Aristotle argues against Anaxagoras as follows: if the elements were at rest in that way, there is no more reason why the separating intellect should arrive at one time rather than at another; therefore, either the intellect separates them at all times or at no time.[57] If Aristotle argues well, then I argue in the same way as follows: the first cause existed from eternity; therefore, if it had to produce something, there is no more reason why it should produce that thing at one time rather than at another; therefore, either it produces it at no time or at all times.

2. Furthermore, Aristotle says that, if that chaos is at rest, there is no more reason why that rest should end at one time rather than at another.[58] I then argue about motion that, if something moves, there is no more reason why it should move at one time rather than at another, and thus there is no reason why it should stop moving at one given time rather than at another.

3. Solution. It must be said that Aristotle does argue well because rest is the privation of motion and motion is prior by nature to rest; therefore, positing that rest is prior to motion is to posit the reversal of that order, and since there is no cause for such a reversal, Aristotle does argue well. But in things that are created non-being precedes being absolutely, and so in this case the order is correct. And this is the cause why they are created then and not at another time. But this cause is nothing other than the will of God, according to Augustine, who raises three questions. The first is why God made the world so small. The second is why God made the world here rather than elsewhere. The third question is why he made the world then rather than before or after. And he says that the first two questions are laughable, but the third one is difficult, and there is no cause for this other than the will of God.[59]

4. To the second argument[60] it must be said that Aristotle's argument is valid about rest, but not about motion. And the reason for this is that motion is determined in three ways, that is, by the power of the mover, by the terminus

[56] Arist. *Phys.* VIII.1, 250b24–26.
[57] Arist. *Phys.* VIII.1, 252a14–19.
[58] Arist. *Phys.* VIII.1, 252a14–19.
[59] The reference to Augustine seems clearly wrong. Note, for example, that the third question implies that time pre-existed the creation of the world, which is against Augustine's view (see, e.g., Aug. *Conf.* XII.15.20, CCSL 27, 225.32–226.52).
[60] See above, n. 2.

una vice magis quam alia terminetur. Et ita patet quod, si aliquid[a] quiescat naturaliter, numquam per solam naturam terminabitur quies eius; si autem moveatur, necessario terminabitur.

<Quaestio 9>
Consequenter [b]quaeritur de hoc quod dicit ibi: "Contraria autem magis his". Ponit tres rationes, et in tertia supponit quod homo moveatur quandoque et quandoque non.

1. Contra: non est aliqua hora quin voluntas alicuius sit ut moveat et quin moveatur illa voluntate; sed eadem est voluntas movendi de loco ad locum in omnibus, quia vult Commentator super tertium *De anima* quod idem est intellectus possibilis in omnibus, et per intellectum possibilem volumus quod volumus. Quod autem intellectus possibilis sit unus numero in omnibus[c] videtur, quia est aliqua forma quae est una numero in toto uno individuo et est alia forma quae est una numero [d]in toto genere rerum, ut causa prima, quae est una numero in omnibus; ergo erit aliqua forma quae erit una forma numero in tota una specie, quae non videtur alia quam anima rationalis; quare etc.

2. Item, sensitiva est absoluta a materia et intellectiva multo magis absolvitur; sed sensitiva propter[e] eius absolutionem est tota et una numero in qualibet parte sui corporis; cum ergo rationalis magis absolvitur a[f] condicionibus materialibus, ipsa propter eius absolutionem erit reperta eadem numero plus quam in uno individuo.

3. Aliter formetur ratio sic: plus excedit rationalis sensitivam quam plura eiusdem speciei excedunt unum secundum speciem, quia rationalis[g] et[h] sensitiva diversificant speciem, plura secundum speciem et unum secundum speciem non. Cum ergo sensitiva sit in uno secundum speciem, tunc rationalis erit in pluribus secundum speciem sive in tota specie, et hoc idem numero.

[a] aliquid] aliquis *ex corr.* O; aliquis L
[b] quaeritur ... his *om.* T
[c] omnibus] eiusdem speciei *add.* T
[d] in ... numero *om. hom.* L
[e] propter] praeter OL, *sed corr.* O
[f] a *suppl. s.l.* O; *om.* L
[g] rationalis *corr. ex* rationales O
[h] et *suppl. s.l.* O

from which and by the terminus towards which. Thus motion is determined by a terminus; and this is the reason why motion ends. But this does not apply to rest, because rest only determines something as its own subject and does not determine something as its own terminus, so that there is no more reason for rest to end on one occasion rather than on another. It is clear then that if something is naturally at rest, its rest will never end by virtue of nature alone, but if it is in motion, its motion will necessarily end.

<Question 9>
The next question is about Aristotle's statement here: "Rather the contraries to these".[61] He posits three arguments, and in the third he assumes that a man sometimes moves and sometimes does not.[62]

1. Against this: there is no time at which someone has the will to move without being moved by that will; but the will to move from a place to a place is the same for all men, because in his commentary on *On the Soul* III the Commentator wants the possible intellect to be the same in all men,[63] and it is through the possible intellect that we will what we will. That the possible intellect is numerically one in all men is shown as follows: there is a form that is numerically one in one whole individual, and there is another form that is numerically one in the whole genus of things, that is, the first cause, which is numerically one in all things; therefore, there will be a form that is numerically one in one whole species, and this form does not seem to be other than the rational soul; hence, <the possible intellect is one in all men>.

2. Furthermore, the sensitive soul is released from matter and the intellectual soul <is released> to a much greater extent; but the sensitive soul is in any part of its body as a whole and as numerically one in virtue of its release from matter; therefore, since the rational soul is released from material conditions to a greater extent, because of this release it will be found as numerically the same in more than one individual.

3. An alternative formulation of this argument is the following: the rational soul exceeds the sensitive soul to a greater extent than many things belonging to the same species exceed a single member of that species, because the rational soul and the sensitive soul diversify a species, whereas many things belonging to one species and one thing of that species do not diversify that species. Therefore, since the sensitive soul is in one thing belonging to one species, the rational soul will be in many things belonging to one species or in the species as a whole, and it will remain the same in number.

[61] Arist. *Phys*. VIII.2, 252b7.
[62] Arist. *Phys*. VIII.2, 252b17–21.
[63] Aver. *In De anima* III, t.c. 5, ed. Crawford, 392.158–413.758.

4. Item, ad principale: si animal aliquando movetur et aliquando non movetur, tunc, quando movetur de potentia ad actum, et hoc per aliquid excitans [a]extra, tunc excitans fit de non excitante excitans; ergo per aliquid extra fit[b] excitans[c], et ita est procedere in infinitum aut est stare in principio, et dicendum quod semper movetur aut quod non movetur.

5. Solutio. Dicendum quod Aristoteles verum supponit.

6. Unde dicendum argumentis et primo ultimo argumento quod[d], si loquamur de actione causae primae, cum fiat de non agente agens, non educitur in actionem per aliquid extra, sed se ipso exserit se in suam actionem. Sed in homine ita est quod, si fiat de non agente agens, necesse est quod moveatur specie appetibili vel tristabili. Et non sequitur quod illud quod movet[e] ipsum moveatur, quia aliquid movere aliud est dupliciter: vel ut finis et desideratum, et quod sic movet non movetur; aut ut efficiens, et tale necessario[f] transmutatur.

7. Ad aliud, cum dicit quod idem est intellectus possibilis in omnibus, haec propositio non habet veritatem nisi talem: [g]intellectus possibilis est idem numero, hoc est assimilatur ei quod est idem numero, in omnibus hominibus, ut materiae primae, quae[h] est eadem numero per privationem cuiuslibet actualitatis formae et distinctionis et est in potentia ad omnes formas. Et ita est de possibili, quia in prima sui unione, antequam recepit aliquid ab extra, fuit essentialiter in quolibet numeratus, actualiter autem non fuit numeratus, quia non recepit phantasmata ab extra, per quae numeratur actualiter[i], et ita in prima unione est in potentia ad actualem numerationem.

8. Ad argumentum dicendum quod non est uniformiter **[O 114va]** aliquid unum numero in [j]genere rerum et aliquid unum numero in uno individuo, quia, si sic, tunc sequeretur[k]. Est enim sensitiva in uno individuo ut pars essentialis et ut perfectio, prima causa est in uno genere quia est ubique praesentialiter, potentialiter et essentialiter.

9. Ad aliud dicendum quod anima rationalis plus absolvitur quam sensitiva non quia sit in tota specie quando aliud est in toto individuo, sed[l]

[a] extra, tunc excitans *om. hom.* T
[b] fit *suppl. s.l.* O; quod est T
[c] excitans *corr. ex* excitatur O
[d] quod] quia OT
[e] movet *corr. ex* movetur O; movetur L
[f] necessario] movetur vel *add.* T
[g] intellectus … numero *om.* T
[h] quae *corr. ex* et O; et LT
[i] actualiter *ex corr.* O; accidentaliter LT
[j] genere ... uno *om.* L
[k] sequeretur] sequere O
[l] sed *corr. ex* et O

4. Furthermore, <a reply> to the main claim: if an animal is sometimes in motion and is sometimes not in motion, then, when it moves from potency to act, and this in virtue of something stimulating it from outside, the thing that stimulates it passes from being non-stimulating to being stimulating; therefore, that thing becomes stimulating in virtue of something external to it, and thus there will be an infinite regress. Alternatively one must stop at the first step and say that an animal is either always in motion or always not in motion.

5. Solution. It must be said that the assumption of Aristotle is true.

6. Thus, in reply to the arguments and first of all to the last one,[64] it must be said that, if we speak of the action of the first cause, when it passes from being a non-agent to being an agent, the first cause is not brought into action in virtue of something external, but by itself brings itself into action. In the case of man, however, if he passes from being a non-agent to being an agent, it is necessary for him to be moved by the species of something desirable or something saddening. And it does not follow that the thing that moves him is moved, because something moves something else in two ways: either as an end and something desired, and what moves in this way is not moved; or as an efficient cause, and this necessarily changes.

7. To the other argument,[65] when he says that the possible intellect is the same in all men, this proposition is true only in the following sense: the possible intellect is numerically one, that is, it is assimilated to what is numerically one in all men, that is, to prime matter, which is numerically one by virtue of the privation of any actuality of a form and of any distinction, and it is in potency to all forms. And so it is in the case of the possible intellect, because in its first union, before it had received anything from outside, the possible intellect was essentially numbered in every man, but it was not actually numbered, because it had not received from outside the phantasms by which it is actually numbered, and thus in its first union it is in potency to actual numeration.

8. To the argument[66] it must be said that the one in number does not exist uniformly in a genus of things and in one individual, because if it did then the conclusion <that the possible intellect is numerically one in all men> would follow. The sensitive soul is in one individual as an essential part and as a perfection, whereas the first cause is in one genus because it is everywhere by its presence, its potency, and its essence.

9. To the other argument[67] it must be said that the rational soul is released <from matter> to a greater extent than the sensitive soul not because the

[64] See above, n. 4.
[65] See above, n. 1.
[66] See above, n. 1.
[67] See above, n. 2.

quia minus proprie adhaeret corpori, et quia sensitiva operatur per instrumentum corporeum et intellectiva non, et quia sensitiva est generabilis et corruptibilis et[a] intellectiva non. Et ita non valet ultima consequentia.

1*. Aristoteles quaerit consequenter, quando corda continue percutitur, utrum sit sonus unus aut non.

<Quaestio 10>
Propter hoc quaeritur primo an sonus simpliciter sit accidens unum.

1. Et videtur quod non, quia omne accidens simpliciter[b] unum numero est tantum [c]in uno subiecto numero[d]; sed sonus, cum generetur ex duobus percussis, erit in ipsis ut in subiecto; quare non est accidens unum numero.

2. Quod autem sonus generatus ex continua et iterata percussione cordae non sit unus videtur, quia dicit Aristoteles quod ad hoc quod motus sit unus oportet quod mobile sit unum et movens unum, et in utroque exigitur continuitas. Cum ergo tanta[e] sit diversitas duorum generantium sonum quanta est duorum generantium motum, tunc ad hoc quod sonus sit unus, exigitur continuitas percutientis et percussi sine diversa iteratione, sicut fuit in motu.

3. Contra: sensus non decipitur circa proprium obiectum; cum ergo in auditu sit continuitas et unitas, tunc in sono erit similiter.

4. Ad primum dicunt quidam quod est accidens unum et est in utroque ut in subiecto. Et hoc quia conveniunt in hoc quod in generatione soni est necesse quod utrumque, scilicet et percutiens et percussum, sint percussa et in ratione percussi, et ita sunt ut subiectum unum, sicut est si duo tangant se, tactus non est in altero, sed in utroque tangentium; unde utrumque tactum est. Et etiam numerus quinarius in quinque, unde numerata sunt.

5. Aliter dicitur et melius quod, quando generatur sonus inter percutiens et percussum, non est sonus[f] unus neque tinnitus[g] unus, sed est ibi confusio diversorum sonorum, et ita absolute non est sonus unus, sicut est de luce quod plures luces confundunt se in una parte medii et tamen non est lux una numero.

[a] et *om.* LT
[b] simpliciter *lectio incerta* O
[c] in uno subiecto] unum subiectum T
[d] numero] quare non est accidens unum numero *scrips. sed del.* O
[e] tanta *corr. s.l. ex* causa O
[f] sonus] neque *add. sed del.* O
[g] tinnitus *scripsimus*; tumultus OL; timitus (*lectio incerta*) T

rational soul is in the whole species whereas the sensitive soul is in the whole individual, but because the rational soul adheres to the body less properly, and because the sensitive soul acts through a corporeal instrument and the intellectual soul does not, and because the sensitive soul is generable and corruptible, and the intellectual soul is not. And thus the last inference is not valid.

1*. Aristotle next asks whether the sound is one or not when a string is continuously struck.[68]

<Question 10>

The first question arising from this is whether sound is absolutely one accident.

1. Arguments for the negative: every accident that is absolutely one in number is only in numerically one subject; but when a sound is generated by two bodies that are struck, it will be in both of them as its subject; hence, sound is not numerically one accident.

2. That the sound produced by the continuous and repeated striking of a string is not one is shown because Aristotle says that in order for motion to be one it is necessary for the mobile thing to be one and for the mover to be one, and continuity in each of them is required.[69] Therefore, since there is as much diversity between two things generating a sound as there is between two things generating a motion, then for sound to be one the continuity of the striking thing and of the struck thing without a different repetition is required, as in the case of motion.

3. Against this: a sense is not deceived about its proper object; therefore, since there is continuity and unity in hearing, then the same will apply to sound.

4. To the first argument[70] some people say that sound is numerically one accident and it is in each of the two things as its subject. And this happens because the two things agree in that in the production of sound it is necessary for each of them, that is, both the striking and the struck thing, to be struck and to have the *ratio* of being struck, and thus they are like one subject, in the same way that if two things touch each other, touch is not just in one of them, but it is in both of the things that are touching; therefore, each of them is touched. Also, the number five is in five things in so far as they are numbered.

5. Another and better reply is that, when a sound is produced between a striking and a struck thing, the sound is not one nor is the ringing one, but there is a confusion of different sounds, and thus sound is not one absolutely, just as in the case of light many lights mingle in one part of the medium, but light is not one in number.

[68] Arist. *Phys.* VIII.2, 252b33–35.
[69] This probably is a reference to Aristotle's discussion of the numerical unity of a motion in *Phys.* V.4, 227b21–228a3, although Aristotle does not mention the mover there.
[70] See above, n. 1.

6. Ad secundum dicendum quod non est sonus unus, sed est sonorum confusio, et ita, si auditus simul et semel audiat, non tamen est hoc una immutatione.

7. Ad ultimum dicendum quod sonus est proprium obiectum auditus, continuatio eius sive numeratio eius non, sicut color est obiectum visus nec decipitur visus circa ipsum, tamen circa numerum coloratorum decipitur, ut patet: si oculus amoveatur a situ suo proprio, apparebit unum duo.

<Quaestio 11>
Quaeritur consequenter utrum sonus idem numero[a] deferatur a percusso usque ad auditum.

1. Et videtur quod non, quia una species numero non facit se usque ad visum, sed quaelibet gignit aliam speciem; ergo sic erit in sono.

2. Contra: lux protensa[b] et reflexa est eadem numero; ergo multo fortius lux protensa est[c] eadem numero, et ita est[d] de sono. Probatio maioris: lux reflexa prohibet lucem protendi; sed si Marchus in theatro prohibeat [e]ipsum esse in foro, hoc non est nisi quia sunt idem numero; ergo lux reflexa non prohibet [f]lucem protendi nisi quia sunt idem numero.

3. Dico quod[g] non est idem sonus numero nec est eadem lux protensa et reflexa; unde obstaculum prohibet per se lucem protendi et accidit luci ut reflectatur. Hoc autem est per se, quod[h] Marchus in theatro prohibet ipsum esse in foro, et ita hic sequitur identitas numeralis, ibi autem non.

<Quaestio 12>
[i]Quaeritur consequenter de hoc quod dicit quod animal solum movetur localiter ex se.

1. Contra: movetur motu augmentationis et nutritionis.

2. [j]Item, quare est quod elementum non movetur ex se, [k]animal tamen movetur ex se, et hoc in omni suo motu?

[a] numero *ex corr.* O
[b] protensa *scripsimus*; pertenssa *corr. ex* percussa O; percussa LT
[c] est] erit LT
[d] est] erit LT
[e] ipsum ... foro] Tullium esse T
[f] lucem protendi *om.* L
[g] quod *suppl. s.l.* O
[h] quod *om.* LT
[i] Quaeritur ... quod] consequenter T
[j] Item ... elementum *corr. in* Contra quod est O
[k] animal … se *om. hom.* O

6. To the second argument[71] it must be said that sound is not one, but there is a confusion of sounds, and thus even if hearing hears at one and the same time, this does not, however, come about by one change.

7. To the last argument[72] it must be said that sound is the proper object of hearing, but its continuity or its number is not, just as colour is the object of sight and sight is not deceived about colour, but it is deceived about the number of coloured things, as is clear, for if an eye is displaced from its proper location, one thing will appear as two.

<Question 11>
The next question is whether a sound that is numerically one is carried from the struck thing to the hearing.

1. Arguments for the negative: a species that is numerically one does not carry itself to sight, but every species generates another one; therefore, the same will apply to the case of sound.

2. Against this: a direct and a reflected light are the same in number; therefore, there is a much stronger reason to say that a direct light is the same in number; and so it is in the case of sound. Proof of the major premise: a reflected light prevents light from extending further; but if the fact that Mark is in the theatre prevents him from being in the market, this is only because Mark in the theatre and Mark in the market are the same in number; therefore, a reflected light prevents light from extending further only because a direct light and a reflected light are the same in number.

3. I say that sound is not numerically one, nor are a direct and a reflected light the same; thus, what per se prevents light from extending further is an obstacle and it is accidental to light that it is reflected. But the fact that Mark's presence in the theatre prevents him from being in the market is per se, and thus in this case numerical identity follows, but not in the other case.

<Question 12>
The next question is about Aristotle's claim that an animal is moved by itself only in locomotion.[73]

1. Against this: an animal is moved <by itself> in the motion of growth and of nutrition.

2. Furthermore, why is it the case that an element is not moved by itself, whereas an animal is moved by itself in all its motions?

[71] The argument is missing.
[72] See above, n. 3.
[73] Arist. *Phys.* VIII.2, 253a14–15.

3. Sed[a] quod movetur ex se potest quiescere quando vult; sed in motu augmentationis aut nutritionis non; quare etc.

4. Dicitur quod ad hoc quod aliquid moveatur ex se proprie oportet quod respectu illius motus ordinetur ut voluntas; et quia respectu motus augmentationis et nutritionis non ordinatur ut voluntas, sed ut natura, non proprie movetur animal tali motu ex se.

5. Vel sic: in motu nutritionis sunt virtutes caelestes operantes, anima autem est solum regens, et ideo in tali motu non movetur ex se. In mobilibus autem voluntariis est ita quod anima est principale agens et ita quod, sicut movetur, movetur proprie ex se.

<Quaestio 13>
[b]Quaeritur de hoc quod dicit consequenter quod inter quoslibet[c] duos motus contrarios cadit quies media.

1. Et videtur quod non dicat verum, quia, si sic esset, tunc, si faba proiciatur sursum et obviet in suo motu cuidam lapidi molari, tunc antequam [d]motu lapidis descendat, faba necessario quiescet, et ita sustinebit magnum lapidem per aliquod tempus, quod est impossibile; quare etc.

2. Item, si aliquis moveatur secundum sanitatem, cum sanitas consistat in indivisibili, tunc in instanti fiet acquisitio termini motus; tunc possibile est quod in illo **[O 114vb]** eodem instanti incipiat[e] transitio ad aegritudinem, et isti motus sunt contrarii; quare possibile est quod non intercidat quies media inter motus contrarios.

3. Ad oppositum: inter omnes quietes contrarias, qualiscumque ponatur casus, intercidit motus medius; ergo, per oppositum, inter quoslibet motus contrarios, [f]qualiscumque sit casus, intercidet quies media.

4. Solutio. Dicendum quod Aristoteles verum dicit, quia inter quoslibet motus contrarios et a causis contrariis cadit quies media, ut, si lapis proiciatur sursum violenter et in fine motus violenti descendat naturaliter, necessario intercidet quies media. Et hoc quia maioris virtutis est pellere et sustinere quam sustinere tantum; ergo in fine motus, quando virtus violenta non potest

[a] Sed *corr. in* quia O
[b] Quaeritur ... quod *om.* T
[c] quoslibet *corr. ex* quolibet O
[d] motu lapidis] lapis (*ex corr.*) O; *inv.* L
[e] incipiat] incipiet OL
[f] qualiscumque ... contrarios (n. 4) *om. hom.* T

3. But what is moved by itself can be at rest when it wants; but in the motion of growth or of nutrition an animal cannot do that; hence, <an animal is not moved by itself in the motion of growth and of nutrition>.

4. It is said that in order for something to be moved by itself in the proper sense it is necessary for it to be ordered to that motion by will; and because an animal is not ordered with respect to the motion of growth and of nutrition by will, but by its nature, it is not moved by itself in the proper sense in that motion.

5. Another reply is that there are celestial powers acting in the motion of nutrition, but the soul only directs, and therefore an animal is not moved by itself in that motion. But in voluntary mobile things, the soul is the principal agent so that, when something of that kind is moved, it is moved by itself in the proper sense.

<Question 13>
The question here is about what Aristotle says next, namely that between any two contrary motions an intermediate rest intervenes.[74]

1. Arguments that Aristotle's claim is not true: if this were the case, then, if a bean were projected upward and in its motion met a millstone, then before it descended through the motion of the millstone, the bean would necessarily be at rest, and thus it would support a big stone for some moment of time, which is impossible; hence, <there is no intermediate rest between two contrary motions>.

2. Furthermore, if something moves with respect to health, then since health consists in an indivisible, the acquisition of the terminus of motion happens at an instant; it is then possible that at that same instant a transition to illness begins, and these motions are contrary; hence, it is possible that an intermediate rest does not intervene between contrary motions.

3. To the contrary: between all contrary rests, whatever case is posited, an intermediate motion intervenes; therefore, by an argument from opposites, between any contrary motions, whatever case is posited, an intermediate rest will intervene.

4. Solution. It must be said that Aristotle's claim is true, because between any two contrary motions that take place from contrary causes an intermediate rest intervenes, for example, if a stone is projected upwards by violence and at the end of the violent motion it naturally moves downwards, then an intermediate rest will necessarily intervene. The reason for this is that it requires a greater power to push and to support than only to support; therefore, at the end of the motion, when the violent power cannot do both,

[74] Arist. *Phys.* VIII.7, 261b5–7.

super utrumque, alterum consequetur. Sed pellere sursum non potest facere nisi sustineat, et[a] sustinere potest quando non pellit; ergo aliquo tempore sustinebit quando non pellet. Et tunc erit quies; quare etc.

5. [b]Et sic patet ad primum.

6. Ad secundum dicendum quod impossibile est quod in eodem instanti in quo acquiritur sanitas incipiat transitio in aegritudinem.

<Quaestio 14>

[c]Quaeritur de hoc quod supponit Aristoteles consequenter[d] quod motus contrarii naturales[e] fiunt in aequalibus temporibus.

1. Et videtur quod sic: secundum Aristotelem in secundo *De generatione*, aequale tempus respondet augmentationi et diminutioni[f].

2. Item, aequale est tempus generationis et corruptionis, et augmentum[g] proportionatur generationi et decrementum corruptioni; quare etc.

3. Item, tempus accessus solis et recessus sunt aequalia et per medium annum accedit et per medium recedit. Sed in accessu est causa generationis, quia tunc generantur plantae et virentia omnia, et in recessu corrumpuntur. Et ita aequale est tempus generationis et corruptionis; quare etc.

4. Ad oppositum: si aliquid potest in [h]duas actiones, et in unam[i] non potest nisi per extrinsecum et in[j] aliam[k] potest sine extrinseco, tunc citius faciet illud in quod[l] potest sine extrinseco quam in quod potest solum cum extrinseco. Sed natura non potest in augmentationem sine extrinseco et potest in diminutionem sine extrinseco; quare minus tempus respondet diminutioni quam augmentationi.

5. Item, elementa de se tendunt ad corruptionem, et non tendunt ad augmentationem nisi per extrinsecum adiuvans; quare citius possunt in decrementum quam in augmentum.

6. Item, illud quod[m] quandoque capit interpolationem agit in maiori tempore quam illud quod non capit interpolationem; [n]sed motus

[a] et] vel *scrips. sed corr. s.l. in* sed O
[b] Et ... primum *om.* L
[c] Quaeritur ... consequenter] consequenter supponit T
[d] consequenter *ex corr.* O
[e] naturales] aequales OL (*corr. s.l. ex* quales O) *et* sunt et *add. s.l.* O
[f] diminutioni *ex corr.* O
[g] augmentum] augmentatum O
[h] duas actiones *corr. ex* duabus actionibus O; duabus actionibus LT
[i] unam] una LT
[j] in *suppl. s.l.* O
[k] aliam] alia LT
[l] quod] quo OL
[m] quod *suppl. s.l.* O
[n] sed ... interpolationem *om. hom.* T

only one of the these two states will follow. But it cannot push upwards unless it supports, and it can support when it does not push; therefore, for some moment of time it will support without pushing. And then there will be rest; hence, <an intermediate rest does intervene between contrary motions>.

5. With this the reply to the first argument[75] is clear.

6. To the second argument[76] it must be said that it is impossible for the transition to illness to begin at the same instant in which health is acquired.

<Question 14>
The question here is about what Aristotle assumes next, namely that contrary natural motions occur in equal times.[77]

1. Arguments for the affirmative: according to Aristotle in *On Generation* II, an equal time corresponds to increase and to diminution.[78]

2. Furthermore, the times of generation and of corruption are equal, and increase is proportionate to generation, and decrease to corruption; hence, <an equal time corresponds to increase and decrease>.

3. Furthermore, the times of the accession and recession of the sun are equal, and for half the year the sun accedes and for the other half it recedes. But in its accession the sun is the cause of generation, because then plants and vegetables are generated, and in its recession they are corrupted. Thus, the times of generation and corruption are equal; hence, <the times of increase and of diminution are equal>.

4. To the contrary: if something can perform two actions, and it can perform one of them only through something extrinsic and the other without something extrinsic, then it will perform the one which it can perform without something extrinsic faster than the one which it can only perform with something extrinsic. But nature cannot produce an increase without something extrinsic and it can produce a diminution without something extrinsic; therefore, a shorter time corresponds to the diminution than to the increase.

5. Furthermore, the elements have a tendency to corruption by themselves, and they do not have a tendency to increase except with the aid of something extrinsic; hence, they can undergo a decrease faster than an increase.

6. Furthermore, the thing that is sometimes interrupted acts for a longer time than the thing that is not interrupted; but the motion of increase is

[75] See above, n. 1.
[76] See above, n. 2.
[77] Arist. *Phys.* VIII.8, 262b9–15.
[78] Arist. *De gen. et corr.* II.10, 336b16–18.

augmentationis quandoque capit interpolationem, cum non fiat sine extrinseco adveniente, et motus decrementi non; quare etc.

7. Dicendum quod possumus loqui de augmento et diminutione dupliciter: aut secundum quod continue et per naturam consequuntur actiones suas aut secundum quod aliquo modo propter impedimentum fit interpolatio. Isto ultimo modo potest unum consequi actionem suam in minori tempore quam aliud, primo modo non.

8. Ad primum dicendum quod, sicut calor naturalis agens et anima regens exigunt aliquid extrinsecum ad hoc quod consequantur actionem augmentationis, sic in tempore decrementi exigunt[a] alimentum. Sed differenter, quia in tempore augmentationis exigunt aliquid ordinatum ad augmentationem, in tempore decrementi exigunt aliquid quod non est ordinatum[b] [c]ad decrementum, sed ad nutritionem.

9. Vel sic: esto quod non sic exigatur extrinsecum ad diminutionem tantum. Et quia natura plus[d] nititur ad esse quam ad non esse, licet in augmentatione requiratur extrinsecum et in diminutione non, natura tamen[e] retardat diminutionem et corruptionem in quantum potest, quatenus tempus diminutionis adaequetur tempori augmentationis.

<Quaestio 15>

[f]Quaeritur consequenter de hoc quod dicit quod partes cavationis non respondent partibus casus guttae, sed vult quod tota cavatio fiat in ultimo temporis.

1. Contra: cavatio est motus; quare etc.

2. Solutio[g]. Dicendum quod [h]intendit quod lapis non cavatur complete quousque ultimus casus ultimam fecerit cavationem, et isto eodem modo loquitur de alteratione.

3. Commentator tamen exponit illud[i] pro prima mutatione in alterationem.

[a] exigunt] *illegibile add. sed del.* O
[b] ordinatum] per se *add.* LT
[c] ad decrementum *om. sed suppl. mg.* ad augmentationem O
[d] plus] utitur *add. sed del.* O; *om.* T
[e] tamen *suppl. s.l.* O; *om.* L
[f] Quaeritur ... quod] consequenter T
[g] Solutio *om.* LT
[h] intendit quod *om.* L
[i] illud *om.* LT

sometimes interrupted, since it does not occur without something coming from outside, whereas the motion of decrease is not interrupted; hence, <increase takes more time than decrease>.

7. It must be said that we can speak of increase and diminution in two ways: either (i) in so far as they perform their actions continuously and in accordance with nature or (ii) in so far as an interruption occurs because of an obstacle. In case (ii) one of the two can perform its action in less time than the other, but not in case (i).

8. To the first argument[79] it must be said that, just as an active natural heat and a directing soul require something extrinsic in order to perform the action of increase, so they require food in the time of decrease. But <they require these things> in a different way, because in the time of increase they require something ordered to increase, whereas in the time of decrease they require something which is not ordered to decrease, but to nutrition.

9. Another reply is the following. Let it be conceded that something extrinsic is not required in this way only for diminution. Since nature strives more for being than for non-being, even though something extrinsic is required in increase and not in diminution, nonetheless nature delays diminution and corruption as much as it can, so that the time of diminution becomes equal to that of increase.

<Question 15>

The next question is about Aristotle's statement that the parts of an erosion do not correspond to the parts of the falling of a drop <of water>, but he wants the whole erosion to occur in the last instant of time.[80]

1. Against this: erosion is a motion; hence, <it does not occur in an instant>.

2. Solution. It must be said that Aristotle's claim means that a stone is not completely eroded until the last drop to fall produces the final erosion, and it is in this way that he talks about alteration.

3. The Commentator, however, interprets Aristotle's claim as referring to the first change in an alteration.[81]

[79] See above, n. 4.
[80] Arist. *Phys.* VIII.3, 253b14–21.
[81] Aver. *In Phys.* VIII, t.c. 23, fol. 359vb.

<Quaestio 16>

[a]Quaeritur consequenter de hoc quod dicit quod elementa quiescunt naturaliter in suis locis et ibi non moventur nisi violenter.

1. Contra: videmus de partibus terrae quod, si amoveantur partes inferiores, descendunt naturaliter; ergo quieverunt violenter.

2. Item, quod aqua non quiescat naturaliter[b] videtur, quia, secundum Aristotelem in *De caelo et mundo* aqua tendit ad inferiorem locorum et ita in superiori quiescit violenter.

3. Et iterum, si amoveatur[c] terra, descendit naturaliter usque ad centrum.

4. Et iterum, secundum Aristotelem, cum corpora caelestia moventur per se, aer et ignis moventur per accidens, et ita non quiescunt naturaliter.

<Quaestio 17>

[d]Quaeritur consequenter de hoc quod dicit quod in omni motu est corruptio termini a quo et generatio termini in quem, et ita omnis motus est quodam modo generatio et quodam modo corruptio.

1. Contra: licet hoc sit verum in tribus speciebus motus, tamen in[e] loci mutatione falsum est.

<Quaestio 18>

Consequenter [f]quaeritur de hoc quod arguit Aristoteles contra opinionem unam sic: aestimaverunt motum non esse; cum ergo omnis aestimatio motus sit, tunc ponunt oppositum sui. Propter hoc quaeritur utrum aestimatio sit motus.

1. Et videtur quod non, **[O 115ra]** quia impartibile non movetur; sed anima secundum omnes suas potentias est impartibilis, et ita in aestimando non movetur.

2. Item, si aestimatio sit motus, non est nisi alteratio; sed hoc non contingit, quia dicit Aristoteles in septimo huius quod circa intellectivam partem non cadit alteratio.

3. Ad oppositum: in primo *De anima* scribitur quod est motus ab anima et quod est motus ad animam; ab anima, ut in somno, quando species repositae

[a] Quaeritur ... quod] consequenter T
[b] naturaliter *om.* O
[c] amoveatur] amovetur O
[d] Quaeritur ... quod] consequenter T
[e] in *ex corr.* O
[f] quaeritur ... Aristoteles] arguit T

<Question 16>

The next question is about Aristotle's statement that the elements are naturally at rest in their places, and they move there only by violence.[82]

1. Against this: we see that parts of earth naturally move downwards when lower parts are removed; therefore, they were at rest by violence.

2. Furthermore, that water is not naturally at rest is shown because, according to Aristotle in *On the Heavens*, water tends towards a lower place, and thus it is at rest in a higher place by violence.[83]

3. Again, if the earth is displaced, it naturally moves down to the centre.

4. Also, according to Aristotle, since the celestial bodies move per se, whereas air and fire move accidentally, the latter are thus not naturally at rest.[84]

<Question 17>

The next question is about Aristotle's statement that in every motion there is a corruption of the terminus from which and a generation of the terminus towards which, and thus every motion is in some way a generation and in some way a corruption.[85]

1. Against this: although this is true for three species of motion, it is false, however, for locomotion.

<Question 18>

The next question is about the following argument of Aristotle against one of the other views: those who hold it judge that motion does not exist; therefore, since every judgment is a motion, they posit the opposite of their own position.[86] **The question then is whether judgment is a motion.**

1. Arguments for the negative: an indivisible thing does not move; but the soul is indivisible in all its powers, and thus it does not move when it judges.

2. Furthermore, if judgment is a motion, it can only be an alteration; but this cannot be, because in *Physics* VII Aristotle says that no alteration occurs in the intellect.[87]

3. To the contrary: in *On the Soul* I Aristotle writes that there is a motion from the soul and a motion towards the soul: (i) from the soul, for instance in sleep, when the species stored in the sensitive soul offer themselves to the particular senses; (ii) towards the soul, that is, when the species are carried to

[82] Arist. *Phys.* VIII.3, 253b33–35.
[83] Arist. *De caelo* II.4, 287b4–6.
[84] It is not clear where Aristotle may have suggested this.
[85] Arist. *Phys.* VIII.3, 254a11–14.
[86] Arist. *Phys.* VIII.3, 254a24–30.
[87] Arist. *Phys.* VII.3, 247b1–2.

in sensitiva offerunt se sensibus particularibus, ad animam, ut quando species ab extra deferuntur ad animam. Et ita vult quod sit motus in virtutibus.

4. Item, Commentator dicit quod, si de vero aestimante fiat falso aestimans, fit aliquis motus.

5. Solutio. Dicendum quod ad aestimationem exigitur motus, quia prius exigitur immutatio sensus et huiusmodi, et ita circa condiciones sive circumstantias aestimationis exigitur motus, ipsa tamen aestimatio fit in instanti.

6. Ad primum dicitur quod impartibile non movetur. Hoc est verum localiter.

7. Vel sic: huiusmodi operationes non sunt animae[a] absolute, sed corporis per animam. Et similiter intelligitur ista 'circa intellectivam [b]non cadit alteratio'.

8. Aut distinguendum est alterationem, quia, si sumatur alteratio naturalis, quae est inter qualitates naturales, verum dicit, si alteratio spiritualis, falsum dicit.

<Quaestio 19>
Consequenter quaeritur utrum anima componat[c] in tempore.

1. Et videtur quod sic, quia informatur in tempore, secundum quod dicit [d]Commentator, quia, si intelligibile sit divisibile, intelligitur in tempore divisibili, si indivisibile, intelligitur in indivisibili; et ita est informatio aliqua in tempore, ergo et compositio.

<Quaestio 20>
[e]Quaeritur de hoc quod dicit consequenter quod gravia et levia non moventur ex se.

1. [f]Contra: grave et leve moventur a natura; sed natura est principium etc.; ergo moventur a principio intra; [g]sed quod movetur a principio intra movetur ex se; quare etc.

[a] animae] per se et *add.* T
[b] non cadit *om.* T
[c] componat *corr. ex* non ponat O
[d] Commentator, quia *om.* L
[e] Quaeritur ... levia] consequenter dicit quod grave et leve T
[f] Contra ... leve] et videtur quod sic quia T
[g] sed ... etc.] et tale quid movetur ex se T

the soul from outside.[88] Thus, Aristotle wants there to be motion in the powers <of the soul>.

4. Furthermore, the Commentator says that, if someone passes from a true judgment to a false one, a motion occurs.[89]

5. Solution. It must be said that a motion is required for a judgment, because a change in the senses and in things of this kind is first required, and thus a motion is required where the conditions or circumstances of a judgment are concerned, but the judgment itself is made at an instant.

6. To the first argument[90] it is said that the claim that an indivisible does not move is true only about locomotion.

7. Another reply is that such operations do not belong to the soul absolutely, but to the body in virtue of the soul. And the proposition 'an alteration does not occur in the intellect'[91] should be understood in the same way.

8. Alternatively, one must draw a distinction about alteration, because in the case of natural alteration, which involves natural qualities, the proposition is true, but in the case of spiritual alteration, the proposition is false.

<Question 19>
The next question is whether the soul produces a composition in time.

1. Argument for the affirmative: the soul receives some form in time, according to the Commentator's statement that, if an intelligible thing is divisible, then it is understood in a divisible time, if it is indivisible, it is understood in an indivisible time;[92] and thus some reception of form takes place in time; therefore, some composition also <takes place in time>.

<Question 20>
The question here is about what Aristotle says next, namely that heavy and light things are not moved by themselves.[93]

1. Against this: heavy and light things are moved by nature; but nature is a principle <of motion of the thing in which it exists>; therefore heavy and light things are moved by an internal principle; but what is moved by an internal principle is moved by itself; hence, <heavy and light things are moved by themselves>.

[88] Arist. *De anima* I.4, 408b15–18.
[89] Aver. *In Phys.* VIII, t.c. 26, fol. 363rb.
[90] See above, n. 1.
[91] See above, n. 2.
[92] Aver. *In De anima* III, t.c. 23, ed. Crawford, 458.13–459.58.
[93] Arist. *Phys.* VIII.4, 255a5–6.

2. Item, per illam eandem virtutem[a] per quam aliquid quiescit in suo loco, [b]per eandem virtutem movetur ad suum locum; sed grave et leve[c] quiescunt naturaliter et per suas formas proprias in locis suis; ergo moventur[d] naturaliter et per suas formas proprias ad loca sua, et ita a principio [e]intra; ergo ex se; quare etc.

3. Item, causa Aristotelis quare moventur a principio extrinseco et non ex se est quia moventur a[f] generante vel[g] a solvente prohibens. Si ita sit [h]quod a solvente prohibens, tunc quaero, quando solvens prohibens solverit[i], post amotionem istius prohibentis a quo sit[j] motus, scilicet aut a[k] gravi descendente aut ab aliquo[l] extrinseco. Si a gravi descendente, ergo ex se. Si ab alio, aut ergo a medio aut a virtute supracaelesti. Non a medio, tum quia medium est immobile et non movet[m] nisi moveatur—tunc esset ulterius quaerere de illo medio quod movet a quo moveatur, et tandem erit processus in infinitum, quod est inconveniens dicere. Tum quia non est maior ratio quare moveatur [n]ad unam partem medii quam [o]ad aliam. Medium enim non plus movet in[p] unam partem quam aliam; quare etc. Si a generante, tunc generans facit de potentia gravi actu grave, et ita dat formam. Sed dicit Aristoteles quod, quantum[q] aliquid dat alicui de forma, tantum dat ei de loco. Tunc generans [r]motum[s] gravis, in dando ei formam, dat ei locum et aliquid de loco. [t]Cum ergo[u] hoc fuerit a principio intra, tunc, ut videtur, movetur ad locum suum a principio intra, et ita ex se.

[a] virtutem] movetur ad suum locum *add. sed del.* O; naturam T
[b] per eandem virtutem] et T
[c] leve] movetur *add. sed del.* O
[d] moventur] movetur O
[e] intra; ergo] et tale movetur T
[f] a *ex corr.* O
[g] vel] et LT
[h] quod ... prohibens *om.* T
[i] solverit] et *add.* LT
[j] sit] fit LT
[k] a *om.* O
[l] aliquo] alio O; illo L
[m] movet] a quo moveatur et tandem erit processus in infinitum quod est inconveniens dicere *add. sed del.* O; *corr. s.l. ex* oportet L
[n] ad unam partem *corr. ex* ab una parte O; ab una parte LT
[o] ad aliam] ab alia LT
[p] in *suppl. s.l.* O; *om.* LT
[q] quantum] quanto OL
[r] motum ... ei] dans gravi T
[s] motum *ex corr.* O
[t] Cum ergo] sed cum T
[u] ergo *suppl. s.l.* O; *om.* L

2. Furthermore, something is moved towards its place by the same power by which it is at rest in its place; but heavy and light things are naturally at rest in their places and in virtue of their proper forms; therefore, they are naturally moved towards their places and in virtue of their proper forms, and so by an internal principle; therefore, by themselves; hence, <heavy and light things are moved by themselves>.

3. Furthermore, the reason given by Aristotle to explain why heavy and light things are moved by an extrinsic principle and not by themselves is that they are moved by their generator or by what removes an obstacle to their movement.[94] (i) If they are moved by what removes an obstacle, then I ask: when the thing that removes the obstacle has removed it, what is the motion caused by after the removal of this obstacle, that is, <is it caused> by the heavy thing moving downwards or by something extrinsic to it? (ia) If it is caused by the heavy thing moving downwards, then the heavy thing is moved by itself. (ib) If it is caused by something else, then it is caused either (ib') by the medium or (ib") by a celestial power. It is not caused by the medium, for two reasons. The first is that the medium is immobile and moves only if it is moved. That would then raise a further question about the medium that moves, namely what the medium is moved by, and then there would be an infinite regress, which is absurd. The second is that there is no more reason why the heavy thing is moved towards one part of the medium rather than towards another; for the medium does not move more towards one part rather than another; hence, <the heavy thing is not moved by the medium>. (ii) If the heavy thing is moved by its generator, then the generator brings the heavy thing into act out of the heavy thing in potency, and thus it gives the heavy thing its form. But Aristotle says that whatever amount of form one thing gives to another, it gives it a similar amount of place.[95] Thus, the generator of the motion of the heavy thing, in giving the heavy thing its form, also gives it place and something pertaining to place. Therefore, since place derives from an internal principle, then, it seems, the heavy thing is moved towards its place by an internal principle, and so by itself.

[94] Arist. *Phys.* VIII.4, 255b35–256a2.
[95] Arist. *De caelo* IV.3, 310a33–b1.

4. Ad oppositum: omne quod movetur ex se dividitur in duo quorum unum est per se movens et aliud per se motum, quae quidem sunt [a]distincta per essentiam, ut animal sive homo; sed grave et leve non sic possunt dividi. Probatio, quia, si dividatur grave, non potest dividi nisi in materiam et formam; sed materia non est sufficiens motum per se, et sic impossibile est resolvere grave in talia duo quorum unum est sufficiens motum et aliud sufficiens movens.

5. Et praeterea, motus localis tantum est perfectorum; quare, cum materia per se sit imperfecta, tunc per se non poterit esse sufficiens motum.

6. Item, grave et leve non moventur naturaliter; ergo non moventur ex se. Probatio praemissae: si grave et leve [b]moverentur naturaliter, tunc velocius moverentur[c] in fine motus quam in principio, quod falsum est. [d]Probatio: imaginetur linea una super quam moveantur duo gravia aequalia in magnitudine quorum unum est propinquius centro [e]quam aliud; tunc, cum super idem spatium et aequale ista duo gravia moveantur et in aequali tempore, simul terminabunt suos motus. Et ita videtur quod mobile quod est propinquius centro non citius movetur; quare non moventur naturaliter.

7. Item, non moventur[f] naturaliter; ergo non ex se. Probatio assumpti: dicit Aristoteles in septimo *Physicorum* et dat talem regulam, quod, si aliquod movens naturaliter moveat aliquod mobile per aliquod spatium in tempore aliquo, medietas virtutis ipsius moventis movebit medietatem ipsius mobilis per tantum spatium in tanto tempore; sed hoc deficit in [g]gravi et levi; ergo non moventur naturaliter. Probatio minoris: sumamus aliquam glebam terrae; ista gleba movetur a tota virtute manus[h] in aliquo tempore. Dividatur illa gleba in athomos; tunc manus movet illum athomum in maiori tempore et non in aequali. Et praeter hoc non est ibi medietas[i] virtutis **[O 115rb]** manus.

8. Item, in omni motu ubi plures partes moventur violenter quam naturaliter, talis motus est violentus; et[j] sic moventur grave et leve; quare etc. Probatio minoris: moveatur aliquod mobile ad centrum; tunc naturaliter movetur per lineas brevissimas. Sed hoc non[k] contingit nisi[l] unicus punctus descendat ad centrum linea brevissima, et sic omnes aliae partes moventur violenter; quare etc.

[a] distincta per essentiam] distantia O
[b] moverentur ... velocius *om. hom. sed* velocius *suppl. s.l.* O
[c] moverentur] moventur O
[d] Probatio: imaginetur] ponatur L
[e] quam ... centro *om. hom.* L
[f] moventur] movetur LT
[g] gravi et levi *corr. ex* grave et leve O
[h] manus] et *add.* O
[i] medietas] indivisibilitas (*lectio incerta*) O
[j] et] sed LT
[k] non *suppl. s.l.* O
[l] nisi] quod *add. s.l.* O

4. To the contrary: everything that is moved by itself is divided into two parts, one of which moves per se and the other of which is moved per se, and these parts are distinct in essence, as in the case of an animal or a man; but heavy and light things cannot be divided in this way. Proof: if a heavy thing is divided, it can only be divided into matter and form; but matter is not sufficient to serve as something moved per se, and so it is impossible to resolve a heavy thing into two parts one of which is sufficient to serve as the moved and the other of which is sufficient to serve as the mover.

5. Moreover, locomotion only belongs to perfect things; hence, since matter is imperfect per se, then matter cannot by itself be sufficient to serve as the moved thing.

6. Furthermore, heavy and light things are not moved naturally; therefore, they are not moved by themselves. Proof of the premise: if heavy and light things were moved naturally, then they would be moved faster at the end of the motion than at the beginning, which is false. Proof: imagine a line over which two heavy things equal in magnitude are moved, one of which is closer to the centre than the other; then, since these two heavy things are moved over the same equal space and in equal time, they will end their motions simultaneously. Thus, it seems that the mobile thing that is closer to the centre is not moved faster; hence, they are not moved naturally.

7. Furthermore, heavy and light things are not moved naturally; therefore, they are not moved by themselves. Proof of the assumption: in *Physics* VII Aristotle gives the rule 'if something that moves naturally moves a mobile thing over some space in some time, half of the power of that mover will move half of the mobile thing over so much space in so much time';[96] but this rule is not valid in the case of heavy and light things; therefore, they are not moved naturally. Proof of the minor premise: let us take a lump of earth. This lump is moved by the total power of a hand in some time. Let us divide that lump into atoms; then the hand moves a given atom in a longer time and not in an equal time. Also, in this case we do not find half the power of a hand.

8. Furthermore, every motion such that more parts are moved by violence than by nature is violent; but heavy and light things are moved in this way; hence, <their motion is by violence>. Proof of the minor premise: let a mobile thing be moved towards the centre; then it is naturally moved through the shortest lines. But this does not happen unless there is just one point that moves downwards to the centre through the shortest line, and hence all the other parts are moved by violence; hence, <more parts of a heavy thing are moved by violence>.

[96] Arist. *Phys*. VII.5, 250a6–7.

9. [a]Ad quaestionem dicunt[b] sic quod non moventur ex se, quia non moventur[c] a principio intra quod est principium elementare, sed aliqua virtute immateriali quae est praesens et naturalis.

10. Sed contra: dicit Aristoteles quod motus contrarii in elementis sunt a formis contrariis; sed huic formae immateriali nihil est contrarium. Probatio: contraria nata sunt fieri circa idem[d] et eandem materiam, et sic omnis forma quae [e]aliquam habet contrariam est materialis, cum [f]esse habet in materia; cum igitur forma immaterialis non sit in materia, non habebit contrarium; quare etc.

11. [g]Isti quaestioni respondet Commentator quod grave et leve non moventur ex se per se, moventur tamen ex se per accidens, quia grave motum movet medium et medium motum movet grave quod est in eo, et sic grave movet se per accidens, [h]et ita dicuntur moveri ex se per accidens.

12. Sed istud habet dubitationem, quia non differret tunc huiusmodi motus a motu animae, quia anima, movendo corpus, movet se per accidens, et est exemplum Aristotelis, libro *De anima*, de nauta et navi.

13. Item, nullum corpus movet nisi moveatur; cum igitur grave sit corpus, non movet nisi moveatur, et sic, ut videtur, falsum dicit Commentator.

14. Item, si anima, cum sit forma, non movet nisi moveatur et excitetur, ergo multo fortius grave et leve non movent nisi moveantur et excitentur[i].

15. Quaestioni puto sic esse dicendum quod nulla forma materialis movet nisi moveatur a virtute immateriali quae est forma immaterialis, quae est virtus caelestis diffusa per totum orbem et continue generata, sine qua impossibile est aliquam formam materialem exire in suam operationem. Operatio autem propria ipsius formae materialis[j] est ut perficiat suam materiam. Et etiam alia est operatio ipsius agere in materiam inducendo in eandem formam consimilem in specie et corrumpendo formam contrariam. Est etiam tertia operatio, quae est ut acquirat sibi locum in quo possit salvari. Unde motor gravis sufficiens est forma substantialis ipsius gravis rememorata a virtute caelesti mediantibus suis accidentibus quae sunt [k]gravitas et levitas.

[a] Ad questionem *om.* T
[b] dicunt] quidam *add.* L; quidam *praem.* T
[c] moventur] movetur OL
[d] idem] subiectum *add.* T
[e] aliquam habet contrariam] habet aliquod contrarium habet materiam vel T
[f] esse habet *scripsimus*; se habet OL; sit T
[g] Isti quaestioni] aliter T
[h] et ... accidens *om. hom.* L
[i] excitentur] excitantur OL
[j] materialis *scripsimus*; naturalis OLT
[k] gravitas et levitas] grave et leve LT

9. To this question some people reply that heavy and light things are not moved by themselves because they are not moved by an internal elementary principle, but by some immaterial power, which is present and is natural.

10. Against this: Aristotle says that contrary motions in the elements derive from contrary forms;[97] but to this immaterial form nothing is contrary. Proof: contraries are apt by nature to exist in the same thing and in the same matter, and so every form that has a contrary is material, since it exists in matter; therefore, since an immaterial form does not exist in matter, it will not have a contrary; hence, <contrary motions in the elements do not derive from an immaterial form>.

11. To this question the Commentator replies that heavy and light things are not moved by themselves per se, but they are moved by themselves accidentally, because a heavy thing once moved moves the medium, and the medium once moved moves the heavy thing that is in it, and thus the heavy thing moves itself accidentally, and so heavy and light things are said to be moved by themselves accidentally.[98]

12. But this reply is doubtful because, if this were the case, the motion of heavy and light things would not be different from the motion of the soul, because the soul in moving the body moves itself accidentally, and the example Aristotle gives in *On the Soul* is that of a sailor and a ship.[99]

13. Furthermore, no body moves unless it is moved; therefore, since a heavy thing is a body, it does not move unless it is moved, and so, it seems, the reply of the Commentator is false.

14. Furthermore, if a soul, since it is a form, does not move unless it is moved and stimulated, there is a much stronger reason to say that heavy and light things do not move unless they are moved and stimulated.

15. To this question, in my view, it must be replied that no material form moves unless it is moved by an immaterial power, which is an immaterial form, and this form is the celestial power spread throughout the whole orb and continuously generated, without which it is impossible for a material form to perform its operations. An operation proper to a material form is that of giving perfection to its matter. Another operation proper to it is that of acting on matter and inducing in matter a form similar to itself in species while destroying the contrary form. A third operation is that of making the thing acquire a place in which it can be preserved. Accordingly, the sufficient mover of a heavy thing is the substantial form of the heavy thing, as recalled[100] by the celestial power, by means of its accidents, which are heaviness and lightness.

[97] Arist. *De caelo* I.3, 270a17.
[98] Aver. *In De caelo* III, t.c. 28, ed. Carmody, II, 557.184–192.
[99] Arist. *De anima* I.3, 406a5–8.
[100] In the *Auctoritates Aristotelis* a similar claim ("Natura nihil facit nisi rememorata a causis superioribus quae sunt deus et intelligentiae") is ascribed to Averroes in his commentary on *Metaph.* XII, although the editor could not locate this claim. See Hamesse, 1974, 138, (281).

16. Unde una condicio eius quod [a]est [b]ex se moveri est quod illud[c] dividatur in talia duo quorum unum est movens per se [d]et aliud motum per se, quae quidem fuerint distincta per essentiam. Et etiam alia condicio est quod, quando aliquid movetur ex se, quod possit resistere suo motori[e] et quiescere quando voluerit. Et istas condiciones impossibile est reperire in gravi et levi. Isto modo non moventur gravia et levia ex se.

17. Alio modo dicitur[f] moveri ex se, hoc est ab aliquo existente in ipso quod est principium intrinsecum. Et sic dicitur quod gravia et levia moveantur ex se. Tamen Aristoteles dicit quod moventur per accidens propter illas condiciones supradictas quae deficiunt in illis.

18. Quod autem dicit Aristoteles quod motus[g] contrarii fiunt a formis contrariis, intelligendum est de formis propriis contrariis quae sunt in materia.

19. Ad rationes in contrarium dicendum. Ad primum sic, quod haec est per se, quod omne grave motum deorsum, quanto plus pertransit, tanto velocius movetur, [h]et haec per accidens 'quanto propinquius est centro, tanto velocius movetur'. Stant igitur ista duo simul, quod ista duo gravia in eodem et aequali tempore moveantur et tamen quod moventur[i] naturaliter.

20. Ad aliud dicendum quod regula Aristotelis in septimo debet sic condicionari, quod, si aliquod movens moveat aliquod mobile, tunc medietas virtutis moventis movet medietatem ipsius mobilis in eodem tempore[j] et[k] in eodem [l]spatio, quod, sicut dividitur virtus, sic per intellectum abstrahatur densitas a medio, et sic tenebit regula Aristotelis in motu [m]gravis et levis.

1*. Quaeritur consequenter utrum anima moveat corpus, secundo utrum ex se, tertio utrum anima moveat corpus absque quod informetur specie aliqua determinata, quarto utrum moveat corpus secundum omnem speciem motus localis.

[a] est ex se *om.* L
[b] ex se *suppl. s.l.* O
[c] illud *fortasse corr. ex* ibi O; ibi LT
[d] et ... se *om. hom. sed* et aliud motum *suppl. s.l.* O
[e] motori *ex corr.* O; vel motui *add.* T
[f] dicitur] dicuntur LT
[g] motus *suppl. s.l.* O; *om.* L
[h] et ... movetur *om. hom.* T
[i] moventur] movetur OT
[j] tempore] et in tanto tempore *add.* L
[k] et] in tanto et *add.* T
[l] spatio ... medio *om.* L
[m] gravis et levis] gravi et levi O

16. Thus, one condition which something that is moved by itself must satisfy is that it is divided into two things one of which is a per se mover and the other of which is per se moved, and that these things are distinct in essence. Another condition is that, when something is moved by itself, it can offer resistance to its mover and be at rest when it wants to. But it is impossible for these conditions to be satisfied in the case of heavy and light things. Thus, heavy and light things are not moved by themselves.

17. Something is also said to be moved by itself in another way, that is, if it is moved by something existing in it and which is an intrinsic principle. And thus it is said that heavy and light things are moved by themselves. Aristotle says, however, that they are moved accidentally because of the aforementioned conditions, which heavy and light things do not satisfy.[101]

18. Aristotle's claim that contrary motions derive from contrary forms[102] must be understood as applying to the proper contrary forms existing in matter.

19. Reply to the arguments to the contrary. To the first argument[103] it must be said that the proposition 'for every heavy thing moved downwards, the more space it traverses, the faster it is moved' is per se, but the proposition 'the closer it is to the centre, the faster it is moved' is accidental. Therefore, these two facts are compatible, that is, that these two heavy things are moved in the same equal time and that they are moved naturally.

20. To the other argument[104] it must be said that Aristotle's rule in *Physics* VII that 'if a mover moves a mobile thing, then half of the power of the mover moves half of the mobile thing in the same time and over the same space' must be understood in a qualified sense as follows: the intellect abstracts density from the medium in accordance with the division of the power. And understood in this way Aristotle's rule will be valid when applied to the motion of heavy and of light things.

1*. The next questions are first, whether the soul moves the body, second whether the soul moves the body by itself, third whether the soul moves the body without being in-formed by a determinate species, and fourth whether the soul moves the body in accordance with every species of locomotion.

[101] Arist. *Phys*. VIII.4, 255b24–29.
[102] See above, n. 10.
[103] See above, n. 6.
[104] See above, n. 7.

<Quaestio 21>

<Quaestio 21.1>

1. De primo sic, primo ad partem negativam: movens et motum sunt proportionalia. Aliter enim impossibile esset quod movens moveat aliquid nisi[a] virtus sua proportionatur. Sed anima et corpus non proportionantur, quia proportio est rerum existentium in eodem genere; sed anima et corpus in nullo eodem genere sunt. Corruptibile enim et incorruptibile [b]in nullo eodem sunt per Aristotelem.

2. Item, omnis motor unus movet motum tantum unico motu et non simul et semel oppositis motibus; sed [c]anima movet corpus oppositis motibus, sicut patet: una manus potest movere sursum et alia deorsum; quare etc.

3. Item, movens et motum sunt simul etc. Cum ergo tota moveat caput, tunc tota anima erit simul cum capite, quia anima est substantia indivisibilis. Sed totum est extra quod nihil, per Aristotelem; ergo, ut videtur, anima non potest movere manum. Sed in motu locali quaelibet pars movetur[d]; quare etc.

4. Item, sit A una pars, B alia; anima, cum movet A, non movet B, **[O 115va]** quia non potest simul esse in diversis partibus; quare etc.

<Quaestio 22>

<Quaestio 22.1>

1. De secundo sic ad partem negativam: animal non dividitur in duo quorum unum est movens per se et aliud per se motum; quare etc. Probatio praemissae: actus et potentia non diversificant essentiam; sed sic se habent corpus et anima. Corpus enim est in potentia respectu animae, quae est actus eius. Quare etc.

2. Item, anima et corpus sunt unius generis, quia perfectio et illud cuius est perfectio; ergo unius essentiae.

3. Dicitur quod unius essentiae generalis et incompletae.

[a] nisi *ex corr.* O
[b] in ... Aristotelem] plus differunt quam secundum genus T
[c] anima ... sicut *om.* L
[d] movetur *scripsimus*; movet OLT

<Question 21>

<Question 21.1>

1. As to the first question, we begin with the arguments for the negative: the mover and the moved are proportionate; for it would be impossible for the mover to move something if its power were not proportionate to it. But the soul and the body are not proportionate, because a relation only holds between things of the same genus; but there is no one and the same genus common to body and soul, since, according to Aristotle, there is no one and the same genus common to the corruptible and to the incorruptible.[105]

2. Furthermore, every single mover moves what is moved with only one type of motion and not with opposite types of motion at one and the same time; but the soul moves the body with opposite types of motion, as is clear: one hand can move upwards and the other downwards; hence, <the soul does not move the body>.

3. Furthermore, the mover and the moved are in the same place. Therefore, since the whole soul moves the head, the whole soul will be in the same place as the head, because the soul is an indivisible substance. But the whole is that outside which there is nothing, according to Aristotle;[106] therefore, it seems, the soul cannot move the hand. But in locomotion every part is moved; hence, <the soul does not move the body>.

4. Furthermore, let A be one part <of the body> and B another part; when the soul moves A, it does not move B, because it cannot be simultaneously in different parts; hence, <the soul does not move the body>.

<Question 22>

<Question 22.1>

1. As to the second question, we begin with the arguments for the negative: an animal is not divided into two things one of which is a per se mover and the other of which is per se moved; hence, <an animal does not move by itself>. Proof of the premise: act and potency do not diversify essence; but body and soul are related in this way; for the body is in potency to the soul, which is its act. Hence, <an animal is not divided into a per se mover and a per se moved>.

2. Furthermore, the soul and the body belong to one genus, because they are respectively a perfection and the thing to which that perfection belongs; therefore, they have one essence.

3. It is said that they have one general and incomplete essence.

[105] Arist. *Metaph.* X.10, 1058b26–29.

[106] Arist. *Phys.* III.6, 207a8–10.

4. Contra: [a]probatio quod, si sint[b] eiusdem essentiae generalis, quod et specialis probatio: differentia nullam novam [c]naturam nec essentiam supra genus addit[d]; eadem ergo essentia non diversificata est in una specie et in alia[e], quia eadem fuit essentia generis; quare etc.

5. Praeterea, essentia generis fit species et non sub specie; cum ergo actus et potentia non diversificant essentiam et differentiae potestate sunt in genere, tunc, ut videtur, conveniunt in eadem essentia speciali; quare etc.

6. [f]Ad principale: anima sensitiva egreditur ex principiis corporis; eadem tunc sunt principia corporis et animae sensitivae. Sed si principia eadem, et principiata sunt eadem; quare etc.

7. Item, si anima sensitiva corrumpatur, resolvitur in materiam et formam; ergo ad materiam primam. Sed materia prima est eadem per essentiam cum corpore; quare etc.

<Quaestio 23>

<Quaestio 23.1>

1. De[g] tertio sic, primo ad partem affirmativam sic: virtus appetitiva appetit per speciem apprehensam consequi delectabile [h]et fugere tristabile, secundum Aristotelem in libro *De anima*; suus ergo appetitus est propter informationem specierum aliquarum; sed numquam movet corpus nisi propter aliquem appetitum et quando excitatur[i] ab aliquo; quare etc.

2. Ad oppositum: aliquis exiens domum vadit, non tamen percipit se ire huc vel illuc propter speciem aliquam; quare etc.

3. Item, sicut est in maiori mundo, sic erit in minori, per Aristotelem in octavo; sed ita est in maiori mundo quod motor ibi movet continue sine specie extrinseca vel sine speciei extrinsecae apprehensione; ergo similiter erit in minori mundo; quare etc.

[a] probatio quod *om.* T
[b] sint *ex corr.* O
[c] naturam ... addit] essentiam supra genus addit neque novam naturam T
[d] addit] in *add.* O
[e] alia] quare eadem *add.* L
[f] Ad principale] Item L; Item ad principale T
[g] De] *illegibile add. sed del.* O
[h] et fugere] vel OL
[i] excitatur] exercetur *ex corr.* O

4. Against this: proof that, if they have the same general essence, they also have the same specific essence: a difference does not add a new nature or essence to a genus; therefore, one and the same non-diversified essence is in different species, because the essence of the genus was the same; hence, <if body and soul have the same general essence, then they also have the same specific essence>.

5. Furthermore, the essence of a genus becomes a species and is not put under a species; therefore, since act and potency do not diversify essence and differences are virtually contained in a genus, it seems that act and potency do agree in the same specific essence; hence, <if body and soul have the same general essence, then they also have the same specific essence>.

6. Argument for the main claim: the sensitive soul emerges from the principles of the body; thus, the principles of the body and of the sensitive soul are the same. But if the principles are the same, then the things derived from the principles are also the same; hence, <the sensitive soul and the body are the same>.

7. Furthermore, if the sensitive soul is corrupted, it is resolved into matter and form; therefore, < it is resolved> into prime matter. But prime matter is the same in essence with the body; hence, <the sensitive soul is the same in essence with the body>.

<Question 23>

<Question 23.1>

1. As to the third question, let us first consider the arguments for the affirmative: the appetitive power desires to attain what is pleasant and to flee from what is saddening in virtue of the species apprehended, according to Aristotle in *On the Soul*;[107] therefore, its appetite occurs in virtue of its being in-formed by some species; but the soul never moves the body except in virtue of some appetite and only when it is stimulated by something; hence, <the soul moves only in virtue of a species>.

2. To the contrary: someone leaving a house goes, but he does not perceive that he is going here or there in virtue of some species or other; hence, <the soul does not move the body in virtue of a species>.

3. Furthermore, as things are in the greater world, so they will be in the lesser world, according to Aristotle in *Physics* VIII;[108] but in the greater world a mover moves continuously without any extrinsic species or without the apprehension of any extrinsic species; therefore, the same will happen in the lesser world; hence, <the soul does not move the body in virtue of a species>.

[107] Arist. *De anima* III.7, 431b2–9.
[108] Arist. *Phys.* VIII.2, 252b26–27.

4. Item, dicit Aristoteles quod quaedam animalia[a] amota a suis locis moventur[b] infinite, et tamen non formantur specie aliqua, sicut vermes et consimilia.

<Quaestio 24>

1. De quarto sic, et primo ad partem affirmativam sic: anima non magis determinatur ad unam partem [c]secundum speciem quam aliam motus localis; ergo indifferenter[d], quantum est de se, potest movere sursum et deorsum.

2. Item, anima, in vivificando corpus, una vita vivificat totum corpus et quamlibet eius partem; sed per illam vitam facit anima partem movere sursum; ergo eadem ratione et totum, cum eadem vita vivificentur.

3. Item, si aliqua virtus possit super aliquod mobile [e]in uno tempore, aequalis virtus poterit super aequale mobile in tanto tempore. Sed corpus totum et aliqua eius pars, ut manus, sunt aequalia, quia anima ubique tota est in corpore; cum ergo tota anima sit in manu et tota anima in toto corpore, tunc manus et corpus sunt aequalia, quia movens et motum proportionantur, sicut praedictum est. Sed anima sive virtus in manu potest elevare manum sursum; quare et totum corpus.

4. Item, minoris virtutis est aliquid depulsum sustinere quam ipsum[f] sustinere et elevare sive propellere[g], sicut in motu violento sive in motu proiectorum. Aer enim potest sustinere per aliquod tempus lapidem propulsum, non tamen potest ipsum ulterius propellere. Cum ergo anima possit corpus elevare in aere, ut patet in saltu, tunc multo fortius poterit ipsum ibidem[h] sustinere; quare etc.

<Quaestio 21.2>

1. Ad primam quaestionem dicendum cum Aristotele quod anima movet corpus.

[a] animalia] alia (*lectio incerta*) O
[b] moventur *scripsimus*; movent OLT
[c] secundum speciem *om.* L
[d] indifferenter *lectio incerta* O; infinitum T
[e] in uno tempore *om.* L
[f] ipsum *iter. sed corr.* O
[g] propellere] pellere LT
[h] ibidem] ibi LT

4. Furthermore, Aristotle says that some animals, such as worms and the like, move endlessly when they are removed from their places, yet they are not in-formed by a species.[109]

<Question 24>

1. As to the fourth question, let us first consider the arguments for the affirmative: the soul is no more determined to one specific part of locomotion than to another; therefore, in so far as it depends on itself it can indifferently move upwards and downwards.

2. Furthermore, in giving life to the body, the soul gives life to the whole body and to every one of its parts in virtue of one life; but in virtue of that life the soul makes a part move upwards; therefore, for the same reason it will also make the whole move upwards, since a part and its whole are made alive by the same life.

3. Furthermore, if one power can move a mobile thing in one timespan, an equal power can move a mobile thing equal to the other mobile thing in an equal timespan. But a whole body and a part of it, such as a hand, are equal because the soul is everywhere as a whole in a body; therefore, since the whole soul is in the hand and the whole soul is in the whole body, then the hand and the body are equal, because the mover and the moved are proportionate, as has been said before.[110] But the soul or the power in the hand can raise the hand; therefore, it should also <be able to> raise the whole body.

4. Furthermore, it requires less power to support something that has been pushed than to support it and <at the same time> to raise or propel it, as happens in a violent motion or in the motion of projectiles; for the air can support a stone that has been propelled for some time, but it cannot propel it further. Therefore, since the soul can raise the body in the air, as is clear when someone jumps, there is a much stronger reason to say that the soul will be able to support the body in the air; hence, <the soul can move the body with every species of locomotion>.

<Question 21.2>

1. In reply to the first question[111] one should say, following Aristotle, that the soul does move the body.[112]

[109] Arist. *De anima* III.11, 433b31–434a5.
[110] See above, q. 21.1, n. 1.
[111] That is, q. 21.
[112] Arist. *De anima* II.4, 415b8–10.

2. Ad primum in oppositum patet qualiter est respondendum multotiens alibi.

3. Ad secundum dicendum quod ‘unicus motor non movet nisi in unicam partem’, hoc est intelligendum quando motor est in una parte; sed anima est in omni parte; et ideo potest movere secundum oppositos motus, ut deprimere unam manum et elevare aliam.

4. Ad tertium dicendum sic[a] secundum quosdam qui ponunt animam esse ubique in corpore per naturam lucis. Et dicunt quod triplex est [b]lux. Est enim una lux quae est caeli empyrei, et ista lux, ut dicunt, respondet animae intellectivae. Est autem alia lux quae est caeli aquei sive cristallini secundum theologos, et non dicitur aqueum quia ibi est aqua, sed propter transparentiam et diaphaneitatem quae in aqua est, cuiusmodi diaphaneitas et transparentia est in illo caelo. Et nota quod, licet aer sit transparens et corpus diaphanum magis quam aqua vel aequaliter, non tamen possumus videre imagines[c] neque similitudines nostras in aere sicut in aqua. Et huius causa est quia aer est solum receptivus et non terminativus, aqua vero est receptiva et etiam terminativa. Et ideo, quia aqua terminativa est et receptiva est, de[d] natura aquae est ut repraesentet imagines et similitudines rerum, aer autem non[e]. Et ista lux respondet animae sensitivae. Est etiam tertia lux, quae est ipsius caeli stellati, et ista lux respondet animae vegetativae, quia per illam lucem crescunt haec inferiora. Mediantibus autem istis lucibus dixerunt ipsi[f] animam multiplicare[g] se ubique in corpore et mediante medio corporeo, scilicet sensitiva et vegetativa.

5. Sed contra: lux multiplicat se secundum numerum, et ita essent infinitae animae in corpore sicut et infinitae luces.

6. Ideo aliter dicendum est et sic quod per rationem [h]talem quae dicta est et consi **[O 115vb]** milem in aliis debemus hoc intelligere in proposito. Videmus enim quod ita est de puncto quod punctus est in linea non semper[i] [j]unus et idem, eo quod habet positionem in linea [k]et eo quod quantus[l] est. Circumscribamus tunc per intellectum positionem a puncto in linea et suam

[a] sic *om.* LT
[b] lux. Est enim *om.* T
[c] imagines] imaginationes O
[d] de] quod OL; ita (*lectio incerta*) *praem. s.l.* O
[e] non] illa *add. sed del.* O
[f] ipsi] corpus *add. sed del.* O; ipsius corpus L
[g] multiplicare] intelligere O
[h] talem ... et *scrips. sed del.* O (quaedam *pro* quae dicta)
[i] semper] et ubique *add.* T
[j] unus et idem *scripsimus*; una et eadem OLT
[k] et eo quod] quae T
[l] quantus *scripsimus*; quanta OLT

2. How one should reply to the first argument to the contrary[113] is made clear many times elsewhere.[114]

3. In reply to the second argument[115] one must say that the claim that a single mover only moves towards a single part must be understood <as applying> when the mover is only found in one part; but the soul is in every part; and therefore the soul can move with opposite types of motion; for example, it can move one hand down and raise the other.

4. In accordance with some people who posit that the soul is everywhere in the body in virtue of the nature of light, one should reply to the third argument[116] as follows. They say that there are three kinds of light. (i) One is the light that belongs to the empyreal heavens, and this light, according to them, corresponds to the intellectual soul. (ii) Another kind of light is that which belongs to the watery heavens (or what the theologians call the crystalline heavens), and these heavens are not said to be watery because there is water in them, but because the kind of transparency and diaphaneity that is found in water is also found in these heavens. Note that, although air is a transparent and diaphanous body to an extent that is greater than or similar to that of water, we cannot, however, see our images or likenesses in air as we do in water. And the reason for this is that air is only receptive and not terminative, whereas water is both receptive and terminative. Therefore, since water is terminative and receptive, it is in the nature of water to represent the images and likenesses of things, but this is not in the nature of air. And this kind of light corresponds to the sensitive soul. (iii) There is also a third kind of light, which belongs to the starry heavens, and this kind of light corresponds to the vegetative soul, because it is in virtue of that light that the things here below grow. These people maintained that it is by means of these kinds of light that the soul multiplies itself everywhere in the body and, in the case of the sensitive and vegetative souls, by means of the corporeal medium.

5. Against this: light multiplies itself numerically, and thus there would be infinitely many souls in a body just as there are infinitely many lights.

6. Therefore, a different reply must be given, to the effect that what happens in the present case must be understood through an appeal to the reason given above[117] and to a similar one given in other cases; for in the case of a point we see that the point is not always one and the same in a line, in that it has a position in the line and in that it is <also> a quantum. Let us then conceptually separate the position from the point in a line and its quantity. We will get the

[113] See above, q. 21.1, n. 1.
[114] See, for example, Book I, q. 59, nn. 6, 16–26, 31–33.
[115] See above, q. 21.1, n. 2.
[116] See above, q. 21.1, n. 3.
[117] See above, n. 3.

quantitatem; habebimus oppositum effectum, scilicet quod ubique erit totus et unus idem punctus in linea. Per istam rationem consimilem de puncto possumus videre istud idem de anima multo fortiori ratione. Anima enim nullam positionem in corpore nec aliquam extensionem habet. Non enim est anima quanta per se vel per accidens. Isto tunc modo possumus per rationem ad hoc intelligendum devenire, non autem per imaginationem.

7. Per exemplum possumus istud rationabiliter videre. Aeternitas enim tota est in qualibet parte temporis[a], secundum quod vult Boethius, dicens quod aeternitas est interminabilis vitae possessio tota simul; multo fortius ergo anima.

8. Ad argumentum in oppositum dicendum quod tota anima est in capite[b]. Et cum dicit quod totum est extra quod nihil, dicendum quod verum est, nihil scilicet[c] quod sit pars eius. Unde non pono illam animam quae est in capite esse partem animae[d] quae est in toto corpore nec animam quae est in manu esse partem animae[e] in capite, quia non pono animam habere partem, cum sit substantia indivisibilis. Sed pono totam esse hic et totam esse ibi, id est perfectam.

<Quaestio 22.2>

1. Ad secundam quaestionem dicendum quod animal movetur ex se et quod dividitur in duo distincta per essentiam.

2. Ad primum argumentum in oppositum dicendum quod duplex est potentia: est quaedam potentia quae numquam fit actus, sed solum est receptiva actus, et in talibus bene possunt esse distincta per essentiam, verbi gratia materia est in potentia ad formam non ita quod materia fit[f] forma, sed ita quod materia est receptiva formae et est sub forma, et ideo materia et forma sunt distincta inter se per essentiam, licet non in composito. Consimiliter est de anima [g]et corpore. Corpus enim est in potentia et anima in actu non ita in potentia quod corpus fit anima, sed ita quod corpus est receptivum animae; et ideo possunt esse distincta per essentiam. Est alia potentia quae aliquando fit actus, et in talibus verum est quod non differunt [h]per essentiam, sicut

[a] temporis *lectio incerta* (*ex corr.*) O
[b] capite *ex corr.* O
[c] scilicet *om.* LT
[d] animae] in capite *add. sed del.* O
[e] animae] quae est *add.* LT
[f] fit] sit LT
[g] et ... essentiam *om.* T
[h] per essentiam *om.* T

opposite effect, that is, that one and the same point as a whole will be everywhere in the line. On the basis of an argument similar to this one about the point there is a much stronger reason to reach the same conclusion about the soul; for the soul does not have any position in the body nor <does it have> any extension, because the soul is not a quantity either per se or accidentally. In this way we can reach an understanding through reason but not through the imagination.

7. We can also grasp this rationally through an example. Eternity is as a whole in any part of time, in accordance with what Boethius meant when he said that eternity is the simultaneous possession as a whole of an endless life;[118] therefore, there is a much stronger reason to say that the soul is like that.

8. In reply to the contrary argument[119] it must be said that the whole soul is in the head. And when Aristotle says that the whole is that outside which there is nothing, it must be said that this is true in the sense that nothing that is a part <is outside its whole>. Thus, I do not posit that the soul that is in the head is a part of the soul that is in the whole body nor that the soul that is in a hand is a part of the soul that is in the head, because I do not posit that the soul has parts, since the soul is an indivisible substance. But I posit that the whole soul is here and the whole soul is there, that is, that the soul is perfect.

<Question 22.2>

1. To the second question[120] it must be said that an animal is moved by itself and that it is divided into two things distinct in essence.

2. In reply to the first contrary argument[121] it must be said that there are two kinds of potency: (i) one kind of potency is that which never comes into act, but is only receptive of an act; and in the things <with this kind of potency potency and act> can indeed be distinct in essence. For instance, matter is in potency to a form not in such a way that matter becomes that form, but in such a way that matter is receptive of that form and is under that form, and therefore matter and that form are distinct one from the other in essence, although not in so far as they exist in a composite. The case of soul and body is similar; for the body is in potency and the soul is in act not in such a way that the body becomes the soul, but in such a way that the body is receptive of the soul; therefore, the body and the soul can be distinct in essence. (ii) Another kind of potency is that which sometimes comes into act; and in the

[118] Boethius, *De cons.* V.6.4, CCSL 94, 102.8–9.
[119] See above, q. 21.1, n. 3.
[120] That is, q. 22.
[121] See above, q. 22.1, n. 1.

[a]manifeste patet de genere et[b] specie. Essentia enim generis fit species, et ideo genus et species non differunt per essentiam.

3. Ad secundum dicendum sicut dictum est.

4. Ad argumentum in oppositum dicendum quod duplex est fieri, scilicet rationis et naturae. Cum vero dicimus quod genus fit species, hic sumitur fieri rationis; cum vero dicimus quod ex igne fit aer, hic sumitur fieri naturae. Unde, si loquamur proprie de essentia, [c]male dicitur 'essentia generis fit essentia speciei', quia essentia non fit. Et hoc patet per Aristotelem in secundo *Posteriorum*, cum dicit quod, etsi nullus esset specialis triangulus, tamen esset demonstrare de triangulo habere tres. Iste triangulus de quo loquitur Aristoteles est essentia pura ipsius trianguli, et ista essentia non est in fieri. Speciales[d] vero trianguli, quorum est ista essentia, sunt in fieri[e]. Unde improprie loquitur, [f]nisi sic dicamus: 'essentia, hoc est forma, generis fit species'. Et tunc dicendum est sic quod, sicut est in fieri naturae, scilicet quod, si ex igne debeat fieri aer, in igne tunc est essentia aeris, sed incomplete et diminute[g], similiter dicimus et ponimus esse in via rationis[h] quod, si ex genere debeat fieri species, tunc in genere est essentia speciei, [i]incomplete tamen, quia, si in genere esset tota essentia speciei, tunc falsum diceret Aristoteles cum dicit quod genus non est nisi pars materialis, differentia vero pars formalis et completiva. Et praeter hoc, tunc non necesse esset ponere in definitione speciei nisi solum genus[j], contra omnes philosophos. Unde dicendum quod, si ista essentia generis quae fit species secundum talem modum loquendi esset[k] completa et non diminuta, verum concluderent rationes. Sed quia ista essentia est incompleta et diversimode completur, ideo est duplex essentia completa et diversa in speciebus.

5. Sciendum etiam quod in genere est una potentia generalis et communis et praeter hoc sunt ibi multae potentiae speciales, ut duae, loquendo de

[a] manifeste patet] manifestum est T
[b] et *lectio incerta* (*ex corr.*) O
[c] male ... fit essentia *om. hom.* T
[d] Speciales *ex corr.* O
[e] fieri] et ob hoc dicit (dicunt T) quod contingit demonstrare etc. *add.* LT
[f] nisi sic dicamus] si sic dicatur T
[g] diminute] quod *add.* OL
[h] rationis] naturae OL
[i] incomplete ... speciei *om. hom.* T
[j] genus] quod est *add.* LT
[k] esset *ex corr.* O; essentia *add.* LT

things <with this kind of potency>, it is true that act and potency do not differ in essence, as is absolutely clear in the case of genus and species; for the essence of the genus becomes the species, and therefore the genus and the species do not differ in essence.

3. We must reply to the second argument[122] as before.[123]

4. In reply to the contrary argument[124] it must be said that there are two kinds of becoming: the becoming of reason and that of nature. When we say that the genus becomes the species, in this case we speak of the becoming of reason; but when we say that air comes to be from fire, in this case we speak of the becoming of nature. Thus, if we speak of essence in the proper sense, it is wrong to say that the essence of the genus becomes the essence of the species, because an essence does not become. And this is made clear by Aristotle in the *Posterior Analytics* II when he says that, even if there were no specific triangle, it would still be possible to prove that a triangle has three <angles equal to two right angles>.[125] The triangle of which Aristotle speaks here is the pure essence of triangle, and this essence does not become, but the specific triangles, to which this essence belongs, do become. Accordingly, we speak improperly unless we say 'the essence, that is, the form, of the genus, becomes the species'. And so we must say that, just as in the becoming of nature, if air is to come to be from fire, the essence of air must be in fire, but in an incomplete and diminished way, we must similarly say and posit that, in accordance with the approach of reason, if the species is to come to be from the genus, then the essence of the species is in the genus, but in an incomplete way. This is because, if the whole essence of the species were in the genus, then Aristotle would say something false when he says that the genus is only the material part <of the species>, whereas the difference is the formal and completing part.[126] Moreover, in that case it would not be necessary to posit anything apart from the genus in the definition of the species, which is contrary to all the philosophers. Accordingly, it must be said that, if this essence of the genus, which becomes the species according to this way of speaking, were complete and not diminished, then the conclusion of the arguments would be true. But because this essence is incomplete and is completed in different ways, therefore in the species there is a twofold essence which is complete and different.

5. We should also know that in a genus there is one general and common potency and in addition to this there are there many specific potencies, for example, two in the case of animal, and the specific potencies are diverse and

[122] See above, q. 22.1, n. 2.
[123] See above, q. 22.1, n. 3.
[124] See above, q. 22.1, n. 5.
[125] Arist. *An. post.* II.3, 91a3–5.
[126] Arist. *Metaph.* VII.12, 1038a5–8.

animali, quae potentiae diversae sunt et distinctae per essentiam, [a]licet potentia generalis sit una et eadem. Et ideo, quia illae potentiae sunt distinctae per essentiam, et species ob hoc sunt distinctae et differunt essentialiter.

6. Ad aliam rationem dicendum quod duplex est ibi positio circa inductionem animae sensitivae in esse. Quidam ponunt ita quod in animali bruto[b], [c]sicut corpus descinditur[d] a corpore quodam modo, [quod[e]] similiter anima sensitiva generati descindatur ex anima sensitiva patris et quod una sensitiva[f] possit cedere in duas, sicut unus punctus in duo puncta.

7. Alia est positio talis. Quidam dicunt quod in animali est quaedam vita sopita in toto corpore, quae[g], posita in matrice, per virtutes naturales et virtutes caelestes consolidat[h] materiam[i] et organizat[j] membra sive facit[k] membra organica, et post exit ista vita quae est ibi sopita et in potentia in actum, [l]et est consimilis in potentia qualis est in actu post consolidationem.

8. Ad aliud dicendum quod supponit falsum, cum dicit quod anima sensitiva componitur[m]. Quidam [n]enim[o] dicunt quod est sola forma.

<Quaestio 23.2>

1. Ad tertiam quaestionem dicendum quod animam movere corpus est dupliciter, sicut [p]dicit Aristoteles et innuit in secundo[q] *De anima*: aut finite aut infinite. Si primo modo, sic semper informatur aliqua [r]specie vel vere[s] accepta vel specie ficta et imaginatione. Et licet homo non percipiat[t] se aliquando specie aliqua informari, tamen[u], quandocumque movetur, movetur[v] propter informationem **[O 116ra]** alicuius speciei qualitercumque. Si secundo modo,

[a] licet ... essentiam *om. hom.* L
[b] bruto] quod *add.* OT
[c] sicut corpus *om.* L
[d] descinditur *ex corr.* O
[e] quod *delevimus*
[f] sensitiva] non *add. sed del.* O; in ipso *add.* L; in potentia *add.* T
[g] quae *scripsimus*; qua OLT
[h] consolidat *corr. ex* consolidant O; consolidant LT
[i] materiam] naturam LT
[j] organizat *corr. ex* organizant O; organizant LT
[k] facit *ex corr.* O; faciunt LT
[l] et ... actu *om.* T
[m] componitur] vel corrumpitur *add. s.l.* O
[n] enim dicunt *ex corr.* O
[o] enim *om.* T
[p] dicit ... innuit] innuit Aristoteles LT
[q] secundo] *recte* tertio
[r] specie ... specie] specialis natura vel tempore sita L
[s] vere *ex corr.* O
[t] percipiat *corr. ex* participat O; participat T
[u] tamen *ex corr.* O
[v] movetur *suppl. mg.* O

distinct in essence, although the general potency is one and the same. Therefore, it is because the specific potencies are distinct in essence that the species are distinct and differ essentially.

6. To the other argument[127] it must be said that there are two positions about the coming into being of the sensitive soul. (i) Some people posit that in a brute animal, just as one body is in some way separated from another body, the sensitive soul of the animal generated is similarly separated from the sensitive soul of the father, and that one sensitive soul can yield two, just as one point yields two points.

7. (ii) Another position is the following. Some people say that in an animal there is a life that is <as if> asleep throughout the whole body. And when this life is located in the mother, in virtue of natural and celestial powers it consolidates matter and gives organization to the members <of the new body> or makes those members organic, and afterwards this life, which is asleep there and in potency, comes into act, and it is as similar in potency as it is in act after this consolidation.

8. To the other argument[128] it must be said that it is based on the false assumption that the sensitive soul is composed; for some people say that the sensitive soul is only a form.

<Question 23.2>

1. To the third question[129] it must be said that the soul moves the body in two ways, as Aristotle says and suggests in *On the Soul* II: either (i) in a finite way or (ii) endlessly.[130] In case (i), the soul is always in-formed by a species, either a species that is truly received or a fictional species or a product of imagination. Although a man sometimes does not perceive that he is in-formed by a species, nonetheless whenever he is in motion he is in motion because he is in-formed by a species in one way or another. In case (ii), given that the

[127] See above, q. 22.1, n. 6.
[128] See above, q. 22.1, n. 7.
[129] That is, q. 23.
[130] The correct reference is Arist. *De anima* III.11, 433b31–434a21.

sic non est necesse, [a]sicut patet per Aristotelem ibidem in libro *De anima*, eo quod continue moventur.

<Quaestio 25>
Quaeritur de hoc quod dicit Aristoteles consequenter, scilicet quod infinitorum non est primum.

1. Contra[b]: accipiamus aliquam lineam rectam. Ibi sunt infinita puncta, et tamen sunt ibi duo puncta prima, videlicet punctus a quo et punctus in quem.

2. Item, dicit Aristoteles: "Infinitae sunt scientiae"; et tamen est aliqua scientia prima, ut illa quae est de primo scibili sive de causa prima.

3. Item, sequitur 'si est ultimum, ergo et primum' et non econverso; [c]sequitur ergo, si est accipere ultimum, quod sit accipere primum; sed in infinitis est accipere ultimum; ergo et primum. Probatio minoris: [d]revolutiones praecedentes revolutionem hodiernam sunt infinitae secundum Commentatorem, et tamen est ultima revolutio, ut quae est hodie.

4. Huic respondet Commentator dicens quod possibile est pertransire infinita per accidens, non tamen per se. Et vocat infinita per se ubi sunt causae agentes et essentiales, verbi gratia animal movetur[e] a virtute sensitiva, virtus sensitiva movetur a virtute supracaelesti, virtus[f] caelestis a corpore caeli, corpus caeli ab intelligentia et illa intelligentia a causa prima et causa prima a nullo alio. Ecce secundum Commentatorem quod in causis essentialibus efficientibus est sumere primum[g]. Infinita pertransire per accidens vocat ut pater patris mei respectu mei[h] et similiter revolutio praecedens esternam[i] revolutionem respectu revolutionis hodiernae, et hoc modo secundum ipsum contingit infinita pertransire.

5. Contra: transitus infiniti prohibetur propter suam dimensionem veram vel proportionalem; cum ergo tanta sit divisio sive dimensio rerum per accidens, eo modo quo Commentator sumit [j]per accidens, quanta est dimensio[k] rerum per se, eo modo quo Commentator sumit per se, tunc tanta[l] erit impossibilitas pertransire infinita per accidens sicut et per se.

[a] sicut ... anima *om.* T
[b] Contra *om.* O
[c] sequitur ... primum *om.* T
[d] revolutiones praecedentes] revolutione simul recedentes L
[e] movetur] movet OL
[f] virtus *om.* O
[g] primum] et non est *add.* OLT *sed del.* O
[h] mei] sui T
[i] esternam] hodiernam T
[j] per ... sumit *om. hom.* T
[k] dimensio *scripsimus*; divisio OL
[l] tanta *corr. s.l. ex* tam O

body is continuously in motion, it is not necessary <for the soul to be in-formed by a species>, as is made clear by Aristotle in the same passage of *On the Soul*.

<Question 25>
This question is about what Aristotle says next, namely that there is not a first among infinitely many things.[131]

1. Against this: let us take a straight line. There are infinitely many points there, but there are nonetheless two first points there, that is, the point from which and the point towards which.

2. Furthermore, Aristotle says: "There are infinitely many sciences";[132] but there is nonetheless a first science, namely the one which is about the first knowable thing or the first cause.

3. Furthermore, the inference 'if there is a last, therefore, there is also a first' is valid, but the converse inference is not; therefore, it follows that, if it is possible to accept a last, it is also possible to accept a first; but in infinites it is possible to accept a last; therefore, <it is> also <possible to accept> a first. Proof of the minor premise: the revolutions that precede today's revolution are infinitely many, according to the Commentator,[133] and yet there is a last revolution, namely the one that occurs today.

4. The Commentator replies to this argument by saying that it is possible to traverse accidental infinites, but not per se infinites. And he calls an infinite 'per se' where there are efficient and essential causes, for example, an animal is moved by the sensitive power, and the sensitive power is moved by a celestial power, and the celestial power is moved by the body of the heavens, and the body of the heavens is moved by an intelligence, and that intelligence is moved by the first cause, and the first cause is moved by nothing else. Thus it is that according to the Commentator there is a first in the <series of> essential efficient causes. But when he talks of traversing accidental infinites the Commentator has in mind cases like that of the father of my father with respect to me, and similarly the revolution preceding the revolution of yesterday with respect to the revolution of today, and according to him it is possible to traverse the infinites in this way.[134]

5. Against this: the traversing of an infinite is prevented by its dimension, either true or proportional; therefore, since there is as much division or dimension in accidental things, given the way the Commentator interprets 'accidental', as there is dimension in per se things, given the way the Commentator interprets 'per se', it will be as impossible to traverse an accidental infinity as it is to traverse an infinity per se.

[131] Arist. *Phys.* VIII.5, 256a18–19.
[132] Arist. *An. post.* I.32, 88b7.
[133] Aver. *In Phys.* VIII, t.c. 9, fol. 345rb–vb .
[134] Aver. *In Phys.* VIII, t.c. 15, fol. 350ra-b; t.c. 47, fol. 388va–b.

6. Item, accipiamus A, B, C aequivoce et diversimode: uno modo quod quodlibet[a] sit[b] causa alterius, ut A causa de B et B causa de C, et alio modo ita quod nullum sit causa alterius. Quaero tunc an sit maior difficultas transire illa quorum quodlibet[c] est causa alterius quam illa quorum nullum est causa alterius. Constat quod non. Apponamus infinita ex utraque parte. Tunc, ut videtur, non erit maior difficultas pertransire haec quam illa. Falsum ergo dicit Commentator.

7. Et hoc verum est. Tamen dixit ad sustinendum illud falsum quod non fuit primus homo et quod tempus non incepit. Et narravit tunc illud falsum probabile ad sustinendum illud falsum.

8. Ideo dicendum est aliter et melius sic, quod in infinitis per se sive actu non est primum, in infinitis tamen potentia, ut in linea, ubi[d] sunt infinita puncta in potentia, est primum. Sed duplex est primum, videlicet ordinationis et inceptionis. Ibi vero est primum inceptionis, primum tamen ordinationis non, quia quilibet punctus potest esse aequaliter primum in ordinatione, et ideo non est ibi aliquod primum.

9. Et per hoc patet solutio ad primum argumentum.

10. Ad secundum dicendum quod Aristoteles non dicit quod infinitae sunt scientiae nisi[e] sic glossetur: infinitae, id est innumerabiles, et hoc quia loquitur[f] ibi de scientiis quantum ad nos et quae sunt de compositione nostra. Si tamen sumamus scientias [g]pro ideis rerum quae sunt in causa prima, sic forte scientiae sunt infinitae. De scientiis tamen nostris, quae sunt de nostra compositione, loquendo, non est verum dicere quod sunt infinitae, et hoc quia defectus est a parte nostra. Unde sciendum quod non est defectus a parte scibilium, quia in scibilibus est scientia infinita potentia, quia, cum cuilibet numero praeterquam binario correspondeat sua figura, tunc de quolibet numero contingit[h] demonstrare suam passionem mediante figura ei correspondente, et ita patet quod[i] a parte scibilium est scientia infinita in potentia.

[a] quodlibet *scripsimus*; quaelibet OLT
[b] sit *om.* O
[c] quodlibet *scripsimus*; quaelibet OLT
[d] ubi *suppl. s.l.* O; *om.* L; sed T
[e] nisi *ex corr.* O
[f] loquitur] quantum ad nos et quae sunt de compositione *add. sed del.* O
[g] pro ideis *om.* L
[h] contingit *suppl. s.l.* O; *om.* L
[i] quod *suppl. s.l.* O

6. Furthermore, let us take A, B, and C equivocally and in different ways: in one way so that each is the cause of another, for instance, A is the cause of B and B the cause of C, in another way so that none of them is the cause of any other. I then ask whether it is more difficult to traverse those things each of which is the cause of another than those things none of which is the cause of another. It is clear that it is not more difficult. Let us take infinitely many things in each of the two cases. Then, it seems, it is not more difficult to traverse these things rather than those. Therefore, the Commentator says something false.

7. And this conclusion is true. However, in support of his false claim, the Commentator said that there was no first man and that time did not begin.[135] And he made this plausible but false claim in support of the above false claim <that is, that it is possible to traverse accidental infinites >.

8. Therefore, it must be said in a different and better way that in the infinites per se or in act there is not a first, but in the infinites in potency, as in a line, where there are infinite points in potency, there is a first. But something can be first in two ways: (i) first as to the ordering and (ii) first as to the beginning. In the present case there is a first as to the beginning, but there is not a first as to the ordering, because any point can be equally first in the ordering, and therefore there is not anything first there.

9. And with this the solution to the first argument[136] is clear.

10. In reply to the second argument[137] it must be said that Aristotle does not say that sciences are infinitely many unless this claim is glossed as follows: "infinite' means 'innumerable' here', and this is because he is speaking there of the sciences in relationship to us and as dependent on our action of composition. However, if we take sciences to be the ideas of the things which exist in the first cause, the sciences are perhaps infinitely many in this sense. Speaking of our sciences, however, which depend on our action of composition, it is not true to say that they are infinitely many, and this because of a defect on our part. Thus, we should know that there is no defect on the side of knowable things, because so far as knowable things are concerned a science is potentially infinite; for, given that to every number except the number two there corresponds its own geometrical figure, it is possible to prove a passion of each number by means of the geometrical figure corresponding to it, and thus it is clear that science is potentially infinite on the side of knowable things.

[135] Ibid..
[136] See above, n. 1.
[137] See above, n. 2.

<Quaestio 26>

Aristoteles consequenter arguit sic: est reperire movens et motum simul, et motum per se; ergo et movens[a]. Quaeritur de isto modo arguendi.

1. Et videtur quod fallat, quia est reperire substantiam et accidens simul, et est reperire substantiam per se, et tamen non sequitur quod sit reperire accidens per se.

2. Item, est reperire materiam et formam simul, et est reperire formam per se, ut causam primam, [b]et tamen non sequitur quod sit reperire materiam per se.

3. Item, est reperire corpus et animam sensitivam simul, et est reperire corpus per se, ut post mortem, et tamen non sequitur quod sit reperire sensitivam per se.

4. Ad omnes istas instantias dicendum quod regula intelligenda est quando illa duo quae simul sunt ita se habent quod unum non plus requirit aliud ad sui existentiam quam econverso. Et hoc deficit in omnibus instantiis praedictis[c].

1*. Aristoteles consequenter habet pro impossibili quod aliquid[d] movens moveatur eadem specie motus qua movet, et [e]etiam habet pro impossibili quod docens doceatur.

<Quaestio 27>

1. Contra primum sic: omne quod alterat alterat per qualitates activas et passivas; sed quod[f] per [g]qualitates activas et passivas agit quodam modo patitur; quare etc.

2. Solutio. Dicendum quod Aristoteles intelligit sic quod alterans non alteratur primo et per se, per accidens tamen potest alterari, sicut concludit ratio. Et similiter est de docente et docto.

1*. Iuxta secundum sive iuxta hoc quod habet pro impossibili quod docens doceatur duo possunt quaeri. Primum est an aliquis possit addiscere, sed hoc ad praesens dimittatur.

[a] movens] per se *add.* T
[b] et ... se] ergo et accidens quod est falsum T
[c] praedictis] et etiam respectu tertii fallit *add.* T
[d] aliquid] aliquod LT
[e] etiam ... impossibili] similiter T
[f] quod *suppl. s.l.* O
[g] qualitates ... passivas] huiusmodi T

<Question 26>

Aristotle next argues as follows: one can find a mover and a moved thing together, and one can find a moved thing by itself; therefore, one can also find a mover <by itself>.[138] The question here is about this way of arguing.

1. Arguments that this way of arguing is fallacious: one can find a substance and an accident together, and one can find a substance by itself; however, it does not follow that one can find an accident by itself.

2. Furthermore, one can find matter and form together, and one can find a form by itself, for instance, the first cause; however, it does not follow that one can find matter by itself.

3. Furthermore, one can find a body and a sensitive soul together, and one can find a body by itself, that is, after death; however, it does not follow that one can find a sensitive soul by itself.

4. One should reply to all these counterexamples[139] that Aristotle's rule is to be understood as applying when the two things that are together are related in such a way that one of them no more requires the other for its existence than the other way around. And this condition is missing in all the counterexamples above.

1*. Aristotle next takes it to be impossible that a mover should be moved by the same species of motion with which it moves, and he also takes it to be impossible that a teacher should be taught. [140]

<Question 27>

1. Argument against the first claim:[141] everything that produces an alteration does this in virtue of active and passive qualities; but what acts in virtue of active and passive qualities is in some way subject to a passion; hence, <a mover can be moved by the same species of motion with which it moves>.

2. Solution. It must be said that what Aristotle means is that what produces an alteration is not subject to an alteration per se and primarily, but it can be subject to an alteration accidentally, as the argument establishes. And the same applies to the case of the teacher and the person taught.

1*. As to the second claim, that is, that it is impossible for a teacher to be taught, there can be two questions. The first is whether someone can learn, but let this question be dismissed for the time being.

[138] Arist. *Phys*. VIII.5, 256b20–24.
[139] See above, nn. 1–3.
[140] Arist. *Phys*. VIII.5, 256b27–257a2.
[141] That is, that it is impossible that a mover should be moved by the same species of motion with which it moves.

<Quaestio 28>
Secundum est utrum scientiae acquisitae in discente sint una qualitas animae vel plures.

1. Et videtur quod sint una per simile sic: color, cum multas habeat[a] species, in eodem tamen subiecto numero sive in una parte non est nisi unus color; ergo a simili, cum scientia habuerit multas species, in eodem tamen subiecto, ut in anima, quae in **[O 116rb]** divisibilis [b]est, non habebit nisi unam qualitatem nec[c] erit nisi una qualitas.

2. Item, quanto subiecta sunt materialiora[d] tanto accidentia numerabiliora. Verbi gratia, plura sunt accidentia in corpore hic inferius quam in corpore caeli, et hoc est propter materialitatem. Sed anima[e] est minus materiale quam corpus; ergo accidentia corporis erunt numerosiora quam accidentia animae. Sed numerositas accidentium corporis non facit idem accidens quod est genus proximum multiplicari per multas species, [f]sicut patet de colore. Color enim, quod est genus proximum, non [g]multiplicat se per suas species in eadem parte. Genus tamen quod est remotum, ut qualitas, potest se multiplicare, quia plures possunt esse qualitates in eadem parte. Ergo numerositas accidentium animae multo fortius non faciet idem accidens quod est genus proximum, ut scientia, multiplicare se per plures species. Scientia tunc in anima est una scientia et non plures.

3. Item, lumen est perfectio medii et habitus eius sicut scientia est perfectio animae et habitus eius. Probatio similitudinis. Sicut enim aer est aliquid in se post absentiam luminis et circumscripto lumine, similiter anima est aliquid in se circumscripta scientia. Unde, sicut lumen non est perfectio aeris quantum ad positionem in specie, sed solum in esse quodam, similiter scientia est perfectio animae; quare etc. Sed duo lumina in eadem parte medii sunt unum. Probatio: albedo addita albedini et habitus realis additus habitui reali non facit numerositatem, sed intensionem. Non enim dicimus quod ibi sunt duae albedines, sed ibi est albedo intensior. Econverso enim modo est de habitu reali et de habitu intentionali, quia habitus intentionalis facit numerum, quia dicimus quod in medio sunt duae species et non dicimus quod in corpore sunt duae albedines. Sed lumen est[h] habitus realis et non intentionalis in medio; quare etc. Probatio minoris: sola species non denominat secundum suum

[a] habeat] qualitates *add. sed del.* O
[b] est ... qualitatem *om.* T
[c] nec] non T
[d] materialiora] quam accidentia animae *add.* L
[e] anima *ex corr.* O
[f] sicut ... species *om. hom.* L
[g] multiplicat se] facit se multiplicare T
[h] est *ex corr.* O

<Question 28>

The second question is whether the sciences acquired by the learner are one or more qualities of his soul.

1. Arguments that it is one: this is argued on the basis of a similarity with the case of colour. Although colour has many species, there is however only one colour in a subject that is numerically one or in one part ; therefore, by an argument from similarity, although science has many species, in one and the same subject, that is the soul, which is indivisible, science will have only one quality and will be only one quality.

2. Furthermore, the more material the subjects are, the more numerous the accidents are. For example, there are more accidents in a body here below than in a celestial body, and this because of its materiality. But the soul is less material than the body; therefore, the accidents of the body will be more numerous than the accidents of the soul. But the plurality of the accidents of the body does not cause one and the same accident that is the proximate genus to be multiplied through many species, as is clear in the case of colour; for a colour, which is a proximate genus, does not multiply itself through its species in the same part of the body. A remote genus, however, such as quality, can multiply itself, because there can be many qualities in the same part of the body. Therefore, there is a much stronger reason to say that the plurality of the accidents of the soul will not make the same accident that is the proximate genus, for instance, science, multiply itself in many species. Science in the soul then is one science and not many.

3. Furthermore, light is a perfection and a habit of the medium just as science is a perfection and a habit of the soul. Proof of the similarity. Just as air is something in itself when light is absent and when light is removed, similarly the soul is something in itself when science is removed. Thus, just as light is not a perfection of air with respect to the placing of air in a species <of a genus>, but only with respect to the kind of existence air has, so science is a perfection of the soul in the same way; hence, <the similarity holds>. But two lights in the same part of the medium are one. Proof: a whiteness added to a whiteness and a real habit added to a real habit do not produce plurality, but intension; for we do not say that there are two whitenesses there but that there is a more intense whiteness. The case of a real habit and that of an intentional habit are opposite, because an intentional habit does produce a number, because we say that there are two species in the medium but we do not say that there are two whitenesses in the body. But light is a real and not an intentional habit in the medium; hence, <there are not two lights in the

nomen[a], licet aliquo modo denominet, ut secundum immutationem. Verbi gratia, non dicimus quod species albi in medio faciat medium esse album. Sed lux denominat medium. Possumus enim dicere quod medium est illuminatum. Quare etc. Ergo a simili, duae scientiae in anima erunt una scientia seu[b] qualitas et non plures.

4. Ad oppositum: quaecumque numerantur secundum speciem, et secundum numerum. Sed in eadem anima sunt[c] diversae scientiae secundum speciem; ergo et secundum numerum.

5. Item, in anima sunt diversae virtutes secundum numerum et in ea est numerositas virtutum; ergo eadem ratione et scientiarum.

6. Solutio. Ad quaestionem in se dicendum [d]quod sunt ibi ut plures scientiae et non ut una.

7. Ad primum in oppositum dicendum quod est accidens dupliciter. Est enim quoddam accidens quod absolute egreditur de principiis subiecti, ut color, quia tu non es dictus albus respectu parietis. Quoddam vero est accidens ab extra et per relationem ad aliud, [e]ut scientia, nec[f] dicitur accidens per se nisi in comparatione ad scibile. De accidente primo modo dicto verum est quod non multiplicat se, si fuerit genus proximum, nisi fuerit accidens quod sit genus remotum, sicut dictum est. De accidente autem secundo modo fallit, et hoc quia tale accidens numeratur in comparatione ad ea quae sunt extra, sicut scientia in comparatione ad scibilia; et ideo, quia plura possunt esse scibilia, ideo et plures scientiae. Unde dicendum quod non est simile, et causa patet.

8. Ad secundum dicendum quod verum concludit, nisi esset prohibitio ex alia parte, ut a parte scibilium. Huiusmodi enim accidens est comparatum. Et ideo, nisi esset prohibitio a parte illius ad quod comparatur, verum concluderet.

9. Ad tertium dicendum quod quaedam sunt dispositiones sive[g] habitus quae, praeter hoc quod dependent, sunt in se hoc aliquid, cuiusmodi est lux, et quaedam sunt dispositiones quae, praeter hoc quod dependent, non sunt

[a] nomen] suum subiectum simpliciter *add.* T
[b] seu *ex corr.* O; sive T
[c] sunt *lectio incerta* (*ex corr.*) O
[d] quod ... est (n. 7) *om.* T
[e] ut scientia *om.* T
[f] nec *ex corr.* O
[g] sive *om.* O

medium>. Proof of the minor premise: the name of a species alone cannot be applied to the medium as such, although it can be applied in some way, that is, in accordance with an intentional change. For example, we do not say that a species of white in the medium makes the medium white. But the name of light can be used of the medium, because we can say that the medium is illuminated. Hence, <light is a real and not an intentional habit in the medium>. Therefore, by an argument from similarity, two sciences in the soul will be one science or one quality and not many.

4. To the contrary: any things that are many in species are also many in number. But in the same soul there are different species of science; therefore, they are also different in number.

5. Furthermore, in the soul there are powers that are different in number and so there is a plurality of powers in the soul; therefore, for the same reason there is also a plurality of sciences.

6. Solution. To the question in itself one must reply that the sciences <acquired by the learner> are in the soul as many and not as one.

7. To the first argument to the contrary[142] it must be said that there are two kinds of accident. (i) One kind of accident is such that it emerges from the principles of its subject absolutely, as does a colour, because you are not said to be white in relation to the wall. (ii) Another kind of accident is that which comes from outside and in relation to something else, as does science, which is not said to be an accident per se except in relation to the knowable. For an accident of kind (i) it is true that it does not multiply itself, if it is a proximate genus, unless it is an accident that is a remote genus, as has been said. For an accident of kind (ii) this fails to be true, and this is because such an accident is numbered in relation to external things, in the way that science is numbered in relation to knowable things. Therefore, since knowable things can be many, sciences too can be many. Thus, it must be said that the case of science is not similar <to that of colour>, and the reason is clear.

8. To the second argument[143] it must be said that it would reach a true conclusion, if there were not a prohibition on the other side, that is, on the side of the knowable things; for this accident <namely science> is relative. Therefore, the argument would reach a true conclusion if there were not a prohibition on the side of the thing to which science is compared.

9. To the third argument[144] it must be said that (i) some dispositions or habits are such that, in addition to being dependent, they are a determinate thing in themselves, and light is a disposition of this kind, and (ii) other dispositions are such that they are not a determinate thing in themselves in

[142] See above, n. 1.
[143] See above, n. 2.
[144] See above, n. 3.

hoc aliquid in se, sed solum prout disponunt subiectum in quo sunt sive solum sunt habitus. De istis dispositionibus ultimo modo dictis verum est quod non sunt duae, sed est habitus sive dispositio intensior. De dispositionibus autem primo modo dictis falsum est, immo debent[a] dici duae[b], sicut vult auctor *Perspectivae*. Et probat hoc per experimentum tale. Sit in pariete aliquod foramen et extra parietem sit[c] corpus opacum; affigantur duae candelae vel[d] plures e directo contra foramen. Illae duae candelae ad diversas partes corporis opaci tendunt radios suos, et sensibiliter videtur qualiter unus radius subtrahitur [e]si subtrahatur una candela inter decem candelas. Hoc autem non posset esse nisi essent ibi duo lumina. Quod autem lux non sit omnino dependens, sed sit hoc aliquid in se patet, quia, ponamus per impossibile[f] quod caelum staret et quod [g]sol currit et quod radius solis non multiplicet se, sed [h]sit unus[i] et idem[j]; tunc partes aeris diversae numero transeunt, et tamen remanet eadem lux numero. [k]Patet ergo sic quod non est dependens.

1*. Consequenter dicit Aristoteles quod motor est in actu et mobile in potentia, et dicit similiter quod generans habet speciem.

<Quaestio 29>

1. Contra primum sic: Aristoteles, cum dicit quod motor est in actu et mobile in potentia, aut intendit sic quod[l] motor habet in actu illud ad quod mobile est in potentia, ut terminum in quem, et sic falsum intendit, quia anima non habet actu terminum ad quem movetur corpus. Aut intendit sic quod motor est ens actu et[m] mobile est in potentia ens, et sic adhuc intendit falsum, quia loquitur ibi de motu locali, et illud quod debet esse mobile in motu locali debet esse perfectum; non ergo est in potentia. Tunc, si exponatur primo modo, dicit falsum a parte moventis; si secundo modo, dicit falsum a parte moti; quare etc.

[a] debent] debet OT
[b] duae] quod duae sunt T
[c] sit *ex corr.* O
[d] vel *lectio incerta* (*ex corr.*) O
[e] si subtrahatur *om.* T
[f] impossibile *corr. s.l. ex* possibile O
[g] sol ... quod *om. hom.* LT
[h] sit ... remanet] semper una et T
[i] unus *corr. ex* una O; una L
[j] idem *corr. ex* eadem O; eadem L
[k] Patet ... dependens *om.* L
[l] quod] *illegibile add. sed del.* O
[m] et] non *add. sed del.* O

addition to being dependent, but only in so far as they give a disposition to the subject in which they exist, that is, they are merely a habit. About dispositions of kind (ii) it is true to say <in the example above> that they are not two but that the habit or disposition is more intense. About dispositions of kind (i), however, this claim is false — on the contrary, they must be said to be two, as the author of the *Perspective* wants. And he proves this by the following experiment. Suppose that there is a hole in the wall and that outside the wall there is a dark body. Let two or more candles be placed directly in front of the hole. Those two candles direct their rays towards different parts of the opaque body, and we can experience through our senses that one ray disappears if one of the ten candles is taken away. But this could not happen unless there were <at least> two lights there.[145] That that light is not absolutely dependent, but a determinate thing in itself, is clear, because if we posit the impossible case that the heavens stop moving and that the sun continues to move and that a ray of sun does not multiply itself but is one and the same, then numerically distinct parts of air will pass by, but light will remain the same in number. This makes it clear, then, that light is not dependent.

1*. Aristotle next says that the mover is in act and the mobile thing is in potency, and he similarly says that the generator has the <same> species <as the thing generated by it>.[146]

<Question 29>

1. Arguments against the first claim:[147] when Aristotle says that the mover is in act and the mobile thing is in potency, either (i) he means that the mover has in act that to which the mobile thing is in potency, as the terminus towards which, and in this case he says something false, because the soul does not have in act the terminus towards which the body is moved. Or (ii) he means that the mover is a being in act and the mobile thing is a being in potency, and in this case too he says something false, because he is speaking there of locomotion, and a thing that has to move by locomotion must be perfect; therefore, it is not in potency. Thus, if Aristotle's claim is explained as in (i), Aristotle says something false about the mover; if it is explained as in (ii), he says something false about what is moved; hence, <Aristotle's claim is false>.

[145] Alhacen, *De aspectibus* I.7, ed. Smith, 57.146–157.
[146] Arist. *Phys*. VIII.5, 257b6–10.
[147] That is, that the mover is in act and the mobile thing is in potency.

<Quaestio 30>

1. Contra secundum sic: mulus generatur, et tamen generantia non habent speciem, eo quod non sunt in eadem specie[a]. **[O 116va]**

2. Item, in generatis per putrefactionem ita est quod virtutes caelestes sive sol cum suo calore[b] sunt generantes, sicut dicit Commentator super [c]septimum huius, et tamen virtutes caelestes non sunt in eadem specie cum putrefacto; quare etc.

3. Item, Commentator in undecimo dicit quod corrigia agunt dolorem, et tamen non sunt dolor.

4. Item, illud videtur adhuc[d] fallere in accidentibus, quia frigiditas facit nigrum et caliditas similiter, et tamen [e]frigiditas et caliditas non sunt in eadem specie [f]cum nigredine; [g]quare etc.

5. Ad oppositum: si illud agens quod maxime distat ab acto habeat veram speciem acti, ut causa prima, secundum Augustinum, qui dicit quod causa prima habet ideas omnium rerum factarum, quae ideae sunt verae rationes causales, et Commentator super undecimum quod omnes formae sunt actu in motore primo[h], ergo multo fortius et agens proximum sive quod magis convenit cum eo, ut motor sive agens hic inferius, habebit veram speciem acti.

6. Solutio. Quidam dicunt quod hoc quod Aristoteles dicit, quod generans habet speciem, intelligendum est in[i] generatis univoce, ubi scilicet generans et generatum sunt in eadem specie specialissima.

7. Sed contra: tunc non esset sermo Aristotelis universalis.

8. Item, tunc nihil aliud esset dicere ‘generans habet speciem’, nisi quod generans habens speciem habet speciem, quod nihil esset dicere, [j]quod concedendum est.

9. Unde dicendum quod sermo Aristotelis est universalis in omnibus. Sed distinguendum est quod[k] habere speciem, [l]hoc est duobus modis: aut speciem causalitatis aut speciem praedicationis. [m]Et voco speciem[n] causalitatis quando

[a] specie] cum generato *add.* T
[b] calore *corr. ex* colore O; colore T
[c] septimum huius] *recte* undecimum Metaphysicae
[d] adhuc *om.* LT
[e] frigiditas et caliditas *om.* T
[f] cum nigredine *om.* L
[g] quare etc. *om.* T
[h] primo] quae sunt in potentia etc. *add.* T
[i] in] generantibus et *add.* LT
[j] quod concedendum est *om.* T
[k] quod *om.* LT
[l] hoc est *om.* T
[m] Et ... praedicationis *om. hom.* L
[n] speciem] praedicationis *add. sed del.* O

<Question 30>

1. Against the second claim:[148] a mule is generated, yet its generators do not have its species, because they are not in the same species.

2. Furthermore, in the case of things generated by putrefaction the celestial powers or the sun with its heat are the generators, as the Commentator says in his comment on *Physics* VII,[149] the celestial powers, however, are not in the same species as the putrefied thing; hence, <Aristotle's claim is false>.

3. Furthermore, in *Metaphysics* XI the Commentator says that sandal straps produce pain;[150] however, they are not a pain.

4. Furthermore, Aristotle's claim also seems to be false when said of accidents, because coldness produces something black, as does heat, yet coldness and heat are not in the same species as blackness; hence, <Aristotle's claim is false>.

5. To the contrary: if the agent that is maximally distant from the thing produced by it has the true species of that thing, as does the first cause, according to Augustine, who says that the first cause has the ideas of all things produced, which ideas are true causal principles,[151] and <also according to> the Commentator, who in his comment on *Metaphysics* XI says that all forms are in act in the first mover,[152] it follows that there is a much greater reason to hold that the proximate agent or the agent that is more in agreement with the thing produced by it, as is the mover or agent here below, will also have the true species of that thing.

6. Solution. Some people say that Aristotle's claim that a generator has the species <of the thing generated by it>[153] must be understood as applying to the case of things generated univocally, that is, to the case in which the generator and the generated are in the same most special species.

7. Against this: if that is so, then the claim of Aristotle would not be universal <in its application>.

8. Furthermore, if that is so, then saying 'the generator has the species' would amount to nothing else but saying that the generator having the species has the species, which would be to say nothing at all. This must be conceded.

9. Thus, it must be said that Aristotle's claim does apply to all cases. But a distinction must be drawn about two ways in which a generator can have the species <of the thing generated by it>: either it has (i) the species of causality

[148] That is, that the generator has the same species as the thing generated by it .

[149] The correct reference is Aver. *In Metaph.* XII, t.c. 13, fol. 299va–b; t.c. 18, fol. 305ra–b.

[150] Aver. *In Metaph.* XII, t.c. 24, fol. 309va.

[151] Aug. *De genesi ad litt*. VI.15, CSEL 28.1, 189.20–190.6; *De diversis quaestionibus octoginta tribus*, q.46, CCSL 44A, 72.41–73.64.

[152] Aver. *In Metaph.* XII, t.c. 18, fol. 305va.

[153] Arist. *Phys*. VIII.5, 257b10.

aliquis generans habet speciem aliquam penes se mediante qua est[a] causa rei fiendae. Et voco speciem praedicationis quando aliqua conveniunt in aliqua forma una praedicabili, ut in forma speciei specialissimae. In quibusdam vero est species utroque modo dicta, in quibusdam autem non, [b]sed prima tantum. In omnibus autem est species primo modo dicta, quae est species causalitatis. Quando enim equus generat equum vel homo hominem, ibi habet generans speciem causalitatis—est enim [c]in eo aliquid mediante quo etc.—, et etiam speciem praedicationis, quia sunt in eadem specie. Quando autem sol sive virtutes caelestes generant res putrefactas, ibi generans habet speciem causalitatis. Est enim aliquid in virtutibus caelestibus mediante quo agunt in haec inferiora.

10. Et non valet: 'sol est potentior generans quam equus vel homo, quia homo generat hominem et sol, [d]et dicit Commentator quod maxime sol; ergo multo fortius habebit utramque speciem', quia quod generans habeat utramque [e]speciem, hoc est propter eius impotentiam. Verbi gratia, si faber faciat mediante instrumento cultellum, si quaererem an faber[f] vel instrumentum est magis agens, diceremus quod faber, licet tamen non possit agere sine instrumento, et tamen instrumentum et cultellum plus conveniunt in communi praedicabili [g]sive in specie praedicationis[h], et faber et cultellus in specie causalitatis. Est enim aliquid in fabro mediante etc. Sol vero non habet speciem praedicationis in generando res istas inferiores. Similiter artifex agens domum[i] non[j] habet speciem domi secundum praedicationem, nisi solum secundum causalitatem. Est enim aliquid sive aliqua species in artifice mediante qua facit domum.

11. Iam patet quod [k]instantiae perimuntur. In omnibus enim est species causalitatis, licet non in omnibus sit species praedicationis, [l]ut declaratum est [m]in exemplis praedictis.

[a] est *suppl. s.l.* O
[b] sed prima tantum *om.* T
[c] in ... enim *om. hom.* T
[d] et ... sol *om. hom.* T
[e] speciem ... eius *om.* T
[f] faber] ferrum OL
[g] sive ... praedicationis *om.* T
[h] praedicationis *corr. mg. ex* causalitatis O; causalitatis L
[i] domum] et ideo *add.* O
[j] non *suppl. s.l.* O
[k] instantiae perimuntur] omne perimit L
[l] ut ... praedictis *om.* T
[m] in exemplis praedictis] ex praedictis L

or it has (ii) the species of predication. By (i) I mean the case in which a generator possesses a species by means of which it is the cause of the thing to be produced, and by (ii) I mean the case in which some things agree in one predicable form, such as the form of the most special species. In some generators, however, we find both kinds of species, whereas in other generators this is not so, but we only find the first kind. But all generators possess the first kind of species, that is, the species of causality. When a horse generates a horse or a man generates a man, the generator there has the species of causality—for there is something in it by means of which <it is the cause of the thing produced>—and it also has the species of predication, because the generator and the thing generated by it are in the same species. When the sun or the celestial powers, however, generate things by putrefaction, the generator there <only> has the species of causality; for there is something in the celestial powers by means of which they act on the things here below.

10. Thus the following inference is not valid: 'the sun is a more powerful generator than a horse or a man because a man and the sun together generate a man, and the sun <does so> maximally, according to the Commentator;[154] therefore, there is a much stronger reason to hold that the sun will have both species <that is, that of causality and that of predication>'. The reason why this inference is not valid is that it is because of the impotence of a generator that it has both species. For example, if a blacksmith produces a knife by means of an instrument, if I were to ask whether the blacksmith or the instrument is more of an agent, we would reply that the blacksmith is, although the blacksmith cannot act without the instrument; the instrument and the knife, however, agree more in a common predicable or in the species of predication, and the blacksmith and the knife agree more in the species of causality; for in the blacksmith there is something by means <of which he is the cause of the knife>. The sun does not have the species of predication when it generates the things here below. Similarly, the artisan who produces a house does not have the species of predication of the house but only the species of causality; for there is something or some species in the artisan by means of which he produces a house.

11. It is now clear that all the counterexamples are destroyed, because all those cases involve the species of causality, although they do not all involve the species of predication, as has been declared in the examples above.

[154] Aver. *In Phys.* II, t.c. 26, fol. 59ra; *In Metaph.* XII, t.c. 18, fol. 305ra–b.

1*. Aristoteles consequenter, ibi: *Dubitationem* [a]*autem habet*, movet talem dubitationem[b]: quando movens et motum dividuntur, utrum pars moventis[c] possit movere partem moti.

<Quaestio 31>

Iuxta hoc quaeritur utrum anima plus moveat corpus[d] in[e] una parte quam in alia sive utrum plus animae sit in una parte corporis quam in alia.

1. Et videtur quod non: a substantia egreditur virtus; ubi ergo est tota substantia alicuius rei, ibi erit tota virtus; sed in qualibet parte corporis est tota[f] substantia animae; ergo et tota virtus.

2. Item, nobilius non determinatur per ignobilius nec perfectio per perfectibile; sed anima est perfectio corporis; ergo non determinabitur per corpus; ergo nec virtus eius determinabitur per diversitatem partium corporis. Et si hoc, tunc in qualibet [g]parte corporis erit tota virtus animae et tota substantia.

3. Item, corrupto oculo, adhuc manet virtus visiva, per Aristotelem in libro *De anima*. [h]Dicit enim Aristoteles quod, si oculus senis corrumperetur et senex accipiat oculum iuvenis, utique videret. Hoc autem non posset esse nisi virtus visiva maneret in potentia; quare etc. Et similiter est de aure et de ceteris sensibus[i]. Cum ergo illa virtus maneat adhuc et non manet in illa parte, tunc manebit [j]in alia. Et ita videtur quod virtus quaelibet ipsius animae sit in qualibet parte corporis essentialiter.

4. Ad oppositum: sicut est in maiori mundo, similiter erit in minori. Sed ita est in maiori, secundum Aristotelem, quod motor caeli non determinat sibi locum in qualibet parte[k] circuli sed in circumferentia tantum; determinat ergo motor caeli sibi suam existentiam plus in una parte quam in alia. Ergo, a simili, sic erit in minori mundo quod anima, quae est ibi motor, determinabit sibi existentiam plus in una[l] parte corporis quam in alia.

5. Item, ubicumque est operationum maior manifestatio, ibi est maior virtutum existentia; sed in una parte corporis, ut in capite, est maior

[a] autem habet *inv. sed corr.* O
[b] dubitationem] utrum *add.* OL
[c] moventis *corr. s.l. ex* movens O; movens T
[d] corpus] quam *add. sed del.* O; *om.* L
[e] in] alia *add. sed del.* O
[f] tota] anima vel *add.* T
[g] parte corporis *om.* T
[h] Dicit ... videret] ibi si senex acciperet oculum iuvenis T
[i] sensibus] similibus OL
[j] in alia *suppl. s.l.* O; et illa L; alibi T
[k] parte] caeli vel *add.* T
[l] una *corr. s.l. ex* sua O

1*. In the same place, Aristotle next raises the following difficulty with the words *A difficulty arises*: when the mover and the moved are divided, can a part of the mover move a part of the moved?[155]

<Question 31>
In connection with this, we must ask whether the soul moves the body more in one part than in another or whether more of the soul is in one part of the body than in another.

1. Arguments for the negative: power emerges from substance; therefore, where there is the whole substance of a thing, the whole power is also there; but the whole substance of the soul is in every part of the body; therefore, the whole power is also there.

2. Furthermore, the more noble is not determined by the more ignoble nor is a perfection determined by the perfectible; but the soul is the perfection of the body; therefore, it will not be determined by the body; therefore, neither will its power be determined by the diversity of the parts of the body. And if so, then the whole power of the soul and its whole substance will be in every part of the body.

3. Furthermore, when an eye is corrupted, the visual power still remains, according to Aristotle in *On the Soul*. Aristotle says that, if the eye of an old man were corrupted and the old man received the eye of a young man, the old man would indeed see; but this could not happen unless the visual power remained in potency; hence, <the visual power remains>.[156] And the same applies to the ear and the other senses. Therefore, since that power still remains and does not remain in that part of the body, then it will remain in another part. It seems, then, that any power of the soul is essentially in any part of the body.

4. To the contrary: as it is in the greater world, so it will be in the lesser world. But according to Aristotle, in the greater world it is the case that the mover of the heavens does not determine a place for itself in any part of the circle but only in the circumference;[157] therefore, the mover of the heavens determines its own existence in one part more than in another part. Therefore, by an argument from similarity, in the lesser world it will be the case that the soul, which is the mover there, will determine its own existence more in one part of the body than in another.

5. Furthermore, wherever there is a greater manifestation of operations there is a greater existence of powers; but in one part of the body, namely the

[155] Arist. *Phys.* VIII.5, 258a27–30.
[156] Arist. *De anima* I.4, 408b21–22.
[157] Arist. *Phys.* VIII.10, 267b6–9.

manifestatio operationum virtutum mediantibus organis sitis in capite; ergo et maior existentia virtutum; quare etc.

6. Item, si in qualibet parte corporis esset tota virtus, tunc in pede esset virtus visiva. Et tunc ponamus per impossibile quod oculus esset in pede; sequetur quod pes possit videre, quod falsum est. Et hoc sequitur ex hypothesi; quare etc.

7. Item, per auctoritates **[O 116vb]** patebit illud. Primo per medicos. Dicunt enim medici quod illud membrum in quo plus viget anima est [a]cerebrum. Patet ergo secundum medicos quod plus est in una parte quam in alia.

8. Item, secundo per Aristotelem patet illud, qui dicit quod illud membrum in quo plus viget anima est cor, et ita, [b]secundum ipsum Aristotelem, plus determinat sibi unam partem corporis quam aliam.

9. Ad quaestionem respondent quidam distinguendo sic quod quaedam virtutes [c]sunt[d] quae sunt animae per totam essentiam et substantiam eius, non in comparatione[e] ad instrumenta aliqua, cuiusmodi sunt animare et vivificare, et istae virtutes sunt in omni parte corporis. Quaedam [f]sunt animae virtutes quae[g] uniuntur instrumento determinato, cuiusmodi sunt virtus visiva, auditiva et consimilia. Et [h]de istis non est verum secundum illos quod sunt in omni parte corporis, sed[i] solum in partibus determinatis[j] ubi[k] instrumenta eorum situantur, et hoc secundum apparitionem et manifestationem, radicaliter tamen origo istarum virtutum est in corde, a quo fluunt et procedunt illae virtutes originaliter.

10. Sed contra istam responsionem est penultima ratio sumpta ab Aristotele in libro *De anima*, ubi dicit quod, si senex acciperet, etc.

11. Ideo dicendum est aliter et melius, ut mihi videtur, quod tota virtus [l]animae est in qualibet parte corporis et quod virtus visiva est in manu. Et tamen dico quod non sequitur, ergo, si oculus esset ibi, quod posset videre. Hoc tamen concedere, si oculus esset ibi per naturam sicut est in capite, non

[a] cerebrum ... est *om. hom.* T
[b] secundum ... aliam] ut prius T
[c] sunt quae sunt] insunt T
[d] sunt *suppl. s.l.* O; *om.* L
[e] comparatione *corr. ex* corpore O; tempore (*lectio incerta*) L
[f] sunt ... determinato] autem insunt animae secundum partem essentiae eius et per determinatum respectum ad partes corporis T
[g] quae *suppl. s.l.* O; insunt animae vel secundum partem essentiae eius vel secundum quod illae virtutes *add.* L
[h] de ... quod] huiusmodi non T
[i] sed *corr. s.l. ex* non O; nisi L
[j] determinatis] vel *add.* T
[k] ubi *ex corr.* O
[l] animae ... virtus *om. hom.* T

head, there is a greater manifestation of the operations of powers by means of the organs located in the head; therefore, the powers have a greater existence there; hence, <the soul determines its own existence more in one part of the body than in another>.

6. Furthermore, if the whole power of the soul were in any part of the body, then the visual power would be in a foot. Let us then posit the impossible case that the eye is in the foot; it would follow that the foot could see, which is false. And this follows from the hypothesis; hence, <the hypothesis that the whole power of the soul is in any part of the body is false>.

7. Furthermore, that the soul exists more in one part of the body than in another will be made clear by authoritative texts. First, by the medical doctors; for they say that the member on which the soul is more influential is the brain. It is clear, therefore, that according to the doctors the soul is more in one part of the body than in another.

8. Furthermore, in the second place that conclusion is made clear by Aristotle, who says that the member in which the soul has more power is the heart,[158] and thus according to Aristotle the soul determines its own existence more in one part of the body than in another.

9. Some people reply to this question by making the following distinction: (i) some powers belong to the soul throughout its whole essence and substance, and not in comparison with any instruments. These powers include the power to animate and to make alive, and they are in every part of the body. (ii) Other powers are linked to a determinate instrument, such as the visual power, the hearing power, and the like. And, according to these people, it is not true of these powers that they are in every part of the body, but they are only in determinate parts, those where their instruments are situated. This applies to their appearance and manifestation, but the origin of these powers is fundamentally in the heart, from which those powers originally flow and proceed.

10. But against this reply there is the penultimate argument taken from Aristotle's *On the Soul,* where he says that, if an old man received <the eye of a young man, he would see>.[159]

11. Therefore, one must give another reply that is better in my view, namely that the whole power of the soul is in any part of the body and that the visual power is indeed in a hand. I say, however, that this does not follow: 'therefore, if the eye were in a hand, the eye could see'. But it would not be absurd to grant this conclusion if the eye were in a hand by nature just as it is in the

[158] Arist. *De motu anim.* 10, 703a11–16.
[159] See above, n. 3.

est inconveniens. Dico tamen quod [a]non[b] sequitur quod manus videret. Et huius causa est quia non est in debito situ, quia omne sensitivum est sensitivum calore cordis et mediante spiritu; si vero oculus sit in manu, tunc deficit proportio visus in debito situ et respectu in comparatione[c] ad cor, [d]a quo corde fluunt et procedunt spiritus et calores deservientes ipsi animae sensitivae. Unde dicendum quod oculus non[e] videt, et tota causa est quia deficit proportio caloris ipsius cordis respectu visus. Praeterea, visus ordinatur in capite, et hoc prope[f] cerebrum, ut accipiat species rerum visibilium et postea deferat per media et media usque quo deveniat ad intelligentiam[g]. Hoc autem non posset facere[h] si esset in manu.

<Quaestio 32>
Consequenter dicit Aristoteles quod in motu locali utrumque tangit [i]utrumque aut alterum alterum.

1. Contra: incorporeum non potest tangere corporeum; sed anima est incorporea, et corpus est corporeum[j].

2. Item, contacta sunt quorum ultima sunt simul; quae ergo non habent ultima non sunt contacta. Sed anima non habet ultimum, licet corpus habeat, [k]quia, ubi est ultimum, ibi est primum et medium; sed ubi[l] sunt ista tria, ibi[m] est divisio; anima autem omnino indivisibilis est; quare etc.

3. Solutio. Dicendum quod, cum duplex sit tactus, scilicet naturalis et mathematicus[n], naturalis dupliciter: aut corporalis aut virtualis, quod[o] Aristoteles non intelligit, cum dicit: "Utrumque tangit utrumque etc.", de tactu primo modo dicto, scilicet de tactu mathematico, quia, ut dicit Commentator super quintum huius, [quod[p]] non est tactus in mathematicis nisi solum continuitas, nec de tactu secundo modo dicto sive de tactu naturali [q]primo modo, quia contra istud procedit ratio prima, sed intelligit Aristoteles

[a] non ... manus] manus non L
[b] non *suppl. s.l.* O
[c] comparatione] corpore O
[d] a ... visus *om.* T
[e] non] in pede in L
[f] prope] praeter OL
[g] intelligentiam] intellectum T
[h] facere *corr. ex* patere O
[i] utrumque aut alterum *om.* L
[j] corporeum] quare etc. *add.* T
[k] quia, ubi est *om.* T
[l] ubi] non *add.* L
[m] ibi] non *add.* L
[n] mathematicus *corr. ex* metaphysicus O
[o] quod] aut *add. sed del.* O; quia L; unde T
[p] quod *delerimus*
[q] primo ... naturali *om. hom.* T

head. I say, however, that it does not follow that the hand would see. And the cause of this is that the eye is not in the appropriate place, because everything capable of sensation is such in virtue of the heat of the heart and by means of the spirit. But if the eye were in a hand, then the relation that sight in its appropriate place and position has to the heart is missing, and it is from the heart that the spirits and the hot fluids that serve the sensitive soul flow and proceed. Thus, it must be said that the eye does not see, and the whole cause of this is that the relation of the heat of the heart to sight is missing. Moreover, sight is located in the head next to the brain in order to receive the species of visible things and afterwards to carry them through different media until the species get to the intelligence. But sight could not do this if it were in a hand.

<Question 32>
Aristotle next says that in locomotion each of the two <that is, the mover and the moved> is in contact with the other or alternatively that one of the two is in contact with the other.[160]

1. Against this: what is incorporeal cannot be in contact with what is corporeal; but the soul is incorporeal, and the body is corporeal.

2. Furthermore, things in contact are those whose extremities exist in the same place; therefore, the things that do not have extremities are not in contact. But the soul does not have an extremity, although the body has, because wherever there is an extremity, there is also a first point and a mid-point; but wherever there are these three things there is a division; but the soul is absolutely indivisible; hence, <the soul is not in contact with the body>.

3. Solution. It must be said that there are two kinds of contact, that is, (i) natural contact and (ii) mathematical contact, and there also are two kinds of natural contact, that is, (ia) corporeal contact and (ib) virtual contact. And when Aristotle says: "Each of the two is in contact with the other <or alternatively one of the two is in contact with the other>",[161] he does not refer to contact of kind (ii), that is, to mathematical contact, because, as the Commentator says in his comment on *Physics* V, there is no contact in mathematical things other than continuity.[162] Nor does Aristotle refer to contact of kind (ia), because the first argument is against this, but Aristotle

[160] Arist. *Phys*. VIII.5, 258a20–21.
[161] Arist. *Phys*. VIII.5, 258a20–21.
[162] Aver. *In Phys*. V, t.c. 22, fol. 223ra–b.

de tactu naturali secundo modo dicto, sumendo naturam secundum proportionem. Unde possumus dicere quod alio modo est tactus metaphorice[a] dictus, sicut dicimus quod appetibile tangit appetitum et similiter [b]quod anima tangit[c] corpus. Et de isto tactu intelligit Aristoteles.

4. Et per hoc patet solutio secundi argumenti. [d]Cum enim dicitur quod contacta sunt quorum ultima sunt simul, intelligit de tactis proprie sumendo tactum, non autem metaphorice vel secundum proportionem.

<Quaestio 33>
Sed quaeritur quare dicit: "Utrumque utrumque[e] aut[f] alterum alterum".

1. Et dicendum quod quia duplex est motor [g]per se. Est unus motor [h]per se, ut motor primus, qui movet et nullo modo movetur, nec per se nec per accidens. Et propter talem motorem dicit "aut alterum alterum". Est alius per se qui, licet non movetur [i]per se, movetur tamen per accidens, ut anima, quae est motor corporis. Et propter talem motorem dixit "utrumque utrumque".

1*. Quaeritur de caelo. Et primo de caelo in se, secundo de suis partibus, prout est motum per se, [j]ut de intelligentia, quae movet ipsum, et de corpore caeli, quod movetur[k].

<Quaestio 34>

1. [l]Quod autem caelum non moveatur ex se probatio: non dividitur in talia duo quorum unum est per se movens et aliud per se motum; quare etc. Probatio praemissae: dicit Aristoteles in primo *Caeli et mundi* quod motus naturalis vel est rectus, et hoc a medio vel ad medium, vel circularis, ut motus caeli. Cum ergo ista divisio sit univoca, tunc, si ita est in uno quod non[m] sic

[a] metaphorice *fortasse corr. ex* mathematice O; mathematice LT
[b] quod anima tangit] anima T
[c] tangit] tangat O
[d] Cum ... proportionem] quia procedit de contactu proprio et non metaphorice sumpto T
[e] utrumque] scilicet utrumque tangit utrumque T
[f] aut *om.* O
[g] per se *om.* T
[h] per se *om.* L
[i] per se, movetur *om. hom.* L
[j] ut ... movetur *om.* L
[k] movetur] Circa primam partem ut circa intelligentiam *add.* O; Primo quaeratur utrum caelum moveatur ex se *add.* T
[l] Quod ... probatio] Et videtur quod non quia T
[m] non *suppl. s.l.* O; *om.* LT

refers to contact of kind (ib), and he takes 'nature' analogically. Thus, we can say that there is another kind of contact, contact in the metaphorical sense, as when we say that the desirable is in contact with the appetite and similarly that the soul is in contact with the body. And Aristotle refers to this kind of contact.

4. And with this the solution to the second argument[163] is clear; for when it is said that the things in contact are those whose extremities exist in the same place, this refers to contact taken properly, not metaphorically or by analogy.

<Question 33>

The question, however, is why Aristotle says: "Each of the two <is in contact> with the other or one of the two <is in contact> with the other".[164]

1. It must be said that the reason for this is that there are two kinds of per se mover. (i) One is such that it moves but is in no way moved, either per se or accidentally, that is, the first mover. And because of this mover Aristotle says "or one of the two <is in contact> with the other". (ii) Another kind of mover is such that, although it is not moved per se, it is, however, moved accidentally, such as the soul, which is the mover of the body. And because of this mover Aristotle said "each of the two is in contact with the other".

1*. Questions about the heavens. And the first question is about the heavens in themselves, the second is about their parts, in so far as the heavens are moved per se, that is, about the intelligence that moves them, and about the body of the heavens which is moved.

<Question 34>

1. Arguments that the heavens are not moved by themselves: the heavens are not divided into two things one of which is a per se mover and the other of which is per se moved; hence, <the heavens are not moved by themselves>. Proof of the premise: in *On the Heavens* I Aristotle says that natural motion is either rectilinear, and this occurs either from the middle or towards the middle, or circular, such as the motion of the heavens.[165] Therefore, since this division is univocal, if one of the two kinds of natural motion is not divided

[163] See above, n. 2.
[164] Arist. *Phys.* VIII.5, 258a20–21.
[165] Arist. *De caelo* I.2, 268b17–24.

dividatur etc., similiter erit in reliquo[a]. Sed [b]motus naturalis rectus non dividitur in talia duo, per Aristotelem. Dicit enim Aristoteles quod gravia et levia non moventur ex se; ergo similiter nec [c]motus circularis [d]vel ergo multo fortius nec motus circularis, eo quod minus naturalis est [e]quam motus rectus.

2. Item, ad idem: si caelum divideretur sic etc., tunc esset dicere quod motor caeli esset in corpore caeli; sed hoc est falsum. Probatio, quia omne quod est in alio movetur per accidens ad motum illius, per Porphyrium[f] dicentem quod, moventibus nobis, moventur [g]et omnia quae in nobis sunt per accidens. Tunc, si intelligentia esset in corpore caeli, moveretur[h] per accidens; [i]hoc autem est falsum[j], secundum Aristotelem.

3. Solutio. Dicendo secundum theologos, dicemus quod non necesse est ponere caelum moveri ex se [k]nec est necesse secundum eos et secundum rei veritatem forte ponere intelligentiam esse motorem ibi, sed solum quod moveatur a sua forma corporali. Et hoc satis ultima ratio posita ad hanc partem convincit.

4. [l]Item, natura operatur in his inferioribus breviori via qua potest propter compendium, ita scilicet quod non ponit alium motorem hic inferius nisi solum [m]formam corporalem, ut patet in levi et gravi; ergo multo fortius natura operatur in natura superiore brevi **[O 117ra]** ori via qua potest propter suum compendium, ita scilicet quod non[n] ponet ibi motorem nisi solum [o]formam corporalem et quod solum movebitur a forma corporali. Ergo non movebitur ab intelligentia, quae est forma spiritualis; quare etc.

5. Dicendo tamen secundum philosophos, debemus hoc ponere. Philosophi enim, quia noluerunt[p] attribuere motum minus nobilem caelo quam alicui[q] corpori hic inferius, dixerunt caelum moveri ex se.

[a] reliquo *corr. s.l. ex* recto O; recto L; alio T
[b] motus naturalis rectus] motum motu recto T
[c] motus circularis] motum motu circulari T
[d] vel ... circularis *om. hom.* LT
[e] quam motus rectus *om.* T
[f] Porphyrium] *recte* Philosophum
[g] et ... moveretur *om.* L
[h] moveretur] movetur O
[i] hoc ... secundum] hoc est contra L; quod est contra T
[j] falsum *suppl. s.l.* O
[k] nec ... convincit *om.* T
[l] Item ... etc. *post* etc. (n. 10) *transp.* LT, *fort. recte*
[m] formam corporalem *corr. ex* forma corporalis O; forma corporalis L
[n] non *suppl. s.l.* O; *om.* L
[o] formam corporalem *corr. ex* forma corporalis O; forma corporalis L
[p] noluerunt *corr. ex* voluerunt O; voluerunt T
[q] alicui *ex corr.* O

<into a per se mover and a per se moved>, neither will the other kind be so divided. But rectilinear natural motion is not divided into two such things, according to Aristotle; for Aristotle says that heavy and light things are not moved by themselves.[166] Therefore, by an argument from similarity, neither is circular motion so divided; or alternatively, therefore there is a much stronger reason to say that circular motion is not so divided, because it is less natural than rectilinear motion.

2. Furthermore, another argument for the same conclusion is this: if the heavens were divided in this way, then it would be possible to say that the mover of the heavens is in the body of the heavens; but this is false. Proof: everything that is in another thing is moved accidentally with the motion of that thing, in accordance with Porphyry's statement that, when we are in motion, all the things that are in us are also moved accidentally.[167] Thus, if the intelligence were in the body of the heavens, it would be moved accidentally; but this is false, according to Aristotle.[168]

3. Solution. If we follow the theologians, we should say that it is not necessary to posit that the heavens are moved by themselves, nor is it necessary, according to them and perhaps according to the truth of the matter, to posit that an intelligence is the mover there, but only <to posit> that the heavens are moved by their corporeal form. And the last argument for this part of the question[169] is enough to prove this conclusion.

4. Furthermore, nature acts on the things here below by the shortest possible way for the sake of economy, that is, in such a way that nature does not place another mover here below apart from the corporeal form, as is clear in the case of heavy and light things; therefore, there is a much stronger reason to hold that nature acts on the higher natural realm by the shortest possible way for the sake of economy, that is, in such a way that nature will not place a mover there apart from the corporeal form and so the heavens are only moved by their corporeal form. Therefore, the heavens will not be moved by an intelligence, which is a spiritual form; hence, <the heavens are only moved by their corporeal form>.

5. If we follow the philosophers, however, we should posit that the heavens are moved by themselves; for, since the philosophers did not want to ascribe a less noble motion to the heavens than to a body here below, they said that the heavens are moved by themselves.

[166] Arist. *Phys*. VIII.4, 255a5–6.
[167] The correct reference is Arist. *Top*. II.7, 113a29–30.
[168] Arist. *Phys*. VIII.6, 258b13–16.
[169] See above, n. 2.

6. Ad primum argumentum in oppositum dicendum quod duplex est motus naturalis: vel a forma corporali naturali vel a natura dicta secundum proportionem. Caelum autem movetur utroque motu. Movetur enim a quadam forma naturali corporali. Ponimus enim quod corpus caeli habet inclinationem ad talem motum qui circularis est. Movetur etiam a quadam natura dicta secundum proportionem, cuiusmodi est intelligentia, sicut apprehendens vel appetens aliquid natura est. Intelligentia enim est appetens et volens movere caelum. Et iste appetitus quodam modo naturalis est, communiter et proportionaliter sumendo naturam.

7. Ad aliud argumentum patebit super quaestionem sequentem.

8. Ad principale adhuc probatio quod non dividitur etc., quia, si divideretur sic, tunc intelligentia etc.[a]; sed non est in caelo; quare etc. Probatio minoris: in genere animalium propter imperfectionem animae est quod non movet animal ad certum locum quietis et propter eius perfectionem est quod moveat[b] animal ad certum locum quietis, sicut dicit Aristoteles in tertio *De anima*, capitulo de movente, ubi dicit quod quaedam animalia moventur infinite, et hoc propter eorum imperfectionem [c]sive animae eorum, et quaedam finite, et hoc propter eorum[d] perfectionem[e], ut homo; unde anima movet corpus ad certum locum quietis propter suam perfectionem. Cum ergo intelligentia sit multo perfectius motor quam anima, tunc movebit suum mobile in partem determinatam et ad locum quietis determinatum; sed hoc non potest esse, eo quod caelum movetur secundum quamlibet suam[f] partem et semper et continue; necesse est ergo ponere quod intelligentia non[g] sit ibi motor.

9. Item, ita est de motore hic inferius quod non movet suum mobile immediate, sed mediate, ut mediante spiritu, [h]et hoc est propter suam nobilitatem. Cum ergo intelligentia sit multo nobilius motor quam anima, tunc[i] movebit suum mobile sive caelum mediante spiritu. Sed in caelo non sunt spiritus, quia caelum non est corpus complexionatum, et spiritus proveniunt ex complexione corporis; quare [j]intelligentia non potest movere caelum; et ita etc.

[a] etc. *suppl. s.l.* O; esset in caelo L; esset in caelo et motor eiusdem T
[b] moveat] aliquod *add. s.l.* O
[c] sive ... perfectionem *om. hom.* T
[d] eorum *suppl. s.l.* O; *om.* L
[e] perfectionem] cum ergo intelligentia sit multo perfectius *add. sed del.* O
[f] suam] sui LT
[g] non *suppl. s.l.* O
[h] et ... spiritu *om. hom.* T
[i] tunc] non *add. sed del.* O; non *add.* L
[j] intelligentia ... ita *om.* T

6. In reply to the first argument to the contrary[170] it must be said that there are two kinds of natural motion: (i) one that derives from a natural corporeal form and (ii) the other that derives from a nature said by analogy. The heavens are moved with each of these two kinds of motion; for they are moved by a natural corporeal form, since we posit that the body of the heavens has an inclination towards circular motion. The heavens are also moved by a nature taken in an analogical sense, in the way in which something such as an intelligence which apprehends or desires is called a nature; for the intelligence <in question> has the appetite and the will to move the heavens. And this appetite is in some sense natural, if we take nature in a broad and analogical sense.

7. The reply to the other argument[171] will be made clear in the following question.[172]

8. Still another argument to prove the main claim that the heavens are not divided <into a per se mover and a per se moved> is this: if they were divided in this way, then an intelligence <would be the per se mover>; but an intelligence is not <the per se mover> in the heavens; hence, <the heavens are not divided in this way>. Proof of the minor premise: in the genus of animals it is because of the imperfection of the soul that a soul does not move an animal towards a determinate place of rest and it is because of its perfection that it moves an animal towards a determinate place of rest, as Aristotle says in *On the Soul* III, in the chapter about the mover, where he says that some animals are moved endlessly, and this because of their imperfection or the imperfection of their soul, and some are moved in a finite way, and this because of their perfection, as in the case of man;[173] thus, the soul moves the body to a determinate place of rest because of its perfection. Therefore, since an intelligence is a much more perfect mover than a soul, it will move its mobile body towards a determinate part and towards a determinate place of rest; but this cannot be the case because the heavens are always and continuously moved in all their parts ; therefore, it is necessary to posit that an intelligence is not the mover in the heavens.

9. Furthermore, a mover here below is such that it does not move its mobile body immediately, but mediately, that is by means of a spirit, and this is because of its nobility. Therefore, since an intelligence is a much nobler mover than a soul, it will move its mobile body or the heavens by means of a spirit. But in the heavens there are no spirits because the heavens are not a composite body, whereas spirits derive from the composition of the body; hence, an intelligence cannot move the heavens; and so <the heavens are not divided into a per se mover and a per se moved>.

[170] See above, n. 1.
[171] See above, n. 2.
[172] See below, q. 35, nn. 16–21.
[173] Arist. *De anima* III.11, 433b31–434a21.

10. Item, Augustinus dicit quod intelligentia et anima sunt natura pares, officio autem impares; sed officium animae est movere corpus; ergo hoc non erit officium intelligentiae; quare etc.

11. Ad primum istorum argumentorum [a]ultimo positorum dicendum quod in omnibus eis quae moventur de loco ad locum totaliter, sicut est in motu animalium, finis motus est quies. Et sic est in istis inferioribus. In superioribus vero, ubi fit motus in loco et non de loco[b], ibi finis non est[c] quies, sed aliquid extrinsecum. Sed ille finis qui est aliquid extrinsecum assignatur dupliciter. Quidam enim dicunt quod ille finis est quatenus assimiletur causae primae hoc modo. Quia enim Deus, secundum Augustinum, praesentialiter, potentialiter et essentialiter est ubique, ideo caelum appetit similiter esse; sed quia caelum est corpus, ideo non potest esse[d] ubique[e]; appetit igitur ut sit in uno loco successive post alium.

12. Sed iste finis non est verus finis, ut mihi videtur, quia in fine saeculi tunc caelum magis assimilabitur causae primae, cum corpora fuerint[f] glorificata et quando caelum stabit. Suus ergo finis, dum movetur, non est ut assimiletur causae primae, eo quod magis assimilabitur causae primae post suum motum, ut quando quiescit[g].

13. Ideo est quidam alius[h] finis melior, ut mihi videtur, et assignatur[i] ab Aristotele in fine[j] *Caeli et mundi*, ubi Aristoteles quaerit quare multae sunt transmutationes in caelo. Et respondet dicendo quod hoc est propter multitudinem corporum hic inferiorum sive propter eorum diversam generationem. Eodem modo dicendum est a parte ista[k] quod finis motus caeli sive corporum supracaelestium est ut faciat generationem in istis corporibus inferioribus. Et propterea motus stellae sunt[l] ut illuminent visum nostrum[m] et etiam ut salvent generationem corporum hic inferiorum. Dicendum tunc quod haec est causa[n] praecisa.

14. Et per istam narrationem patet manifeste solutio ad argumentum primum.

[a] ultimo positorum *om.* T
[b] loco] ad locum et *add.* T
[c] est] finis *add. sed del.* O
[d] esse] simul *add. mg.* O
[e] ubique *corr. ex* ubi O
[f] fuerint *corr. ex* fuit (*lectio incerta*) O; sint T
[g] quiescit *lectio incerta* (*ex corr.*) O
[h] alius] motus *add. sed del.* O
[i] assignatur *corr. ex* assumatur O
[j] fine] *recte* secundo
[k] ista *suppl. s.l.* O
[l] sunt] et *add.* O
[m] nostrum] et visui nostro facissiant (*sic*) *add. sed del.* O; et visui nostro facissierant (*sic*) *add.* L
[n] causa *suppl. s.l.* O

10. Furthermore, Augustine says that an intelligence and a soul are equal by nature, but unequal by duty;[174] but the duty of a soul is to move the body; therefore, this will not be the duty of an intelligence; hence, <the heavens are not moved by an intelligence>.

11. To the first argument in this last series[175] it must be said that for all the things that are moved from place to place as a whole, as happens in the motion of animals, the end of their motion is rest. And this is the case for the things here below. In things above, however, where motion occurs in a place and not from a place, the end of their motion is not rest, but something extrinsic. But the extrinsic end of a motion is specified in two ways. (i) Some people say that that end is the mobile thing's being assimilated to the first cause in this way. Since, according to Augustine, God is everywhere by his presence, his power, and his essence,[176] the heavens desire to exist in the same way, but since the heavens are a body, they cannot be everywhere; therefore, they desire to be successively in one place after another.

12. It seems to me, however, that this kind of end is not the true end, because at the end of time the heavens will be more fully assimilated to the first cause, when bodies will be glorified and the heavens will cease to be in motion. Therefore, the end of the heavens, as long as they are moved, is not to be assimilated to the first cause, given that they will be more fully assimilated to the first cause after their motion <has ceased>, that is, when they will be at rest.

13. (ii) Therefore, there is another and better end, in my view, and this is specified by Aristotle at the end of *On the Heavens*, where Aristotle asks why there are many changes in the heavens.[177] His reply is that this happens because of the multitude of bodies here below or because of their different generations. In this case then one must say that the end of the motion of the heavens or of the celestial bodies is to produce generation in the bodies here below. Consequently, the motions of the stars occur in order to illuminate our sight and in order to preserve the generation of the bodies here below. It must be said, then, that this is the precise cause.

14. And with this explanation the solution of the first argument[178] is absolutely clear.

[174] Aug. *De lib. arb.* III.11, CCSL 29, 294.15–16.

[175] See above, n. 8.

[176] The correct reference is Peter Lombard, *Sententiae* I, dist. 37 c.1 § 5, ed. Coll. S. Bonav. IV, 265.2–4; based on *Glossa ord. in Cant*. 5.17, ibid., § 3, 264.6–8 and critical apparatus .

[177] The correct reference is Arist. *De caelo* II.4, 286b6–9.

[178] See above, n. 8.

15. Ad secundum dicendum quod causa quare anima movet mediante spiritu et non intelligentia caelum est quia corpus quod[a] movetur ab anima componitur ex partibus differentibus, ut ex quattuor elementis, et, post compositionem suam, habet partes diversimode organizatas. Propter ergo diversitatem partium componentium corpus et propter diversitatem partium corporis organicarum oportet ponere spiritum in corpore, qui spiritus habeat convenientiam cum omnibus partibus corporis et ut mediante illo spiritu possit anima movere quamlibet partem corporis et totum corpus secundum quamlibet speciem. Caelum vero non est corpus compositum sic ex diversis partibus, ut ex quattuor elementis, nec habet partes ita diversimode organizatas; et ideo non necesse est ponere ibi spiritum quatenus intelligentia moveat totum caelum mediante illo spiritu. Per istam causam patet responsio.

16. Ad tertium dicendum quod Augustinus non negat eos convenire in aliquo officio generali, sicut in eo quod est movere corpus, verbi gratia non negat eos convenire[b] in hoc quod utrumque debet contemplari suum creatorem, sed negat eos convenire in specialibus officiis quibusdam, verbi gratia anima movet corpus ad quietem, intelligentia vero non movet caelum ad quietem; praeterea, anima est perfectio corporis, intelligentia vero non est perfectio caeli.

17. Ad ultimum dicendum quod propter compendium est ut grave **[O 117rb]** et leve solum moveantur a forma corporali [c]et quod caelum moveatur et a forma corporali et a forma spirituali. Et huius causa est vilitas et nobilitas. Propter nobilitatem enim est quod caelum moveatur ita a quadam forma spirituali quatenus caelum sit motum ex se et dividatur in talia duo quae etc. Et hoc non posset esse nisi poneremus intelligentiam esse motorem[d]. Quatenus ergo caelum moveatur [e]ex se et [f]hoc propter sui nobilitatem, eo quod non est minus nobile[g] quam corpus hic inferius, quod movetur de se, ponimus ipsum esse[h] motum et a forma corporali et a forma spirituali. Propter vilitatem vero ipsius gravis est quod non movetur ex se; et ita propter eius vilitatem [i]non debemus ponere ipsum moveri a forma spirituali, quia, quod aliquid moveatur [j]ab aliqua forma spirituali, hoc est propter eius nobilitatem, et ita non propter vilitatem. Per [k]hanc causam patet responsio.

[a] quod *ex corr.* O
[b] convenire *corr. mg. ex* movere O
[c] et ... corporali *om. hom.* T
[d] motorem *ex corr.* O
[e] ex se *om.* T
[f] hoc propter *inv.* O
[g] nobile] nobilis OL
[h] esse *om.* O
[i] non debemus ponere *om.* L
[j] ab aliqua forma] a forma aliqua T
[k] hanc causam] hoc T

15. In reply to the second argument[179] it must be said that the reason why a soul moves <a body> by means of a spirit but the intelligence does not move the heavens <by means of a spirit> is that a body moved by a soul is composed of different parts, such as the four elements, and after its composition, it has parts organized in different ways. Therefore, because of the diversity of the composing parts and because of the diversity of the organic parts of the body it is necessary to posit a spirit in the body which agrees with all the parts of the body, so that by means of that spirit the soul can move both any part of the body and the body as a whole with any species <of motion>. The heavens, however, are not a body composed in this way of different parts, such as the four elements, nor do they have parts differently organized in this way; therefore, it is not necessary to posit a spirit in order for an intelligence to move the whole heavens by means of that spirit. With this explanation the reply is clear.

16. In reply to the third argument[180] it must be said that Augustine does not deny that a soul and an intelligence agree in some general duty, such as the role of moving the body—for example, he does not deny that they agree in that each of the two must contemplate its creator; but he denies that they agree in some special duties—for example, a soul moves a body towards rest, whereas an intelligence does not move the heavens towards rest; moreover, a soul is the perfection of a body, whereas an intelligence is not the perfection of the heavens.

17. To the last argument[181] it must be said that it is for the sake of economy that heavy and light things are only moved by a corporeal form and that the heavens are moved both by a corporeal form and by a spiritual form. And the cause of this is their ignobility and nobility respectively. It is because of their nobility that the heavens are moved by a spiritual form with the result that the heavens are moved by themselves and are divided into two things such that <one is a per se mover and the other is per se moved>. And this could not happen unless we posited that an intelligence was their mover. Therefore, in order for the heavens to be moved by themselves, and this because of their nobility, given that they are not less noble than a body here below, which is moved by itself, we posit that the heavens are moved both by a corporeal form and by a spiritual form. It is instead because of the ignobility of a heavy thing that a heavy thing is not moved by itself; and thus it is because of its ignobility that we do not have to posit that a heavy thing is moved by a spiritual form, given that something is moved by a spiritual form because of its nobility and not because of its ignobility. With this explanation the reply to the argument is clear.

[179] See above, n. 9.
[180] See above, n. 10.
[181] See above, n. 4.

1*. Habito quod caelum ex se moveatur, quaeritur de suis partibus.

<Quaestio 35>
Et primo de intelligentia utrum ipsa sit ubique tota in caelo.

1. Et videtur quod sic, quia inter omnia moventia et mota intelligentia maxime assimilatur causae primae et, praeter hoc, in qualibet parte ipsius caeli indifferenter potest motus incipere. [a]Sed sic non est in his corporibus inferioribus, quia non indifferenter potest motus incipere hic inferius, quia, per Aristotelem, incipit a dextra parte. Et similiter ista corpora inferiora, quae sunt mobilia, et animae rationales, quae sunt motores, sunt remotiores a causa prima quam ipsum caelum et intelligentia movens ipsum. Ergo ista corpora inferiora et motores hic inferius [b]minus assimilantur causae primae; sed motores hic inferius assimilantur in tantum quod sunt ubique in suis corporibus motis; ergo et[c] multo fortius intelligentiae in tantum assimilantur ipsi primo quod etc.

2. Item, quaelibet pars movetur in loco [d]et est in loco per motum totius; cum ergo quaelibet pars moveatur, constat quod non ab alio motore [e]quam a motore movente totum, et hoc est ab intelligentia. Sed movens et motum sunt simul etc.; ergo intelligentia est in qualibet parte tota, cum ab ipsa tota quaelibet pars moveatur.

3. Item, forma situalis, eo quod situalis est, non est ubique in toto corpore, ut patet de albedine et nigredine; ergo forma non situalis erit ubique in toto corpore. Sed intelligentia est maxime tale, scilicet non situale; quare etc.

4. Item, anima sensitiva, quae materialis est et operatur per instrumenta corporea, est ubique tota in corpore; sed intelligentia nec materialis est nec operatur per instrumenta corporea; quare multo fortius intelligentia[f] erit ubique [g]tota in corpore.

[a] Sed ... incipere *om. hom.* T
[b] minus ... inferius *om. hom.* T
[c] et *om.* LT
[d] et ... loco *om. hom.* T
[e] quam a motore *om. hom.* L
[f] intelligentia] intellectus L; in aliquo T
[g] tota in] in toto L

1*. Now that it has been established that the heavens are moved by themselves, questions are raised about their parts.

<Question 35>
And the first question is about the intelligence, whether it is everywhere as a whole in the heavens.

1. Arguments for the affirmative: of all the things that are movers and those that are moved, an intelligence is maximally similar to the first cause, and in addition to this, the motion of the heavens can begin in any of their parts indifferently. But this is not the case in the bodies here below because a motion here below cannot begin from any part indifferently, given that, according to Aristotle, it begins from the part on the right.[182] Similarly the bodies here below, which are mobile things, and the rational souls, which are their movers, are more remote from the first cause than are the heavens themselves and the intelligence that moves them. Therefore, the bodies here below and the movers here below are less similar to the first cause; but the movers here below are similar <to the first cause> in so far as they are everywhere in the bodies moved by them; therefore, there is a much stronger reason to say that intelligences are similar to the first cause to such an extent that <they are everywhere in the things moved by them>.

2. Furthermore, each and every part <of the heavens> is moved in a place and is in a place in virtue of the motion of the whole; therefore, since each and every part is moved, it is manifest that no part is moved by a mover other than the mover that moves the whole, and this mover is an intelligence. But the mover and the moved exist simultaneously; therefore, the intelligence is in each and every part as a whole, since each and every part is moved by the whole intelligence.

3. Furthermore, a form that has a position in so far as it has a position is not everywhere in the whole body, as is clear in the case of whiteness and blackness; therefore, the form that does not have a position will be everywhere in the whole body. But an intelligence is maximally of this kind, that is, without a position; hence, <the intelligence is everywhere in the heavens>.

4. Furthermore, the sensitive soul, which is material and acts through corporeal instruments, is everywhere as a whole in the body; but an intelligence is not material nor does it act through corporeal instruments; hence, there is a much stronger reason to say that an intelligence is everywhere as a whole in the body.

[182] Arist. *De progr. anim.* 4, 705b15–706a26.

5. Item, Commentator[a] in capitulo *De substantia orbis* dicit quod caelum est movens et intelligens, et in primo *Caeli et mundi* dicit Aristoteles quod caelum vivit vita nobilissima, et etiam illud dicit in libro *De vegetabilibus*. Sed partes vivunt vita totius, totum vero non vivit vita partium; ergo non debemus ponere aliquam partem determinatam quam movet intelligentia, ut per eius motionem sive vitam moveatur totum caelum; debemus ergo ponere intelligentiam esse ubique et in qualibet parte[b] movens totaliter.

6. Item, sit A illa pars determinata caeli in qua est intelligentia; ista pars, cum sit continua, est divisibilis in infinitum; quaero tunc an intelligentia sit in aliqua parte continua ipsius A vel in parte indivisibili. Si in parte continua, tunc, cum illa pars sit divisibilis, erit tunc in qualibet parte illius, et ita non in aliqua parte determinata. [c]Dato tunc quod in aliqua parte determinata, sequitur oppositum, scilicet quod non in aliqua parte determinata. Si in parte indivisibili, tunc, cum intelligentia solum [d]vivificet et moveat illud in quo est sicut anima, tunc solum vivificabit unam partem tantum, et ita non totum caelum, quod est inconveniens.

7. Item, si in uno puncto indivisibili esset intelligentia et non est maior ratio quare sit in uno quam in alio, tunc erit aut in quolibet[e] aut in nullo. Sed non in nullo, [f]per suppositionem nostram; ergo in quolibet[g], et ita sequitur oppositum suppositi, scilicet quod non in aliqua parte determinata. Quod vero non sit maior ratio quare sit in uno quam in alio, probatio: ille punctus, cum sit indivisibilis, est mathematicus. Punctus enim naturalis est divisibilis. Sed in mathematicis non est bonum; ergo, cum intelligentia sit bonum et nullus punctus indivisibilis est bonus, tunc non erit maior ratio quare erit in uno quam in alio.

8. Item, sit A illa pars in qua fuerit intelligentia; moveatur tunc et elevetur ista pars ab oriente in occidens; sequetur intelligentiam moveri per accidens, et hoc est inconveniens.

9. Item, est intelligentia suprema incausata, ut causa prima, quae movet[h] et est ubique, et est intelligentia causata et infima, ut anima, quae etiam est ubique in suo corpore quod movet; ergo, cum medium participet naturam extremorum, tunc intelligentia media, quae est motor caeli, erit ubique in caelo.

[a] Commentator] Aristoteles OL
[b] parte *om.* O
[c] Dato ... determinata *om. hom.* T
[d] vivificet et *om.* T
[e] quolibet *scripsimus*; qualibet OLT; et ita sequitur *add. sed del.* O
[f] per suppositionem nostram *om.* T
[g] quolibet *scripsimus*; qualibet OLT
[h] movet] movetur OL

5. Furthermore, in the chapter *On the Substance of the Orb* the Commentator says that the heavens are moving and understanding,[183] and in *On the Heavens* I Aristotle says that the heavens live the noblest life,[184] and he also says that in *On Plants*.[185] But parts live the life of their whole, whereas a whole does not live the life of its parts; therefore, we do not have to posit a determinate part which the intelligence moves, so that in virtue of the motion or life of this part the whole heavens are moved; therefore, we have to posit that the intelligence is everywhere and moving in every part as a whole.

6. Furthermore, let A be that determinate part of the heavens in which the intelligence is; since this part is continuous, it is divisible to infinity; then I ask whether the intelligence is (i) in a continuous part of A or (ii) in an indivisible part. (i) If it is in a continuous part, then, since that part is divisible, the intelligence will be in any part of it, and so it will not be in a determinate part. Granted that the intelligence is in a determinate part, then, the opposite conclusion follows, that is, that the intelligence is not in a determinate part. (ii) If the intelligence is in an indivisible part, then, since the intelligence only gives life to and moves the thing in which it is, as does the soul, it will give life to only one part, and so it will not give life to the whole heavens, which is absurd.

7. Furthermore, if the intelligence were in one indivisible point, then, since there is no more reason why it should be in one point rather than in another, it will be either in any point or in none. But it is not in none, by our assumption; therefore, it is in any point, and thus the opposite of the assumption follows, that is, that the intelligence is not in a determinate part. Proof that there is no more reason why the intelligence should be in one point rather than in another: since that point is indivisible, it is a mathematical point, given that a natural point is divisible. But the good does not exist in mathematical things; therefore, since an intelligence is a good, and no indivisible point is a good, there will be no more reason why the intelligence should be in one point rather than in another.

8. Furthermore, let A be the part in which the intelligence is found; let this part be moved and raised from east to west; it follows that the intelligence is moved accidentally, and this is absurd.

9. Furthermore, there is a supreme uncaused intelligence, that is, the first cause, which moves and is everywhere, and there is a caused and lowest intelligence, that is, the soul, which is also everywhere in the body which it moves; therefore, since what is intermediate participates in the nature of the extremes, then the intermediate intelligence, which is the mover of the heavens, will be everywhere in the heavens.

[183] Aver. *De sub. orb.* 1, fol. 5va.
[184] Arist. *De caelo* I.9, 279a20–22.
[185] Ps. Arist. *De plantis* I.1, 816a32–33.

10. Ad oppositum: si intelligentia sit ubique tota et in qualibet parte et vivificet quamlibet partem, tunc, cum suum vivificare sit intelligere, quaelibet pars intelligeret; hoc autem falsum est. Et idem est argumentum de anima rationali.

11. Item, anima sensitiva, existens ubique in toto corpore, vivificat ita totum corpus quod facit quamlibet partem sentire; ergo a simili anima rationalis et similiter intelligentia vivificant totum corpus ita quod faciunt quamlibet partem intelligere, dato quod sint ubique in suis corporibus. Sed hoc est falsum. Totum enim intelligit et non pars.

12. Item, si sit in qualibet parte, tunc est in B; elevetur tunc ista pars ab oriente in occidens; sequetur intelligentiam moveri[a] per accidens, quod est inconveniens.

13. Item, tota intelligentia est in A, et totum est extra quod nihil; [b]quare etc.

14. Solutio. Dicendum quod[c] intelligentia **[O 117va]** est ubique tota in caelo. Sed aliquid esse ubique, hoc est dupliciter, aut ita quod respiciat secundum suas operationes quamlibet partem et non solum totum, et sic est de anima sensitiva. Anima enim sensitiva est ita ubique quod in qualibet parte secundum suam operationem et secundum speciales operationes in specialibus partibus, ut secundum operationem sensitivam visivam in oculo, et sic de aliis partibus. Aut ita quod secundum suas operationes solum respiciat totum et non partes. Sed hoc potest esse dupliciter: aut ita quod sit eius motor et perfectio, et sic anima rationalis est ubique tota, aut ita quod sit motor tantum, et sic est intelligentia ubique in caelo.

15. Per haec dicta patet responsio ad duas primas rationes.

16. [d]Et ad tertium secundum quosdam. Dicunt enim quidam quod causa quare intelligentia non movetur per accidens hac parte B mota ab oriente in occidens est[e] quia tantum est eius motor et non perfectio.

17. [f]Alii autem dicunt sic, quod, quia intelligentia respicit totum primo et per se et non partes, et totum quiescit et numquam movetur nec per se nec per

[a] moveri] movere OL
[b] quare etc.] ergo nihil sui est in B T
[c] quod *suppl. s.l.* O
[d] Et ... enim] Ad tertium dicunt T
[e] est *suppl. s.l.* O; *om.* L
[f] Alii autem dicunt] alii dicendum *scrips.* O *sed corr. s.l. in* aliis videtur esse dicendum

10. To the contrary: if the intelligence is everywhere as a whole and in each and every part, and gives life to each and every part, then, since the life which an intelligence gives is understanding, each and every part would understand; but this is false. And the same argument holds for the rational soul.

11. Furthermore, the sensitive soul, which exists everywhere in the whole body, gives life to the whole body in such a way that it makes each and every part of the body have sense perception; therefore, by an argument from similarity, the rational soul and likewise an intelligence give life to the whole body in such a way that they make each and every part understand, given that they are everywhere in their bodies. But this is false, because it is the whole that understands and not a part.

12. Furthermore, if the intelligence is in each and every part, then it is in part B; then let this part be raised from east to west; it follows that the intelligence is moved accidentally, which is absurd.

13. Furthermore, <suppose that> the whole intelligence is in A, and a whole is that outside which there is nothing; hence, <the intelligence is not in a part distinct from A>.

14. Solution. It must be said that the intelligence is everywhere as a whole in the heavens. But something can be everywhere in two ways: either (i) in such a way that it is related through its operations to each and every part and not only to the whole. And this is the case of the sensitive soul, because the sensitive soul is everywhere in such a way that it is in each and every part in accordance with its <general> operation and in special parts in accordance with its special operations, for instance, it is in the eye in accordance with the sensitive visual operation, and the same applies to the other parts. Or (ii) <something can be everywhere> in such a way that it is related through its operations only to the whole and not to the parts. And this case is twofold: either (iia) <it is related> in such a way that it is the mover of the whole and also its perfection, and in this way the rational soul is everywhere as a whole; or (iib) <it is related> in such a way that it is only the mover, and in this way the intelligence is everywhere in the heavens.

15. From what has been said the reply to the first two arguments[186] is clear.

16. And the reply to the third argument[187] is also clear, according to some people, who say that the reason why the intelligence is not moved accidentally when part B is moved from east to west is that the intelligence is only that part's mover and not its perfection.

17. Other people, however, think that one should reply to the third argument by saying that since the intelligence is primarily and per se related to the whole and not to the parts, and the whole is at rest and is never moved

[186] See above, nn. 10–11.
[187] See above, n. 12.

accidens, ideo intelligentia, quia primo et per se respicit totum, [quod[a]] nullo modo[b] movetur. Et si ex consequenti respiciat partes et moveat ipsas, non tamen debet moveri.

18. Contra: eadem ratione non moveretur anima rationalis per se nec per accidens, cum ipsa primo et per se respiciat corpus totum et ex consequenti partes, quod falsum est.

19. Sed dicendum quod bene argueret si totum corpus hic inferius non moveretur[c], sed quiesceret, sicut est a parte caeli, sed sic non est.

20. Aliter dicitur sic quod est aliqua forma quae propter suam materialitatem et inseparabilitatem non potest esse in toto ita quod in qualibet[d] parte, cuiusmodi sunt albedo et nigredo et similia; et est alia forma quae propter sui immaterialitatem est in toto et in qualibet parte; ergo, ut videtur, erit reperire tertiam formam quae erit ita in toto quod et in nulla parte. Et talem formam ponimus intelligentiam.

21. Et sic patet qualiter possumus ponere quod intelligentia sit ubique in caelo, ita tamen quod [e]non in qualibet parte, aliter quam praedictum est.

<Quaestio 36>
Quaeritur utrum intelligentia moveatur per se vel per accidens.

1. Et videtur quod sic[f], quia omne quod est in alio, ipso moto per se, movetur per accidens; sed intelligentia est in caelo; ergo, ipso caelo moto per se, movetur per accidens suus motor sive intelligentia. Probatio minoris: dicit Aristoteles quod primum motum est motum ex se; ergo dividitur in talia duo quorum unum est per se movens et aliud per se motum, ut in ipsum caelum et intelligentiam; [g]sed nihil dividitur in illud quod est extra ipsum; cum ergo primum motum dividatur in caelum et intelligentiam, tunc intelligentia erit in ipso caelo, [h]et tunc de necessitate oportet ponere quod moveatur per accidens.

2. Praeterea, Aristoteles in fine octavi, determinans istam quaestionem 'Ubi est primus motor?', dicit quod in circumferentia; sed circumferentia est in caelo; ergo et intelligentia, et ulterius ut prius.

[a] quod *delevimus*
[b] modo *suppl. s.l.* O; *om.* L
[c] moveretur] moveret OL
[d] qualibet] aliqua O
[e] non ... est] in aliqua parte T
[f] sic] per accidens T
[g] sed ... intelligentiam *om. hom.* T
[h] et ... accidens] quare etc. T

either per se or accidentally, therefore, since it is primarily and per se related to the whole, the intelligence is in no way moved. And even if the intelligence is secondarily related to the parts and moves them, it does not follow that it is moved.

18. Against this: by the same reasoning, since the rational soul is primarily and per se related to the whole body and secondarily to the parts, it would not be moved per se or accidentally, which is false.

19. It must be said, however, that the argument would be valid if the whole body here below were not moved but were at rest, as happens in the case of the heavens, but this is not so in this case.

20. Another reply is that there is a kind of form that because of its materiality and inseparability cannot be in the whole in such a way that it is in every part, and whiteness, blackness and the like are of this kind; there is another kind of form that because of its immateriality is in the whole and in every part; therefore, it seems, there will be a third kind of form that is in the whole in such a way that it is in no part. And we posit that this kind of form is an intelligence.

21. And thus it is clear how we can posit that the intelligence is everywhere in the heavens, but in a way different from that posited above, for it is not in any of their parts.

<Question 36>
The question is whether the intelligence is moved either per se or accidentally.

1. Arguments for the affirmative: everything that is in another thing is moved accidentally, when that other thing is moved per se; but the intelligence is in the heavens; therefore, when the heavens are moved per se, their mover or the intelligence is moved accidentally. Proof of the minor premise: Aristotle says that the first moved thing is moved by itself;[188] therefore, it is divided into two things such that one is a per se mover and the other is per se moved, that is, into the heavens themselves and the intelligence; but nothing is divided into something that is outside it; therefore, since the first moved thing is divided into the heavens and the intelligence, the intelligence will then be in the heavens themselves, and thus it is necessary to posit that it is moved accidentally.

2. Furthermore, at the end of *Physics* VIII in replying to the question 'where is the first mover?', Aristotle says that it is in the circumference;[189] but the circumference is in the heavens; therefore, the intelligence is also in the heavens; and then one can go on to infer as before <that the intelligence is moved accidentally>.

[188] Arist. *Phys.* VIII.6, 259a32–33.
[189] Arist. *Phys.* VIII.10, 267b6–9.

3. Item, intelligentia aut est in una parte caeli aut in qualibet; sed sive sic sive sic, semper sequitur quod intelligentia per accidens moveatur per motum talis partis vel per motum[a] cuiuslibet.

4. Item, dicit Aristoteles in libro *De animalibus* quod intelligentia movens caelum est in caelo sicut spiritus movens semen est in semine; [b]sed spiritus movens semen ita est in semine quod, moto isto semine per se, movebitur et ista virtus per accidens; ergo intelligentia movens caelum ita erit in caelo [c]quod, caelo moto per se, movebitur intelligentia per accidens.

5. Item, sicut est in minori mundo, sic[d] erit in maiori; sed ita est in minori mundo quod, si corpus animalis moveatur per se, suus motor movetur per accidens; ergo similiter erit in maiori mundo quod, moto caelo per se, movebitur intelligentia per accidens.

6. Item, Aristoteles dicit in octavo quod motores orbium inferiorum moventur per accidens. Et Commentator exponit Aristotelem dicens quod huius causa est propter motum orbis superioris sive propter motum magni firmamenti, [e]quia per motum magni firmamenti moventur motores[f] orbium inferiorum ab oriente in occidens. Cum ergo orbis primus per se moveatur in illam partem eandem in quam movetur orbis inferior per accidens, secundum Commentatorem, quia per motum magni firmamenti, tunc multo fortius, ut videtur, motor primi orbis [g]movebitur per accidens quam motor orbis inferioris.

7. Item, qualitercumque animal moveatur, sive per se sive per accidens, nihil facit ad hoc quod motor eius moveatur per accidens, verbi gratia, sive animal moveatur ex se sive feratur ab alio, semper motor eius movetur per accidens. Ergo sic est a parte primi mobilis.

8. Ad oppositum: Aristoteles probat quod nec movetur per se nec per accidens.

9. Item[h], nihil quod movetur per accidens movetur continue et aeternaliter; cum igitur motor primi orbis moveatur continue et eternaliter et[i] de necessitate, non per accidens movebitur.

[a] motum] corporali *add. sed del.* O
[b] sed ... semen *om.* L
[c] quod ... accidens] etc. T
[d] sic] similiter LT
[e] quia ... firmamenti *om. hom.* L
[f] motores] hic *add. sed del.* O
[g] movebitur ... orbis *om. hom.* T
[h] Item] et O
[i] et *om.* LT

3. Furthermore, the intelligence is either in one part of the heavens or in each and every part; but in either case it follows that the intelligence is moved accidentally either because of the motion of that part or because of the motion of every part.

4. Furthermore, in *On Animals* Aristotle says that the intelligence moving the heavens is in the heavens just as the spirit moving the semen is in the semen;[190] but the spirit moving the semen is in the semen in such a way that, when the semen is moved per se, that power will also be moved accidentally; therefore, the intelligence moving the heavens will be in the heavens in such a way that, when the heavens are moved per se, the intelligence will be moved accidentally.

5. Furthermore, as it is in the lesser world, so it will be in the greater world; but in the lesser world it is the case that, if the body of an animal is moved per se, its mover is moved accidentally; therefore, the same will apply in the greater world, namely that, when the heavens are moved per se, the intelligence will be moved accidentally.

6. Furthermore, in *Physics* VIII Aristotle says that the movers of the inferior orbs are moved accidentally.[191] And the Commentator explains Aristotle's claim by saying that this happens because of the motion of the superior orb or because of the motion of the great firmament, given that the movers of the inferior orbs are moved from east to west because of the motion of the great firmament.[192] Therefore, since, according to the Commentator, the first orb is moved per se towards that part towards which an inferior orb is moved accidentally, and given that this happens because of the motion of the great firmament, then, it seems, there is a much stronger reason to say that the mover of the first orb will be moved accidentally than <to say this of> the mover of the inferior orb.

7. Furthermore, in whatever way an animal is moved, either per se or accidentally, this does not in any way affect the fact that its mover is moved accidentally. For example, whether an animal is moved by itself or whether it is carried by another thing, its mover is always moved accidentally. Therefore, the same happens in the case of the first mobile thing.

8. To the contrary: Aristotle proves that the intelligence is moved neither per se nor accidentally.[193]

9. Furthermore, nothing that is moved accidentally is moved continuously and eternally; therefore, since the mover of the first orb is moved continuously and eternally and necessarily, it will not be moved accidentally.

[190] Not found.
[191] Arist. *Phys.* VIII.6, 259b 28–31.
[192] Aver. *In Phys.* VIII, t.c. 52, fols. 392vb–393ra.
[193] Arist. *Phys.* VIII.6, 259b20–26.

10. [a]Item, Commentator dicit quod motores animalium inferiorum moventur per accidens, quia determinantur et constituuntur per sua subiecta; cum ergo motor primi orbis non constituitur per suum subiectum, tunc, per Commentatorem, non movebitur per accidens.

11. [b]Quaestioni de motore primo utrum moveatur etc. dicunt quidam sic, quod duplex est motor: coniunctus, ut intelligentia, et talis motor potest moveri per se vel per accidens, nec negat hoc Aristoteles; alius est motor separatus, ut causa prima, et de hac dicit Aristoteles quod non movetur per se vel per accidens.

12. Sed quod[c] ista responsio sit contra mentem Aristotelis et Commentatoris probatio, quia Aristoteles dicit quod caelum est motum ex se; ergo dividitur in etc.; ergo, ut videtur, Aristoteles intelligit de [d]motore in quod dividitur sive de motore coniuncto, de quo dicit statim quod neque movetur **[O 117vb]** per se nec per accidens.

13. Item, si hoc esset verum, tunc adeo bene diceret Commentator quod motor primus primi orbis per accidens moveatur sicut et motores orbium inferiorum.

14. Patet tunc plane quod ista responsio est[e] contra Commentatorem et[f] [g]etiam contra Aristotelem, et ideo non est approbanda.

15. Ideo, ut mihi videtur, dicendum [h]est aliter et sic, quod motor coniunctus, ut intelligentia, non movetur per se nec per accidens. Et huius causa est duplex: una quae praetacta est, scilicet quia motor caeli respicit ipsum totum caelum unde totum est, et totum unde totum est immobile; et ideo non debet moveri per se nec per accidens, quia suum mobile quod respicit non movetur per se nec per accidens, et ideo nec eius motor.

16. Alia causa talis est, secundum Commentatorem, quod, quia motores[i] orbium inferiorum constituuntur per sua subiecta, [j]hoc est definiuntur per sua subiecta, ut patet—anima enim est actus corporis physici organici—, ideo oportet ponere quod isti[k] motores moveantur per accidens. Non est autem ita in motore primi mobilis[l], quia non constituitur nec substentatur per suum subiectum, sed potius econverso. Pono enim quod intelligentia substentat

[a] Item ... movebitur per accidens *om.* L
[b] Quaestioni ... etc. *om.* T
[c] quod *corr. ex* quia O; quia L
[d] motore ... de] tali T
[e] est] communis (*lectio incerta*) *add. sed del.* O
[f] et *iter.* O
[g] etiam ... approbanda] Aristotelem T
[h] est ... sic *om.* T
[i] motores *ex corr.* O
[j] hoc ... subiecta *om. hom.* T
[k] isti *corr. ex* iste O; iste L
[l] mobilis] vel orbis *add.* T

10. Furthermore, the Commentator says that the motors of the animals here below are moved accidentally, because they are determined and constituted by their subjects;[194] therefore, since the mover of the first orb is not constituted by its subject, then, according to the Commentator, it will not be moved accidentally.

11. In reply to the question about the first mover, that is, whether it is moved <per se or accidentally>, some people say that there are two kinds of mover: (i) a conjoined mover, such as an intelligence; and this mover can be moved either per se or accidentally, and Aristotle does not deny this; (ii) a separated mover, such as the first cause; and of the first cause Aristotle says that it is moved neither per se nor accidentally.

12. Proof that this reply is against Aristotle's and the Commentator's intention: Aristotle says that the heavens are moved by themselves;[195] therefore, they are divided <into a per se mover and a per se moved>; therefore, it seems, Aristotle is referring to the mover into which the heavens are divided, that is, the conjoined mover, about which he immediately says that it is moved neither per se nor accidentally.[196]

13. Furthermore, if this reply were true, then the Commentator could equally well say that the first mover of the first orb is moved accidentally just as <he said that of> the movers of the inferior orbs.

14. It is absolutely clear then that this reply is contrary to the Commentator and also contrary to Aristotle, and therefore we should not accept it.

15. Therefore, in my view, this question must receive another reply as follows: a conjoined mover, such as the intelligence, is moved neither per se nor accidentally. And there are two reasons for this: (i) one, which has been touched upon before, is that the mover of the heavens is related to the whole heavens in so far as they are a whole, and the whole in so far as it is a whole is immobile; therefore, the mover of the heavens cannot be moved either per se or accidentally, since the mobile thing to which it is related is not moved either per se or accidentally, and therefore neither is its mover.

16. (ii) The other reason, according to the Commentator, is that since the movers of the inferior orbs are constituted by their subjects, that is, they are defined by their subjects, as is clear—for the soul is the act of an organic natural body—it is therefore necessary to posit that these movers are moved accidentally.[197] But this is not the case for the mover of the first mobile thing,

[194] Aver. *In Phys*. VIII, t.c. 52, fols. 392vb–393rb.
[195] This is probably a reference to Arist. *Phys.* VIII.6, 259a32–33 .
[196] Arist. *Phys*. VIII.6, 258b13–16.
[197] See above, n. 10.

caelum et non econverso; unde pono quod intelligentia sit in caelo, tamen non substentatur nec defertur[a] per caelum sicut anima substentatur per corpus. Et ideo pono quod possit esse eius motor coniunctus, videlicet quod sit in caelo, et tamen quod nec moveatur per se nec per accidens.

17. Et illud idem est quod dicitur sub aliis verbis quod est duplex motor: vel ut perfectio et motor, [b]et sic est anima motor, et de tali motore oportet quod moveatur per accidens; vel ut motor tantum [c]et non ut perfectio. Et[d] hoc est dupliciter. Aut enim defertur[e] et substentatur per suum mobile, sicut est de nauta in navi—licet enim nauta non sit nisi motor tantum et non perfectio, quia non ponit navem in esse, tamen substentatur et defertur[f] per motum navis—, et de tali adhuc est verum quod [g]movetur per accidens. Aut enim non defertur nec substentatur[h] per suum mobile, et sic est intelligentia motor caeli, et de tali non est verum quod moveatur per se vel per accidens.

<Quaestio 37>

Quia suppositum est in ista responsione quod motor primi orbis sit motor tantum et non perfectio, ideo quaeratur utrum sit eius perfectio.

1. Et videtur quod sic: nobiliori perfectibili nobilior debetur perfectio; sed caelum est corpus nobilissimum; ergo ei debetur perfectio nobilissima. Sed aliquod corpus vilius, ut corpus hic inferius[i], perficitur anima, [j]quae vita est; ergo[k] corpus caeli[l] non perficitur anima[m]; oportet ergo ponere quod intelligentia, quae vita est[n].

2. Item, motor tantum motor non vivificat suum mobile, quia, si sic, tunc navis diceretur[o] viva[p], quod falsum est; cum ergo caelum dicatur vivum, oportet[q] ponere quod motor eius non sit tantum motor, sed etiam perfectio.

[a] defertur *corr. ex* definitur O
[b] et ... motor] ut anima est motor etc. T
[c] et ... perfectio *om.* T
[d] Et] motor *add. sed del.* O
[e] defertur *corr. ex* definitur O
[f] defertur *corr. ex* definitur O
[g] movetur per accidens *ex corr.* O
[h] substentatur] substantur (*corr. ex* substatur) O
[i] inferius *ex corr.* O
[j] quae … anima *iter hom.* L (quod est caelum *pro* caeli)
[k] ergo] sed *ex corr.* O
[l] caeli *ex corr.* O
[m] anima] sed aliquo nobiliori *add.* T
[n] est] quare etc. *add.* T
[o] diceretur] dicetur O
[p] viva *ex corr.* O
[q] oportet *ex corr.* O

because it is not constituted or upheld by its subject, but rather the other way around; for I posit that the intelligence upholds the heavens and not the other way around. Thus, I posit that the intelligence is in the heavens, yet it is not upheld or carried by the heavens in the way that the soul is upheld by the body. And therefore I posit that the intelligence can be the conjoined mover of the heavens, that is, that the intelligence is in the heavens, and yet it is moved neither per se nor accidentally.

17. This same position can be formulated in different words by saying that there are two kinds of mover: (i) one which is both a perfection and a mover; and the soul is a mover of this kind, and of this kind of mover it is necessary to say that it is moved accidentally; (ii) another which is only a mover and not a perfection. And case (ii) is twofold: either (iia) that mover is carried and upheld by its mobile thing, as happens to the sailor in the ship—for, although the sailor is only a mover and not also a perfection, because he does not bring the ship into being, he is, however, upheld and carried by the motion of the ship—and of this kind of mover it is still true to say that it is moved accidentally. Or (iib) that mover is not carried or upheld by its mobile thing; and in this way the intelligence is the mover of the heavens; and of this kind of mover it is not true to say that it is moved either per se or accidentally.

<Question 37>

Since it has been assumed in this reply that the mover of the first orb is only its mover and not its perfection, the question now is whether it is its perfection.

1. Arguments for the affirmative: to a nobler perfectible thing a nobler perfection is due; but the heavens are the noblest body; therefore, the noblest perfection is due to them. But a more ignoble body, such as a body here below, is perfected by the soul, which is life; therefore, the body of the heavens is not perfected by a soul; therefore, it is necessary to posit that it is perfected by an intelligence, which is life.

2. Furthermore, the mover that is only a mover does not cause its mobile thing to be alive, because, if so, then the ship would be said to be alive, which is false; therefore, since the heavens are said to be alive, it is necessary to posit that their mover is not only a mover, but also their perfection.

3. Item, nihil dicitur moveri ex se quod habet motorem tantum qui non est eius perfectio; cum igitur caelum moveatur ex se, [a]ut dicit Aristoteles, tunc motor eius, ut intelligentia, non erit motor tantum, sed[b] perfectio.

4. Item, verior est unio intelligentiae cum corpore caeli quam animae rationalis cum corpore hominis; sed verior est unio eius quod unitur ut perfectio et motor quam eius quod[c] unitur ut motor tantum; ergo intelligentia unietur corpori caeli ut motor et perfectio. Probatio maioris: verior est unio eius quod unitur [d]sine medio quam eius quod unitur per multa media; sed anima rationalis unitur corpori mediante spiritu et[e] mediantibus multis transmutationibus; intelligentia vero unitur caelo sine omni medio; quare etc.

5. Item, est reperire intellectum unitum corpori ut motor et perfectio; si ergo contingat reperire unum per se[f], et alterum; tunc, si in caelo esset intelligentia tantum motor, esset reperire alicubi ubi tantum esset perfectio; hoc autem est falsum; quare etc.

6. Item, in genere sensitivae est reperire quod multae sensitivae secundum speciem diversae unitae sunt ut perfectiones et motores eorum[g] quorum sunt; similiter ergo[h] erit reperire in genere intellectivae quod multae intellectivae diversae secundum speciem uniantur ut motores et ut perfectiones[i]; sed hic inferius non sunt istae diversae animae secundum speciem; oportet ergo ponere quod intelligentia sit aliqua illarum; quare etc.

7. Item, in genere formarum ita est quod, quanto forma est[j] magis materialis, tanto minus perficit[k], ut patet in formis elementaribus; sed intelligentia est forma maxime immaterialis; ergo maxime perficiet suum mobile; quare etc.

8. Ad oppositum: Aristoteles in libro *De animalibus* dicit quod intelligentia est in caelo sicut motor in semine ante membrorum organizationem; sed spiritus[l] tunc existens est ibi tantum ut motor et non ut[m] perfectio; [n]ergo intelligentia erit similiter [op]in caelo tantum ut motor et non ut perfectio.

[a] ut dicit Aristoteles *om.* T
[b] sed] et *add. s.l.* O
[c] quod *om.* O
[d] sine … unitur] cum medio vel T
[e] et] in *add. sed del.* O
[f] se *ex corr.* O
[g] eorum] in quibus sunt vel *add.* T
[h] ergo *suppl. s.l.* O
[i] perfectiones] perfectiores OL
[j] est *suppl. s.l.* O; *om.* LT
[k] perficit] et e contrario *add. s.l.* O
[l] spiritus] ibi *add. s.l.* O
[m] ut *suppl. s.l.* O; *om.* L
[n] ergo … perfectio *iter. hom.* L (in caelo *om.*)
[o] in caelo *suppl. s.l.* O; *om.* L
[p] in ... perfectio *om.* T

3. Furthermore, nothing is said to be moved by itself when it has a mover that is only its mover and not its perfection; therefore, since the heavens are moved by themselves, as Aristotle says,[198] then their mover, that is, the intelligence, will not be only a mover, but also a perfection.

4. Furthermore, the union of the intelligence with the body of the heavens is truer than the union of a rational soul with the body of a man; but the union of something that is united as a perfection and a mover is truer than the union of something that is united only as a mover; therefore, the intelligence is united to the body of the heavens as their mover and perfection. Proof of the major premise: the union of something that is united without a medium is truer than the union of something that is united through many media; but a rational soul is united to a body by means of a spirit and by means of many changes, whereas the intelligence is united to the heavens without any medium; hence, < the union of the intelligence with the heavens is truer than the union of a rational soul with a body>.

5. Furthermore, one can find an intellect united to a body as its mover and perfection; therefore, if one can find one of the latter two by itself, then one can also find the other by itself; hence, if the intelligence were in the heavens only as a mover, one would be able to find something in which the intelligence was only a perfection; but this is false; hence, <the intelligence is not only the mover of the heavens>.

6. Furthermore, in the genus of the sensitive soul one can find that many sensitive souls different in species are united as the perfections and movers of the things to which they belong; therefore, similarly, in the genus of the intellectual soul one would be able to find that many intellectual souls different in species were united as movers and as perfections; but here below there are no intellectual souls that are different in species; therefore, it is necessary to posit that the intelligence is one of them; hence, <the intelligence is united to the heavens as their mover and perfection >.

7. Furthermore, in the genus of forms it is the case that, the more material a form is the less perfection it gives, as is clear in the elementary forms; but an intelligence is a maximally immaterial form; therefore, it maximally gives perfection to its mobile thing; hence, <the intelligence is united as a perfection to the heavens>.

8. Arguments to the contrary: in *On Animals* Aristotle says that the intelligence is in the heavens in the way that the mover is in the semen before the organization of the <body's> members;[199] but the spirit that exists in the semen at that time is there only as a mover and not as a perfection; therefore, the intelligence will similarly be in the heavens only as a mover and not as a perfection.

[198] This is probably a reference to Arist. *Phys.* VIII.6, 259a32–33 .
[199] Not found.

9. Item, forma quae magis est abstracta minus unitur; sed intelligentia magis abstrahitur quam intellectiva hominis; [a]et intellectiva hominis unitur ut perfectio et motor; ergo intelligentia minus, et ita ut motor tantum.

10. Solutio. Dicendum quod est in caelo ut motor tantum.

11. Ad primum in oppositum[b] quod ei debetur nobilior perfectio in suo genere, et illa perfectio est forma corporalis et non spiritualis[c].

12. Vel dicendum, secundum quosdam, quod[d] minor falsa est, ista scilicet, quod caelum est corpus nobilius[e] quam corpus hominis, quia quidam dicunt quod incorruptibilitas nihil facit ad nobilitatem, quia dicunt quod lapis tunc esset nobilissima creatura. Unde ipsi dicunt quod, considerando corpus humanum quoad [f]nobilem complexionem et harmoniam quae additur super corpus caeli, sic est corpus nobilissimum [g]sive considerando condiciones corporis humani quae adduntur super corruptibilitatem et considerando illas invenimus ipsum corpus humanum nobilissimum.

13. Ad alias **[O 118ra]** rationes dicendum quod aliquid potest esse motor tantum dupliciter. Aut enim, praeter hoc quod movet mobile, regit ipsum in existendo, et sic intelligentia est motor tantum; alius est motor tantum qui non substentat suum mobile nec regit ipsum [h]in existendo, sed solum regit ipsum in movendo, ut nauta in navi. De primo motore potest dici quod quoquo modo vivificat suum mobile, de secundo motore non. Et [i]de hoc secundo motore[j] concludunt rationes.

<Quaestio 38>
Quaeritur utrum intelligentia movens caelum moveat ipsum cum tota sua potentia.

1. Et videtur quod non, quia, si corpus caeli esset maius[k] quam nunc est, adhuc intelligentia moveret ipsum; ergo non movet ipsum nunc secundum suam totam potentiam. Probatio praemissae: ita est in istis inferioribus quod quodlibet moventium inferiorum in animalibus, etsi corpus motum esset maius quam nunc est, dummodo esset bene dispositum sive complexionatum, adhuc moveret ipsum; ergo similiter erit in superioribus etc.

[a] et intellectiva hominis *om. hom.* T
[b] oppositum] dicendum *add.* LT
[c] spiritualis] propter sui nobilitatem *add.* OLT *sed del.* O
[d] quod *ex corr.* O
[e] nobilius *ex corr.* O
[f] nobilem complexionem] nobilitatem complexionum T
[g] sive ... nobilissimum *scrips. sed del.* O
[h] in ... movendo] nisi in movendo tantum in modo L
[i] de ... motore] hoc T
[j] motore *ex corr.* O
[k] maius *ex corr.* O

9. Furthermore, a form that is more abstract is less united <to its perfectible thing>; but an intelligence is more abstract than the intellectual soul of a man; and the intellectual soul of a man is united <to a man> as a perfection and a mover; therefore, the intelligence is less united <to the heavens>, and thus it is united to them only as a mover.

10. Solution. It must be said that the intelligence is in the heavens only as a mover.

11. To the first argument to the contrary[200] it must be said that a nobler perfection is due to the heavens within their genus, and the perfection in question is a corporeal form and not a spiritual one.

12. Alternatively, according to some people we should say that that minor premise is false, that is, that the heavens are a nobler body than the body of a man. They say that this is because incorruptibility contributes nothing to nobility, on the grounds that, if it did, then stones would be the most noble creatures. Hence, these people say that, if one considers the human body with respect to its noble constitution and harmony, which are properties added over and above those belonging to the body of the heavens, the human body is the most noble body, or if one considers the conditions of the human body that are added over and above its corruptibility, one finds that the human body is the most noble.

13. In reply to the other arguments[201] one should say that something can be only a mover in two ways: either (i) when, in addition to moving the mobile thing, it directs the mobile thing in its existence; and in this way the intelligence is only a mover; or (ii) when it does not uphold its mobile thing or direct it in its existence, but only directs it in its motion, as does the sailor in the ship. The claim that a mover in some way gives life to its mobile thing is true of a mover of kind (i), but not of a mover of kind (ii). And the arguments are conclusive about a mover of kind (ii).

<Question 38>
The question is whether the intelligence that moves the heavens moves them with the whole of its power.

1. Arguments for the negative: if the body of the heavens were bigger than it is now, an intelligence would still move it; therefore, an intelligence does not now move it with the whole of its power. Proof of the premise: in the things here below any mover of an animal would still move it even if the moved body were bigger than it is now, as long as the body was well disposed or constituted; therefore, the same will apply to the superior things.

[200] See above, n. 1.
[201] See above, nn. 2–7.

2. Item, si corpus caeli habet inclinationem ad tantam velocitatem quantam nunc habet, tunc motor nihil apponit de sua potentia, ut, si lapis descendat deorsum secundum tantam velocitatem quantam habet ex cursu naturali sive motu naturali, si ei coniungatur manus aliqua, illa manus non confert ad eius motum deorsum. Si autem non habeat inclinationem ad tantam velocitatem quantam nunc habet, tunc resisteret suo motori, et tunc accideret fatigatio in eius motu, quia dicit Aristoteles quod animalia patiuntur fatigationem in suo motu, quia mobile resistit suo motori. Sed non est ponere secundum membrum, scilicet quod caelum non habeat inclinationem etc., quia tunc ibi esset fatigatio, quod inconveniens est ponere. Debemus tunc ponere primum membrum; relinquitur ergo quod intelligentia nihil apponat de sua potentia.

3. Item, ita est in motoribus animatis et vivis, qui movent corpus animatum et vivum, quod, ubi[a] moventur corpora[b] cum tota potentia motoris, ibidem de[c] necessitate fit labor et taedium; ergo similiter erit in motoribus supracaelestium[d] quod, si intelligentia moveret caelum cum tota sua potentia, accideret ei labor et taedium, quod inconveniens est [e]ponere; quare etc.

4. Item, intelligentia movens caelum potest movere ipsum citius quam movet; ergo non movet ipsum cum tota sua potentia. Probatio praemissae: ponamus quod rota aliqua fuerit circa caelum a centro ad circumferentiam. Moveatur ita tarde quod compleat suum motum in uno die naturali; tunc motus caeli et motus istius [f]rotae materialis sunt aequales, cum sint termini eiusdem [g]continui. Sed caelum potest citius moveri[h] qualibet re materiali[i]; ergo intelligentia potest movere ipsum citius[j] quam movet, ut patet per istam imaginationem; quare etc.

5. Item, imaginemur[k] circulum magnum in medio caeli et circulum alium parvum prope iuxta polum. Isti circuli, cum sint termini eiusdem continui, aequaliter et aeque velociter moventur; sed constat quod circulus maior velocius movetur, quia maius[l] spatium pertransit; quare etc.

[a] ubi] mota *add. s.l.* O
[b] corpora *om.* OL
[c] de *suppl. s.l.* O
[d] supracaelestium] supracaelestibus T
[e] ponere; quare etc. *om.* T
[f] rotae materialis] potentiae naturalis T
[g] continui. Sed *ex corr.* O
[h] moveri *scripsimus*; movere OLT
[i] materiali] naturali LT
[j] citius *corr. ex* totius O
[k] imaginemur *scripsimus*; imaginamur OT; imaginatur L
[l] maius] magis OL

2. Furthermore, if the body of the heavens has the inclination to be moved with as much velocity as that with which it is now moved, then its mover does not contribute anything of its own power, just as, if a stone were being moved downward with as much velocity as that which it has from its natural course or natural motion, then if a hand were joined to it, that hand would not contribute to the downwards motion of that stone. If the body of the heavens, however, did not have the inclination to be moved with as much velocity as that with which it is now moved, then it would offer resistance to its mover, and thus weariness would occur in its motion, because Aristotle says that animals suffer fatigue in their motion because a mobile thing offers resistance to its mover.[202] But this second hypothesis, that is, that the heavens do not have the inclination <to be moved with as much velocity as that with which they are now moved> cannot be posited, because otherwise they would experience fatigue, which would be an absurd supposition. We must then posit the first hypothesis; therefore, we are left with the conclusion that the intelligence contributes nothing of its power.

3. Furthermore, where animate and living movers, which move an animate and living body, are concerned, it is the case that when bodies are moved with the whole of power of the mover, fatigue and weariness necessarily occur; therefore, the same would apply to the movers of the celestial bodies so that, if the intelligence moved the heavens with the whole of its power, the heavens would be subject to fatigue and weariness, which it is absurd to posit; hence, <the intelligence does not move the heavens with the whole of its power>.

4. Furthermore, the intelligence moving the heavens is able to move them faster than it does now; therefore, it does not move them with the whole of its power. Proof of the premise: let us posit that there is a wheel in the heavens extending from the centre to the circumference. Let that wheel be moved so slowly that it completes its motion in one natural day. In this case the motion of the heavens and the motion of the material wheel are equal, since they are termini of the same continuum; but the heavens can move faster than any material thing; therefore, the intelligence can move the heavens faster than it does now, as this imaginary example makes clear; hence, <the intelligence does not move the heavens with the whole of its power>.

5. Furthermore, let us imagine a big circle in the middle of the heavens and another small circle close to the pole. Since these two circles are termini of the same continuum, they are moved equally and equally fast; but it is apparent that the bigger circle moves faster, because it traverses more space; hence, <the intelligence does not move the heavens with the whole of its power>.

[202] Arist. *De caelo* II.1, 284a26–33.

6. Item, ignobiliora[a] semper movent cum plus potentiae et cum tota potentia, ut patet de [b]gravi, nobiliora vero movent cum voluerint, ut patet[c] de homine. Eo enim quod nobilius est lapide, ideo movet [d]quando voluerit et cum plus et minus suae potentiae. Ergo, cum caelum fuerit corpus nobilissimum, tunc maxime movebit cum voluerit et cum plus et minus suae potentiae; quare etc.

7. Item, Aristoteles in octavo dicit quod motor primus est infinitae potentiae; ergo movet secundum infinitam potentiam; sed omnis talis motus est in instanti; sequitur tunc quod, si caelum moveatur a suo motore secundum totam suam potentiam, quod caelum movetur in instanti, quod est inconveniens; quare etc.

8. Ad oppositum: ita est in istis inferioribus quod, quando aliquid movet secundum[e] plus et minus, ibi est fatigatio; ergo similiter erit in superioribus. Necesse est tunc ponere quod secundum totam suam potentiam etc.[f]

9. Item, omne voluntarium, si moveat secundum minus suae potentiae, potest movere secundum totam suam potentiam; cum ergo intelligentia sit voluntarium agens, tunc, si moveat secundum minus suae potentiae, movebit et secundum totam; quare etc.

10. Solutio. Dicendum est quod nec secundum plus nec secundum minus suae potentiae proprie loquendo movet intelligentia caelum, quia, ubi proprie est appositio potentiae secundum plus et minus, ibi est motor in pellendo. [g]Intelligentia vero movens caelum non est motor in pellendo, sed solum est motor in regendo. Unde dicendum quod duplex est motor: in pellendo et regendo; de motore in pellendo concludunt rationes, de motore in regendo, cuiusmodi est intelligentia, non.

11. [h]Per istam solutionem patet qualiter respondendum est argumentis.

12. Aliter dicendum est sic quod intelligentia apponit totam potentiam suam ad movendum caelum. Sed apponere totam potentiam, hoc est dupliciter: [i]aut ita quod totam essentiam potentiae, non tamen totam intensionem, aut ita quod et totam essentiam et totam intensionem. Primo modo verum est quod intelligentia apponit totam potentiam, communiter loquendo, ad movendum caelum, secundo modo falsum est. Non enim

[a] ignobiliora *ex corr.* O
[b] gravi ... de *om. hom.* T
[c] patet *suppl. s.l.* O
[d] quando voluerit *om.* L
[e] secundum *suppl. s.l.* O
[f] etc. *suppl. s.l.* O; *om.* L; moveat T
[g] Intelligentia ... pellendo *om. hom.* L
[h] Per ... argumentis] Et ita patet argumentis T
[i] aut ... intensionem *om. hom.* T

6. Furthermore, the things that are more ignoble always move with more power and with their whole power, as is clear in the case of heavy bodies, whereas nobler things move when they want to, as is clear in the case of a man; for it is because a man is nobler than a stone that he moves when he wants to and with more or less of his power. Therefore, since the heavens are the noblest body, they will move when they want to and with more or less of their power; hence, <the intelligence does not move the heavens with the whole of its power>.

7. Furthermore, in *Physics* VIII Aristotle says that the first mover has infinite power;[203] therefore, it moves with infinite power; but every such motion is instantaneous; it follows then that, if the heavens are moved by their mover with the whole of its power, the heavens will be moved instantaneously, which is absurd; hence, <the intelligence does not move the heavens with the whole of its power>.

8. Arguments to the contrary: in the things here below it is the case that, when something moves with more or less power, fatigue will be involved; therefore, the same will apply to the superior things. Hence it is necessary to posit that <the intelligence moves the heavens> with the whole of its power.

9. Furthermore, if a voluntary mover moves with less of its power, it can <also> move with the whole of its power; therefore, since the intelligence is a voluntary agent, then, if it moves with less of its power, it will also move with the whole of it; hence, <the intelligence moves the heavens with the whole of its power>.

10. Solution. It must be said that, properly speaking, the intelligence moves the heavens neither with more of its power nor with less, because if something moves with more or less of its power in the proper sense, it moves by pushing. The intelligence that moves the heavens, however, does not move them by pushing them, but only by directing them. Accordingly, it must be said that there are two kinds of mover: (i) a mover that moves by pushing and (ii) a mover that moves by directing. The arguments above[204] are conclusive about the mover that moves by pushing, but not about the mover that moves by directing, and the intelligence is of this second kind.

11. Given this solution it is clear how to reply to the arguments.[205]

12. Another solution is that the intelligence applies the whole of its power in moving the heavens. But something can apply the whole of its power in two ways: either (i) in such a way that it applies the whole essence of its power, but not the whole intension, or (ii) in such a way that it applies both the whole essence and the whole intension of its power. The claim that the intelligence

[203] Arist. *Phys.* VIII.10, 267b17–24.
[204] See above, nn. 1–7.
[205] See above, nn. 1–7.

apponit totam intensionem[a] potentiae suae. Et huius causa est propter ordinationem[b] et congruentiam universi, quia determinatum est in natura et congruentia universi quam citius caelum debet moveri, quia, secundum motum suum nunc et in tanta intensione[c] fiunt generationes animalium hic inferius. Unde forte, si suus motus esset intensior, non fierent generationes hic inferius sicut nunc fiunt. Unde [d]sciendum quod propter congruentiam universi est quod caelum tali motu movetur et non velociori.

13. Et ista eadem causa assignanda est ad istam quaestionem, cum[e] quaeritur quare Deus non fecit plures mundos. [fg]Dicendum enim quod hoc est propter congruentiam universi et ordinationem naturae. Si enim fecisset plures mundos, universa non possent esse in uno. Ut ergo universa essent in uno mundo, **[O 118rb]** non fecit Deus nisi unum mundum.

1*. Quaeritur de motoribus caeli in se et absolute.

<Quaestio 39>
Et primo quaeratur an sint in genere.

1. Et videtur quod sic. Et quia dicitur quod non sunt in genere propter hoc quod sunt incorruptibilia[h], contra hoc sic: corruptibile et incorruptibile sunt contraria, et[i] contraria sunt [j]quae sunt in eodem genere; ergo etc.

2. Si dicatur quod non sunt contraria, sed potius privativa, contra: omnia privativa aut sunt circa idem genere aut circa idem specie, [k]per Aristotelem: circa idem specie, ut caecitas et visio circa hominem, circa idem genere, ut est de talpa.

3. Item, species una in uno [l]et eodem genere non potest esse in diversis; [m]sed homo est species[n] una in uno genere; ergo non potest esse in diversis. Sed homo componitur ex corpore et anima; quare etc.

[a] intensionem *corr. ex* intentionem O; intentionem T
[b] ordinationem] ordinem LT
[c] intensione *corr. ex* intentione OT
[d] sciendum quod *om.* T
[e] cum *ex corr.* O
[f] Dicendum ... mundos *om. hom.* L
[g] Dicendum enim *om.* T
[h] incorruptibilia] vel incorruptibiles *add. s.l.* O
[i] et] contra sic *add. sed del.* O
[j] quae sunt *om.* T
[k] per ... specie *om. hom.* T
[l] et eodem genere *corr. ex* genere et eodem O
[m] sed ... diversis *iter. hom.* T *et* generibus *add.*
[n] species] in *add. sed del.* O

applies the whole of its power, in a broad sense, in moving the heavens is true in way (i), but it is false in way (ii), since the intelligence does not apply the whole intension of its power. And the reason for this is the ordering and the harmony of the universe, because the nature and the harmony of the universe determine how fast the universe must be moved, given that it is in accordance with its present motion and its present degree of intension that the generations of animals here below take place. Thus, it might be the case that if the motion of the universe were more intense, the generations here below would not take place as they do now. Hence we should know that it is because of the harmony of the universe that the heavens are moved with their present motion and not with a faster one.

13. And the same reason must be given in reply to the question of why God did not make more worlds. It must be said that this is because of the harmony of the universe and the ordering of nature; for if he had made more worlds, then the totality of things could not exist in just one world. In order for the totality of things to exist in just one world, then, God only made one world.

1*. Questions about the movers of the heavens in themselves and absolutely.

<Question 39>
The first question is whether they are in a genus.

1. It seems that they are. And when it is objected that the movers of the heavens are not in a genus because they are incorruptible, one can argue against this as follows: corruptible and incorruptible are contraries; and contraries are in the same genus; therefore, <the movers of the heavens are in a genus>.

2. If it is said that corruptible and incorruptible are not contraries, but rather privatives, one can argue against this as follows: all privatives are either in the same genus or in the same species, according to Aristotle — that is, in the same species, in the way that blindness and vision are in man; or in the same genus, in the way that blindness and vision <are in a man and> in a mole.[206]

3. Furthermore, one species in one and the same genus cannot be in different <genera>; but man is one species in one genus; therefore, man cannot be in different genera. But man is composed of body and soul; hence, <body and soul are in the same genus>.

[206] Arist. *Metaph.* V.22, 1022b24–27.

4. Dicunt quidam quod anima absolute non est perfectio hominis, sed mediante quodam esse comparato ad imaginationem, et illud esse est corruptibile.

5. Contra: verae rei vera forma est perfectio; sed homo est res vera; ergo eius perfectio est forma vera; non ergo esse aliquod, [a]ut tu dicis.

6. Item, illud esse aut est substantiale aut accidentale. Non accidentale, quia[b] nullum accidens est perfectio substantiae. Oportet igitur ponere quod sit substantiale. Sed tale esse est incorruptibile; ergo redit idem quod prius, quod incorruptibile[c] est in genere.

7. Item, omne commune praedicabile de multis est in genere[d]; sed intelligentia est praedicabilis de multis; ergo[e] etc.

8. Item, quaecumque sunt univoca sunt in praedicamento; sed sol et corpora hic inferiora sunt univoca; ergo sunt in praedicamento, et tamen sol est incorruptibilis sicut intelligentia. Probatio minoris: [f]si passio univoca, et subiectum univocum; sed passio eorum est univoca; ergo et ipsa subiecta sunt univoca. Probatio minoris huius[g]: passio solis et corporum supracaelestium est motus et corporum hic inferiorum; sed motus est univocus hinc[h] inde, quia tempus[i] est mensura <motus[j]> caeli et corporum inferiorum[k], et mensura et mensuratum sunt unigenea; cum ergo istae passiones immediate egrediantur a substantia et sunt univocae, necessario sequitur quod [l]subiecta sunt univoca, et ita in genere[m]. Item, quantitas solis et terrae sunt comparabiles, et omnis comparatio est in univoco; quare etc. [n]Et ulterius ut prius. Probatio maioris: in libro *De anima* comparat quantitates eorum ad invicem. Dicit enim [o]quod quidam, ut qui sequuntur sensum, dicunt quod terra est maior sole, quia sensus dicit esse ita, ratio tamen contradicit huic. Dicit enim ratio quod sol est maior terra. In hoc tunc facit Aristoteles comparationem.

[a] ut tu dicis *om.* T
[b] quia *suppl. s.l.* O
[c] incorruptibile *ex corr.* O
[d] genere] vel ipsum genus *add.* T
[e] ergo] quare LT
[f] si ... minoris *om. hom.* L
[g] huius *ex corr.* O
[h] hinc *ex corr.* O
[i] tempus *scripsimus*; motus OLT
[j] motus *supplevimus*
[k] inferiorum *scripsimus*; supracaelestium OLT
[l] subiecta sunt univoca] substantiae sunt univocae T
[m] genere] paratio est in univoco quare etc. et ulterius ut prius *add. sed del.* O
[n] Et ... prius *om.* T
[o] quod ... enim *om. hom.* L

4. Some people say that the soul is not the perfection of man absolutely, but by means of some kind of being relative to the imagination, and that being is corruptible.

5. Against this: the perfection of a genuine thing is a genuine form; but a man is a genuine thing; therefore, his perfection is a genuine form; therefore, his perfection is not some kind of being, as you claim.

6. Furthermore, that being is either substantial or accidental. It is not accidental, because no accident is a perfection of a substance. Therefore, it is necessary to posit that that being is substantial. But the substantial being <of the soul> is incorruptible; therefore, we get back to the same conclusion, namely that what is incorruptible is in a genus.

7. Furthermore, every common term predicable of many is in a genus; but intelligence is predicable of many; therefore, <it is in a genus>.

8. Furthermore, all univocal things are in a category; but the sun and the bodies here below are univocal; therefore, they are in a category, yet the sun is incorruptible just as an intelligence is. Proof of the minor premise: if a passion is univocal, its subject is also univocal; but a passion of the sun and of the body here below is univocal; therefore, the subjects themselves of that passion are also univocal. Proof of the minor premise of this argument: motion is a passion of the sun and the <other> celestial bodies and of the bodies here below; but motion is univocal in these two kinds of thing, because time is the measure of the motion of the heavens and of the bodies here below, and the measure and the measured are of the same nature; therefore, since these passions emerge directly from substance and are univocal, it necessarily follows that their subjects are univocal, and hence they are in a genus. Furthermore, the quantities of the sun and of the earth are comparable, and every comparison is between univocal things; hence, <the quantities of the sun and of the earth are univocal>. And we can conclude as before <that their subjects are univocal>. Proof of the major premise: in *On the Soul* Aristotle compares the quantity of the sun to that of the earth, since he says that some people, who follow their senses, say that the earth is bigger than the sun, because their senses say so, yet reason contradicts this, given that reason tells us that the sun is bigger than the earth.[207] In saying this, then, Aristotle offers a comparison.

[207] Arist. *De anima* III.3, 428b2–4.

9. Ad oppositum: omne quod est in genere egreditur [a]de potentia generis; sed incorruptibile non potest de potentia alicuius educi. Educi enim de potentia alicuius repugnat naturae incorruptibilitatis; quare etc.

10. Item, quorum genus idem et materia eadem; si ergo istae formae quae sunt corruptibilitas et incorruptibilitas habeant genus idem aliquod, tunc materia manet eadem illarum formarum, et habebit duplicem potentiam. Inducatur[b] forma incorruptibilis; ista forma [c]inducta[d] ex potentiis illius materiae est posterior illis potentiis. Sed quod posterius est non tollit [e]illud quod prius est; ergo post inductionem huius formae incorruptibilis manet potentia respectu formae corruptibilis, quod falsum est. Hoc autem sequitur dato quod sint in genere; quare [f]non possumus ponere quod incorruptibile sit in genere.

11. Item, Aristoteles[g]: "Corruptibile et incorruptibile plus differunt[h] quam secundum genus"; ergo non sunt in eodem genere; quare etc.

12. Item, Commentator: "'corpus' dicitur aequivoce de corporibus supracaelestibus et corporibus hic inferius".

13. [i]Item, Commentator: "'natura' dicitur aequivoce in corporibus supracaelestibus et in corporibus hic inferius"; sed aequivoca non sunt in genere; quare etc.

14. Item, Aristoteles[j]: "Non habentia speciem in materia agunt impassibiliter entia". Ista propositio esset falsa, ut patuit prius, nisi[k] sic intelligeret quod corpora supracaelestia et corpora inferiora non habent speciem in materia univoce sed aequivoce; aequivoca vero in genere non sunt; [l]quare etc.

15. Solutio. Dicendum quod sunt in genere.

16. Ad primum in oppositum dicendum quod incorruptibile educitur de potentia ipsius generis. Et non est intellectus huius propositionis quod [m]species[n] debeant educi de potentia generis ita quod generentur ex potentia generis, sed quod debeant educi de potentia generis per intellectum hoc modo:

[a] de potentia generis *suppl. s.l.* O; *om.* L
[b] Inducatur *ex corr.* O
[c] inducta ex potentiis] inducitur ex potentia LT
[d] inducta *ex corr.* O
[e] illud ... ergo *om.* T
[f] non ... genere] etc. T
[g] Aristoteles] in decimo *add.* T
[h] differunt *lectio incerta* O
[i] Item ... inferius *om. hom.* T
[j] Aristoteles] dicit quod *add.* T
[k] nisi *ex corr.* O
[l] quare etc. *om.* T
[m] species ... generis] educatur T
[n] species *suppl. s.l.* O; *om.* L

9. Arguments to the contrary: everything that is in a genus emerges from the potency of that genus; but what is incorruptible cannot be educed from the potency of something, since to be educed from the potency of something is inconsistent with the nature of incorruptibility; hence, <what is incorruptible is not in a genus>.

10. Furthermore, things with the same genus also have the same matter; therefore, if these forms—corruptibility and incorruptibility—had the same genus, then the matter of those two forms would remain the same and would have two kinds of potencies. Let the incorruptible form be induced <in matter>. This form induced from the potencies of that matter is posterior to those potencies. But what is posterior does not remove what is prior; therefore, after the induction of this incorruptible form, the potency <of that matter> with respect to the corruptible form remains, which is false. But this claim follows from the assumption that <the forms of corruptibility and of incorruptibility> are in a genus; hence, we cannot posit that what is incorruptible is in a genus.

11. Furthermore, Aristotle <says>: "Corruptible and incorruptible things differ more than in genus";[208] therefore, they are not in the same genus; hence, <an incorruptible thing is not in a genus>.

12. Furthermore, the Commentator <says>: "'body' is said equivocally of the celestial bodies and of the bodies here below".[209]

13. Furthermore, the Commentator <says>: "'nature' is said equivocally of the celestial bodies and of the bodies here below";[210] but equivocal things are not in a genus; hence, <incorruptible and corruptible things are not in the same genus>.

14. Furthermore, Aristotle <says>: "Things such that their species is not in matter act without being subject to a passion".[211] But this proposition would be false, as was made clear before,[212] if Aristotle did not mean that the celestial bodies and the bodies here below do not have their species in matter univocally but equivocally; but equivocal things are not in a genus; hence, <incorruptible and corruptible things are not in a genus>.

15. Solution. It must be said that the movers of the heavens are in a genus.

16. To the first argument to the contrary[213] it must be said that what is incorruptible is educed from the potency of the genus itself. And the sense of this proposition is not that species must be educed from the potency of the genus in such a way that they are generated from the potency of the genus, but

[208] Arist. *Metaph.* X.10, 1058b26–29.
[209] Aver. *In Metaph.* X, t.c. 26, fol. 276va.
[210] Aver. *In Phys.* II, t.c. 1, fol. 48rb; *In Metaph.* II, t.c. 16, fol. 36ra.
[211] Arist. *De gen. et corr.* I.7, 324b4–5 .
[212] See above, Book I, q. 1, n. 2.
[213] See above, n. 9.

primo respicere genus, secundo invenire differentiam et addere generi, et ita ultimo per intellectum educere species de potentia generis. Et sic possumus facere a parte incorruptibilitatis.

17. Per hoc patet solutio ad secundum.

18. Ad alia patuit prius.

<Quaestio 40>
Quaeritur utrum intelligentiae differant specie.

1. Et videtur quod sic, quia dicit Aristoteles in undecimo [a]*Philosophiae primae*, ubi probat numerum intelligentiarum: motores caeli differunt secundum prius et posterius, nobilius et ignobilius; sed quae sunt eiusdem speciei, neutrum est prius nec nobilius altero; quare etc.

2. Item, Aristoteles in primo *Caeli et mundi* probat mundum tantum esse unum per hoc quod forma eius occupat totam suam materiam; cum ergo quaelibet intelligentia occupet totam suam materiam, tunc non erit nisi una intelligentia[b] [c]una specie, et ita non erit nisi una intelligentia in una specie.

3. Item, Aristoteles in secundo *De anima*, capitulo de potentia vegetativa, dicit quod, quia omne individuum est corruptibile nec potest salvare [d]se in se, propter hoc dedit[e] natura eis virtutem generativam quatenus salvent se in suo simili et ut perpetuetur esse dividuum. [fg]Tunc per Aristotelem, videtur quod multitudo individuorum sub specie est ut perpetuetur esse dividuum. Sed in qualibet intelligentia potest esse[h] dividuum salvari, cum quaelibet sit incorruptibilis; ergo in nulla specie intelligentiarum erit nisi una intelligentia tantum; [i]quare etc.

4. Item, ita est in partibus essentialibus mundi, quae sunt incorruptibiles, quod[j] in una specie est tantum unicum individuum; ergo, cum intelligentia sit multo incorruptibilior, tunc multo fortius in una specie non erit nisi[k] unicum individuum.

[a] Philosophiae primae *om.* T
[b] intelligentia] sub *add. s.l.* O
[c] una ... intelligentia *om. hom.* T
[d] se in se *ex corr.* O; suum esse LT
[e] dedit] in dicit O
[f] Tunc ... potest esse dividuum *om. hom.* L
[g] Tunc ... dividuum *om. hom.* T
[h] esse *ex corr.* O
[i] quare etc. *om.* T
[j] quod *corr. ex* quia O
[k] nisi] est *add.* O

that they must be educed from the potency of the genus through the work of the intellect as follows: first by considering the genus, secondly by finding the difference and by adding it to the genus, and thus finally by educing the species from the potency of the genus through the work of the intellect. And we can apply this process to incorruptibility.

17. With this the solution to the second argument[214] is clear.

18. The reply to the other arguments[215] was made clear before.[216]

<Question 40>
Whether the intelligences differ in species.

1. It seems that they do, because in *First Philosophy* XI where he establishes the number of intelligences, Aristotle says that the movers of the heavens differ with respect to before and after, and to the more noble and the more ignoble;[217] but things of the same species are not such that one is prior to or more noble than another; hence, <the intelligences differ in species>.

2. Furthermore, in *On the Heavens* I Aristotle proves that there is only one world because its form occupies the whole of its matter;[218] therefore, since any intelligence occupies the whole of its matter, then there is only one intelligence of one species, and thus there is only one intelligence in one species.

3. Furthermore, in *On the Soul* II, in the chapter on the vegetative power, Aristotle says that since every individual is corruptible and cannot preserve itself in itself, this is why nature provided individuals with generative power so that individuals can preserve themselves in something similar to them and so that specific being can be perpetuated.[219] According to Aristotle, then, it seems that the multitude of individuals under a species is for the sake of perpetuating specific being. But specific being can be preserved in any intelligence, since any intelligence is incorruptible; therefore, in no species of intelligence is there more than one intelligence; hence, <the intelligences differ in species>.

4. Furthermore, the same applies to the essential parts of the world, which are incorruptible, namely, that in one species there is only one individual; therefore, since an intelligence is much more incorruptible <than these essential parts>, there is a much stronger reason to say that in one species <of intelligence> there is only one individual.

[214] See above, n. 10.
[215] See above, nn. 11–14.
[216] See above, Book I, q. 59, nn. 31–36.
[217] Arist. *Metaph.* XII.8, 1073b1–3.
[218] Arist. *De caelo* I.9, 278a25–27.
[219] Arist. *De anima* II.4, 415a26–b7.

5. Item, dicit Commentator **[O 118va]** quod, si [a]in aliqua specie fuerit unicum individuum incorruptibile, impossibile est quod sit nisi unum individuum in illa specie.

6. Ad oppositum: [b]unitum et non unitum non sunt differentiae essentialiter facientes differre in substantia; sed anima rationalis et intelligentia solum differunt penes unitum et non unitum; ergo non differunt essentialiter. Sicut ergo est in anima rationali quod omnes animae rationales sunt eiusdem speciei, quia idem[c] specie perficiunt[d] omnes, tunc a simili vel multo fortius spiritus superiores, ut intelligentiae, erunt omnes eiusdem speciei.

7. Item, maior est nobilitas et evidentia nobilitatis in natura ubi nobilior species est multiplicata per multa individua quam ubi ignobilior, verbi gratia maior nobilitas est quando rosa bene se multiplicat quam quando urtica. Cum ergo intelligentia sit species nobilissima, tunc maxime congruit naturae ut talis species multiplicetur per plura individua. [e]Tunc, ut videtur, intelligentiae omnes conveniunt specie et non differunt specie, sicut multa individua conveniunt in specie et non differunt.

8. Item, nobilissima operatio est eadem in omnibus intelligentiis, ut intelligere; et operatio procedit a virtute, et virtus a forma nobilissima; ergo virtus erit eadem in omnibus, [f]et similiter substantia sive forma nobilissima erit eadem in omnibus. Sed forma nobilissima in unoquoque est sua forma completiva et perfectiva; ergo eadem est forma completiva omnium. [g]Necessario tunc conveniunt in specie omnes sicut individua in specie.

9. Item, est unum quid simplicissimum omnino indivisibile per Aristotelem, ut causa prima, ad quod omnes res habent ordinem; quod ergo magis appropinquatur[h] illi simplicissimo minus erit divisibile secundum formam et quod maxime ei appropinquatur minime erit divisibile secundum formam. [i]Sed intelligentia maxime appropinquatur causae primae; ergo minime erit divisibile secundum formam et, si hoc, oportet quod omnes conveniant in una forma, et ita [j]quod conveniant in specie.

10. Item, dicit Augustinus quod intelligentia et anima sunt natura pares, officio autem impares; sed talis est natura in animalibus quod omnes animae

[a] in aliqua specie *om.* L
[b] unitum ... unitum *om.* L
[c] idem *corr. in* eadem O
[d] perficiunt *corr. in* perficiunur O
[e] Tunc ... differunt *om.* T
[f] et ... omnibus *om. hom.* LT *sed suppl. mg.* L
[g] Necessario ... individua in specie] et ita omnes sunt eiusdem speciei T
[h] appropinquatur *corr. ex* appropinquat O; propinquatur T
[i] Sed ... formam *om. hom.* T
[j] quod conveniant *om.* T

5. Furthermore, the Commentator says that, if there is only one incorruptible individual in some species or other, it is impossible for there not to be only one individual in that species.[220]

6. Arguments to the contrary: being united and being non-united are not differences that produce an essential difference in a substance; but a rational soul and an intelligence only differ with respect to being united and non-united; therefore, they do not differ essentially. Therefore, just as in the case of the rational soul all rational souls are of the same species, because they are perfections of things of the same species, there is a similar or much stronger reason to hold that superior spirits, that is the intelligences, will all be of the same species.

7. Furthermore, there is more nobility and more evidence of nobility in nature when a nobler species is multiplied in many individuals than when a more ignoble species is <multiplied>; for example, there is more nobility when the rose multiplies itself than when the nettle does. Therefore, since intelligence is the noblest species, it is maximally appropriate to nature that this species should be multiplied in many individuals. Thus, it seems, all intelligences agree in species and do not differ in species, just as many individuals agree in species and do not differ.

8. Furthermore, the noblest operation, that is, understanding, is the same in all intelligences; and the operation proceeds from a power, and the power proceeds from the noblest form; therefore, the power will be the same in all intelligences, and similarly substance or the noblest form will be the same in all of them. But the noblest form in any thing is the completing and perfecting form of that thing; therefore, the completing form of all intelligences is the same. All intelligences, then, necessarily agree in species, as do individuals in a species.

9. Furthermore, according to Aristotle, there is one being that is the simplest being and is an absolutely indivisible being, namely the first cause, to which all things are ordered;[221] therefore, whatever is closer to that simplest thing will be less divisible in form and whatever is maximally close to it will be minimally divisible in form. But an intelligence is maximally close to the first cause; therefore, it will be minimally divisible in form, and if so, it is necessary for all intelligences to agree in one form, and thus for them to agree in species.

10. Furthermore, Augustine says that an intelligence and a soul are equal by nature, but unequal by duty;[222] but the nature of animals is such that all

[220] Aver. *In De caelo* I, t.c. 94, ed. Carmody, I, 176.46–177.74.
[221] Arist. *Phys.* VIII.10, 266a10–12; *Metaph.* XII.10, 1075a11–15 .
[222] Aug. *De lib. arb.* III.11, CCSL 29, 294.15–16.

rationales sunt eiusdem speciei et in specie conveniunt; ergo talis erit natura in intelligentiis quod omnes erunt eiusdem speciei et in una specie convenient.

11. Quaestioni, secundum Commentatorem et secundum Aristotelem, ut mihi videtur, dicendum est quod differunt specie, quia dicit Aristoteles quod motores differunt sicut et mobilia[a], mobilia[b] vero specie differunt, et ita, per Aristotelem, motores eorum [c]differrent specie.

12. Commentator etiam dicit quod, si sit aliquod individuum incorruptibile sub una specie, impossibile est quod sit nisi unicum individuum in illa specie. Sic autem est de intelligentia.

13. Ponendo tamen secundum theologos et secundum rei veritatem, dicemus quod conveniunt in specie [d]et quod omnes sunt in una specie, et ponemus etiam quod quaedam intelligentiae sunt nobiliores et priores.

14. Hoc modo ultimo ponendo, dicendum est ad primum argumentum in oppositum, cum dicit 'quae sunt sub eadem specie, neutrum est posterius nec ignobilius' etc., quod hoc est intelligendum sic quod neutrum est [e]nobilius nec prius altero quantum ad condiciones ipsius speciei, ut quoad condiciones praedicationis. Aequaliter enim et aeque primo praedicatur species de omnibus individuis. Quantum tamen ad condiciones individuales, falsum est, sicut patet de corpore et similiter de anima. Videmus enim quod aliquod corpus est nobilius alio, ut corpus melius dispositum et complexionatum. Similiter videmus et dicimus quod quaedam anima est nobilior alia, et dicimus etiam hoc esse verum in prima sui creatione, quod quaedam anima creatur nobilior alia, licet omnes sint aeque natae ad suscipiendum bonitatem divinam. Patet tunc qualiter respondendum est textui Aristotelis. Dicendum enim est quod intelligentiae omnes sunt aeque nobiles quoad proprietates et condiciones speciales, non autem quoad condiciones individuales. Et hoc non est inconveniens. Ita enim est in his corporibus inferioribus.

15. Ad secundum dicendum quod aliquod individuum occupare totam materiam, hoc est dupliciter: aut totam materiam suae speciei deputatam, et sic caelum, mundus, sol et luna et consimilia occupant totam suam materiam —tota enim materia illius speciei non multiplicatur nec extenditur nisi in uno;

[a] mobilia *ex corr.* O; differunt *add.* LT
[b] mobilia *suppl. s.l.* O
[c] differrent specie *om.* T
[d] et ... specie *om. hom.* T
[e] nobilius nec prius] posterius nec ignobilius T

rational souls are of the same species and agree in species; therefore, the nature of intelligences will be such that they are all of the same species and agree in one species.

11. It seems to me that, according to the Commentator and Aristotle, one should reply to the question by saying that the intelligences differ in species, because Aristotle says that movers differ in the way that mobile things do;[223] but mobile things differ in species, and thus according to Aristotle their movers would differ in species.

12. The Commentator also says that, if there is an incorruptible individual in one species, it is impossible for there not to be only one individual in that species.[224] But this is the case of intelligences.

13. Following both the theologians and the truth of the matter, however, we shall say that the intelligences agree in species and that they are all in one species, and we shall also posit that some intelligences are more noble than and prior to others.

14. Following this second position, we must reply to the first argument to the contrary[225] by saying that the claim 'among things in the same species none is posterior to or more ignoble <than another>' must be understood in the sense that none is more noble than or prior to another with respect to the conditions of the species itself, for instance, with respect to the conditions of predication; for a species is equally and equally primarily predicated of all its individuals. With respect to the individual conditions, however, this claim is false, as is clear in the case of the body and similarly in that of the soul; for we see that one body is more noble than another, namely the body that has a better disposition and constitution. Similarly, we see and say that one soul is more noble than another, and we also say that it is true that at the first creation of a soul one soul is created as more noble than another, even though all souls are equally apt by nature to receive divine goodness. It is clear, then, how one should reply to the words of Aristotle. That is, one must say that all intelligences are equally noble with respect to the specific properties and conditions, but not with respect to individual conditions. And this is not absurd; for this is the case for bodies here below.

15. One must reply to the second argument[226] by saying that the claim that an individual occupies the whole matter can be understood in two ways: either (i) it occupies the whole matter assigned to its species; and in this way the heavens, the world, the sun, the moon, and the like each occupy the whole of their matter—for the whole matter of that species is only multiplied or

[223] This is probably a reference to Arist. *Metaph.* XII.8, 1073b1–3.

[224] Aver. *In De caelo* I, t.c. 94, ed. Carmody, I, 176.46–177.74.

[225] See above, n. 1.

[226] See above, n. 2.

aut totam materiam suo[a] individuo deputatam. Et sic anima rationalis et intelligentia occupant totam suam materiam. Et ita patet quod, licet argumentum teneat in mundo, [b]non tamen tenet in intelligentiis, et hoc quia aequivocatur occupare totam[c] materiam.

16. Ad tertium dicendum quod haec [d]non est causa quare multa individua sunt sub una specie absolute et simpliciter, quia individuum est generabile et corruptibile, sed haec est causa praecisa quare multa sunt individua sub una specie per generationem et corruptionem.

17. Ad alias patet per iam dicta.

1*. Quaeritur de motoribus [e]in comparatione.

<Quaestio 41>

[f]Et primo in comparatione ad animam rationalem, inquirendo[g] utrum intelligentia et anima[h] rationalis sint eadem specie [i][j]vel non.

1. Et videtur quod sic, quia nobilissima operatio est eadem specie in illis, ut intelligere; et operatio [k]egreditur a virtute; ergo eadem est virtus in illis. Et virtus a substantia sive a forma nobilissima quae completiva est; ergo conveniunt in eadem substantia sive forma completiva; ergo in specie.

2. Item, dicit Augustinus quod intelligentia et anima rationalis sunt natura pares, officio vero impares; sed si sunt pares natura, tunc conveniunt in natura aliqua, et ita in specie.

3. Item, unitum et non unitum non sunt differentiae facientes essentialiter[l] differre in substantia, quia solum sunt accidentales proprietates, quia dicunt relationem; sed intelligentia et anima solum differunt sicut unitum et non unitum; ergo non differunt essentialiter, sed solum accidentaliter; quare etc.

4. Si dicatur quod anima rationalis et intelligentia non differunt penes unitum et non unitum, sed penes unibile et non unibile, et istae sunt differentiae facientes differre essentialiter in substantia, contra[m]: unibile et non unibile

[a] suo *corr. in* sibi O; sibi T
[b] non tamen *corr. in* tamen non O; *inv.* L
[c] totam] suam *add. s.l.* O
[d] non est causa *corr. ex* est causa non O
[e] in comparatione *om.* T
[f] Et ... comparatione *om. hom.* L
[g] inquirendo] et primo T
[h] anima *suppl. s.l.* O
[i] vel ... specie (n. 1) *om. hom. sed suppl. mg.* L
[j] vel non *om.* T
[k] egreditur ... completiva; ergo] a virtute et virtus a substantia quae est forma nobilissima et completiva et ita T
[l] essentialiter] sed solum accidentaliter *add. sed del.* O
[m] contra *lectio incerta* O; si *add.* L

extended in one individual; or (ii) it occupies the whole matter assigned to that individual; and in this way a rational soul and an intelligence occupy the whole of their matter. And thus it is clear that, although the argument is valid for the world, it is not valid for the intelligences, and this is because 'occupying the whole matter' is taken equivocally.

16. One must reply to the third argument[227] by saying that this is not absolutely and simply the cause of there being many individuals in one species, that is, that this happens because an individual is generable and corruptible, but rather that this is the precise cause of there being many individuals in one species through generation and corruption.

17. The reply to the other arguments[228] is clear from what has already been said.

1*. Questions about the movers <of the heavens> when they are compared <to something>.

<Question 41>

The first question is about the movers as compared to a rational soul. We ask whether an intelligence and a rational soul are the same in species or not.

1. Arguments for the affirmative: the noblest operation, namely understanding, is the same in species in an intelligence and in a rational soul; and the operation emerges from a power; therefore, the power is the same in them both. And the power emerges from the substance or the noblest form, which is the completive form; therefore, an intelligence and a rational soul agree in the same substance or completive form; therefore, they agree in species.

2. Furthermore, Augustine says that an intelligence and a rational soul are equal by nature, and unequal by duty;[229] but if they are equal by nature, then they agree in some nature and so they agree in a species.

3. Furthermore, being united and being non-united are not differences that produce an essential difference in a substance, because they are simply accidental properties, since they express a relation; but an intelligence and a soul only differ as being united and non-united; therefore, they do not differ essentially, but only accidentally; hence, <they are in the same species>.

4. If it is said that an intelligence and a rational soul do not differ with respect to being united and being non-united, but with respect to being able to be united and not being able to be united, and that the latter are differences that cause one substance to differ essentially <from another>, one can argue

[227] See above, n. 3.
[228] See above, nn. 4–5.
[229] Aug. *De lib. arb.* III.11, CCSL 29, 294.15–16.

dicunt habitudines accidentales, quia **[O 118vb]** dicunt relationem similiter sicut et illae differentiae —unitum et non unitum—et ita, sicut non differunt essentialiter per has—unitum etc. — similiter nec per alias.

5. Item[a], anima intellectiva[b] et intelligentia, quantum est de se, habent eandem mensuram, quia utraque[c] mensuratur aevo; cum ergo mensura et mensuratum sint unigenea[d], tunc, [e]cum ista mensura sit utriusque, erit unigenea cum utroque mensurato. Et quaecumque sunt eadem respectu alicuius tertii ipsa inter se sunt eadem; ergo anima rationalis et intelligentia sunt eadem.

6. Item, dicit Commentator quod primus motor[f] est causa diversorum motuum non de se, sed mediante [g]moto primo; quae igitur immediate procedunt a [h]causa prima maxime conveniunt[i]; [j]ergo, quantum possibile est ipsa convenire conveniunt. Sed multa sunt quae conveniunt et genere et specie; ergo de necessitate oportet nos ponere ipsa convenire in genere et specie; quare etc.

7. Ad oppositum: plus conveniunt anima et anima quam anima et intelligentia; sed anima et anima differunt numero; ergo [k]oportet quod anima et intelligentia plus differant quam numero; sed plus differre quam secundum numerum non potest esse nisi differre specie; quare oportet ponere quod anima et intelligentia differant specie.

8. Item, si aliqua duo sint in eadem specie, si unum illorum faciat sive[l] constituat unum per essentiam cum quodam alio consimili, tunc aliud cum quodam alio sibi consimili faciet unum per essentiam. Si ergo anima et intelligentia [m]sint convenientes secundum speciem, tunc cum anima cum quodam alio, ut cum corpore, constituat unum per essentiam, ut hominem, ergo intelligentia cum quodam alio constituet unum per essentiam. Hoc autem falsum est.

[a] Item *om.* O
[b] intellectiva] quantum est de se *add. sed del.* O
[c] utraque] mensura *add.* O; utrumque T
[d] unigenea] per Aristotelem in quinto Metaphysicae capitulo de uno *add.* T
[e] cum ... mensurato] aevum erit unigeneum cum utroque T
[f] motor] motus T
[g] moto primo *scripsimus*; causa prima OLT
[h] causa prima *sc.* primus motor
[i] conveniunt *ex corr.* O
[j] ergo ... conveniunt *om. hom.* T
[k] oportet ... differant specie] anima et intelligentia plus differunt et ita specie T
[l] sive *suppl. s.l.* O; et LT
[m] sint ... est] conveniant specie tunc sicut anima cum corpore facit unum per essentiam, ut hominem, sic intelligentia cum alio corpore faciet alia per essentiam. Sed hoc est falsum; ergo primum T

against this as follows: being able to be united and not being able to be united express accidental relationships, because they express a relation in a similar way to the differences 'being united', 'being non-united', and thus, given that an intelligence and a rational soul do not differ essentially because of the differences 'being united', <'being non-united'>, they similarly fail to differ because of these other differences.

5. Furthermore, an intellectual soul and an intelligence, in so far as it depends on them themselves, have the same measure, because each of them is measured by the *aevum*; therefore, since a measure and what is measured by it are of the same nature, then, since the same measure belongs to each of them, this measure will be of the same nature as each of the two measured things. And any two things that are the same as a third thing are the same as each other; therefore, a rational soul and an intelligence are the same.

6. Furthermore, the Commentator says that the first mover is the cause of different motions not by itself, but by means of the first moved thing;[230] therefore, the things that immediately proceed from the first cause are maximally in agreement; therefore, they agree as much as is possible for them. But there are many things that agree both in genus and in species; therefore, it is necessary for us to posit that an intelligence and a rational soul agree in genus and in species; hence, <they are of the same species>.

7. Arguments to the contrary: two souls agree more than do a soul and an intelligence; but two souls differ in number; therefore, it is necessary for a soul and an intelligence to differ in more than in number; but to differ in more than in number can only be to differ in species; therefore, it is necessary to posit that a soul and an intelligence differ in species.

8. Furthermore, if two things <A and B> are in the same species, if one of them, <say A>, in conjunction with another thing <C> similar to itself, produces or constitutes something that is one in essence, then the other thing <B> in conjunction with another thing <D> similar to itself, will produce something that is one in essence. Therefore, if a soul and an intelligence agree in species, since a soul in conjunction with another thing, namely a body, constitutes something that is one in essence, namely a man, then an intelligence in conjunction with some other thing will constitute something that is one in essence. But this is false.

[230] Aver. *In Metaph.* XII, t.c. 37, fol. 320va.

9. Item, quod non est species [a]non convenit cum alia specie in specie; sed anima non est species; [b]ergo etc. Probatio [c]minoris: dicit Porphyrius quod duae species non conveniunt ad constituendum tertiam speciem; sed corpus est species et anima convenit cum ipso ad constituendum hominem; ergo [d]oportet ponere quod anima non sit species; quare etc.

10. Solutio. Quidam dicunt quod anima intellectiva et intelligentia non sunt eiusdem speciei, sed diversae.

11. Ad primam rationem in oppositum dicunt sic quod talis forma arguendi fallit quia diversae formae per esse consimilia quae habent in materia possunt esse causae eiusdem effectus sive operationis, sicut eadem forma per esse diversa potest esse causa diversorum effectuum. [e]Verbi gratia de primo, ut cum dicitur quod canis est albus et quod panis sive homo similiter, et tamen istae formae substantiales differunt secundum speciem. Verbi gratia de secundo, ut cum dicitur quod idem homo per esse diversa potest esse albus et niger. Et dicunt similiter esse de hac operatione, intelligere[f], sicut est de albedine in cane et pane[g], quantum ad hoc.

12. Aliter dicitur sic, quod non est idem intelligere[h] intelligentiae et animae rationalis, quia pono quod intelligentia solum intelligat per species innatas, [i]anima autem rationalis intelligit et per species innatas et per species receptas; unde possumus dicere quod non est omnino idem[j] intelligere utrobique.

13. Alia opinio est quae ponit quod non sunt eiusdem speciei nec diversae. Et[k] ponit quod non differunt sicut unibile et non unibile nec sicut unitum et non unitum, sed differunt per formas quae subsunt istis habitudinibus, quae formae sunt habere inclinationem ad corpus, [l]non habere [m]inclinationem ad corpus.

14. Et illi qui sunt huius opinionis respondent ad rationem sequentem, quae est de mensura etc., dicendo quod duplex est mensura: intrinseca et extrinseca. Mensura intrinseca est ut cum dicimus quod homo mensurat

[a] non ... specie] cum aliquo non convenit T
[b] ergo etc. *om.* T
[c] minoris ... quod] per Porphyrium T
[d] oportet ... species] anima non est species T
[e] Verbi ... niger *om.* T
[f] intelligere] quod *add.* O
[g] pane *ex corr.* O
[h] intelligere] et *add.* O
[i] anima ... innatas *om. hom.* T
[j] idem *om.* LT
[k] Et *ex corr.* O
[l] non ... corpus *om. hom.* L
[m] inclinationem ad corpus *om.* T

9. Furthermore, what is not a species does not unite with another species in a species; but soul is not a species; therefore, <a soul does not unite with an intelligence in a species>. Proof of the minor premise: Porphyry says that two species do not unite in order to constitute a third species;[231] but body is a species and soul unites with it in order to constitute man; therefore, it is necessary to posit that soul is not a species; hence, <soul is not a species>.

10. Solution. Some people say that an intellectual soul and an intelligence are not of the same species, but of different species.

11. These people reply to the first argument to the contrary[232] by saying that this form of arguing fails because diverse forms in virtue of the similar kind of being that they have in matter can be the cause of the same effect or of the same operation, just as the same form in virtue of different kinds of being can be the cause of different effects. An example of the first case is when we say that a dog is white and similarly that bread or a man is white, yet their substantial forms are different in species. An example of the second case is when we say that the same man in virtue of different kinds of being can be white and black. And these people say that the case of the operation of understanding is in this respect like that of whiteness in a dog and in bread.

12. Another reply is that the understanding of an intelligence and of a rational soul are not the same, because I posit that an intelligence only understands in virtue of innate species, but that a rational soul understands both in virtue of innate species and in virtue of received species; thus, we can say that understanding is not exactly the same in the two cases.

13. Another opinion posits that an intelligence and a rational soul are neither of the same species nor of different species. It also posits that they differ neither as being able to be united and not being able to be united nor as being united and being non-united, but they differ in virtue of the forms that underlie these relations, and these forms are 'to have an inclination towards the body' and 'not to have an inclination towards the body'.

14. And the supporters of this opinion reply to the argument that followed, which is about the measure <and the thing measured>,[233] by saying that there are two kinds of measure: (i) an intrinsic measure and (ii) an extrinsic measure. (i) An intrinsic measure is what we refer to when we say that a man measures

[231] Porph. *Isag.*, AL I.6–7, 28.5–6.
[232] See above, n. 1.
[233] See above, n. 5.

hominem, equus equum; et de hac mensura intelligendus[a] est textus Aristotelis cum dicit "Mensura et mensuratum sunt unigenea". Sed talem mensuram non habent intelligentia et anima rationalis eandem, [b]sed diversam. Hoc enim modo anima mensurat animam et intelligentia intelligentiam. Mensura extrinseca est ut cum dicimus quod aeternitas sive aevum mensurat [c]durationem[d] intelligentiae et etiam durationem[e] animae rationalis. Et de[f] ista mensura extrinseca non intelligendus est textus Aristotelis, et secundum hanc mensuram procedit ratio illa.

15. Quidam dicunt tertio modo adhuc quod anima rationalis non differt secundum speciem ab intelligentia. Et ratio [g]eorum est eo quod anima non est species. [h]Et dicunt quod anima non est species propter hoc quod anima rationalis est pars[i], species autem debet esse totum. Cum ergo anima sit pars et non totum, [j]nullo modo potest esse species vera.

16. Sed contra: anima potest esse non pars, sed aliquid in se, sicut dicit Aristoteles in libro *De anima,* [k]cum dicit quod anima separatur a corpore sicut perpetuum a corruptibili. Post istam separationem, tunc, cum anima non est pars, sed hoc aliquid in se, potest esse species. Falsum ergo dixerunt dicentes animam non esse speciem. Et similiter patet quod ratio eorum fuit nulla.[l]

[a] intelligendus *corr. ex* intelligendum O
[b] sed diversam *om.* T
[c] durationem ... durationem *om.* L
[d] durationem *scripsimus*; dum OT
[e] durationem *scripsimus*; dum OT
[f] de *ex corr.* O
[g] eorum est *corr. in* est eorum O; *inv.* L
[h] Et ... species *om. hom.* T
[i] pars] speciei *add.* T
[j] nullo modo] non T
[k] cum dicit *om.* T
[l] nulla] *Hic aliam quaestionem add.* LT. *Cf. Appendix, Liber VIII, q. 42*

a man, and a horse measures a horse; and the text of Aristotle saying "the measure and what is measured by it are of the same nature"[234] must be understood with reference to this kind of measure. But an intelligence and a rational soul taken together are not subject to this kind of measure, but to a different kind; for by this kind of measure a soul measures a soul and an intelligence measures an intelligence. (ii) An extrinsic measure is what we refer to when we say that eternity or the *aevum* measures the duration of an intelligence and also measures the duration of a rational soul. And the text of Aristotle must not be understood with reference to this kind of measure, whereas the argument in question does refer to this kind of measure.

15. Some people have a third view to the effect that a rational soul does not differ in species from an intelligence. And their argument is that soul is not a species. And they say that soul is not a species because the rational soul is a part, whereas a species must be a whole. Therefore, since the soul is a part and not a whole, it can in no way be a genuine species.

16. But <one can argue> against this: the soul can be not a part but something in itself, as Aristotle observes in *On the Soul* when he says that the soul is separated from the body in the way that what is perpetual is separated from what is corruptible.[235] After this separation, then, when the soul is not a part, but a determinate thing in itself, the soul can be a species. Therefore, those who said that soul is not a species said something false. And similarly it is clear that their argument was null and void.

[234] Arist. *Metaph.* X.1, 1053a24–25.
[235] Arist. *De anima* II.2, 413b26–27.